H. Crapo
D. Senato
(Eds)

Algebraic Combinatorics and Computer Science

A Tribute to Gian-Carlo Rota

Springer

Springer-Verlag Italia Srl.

This volume is dedicated to Gian-Carlo Rota,
one of the great mathematicians of the 20th century.
A tribute by friends, pupils and admirers

H. CRAPO
E.H.E.S.S. – C.N.R.S.
Centre d'Analyse
et Mathematique Sociales
Paris, France

D. SENATO
Dipartimento di Matematica
Università della Basilicata
Potenza, Italy

Originally published by Springer-Verlag Italia, Milano in 2001
Softcover reprint of the hardcover 1st edition 2001

ISBN 978-88-470-2159-4 ISBN 978-88-470-2107-5 (eBook)
DOI 10.1007/978-88-470-2107-5

Library of Congress Cataloging-in-Publication Data: Applied for

Cover design: Simona Colombo, Milan
Typesetting: Bürosoft/Text- und DTP-Service, Berlin/Heidelberg

SPIN: 10781666

Foreword

This book, dedicated to the memory of Gian-Carlo Rota, is the result of a collaborative effort by his friends, students and admirers. Rota was one of the great thinkers of our times, innovator in both mathematics and phenomenology. I feel moved, yet touched by a sense of sadness, in presenting this volume of work, despite the fear that I may be unworthy of the task that befalls me. Rota, both the scientist and the man, was marked by a generosity that knew no bounds. His ideas opened wide the horizons of fields of research, permitting an astonishing number of students from all over the globe to become enthusiastically involved. The contagious energy with which he demonstrated his tremendous mental capacity always proved fresh and inspiring. Beyond his renown as gifted scientist, what was particularly striking in Gian-Carlo Rota was his ability to appreciate the diverse intellectual capacities of those before him and to adapt his communications accordingly. This human sense, complemented by his acute appreciation of the importance of the individual, acted as a catalyst in bringing forth the very best in each one of his students. Whosoever was fortunate enough to enjoy Gian-Carlo Rota's longstanding friendship was most enriched by the experience, both mathematically and philosophically, and had occasion to appreciate son cotè de bon vivant.

The book opens with a heartfelt piece by Henry Crapo in which he meticulously pieces together what Gian-Carlo Rota's untimely demise has bequeathed to science. Mathematics is a discipline which does not follow straight lines of progression, but which undergoes sudden interruptions, expansions and contractions. Mathematicians are lured by contemporary trends, prolifically writing articles each of which unveils some aspect of the truth. The more the subject grows, the more complex the language becomes. As a consequence, it is increasingly difficult, even for experts in the field, to determine the importance of new results and the potential avenues which they open before them. Gian-Carlo Rota's most outstanding talent, was his all-encompassing vision of the manifold aspects of mathematics, his ability to distil the 'underlying ideas' of each domain and precisely to state the formulation which would most naturally lend itself to limpidity of results. This is borne witness to by David Buchsbaum's salient article, which introduces the second chapter which deals with a number of themes of research which were particularly dear to Rota himself or which formed part of his work in progress.

Rota would take no "accepted truth" for granted and reconsidered and discussed issues unremittingly. In one of his most recent essays, "Twelve problems in probability no one likes to bring up", Rota bravely reconsidered the question of Kolmogorov's elegant and neat axiomatization, and pointed to the possibility of a deeper level of understanding. The essay in question forms part of the highly regarded Fubini lectures, which he gave in 1998 at the Institute for Scientific Interchange and which are published here as the first chapter of this book, with a refined preface to the lectures by Edoardo Vesentini.

The third chapter contains two thought-provoking expository articles, by Martin Aigner and Dominique Perrin, presented by Marilena Barnabei and Flavio Bonetti. The first exposition takes an in-depth look at the properties of one of the most profoundly studied sequences of combinatorial counting numbers, the Catalan Numbers. In the second exposition, Perrin presents an overview of the field of generating series for formal languages and rational series. Chapter 4, which opens with a piece by Alain Lascoux (in which he describes the impact of "On the Foundation of Combinatorial Theory. VIII. Finite Operator Calculus"), provides the reader with a selection of texts, representative of some of the most important lines of research in the fields of Algebraic Combinatorics and Theoretical Computer Science, fields profoundly influenced by Rota's thinking. Over and above the fact that many of these essays contain new results, they also furnish us with an up-to- date, self-contained overview of these areas of ongoing research. Gian-Carlo Rota has left us an invaluable wealth of ideas, which we shall do our utmost to keep alive. This volume is the first step. It would have been impossible without the generous support of the Department of Mathematics of the University of Bologna and of the University of Basilicata and without Elvira Di Nardo's careful editing.

November 2000 Domenico Senato

List of Contributors

- **Martin Aigner**, Institut für Mathematik II, Arnimallee 3, 14195 Berlin, Germany
- **Marilena Barnabei**, Dipartimento di Matematica, Università di Bologna, Piazza di Porta San Donato 5, 40126 Bologna, Italy
- **Flavio Bonetti**, Dipartimento di Matematica, Università di Bologna, Piazza di Porta San Donato 5, 40126 Bologna, Italy
- **Paolo Bravi** Dipartimento di Matematica, Università degli Studi di Roma "La Sapienza", Piazzale Aldo Moro 2, 00185 Roma, Italy
- **Andrea Brini**, Dipartimento di Matematica, Università di Bologna, Piazza di Porta San Donato 5, 40126 Bologna, Italy
- **Thomas Britz**, Department of Mathematical Sciences, University of Aarhus, 8000 Århus, Denmark
- **David A. Buchsbaum**, Department of Mathematics, Brandeis University, Waltham, MA 02154, USA
- **Ferdinando Cicalese**, Dipartimento di Informatica ed Applicazioni, Università di Salerno, via S. Allende, 84081 Baronissi Salerno, Italy
- **Bruno Codenotti**, Istituto di Matematica Computazionale del CNR, Area della Ricerca di S. Cataldo, Via Alfieri 1, 56010 Ghezzano Pisa, Italy
- **Henry Crapo**, E.H.E.S.S. – C.N.R.S. – Paris IV, Centre d'Analyse et de Mathématique Sociales, 54 boulevard Raspail, porte 201, 75270 Paris Cedex 06, France
- **Alessandro Di Bucchianico**, Eindhoven University of Technology, Section Quality of Products & Processes, Paviljoen C11, P.O. Box 513, 5600 MB Eindhoven, The Netherlands
- **Elvira Di Nardo**, Dipartimento di Matematica, Università della Basilicata, Via N. Sauro 85, 85100 Potenza, Italy
- **Adriano M. Garsia**, Department of Mathematics, University of California, San Diego, La Jolla, CA 92093, USA
- **Antonio Giambruno**, Dipartimento di Matematica ed Applicazioni, Università di Palermo, Via Archirafi 34, 90123 Palermo, Italy
- **Jacques Justin**, LIAFA, 2 place Jussieu, 75251 Paris Cedex 05, France
- **Gerardo Soto y Koelemeijer**, Delft University of Technology, Applied Mathematical Analysis Department, P.O. Box 5031, 2600 GA Delft, The Netherlands
- **Alain Lascoux**, CNRS, Institut Gaspard Monge, Université de Marne-la-Vallée, 77454 Marne-la-Vallée Cedex, France
- **Matteo Mainetti**, Department of Mathematics, Massachusetts Institute of Technology, Cambridge, MA 02139, USA
- **Laura B. Montefusco**, Dipartimento di Matematica, Università di Bologna, Piazza di Porta San Donato 5, 40126 Bologna, Italy

- **Daniele Mundici**, Dipartimento di Informatica, Università di Milano, Via Comelico 39–41, 20135 Milano, Italy
- **Dominique Perrin**, Institut Gaspard Monge, Université de Marne-la-Vallée, 77454 Marne-la-Vallée Cedex 2, France
- **Giuseppe Pirillo**, Istituto per le Applicazioni della Matematica e dell'Informatica, CNR Viale Morgani 67/A, 50134 Firenze, Italy
- **Luigi Pezzoli**, Centro Interdipartimentale di Ricerca per le Applicazioni della Matematica, 40123 Bologna, Italy
- **Francesco Regonati**, Dipartimento di Matematica, Università di Bologna, Piazza di Porta San Donato 5, 40126 Bologna, Italy
- **Giovanni Resta**, Istituto di Matematica Computazionale del CNR, Area della Ricerca di S. Cataldo, Via Alfieri 1, 56010 Ghezzano Pisa, Italy
- **Domenico Senato**, Dipartimento di Matematica, Università della Basilicata, Via N. Sauro 85, 85100 Potenza, Italy
- **Joel A. Stein**, Department of Mathematics, California State University, 5500 University Pkwy, S. Bernardino, CA 92407, USA
- **Antonio G. B. Teolis** Dipartimento di Matematica, Università di Bologna, Piazza di Porta San Donato 5, 40126 Bologna, Italy
- **Ugo Vaccaro**, Dipartimento di Informatica ed Applicazioni, Università di Salerno, via S. Allende, 84081 Baronissi Salerno, Italy
- **Edoardo Vesentini**, Accademia Nazionale dei Lincei, Palazzo Corsini, Via della Lungara 10, 00165 Roma, Italy
- **Mike Zabrocki**, Centre de Recherche Mathématiques, Université de Montréal/LaCIM, Université du Québec à Montréal, Montreal QC H3C 3J7, Canada

Contents

Introduction

Ten abandoned gold mines

H. Crapo

1 Introduction

I thank Domenico Senato and his fellow organizers for the kind invitation to participate in this gathering of friends of Gian-Carlo Rota – to grieve with you his passing and to search with you for a renewal of the work he laid out for us. Mimmo and Elvira have asked me to speak about Gian-Carlo's mathematical work – and I do so in keen recollection of the happy evenings the four of us spent together at a café on the *corso* in Cortona just in the summer of 1998. Gian-Carlo had a way of creating order in his life, imposing patterns on time so as to be able to concentrate his energies, and to plan for discussions with innumerable people. I like to compare this practice with the establishment of rules in the monastic orders. That summer's *Regola Cortoniense* starts, as usual, not with the *angelus* but with lunch (no pasta, but with a big bowl of fruit to take to his room in prospect of an evening and morning without supplies). Then the combinatorial seminars, followed by a carefully scheduled series of tête-à-têtes with individual students and visitors. At 17h30 sharp, we climb into *la Macchina* Senato, for the short drive uphill from the Palazzone to the town. There, always at that table just outside the door of his favorite café, Gian-Carlo orders his evening meal: three scoops of *gelato al cioccolato* covered with a rich chocolate sauce, topped off with those tiny but ubiquitous Japanese parasols, which are promptly distributed as offerings. Discussion begins immediately on the umbral calculus, and lasts until after dark. A few pleasantries off the subject, to relax in the evening calm, and Mimmo and Elvira head for their lodging. Gian-Carlo and I return on foot (downhill) to the Palazzone. The day is complete. These were such happy hours for us all.

I've decided, under the title "Ten abandoned gold mines", to pinpoint areas where Gian-Carlo has sketched routes that he himself had not the possibility to follow. There's a romance about abandoned mines, a romance which I will try to nourish.

Where should we look for such mines? Gian-Carlo's early studies in combinatorial theory are now shelved with the classics. I went through much of this work in preparing my introduction to Volume I of his collected works, where I point out a few matters I feel should be pursued. Since I don't want to repeat myself, my references today to the earlier and better-known combinatorial subjects will be slight. No offense intended.

The 30-year program Gian-Carlo undertook to modernize invariant theory was already reaching fulfillment when he died: a major collaborative work with Joel Stein on resultants was ready for the *Advances* (and will be the subject of Joel's intervention here in Maratea). The ultimate realization of Gian-Carlo's dreams for the umbral operator in invariant theory remains in competent hands: Andrea Brini, Francesco Regionati, Antonio Teolis. I have to scold them for not talking at this meeting, but can recommend their beautiful paper [3]: in this week's *Advances*, as a first installment in their fascinating tale of umbral operators and duality in geometry.

Approaching a turning point in (and, in retrospect, the conclusion of) his life's work, Gian-Carlo seemed to adopt a new strategy, devoting an increasingly large proportion of his time and energy to the preparation of addresses to large audiences: witness the memorial lecture for Garrett Birkhoff, the MIT Killian lectureship, the Fubini lectures in Torino, the upcoming birthday lecture for Dan Kleitman and several major addresses to the AMS, including the prestigious Colloquium Lectures last year in Baltimore. These lectures often bore titles beginning with a digit: "10 things I wish I had learned ...", "12 problems no one wants to bring up ...", or, sadly, "10 mathematical problems I will never solve". Since it will be he who writes the majority of my talk today, I let him continue in this tradition. I have been able to draw much of today's list of 10 gold mines from these recent lectures, or even from the remarkable (and prize-winning) *Indiscrete Thoughts*, a book in which he took an evident pride. These texts have the particularity that they concentrate the expression of Gian-Carlo's desires for future developments in mathematics. I will let Gian-Carlo speak for himself, in direct quotation (displayed in narrower paragraphs, with specific reference to source). I hope in this way to be able to lead you to texts of which you may not yet be aware, and perhaps also to provide occasions for you to recall with pleasure and sadness the tenor of his words[1]. I'm sure Gian-Carlo would like to apologize for not being present to discuss these questions with you in person.

2 Gold mine 1. From probabilistic proofs to algorithms

The first two gold mines are from the area of probability theory. Gian-Carlo was always on the lookout for novel *logics*, methods of reasoning that could find service in areas where traditional methods failed. He felt, in particular, that probabilistic arguments deserved to find a place as part of formal logic, even in the algorithmic constructions so essential to computer science.

> The idea of proving the existence of a mathematical object by proving that the probability of its existence is positive is one of the most fertile ideas to come out of the great Erdös. There is a large number of objects whose existence can only be proven probabilistically, and for which no construction is known.
>
> Professional logicians should become interested in this method of proving existence theorems. There ought to be a correspondence principle allowing us to translate probabilistic algorithms into combinatorial algorithms or, lacking such a translation, a new logic associated with probabilistic reasoning. Unfortunately, those people who are competent to carry out such a program have better fish to fry, and philosophers of mathematics are too

---

[1] As the personnage Alexandre, played by Jean-Pierre Léaud, says in Jean Eustache's film *La Maman et la Putain* (1973): "Parler avec les mots des autres, voilà ce que je voudrais. Ça doit être ça, la liberté".

> incompetent to deal with a problem which ought to be their bailiwick. [19, Chap. XX. A mathematician's gossip, p. 225]

Recall this was an approach favored also by W. T. Tutte, who wrote a series of deep papers on enumeration (or "census") of planar maps, doubtless with a view to disproving the four-color conjecture by showing that the probability of choosing a non-four-colorable planar map was not equal to zero.

3 Gold mine 2. New concept of sample space

The second gold mine I propose from probability theory is Gian-Carlo's program for reshaping the fundamental concept of sample space. He approached this subject from several directions, recognizing in the algebra and combinatorics of multisets an appropriate departure from the familiar landscape of set theory, and recognizing in the notion of probability amplitude in quantum mechanics the need for a notion of sample space with more internal structure. It is in rewriting the foundations of quantum mechanics that new concepts of sample spaces would find their test of adequacy.

On several occasions, and as early as the 1960's, Gian-Carlo discussed with me the connections between elementary combinatorics, probability theory and various statistics of physical theories. Hints in this direction are in the courses he taught in both combinatorics and probability. I always encouraged him to write this all out in detail; his students tell me he did, and that I will find it in his recent course material.

Theme 2.1. Bottom lines

> To combinatorialists, the bottom lines are the Yang–Baxter equation, the representation theory of the classical groups, and the Schensted algorithm. To some algebraists and combinatorialists of the next ten or so years, the bottom line may be elimination theory.
> I will shamelessly tell you what my bottom line is. It is placing balls into boxes, or, as Florence Nightingale David put it with exquisite tact in her book *Combinatorial Chance*, it is the theory of distribution and occupancy. [19, Chap. III. Combinatorics, representation theory and invariant theory, pp. 51–52]

This is a first hint as to how a reformulation of sample space might take form. It would be based on a notion of distribution and occupancy with a different combinatorics; say, placing distinguishable or indistinguishable balls into distinguishable or indistinguishable boxes, with or without additional internal structure. Second, the familiar calculation of expectation should be replaced by the more general linear algebraic calculation of trace.

Theme 2.2. Sample spaces

> I will lay my cards on the table: a revision of the notion of sample space is my ultimate concern.

> ...a random variable is better described by its cumulants than by its moments or its characteristic function.
> What remains to be defined is the quantum analog of the multivariate normal distribution. Forget the old platitude that joint distributions do not exist in the quantum case. We present here an as yet unpublished proposal ... [a] multivariate distribution [we call] the Clifford distribution. [22]

Is quantum probability combinatorial? The search for an appropriate combinatorial theory is underway, concurrent with a fully developed sentiment that quantum phenomena are ultimately random, that to search for "hidden structure" is a counterproductive endeavor. Witness the illuminating one-sided debate in the recent book *The Feynmann Processor*[2]. I can't help feeling there is a middle ground here. It would be natural if the new notion of sample space pointed to by Gian-Carlo were to involve ideas from linear algebra, as opposed to manipulations of finite sets. Signs of this are already visible in Peter Shor's work on quantum computing.

Theme 2.3. Probability amplitude

> The Feynman path integral is the mathematicians' *pons asinorum*. Attempts to put it on a sound footing have generated more mathematics than any subject in physics since the hydrogen atom. To no avail. The mystery remains, and it will stay with us for a long time.
> The Feynman integral, one of the most useful ideas in physics, stands as a challenge to mathematicians. While formally similar to Brownian motion, and while admitting some of the same manipulations as the ones that were made rigorous long ago for Brownian motion, it has withstood all attempts at rigor. Behind the Feynman integral there lurks an even more enticing (and even less rigorous) concept: that of an amplitude which is meant to be the quantum-mechanical analog of probability (one gets probabilities by taking the absolute values of amplitudes and squaring them: hence the slogan "quantum mechanics is the imaginary square root of probability theory"). A concept similar to that of a sample space should be brought into existence for amplitudes and quantum mechanics should be developed starting from this concept. [19, Chap. XX. A mathematician's gossip, p. 229]

The final three quotes I extracted on this subject deal with a substitute for "sets" in the definition of sample space, a substitute needed to build a probability theory more adequate to the needs of quantum theory. The basic notion is that of "multiset", a concept Gian-Carlo explains is dual to that of set, via the notion of "support set" of a vector relative to a basis.

Theme 2.4. Multisets

> An instance of a semantics without syntax is the theory of multisets. A multiset of a set S is a generalization of a subset of S, where elements are allowed

[2] Milburn, G. J. (1998): The Feynman Processor. Perseus Books, Reading, MA.

to occur with multiplicities. Multisets can be added and multiplied; however, a characterization by algebraic operations of the family of multisets of a set S – an analog of what Boolean algebra is for sets – is not known at present. The problem is of more than academic interest. There is a deep duality between the algebra of sets and the algebra of multisets that a syntactical description may well elucidate. [19, Chap. XII. Syntax, semantics, and the problem of identity of mathematical items, pp. 154–155]

Theme 2.5. Fourier–Gelfand transform

A multiset can be described either by listing its elements in any order, each element repeated as often as its multiplicity. Or else, by first giving a set – the support of the multiset – together with a multiplicity assigned to each element of the support. Our Fourier–Gelfand transformation is the recognition that this double notation is an instance, perhaps the simplest instance, of harmonic analysis ("spec" analysis) on a commutative ring. We believe that the duality between support and multiplicity, i.e., between set and multiset (elsewhere thought of as the duality between the commutative and the anti-commutative), far from being just a notational quirk, is instead by its very simplicity, the primordial case which may throw light on the foundations of harmonic analysis. [14]

Theme 2.6. Placings and reluctant functions

To get a combinatorial interpretation for the complete homogeneous functions, one needs to look at *placings*, that is, distributions of balls into boxes in which the balls in each box are "placed in some configuration". Placings are closely related to reluctant functions and are reminiscent of MacMahon's distributions into groups. ... The notion of placings or a reluctant function can be naturally described using category theory and led to the notion of species (Bonnetti, Rota, Senato, and Venezia, 1985, 1990.)
Missing from the roster of symmetric functions in Doubilet (Foundations VII) are the Schur functions. Although there are combinatorial interpretations of the Schur functions, they do not seem to fit into the theory of distribution and occupancy. The most intriguing and mysterious problem area suggested by Doubilet is to find such an interpretation and to use it to obtain a purely combinatorial approach to representations of the symmetric group. (See Doubilet, Rota, Fox, 1980) [13]

I should also add here a reference to one of my favorites [15] among Gian-Carlo's early papers, where reluctant functions were introduced and described concretely.

4 Gold mine 3. Entropy

Gian-Carlo made a few attempts at seeing how the appealing notion of entropy could find a more natural role in mathematics. I select two examples: his suggestion that entropy provides a natural proof of the central limit theorem, and his idea that σ-subalgebras could be made to form a universal entropy group.

Theme 3.1. Maximum entropy principle

> Shannon entropy is maximized for atomic Boolean σ-algebras where all atoms have the same probability. There is a continuous analog of this fact, first stated by Kampé de Fériet. Among all random variables with finite variance, those having the normal distribution have maximum Boltzmann entropy. Kampé de Fériet's discovery led to the development of an information theory for Gaussian processes, which is now the bread and butter of signal processing.
> ... the complete and utter lack of justification of the maximum entropy principle is nothing short of a scandal. On learning that a normally distributed random variable of finite variance has maximum entropy, it is natural to ask for an intuitive proof of the central limit theorem that relies on maximum entropy; such a proof has never been given, ... although several mathematicians attempted it, among them Linnik and Rényi. [22]

Theme 3.2. Entropy as a lattice concept

> I should like to submit to you a proposal for a definition of entropy on the lattice of σ-subalgebras. We follow an idea that gained favor in the latter half of this century and that was brilliantly exploited by Grothendieck. Accordingly, let us see how a "universal entropy group" might be defined. We begin by taking the free Abelian group generated by all those σ-subalgebras; call this free Abelian group F. We then impose on this group the identity $\pi \wedge \pi' = \pi + \pi'$ whenever π and π' are independent σ-subalgebras.
> The quotient group thus obtained may be denoted by F'. The quotient group F' is the universal object that any entropy function must factor through, as is evident from the construction.
> An abyss of ill-understood ideas separates Shannon entropy from Boltzmann entropy. No one to this day has seen through the dilemma of choosing between entropy as a measure of randomness of probability densities of random variables and entropy as a measure of randomness of σ-subalgebras. [22]

5 Gold mine 4. Intrinsic volumes

The notion of a valuation on a distributive lattice is a recurrent theme in Gian-Carlo's work. The simplest valuation is the binary-valued function ("truth value") τ_x defined on the Boolean algebra of subsets of a set S, and, for any element $x \in S$, by

$$\tau_x(A) = \begin{cases} 1 & \text{if } x \in A \\ 0 & \text{otherwise.} \end{cases}$$

He wrote an early paper on the subject, which is well-known, but the full sweep of his ambitions for the subject, from Minkowski's mixed volumes to applications to quantifiers in formal logic, becomes visible only in the little-known volume of conference proceedings[3]. [16]

I think it was only after he read Steve Schanuel's charming paper [23] on geometric measures, and started writing his book [12] on the subject with Dan Klain that Gian-Carlo realized what a good subject this made for public lectures. An *intrinsic volume* has four basic properties. It is a real-valued valuation defined on a specified class of subsets of space. It has convenient values on rectangular subsets and on the empty set, and is a Euclidean invariant:

$$\begin{array}{rl} \text{initial:} & \mu(\emptyset) = 0 \\ \text{valuation:} & \mu(A) + \mu(B) \;=\; \mu(A \cup B) + \mu(A \cap B) \\ \text{invariant:} & \text{under translation and rotation of space.} \\ \text{normalization:} & \mu(A \times B \times \cdots \times C) \;=\; e_k(a, b, \cdots c), \end{array}$$

where e_k is Euclidean volume.

Theme 4.1. Generalize the isoperimetric inequalities

> The $n+1$ intrinsic volumes $\mu_0, \mu_1, \cdots, \mu_n$ are a basis for the space of all continuous invariant measures defined on all finite unions of compact convex sets. ... The intrinsic volumes are independent of each other, except for certain inequalities they satisfy. These inequalities generalize the classical isoperimetric inequality that relates volume to area. At present, we know very little about the intrinsic volumes. We do not even know the formula for the intrinsic volumes of an n-simplex. [20]

6 Gold mine 5. Symmetry classes of tensors

The next four gold mines are located in the domain Gian-Carlo selected for his life's work: invariant theory.

Hermann Weyl's writings in invariant theory held forth two principles Gian-Carlo found to be compelling: (a) all geometric facts are expressed by the vanishing of invariants, and (b) all invariants are invariants of tensors. These principles lead naturally to a program for geometry: (1) to decompose tensor algebra into irreducible components, and (2) to devise efficient notation for invariants of each irreducible component.

[3] This text is hard to come by, but I'd be pleased to send you a copy.

Irreducible components are symmetry classes, a subject Gian-Carlo approached in a number of papers with Nick Metropolis. There is a nice review of this material for a general audience in the Colloquium Lectures, as cited below. There are only two symmetry classes of functions in two variables, the symmetric and skew symmetric. For three variables, there is one new class, which he terms "cyclic". For four variables, there are five classes; for five variables, seven classes. The set of symmetry classes of functions with n variables is in bijection with the set of standard tableaux with n boxes, filled with $1, \cdots, n$.

> To this day, only two symmetry classes of tensors have been studied in any detail. ... Tensors belonging to symmetry classes other than the classes of symmetric and skew symmetric tensors also occur in geometry and physics. However, these symmetry classes have been studied very little, and they are a long way from being understood. [21]

7 Gold mine 6. Invariant theory and superalgebra

We continue a discussion of Gian-Carlo's work in invariant theory with some indications of what he thought were his main contributions: the clarification of the classical symbolic method, and its superalgebraic expression, with commutation rules derived from a systematic use of signed and unsigned alphabets.

Theme 6.1. The symbolic method

> In mathematics, it is extremely difficult to tell the truth. The formal exposition of a mathematical theory does not tell the whole truth. The truth of a mathematical theory is more likely to be grasped while we listen to a casual remark made by some expert that gives away some hidden motivation, when we finally pin down the typical examples, or when we discover what the real problems are that are stored behind the showcase problems.
> The hidden purpose of the symbolic method in invariant theory was not simply that of finding easy expressions for invariants. A deeper faith was guiding this method. It was the expectation that the expression of invariants by the symbolic method would eventually guide us to single out the "relevant" or "important" invariants among an infinite variety. It was the hope that the significance of the vanishing of an invariant could be gleaned from its umbral expression. The vanishing of this faith was the main reason for the demise of classical invariant theory, and the revival of this faith is the reason for its present rebirth. [21]

Theme 6.2. Kinds of coordinates

> The realization that the classical system of Cartesian coordinates can and should be enriched by the simultaneous use of two kinds of coordinates, some of which commute and some of which anticommute, has been slow

> in coming; its roots, like the roots of other overhaulings of our prejudices about space, go back to physics, to the mysterious duality that is found in particle physics between identical particles that obey or do not obey the Pauli exclusion principle.
> Both anticommuting variables (exterior algebra) and commuting variables (ordinary cartesian coordinates) have long been used in geometry. However, the interactive role of the two kinds of variables has been far from clear; glimmerings of an understanding of the need for both kinds of variables can be gleaned from homological algebra, from the de Rham and Koszul complexes. [4]

Gian-Carlo made a conscious decision 25 years ago to devote his best efforts to reviving and rewriting invariant theory. As I try to sort out the various shafts leading to this abandoned mine, I am led invariably to work ultimately written up by his graduate students and close colleagues, either in collaboration with Gian-Carlo, or independently. Gian-Carlo's early work with Joel Stein, Peter Doubilet, Joseph Kung, Andrea Brini, Marilena Barnabei et al. on invariants of symmetric and of antisymmetric tensors was first unified in a superalgebraic setting at the AMS Regional Conference in West Chester, 1985. [6]

They made several attempts to polish and extend this document; the results are best seen in the Curaçao conference proceedings. [4]

There, they use covariants to express the Grassmann conditions for the decomposibility of a skew symmetric tensor. The clean, perhaps definitive, version is now in the introductory sections of Wendy Chan's excellent presentation [5] although for the proofs we must still refer back to the earlier papers.

Theme 6.3. Duality

The full power of the superalgebraic approach to invariant theory appears in duality theorems summarizing the effect of repeated polarization operators, and expressing the dual exterior product (the *meet* operator that launched the article by Doubilet, Rota, Stein) in an appropriate superalgebraic context. This is now the work of Andrea Brini and his colleagues; their most recent article is cited above, in the introduction. This is important work, central to the fruition of the research program so actively promoted by Gian-Carlo. See also [2].

8 Gold mine 7. Synthetic geometry

After the years Gian-Carlo spent developing and promoting the superalgebraic approach to invariant theory, he became capable of saying some unkind things about a field that had earlier on held great attraction for him: synthetic geometry.

> Synthetic projective geometry in the plane held great sway between 1850 and 1940. It is an instance of a theory whose beauty was largely in the eyes of its beholders. Numerous expositions were written of this theory by English and Italian mathematicians (the definitive one being the one given by

> the American mathematicians Veblen and Young). These expositions vied with one another in elegance of presentation and in cleverness of proof; the subject became required by universities in several countries. In retrospect, one wonders what all the fuss was about. Nowadays, synthetic geometry is largely cultivated by historians, and the average mathematician ignores the main results of this once flourishing branch of mathematics. The claim that has been raised by defenders of synthetic geometry, that synthetic proofs are more beautiful than analytic proofs, is demonstrably false. Even in the nineteenth century, invariant-theoretic techniques were available that could have provided elegant, coordinate-free analytic proofs of geometric facts without resorting to the gymnastics of synthetic reasoning and without having to stoop to using coordinates. [19, Chap. X. The phenomenology of mathematical beauty, p. 125]

But this quote does not tell the whole story. Rosa Huang and Michael Hawrylycz wrote their doctoral theses with Gian-Carlo on themes that tended to bring the superalgebra invariant theory back into significant contact with the geometry it was designed to study, a geometry whose intuitive access is through a synthetic approach involving flats (subspaces) and operators (join and meet, even complementation).

Theme 7.1. Four subspaces

Rosa Huang set out to classify configurations of projective subspaces, the first difficult case being that of four subspaces. She settled the problem for configurations of four, then of n, lines [8, 9], then went on to study the invariants of configurations of four subspaces of arbitrary rank. Her thesis [7] contains most of her work to classify configurations of four projective planes in projective 5-space (subspaces of rank 3 in a space of rank 6). Her untimely death brought this project to a halt: her colleague Roger Howe assured the publication of their joint work [11] for four subspaces in projective space.

The essential mine-shaft, to my mind, would be reopened if we were to complete Rosa's proposed concordance between the catalogue of invariant properties of subspace configurations with the Gelfand–Ponomarev construction of configurations irreducible with respect to direct sum. In the latter one finds eight denumerably infinite families of irreducible configurations, one of these (the most interesting and time-consuming) having an additional continuous variation with respect to a choice of cross-ratio.

Theme 7.2. Analytic trivialities

Of course the aim of capturing synthetic geometry in a superalgebra is not to make a zoo of special configurations.

> The ideal of *all* science, not only of mathematics, is to do away with any kind of synthetic *a posteriori* statement and to leave only analytic trivialities in its wake. [19, Chap. IX. The phenomenology of mathematical truth, p, 119]

The superalgebra is an organized, hierarchical language for the description of the truths of geometry. Now we should get busy developing those truths and learning to read those descriptions!

9 Gold mine 8. The search for a universal geometry

Significant advances in geometrical theory are founded on simplification and universality of methods and calculation. It was Felix Klein's realization that projective geometry is universal geometry that permitted the formulation of the Erlangen program relative to a hierarchy of groups of possible symmetry. Gian-Carlo's *leit-motif* was his concern that the methods and proofs of geometry and its theory of invariants be characteristic free, that is, independent of choice of field of coordinates. For geometry to have methods of calculation that are characteristic free, one must build an appropriate algebra from the material of the configuration itself, as in von Staudt's algebra of throws or Emil Artin's *Geometric Algebra.* Gian-Carlo pinned his hopes on some extension of Neil White's work on the bracket ring. He came back to this question often in his recent writings. I hope he would also have enjoyed my recent work with Bill Schmitt (J. Combin. Theory Ser. A, to appear), developing an analogue of the tensor algebra of an exterior algebra for any matroid. Gian-Carlo chose the name *Whitney algebra* for this concept; there is at least a chance that it will turn out to be the closest thing to a universal coordinatizing algebra for combinatorial geometric configurations.

Theme 8.1. A ring for universal geometry

> In the latter part of this century, algebra has so overwhelmed geometry that geometry has come to be viewed as a mere "façon de parler". Sooner or later, geometry in the synthetic vein will reassert its rights, and the lattice theory of ideals will be its venue. We intuitively feel that there is a geometry, projective, algebraic, or whatever, whose statements hold independently of the choice of a base field. Desargues's theorem is the simplest theorem of such "universal" geometry. A new class of commutative rings remains to be discovered that will be completely determined by their lattice of ideals. Von Neumann found a class of noncommutative rings that are determined by their lattices of ideals, ... but the problem for commutative rings seems much more difficult. A first step in this direction was taken by Hochster. Algebraic geometry done with such rings might be a candidate for "universal geometry". [18]

Theme 8.2. Universal vector space

> Our first step is to establish the existence of a "universal vector space" over a "universal field", in which all facts of projective geometry that are true irrespective of the choice of field will hold, and conversely. Such a universal vector space, which we call the White module, must be constructed over a

> field that is in some way "intrinsic" to projective geometry. It is absurd to do geometry over a field of scalars that is prescribed in advance. The theorems of geometry are ontologically prior to the choice of field, and if there is a field over which we will decide to coordinatize, such a field should be dictated directly by those facts of projective geometry that are true irrespective of the field of scalars. [4]

Theme 8.3. The White module

> The White module is the module through which all such maps factor; speaking somewhat loosely, the White module is the universal module for all representations of the bracket ring into integral domains. Thus, it suffices to consider models taking values in the White module, in order to have the completeness theorem. Thus, we expect that any fact of projective geometry that is true over every field should be expressible in the White module. That this expectation is fulfilled is the ultimate distillate of the fundamental theorems of classical invariant theory.
> This algebra [i.e., the *umbral algebra* for skew-symmetric tensors] now plays the role of the exterior algebra of the White module, and thus is a "universal model". [4]

It is high time we establish once and for all the universal properties of the superalgebras relative to choice of configuration and choice of coordinatizing field (separating these two issues as far as possible). That is, not only should any calculation in the superalgebra be reflected faithfully in appropriate synthetic (or ultimately coordinate) calculations in any model, but also some sort of converse should hold: if some calculation is valid in a sufficiently general class of models, the supporting calculation in the superalgebra should be valid. The tricky part is always the change of field!

The concept of *models* is treated in [17].

Theme 8.4. Bits and pieces

> There remain a number of offshoots of Gian-Carlo's work in invariant theory that could now be followed up. A few suggestions:
>
> (1) Provide synthetic geometric expressions (in $\vee$ and $\wedge$) for invariant properties. Find closer ties between synthetic geometry and the (super) Hopf algebra.
> (2) Parallel the Klein program in the superalgebraic setting, taking projective geometry as universal, and building other geometries by adjunction of fixed reference elements, such as the imaginary "cyclic points" on the imaginary line at infinity, that permit the introduction of the concept of "circle" in a projective geometry.
> (3) Show how special geometric contexts (geometric specializations) affect the interplay of invariant properties.

(4) Investigate higher order syzygies (and their synthetic conditions for their occurrence in special position) in the umbral module. There is a crucial role here for the "geometry of circuits", analogue of the dual matroid, and for the meet operation in the tensor algebra. To date, the only general theoretical approach to this question is via the Anick–Rota resolution. Here, at least, I plan to take my own advice, and get back to work on it. Gian-Carlo agreed:

> We conjecture that the translation of an excluded minor characterization into a property of the bracket ring can be carried out by use of higher-order syzygies ... Recent work of Dress and his school supports this conjecture. [1]

10 Gold mine 9. Latin squares and the basis conjecture

With what will certainly be an unexpected *virage* toward concreteness, let me point to a gold mine in one of those rare situations where Gian-Carlo formulated and studied what could be called a "problem". In general he didn't like problems, he liked theories. I suppose he chose to state the basis conjecture as a "problem" in order to attract another clientèle, the question having already resisted his best efforts.

Conjecture. Let V be a vector space over an arbitrary infinite field. Suppose that

$$\{a_1, a_2, \cdots, a_n\} \quad \{b_1, b_2, \cdots, b_n\} \quad \cdots \quad \{c_1, c_2, \cdots, c_n\}$$

are n sets, bases for V. Then one can relabel the elements within each basis such that

$$\{a_1, b_1, \cdots, c_1\} \quad \{a_2, b_2, \cdots, c_2\} \quad \cdots \quad \{a_n, b_n, \cdots, c_n\}$$

are also n bases for V.

Stronger conjecture. If $a, b, \cdots, c$ are n positive letters, then the element $[ab \cdots c]^n$ is non-zero in the bracket superalgebra Bracket(L^+).

The stronger conjecture is equivalent to conjectures on Latin squares by Alon and Tarsi, 1986, and by Huang and White, 1991. These in turn imply the basis conjecture of Rota, and an earlier conjecture by Dinitz, 1978, namely: if S_{ij} is an n-element set, for i and j from 1 to n, then there is a partial Latin square whose ij entry is chosen from the set S_{ij}, for all i, j. See [10]

11 Gold mine 10. The phenomenology of perception

Gian-Carlo was optimistic about the future impact of modern science on philosophy. Indeed, he was to be one who would carry that influence into the philosophers' camp, even to meetings of their inner "circles". But this does not mean he was enthusiastic about earlier attempts to claim that philosophy could be reduced to some sort of mathematics.

He used to tell me of his interest in this interaction, and how he felt the philosophers were not yet taking advantage of openings afforded them by mathematics and the physical sciences. It seems they were unable to interpret what is essential to scientific activity. I told Gian-Carlo that there was a philosopher at the University of Waterloo who wrote on the work of Charles Sanders Peirce, and that I could arrange for him to speak to their departmental colloquium on his next visit to Canada. Fortunately, he accepted the challenge. I think it was the first public lecture he gave on philosophy.

Some years later, Gian-Carlo took the bull by the horns and insisted that MIT appoint him to a joint professorship in Mathematics and Philosophy, to the distress of some, to the joy of others. It was not an easy move, but it made all the difference. The reading and teaching of philosophy came in recent years to occupy roughly half of Gian-Carlo's energies. And the influence on his students in philosophy was enormous.

Theme 10.1. Phenomenology: perception

> A phenomenological description of the act of perception will begin by observing that perception depends essentially on a background of past experience, and is oriented towards some future goal. This obvious remark is supplemented by an excruciating analysis of the structure of this double dependency, a task that no philosopher or scientist had previously attempted. Husserl arrives at a rigorous theoretical characterization of the multiple dependencies of every act of perception.
>
> We may apply Husserl's ideas to the raging discussion of reductionism. Some philosophers assert that perception can be entirely accounted for by brain functions; others clamor for an essential distinction between brain functions and perception. Which of the two schools is right?
>
> Phenomenologists observe that whether one holds to one of these views or the other, the phenomenological description has to be taken into account in any dealings with perceptual phenomena, no matter where one believes these phenomena to be located. One will have to admit the dependence of perception on past background and future projection, irrespective of the anatomical basis of perception.
>
> Rota, G.-C. (1998): Ten remarks on Husserl and phenomenology. Provost's Seminar. MIT

Theme 10.2. Phenomenology: computer – brain

> The once abstruse conclusions of Husserlian phenomenology, the exasperating counterexamples in Wittgenstein's *Philosophical Investigations* will be vindicated by the demands of intellectual honesty of the computer scientist who has to make programs that work, and by the brain scientist's search for a theory that will adequately fit his experimental data. The theme of our

time is the improbable but inevitable marriage of philosophy with the new sciences of the mind. [19, Chap VIII. Philosophy and computer science, pp. 106–107]

Rota, G.-C. (1997): Philosophy and computer science. In: Indiscrete Thoughts. Birkhäuser Boston, Boston, MA, Chap. VIII, pp. 106–107

Theme 10.3. Axiomatic method

Every fact of mathematics must be ensconsed in an axiomatic theory and formally proved if it is to be accepted as true. Axiomatic exposition is indispensible in mathematics because the facts of mathematics, unlike the facts of physics, are not amenable to experimental verification.
The axiomatic method of mathematics is one of the great achievements of our culture. However, it is only a method. Whereas the facts of mathematics once discovered will never change, the method by which these facts are verified has changed many times in the past, and it would be foolhardy to expect that changes will not occur again at some future date. [19, Chap. VII. The pernicious influence of mathematics upon philosophy, p. 90]

Is the Appel–Haken proof of the four-color theorem a new method of verification? Yes and no. Many classical proofs (among our least favorites) proceed by exhaustion of a long list of conceivable exceptions. Are the topos-theoretical proofs on localization in ring theory that proceed by intuitionist arguments on global rings a new method of verification? Yes. There you have a technique for proving a difficult theorem by proving that an analogous statement is true about a simpler structure, but proving it with one hand tied behind your back (using a weaker logic, say, without double negation). There's a very readable page of introduction to this point of view on the web: http://math.ucr.edu/home/baez/week68.html

I think the following quotation reveals the reason Gian-Carlo took phenomenology seriously: he felt it to be the only philosophical current that can hope to account for the reality of his mathematical experience.

Theme 10.4. Evidence

Phenomenological evidence is a *logical concept*, on a par with definition or inference. The de-psychologization of evidence is a great achievement of phenomenology.
Primordial evidence, $\alpha\lambda\eta\theta\epsilon\iota\alpha$ to the Greeks, is constitutive of mathematical reasoning. The experience that most mathematicians will corroborate, that a statement "must" be true, is not psychological. In point of fact, it is not an experience. It is the condition of possibility of our experiencing the truth of mathematics.
An experience that is shared by most mathematicians betokens the standing of evidence. After reaching evidence we go back to our train of thought and try to pin down the moment when evidence was arrived at. We pretend to

locate the moment we arrived at evidence in our conscious life. But such a search for a temporal location is wishful thinking. [19, Chap. XIV, Kant and Husserl, pp. 168–169]

Theme 10.5. Phenomenology: rigor

Mathematics is a rigorous theoretical science. The rigor of mathematics enables mathematicians to carry out their work without any need for experiment. The results of mathematics, although expressed in terms that the man in the street might label "imaginary", find striking confirmation and lucrative applications in everyday reality, whether in engineering, in business or in medicine.
Among all theoretical disciplines, mathematics is the only one that is successful. No other theoretical undertaking has enjoyed the success of mathematics, with the possible exception of voodoo.

Rota, G.-C. (1998): Ten remarks on Husserl and phenomenology. Provost's Seminar, MIT

12 Conclusion. In memoria

I close with the words of some who attended the memorial meetings for Gian-Carlo last spring in Cambridge[4]. I'll read four of these statements, with the hope that, in so doing, I will forge some link between the sentiments of gratitude and loss expressed this year on the two sides of the Atlantic. These memorial statements are by Sara Billey, Gian-Carlo's colleague at MIT, by Joe Oliveira, a former Ph.D. student, by an undergraduate student in his course on probability, whose name, unfortunately, I do not know. I close with my own statement on that occasion.

Teaching

Well, I'm taking over Professor Rota's class in probability now. Last Wednesday we postponed the quiz that was supposed to take place and instead had a discussion on what the class and Professor Rota meant to us. I guess word got out and three generations of students showed up, almost two hundred people. In that discussion the students finally told me what I was missing [i.e., in my teaching]: that he spoke to each of them directly, that he connected with them on day one, that he spoke their language quite literally at times: Spanish, French, Italian, whatever; that he had an exquisite taste in mathematics, that he answered every question with respect, and mostly that he believed in them. He believed in us. He believed that we could all learn deeply as he had, he believed we were all smart enough, and that unwavering confidence in our abilities made us want to succeed. After our discussion a student gave me the following quote, "Find a teacher, find a friend."

[4] The Department of Mathematics at MIT has produced audio and video tapes of those memorials. These can be ordered from: dept@math.mit.edu.

I've lost both my mentor and my friend. I am very glad to have known Gian-Carlo Rota.
G.-C., you've done great things with your life and you've helped me so much with mine. You've helped me get this far and I thank you for that. May you rest in peace.

Sara Billey: MIT Memorial Assembly. Cambridge, 30 April 1999

Thesis supervision

[*Joseph Oliveira, Bill Chen, Mike Hawrylycz at Los Alamos*]
Our joint realization was that Rota had in different ways within each of us cut to the core of our being in the world and like a skilled surgeon with his verbal scalpel began to probe and peel back our layers of self-deception and self-doubt. He used his cutting sarcasm not truly to injure us, but to surgically remove our self-imposed fears and the pockets of our minds' psychological infections. With the surgery complete Gianco would then masterfully close the wounds that he had made in our beings with words of support and encouragement. The three of us came to realize that each time Gianco performed surgery on our minds, in our beings, he packed those surgical wounds with a piece of his own being, with a part of his spirit, with a part of his mathematical knowledge, with his self-confidence and courage. Rota showered his healing gifts on each of his students.

Joseph Oliveira: MIT Memorial Assembly. Cambridge, 30 April 1999

Dreams

Well, my last memory of Professor Rota was this morning, when I woke from a dream which he was in. It was an odd dream because he was lecturing about the normal distribution, and I'm used to having normal distribution dreams, so that wasn't the odd part. But I was looking at him and just saying, "Are you sure you should be here teaching 18.313?" Well, OK, it seemed fine he was teaching. And he said some very deep things about the normal distribution, but of course my memory has not served. When I woke up I didn't remember what they were. But it was very comfortable, it was just like old times. Also, we had a class on Wednesday when Professor Billey lectured, and we had that Memorial the Wednesday afterwards, and I had a dream that night, and he was lecturing on the Poisson distribution again, in keeping with the dreams, and I thought, "But he's not here, he's not alive." But I said, "All right, he's lecturing because he's missed Wednesday's lecture, that's it, so he has to do one more make-up lecture because he missed that." He promised he would tell us:

"Next Friday I'll tell you what the blips are."

A student in the probability course 18.313: MIT Student Memorial Vigil. Kresge Oval, Cambridge, 15 May 1999

Symbols

It is good that we can be together to learn from each others' testimony on this tearful spring day of Gian-Carlo Rota's 66th year. My love for Gian-Carlo runs deep. In mourning his departure I try to understand the depth of that love. Just over forty years have passed since that day Kenneth Hoffman tucked me under Gian-Carlo's wing. I know I speak for us all when I say Gian-Carlo has been an unfailing and forgiving friend.

We were the same age (and perhaps *of* the same age – pre-war models), but he was like a grandfather to me, possessed of a gnarly old wisdom. At the same time he was that ten year old you can read in (his sister) Esther Gasperoni's novel *Orage sur le Lac*, the touching story based on the wartime years when their antifascist father found refuge for his family in the mountain village of Rimella.

I'm happy Gian-Carlo had many honors in his time, and in particular that MIT understood and thanked him with the Killian Award and the Norbert Wiener Professorship. I'm relieved also that Franco Gasperoni is undertaking to see a number of uncompleted papers through to publication.

I would have wished a peaceful retirement for Gian-Carlo, writing and lecturing, taking long walks (or making even longer telephone calls) to talk with his friends, or sitting in the Galleria in Milano to reread Proust and Heidegger. Time to stretch and weave his bright colored threads of thought into a rich fabric, time to take delight in the careers of his countless offspring. If perchance, the universe be built that way, I wish him now that peace, that time. But I fear that it is we who must gather those threads, and sing him the song he would have liked to hear. I suppose he felt his time was short. Like Feynman, he worked hard to his last day, and, we pray, went peacefully.

Gian-Carlo interacted with and enriched the lives of four generations, from Stan Ulam to William Feller, Marshall Hall, Richard Rado, Jack Schwartz and Garrett Birkhoff, to his most recent graduate students and to those who wanted to study with him.

I search for the right symbol for this moment. I see the dead fawn we found at the entrance to the narrows of the Green River on the day I heard of Gian-Carlo's death. Symbol beastly and eternal – a delicate arrangement of bones, with a dreadful secret. Then I think of the pilot navigating landfalls on the high seas, identifying those signs, be they averaging operators, Möbius functions, topoi, Hopf algebras, umbra or linear lattices that, through his guidance, have come to light our way. But my thought comes to rest on a figure wrought by Michelangelo Buonarotti – of one who parted the waters (the waves of chaos of our times, the sterile waters of old constituencies) to lead his people through the desert to a land where that new mathematics he dubbed *combinatorial theory* could flourish.

So, no more recipes for risotto, no more hints for doing neckties, no more highs, no more lows.
We're sorry you had to leave us so soon, Gian-Carlo.
We miss you.
In the words of a Cherokee poet[5]:

Not in the time of pleasure
Hope doth set her bow;
But in the sky of sorrow,
Over the vale of woe.

Through gloom and shadow look we
On beyond the years:
The soul would have no rainbow
Had the eyes no tears.

Henry Crapo: MIT Memorial Assembly. Cambridge, 30 April 1999

Now, as I prepare this text for publication, nearly a year has passed since April 1999. I find I still can't peel a garlic without remembering those evenings when we would take a break for supper, and Gian-Carlo would prepare his favorite spaghetti.

References

[1] Anick, D., Rota, G.-C. (1991): Higher order syzygies for the bracket algebra and for the ring of coordinates of the Grassmannian. Proc. Nat. Acad. Sci. USA **88**, 8087–8090
[2] Brini, A., Huang, R.Q., Teolis, A.G.B (1992): The umbral symbolic method for supersymmetric tensors. Adv. Math. **96**, 123–193
[3] Brini, A., Regonati, F., Teolis, A.G.B. (1999): Multilinear algebra over supersymmetric rings. Adv. Math. **145**, 98–158
[4] Chan, W., Rota, G.-C., Stein, J. (1995): The power of positive thinking. In: White, N.L. (ed.) Invariant Methods in Discrete and Computational Geometry. Kluwer, Dordrecht, pp. 1–36
[5] Chan, W. (1998): Classification of trivectors in $6D$-space. In: Sagan, B.E., Stanley, R.P. (eds.) Mathematical Essays in Honor of Gian-Carlo Rota. Birkhäuser Boston, Boston, MA, pp. 63–110
[6] Grosshans, F.D., Rota, G.-C., Stein, J. (1987): Invariant theory and superalgebras. (CBMS Regional Conference Series in Mathematics, **69**). American Mathematical Society, Providence, RI
[7] Huang, R.Q. (1990): Combinatorial methods in invariant theory. Ph.D. Thesis, Massachusetts Institute of Technology
[8] Huang, R.Q. (1990): Invariants of sets of linear varieties. Proc. Nat. Acad. Sci. U.S.A **87**, 4557–4560
[9] Huang, R.Q. (1991): Invariants of sets of lines in projective 3-space. J. Algebra **143**, 208–218

[5] Chesnay, John Vance (1892): Tears. The Century **44**, 538.

[10] Huang, R.Q., Rota, G.-C. (1994): On the relations of various conjectures on Latin squares and straightening coefficients. Discrete Math. **128**, 225–236
[11] Howe, R., Huang, R.Q. (1996): Projective invariants of four subspaces. Adv. Math. **118**, 295–336
[12] Klain, D.A., Rota, G.-C. (1997): Introduction to geometric probability. (Lezioni Lincei), Cambridge University Press, Cambridge
[13] Kung, J.P.S. (ed.) (1995): Gian-Carlo Rota on Combinatorics. Birkhäuser Boston, Boston, MA
[14] Metropolis, N., Nicoletti, G., Rota, G.-C. (1981): A new class of symmetric functions. In: Mathematical Analysis and Applications. Part B. Academic Press, New York, pp. 563–575
[15] Rota, G.-C., Mullin, R. (1970): On the foundations of combinatorial theory: III. Theory of binomial enumeration. In: Harris, B. (ed.) Graph Theory and its Applications. Academic Press, New York. pp. 167–213
[16] Rota, G.-C. (1973): The valuation ring of a distributive lattice. In: Faitlowicz, S., Kaiser, K. (eds.) Proceedings of the University of Houston Lattice Theory Conference. Department of Mathematics, University of Houston, Houston, TX, pp. 574–632
[17] Rota, G.-C., Stein, J.A. (1990): Supersymmetric Hilbert space. Proc. Nat. Acad. Sci. USA **87**, 653–657
[18] Rota, G.-C. (1997): The many lives of lattice theory. Notices Amer. Math. Soc. **44**, 1440–1445
[19] Rota, G.-C. (1997): Indiscrete thoughts. Birkhäuser Boston, Boston, MA
[20] Rota, G.-C. (1998): Introduction to geometric probability. (AMS Colloquium Lectures, Baltimore, January 1998) American Mathematical Society, Providence, RI, videocasette; edited version: Math. Intelligencer **20** (4), 11–16
[21] Rota, G.-C. (1998): Invariant theory, old and new. (AMS Colloquium Lectures, Baltimore, January 1998); edited version: Math Intelligencer **21** (1), 20–27
[22] Rota, G.-C. (2001): Twelve problems in probability no one likes to bring up. In: Crapo, H. Senato, D. (eds.) Algebraic combinatorics and computer science. A tribute to Gian-Carlo Rota. Springer Milan, pp. 57
[23] Schanuel, S.H. (1986): What is the length of a potato? An introduction to geometric measure theory. In: Lawrence, F.W., Schanuel, S.H. (eds.) Categories in Continuum Physics. (Lecture Notes in Mathematics, (vol. **1174**) Springer, Berlin, pp. 118–126

The Fubini Lectures

Foreword

E. Vesentini

> *Of all escapes from reality, mathematics is the most successful ever. It is a fantasy that becomes all the more addictive because it works back to improve the same reality we are trying to evade.*
>
> *Gian-Carlo Rota, The Lost Café.*

Guido Fubini (Venice 1879–New York 1943), one of the outstanding mathematicians of the past century, spent thirty years of his life in Turin, teaching in the School of Engineering, until the racial laws enacted by the Fascist government in 1938 expelled him from a community of which he had been one of the most prestigious members, with the enthusiasm and the energy which he put into all his endeavours.

Fubini's contributions to mathematics encompass research fields which, in his times, were traditionally considered as quite separate from one another: measure theory, differential equations, differential geometry and projective differential geometry, discontinuous groups and automorphic functions, and so on. His approach to a mathematical problem, although original and sometimes very sophisticated, always focused on the peculiar aspects of the problem itself, and he was even ready – as he once wrote half-jokingly – to "make the monstrous blunder to introduce symbols not yet defined, or – which is perhaps the same – to revert to a daring intuition". An approach that was at a maximal distance from Bourbaki's thomistic vision and close to Raymond Queneau's description of mathematics:

> "Ce n'est pas à l'architecture, à la maçonnerie qu'il faut comparer la géométrie ou l'analyse, mais à la botanique, à la géographie, aux sciences physiques même. Il s'agit de décrire un monde, de le découvrir et non de le construire ou de l'inventer, car il existe en dehors de l'esprit humain et indépendant de lui."

Fubini's attitude toward mathematics would have pleased Gian-Carlo Rota and it fits in well with the metaphor on baroque music that he used in an interview in 1985:

> "People think that they will be either geniuses like Beethoven, or nothing. But look at the Baroque Age - there were hundreds of little Italians who wrote good music and didn't give a hoot about being creative."
> "There is nothing deadlier for a mathematician than to be placed in a beautiful office and instructed to lay golden eggs. Creativity is never directly sought after. It comes indirectly. It comes while you are complaining about too

> much routine work, and so you decide to spend half an hour on your secret pet project. Those are the occasions when you get good ideas. Or while getting ready to lecture your undergraduates, you realise that the textbook for the course is lousy, and that the subject has never before been properly explained. While you work on explaining some old material, lo and behold, you get a great new idea."

In the second half of the past century, Fubini's image seemed to have somehow faded away as if the undisputed quality of his scientific contributions, actively present in the everyday work of many mathematicians, had relegated more and more to the background his personal history and the memory of his intense and lively relationship with the Italian scientific community. This perception was at the origin of the Guido Fubini Lectures, which were created and are currently organised by the Institute for Scientific Interchange, a research centre located in Turin and chaired by Tullio Regge.

In 1998, Gian-Carlo Rota was one of the first Fubini Lecturers, and the notes which appear after this short introduction were written up by Gian-Carlo himself. As he says, I was rather insistent in my proposal that the five lectures he was supposed to deliver should deal with different topics. I trust that, after reading the text, everyone will agree that this was a good idea. What Gian-Carlo doesn't say is that his first lecture, held at the Accademia delle Scienze in Turin, was preceded by a short, elementary introduction, which he called a "Mickey Mouse Lecture'", but refused to include in the final text. It was a not at all trivial, but rather a very amusing bravura, and I regret having accepted Gian-Carlo's decision.

In fact, the topics of the five lectures were not as different as their titles might suggest. Indeed they share a common connecting theme: probability. This sort of coherent unity is also apparent in the second lecture, formally devoted to invariant theory ("the great romantic story of mathematics"), if one accepts Gian-Carlo's idea of renaming the field of geometric probability the oxymoron 'continuous combinatorics'.

The other four seminars are closer to one another: the first one is devoted to an introduction to measure theory and geometric probability[1], and the last three combine twelve open problems (from von Neumann's pointless probability, to entropy, to quantum probability, ...) with ultimate purpose: "to reopen some questions in the foundations of probability, and thus also a much touchier and much neglected subject, the foundations of statistics."

Gian-Carlo's lectures in Turin were due to have a sequel, and he was scheduled to lecture at the Accademia Nazionale dei Lincei, in Rome, on 17 June 1999. Sadly that was too late. He died on 19 April 1999 in Cambridge, Massachussetts.

---

[1] It summarises the opening chapters of the notes of the Lezioni Lincee delivered by Gian-Carlo in 1986 at the Scuola Normale Superiore in Pisa; see Klain, D. A., Rota, G.-C. (1997): Introduction to geometric probability. Cambridge University Press, Cambridge.

The adventures of measure theory*

G.-C. Rota

I am very happy to be here before you as the Fubini Lecturer for this year, and I feel deeply honored to be given this great opportunity to share with you some of the mathematics we love.

When I received from Edo Vesentini the invitation to be the Fubini Lecturer for this year, I rushed to read the list of previous Fubini Lecturers, and I was surprised to discover that there were only two who preceded me: Dennis Sullivan and my old friend Eli Stein. It is a short list, but a distinguished list, and I feel flattered that my name will be lengthening the list. My immediate temptation was to decline the invitation outright; but Edo Vesentini assured me that no one has ever declined the invitation. So I went back to my fantasies and I began to fantasize on the probable topics that other possible candidates on the list of Fubini lectures, past and future, might have chosen. I imagined a brilliant young mathematician, eager to establish himself as a leader in his field, delivering one single dazzling proof beginning with the first Fubini lecture and lasting all the way to the end of the fifth. Or else, some middle-aged mathematician, anxious to have his latest theory accepted by the mathematical world, delivering to a thrilled audience a five-hour condensation of material that would normally take an entire term in an advanced graduate course.

These fantasies came to an abrupt end when Edo Vesentini informed me in no uncertain terms that he thought it wise (for reasons that you will readily understand) that the five lectures should deal with five independent and unrelated topics, thereby allowing any member of the audience to skip one or more lectures, without missing anything. Edo's recommendation deprived me of all role models among previous lecturers in protracted lecture series. In a state of temporary panic, I again tried to remember other speakers in past lecture series I had listened to. Sure enough, one name immediately came to mind: that of Hermann Weyl.

I hope you will forgive me if I digress with some personal reminiscences.

In the fall of 1950 I enrolled as a freshman at Princeton, an Italian boy of eighteen who hardly knew how to speak English, but who had studied Latin and Greek, unlike most other freshmen at Princeton.

The following year, in November 1951, I listened to my first mathematics lectures. These were the three Vanuxem lectures, delivered by Hermann Weyl and bearing the generic title "Symmetry". These lectures were an unforgettable experience. The lectures took place in the old chemistry auditorium, packed with an expectant public. As I shamelessly sat in the first row trying to guess which of the other persons sitting in the same row was to be the speaker, a hush fell upon the audience: Einstein was

* The sections of these Fubini Lectures entitled *The adventures of measure theory* and *What is invariant theory, really?* are slightly revised versions of two Colloqium Lectures delivered to the Annual Meeting of the American Mathematical Society, Baltimore, 1977. The texts of these talks were first published, respectively, in *The Mathematical Intelligencer*, Vol. 20, Number 4, 1998 and Vol. 21 Number 1, 1999

entering the lecture room. To my disappointment, he sat somewhere in the middle of the auditorium.

The first lecture began with an impressive and lengthy quotation in Greek, which no one in the audience understood except Luther Pfahler Eisenhart. This brilliant start was followed by a display of slides portraying charming women wearing the long brimmed hats fashionable at the time, and later by more slides showing the Alhambra and the Pentagon. Not a word of mathematics. The audience was left wondering where such a sparkling display of "Kultur" was leading up to. Not much more mathematics was mentioned in the second lecture, when more slides were shown of physics experiments, for which the lecturer provided a learned oral commentary. Only in the last lecture did some group theory make a modest appearance. By that time the audience, which had not dwindled, was enthralled with the subject, and did not mind the fact that the speaker had said very little about mathematics; actually he had said very little about anything at all. What is more remarkable, the audience seemed to be thankful to the speaker for making the contents of the three lectures independent of one another, thereby minimizing all memory requirements. I hazard to guess that the success of Hermann Weyl's lectures may be in part attributed to the speaker's astute foresight in making his lectures self-contained, independent and lightweight. As I recall this distant episode, I realize that Edo Vesentini's recommendation about the independence of the present Fubini lectures is a wise one, all the more so when the speaker is not Hermann Weyl.

As I thought about the topics of these lectures, I realized that my age would be of some advantage. Why? It is a relief, both to you and to me, to know right at the start that the speaker does not feel the need to impress you by stating the results of his latest research. Nor do you or I suffer from any lack of exposure to the latest fashions in mathematics. We can therefore afford to spend these five hours on leisurely discussion of some mathematics that may matter to both you and me.

We will cover in these lectures a few items that are not widely known, and that should be better known. I vouch that the first two lectures can be understood by anyone with a B.A. in mathematics. The remaining three lectures will not be devoted to convoluted mathematical argument. They will survey some areas of probability theory that in my view are in dire need of further research. I hope the survey will be accessible to anyone with a minimal background in probability.

The title of this lecture is: "The adventures of measure theory".

About one hundred years ago, the properties that underlie such notions as length, area, volume, as well as the probability of events, were abstracted under the banner of the word "measure". Let us review the definition of measure, since we will be using this definition in an unusual way.

A measure μ is a function defined on a family of subsets of a set S, which takes real values, not necessarily positive. The family of sets on which a measure is defined is closed under unions and intersections, and contains the empty set.

A measure is characterized by two simple axioms. Let us take a minute to review these axioms.

Axiom 1.

$$\mu(\emptyset) = 0,$$

where $\emptyset$ is the empty set.

Axiom 2. If A and B are two measurable sets, then

$$\mu(A \cup B) = \mu(A) + \mu(B) - \mu(A \cap B).$$

The meaning of this second axiom is clear. The axiom states that measure is additive. In particular, if we have two disjoint sets A and B, then

$$\mu(A \cup B) = \mu(A) + \mu(B).$$

More generally, for any finite family F whose members are sets, and for which any two members are disjoint, we have:

$$\mu(\bigcup_{A \in F} A) = \sum_{A \in F} \mu(A).$$

We most emphatically do not assume that a measure is countably additive.

The best-known example of a measure is the volume $\mu_n(A)$ of a solid A in ordinary n-dimensional Euclidean space. The volume $\mu_n(A)$ of a solid A satisfies Axioms 1 and 2 above, but Axioms 1 and 2 do not characterize volume among all possible measures. What additional axioms must we add to the definition of a measure in order to characterize volume? It is possible to characterize volume among all measures by adding to Axioms 1 and 2 two additional intuitive axioms, namely, the following:

Axiom 3. The volume of a set A is independent of the position of A. If a set A in n-dimensional Euclidean space can be rigidly moved onto a set B, then A and B have the same volume.

In other words, volume is invariant under the group of Euclidean motions.

Lastly, we must prescribe a normalization, as physicists say. This is done by taking a parallelotope P with orthogonal sides of lengths $x_1, x_2, \ldots, x_n$, and setting

Axiom 4.

$$\mu_n(P) = x_1 x_2 \cdots x_n.$$

These axioms, together with suitable continuity conditions, uniquely determine the volume of solids in Euclidean n-space. For example, starting from these four axioms, by a limiting process such as one finds in an advanced calculus textbook, one establishes the fact that the volume of a ball S_r of radius r in n-dimensional space is given by the following formulas:

$$\mu_n(S_r) = \frac{\pi^{n/2} r^n}{(n/2)!}$$

if the dimension n is even and

$$\mu_n(S_r) = \frac{2^n \pi^{(n-1)/2}((n-1)/2)! r^n}{n!}$$

if the dimension n is odd.

It is still widely believed that volume is the only invariant measure in Euclidean n-space. But in point of fact there are other invariant measures, defined on all reasonable subsets of Euclidean n-space, which have a notable geometric significance. Our objective is to describe all such invariant measures.

What happens if we keep the first three axioms, but tamper with the fourth axiom, the normalization axiom ? Will we get something interesting, or will we get nothing new? To answer this question, we will appeal to the basic tools of combinatorial mathematics.

The basic tools of combinatorial mathematics are the elementary symmetric functions, to wit, the following polynomials in n variables:

$$\begin{aligned} e_1(x_1, x_2, \ldots, x_n) &= x_1 + x_2 + \cdots + x_n, \\ e_2(x_1, x_2, \ldots, x_n) &= x_1x_2 + x_1x_3 + \cdots + x_{n-1}x_n, \\ &\vdots \\ e_{n-1}(x_1, x_2, \ldots, x_n) &= x_2x_3 \cdots x_n + x_1x_3x_4 \cdots x_n + \cdots + x_1x_2 \cdots x_{n-1}, \\ e_n(x_1, x_2, \ldots, x_n) &= x_1x_2 \cdots x_n. \end{aligned}$$

Observe an interesting coincidence. The last of these n symmetric functions is also the formula for the volume of a parallelotope. Axiom 4 can be rewritten as

$$\mu_n(P) = e_n(x_1, x_2, \ldots, x_n).$$

Let us try an experiment, and replace the n-th symmetric function by the $(n-1)$-st symmetric function. Let us first take $n = 3$, that is, three-dimensional space, so that we can better visualize what is going on. Let us see whether we can define a measure on subsets of 3-dimensional space by keeping three of the above axioms, but by replacing the normalization Axiom 4 by using another symmetric function instead of the symmetric function $e_3(x_1, x_2, x_3)$ which gives the volume. Let us first replace the symmetric function e_3 by the symmetric function e_2, thereby changing Axiom 4 to

Axiom 4′.

$$\mu_2(P) = x_1x_2 + x_1x_3 + x_2x_3.$$

Does this axiom define a measure? Of course it does. The right-hand side is the formula for the surface area of the parallelotope P, divided by 2. Again we will find in any advanced calculus textbook the explanation of the fact that Axioms 1, 2, 3, and 4′, together with some continuity considerations, completely determine an invariant measure which is the surface area of solids in ordinary space. For example, the following well-known formula for the surface area of a ball S_r of radius r in 3 dimensions is obtained from these axioms:

$$\mu_2(S_r) = 4\pi r^2.$$

Let us take the next step.

Emboldened by our success with two symmetric functions, we now replace Axiom 4 by yet another axiom, using another symmetric function. Let us set

Axiom 4″.

$$\mu_1(P) = e_1(x_1, x_2, x_3) = x_1 + x_2 + x_3.$$

The new measure μ_1 will satisfy Axioms 1, 2, and 3, and in addition it will satisfy Axiom 4″. The symmetric function of degree one plays the role that in the previous two examples was played by the other two symmetric functions.

But wait a minute: is this definition consistent?

To realize that the definition of the new measure μ_1 is consistent, that is, that μ_1 as defined by Axioms 1, 2, 3, and 4″ really exists and is not a dream of reason, look at two parallelotopes P_1 and P_2 that have a face in common. The first parallelotope has sides equal to x_1, x_2, x_3 , and the second parallelotope has sides equal to x_1, x_2, y. The two parallelotopes have a common face with sides equal to x_1, x_2. The measure $\mu_1(P_1 \cup P_2)$ of the parallelotope $P_1 \cup P_2$ can be computed in two ways: using the left side of Axiom 2, or using the right side, and the two computations had better yield the same answer, in symbols:

$$\mu_1(P_1 \cup P_2) = \mu_1(P_1) + \mu_1(P_2) - \mu_1(P_1 \cap P_2).$$

Let us check this.

The left side is computed by observing that the parallelotope $P_1 \cup P_2$ has sides equal to x_1, x_2 and $x_3 + y$. Therefore, Axiom 4″ tells us that

$$\mu_1(P_1 \cup P_2) = x_1 + x_2 + x_3 + y.$$

Now let us compute the right side. We have

$$\begin{aligned} \mu_1(P_1) &= x_1 + x_2 + x_3, \\ \mu_1(P_2) &= x_1 + x_2 + y, \\ \mu_1(P_1 \cap P_2) &= x_1 + x_2, \end{aligned}$$

again by Axiom 4″ applied to $P = P_1 \cap P_2$, since one side equals zero when the parallelotope is a flat, that is, a rectangle. Therefore, the right side of Axiom 2 equals

$$\begin{aligned} \mu_1(P_1) + \mu_1(P_2) - \mu_1(P_1 \cap P_2) &= x_1 + x_2 + x_3 + x_1 + x_2 + y - (x_1 + x_2) \\ &= x_1 + x_2 + x_3 + y, \end{aligned}$$

and the two sides of our equations agree, thereby convincing us that the definition may well be consistent.

The preceding argument is convincing, even though it proves nothing.

Actually, the definition of $\mu_1(P)$ for a parallelotope P has a simple geometric interpretation. When multiplied by 4, it equals the perimeter of the parallelotope P, that is, the sum of the lengths of all the edges of the parallelotope P. Just as happens for volume and area, it can be shown by continuity considerations that the measure μ_1 can be extended to all reasonable solids in ordinary space, for example, to all convex sets and to all polyhedra, convex or nonconvex.

But, one may object, $\mu_1(P)$ makes sense for a parallelotope P, because a parallelotope has a well-defined perimeter. What if A is a solid that does not have a well-defined perimeter, a sphere for example? The definition of the measure $\mu_1(A)$ for solids A that may not have a well-defined perimeter flies in the face of common sense.

Einstein wrote: "Common sense is the residue of those prejudices that were instilled into us before the age of seventeen". Common sense must constantly readjust to reality.

The new measure μ_1 that we obtain in this way is called the mean width, a misnomer that has been kept for historical reasons. The mean width of a solid in space is completely characterized by Axioms 1, 2, 3, and 4$''$. In particular, it is invariant, that is, it does not depend on position. For example, the formula for the mean width of a sphere of radius r is computed to be

$$\mu_1(S_r) = 4r.$$

Thus we see that in three dimensions each of the three elementary symmetric functions of three variables leads to an invariant measure that enjoys equal rights with volume. The first two of these measures are well-known, namely, volume and area. The third, the mean width, is at present almost totally unknown. I know of no person who has an intuitive feeling for the mean width, similar to the intuitive feeling we have for volume and area.

Let us conjecture a possible application of the mean width. A potato grower knows that a potato's volume is important, because it determines the nutritional content of the potato. The potato grower also knows that the surface area of a potato is important, because it is rumored that the vitamins in a potato are concentrated in the skin. We may conjecture that, as soon as the potato grower will become aware of the mean width, he or she will find a nutritional interpretation of the mean width of a potato. I am indebted to Steve Schanuel for this example.

A similar kind of reasoning works in n dimensions. We discover n different invariant measures, each of them well-defined on all polyhedra and on all finite unions of compact convex sets. Each of the n elementary symmetric functions of n variables leads to the definition of a new invariant measure which is a different generalization of the notion of volume. These n measures are called the intrinsic volumes. The intrinsic volumes are first defined on an orthogonal polytope P whose sides equal

$x_1, x_2, \ldots, x_n$ by setting

$$\mu_k(P) = e_k(x_1, x_2, \ldots, x_n),$$

where $e_k(x_1, x_2, \ldots, x_n)$ is the k-th elementary symmetric function. Here the subscript k ranges from 1 to n.

One then proceeds to extend the definition of the intrinsic volumes to more general sets, by a technique which we will shortly see.

The intrinsic volumes are independent of each other, except for certain inequalities they satisfy. Mathematicians are presently working on determining these as yet unknown inequalities among the intrinsic volumes. These inequalities generalize the classical isoperimetric inequality that relates volume to area. At present, we know very little about the intrinsic volumes; they have not been around for long and very little research has been done on them. We do not even know the formulae for the intrinsic volumes of an n-simplex.

Now you are thinking: this is all fine and dandy, but how is the extension of the intrinsic volumes from parallelotopes to more general sets carried out? And besides, isn't there any intuitive interpretation we can give the intrinsic volumes?

We will answer both these questions simultaneously. Let us go back to three-dimensional space. You all know that the set of all straight lines in space – not necessarily through the origin – forms a nice algebraic variety, called the Grassmannian. The group of all Euclidean rigid motions acts on the Grassmannian, and there is an invariant measure on the Grassmannian under the action of the group of Euclidean motions. This invariant measure is unique except for a constant factor. A similar statement may be made about the set of all planes, and more generally for the set of all linear varieties of dimension k in Euclidean space of dimension n. Remember that these linear varieties need not pass through the origin.

In the practice of mathematics, computation with invariant measures on Grassmanians is rare; most mathematicians would be hard put even to recall an explicit formula for the invariant measures on Grassmanians. Let us take a few minutes to get a feeling for the invariant measure on the set of all straight lines in three-space. As is customary, we begin by giving this measure a name: let us call it λ_1^3; the upper index 3 stands for three-dimensional space, and the lower index stands for the dimension of a line, namely, one. To repeat, we use the notation λ_1^3 to denote the invariant measure on the set of all straight lines in three-space.

Consider a rectangle R placed anywhere in space, and consider the set of all straight lines that meet the rectangle R. Can we compute the measure of this set of lines without knowing the formula for the invariant measure on the Grassmannian of all lines in three-space? Of course we can. A straight line meets the rectangle R either at a point or not at all; therefore, the value of the measure of the set of all lines meeting R depends only on the area $\mu_2(R)$ of the rectangle R. If we take another rectangle R' whose area is double the area of R, then the measure of the set of all lines meeting R' is double the measure of the set of all lines meeting R. Proceeding along these lines, we get to Cauchy's functional equation, and we infer that the measure of the set of all straight lines meeting a rectangle R equals a constant times the area

$\mu_2(R)$. Since we are at liberty to choose a normalization of the measure, let us agree to set this constant equal to one.

But instead of working with a rectangle we could have worked with any planar figure C whatsoever, placed in an arbitrary position in space. The measure of the set of lines meeting C equals the area $\mu_2(C)$, by the same reasoning. We stress the assumption that C must lie in a plane. To conclude: even without knowing the formula for the invariant measure λ_1^3, we can nevertheless compute the value of such a measure on certain sets of lines.

Let us now take a more sophisticated set of straight lines. We take a set D in three-space that is the union of disjoint sets $C_1, C_2, \ldots, C_n$, where each of the C_i is contained in a different plane, and we ask for the measure of the set of all straight lines meeting D. Such a computation can be carried out, but it is a combinatorial nightmare; so much so that we are forced to do what mathematicians do when confronted with combinatorial nightmares: they change the problem ever so slightly. In this case we take a hint from the way probabilists work. Let $X_D(\omega)$ equal the number of times the straight line ω meets the set D. Instead of computing a measure, let us compute the integral

$$\int X_D(\omega) d\lambda_1^3(\omega),$$

where ω ranges over the Grassmannian, that is, over the set of all straight lines in space. We will see that we can compute this integral without knowing the measure λ_1^3 on the Grassmannian. Since

$$D = \bigcup_{i=1}^{n} C_i,$$

and since the C_i are disjoint, we have

$$\int X_D(\omega) d\lambda_1^3(\omega) = \sum_{i=1}^{n} \int X_{C_i}(\omega) d\lambda_1^3(\omega).$$

But we have chosen each of the sets C_i to lie in a plane, so that a straight line meets C_i either once or not at all. It follows that

$$\int X_{C_i}(\omega) d\lambda_1^3(\omega) = \mu_2(C_i)$$

and therefore

$$\int X_D(\omega) d\lambda_1^3(\omega) = \sum_{i=1}^{n} \mu_2(C_i).$$

What is this identity telling us? The right-hand side equals the area of the surface D. Nothing stops us from passing to the limit, and making the following assertion. Let

E be "any" surface in space, and let $X_E(\omega)$ be the number of times the straight line ω meets the surface E. Then the integral

$$\int X_E(\omega) d\lambda_1^3(\omega)$$

ranging over all straight lines ω, equals the surface area of E. In probabilistic language: the average number of times a randomly chosen straight line meets the surface E equals the surface area of E.

Let us now retrace our steps, and repeat the same reasoning taking the set of all planes in space, instead of the set of all straight lines. The invariant measure on this Grassmannian is denoted by λ_2^3, where again the upper index stands for three-dimensional space, and the lower index for the dimension of a plane. Since a plane meets a straight line segment either at a point or not at all, the same argument shows that the measure of the set of all planes that meet a line segment L equals $\mu_1(L)$, namely, the length of the segment L; more generally, if F is any curve "whatsoever" in space, and if $X_F(\omega)$ equals the number of times the plane ω meets the curve F, then repeating the argument we used for straight lines we infer that the integral

$$\int X_F(\omega) d\lambda_2^3(\omega)$$

equals the length of the curve F. The variable of integration ω now ranges over planes, not over straight lines. Here again we compute an integral without knowing the measure.

We are now very close to getting an intuitive interpretation of the mean width. Recall the parallelotope P with sides equal to x_1, x_2, x_3. To measure the planes meeting the parallelotope P, we first consider a family of *parallel* planes, all sharing the same fixed unit normal u. In other words, consider the set of all planes parallel to the plane $u^\perp$. Without loss of generality, place the parallelotope in space so that one of the vertices of P is at the origin, and such that the vector u lie in the octant of space opposite to the parallelotope P. (We can do this generically.) Denote the edges of P that meet the origin by x_1, x_2, x_3. Given the fixed unit vector u, and its family of normal planes, let us take the curve F to be a path along the edges (line segments)

$$[0, x_1], \ [x_1, (x_1 + x_2)], \ \text{and} \ [(x_1 + x_2), (x_1 + x_2 + x_3)],$$

in that order. A plane parallel to $u^\perp$ meets the parallelotope P if and only if it meets the curve F on the parallelotope at exactly one point. Therefore, the measure of the set of all planes parallel to $u^\perp$ that meet the parallelotope P is proportional to the length of the curve F. Averaging over all unit vectors u (and hence, over all families of parallel planes), we conclude that the measure of the set of all planes meeting a parallelotope equals the mean width of the parallelotope, except for a constant factor which we will again set to be one.

In view of this realization, we can immediately see how to define the mean width of any closed convex set: it equals the measure of the set of all planes that meet the convex set. Thus, we have shown that the mean width may be extended to all closed convex sets in space.

We are now in a position to give a probabilistic interpretation of the mean width of a convex set.

Take two compact convex sets A and B in three-dimensional Euclidean space, and suppose that A is contained in B. Let us begin by belaboring the obvious. Suppose that we take a point at random belonging to the larger set B. What is the probability that the point shall belong to the smaller set A? The answer is clear: such a probability equals the ratio of the volume of A by the volume of B.

Instead of choosing a point at random, let us choose a straight line at random in space. Assuming that such a straight line meets the larger set B, what is the probability that such a straight line will also meet the smaller set A?

We have already computed the answer to this question, albeit implicitly. Such a probability equals the surface area of the set A divided by the surface area of the set B.

You can tell what is coming next. We now take a random plane in space. Assuming that the plane meets the larger set B, what is the probability that it will also meet the smaller set A? The answer is the following: such a probability equals the mean width of A divided by the mean width of B.

In Euclidean n-space, we obtain by much the same reasoning interpretations of the intrinsic volume $\mu_k(C)$ of a compact convex set C as the Grassmannian measure of the set of all linear varieties of dimension $n-k$ that meet the convex set C, and a similar probabilistic interpretation holds.

What comes next? There are at least two questions still open. First, are there any other invariant measures besides the intrinsic volumes, and second, how can the definition of the intrinsic volumes be extended to more general subsets of n-space than convex sets. The answers to both these questions are closely related.

The answer to the first question is negative. We are missing one measure, and to discover it, we will engage for a minute in the kind of mathematical reasoning that physicists find unbearably pedantic, just to show physicists that such reasoning does pay off.

Let us ask ourselves the question: what is the value of the symmetric function of order zero of a set of n variables $x_1, x_2, \ldots, x_n$, say $e_0(x_1, x_2, \ldots, x_n)$? I will give you the answer, and will leave it to you to justify this answer after the lecture is over. The answer is the following: $e_0 = 1$ if $n > 0$, that is, if the set of variables $x_1, x_2, \ldots, x_n$ is nonempty, and $e_0 = 0$ if the set of variables is empty.

We are led to believe that there may exist an invariant measure in n-space associated with the symmetric function of order zero. We set

$$\mu_0(C) = 1$$

if C is any nonempty compact convex set, and of course $\mu_0(\emptyset) = 0$. Does such a measure exist? It does indeed exist, and the fact that it exists is, in my opinion, one of the most remarkable discoveries ever made in mathematics.

We will prove that such a measure is well-defined on any set which is a finite union of compact convex sets. We do this by employing a classical device borrowed from functional analysis: instead of defining a measure, we define a linear functional on all simple functions, that is, on all real functions $f(\omega)$ defined for $\omega \in \mathbf{R}^n$ which are linear combinations of indicator functions of compact convex sets. Let us first begin with the case $n = 1$, that is, let ω range over points on the line. Define a linear functional χ_1 on simple functions as follows:

$$\chi_1(f) = \sum (f(\omega) - f(\omega+)),$$

where the sum ranges over all real numbers ω. The meaning of the plus sign is best gleaned from an example. Let f be the indicator function of the closed segment $[a, b]$. Then $f(\omega) - f(\omega+) = 0$ for all ω except $\omega = b$, because we have $f(b) = 1$ but $f(b+) = 0$. Thus, we see that $\chi_1(f) = 1$ if f is the indicator function of an interval $[a, b]$.

Now let us go over to n dimensions, proceeding by induction. Do not worry, this won't take long. Take a straight line L and for every point ω in L let H_ω be the hyperplane through the point ω perpendicular to the line L. If f is a simple function defined in n-space, and if ω is a point on the straight line L, let f_ω be the restriction of f to the hyperplane H_ω. Define a linear functional χ_n as follows:

$$\chi_n(f) = \sum \chi_{n-1}(f_\omega) - \chi_{n-1}(f_{\omega+}),$$

where the sum ranges over all points ω on the line L. There is only a finite set of $\omega's$ for which the summand is nonzero. When f is the indicator function of a non-empty compact convex set, then an argument similar to the preceding shows that $\chi_n(f) = 1$. Thus, we may define a measure $\mu_0(G) = \chi_n(f)$, where G is any finite union of compact convex sets, and f is the indicator function of the set G. We have thus proved the existence of a measure μ_0 which is defined on all finite unions of compact convex sets, and which takes the value one on all nonempty compact convex sets. This measure has a long history: it is the Euler characteristic.

Now you are thinking: if this is the Euler characteristic, then it is up to you to show that it coincides with what we ordinarily believe to be the Euler characteristic. Let us conclude this lecture by deriving the formula of Euler–Schläfli–Poincaré for polyhedra. As a matter of fact, this formula can be encapsulated into a simpler formula, one that is easy to remember.

Let C be a nonempty compact convex polytope of dimension n, and let int(C) be the interior of C. Then we have the following fundamental formula for the Euler characteristic of int(C):

$$\mu_0(\text{int}(C)) = (-1)^n.$$

Indeed, if f is the indicator function of the set int(C), we have:

$$\mu_0(\text{int}(C)) = \sum \chi_{n-1}(f_\omega) - \chi_{n-1}(f_{\omega+}),$$

where the sum ranges over all points ω on the line L as above. But by induction, we see that every term on the right-hand side equals zero, except when ω is the first point on the line L for which the intersection $C \cap H_\omega$ is not empty. If ω_ℓ is such a first point, then we have

$$\chi_{n-1}(f_{\omega_\ell}) = 0$$

because the point ω_ℓ is on the boundary of C, and

$$\chi_{n-1}(f_{\omega_\ell +}) = (-1)^{n-1}$$

by the induction hypothesis, because $f_{\omega_\ell +}$ is the indicator function of the set $\text{int}(C) \cap H_{\omega_\ell +}$, which is the interior of a convex polyhedron one dimension lower. Putting all this together, we obtain

$$\mu_0(\text{int}(C)) = \sum \chi_{n-1}(f_{\omega_\ell}) - \chi_{n-1}(f_{\omega_\ell +}) = -(-1)^{n-1} = (-1)^n,$$

as desired. We are now in a position to state the famous Euler formula for polyhedra. What is a polyhedron? A polyhedron is a finite union of convex polyhedra. Given a polyhedron, we must define a system of faces (of all dimensions, ranging from dimension 0 (a point) to dimension n). We will say that a set **F** of convex polyhedra is a system of faces for an arbitrary polyhedron K when the elements of **F**, called faces, are nonempty compact convex sets F with disjoint interiors such that

$$K = \bigcup_{F \in \mathbf{F}} \text{int}(F).$$

Caution: the interior of a face of dimension k is to be taken relative to the linear space of dimension k that contains the face, and the interior of a point is a point.

Under these conditions we may take the Euler characteristic of both sides, and, using the fact that any two interiors of faces are disjoint, we obtain (using the fact that the measure of the disjoint union of a family of sets equals the sum of the measures of the individual sets):

$$\mu_0(K) = \sum_{F \in \mathbf{F}} \mu_0(\text{int}(F)) = f_0 - f_1 + f_2 - \cdots + \cdots,$$

where f_i equals the number of faces of dimension i. This is Euler's formula.

We can now answer the second of the questions we had left open: how to extend the definition of the intrinsic volumes from compact convex sets to all finite unions of compact convex sets. If G is such a finite union of compact convex sets, then we set

$$\mu_k(G) = \int \mu_0(G \cap \omega) d\lambda^n_{n-k}(\omega),$$

where ω ranges over all linear varieties of dimension $n - k$ in n-space. The left-hand side defines a measure, and, when G is a compact convex set, it agrees with the definition we have already given. It is therefore the desired extension. The Euler characteristic does all the work for us.

We are now in a position to state the main theorem of geometric probability. We will say that an invariant measure μ on Euclidean n-space, defined on all finite unions of compact convex sets, is continuous, when

$$\lim_{C_n \to C} \mu(C_n) = \mu(C)$$

for all sequences C_n of compact convex sets converging to the compact convex set C. We have the

Main Theorem of Geometric Probability. *The $n+1$ intrinsic volumes $\mu_0, \mu_1, \ldots, \mu_n$ are a basis of the space of all continuous invariant measures defined on all finite unions of compact convex sets.*

The first proof of this theorem is due to Hadwiger; the first elementary proof was published last year by Dan Klain of Georgia Tech.

In closing, let me try to answer the question you are about to ask: what has this got to do with geometric probability, anyway?

I will attempt a sketchy answer. Consider two compact convex sets A and B. We imagine B to be fixed in n-space, and that we "drop" the rigid set A at random. What is the probability that A meets B? We answer this question in three steps. First, we realize that, by keeping B fixed and varying A by the group of Euclidean motions, we define an invariant measure on convex sets B. Second, we apply Hadwiger's theorem, and infer that such an invariant measure equals a linear combination of the $n+1$ intrinsic volumes, with coefficients depending on A and not on B. Third, we determine these coefficients by taking suitable B's. The end result is an identity which is known as the kinematic formula, which has been the object of much research in this century, still going on today.

References

[1] Klain, D.A., Rota, G.-C. (1997): Introduction to geometric probability. Lezioni Lincee. Cambridge University Press, Cambridge

What is invariant theory, really?*

G.-C. Rota

Invariant theory is the great romantic story of mathematics. For one hundred and fifty years, from its beginnings with Boole to the time, around the middle of this century, when it branched off into several independent disciplines, mathematicians of all countries were brought together by their common faith in invariants: in England, Cayley, MacMahon, Sylvester and Salmon, and later, Alfred Young, Aitken, Littlewood and Turnbull. In Germany, Clebsch, Gordan, Grassmann, Sophus Lie, Study; in France, Hermite, Jordan and Laguerre; in Italy, Capelli, d'Ovidio, Brioschi, Trudi and Corrado Segre; in America, Glenn, Dickson, Carus (of the Carus Monographs), Eric Temple Bell and later Hermann Weyl. Seldom in history has an international community of scholars felt so united by a common scientific ideal for so long a stretch of time. In our century, Lie theory and algebraic geometry, differential algebra and algebraic combinatorics are offsprings of invariant theory. No other mathematical theory, with the exception of the theory of functions of a complex variable, has had as deep and lasting an influence on the development of mathematics.

Eventually, invariant theory was to become a victim of its own success: the very term "invariant theory" is nowadays understood in such a wide variety of senses that it has become all but meaningless. It is no wonder that you are baffled by the title of this lecture, and curious to hear what will be said about invariant theory in the next forty-eight minutes.

Like the Arabian phoenix arising from its ashes, classical invariant theory, once pronounced dead, is once again at the forefront of mathematics. The old treatises are being dusted off the shelves of library basements and reread, reinterpreted and presented in a language that meets the standard of rigor of our day. The program of classical invariant theory, that had for some time been given up as hopeless, is again being pursued, and success may at last be within reach.

We will review two turning points in the history of invariant theory. The first, the "new" one, happened around the turn of the century, and its effects are still being felt all over mathematics. The second, the "old" one, happened very early in the game, and led to a serious misunderstanding that lasts to this day.

A pedestrian definition of invariant theory might go as follows: invariant theory is the study of orbits of group actions. Such a definition is correct, but it must be supplemented by a programmatic statement. Hermann Weyl, in the introduction to his book *The Classical Groups*, was the first in this century to give a sweeping overview of the program of invariant theory. He summarized this program in two basic assertions. The first states that "all geometric facts are expressed by the vanishing of invariants", and the second states that "all invariants are invariants of tensors".

Let us briefly comment on these lofty statements. What is a geometric fact? A geometric fact is a fact about space that is independent of the choice of a coordi-

---

* The second Fubini Lecture, delivered at the Villa Gualino, Torino, 2 June, 1998

nate system. Geometric facts are described by means of equations which require a choice of coordinates. In a vector space V of dimension n one chooses a coordinate system $x_1, x_2, \ldots, x_n$. Since Descartes, we have learned to express geometric facts by equations in the coordinates $x_1, x_2, \ldots, x_n$. However, about one hundred years ago, mathematicians and physicists made the shocking discovery that the usual type of equations, that is, equations in the commutative ring generated by the variables $x_1, x_2, \ldots, x_n$, are inadequate for the description of a lot of geometric and physical facts. Motivated by this discovery, they introduced a more general ring. This is the ring of noncommutative polynomials in the coordinates $x_1, x_2, \ldots, x_n$. Homogeneous elements of this ring, that is, homogeneous noncommutative polynomials in the variables $x_1, x_2, \ldots, x_n$, are called tensors. If we believe Hermann Weyl's philosophy, then we will be satisfied that equations in the tensor algebra suffice for the description of any geometric fact we will ever meet. Furthermore, if these equations are to express geometric properties, then they must hold no matter what coordinate system is chosen; in other words, equations that describe geometric facts must be invariant under changes of coordinates. The program of invariant theory, from Boole to our day, is precisely the translation of geometric facts into invariant algebraic equations expressed in terms of tensors.

This program of translation of geometry into algebra was to be carried out in two steps. The first step consisted in decomposing tensor algebra into irreducible components under changes of coordinates. The second step consisted in devising an efficient notation for the expression of invariants for each irreducible component. The first step was successfully carried out in this century; the second was abandoned sometime in the twenties and only recently has it resurfaced.

The decomposition of tensor algebra into irreducible components was discovered around the turn of the century almost simultaneously by Issai Schur and Alfred Young. The gist of this decomposition is one of the great advances in mathematics of all times, and it may be worthwhile to present it in a form that can be made available to undergraduates.

Let us consider functions of three variables, such as $f(x_1, x_2, x_3)$. Two well-known classes of functions of three variables are symmetric functions, defined to satisfy the equations

$$f_s(x_1, x_2, x_3) = f_s(x_{i_1}, x_{i_2}, x_{i_3})$$

for every permutation sending the indices $(1, 2, 3)$ to (i_1, i_2, i_3), and skew-symmetric functions, defined by the equations

$$f_a(x_1, x_2, x_3) = \pm f_a(x_{i_1}, x_{i_2}, x_{i_3}),$$

where the sign is $+1$ or -1 according as the permutation sending the indices $(1, 2, 3)$ to (i_1, i_2, i_3) is even or odd.

It is not true that a function of three variables is the sum of a symmetric function and a skew-symmetric function. A third type of function is required, which is called a cyclic function, which is defined by the equation

$$f_c(x_1, x_2, x_3) + f_c(x_3, x_1, x_2) + f_c(x_2, x_3, x_1) = 0.$$

Every function of three variables can be uniquely written as the sum of a symmetric function, a skew-symmetric function, and a cyclic function; in symbols:

$$f(x_1, x_2, x_3) = f_s(x_1, x_2, x_3) + f_a(x_1, x_2, x_3) + f_c(x_1, x_2, x_3).$$

Each of the three symmetry classes is invariant under permutations; this fact is obvious for symmetric and skew-symmetric functions but not quite so obvious for cyclic functions. These three invariant subspaces play for the group of permutations of a set of three elements a role analogous to the role of the eigenvectors of a symmetric matrix.

For functions $f(x_1, x_2, x_3, x_4)$ of four variables there are five symmetry classes, which are defined as follows.

1. Symmetric functions.
2. Skew-symmetric functions.
3. Cyclic symmetric functions, satisfying the four equations

$$\begin{aligned}
f(x_1, x_2, x_3, x_4) + f(x_1, x_4, x_2, x_3) + f(x_1, x_3, x_4, x_2) &= 0,\\
f(x_1, x_2, x_3, x_4) + f(x_4, x_2, x_1, x_3) + f(x_3, x_2, x_4, x_1) &= 0,\\
f(x_1, x_2, x_3, x_4) + f(x_4, x_1, x_3, x_2) + f(x_2, x_4, x_3, x_1) &= 0,\\
f(x_1, x_2, x_3, x_4) + f(x_3, x_1, x_2, x_4) + f(x_2, x_3, x_1, x_4) &= 0.
\end{aligned}$$

4. Functions satisfying the four equations

$$\begin{aligned}
f(x_1, x_2, x_3, x_4) + f(x_2, x_1, x_3, x_4) + f(x_1, x_2, x_4, x_3)&\\
+ f(x_2, x_1, x_4, x_3) &= 0,\\
f(x_1, x_2, x_3, x_4) + f(x_3, x_2, x_1, x_4) + f(x_1, x_4, x_3, x_2)&\\
+ f(x_3, x_4, x_1, x_2) &= 0,\\
f(x_1, x_2, x_3, x_4) + f(x_1, x_3, x_2, x_4) + f(x_4, x_2, x_3, x_1)&\\
+ f(x_4, x_3, x_2, x_1) &= 0,\\
\sum \operatorname{sign}(\sigma) f(x_{\sigma 1}, x_{\sigma 2}, x_{\sigma 3}, x_{\sigma 4}) &= 0.
\end{aligned}$$

5. Functions satisfying the equations

$$\begin{aligned}
f(x_1, x_2, x_3, x_4) - f(x_2, x_1, x_3, x_4) - f(x_1, x_2, x_4, x_3)&\\
+ f(x_2, x_1, x_4, x_3) &= 0,\\
f(x_1, x_2, x_3, x_4) - f(x_3, x_2, x_1, x_4) - f(x_1, x_4, x_3, x_2)&\\
+ f(x_3, x_4, x_1, x_2) &= 0,\\
f(x_1, x_2, x_3, x_4) - f(x_1, x_3, x_2, x_4) - f(x_4, x_2, x_3, x_1)&\\
+ f(x_4, x_3, x_2, x_1) &= 0,\\
\sum f(x_{\sigma 1}, x_{\sigma 2}, x_{\sigma 3}, x_{\sigma 4}) &= 0.
\end{aligned}$$

Every function of four variables is uniquely expressible as the sum of five functions, each belonging to one of these symmetry classes. Each symmetry class is invariant under permutations.

More generally, every function of n variables $f(x_1, x_2, \ldots, x_n)$ can be uniquely written as the sum of p_n functions, each one belonging to a different symmetry class. Here, p_n equals the number of partitions of the integer n. Each symmetry class is defined by equations which are not difficult to find.

This decomposition holds for tensors as well, after some cosmetic changes of notation. To this day, only two symmetry classes of tensors have been studied in any detail. Symmetric tensors are ordinary commutative polynomials such as we learned to use in analytic geometry. Skew-symmetric tensors are polynomials in the coordinates $x_1, x_2, \ldots, x_n$ provided that the variables are assumed to satisfy the equations $x_i x_j = -x_j x_i$. Tensors belonging to symmetry classes other than the classes of symmetric and skew-symmetric tensors also occur in geometry and physics. However, these symmetry classes have been studied very little, and they are a long way from being understood.

We will describe the most peculiar feature of classical invariant theory, namely, the symbolic or umbral notation. We will consider the simplest group, namely, the group of translations of the line. The unusual features of the symbolic method will already be apparent in this special case.

Let $p(x)$ and $q(x)$ be monic polynomials in the variable x. We write them in the following quaint notation:

$$p(x) = x^n + \binom{n}{1} a_1 x^{n-1} + \binom{n}{2} a_2 x^{n-2} + \cdots + \binom{n}{n-1} a_{n-1} x + a_n,$$

and

$$q(x) = x^k + \binom{k}{1} b_1 x^{k-1} + \binom{k}{2} b_2 x^{k-2} + \cdots + \binom{k}{k-1} b_{k-1} x + b_k.$$

We assume that the polynomial $q(x)$ is of lower degree than the polynomial $p(x)$, that is, that $k \leq n$.

Define the translation operator T^c on a polynomial $p(x)$ as follows:

$$T^c p(x) = p(x+c).$$

Let us write

$$p(x+c) = x^n + \binom{n}{1} p_1(c) x^{n-1} + \binom{n}{2} p_2(c) x^{n-2} + \cdots + \binom{n}{n-1} p_{n-1}(c) x + p_n(c),$$

The j-th coefficient $p_j(c)$ of the polynomial $p(x+c)$ is computed to be

$$p_j(c) = a_j + \binom{j}{1} a_{j-1} c + \binom{j}{2} a_{j-2} c^2 + \cdots + c^j.$$

A polynomial $I(a_1, a_2, \ldots, a_n, b_1, b_2, \ldots, b_k)$ in the variables $a_1, a_2, \ldots, a_n$, $b_1, b_2, \ldots, b_k$ is said to be an invariant of the two polynomials $p(x)$, $q(x)$ when

$$I(a_1, a_2, \ldots, a_n, b_1, b_2, \ldots, b_k) \\ = I(p_1(c), p_2(c), \ldots, p_n(c), q_1(c), q_2(c), \ldots, q_k(c))$$

for all complex numbers c. By abuse of notation, we write $I(p(x), q(x))$ and we speak of I as being an invariant of the polynomials $p(x)$ and $q(x)$. In this abusive notation, a polynomial I is said to be an invariant of the polynomials $p(x)$ and $q(x)$ whenever

$$I(T^c p(x), T^c q(x)) = I(p(x), q(x))$$

for all constants c.

Invariant theory is concerned with the problem of finding all invariants of a given set of polynomials, as well as their significance.

What is meant by the "significance" of an invariant? We will appeal to Hermann Weyl. "Every" property of polynomials which is invariant under the group of translations is expressed by the vanishing of a set of invariants. In other words, "any" set of polynomials which is invariant under translations is the same set as a set of polynomials obtained by setting to zero a set of invariants of such polynomials.

It is impossible to understand the above statement without examples. Let us consider the simplest and oldest example. The property of a quadratic polynomial

$$q(x) = x^2 + 2b_1 x + b_2.$$

of having a double root, is invariant under translations; in other words, if the polynomial $q(x)$ has a double root, so does the polynomial $q(x + c)$ for any constant c. Following Hermann Weyl, we look for an invariant whose vanishing expresses this property. Sure enough, it is easy to check that the discriminant

$$D(b_1, b_2) = b_1^2 - b_2$$

is the desired invariant. This example, due to Boole, was the spark that led to the birth of invariant theory.

One often hears the sentence, "Hilbert killed invariant theory", repeated as an excuse to ignore all that went on in invariant theory after Hilbert. I don't know who made up this infamous sentence. It is not true. Hilbert loved invariant theory, and he went on publishing striking papers in invariant theory well after he proved the theorem that is nowadays called the Hilbert basis theorem, the theorem that is supposed to have killed invariant theory. Some of the most fascinating results in invariant theory were discovered in the first twenty years of this century, a long time after Hilbert proved his basis theorem.

What then is the reason for the temporary demise of invariant theory in this century? One reason is the endemic use of a notation that lacked rigor and that amounted to little more than handwaving in print. This is the symbolic or umbral notation.

Dieudonné wrote that half the success of a piece of mathematics depends on a proper choice of notation. It would be interesting to make a list of unfortunate notations that killed various chapters of mathematics, as well as a list of felicitous notations that promoted the development of other branches of mathematics. The symbolic or umbral notation that was used by invariant theorists through the nineteen twenties was a catastrophe. A number of mathematicians tried to make sense of the symbolic method without success, the three most notable being Hermann Weyl, Eric Temple Bell and Edward Hegeler Carus. Eric Temple Bell failed to properly define umbral notation, and his book *Algebraic Arithmetic* remains to this day the book of seven seals. If Hermann Weyl and Eric Temple Bell had lived fifty years longer, so as to benefit of the development of what was in their time called "modern" algebra, they would undoubtedly have succeeded in properly defining umbral notation.

In our day, it does not take much work to accomplish this task. Do not be alarmed: it will only take a few minutes. Before I start spouting out definitions, let me say what I am not going to say. Umbral notation can be shown to be equivalent, or "cryptomorphic", to use a term invented by my late friend Garrett Birkhoff, to another notation that has gained great notoriety in our day: I mean the notation of Hopf algebras. I will not justify this Sybilline pronouncement, not because it is difficult to do so, but because it would be too boring to do so.

Let us go on to the definition of umbral notation.

Side by side with the polynomials $p(x)$ and $q(x)$, we consider another polynomial algebra $\mathbf{C}[x, \alpha, \beta]$ in three variables x, α and β, together with a linear functional E defined on the underlying vector space $\mathbf{C}[x, \alpha, \beta]$. The definition of the linear functional E is the key point. It is carried out in the following steps.

Step 1. Set

$$E(x^j) = x^j$$

for all nonnegative integers j; in particular $E(1) = 1$. Thus, the range of the linear functional E is $\mathbf{C}[x]$.

Step 2. Set

$$E(\alpha^j) = a_j;$$

in particular, we have $E(\alpha^j) = 0$ if $j > n$.

Step 3. Set

$$E(\beta^j) = b_j;$$

in particular, we have $E(\beta^j) = 0$ if $j > k$.

Step 4. This is the main step. Set

$$E(\alpha^i \beta^j x^\ell) = E(\alpha^i) E(\beta^j) x^\ell.$$

Following Sylvester, we call the variables α and β umbrae. In other words, the linear functional E is multiplicative on distinct umbrae.

Step 5. Extend by linearity.

This completes the definition of the linear functional E.

We next come to the most disquieting feature of umbral notation. Let $f(\alpha, \beta, x)$ and $g(\alpha, \beta, x)$ be two polynomials in the variables α, β, x. We write

$$f(\alpha, \beta, x) \cong g(\alpha, \beta, x)$$

to mean

$$E(f(\alpha, \beta, x)) = E(g(\alpha, \beta, x)).$$

Read $\cong$ as "equivalent to". The "classics" went a bit too far, they wrote

$$f(\alpha, \beta, x) = g(\alpha, \beta, x)$$

that is, they replaced the symbol $\cong$ by ordinary equality. This was an excessive abuse of notation. The "classics" were aware of the error and, while they avoided computational errors by clever artistry, they were unable to settle on a correct notation.

The umbral or symbolic method consists of replacing all occurrences of the coefficients of the polynomials $p(x)$ and $q(x)$ by umbrae and equivalences. For example,

$$p(x) \cong (x + \alpha)^n$$

and

$$q(x) \cong (x + \beta)^k.$$

Let us carefully check the first equivalence. By definition, the equivalence means the same as

$$E(p(x)) = E((x + \alpha)^n).$$

Since $E(x^j) = x^j$ for all nonnegative integers j, this identity can be rewritten as

$$p(x) = E((x + \alpha)^n).$$

Expanding the right-hand side by the binomial theorem, we obtain

$$\begin{aligned} &E((x + \alpha)^n) \\ &\quad = E\left(x^n + \binom{n}{1}\alpha x^{n-1} + \binom{n}{2}\alpha^2 x^{n-2} + \cdots + \binom{n}{n-1}\alpha^{n-1} x + \alpha^n\right). \end{aligned}$$

By linearity this equals

$$x^n + \binom{n}{1}E(\alpha)x^{n-1} + \binom{n}{2}E(\alpha^2)x^{n-2} + \cdots + \binom{n}{n-1}E(\alpha^{n-1})x + E(\alpha^n).$$

Evaluating the linear functional E, we see that this in turn equals

$$x^n + \binom{n}{1} a_1 x^{n-1} + \binom{n}{2} a_2 x^{n-2} + \cdots + \binom{n}{n-1} a_{n-1} x + a_n,$$

as desired.

The expression

$$(x+\alpha)^n$$

is called an umbral representation of the polynomial $p(x)$.

In umbral notation, a complex number r is a root of the polynomial equation $p(x) = 0$ if and only if

$$(r+\alpha)^n \cong 0.$$

Similarly, in umbral notation the polynomial $T^c p(x) = p(x+c)$ may be represented as follows:

$$p(x+c) \cong (x+\alpha+c)^n,$$

and this yields the umbral expression for the coefficients $p_j(c)$ of the polynomial $p(x+c)$, namely

$$p_j(c) \cong (\alpha+c)^j.$$

Let us next see how umbral notation is related to invariants. We assume that the two polynomials $p(x)$ and $q(x)$ have the same degree n. Then an invariant A of the polynomials $p(x), q(x)$ may be defined as follows:

$$A(q(x), p(x)) \cong (\beta-\alpha)^n.$$

The evaluation of the invariant A in terms of the coefficients of $p(x)$ and $q(x)$ proceeds as follows:

$$\begin{aligned} A(q(x), p(x)) = \; & E((\beta-\alpha)^n) = E\biggl(\beta^n - \binom{n}{1}\beta^{n-1}\alpha + \cdots \\ & + (-1)^{n-1}\binom{n}{n-1}\beta\alpha^{n-1} + (-1)^n\alpha^n\biggr) \\ = \; & E(\beta^n) - E\left(\binom{n}{1}\beta^{n-1}\alpha\right) + \cdots \\ & + (-1)^{n-1} E\left(\binom{n}{n-1}\beta\alpha^{n-1}\right) + (-1)^n E(\alpha^n) \\ = \; & E(\beta^n) - \binom{n}{1} E(\beta^{n-1}) E(\alpha) + \cdots \end{aligned}$$

$$+(-1)^{n-1}\binom{n}{n-1}E(\beta)E(\alpha^{n-1})+(-1)^n E(\alpha^n)$$

$$= b_n - \binom{n}{1}b_{n-1}a_1 + \binom{n}{2}b_{n-2}a_2 - \cdots + \cdots$$

$$+(-1)^{n-1}\binom{n}{n-1}b_1 a_{n-1} + (-1)^n a_n.$$

Why is A an invariant? This is best seen in umbral notation:

$$A(T^c q(x), T^c p(x)) \cong (\beta + c - \alpha - c)^n = (\beta - \alpha)^n.$$

The invariant A is called the apolar invariant; two polynomials $p(x)$ and $q(x)$ having the property that $A(q(x), p(x)) = 0$ are said to be apolar. In umbral notation, two polynomials are apolar whenever

$$(\beta - \alpha)^n \cong 0.$$

The concept of apolarity has a distinguished pedigree going all the way back to Apollonius.

What is the "significance" of the apolar invariant? What does it mean for two polynomials to be apolar? This question is answered by

Theorem 1. *Suppose that r is a root of the polynomial $q(x)$, that is, that $q(r) = 0$. Then the polynomials $q(x)$ and $p(x) = (x - r)^n$ are apolar.*

Proof. For $p(x) = (x - r)^n$ we have $\alpha^j \cong (-r)^j$, and hence

$$A(q(x), p(x)) \cong (\beta - (-r))^n = (\beta + r)^n \cong 0,$$

as desired. □

Corollary. *If the polynomial $q(x)$ has n distinct roots $r_1, r_2, \ldots, r_n$, and if the polynomial $p(x)$ is apolar to $q(x)$, then there exist constants $c_1, c_2, \ldots, c_n$ for which*

$$p(x) = c_1(x - r_1)^n + c_2(x - r_2)^n + \cdots + c_n(x - r_n)^n.$$

Proof. The dimension of the affine subspace of all monic polynomials $p(x)$ which are apolar to $q(x)$ equals n. But if the polynomial $q(x)$ has simple roots, then by the above theorem the polynomials $(x - r_1)^n, (x - r_2)^n, \ldots, (x - r_n)^n$ are linearly independent and apolar to $q(x)$. Hence the polynomial $p(x)$ is a linear combination of these polynomials. This completes the proof. □

Thus, we see that apolarity gives a trivial answer to the following question: when can a polynomial $p(x)$ be written as a linear combination of polynomials of the form $(x - r_1)^n, (x - r_2)^n, \ldots, (x - r_n)^n$?

A beautiful theorem on apolarity was proved by the British mathematician John Hilton Grace. We state it without proof:

Grace's Theorem. *If two polynomials $p(x)$ and $q(x)$ of degree n are apolar, then every disk in the complex plane containing every zero of $p(x)$ also contains at least one zero of $q(x)$.*

Grace's Theorem is an instance of what might be called a sturdy theorem. For almost one hundred years it has resisted all attempts at generalization. Almost all known results about the distribution of zeros of polynomials in the complex plane are corollaries of Grace's theorem.

We will next generalize the apolar invariant to the case of two polynomials $p(x)$ and $q(x)$ of different degrees n and k, with $k \leq n$. To this end, we slightly generalize the definition of invariant, as follows.

A polynomial $I(a_1, a_2, \ldots, a_n, b_1, b_2, \ldots, b_k, x)$ in the variables $a_1, a_2, \ldots, a_n$, $b_1, b_2, \ldots, b_k, x$ is said to be an invariant of the polynomials $p(x), q(x)$ when

$$\begin{aligned} &I(a_1, a_2, \ldots, a_n, b_1, b_2, \ldots, b_k, x) \\ &\qquad = I(p_1(c), p_2(c), \ldots, p_n(c), q_1(c), q_2(c), \ldots, q_k(c), x + c) \end{aligned}$$

for all complex numbers c. Sometimes these more general invariants are called covariants.

We define a more general apolar invariant as follows:

$$A(q(x), p(x)) \cong (\beta - \alpha)^k (x - \alpha)^{n-k}.$$

Again, we say that two polynomials $p(x)$ and $q(x)$ are apolar when $A(q(x), p(x))$ is identically zero, that is, zero for all x. Theorem 1 remains valid as stated. That is, if $q(r) = 0$, then the polynomial $p(x) = (x - r)^n$ is apolar to $q(x)$.

Let us consider a special case. Suppose that $q(x)$ is a quadratic polynomial and $p(x)$ is a cubic polynomial:

$$q(x) = x^2 + 2b_1 x + b_2$$

and

$$p(x) = x^3 + 3a_1 x^2 + 3a_2 x + a_3.$$

Then we have, in umbral notation,

$$\begin{aligned} A(q(x), p(x)) &\cong (\beta - \alpha)^2 (x - \alpha) \\ &= (\beta^2 - 2\alpha\beta + \alpha^2)x - \alpha\beta^2 + 2\alpha^2\beta - \alpha^2. \end{aligned}$$

Evaluating the linear functional E, we obtain the following explicit expression for the apolar invariant:

$$\begin{aligned} &A(q(x), p(x)) = E((\beta^2 - 2\alpha\beta + \alpha^2)x - \alpha\beta^2 + 2\alpha^2\beta - \alpha^2) \\ &\quad = E(\beta^2) - 2E(\alpha\beta) + E(\alpha^2)x + E(-\alpha\beta^2) + 2E(\alpha^2\beta) - E(\alpha^2) \\ &\quad = E(\beta^2) - 2E(\alpha)E(\beta) + E(\alpha^2)x + E(-\alpha)E(\beta^2) + 2E(\alpha^2)E(\beta) - E(\alpha^2) \\ &\quad = (b_2 - 2a_1 b_1 + a_2)x - a_1 b_2 + 2a_2 b_1 - a_3. \end{aligned}$$

Thus, a quadratic polynomial $q(x)$ and a cubic polynomial $p(x)$ are apolar if and only if their coefficients satisfy the two equations

$$\begin{aligned} b_2 - 2a_1b_1 + a_2 &= 0 \\ -a_1b_2 + 2a_2b_1 - a_3 &= 0. \end{aligned}$$

Using these equations, we can prove two important theorems.

Theorem 2. *There is in general one quadratic polynomial which is apolar to a given cubic polynomial.*

Proof. Indeed, the above equations may be rewritten as

$$\begin{aligned} b_2 - 2a_1b_1 &= -a_2 \\ -a_1b_2 + 2a_2b_1 &= a_3. \end{aligned}$$

The solutions b_1, b_2 for given a_1, a_2, a_3 are in general unique. □

Theorem 3. *There is always a two-dimensional space of cubic polynomials which are apolar to a given quadratic polynomial.*

Proof. Indeed, given b_1, b_2 we may solve for a_1, a_2, a_3 from the equations

$$\begin{aligned} -2a_1b_1 + a_2 &= -b_2 \\ -a_1b_2 + 2a_2b_1 &= a_3. \end{aligned}$$

These equations always have a double infinity of solutions, as they used to say in the old days. □

Theorems 2 and 3 provide a simple and explicit method for solving a cubic equation. It goes as follows.

Given the cubic polynomial

$$p(x) = x^3 + 3a_1x^2 + 3a_2x + a_3,$$

first, by Theorem 2, we find a unique quadratic polynomial $q(x)$ which is apolar to $p(x)$. In general, such a quadratic polynomial $q(x)$ has two distinct roots r_1 and r_2. By Theorem 1, the cubic polynomials $(x - r_1)^3$ and $(x - r_2)^3$ are apolar to $q(x)$. Second, by Theorem 3, the affine linear space of cubic polynomials apolar to $q(x)$ has dimension two. Since $p(x)$ is apolar to $q(x)$, we conclude that $p(x)$ is a linear combination of $(x - r_1)^3$ and $(x - r_2)^3$. In symbols:

$$p(x) = c(x - r_1)^3 + (1 - c)(x - r_2)^3$$

for some constant c. Observe that c, r_1 and r_2 are computed by solving linear and quadratic equations.

In this way, the solution of the cubic equation $p(x) = 0$ is reduced to the solution of the equation

$$c(x - r_1)^3 = -(1 - c)(x - r_2)^3,$$

and this equation is easily solved by taking a cube root.

This method of solving a cubic equation is the only one I can remember.

Let me digress with a personal anecdote. A few years ago, I was lecturing on this material at a symposium on combinatorics that took place at the University at Minnesota. Persi Diaconis was sitting in the front row, and I could tell as I started to lecture that he was falling asleep; he eventually began to doze off. But the moment I mentioned the magic words "solving a cubic equation" he woke up with a start and said: "Really! How?"

The preceding two theorems are easily generalized.

Theorem 4. *The dimension of the space of all (monic) polynomials of degree k that are apolar to a polynomial of degree n equals* $\max\{0, 2k - n\}$, *in general, when* $k \leq n$.

Theorem 5. *The dimension of the space of all (monic) polynomials of degree n that are apolar to a polynomial of degree k equals k, if* $k \leq n$.

Let us try to solve an equation of degree 5 in much the same way as we solved a cubic equation. For the quintic polynomial

$$p(x) = x^5 + 5a_1x^4 + 10a_2x^3 + 10a_3x^2 + 5a_4x + a_5 = 0,$$

Theorem 4 assures us that there is in general a unique cubic polynomial $q(x)$ which is apolar to $p(x)$. In general, this cubic polynomial has three distinct roots r_1, r_2, r_3. By Theorem 1, the polynomials $(x - r_1)^5$, $(x - r_2)^5$, $(x - r_3)^5$ are linearly independent and apolar to $q(x)$. By Theorem 5, the dimension of the space of all polynomials apolar to $q(x)$ equals 3. But the polynomial $p(x)$ is apolar to $q(x)$. Hence, $p(x)$ can be written in the form

$$p(x) = c_1(x - r_1)^5 + c_2(x - r_2)^5 + c_3(x - r_3)^5$$

for suitable constants c_i. Thus, we see that a generic polynomial of degree 5 can be written as a linear combination of three fifth powers of linear polynomials. These are computed by solving linear, quadratic and cubic equations. This reduction to canonical form of the quintic is as close as we can come to solving a quintic equation by radicals.

At this point, someone in the audience will raise his or her hand and say: "Excuse me, but the umbral method you have introduced is not even good enough to express the discriminant of a quadratic equation! "

Quite right.

The definitions of umbrae and of the linear functional E have an obvious generalization to any array of polynomials, say $p_1(x), p_2(x), \ldots, p_\ell(x)$. One simply considers the space of polynomials

$$\mathbf{C}[x, \alpha_1, \alpha_2, \ldots, \alpha_\ell]$$

and one sets

$$E(\alpha_t^j)$$

to equal the j-th coefficient of the polynomial $p_t(x)$. What is crucial is the fact that the linear functional E is again multiplicative on distinct umbrae:

$$E(\alpha_1^i \alpha_2^j \alpha_3^k \cdots x^\ell) = E(\alpha_1^i)E(\alpha_2^j)E(\alpha_3^k)\cdots x^\ell.$$

Now comes the catch that in the old days was to turn into a notational nightmare: the polynomials $p_1(x), p_2(x), \ldots, p_\ell(x)$ need not be distinct. In fact, the most important case occurs when each of the polynomials $p_1(x), p_2(x), \ldots, p_\ell(x)$ is equal to one and the same polynomial $p(x)$. In this case, the definition of the linear functional E may be simplified as follows:

1. $E(\alpha_i^j) = a_j$ for every i , and
2. $E(\alpha_1^i \alpha_2^j \alpha_3^k \cdots x^\ell) = a_i a_j a_k \cdots x^\ell$ for all nonnegative integers $i, j, k, \ldots \ell$.

Umbrae $\alpha_1, \alpha_2, \ldots, \alpha_\ell$ satisfying 1 and 2 are said to be exchangeable. Thus, for exchangeable umbrae we have

$$(x + \alpha_1)^n \cong (x + \alpha_2)^n.$$

Eric Temple Bell, who wrote $=$ in place of $\cong$, was baffled by the fact that two umbrae could be exchangeable without being equal.

We can now state the main theorem of invariant theory. We will consider a single polynomial.

Theorem 6. *Every invariant of a polynomial $p(x)$ is obtained by evaluating some polynomial in the differences $\alpha_i - \alpha_j$ and $\alpha_i - x$, where α_i and α_j are exchangeable umbrae. Conversely, every polynomial in such differences is equivalent to an invariant of the polynomial $p(x)$.*

The proof is extremely simple, but will be omitted.

Let us review some classical examples.

The discriminant of a quadratic polynomial $p(x) = x^2 + 2a_1x + a_2$ may be umbrally represented as follows:

$$D(p(x)) \cong (\alpha_1 - \alpha_2)^2/2,$$

where α_1 and α_2 are exchangeable umbrae. Indeed:

$$\begin{aligned} E((\alpha_1 - \alpha_2)^2) &= E(\alpha_1^2) - 2E(\alpha_1\alpha_2) + E(\alpha_2^2) \\ &= a_2 - 2a_1^2 + a_2 = 2(a_2 - a_1^2), \end{aligned}$$

as desired.

We next consider a cubic polynomial $p(x) = x^3 + 3a_1x^2 + 3a_2x + a_3$. The discriminant of this polynomial, let us call it $D(p(x))$, equals, as you know, the expression

$$D(p(x)) = \frac{4(a_2 - a_1)^2(a_1a_3 - a_2^2) - (a_3 - a_1a_2)^2}{2}.$$

The umbral expression of the discriminant is easier to remember.

$$D(p(x)) \cong (\alpha_1 - \alpha_2)^2(\alpha_3 - \alpha_4)^2(\alpha_1 - \alpha_4)(\alpha_2 - \alpha_3).$$

As you know, the discriminant vanishes if and only if the cubic equation $p(x) = 0$ has a double root.

The Hessian of a cubic polynomial can be elegantly written in umbral notation as follows:

$$H(p(x)) \cong (\alpha_1 - \alpha_2)^2(\alpha_1 - x)(\alpha_2 - x).$$

The Hessian vanishes if and only if the cubic polynomial is the third power of a polynomial of degree one.

Allow me another digression. On hearing about the vanishing of the Hessian as the condition that a cubic polynomial be a perfect cube, it is natural to ask the general question: which invariant of a polynomial of degree n vanishes if and only if the polynomial is the k-th power of some polynomial of degree n/k? Here k is a divisor of n. For a long time I thought the answer to this question to be beyond reach, until one day, while leafing despondently through the second volume of Hilbert's collected papers, I accidentally discovered that Hilbert had completely solved it. The solution can be elegantly expressed in umbral notation. This is only one of several striking results of Hilbert on invariant theory that have been forgotten.

We consider next an invariant of the quintic. Theorem 3 tells us that a quintic $p(x) = x^5 + 5a_1x^4 + 10a_2x^3 + 5a_3x^2 + 5a_4x + a_5$ has a unique apolar cubic polynomial $q(x)$. The polynomial $q(x)$ is an invariant of $p(x)$. Does it have a simple expression in umbral notation? Indeed it does. The expression is the following:

$$q(x) \cong (\alpha_2 - \alpha_3)^2(\alpha_3 - \alpha_1)^2(\alpha_1 - \alpha_2)^2(\alpha_1 - x)(\alpha_2 - x)(\alpha_3 - x).$$

In the classical literature this invariant is denoted by the letter j. What property will the quintic polynomial $p(x)$ have when the invariant j vanishes? The answer to this question is pleasing. The invariant j of a quintic polynomial is identically equal to zero if and only if the quintic is apolar to some nontrivial polynomial of degree two. But then Theorem 5 tells us that the quintic may be written in the form

$$p(x) = c(x - r_1)^5 + (1 - c)(x - r_2)^5,$$

where r_1, r_2 are the roots of a quadratic equation. Thus, the vanishing of the invariant j is a necessary and sufficient condition that the quintic polynomial $p(x)$ may be written as the sum of two rather than three fifth powers of linear polynomials. When this is the case, the fifth degree equation $p(x) = 0$ can be solved by radicals.

By similar arguments, one can compute all invariants whose vanishing implies that the equation of degree five is algorithmically solvable by radicals. Twenty-three invariants play a relevant role, as Cayley was the first to show.

Hilbert's theorem on finite generation of the ring of invariants can be recast in the language of umbrae, and can be given a simple combinatorial proof that dispenses with the Hilbert basis theorem.

In closing, let us touch upon another reason for the demise of the symbolic method in invariant theory.

In mathematics, it is extremely difficult to tell the truth. The formal exposition of a mathematical theory does not tell the whole truth. The truth of a mathematical theory is more likely to be grasped while we listen to a casual remark made by some expert that gives away some hidden motivation, when we finally pin down the typical examples, or when we discover what the real problems are that were stored behind the showcase problems.

Philosophers and psychiatrists should explain why it is that we mathematicians are in the habit of systematically erasing our footsteps. Scientists have always looked askance at this strange habit of mathematicians, which has changed little from Pythagoras to our day.

The hidden purpose of the symbolic method in invariant theory was not simply that of finding easy expressions for invariants. A deeper faith was guiding this method. It was the expectation that the expression of invariants by the symbolic method would eventually guide us to single out the "relevant" or "important" invariants among an infinite variety. It was the hope that the significance of the vanishing of an invariant could be gleaned from its umbral expression. The vanishing of this faith was the real reason for the demise of classical invariant theory, and the revival of this faith is the reason for its present rebirth.

Whether or not we will succeed this second time where the classics failed is a cliffhanger that will probably be resolved in the next few years. I would not be speaking to you now if I did not believe in success.

References

[1] Kung, J.P.S., Rota, G.-C. (1984): The invariant theory of binary forms. Bull. Amer. Math. Soc. (N. S.) **10**, 27–85

[2] Grosshans, F.D., Rota, G.-C., Stein, J.A. (1987): Invariant theory and superalgebras. (CBMS Regional Conference Series in Mathematics, vol. **69**) American Mathematical Society, Providence, RI

[3] Metropolis, N., Rota, C.-C. (1991): Symmetry classes: functions of three variables. Amer. Math. Monthly **98**, 328–332

[4] Metropolis, N., Rota, G.-C., Stein, J.A. (1991): Theory of symmetry classes. Proc Nat. Acad. Sci. **88**, 8415–8419

[5] Ehrenborg, R., Rota, G.-C. (1993): Apolarity and canonical forms for homogeneous polynomials. European J. Combin. **14**, 157–181

[6] Rota, G.-C., Taylor, B.D. (1994): The classical umbral calculus. SIAM J. Math. Anal. **25**, 694–711

[7] Di Crescenzo, A., Rota, G.-C. (1994): Sul calcolo umbrale. Ricerche Mat. **43**, 129–162
[8] Metropolis, N, Rota, G.-C., Stein, J.A. (1995): Symmetry classes of functions. J. Algebra **171**, 845–866

Twelve problems in probability no one likes to bring up*

G.-C. Rota

The purpose of this lecture, and of the lectures that will follow it, is a controversial one. It is, plainly and simply, to reopen some questions in the foundations of probability, and thus also a much touchier and much neglected subject, the foundations of statistics. There seems to have been a conspiracy of silence on such foundational questions among mathematicians (not so among philosophers), who would like to believe that the subject has been definitively closed by Kolmogorov, and many of whom react unfavorably to any suggestion of revision of Kolmogorov's elegant setup.

Some of the inadequacies of probability theory as it is practiced at the present time will be displayed by short surveys of items belonging to three different kinds. In the first place, I intend to bring up examples of stochastic processes whose phenomenological behavior is not fully matched in their mathematical presentation. In the second place, we will seek out instances of statistical reasoning that are brushed aside in treatises purporting to be rigorous, as if they didn't exist, despite their practical success. In the third place, we will relate the classical picture of probability on which we have made a living since Kolmogorov's *Ergebnisse* tract of 1932, to the new probability theories that have lately been simmering, what logicians might call "deviant" probabilities. These new probability theories, quantum probability and the free probability discovered as recently as 1985 by Voiculescu, issue from relaxing some algebraic aspect of probability, while attempting to keep a similar analytic structure together with a statistical interpretation that will relate these theories to the real world.

Our survey will lead us to reopen a number of questions in probability and statistics that the professionals would rather pass over in silence. I hasten to add that you should not expect a "catalogue raisonné" of paradoxes of probability, such as one finds in those alluring books that manage to find a publisher every few years. I am sorry to disappoint you if you expect these lectures to provide a short intensive in probability theory that will excuse you from any study of the subject. You are more likely to learn from these lectures what probability is not, and to wonder why such elementary questions as the ones I am about to outline have not been worked upon.

I will lay my cards on the table: a revision of the notion of a sample space is my ultimate concern. I hasten to add that I am not about to put forth concrete proposals for carrying out such a revision. We will, however, be guided by a belief that has been a guiding principle of the mathematics of this century. Analysis will play second fiddle to algebra. The algebraic structure sooner or later comes to dominate, whether or not it is recognized when a subject is born. Algebra dictates the analysis.

I feel honored to observe in this distinguished audience the presence of some of the foremost specialists in the fields to be dealt with. I apologize to them in advance

* The Fubini Lectures, Torino, 3–5 June, 1998

for the elementary character of the presentation they are about to hear. If we are to break out of an established mold, we are all but forced to focus on the elementary features of the subject, those that specialists have internalized and have come to consider as inexorably fixed. Hence the emphasis on certain basics that some of you will find cloyingly familiar.

1 Problem one: the algebra of probability

The beginning definitions in any field of mathematics are always misleading, and the basic definitions of probability are perhaps the most misleading of all. In a strictly formal sense, probability is the study of sets of random variables on a sample space, and of their joint probability distributions.

What is a sample space? A sample space is a measure space (Ω, Π, P), where Ω is a set, Π is a σ-subalgebra of subsets, that is, a Boolean σ-algebra of subsets of Ω, and P is a countably additive nonnegative measure such that $P(\Omega) = 1$. The elements of the Boolean σ-algebra Π are called events. An element of the sample space Ω is called a sample point. A random variable is a real-valued function defined on Ω, which is measurable relative to the σ-subalgebra Π.

It has been argued that the notions of sample space and event are redundant, and that all of probability should be done in terms of random variables alone. As a matter of fact, in that variant of classical probability that is now called quantum probability the introduction of random variables, under the name of observables, is preferable, since the quantum analogs of events come to play a secondary role.

How would one introduce probability in terms of random variables alone? This is a subject that has been thoroughly studied, and a brief mention will suffice. One takes an ordered commutative algebra over the reals, and endows it with a positive linear functional $E(X)$. The elements of the algebra will be the random variables and the linear functional is the expectation of a random variable. I will spare you the details.

The alternative definitions of probability, by means of a sample space and by means of an algebra of random variables, are equivalent. Mathematicians opt for the definition in terms of random variables because they do not wish to miss a chance to appeal to Her Imperial Majesty, the Theory of Commutative Rings. However, no practicing statistician has ever felt comfortable with algebras of random variables. The answer to a problem in statistics is likely to be a number that specifies the probability of an event. Events are essential to statistical intuition and cannot be done without.

From the algebra of random variables one can recover the sample space, and vice versa. Recovering the sample space from the algebra of random variables is easy. The probability of an event A can be recovered from the indicator random variable $I_A(\omega)$, namely, the random variable taking one of the two values zero and one according as ω belongs to the event A, by setting $P(A) = E(I_A)$. The Boolean σ-algebra of events can be reconstructed from the ring generated by indicator random variables. This reconstruction, trivial as we consider it, actually lies at the heart of probability.

What is more interesting is the problem of recovering the algebra of random variables when one is given a mere abstract Boolean σ-algebra, not represented as an algebra of sets, that is, when one is forbidden to mention points. This problem was worked upon with passion and solved by two mathematicians in this century, Carathéodory and von Neumann. Carathéodory was motivated by his desire to put the finishing touch to his measure theory, a subject of which he was one of the founders at the beginning of the century. Von Neumann had an altogether different motivation. He was driven by the hope that, once he understood how the algebra of random variables could be formulated in terms of a Boolean algebra of events, without any appeal to points, then the algebra of observables might be similarly constructed in terms of the lattice structure of the quantum mechanical analogs of events. We will see later in these problems how far von Neumann's work has met with success.

Von Neumann never wrote up his algebraization of probability, though he discussed it with his assistants and he taught it in seminars that were mimeographed and distributed to a small circle of initiates at the Institute for Advanced Study in Princeton. Carathéodory, on the other hand, spent the last years of his life writing a book, *Mass and Integral*, in which he explained in full wealth of detail how to reconstruct the algebra of random variables from an abstract Boolean σ-algebra. It was von Neumann, however, who invented for this theory the name that has stuck: pointless probability.

We will review the main idea of pointless probability. To this end, we must review the definition of one of the fundamental notions of mathematics, namely, the notion of a lattice (which we will also need in the statement of other problems). A strange set of historical circumstances has conspired to make lattice theory less well-known than it ought to be.

What is a lattice? A lattice L is a set endowed with two operations, called meet and join and denoted $\wedge$ and $\vee$, which satisfy some but not of all the properties satisfied by "and" and "or" of ordinary logic, namely, each of them is idempotent, associative and commutative. What is more important, for $a, b, c \in L$ the following absorption laws are satisfied:

$$a \vee (b \wedge a) = a$$

and

$$a \wedge (b \vee a) = a.$$

It was discovered by Dedekind in the last century that two operations $\vee$ and $\wedge$, satisfying the identities stated above, uniquely characterize greatest lower bound and least upper bound in a partially ordered set. Thus, a lattice is a partially ordered set in which any two elements have a least upper bound and a greatest lower bound.

Dedekind's discovery went largely unnoticed until the thirties in this century, when Birkhoff, Ore and von Neumann worked on it with passion. Observe that the definition of a lattice does not assume that meet and join satisfy the distributive law $a \wedge (b \vee c) = (a \wedge b) \vee (a \wedge c)$. A theorem of Birkhoff asserts that a lattice satisfying the distributive law, a distributive lattice, is isomorphic to a lattice of sets with union and intersection of sets playing the role of join and meet.

A Boolean σ-algebra can be defined as a distributive lattice satisfying certain additional technical conditions. A theorem of Loomis and Sikorski insures that every Boolean σ-algebra can be represented as the quotient of a Boolean σ-algebra of events in a sample space by a σ-ideal. Pointless probability deals with an abstract Boolean σ-algebra Π and with a real-valued function defined on Π which imitates the definition of probability. The problem is to define an algebra of random variables.

The setup is not as artificial as it may appear. Among probabilists, mention of sample points in an argument has always been bad form. A fully probabilistic argument must be pointless.

To this end, consider two categories: on the one hand, the category of sample spaces and probability-preserving maps between them; on the other hand, the category of abstract Boolean σ-algebras and their σ-homomorphisms. There is a contravariant functor of the first category into the second that is an isomorphism of categories (an isomorphism is an unusual occurrence in category theory). This isomorphism of categories makes it possible to systematically translate set-theoretic concepts expressed in terms of points, sets and functions, in terms of abstract Boolean σ-algebras and their morphisms, thereby getting rid of points. One advantage of this translation is that it does away with events of probability zero, since one can deal with the Boolean σ-algebra of events in a sample space modulo the ideal of events of probability zero.

In the "pointless" version of probability, arrows point in a direction that is opposite to the one in terms of which most mathematicians still think. For example, a random variable is defined not as a function from a sample space to the reals, but as a homomorphism of the Boolean σ-algebra of Borel sets of the real line into the Boolean σ-algebra of events. Let us take a moment to realize how this weird definition relates to the ordinary definition of a random variable X as a measurable function from a sample space to the real numbers. Given the "normal" definition of a random variable as a measurable function, one defines a homomorphism $\hat{X}$ of the Boolean σ-algebra of Borel sets of the reals to the Boolean σ-algebra of events by setting $\hat{X}(B) = X^{-1}(B)$, that is, $\hat{X}(B)$ is the inverse image of the Borel set B by the random variable X. Conversely, every homomorphism of the Boolean algebra of Borel sets to a Boolean σ-algebra of events can be implemented by an ordinary random variable, though such an implementation is fraught with measure-theoretic difficulties having to do with events of probability zero.

The combinatorial construction of sum and product of random variables in terms of Boolean operations is a tour de force that Carathéodory brilliantly carried out. His definitions of sum and product of random variables in terms of lattice joins and meets remain amazing to this day, though we would rather admire them from a distance. Simpler ways have since been found to attain the same objective. I will review the one that is perhaps the simplest, as well as the one that is conceptually clearest. It goes back to some work I did in 1967.

Given a Boolean σ-algebra $\mathcal{B}$, we construct a ring, functorially associated with $\mathcal{B}$, as follows. We begin with the commutative algebra freely generated over the reals by the elements of $\mathcal{B}$, and impose on this free algebra the identities (where A and B $\in \mathcal{B}$):

1. $A^c = \hat{1} - A$, where $\hat{1}$ is the maximal element of $\mathcal{B}$;
2. $A + B = A \wedge B + A \vee B$;
3. $A \cdot B = A \wedge B$.

The quotient algebra of the free algebra obtained by imposing these identities is called the valuation algebra of $\mathcal{B}$.

It may be shown that the valuation algebra is isomorphic to the algebra of simple random variables in some sample space; functional analysis does the rest.

By virtue of this construction of the algebra of random variables in terms of an abstract Boolean σ-algebra of events, probability can dispense with sample points. We will appeal to the pointlessness of probability when we come face to face with the loud ongoing dispute between Bayesian statisticians and frequentist statisticians.

The pointless view has proved beneficial elsewhere, for example, in universal algebra. For example, instead of taking the product $\prod_{t \in T} S_t$ of a family of sets S_t where the parameter t ranges over a set T, one can define a pointless version of this product, where the set T is replaced by an abstract Boolean algebra.

And yet, one can argue against the pointless point of view, by recalling areas of mathematics where the pointless view has failed to take hold, at least so far. Two related instances where the pointless philosophy has not yet succeeded are the following (forgive me for straying far afield for a minute).

First. Suppose that we are given a one-parameter group s^t of probability-preserving transformations on a Boolean σ-algebra endowed with a probability. The parameter t ranges over the reals. Is it possible to redefine this structure in a "pointless" way? One must succeed in redefining such a one-parameter group exclusively in terms of Borel sets of the reals (of nonzero Lebesgue measure) and elements of the σ-algebra upon which the transformations act, without ever mentioning a value of the parameter t. I have posed this question to some category theorists, without ever receiving a satisfactory answer. A positive solution of this problem would spill over elsewhere.

Second. Mackey's theory of virtual groups would be greatly simplified if it were to be rewritten pointlessly. Despite the pioneering work of Arlan Ramsay, Robert Ziemer and others in this direction, a lily-white pointless presentation of Mackey's theory is missing. In the current presentations of Mackey's theory, the difficulties that arise in dealing with sets of null measure (which have not been completely resolved) make it difficult to see at the end of the tunnel the notable simplification in ergodic theory and representation theory that Mackey's ideas have brought about.

To digress again: the pointless philosophy can be carried to dangerous extremes. One can argue that the definition of a small category in terms of arrows should be replaced by an "arrowless" definition in terms of abstract Boolean algebras. I wonder whether any category theorist will take up this challenge.

In developing pointless probability, Carathéodory and von Neumann had their heart in the right place: they foresaw the necessity of bringing out hidden algebraic underpinnings of probability. However, they did not go far enough. It gradually became clear after their work that all stochastic processes could, and perhaps should, be defined by generators and relations on special Boolean σ-algebras. By way of example, we will outline the algebraization of the Poisson process.

What is the Poisson process, really? Let us first review the ordinary definition in terms of sample points, and then go on to the pointless definition. We are given a set S, a family $\mathcal{P}$ of subsets of S closed under unions and intersections, and a finitely additive atomless unbounded measure μ defined on the Boolean algebra $\mathcal{P}$. Important applications demand that this measure be allowed to be finitely additive rather than countably additive.

The sample points of the Poisson process are rare subsets of the set S. A rare subset of S is an infinite subset ω of the set S with the property that for every subset A such that $A \in \mathcal{P}$ and such that $\mu(A) < \infty$ one has $|\omega \cap A| < \infty$.

In the sample space Ω_P consisting of all sample points we define the event $\binom{A}{n}$, where A is a member of the family $\mathcal{P}$, and where n is a nonnegative integer, to be the set of all sample points having n entries in A, that is, the set of all sample points ω such that $|\omega \cap A| = n$. An event will be a member of the Boolean σ-algebra $\mathcal{E}_P$ of all subsets of Ω_P generated by all events $\binom{A}{n}$ as A ranges over family $\mathcal{P}$ and n ranges over all nonnegative integers.

The Poisson process is defined on the Boolean σ-algebra $\mathcal{E}_P$ by setting

$$P(\binom{A}{n}) = \frac{\mu(A)^n}{n!} e^{-\mu(A)},$$

and by demanding that events $\binom{A}{n}$ and $\binom{B}{k}$ be independent whenever the elements A and B of the family $\mathcal{P}$ are disjoint.

In this definition of the Poisson process, the choice of a probability is a "Deus ex machina" which is justified by the exquisitely named "law of rare events", or by plain handwaving.

Now to the ruthlessly pointless definition by generators and relations, first obtained by Kung. We begin with a distributive lattice L on which an atomless unbounded (finitely additive) measure μ is defined. Next, we consider the Boolean algebra freely generated by elements which we denote by $\binom{A}{n}$, as A ranges over the lattice L and n ranges over the nonnegative integers. On this family of generators we impose the following relations:

1. $\binom{A}{n}^c = \bigcup_{j \neq n} \binom{A}{j}$;
2. $\binom{A}{n} \cap \binom{A}{j} = \hat{0}$ if $n \neq j$.
3. $\binom{A \cup B}{n} = \bigcup_{i+j=n} \binom{A}{i} \cap \binom{B}{j}$ whenever the sets A and B are disjoint, where i and j range over the nonnegative integers.

Let us call *Poisson* the Boolean σ-algebra thus generated. The Poisson process may be now defined pointlessly as the unique probability measure on the Boolean σ-

algebra *Poisson*, subject to some technical details. The proof of uniqueness of such a probability is nontrivial (Kung).

The pointless definition of the Poisson process is an example of an algebraic characterization of a stochastic process.

Similar algebraic characterizations in terms of generators and relations on Boolean σ-algebras can be given for Brownian motion, for shot noise and for the still mysterious gamma process. These algebraic characterizations remain to be worked out, to the best of my knowledge.

2 Problem two: densities of random variables

If X is a random variable, we denote by $(X \leq t)$ the event that the random variable X takes values at most equal to t, that is, the set $\{\omega : X(\omega) \leq t\}$.

We are used to associating to every random variable X the cumulative probability distribution $F_X(t) = P(X \leq t)$, and the statistical properties of a random variable are in principle determined by its cumulative probability distribution. For example, the moments of the random variable are defined by the Stieltjes integral

$$E(X^n) = \int_{-\infty}^{\infty} t^n dF_X(t).$$

As a matter of fact, the invention of the Stieltjes integral was motivated by this definition of the moments of a random variable.

Textbooks in probability do not fail to go into great detail on the Stieltjes integral. However, in the practice of probability the Stieltjes integral is avoided like the plague. In practice, one deals mostly with two kinds of random variables: in the first place, random variables for which the cumulative probability distribution is absolutely continuous, or continuous random variables as they are often called. For these random variables, one defines the probability density $\text{dens}(X = t)$ of a random variable X to be the derivative of the cumulative probability density $P(X \leq t)$ relative to the variable t so that the formulas for the moments are more aptly expressed in terms of the density in the form

$$E(X^n) = \int_{-\infty}^{\infty} t^n \,\text{dens}(X = t)\, dt.$$

The cumulative probability distribution is recovered from the density by the formula

$$P(X \leq t) = \int_{-\infty}^{t} \text{dens}(X = t)\, dt.$$

The second class of random variables one frequently deals with are those that take only a countable set of values, most frequently integers. For these discrete-valued random variables, one can still define a density: it is a linear combination of Dirac delta functions. There are however other kinds of random variables that crop up from time to time that are neither continuous nor discrete, and it would be interesting to be able to work with a notion of density that will work for all random variables whatsoever. Laurent Schwartz's theory of distributions is a modest step in this direction. However, this theory was developed largely to meet the needs of the theory of linear partial differential equations. In its present version, Laurent Schwartz's theory of distributions is unsuitable to define a density for all random variables. A modification of Laurent Schwartz's theory of distributions will be needed, but has never been carried out.

Sometime in the fifties, my late friend Alberto Calderón remarked to me that the weakness of Laurent Schwartz's theory of distributions was the fact that the theory was not adapted to making changes of variables. This weakness makes Laurent Schwartz's theory unsuitable for defining the density of an arbitrary random variable. What is needed, and what to the best of my knowledge no one has carried out to this day, is a modification of Laurent Schwartz's theory of distributions that will take changes of variables into account. More specifically, the density, in whatever sense it is to be defined, of a function $f(X)$ of the random variable X is to be related to the density of the random variable X by an explicit formula similar to the formula that is used for positive continuous random variables and for differentiable functions f with a positive derivative, namely,

$$\mathrm{dens}(f(X) = t) = \mathrm{dens}(X = f^{(-1)}(t))\frac{d}{dt}f^{(-1)}(t).$$

I have no idea whether this problem has a solution, or even whether it makes sense. A solution would be desirable, since one gets a feeling for the statistical properties of a random variable from its histogram, namely, by graphing its density. No one can infer any but the most trivial statistical properties of a random variable by staring at a graph of its cumulative probability distribution.

Such a definition of density as a generalized "distribution" would have to pass another test: it would have to allow changes of variables in several random variables. The basic requirement is simple: one has to define a joint density of, say, two random variables X and Y, call it $\mathrm{dens}(X = t, Y = s)$, in such a way as to have an identity of the form

$$P((X \leq t) \cap (Y \leq s)) = \int_{-\infty}^{t}\int_{-\infty}^{s} \mathrm{dens}(X = t, Y = s)\, dt\, ds.$$

However, the problem of changes of variables becomes even more delicate.

3 Problem three: structure theory for Boolean σ-algebras

Every random variable X determines a unique σ-subalgebra π_X of the σ-algebra of events Π, namely, the minimal σ-subalgebra of Π relative to which X is measurable. Such a σ-subalgebra π_X expresses the ability of the random variable X to distinguish points of the sample space. This ability of distinguishing among points is relevant in the interpretation of random variables as the result of a search, in information theory. The finer the σ-subalgebra π_X defined by a random variable, the better the result of a search provided by the random variable X. Rényi proposed the term "information provided by X" to denote the σ-subalgebra π_X determined by a random variable X.

For random variables X taking integer values, the information π_X provided by X is an atomic Boolean algebra, that is, the σ-subalgebra π_X is generated by a disjoint family of events of the sample space called atoms. The set of atoms of π_X defines a partition of the sample space Ω. Thus, for integer random variables an information-theoretic interpretation of the information provided by the random variable X is the following. Consider a search problem, where an unknown sample point ω_o of Ω is to be searched for by measurements which are the values of random variables. The value of a random variable X which is used in the search process will specify which of the blocks of the partition defined by π_X the unknown element ω_o lies in.

In general, several measurements will be needed, that is, the information provided by several random variables will have to be taken into account in order to determine the unknown point ω_o. We are therefore led to develop a notation for the joint information provided by several random variables, and this in turn leads us to the next fundamental notion of probability, namely, the lattice structure on the family of all σ-subalgebras of the Boolean σ-algebra of events Π.

To this end, the family Σ of all σ-subalgebras of the σ-algebra Π is ordered by reverse containment. That is, if π and π' are σ-subalgebras, say that $\pi \leq \pi'$ when $\pi \supseteq \pi'$ as families of sets. We obtain a partially ordered set, and meet and join are defined in the partially ordered set Σ as follows. The join of two σ-algebras π and π' is defined as the set-theoretic intersection of the two σ-subalgebras considered as families of sets, and the meet of π and π' is the σ-subalgebra generated by π and π'.

The lattice Σ of all Boolean σ-subalgebras of Π is one of the most interesting objects of probability. It is also one of probability's most neglected notions. The lattice Σ of all Boolean σ-subalgebras has never been properly structurally characterized. It suffers from some pathological properties that lead to the suspicion that such a structural definition may be difficult. For example, it has been proved that the lattice Σ satisfies no identities whatsoever in meets and joins, other than the trivial ones, and that every lattice whatsoever can be embedded in Σ, subject only to cardinality restrictions.

The lack of intrinsic definitions of the lattice Σ of all Boolean σ-subalgebras of a sample space stands in stark contrast to the multiple definitions of a Boolean σ-algebra in lattice-theoretic terms. In view of the fact that, categorically speaking, the Boolean σ-algebra of events and the lattice Σ of all Boolean σ-subalgebras are dual notions, this unevenness of definition is puzzling.

Not only do we miss a proper definition of the lattice Σ. Our present knowledge of the structure of individual σ-subalgebras is also deficient. Too many elementary questions remain open or ill-understood. For example, it is known since von Neumann that there is one Boolean σ-algebra of events which meets all the needs of probabilistic computations. It is sometimes called the "standard" Boolean σ-algebra. It has a simple definition. It is a separable atomless Boolean σ-algebra. A theorem of von Neumann states that any two such σ-algebras are isomorphic. The proof of this theorem given by von Neumann in his paper *Zur Operatorenmethoden der klassischen Mechanik* has not been appreciably simplified to this day. As a matter of fact, some later proofs are more complicated than the original proof! We miss a two-line proof of von Neumann's theorem.

The difficulty of every attempted proof lies in proving a lemma stating that that in any atomless separable sample space it is possible to find two complementary events A and A^c of equal probability, that is, with $P(A) = P(A^c) = \frac{1}{2}$. Surely there is a one-line proof of this intuitive fact.

The next hole is the classification of σ-subalgebras π of Π to within probability-preserving transformations. We know the facts, but we do not quite understand what these facts mean. Assume even that both π and Π are atomless and separable. A theorem proved almost simultaneously by Maharam and by Rokhlin states that any two such σ-subalgebras π and π' are isomorphic, that is, that there is always a probability-preserving transformation of the underlying sample space Ω that sends one into the other. In simpler terms, the result of Maharam and Rokhlin can be stated as follows. We can visualize the sample space Ω as the unit square $[0, 1] \times [0, 1]$ in the plane, with Lebesgue probability and with Π being the family of Borel sets. Then every atomless σ-subalgebra of Π is isomorphic to the σ-subalgebra of cylinder sets, that is, to the σ-subalgebra generated by all rectangles of the form $A \times [0, 1]$ whose base is an arbitrary measurable subset A of the interval $[0, 1]$.

A short, elegant proof of this theorem has just been published by Catherine Yan in the latest issue of the journal *Advances in Mathematics*.

Let us go on to the next scorching issue relating to the classification of σ-subalgebras, which has remained open since the beginnings of probability theory. Kolmogorov believed this problem to be one of the most important in the theory of stochastic processes, but, to the best of my knowledge, no work on it has been done to date.

In order to properly motivate the problem, I must remind you of some ideas from functional analysis. Every orthogonal operator U in a real Hilbert space H is endowed with a spectral resolution, that is, with a family of orthogonal projections E_t depending "continuously" on the real parameter t, such that if $t \leq t'$ then $E_t \leq E_{t'}$, that is, the range of the projection E_t is contained in the range of the projection $E_{t'}$. A chapter of the spectral theory of self-adjoint operators in Hilbert space, called multiplicity theory, classifies such spectral resolutions to within orthogonal equivalence, and thereby gives a complete set of invariants for orthogonal operators in Hilbert space to within orthogonal equivalence.

The problem stated by Kolmogorov may be viewed as the analog of multiplicity theory for families π_t of Boolean σ-subalgebras of a sample space Ω depending "continuously" on the real parameter t (I will not make precise what is meant by "continuously"), having the property that $\pi_t \leq \pi_{t'}$ if $t \leq t'$. Every martingale defines such a one-parameter family of σ-subalgebras, but the problem of their equivalence to within a probability-preserving transformation has been neglected. In somewhat optimistic terms, the problem asks for an analog of multiplicity theory for Boolean σ-algebras.

You may object that this problem is not as important as the problem that is solved by multiplicity theory in Hilbert space, because multiplicity theory yields a complete set of invariant for orthogonal operators in Hilbert space. What kind of invariants would such a classification of increasing families of σ-algebras provide? We will come back to this question.

4 Problem four: entropy

The lattice Σ of σ-subalgebras plays for information the role that the lattice of events plays for probability. To get an inkling of this role, let us first consider a σ-subalgebra π_X that is σ-generated by a countable set of atoms of positive probability, and that is therefore associated with a partition of the sample space. Recall that, if π_X is the information provided by a random variable X, then, as we have seen, we can view the random variable X as a question whose answer will tell us which of the blocks of the partition π_X an unknown sample point ω_o lies in. Suppose now that $X, X', \ldots, X''$ are random variables generating information $\pi_X, \pi_{X'}, \ldots, \pi_{X''}$, all of them atomic Boolean σ-algebras. Then the joint information provided by all such random variables is the meet of the Boolean σ-subalgebras $\pi, \pi', \ldots, \pi''$. In terms of partitions, such a meet is the Boolean σ-algebra generated by a partition whose blocks are obtained by taking all possible intersections of sets of blocks, one from each of the partitions $\pi_X, \pi_{X'}, \ldots, \pi_{X''}$, in all possible ways, and discarding intersections of probability zero. This interpretation of the meet of σ-algebras, at least when they are all atomic, renders in precise terms the intuitive feeling that the more questions we ask in the process of search of an unknown sample point ω_o, the closer we get to determining ω_o.

I will briefly digress by stating a seemingly irrelevant open question. Is there an information-theoretic interpretation of the join of Boolean σ-algebras? I have never heard a convincing proposal.

Since the lattice of events and the lattice of Boolean σ-subalgebras are dual notions, we expect to find analogies between the two. An important information-theoretic notion is arrived at by analogizing the probability of events. Probability is a measure on the Boolean algebra of events. In deliberately abstruse terms, let us force ourselves to view such a measure as a real-valued function on the lattice of events which enjoys certain algebraic properties expressed in terms of joins and meets of events, that translate into algebra the intuitive idea of the size of a set. By analogy, we may look for a similarly defined real-valued function on the lattice Σ of σ-subalgebras, which will capture some property that will turn out to be for σ-subalgebras what size is to a set. How shall we be led to such a property? We have already an inkling of what it should be: it should be a measure of the information provided by a random variable. Is there a candidate for a measure of the amount of information?

We encounter here one of those shady areas that have steadfastly refused to be mathematically understood. There is indeed a number that is associated with "certain" σ-subalgebras and, we would like to believe, probably to all σ-subalgebras. This number is called entropy. Entropy plays for σ-subalgebras a role analogous to the role of probability for events.

Entropy has been axiomatized, but only on the sublattice of atomic σ-subalgebras. If π is an atomic Boolean σ-subalgebra, the entropy of π, in symbols $H(\pi)$, is a real number, and the function H is subject to the following axioms, plus some additional technical conditions which I deliberately omit:

1. If $\pi \leq \pi'$, then $H(\pi) \geq H(\pi')$.
2. $H(\pi \wedge \pi') \leq H(\pi) + H(\pi')$.
3. If π and π' are independent σ-subalgebras, then

   $$H(\pi \wedge \pi') = H(\pi) + H(\pi').$$

   Recall that two σ-subalgebras π and π' are said to be independent whenever every pair of events A and B, where A belongs to π and B belongs to π', are independent, that is, $P(A \cap B) = P(A)P(B)$.
4. Let s be a probability-preserving automorphism of the Boolean σ-algebra Π. Then s induces an automorphism of the lattice Σ of all Boolean σ-subalgebras, and under these circumstances we will say that two Boolean σ-subalgebras π and π' are isomorphic.
   One assumes that, if π and π' are isomorphic, then $H(\pi) = H(\pi')$.
5. $H(\pi) \geq 0$.

Under these conditions, one can show that there is a function, unique except for normalization, which is defined on atomic Boolean σ-subalgebras π by the well-known formula

$$H(\pi) = \sum_{B} P(B) \log_2 \frac{1}{P(B)}.$$

The sum ranges over all atoms B of the underlying partition of the σ-algebra π.

This function is the Shannon entropy. One normalizes the Shannon entropy by the logarithms in the base two, because the entropy of a σ-subalgebra generated by a partition of Ω into two blocks, each having probability one half, should equal one, or one bit of entropy, following Shannon. A word of warning: even when π is an atomic σ-subalgebra the entropy $H(\pi)$ may be infinite.

Entropy measures the lopsidedness of a Boolean σ-subalgebra. That is, among all atomic Boolean σ-subalgebras generated by n atoms, entropy takes the maximum value on those σ-subalgebras whose partitions have blocks of the same probability. The more uneven the probability of the blocks, the smaller the entropy.

Before going on to the puzzling aspects of entropy, let us recall the basic information-theoretic application of the notion of entropy. Assume that the σ-algebra Π is atomic, and that $\Pi = \pi \wedge \pi' \wedge \cdots \wedge \pi''$. Then, for any set of σ-subalgebras π, $\pi', \ldots, \pi''$, one of the axioms above tells us that

$$H(\Pi) \leq H(\pi) + H(\pi') + \cdots + H(\pi'').$$

It follows that a measurement scheme, that is, any choice of random variables whose answers will help identify an unknown sample point ω_o, cannot work unless the entropy of the meet of the information provided by each of these random variables is at least equal to $H(\Pi)$. We thus obtain "the information-theoretic lower bound" to a search problem, a bound that has proved useful in engineering.

A celebrated theorem of Shannon asserts that in a certain sense the converse of this statement is true. That is, for an information-theoretic bound $H(\Pi)$, it is possible to devise a scheme of measuring by random variables whose joint entropy equals $H(\Pi)$, in other words, which comes as close as possible to performing a successful and maximally economical search for an unknown sample point.

Now to the bad news. A great deal of effort has gone into extending the definition of entropy to σ-subalgebras other than atomic σ-subalgebras. It cannot be said that these efforts have failed since they have resulted in some of the deepest results ever discovered in ergodic theory, in the work of Kolmogorov, Ornstein, Ruelle and others. Neither can it be said that these efforts have been successful. All definitions attempted so far of the entropy of a atomless σ-subalgebra rely on crude approximation techniques whereby an atomless σ-subalgebra is approximated by a sequence of atomic σ-subalgebras.

A great deal of effort has gone into axiomatizing the notion of entropy on the lattice of σ-subalgebras independently of the notion of probability; most of these efforts were carried out in Eastern Europe in the sixties and seventies. They cannot be said to have been successful either; however, they did bring out some interesting features of the lattice Σ and I should like to review one of them.

Consider the family of σ-subalgebras defined as follows. Take an event A of positive probability, and define the σ-subalgebra π_A as follows. It will contain the event A and every event contained in the complement of A, or A^c. It turns out that σ-subalgebras of the form π_A can be uniquely characterized as the only σ-subalgebras π of the lattice Σ having the property that, for every pair π' and π'' of σ-subalgebras, the distributive law holds:

$$\pi \wedge (\pi' \vee \pi'') = (\pi \wedge \pi') \vee (\pi \wedge \pi''),$$

as well as the dual identity obtained by reversing the roles of $\wedge$ and $\vee$. By virtue of this property, one can show that the sublattice generated by all σ-subalgebras of the form π_A, as A ranges over all events of the sample space, is isomorphic to the Boolean algebra of events Π. Thus, suppose an entropy function were to be defined on the lattice Σ. Then, subject to technical assumptions on the sample space Ω, the restriction of such an entropy function to σ-subalgebras of the form π_A allows us to recover the probability! This is a line of thought that has never been completely carried through, because of combinatorial difficulties.

I should like to submit to you a proposal for a definition of entropy on the lattice of σ-subalgebras. We follow an idea that gained favor in the latter half of this century and that was brilliantly exploited by Grothendieck. Accordingly, let us see how a "universal entropy group" might be defined. We begin by taking the free Abelian group generated by all σ-subalgebras; call this free Abelian group F. We then impose on this group the identity $\pi \wedge \pi' = \pi + \pi'$ whenever π and π' are independent σ-subalgebras.

The quotient group thus obtained may be denoted by F'. The quotient group F' is the universal object that any entropy function must factor through, as is evident from the construction.

Next comes a step that will be more difficult to carry out. Denote by $[\pi]$ the canonical image of a σ-subalgebra π in the group F'. Suppose that s is any probability-preserving transformation of the sample space Ω. Take the quotient of F' obtained by identifying $[\pi]$ with $[s(\pi)]$ for all σ-subalgebras π. Call this quotient the entropy group of the sample space Ω. This group cannot be entirely trivial since the entropy of atomic σ-subalgebras factors through it. The big question is: is this group ordered? If so, then the entropy group will be as good a candidate as any for the values of a universal entropy defined on all σ-subalgebras. I conjecture that the entropy group is linearly ordered. If this conjecture were true, then the linearly ordered entropy group will almost certainly not be Archimedean. I hope someone will get interested in working out the properties of the entropy group.

To proceed from Shannon to Boltzmann entropy, we need to recall a refinement of entropy, namely, conditional entropy. If π and π' are atomic Boolean σ-algebras, with $\pi \leq \pi'$, the conditional entropy $H(\pi|\pi')$ is defined as

$$H(\pi|\pi') = \sum_{C \in \pi'} P(C) H(\pi|C),$$

where the conditional entropy $H(\pi|C)$ relative to the event C is defined as

$$H(\pi|C) = \sum_{B \in \pi} P(B|C) \log_2 \frac{1}{P(B|C)}.$$

In order to motivate the definition of Boltzmann entropy, we go back to the Shannon entropy and first define the Shannon entropy of a random variable rather than the entropy of a σ-subalgebra. This is easy if the random variable X provides information which is an atomic σ-subalgebra, say π. When this is the case, one simply defines $H(X)$ to equal $H(\pi)$. For any measurable function f one has $H(f(X)) \leq H(X)$, and this inequality is further confirmation that entropy is a measure of disorder.

The first definition of entropy of a random variable, due to Boltzmann, is superficially similar to the definition of Shannon entropy, but this similarity is only skin-deep. We recall it in the form of conditional entropy, although this is not the way Boltzmann introduced it. Suppose that X and Y are random variables. Define the measure $\mu_Y(A)$ of an event A to equal $E(Y I_A)$, and let $\text{dens}(X = t|Y)$ be the conditional density of the random variable X given the random variable Y (of course we assume that such a conditional density exists and has suitable properties). One defines the conditional entropy $H(X|Y)$ of the random variable X relative to the random variable Y by the formula

$$H(X|Y) = \int_{-\infty}^{\infty} \text{dens}(X = t|Y)) \log \frac{1}{\text{dens}(X = t|Y)} \mu_Y(dt).$$

One can adapt the axioms for the Shannon entropy and obtain by analogy a similar axiomatization of conditional Boltzmann entropy of random variables which are well-behaved towards each other. Conditional Boltzmann entropy is nonnegative, and $H(f(X)|Y) \leq H(X|Y)$ for "any" measurable function f, so that, again, conditional Boltzmann entropy can be taken to be a measure of disorder.

Conditional Boltzmann entropy, thus defined, assumes that the random variable X is absolutely continuous relative to the random variable Y. It is not known, and probably unlikely, that conditional Boltzmann entropy may be defined for more general pairs of random variables. The appearance of absolute continuity in this context injects a nonalgebraic flavor that Shannon entropy has managed to avoid. Bringing the algebra out of this definition is not likely to be an easy task.

Boltzmann entropy is usually defined in another way, which is analytically simpler but conceptually confusing. The classical definition of the Boltzmann entropy of a random variable X is given by the expression

$$H(X) = \int_{-\infty}^{\infty} \mathrm{dens}(X = t) \log \frac{1}{\mathrm{dens}(X = t)}\, dt.$$

This "absolute" Boltzmann entropy, which is Boltzmann's original definition, has the advantage of simplicity. It does however have a major drawback: it is no longer true that $H(X) \geq 0$ for every random variable X. It assumes that the random variable X has a density relative to Lebesgue measure on the real line. Worse yet Boltzmann entropy cannot be obtained as a limit of discrete Shannon entropies by writing the integral as a limit of sums.

The information-theoretic interpretation in terms of search is not easily derived for Boltzmann entropy, as it is for Shannon entropy, when X is a random variable whose information is an atomic σ-subalgebra. Nonetheless, and remarkably, all the main results of Shannon information theory have been carried over to more general random variables by Boltzmann entropy. The proofs can only be guessed by analogy, and have to be painfully redone from scratch. In this and other thermodynamical situations, the passage from the discrete to the continuous has remained difficult and mysterious.

Despite the algebraic attractiveness of Shannon entropy, Boltzmann entropy is the one that dominates in applications to statistical mechanics and to thermodynamics.

An abyss of ill-understood ideas separates Shannon entropy from Boltzmann entropy. No one to this day has seen through the dilemma of choosing between entropy as a measure of randomness of probability densities of random variables and entropy as a measure of randomness of σ-subalgebras. The intractability of entropy of atomless σ-subalgebras suggests that the program of defining the entropy for σ-algebras other than atomic ones may be hopeless, and that the natural definition in the continuous case will inexorably require an order relation among probability densities, which the Boltzmann entropy hints at. Such an order relation will in turn appeal to the singularity of probability densities, rather than the discrepancy between two σ-subalgebras, an unknown area at present, although one which my friend George Mackey warmly recommends for further research.

The difficulties in extending the notion of entropy to arbitrary random variables raise the suspicion that the present Kolmogorov formulation of probability in terms of events and σ-algebras may be inadequate to the phenomenological needs of probability. It is conceivable that the present picture of probability might one day be replaced by a suppler definition where the notions of σ-subalgebra and probability density are not as intractable as they are in present day measure theory. This suspicion deepens when we take into account the fact that Boltzmann's original definition of entropy may be viewed as conditional relative to the Lebesgue measure on the real line. But Lebesgue measure on the line is not the density of any random variable. Does this mean that the very concept of a random variable should be placed on the operating table? We will shortly deal with another problem that further supports this suspicion.

At the risk of ridicule, I should like to go as far as pinpointing where such a revision of the fundamentals of probability should start. The Kolmogorov consistency theorem states that under certain technical conditions any consistent system of joint densities can be realized by a set of random variables in a sample space. To all practical purposes a sample space is completely determined by a preassigned set of consistent joint densities of random variables. The Kolmogorov consistency theorem confirms the realistic view of randomness. What are given are densities, or rather, the histograms of certain densities. One may argue that every probabilistic notion depends on such densities. As a matter of fact, after the publication of Kolmogorov's tract of 1932, there remained some diehard mathematicians who refused to work with random variables, and who insisted that all of probability should be done directly with densities. One of these mathematicians was Aurel Wintner of Johns Hopkins University, who published a set of notes on probability formulated entirely in terms of densities. In these notes Wintner proved some important new results, well ahead of their time; however, because he wrote in terms of densities instead of random variables, his work was ignored, and his discoveries are now attributed to others.

It may well turn out that an extension of the notion of sample space may have to take its lead from Kolmogorov's consistency theorem, by allowing a consistent set of measures which are not necessarily probability densities. The appearance of the Lebesgue measure on the real line in some statistical contexts supports this contention. However, I have no constructive suggestion on how this extension could be carried out, nor what notion may one day replace our sample spaces, if Kolmogorov's consistency theorem is ever extended to include infinite measures. Replacing the sample space by a measure space of infinite measure is too naive a proposal, and it should be discarded a priori. Rényi tried it and failed, as he confessed to me in a conversation I had with him in Budapest just before he died. A completely original idea is called for. I will shamelessly stop at this point.

5 Problem five: the maximum entropy principle

The generalization of the present day notion of a sample space to accommodate entropy is not an academic problem. Behind this problem lurks one of the burning issues that separate mathematicians and physicists, and that is seldom brought up. I mean the problem of the mathematical axiomatization of thermodynamics. Years ago Luigi Radicati told me in no uncertain terms that he did not believe an axiomatization of thermodynamics is possible. I do not know any thermodynamics, and I will therefore restrict myself to a little corner of this big question.

Shannon entropy is maximized for atomic Boolean σ-algebras where all atoms have the same probability. There is a continuous analog of this fact, first stated by Kampé de Fériet . Among all random variables with finite variance, those having the normal distribution have maximum Boltzmann entropy. Kampé de Fériet's discovery led to the development of an information theory for Gaussian processes, which is now the bread and butter of signal processing.

But Kampé de Fériet's discovery was not an isolated one. It was soon followed by the discovery that every stochastic process that has been observed can be characterized by a maximum entropy principle. For example, an exponential random variable may be defined as the nonnegative random variable of maximum Boltzmann entropy among all nonnegative random variables of finite expectation. Similarly, a random variable that is uniformly distributed over a finite interval of the real line may be defined as the random variable of maximum entropy among random variables supported on the same interval.

The maximum entropy principle, which may be gleaned from the preceding examples, states that, in the absence of any further information, the best guess of a random variable satisfying given conditions is the random variable of maximum entropy that satisfies those conditions. Among all mathematical recipes, this is to the best of my knowledge the one that has found the most striking applications in engineering practice. The best techniques of image reconstruction known to date rely on the maximum entropy principle. I have myself been witness to police cases where the correct number plate of an automobile was reconstructed by maximum entropy from a photograph which to the naked eye looked like chaos. Even the solution of overdetermined systems of equations is at present best carried out by maximum entropy computations.

In view of such a variety of lucrative applications, the complete and utter lack of justification of the maximum entropy principle is nothing short of a scandal. On learning that a normally distributed random variable of finite variance has maximum entropy, it is natural to ask for an intuitive proof of the central limit theorem that relies on maximum entropy; such a proof has never been given, to the best of my knowledge, although several mathematicians attempted it, among them Linnik and Rényi.

A great deal of work has been done, notably by Jaynes and his school, in an attempt to justify the maximum entropy principle, but we cannot say we are any closer to such a justification today than Boltzmann or Szilard ever were. Computations that follow the maximum entropy principle are endowed with a statistical correctness that disappears the moment we try to explain what makes such computations work. I am reminded of St. Augustine's quip about time: "Si nemo a me quaerat, scio; si quaerenti explicare velim, nescio".

6 Problem six: conditional probability, conditional expectation and Bayes' law

Of all applications of probability to engineering and physics, the maximum entropy principle and Bayes' law are the two that have yielded the highest dividends. At present, experts at applying entropy and Bayes are to be found among physicists and engineers, not among mathematicians. It seems that mathematicians have made it a point to neglect precisely the two most successful notions in all probability.

Having dealt with the sad state of entropy and the maximum entropy principle, let us now turn our attention to the sad state of Bayes' law. At the risk of giving away my punch line, we want to show how the problem of injecting rigor into the heuristic but valid applications of Bayes' law leads again to reopening Pandora's box of the foundations of probability.

Bayes's law deals with conditional probability, and, in the measure-theoretic presentation of probability, conditional probability is made to look like as a poor relative of conditional expectation. Let us review the uneasy relationship between these two concepts.

The conditional probability of an event A given an event B, in symbols $P(A|B)$, is defined as $P(A \cap B)/P(B)$. This definition of conditional probability is one of the greatest triumphs of mathematical definition of all times. It is amazing to realize that such a simple definition renders the intuitive concept: "the occurrence of event B makes the occurrence of event A more probable". Keisler has justified the definition of conditional probability by showing that it provides a bridge between probability and logic.

Rényi was one of several mathematicians who were star-struck by conditional probability. He did much work aimed at reshaping the foundations of probability in terms of conditional probability, but his ideas have not caught on. They were ahead of their time, as we will see in the next brief digression.

Rényi's main axiom for conditional probability can be recast as follows (I hope Rényi will not turn in his grave for my restatement of his axiom in a form that he might not recognize):

$$P((A|B)|C) = P(A|B \cap C).$$

Cast in this form, Rényi's axiom can be interpreted category-theoretically. If we view $A|B$ as a morphism from an object A to an object B in a category whose objects are the events, then Rényi's axiom is a condition to be satisfied in a category, if such a category is to play the role of a generalization of a sample space where conditional probability still makes sense. The condition that there be only one arrow from A to B could be relaxed. In other words, we propose to generalize the notion of sample space by allowing an event A to be "conditional" on an event B "in several possible ways". Rényi was groping for some such generalization of probability. I hope you will forgive my "chutzpah" in suggesting such a wild generalization.

Kolmogorov was, I believe, the first to shove aside conditional probability in favor of conditional expectation of a random variable relative to a σ-subalgebra, by making ample use of the recently discovered theorems of Fubini and of Radon-Nikodym. The conditional expectation $E_\pi(X)$ of a random variable X relative to an arbitrary σ-subalgebra is defined in two steps. First, one defines the measure $\mu_X(A) = \int_A X dP$ for every event A in π. Second, one observes that this measure is absolutely continuous relative to the probability P, and therefore may be written in the form $\mu_X(A) = \int_A X' dP$, where X' is a random variable which is measurable relative to the σ-subalgebra π. The linear operator $E_\pi(X) = X'$ is the conditional expectation operator relative to the σ-algebra π. I gave an algebraic definition of conditional expectation in 1958. Let T be a linear operator on random variables which satisfies suitable continuity conditions (for example, it must be a contraction) and which satisfies the two algebraic conditions $T1 = 1$, where 1 is the random variable identically equal to one, and $T(X(TY)) = (TX)(TY)$ for any two random variables X and Y. Then T is a conditional expectation operator for some σ-subalgebra, that is, $T = E_\pi$ for some σ-subalgebra π.

If π' is another σ-subalgebra, the conditional expectation operators E_π and $E_{\pi'}$ do not necessarily commute. We will shortly see how requiring commutativity of conditional expectation operators leads from ordinary probability to quantum probability.

Conditional probability is seen to be a special case of conditional expectation by considering an atomic Boolean σ-subalgebra. Suppose that π is an atomic σ-subalgebra. Then the conditional expectation $E_\pi(I_B)$ of an event B relative to π is the random variable X' which is constant on each atom of π, and which on the atom A of π takes the constant value $\frac{P(A\cap B)}{P(A)}$.

Although historically there is a continuity of thought leading from conditional probability to conditional expectation, conditional probability and conditional expectation are birds of different feathers. The divergence between the two concepts becomes clear in the analysis of one of the most controversial statements of statistics, namely, Bayes's law.

The official statement of Bayes's law that is found in most textbooks of statistics is misleading. Bayes' law is stated in the form of a trivial identity for conditional probabilities: for any events A and B we have

$$P(A|B) = \frac{P(B|A)P(A)}{P(B)}.$$

To the reader ignorant of the intended interpretation, the meaning of this statement will remain elusive. Such a reader will become more perplexed after learning that the terms in the above identity are alluded to in a special jargon. The left side is "the posterior", the first term in the numerator on the right is "the likelihood", and the second term in the numerator on the right as "the prior". The denominator is ignored: as a matter of fact statisticians love to state Bayes' law in the form

$$P(A|B) \propto P(B|A)P(A),$$

using the obsolescent symbol $\propto$ to denote proportionality. What is going on here?

In order to see through the mystery one is forced to dig up the original instance of Bayes' law, due to Laplace. Unless one is privy to this example, Bayes' law will remain mysterious.

Suppose that we are given a set of n balls in an urn, of which an unknown subset of balls is red, the others being black. We extract k balls and find that j of the extracted balls are red. What is our estimate of the number of red balls?

The sample space Ω_{MB} of this example is the sample space of Maxwell–Boltzmann statistics. A sample point is an arbitrary subset of k balls, every subset of the sample space is an event, and we set

$$P_{MB}(\omega) = \frac{1}{\binom{n}{k}}$$

for every sample point ω.

Let the random variable X equal the number of red balls in the urn, and let the random variable Y equal the number of red balls in a set of k balls extracted at random. The probability distributions of the random variables X and Y cannot be both arbitrarily prescribed, because combinatorial considerations dictate that

$$P(Y = \ell | X = j) = \frac{\binom{j}{\ell}\binom{n-j}{k-l}}{\binom{n}{k}}.$$

This conditional probability remains fixed irrespective of the choice of the distributions of X and Y. This fixed quantity is what statisticians call the likelihood. Thus, once the probability distribution of one of the two random variables X and Y is chosen, the probability distribution of the other is determined by the likelihood, but we are at liberty to choose either of the two probability distributions. Irrespective of such a choice, we have

$$P(X = j | Y = \ell) = \frac{P(Y = \ell | X = j) P(X = j)}{P(Y = \ell)}.$$

Suppose we now choose the probability distribution of the random variable X arbitrarily. Then the probability distribution of the random variable Y is determined, and we have

$$P(X = j | Y = \ell) = \frac{\binom{j}{\ell}\binom{n-j}{k-\ell} P(X = j)}{\sum_j \binom{j}{\ell}\binom{n-j}{k-\ell} P(X = j)}.$$

We can now proceed to an interpretation. We guessed a probability distribution of the random variable X, in technical terms we guessed a prior. The likelihood is independent of this guess; it is given by the structure of the sample space. The posterior is, in everyday language, the estimate of the evidence of the assumption $(X = j)$, given the evidence $(Y = \ell)$.

The Laplace example shows that Bayes' law is a comparison of σ-algebras, not a comparison of events. What we are dealing with in the preceding example is a comparison between the information π_X and the information π_Y provided by the random variables X and Y.

Having thus recast Bayes' law in terms of σ-algebras, we may take a stab at generalizing Bayes' law to more general σ-algebras than atomic σ-algebras. Such a generalization has been attempted by several mathematicians (Bahadur, Blackwell, Doob, Halmos and others), but it cannot be said to have been successful, and it has been abandoned because of the measure-theoretic sophistication that it requires, and perhaps also because of lack of applications.

We meet once more the situation we met in trying to generalize Shannon entropy. It turns out once more that for random variables whose σ-subalgebras are atomless the generalization of Bayes' law is better carried out in terms of densities rather than in terms of σ-algebras. Given random variables X and Y with continuous densities, one shows, under suitable technical assumptions, that

$$\begin{aligned}\operatorname{dens}(X = t|Y = s) &= \frac{\operatorname{dens}(Y = s|X = t)\operatorname{dens}(X = t)}{\operatorname{dens}(Y = s)} \\ &= \frac{\operatorname{dens}(Y = s|X = t)\operatorname{dens}(X = t)}{\int_{-\infty}^{\infty}\operatorname{dens}(Y = s|X = t)\operatorname{dens}(X = t)\,dt}.\end{aligned}$$

Remarkably, here again we meet the same awkwardness that we met in the definition of the "absolute" Boltzmann entropy. In the practice of statistics, it is universally accepted that the prior density $\operatorname{dens}(X = t)$ may be chosen to be the uniform density over the line, which is not the density of a random variable. One simply argues that, for several choices of the likelihood, notably when the likelihood is a normal density, the integral in the denominator remains finite when $\operatorname{dens}(X = t)dt$ is set equal to dt. No rigorous justification has ever been given of this procedure. We may take this gap in rigor as further evidence for the need of a rethinking of the notion of sample space and of Kolmogorov's consistency theorem along the lines we have proposed.

7 Problem seven: justification of the univariate normal distribution

The problem here is not a lack of justification of the normal distribution; rather, it is the fact that there are too many such justifications, unrelated to one another. The justification by the maximum entropy argument would be excellent, if it could only be used to prove the central limit theorem. The original justification given by Gauss is convincing, but it relies on the maximum likelihood principle, which works like a charm in applications, but which is presently lacking a rigorous foundation. Mark Kac's derivation of the normal distribution by rotational invariance does not relate to the statistical applications of the normal distribution. The justification that appeals to the central limit theorem is ultimately psychological, unless the speed of convergence to the normal distribution in the central limit theorem is specified. For some thirty years, there was a Russian school of statisticians, headed by Petrov, whose sole concern was the estimation of such a speed of convergence, but again, they cannot be said to have made a dent.

The problem of giving a rigorous justification of the normal law that will have the power of conviction splits into two problems: the justification of the univariate normal law, and the justification of the multivariate normal distribution. The former will require an overhaul of the foundations of statistics, such as has seldom been attempted; the latter has led to some good mathematics that makes the multivariate normal distribution extremely plausible, though some nitty-gritty statistical applications, notably factor analysis, have remained controversial. Paradoxically, the multivariate normal distribution is more tractable than the univariate normal distribution.

What is needed in order to justify the univariate normal distribution is an axiomatic of confidence intervals. Once more, we meet with a technique which is of wide application and which has proven to be effective in statistical practice, but which mathematicians have failed to take seriously.

The only rigorous justification of the notion of a confidence interval attempted so far is due to de Finetti, and I will try to recast de Finetti's ideas in an oversimplified manner. Again, I will use the liberty that is granted to a survey to pass over in silence all nasty technical details.

We are given a parametric family of random variables X_s depending on the real parameter s, with known density $\text{dens}(X_s = t)$. Suppose we record k observations of one and the same random variable belonging to the family. What is our best guess of the parameter s?

De Finetti followed a Bayesian approach. He assumed that the parameter s is itself a random variable m with an unknown density $\text{dens}(m = s)$. The random variable X_m now has a density that is computed by Fubini's theorem as

$$\text{dens}(X_m = t) = \int_{-\infty}^{\infty} \text{dens}(X_s = t)\, \text{dens}(m = s)\, ds.$$

Bayes' law in the language of densities states that, given k "observations" $(X_m = t_1), (X'_m = t_2), \cdots, (X''_m = t_k)$, we have

$$\begin{aligned} &\text{dens}(m = s \mid X_m = t_1, X'_m = t_2, \cdots, X''_m = t_k) \\ &\propto \text{dens}(X_m = t_1, X'_m = t_2, \cdots, X''_k = t_k \mid m = s)\, \text{dens}(m = s). \end{aligned}$$

The right-hand side simplifies admirably, because when m is assigned a fixed value s the random variables on the right are independent. We thus obtain

$$\begin{aligned} &\text{dens}(m = s \mid X_1 = t_1, X_2 = t_2, \cdots, X_k = t_k) \\ &\propto \text{dens}(X_s = t_1)\, \text{dens}(X'_s = t_2) \ldots \text{dens}(X''_s = t_k)\, \text{dens}(m = s). \end{aligned}$$

This identity leads to a simple computation of the probability

$$P(a < m < b \mid X_1 = t_1, X_2 = t_2, \cdots, X_k = t_k),$$

and therefore to a rigorous definition of a confidence interval.

It should not be difficult to carry this train of ideas one step further when the random variables X_s are normally distributed with mean s, and to invent a system of axioms for the normal distribution which will derive the normal distribution from axioms relating to the comparison of ratios of confidence intervals (following Gauss, who came to the normal distribution on the basis of comparisons of expectations). To the best of my knowledge this program has not been undertaken.

De Finetti's argument is by no means universally accepted by the community of statisticians, which is split into two camps: the Bayesians and the frequentists, sharply at odds with each other. The Bayesian motto is: there are no constants, there are only random variables, or equivalently: every constant is a random variable with a sharp density.

Frequentists believe otherwise. The value of a parameter s to be estimated is a God-given constant which is to be approximated by statistical measurement. It is hard to take seriously the suggestion that Planck's constant is a random variable with a peaked density. Frequentists resort to handwaving, and to loud appeals to the central limit theorem, when it comes to giving a mathematically acceptable definition of confidence intervals.

The frequentist point of view seems less rigorous than the Bayesian. Only when one performs actual computations does one realize that frequentist computations are often speedier, and perhaps more sensible. To make things worse, Bayesians often inject into their computations their own never fully justified step. In deciding the choice of a prior, they often opt for the uniform prior, corresponding to no random variable whatsoever. When asked to provide a rigorous justification for this step, Bayesian statisticians raise their voices.

Can the frequentist and the Bayesian points of view be reconciled? I should like to close this problem with a modest suggestion. Carathéodory and von Neumann have shown beyond reasonable doubt by their algebraization of probability that in probabilistic reasoning the notion of a sample point is a psychological prop. To think probabilistically means to think in terms of events in a σ-algebra. To think probabilistically is to think pointlessly. Once this inexorable demand of the algebra is met, then one is all but forced to treat every parameter as a random variable.

8 Problem eight: probability-preserving transformations

This problem has been open for a long time, and little progress has been made on it. Nevertheless, the payoff would be big.

One can make a list of probabilistic analogs of Hilbert space notions. For example, conditional expectation is the probabilistic analog of an orthogonal projection in Hilbert space. A probability preserving transformation is the probabilistic analog of an orthogonal operator in Hilbert space, and so on.

However, whereas Hilbert space theory has been successful in obtaining a structure theory for orthogonal operators, there is at present no analogous structure theory for probability-preserving transformations of a sample space. A long-standing conjecture states that a subfamily of tractable probability-preserving transformations can be singled out, to which one can associate an increasing family of Boolean σ-algebras, which would give a complete set of invariants of a probability-preserving transformation under isomorphisms of the sample space. Such a complete set of invariants would be provided by the probabilistic analog of multiplicity theory, which was stated in a previous problem. There are symptoms that indicate that the solution to this problem ought to be affirmative. Again, to the best of my knowledge no one has seriously worked on this problem. After working on this problem off and on for forty years, my tentative conclusion is that the intractability of the problem stems again from the inadequacy of the notion of a σ-algebra in the formulation of probability. I make this statement with reluctance, since I have no alternative to propose. Nor is it likely that an overhaul of the foundations of probability will soon take place. Mathematicians are conservative when it comes to such overhauls. Nonstandard analysis, a technique which has proved beneficial and effective in innumerable problems, is a case in point. After some forty years, nonstandard analysis is still an outcast.

9 Problem nine: from ordinary probability to quantum probability

All deviant versions of probability theory that have been attempted so far take their lead from a simplification of the structure of the lattice of σ-subalgebras. Let us see how one takes the lead from such a simplification to arrive at quantum probability. Our motivation does not follow the historical development of quantum probability, which originated in quantum mechanics. It may instead be taken as an ideal history of how quantum probability ought to have developed. The "Leitfaden" is a hard look at the concept of independence of σ-algebras, with a view to discovering a hidden inadequacy in this concept, which will lead to an enlarged and hopefully more natural definition.

Recall that two σ-subalgebras π and π' are said to be independent when every pair of events A, B, where $A \in \pi$ and $B \in \pi'$, is independent. When π and π' are independent σ-subalgebras, the conditional expectation operators E_π and $E_{\pi'}$ commute: $E_\pi E_{\pi'} = E_{\pi'} E_\pi$. The converse of this statement is not true: two conditional expectation operators may commute even when the corresponding σ-subalgebras are not independent. We will say that two σ-subalgebras stochastically commute when their corresponding conditional expectation operators commute. It has been shown by Catherine Yan (in the latest issue of *Advances in Mathematics*) that E_π and $E_{\pi'}$ stochastically commute if and only if any two random variables X and Y providing information π and π' respectively are conditionally independent, that is, they are independent conditional upon the σ-algebra $\pi \vee \pi'$, in symbols $E(XY|\pi \vee \pi') = E(X|\pi \vee \pi')E(Y|\pi \vee \pi')$. There are other, combinatorial characterizations of stochastic commutativity of σ-subalgebras (also due to Catherine Yan) that show that stochastic commutativity is a generalization of statistical independence that dovetails with current views of nonstandard logics.

To any stochastic process one may associate a sublattice of the lattice of all σ-subalgebras of the sample space, namely, the lattice generated by all information σ-subalgebras provided by all random variables belonging to the stochastic process. For example, the lattice of σ-subalgebras associated to a martingale is isomorphic to a linearly ordered set. To the best of my knowledge, no stochastic process other than martingales has been studied whose associated lattice of σ-subalgebras is a lattice of stochastically commuting σ-subalgebras.

It would be interesting to discover such a stochastic process arising from some natural considerations.

Among all lattices of stochastically commuting σ-subalgebras, there is one that deserves particular attention. It is a lattice in which every σ-subalgebra π has a complement, that is, to every σ-subalgebra π one may associate one (and in general many) σ-subalgebras π' such that $\pi \vee \pi' = \hat{1}$ and $\pi \wedge \pi' = \hat{0}$, where $\hat{0}$ and $\hat{1}$ are the bottom and top elements of the lattice. Such a lattice, satisfying some additional technical assumptions that will insure that the lattice is large enough, can be shown to be isomorphic to the lattice of all closed subspaces of a Hilbert space belonging to a hyperfinite factor. Let us call it the quantum lattice.

The quantum lattice plays the role of the lattice of events in quantum probability. It is endowed with a real-valued function taking values between 0 and 1, which satisfies some but not all of the axioms of probability, but which satisfies sufficiently many such axioms to reassure us that we are dealing with some sort of generalization of probability. Such a real-valued function is called the trace, for reasons that will be shortly apparent. The quantum lattice does not satisfy the distributive law, but it satisfies another identity in meets and joins, the modular identity, which we need not recall. Suffice it to mention that in their joint paper that gave birth to quantum probability, Birkhoff and von Neumann believed the modular identity to play an essential role. They did not know that the quantum lattice could be realized as a lattice of stochastically commuting σ-subalgebras; as a matter of fact the modular identity hypnotized mathematicians for half a century and distracted them from the fundamental property, which is commutativity of conditional expectations.

The quantum analog of a random variable is called an observable. As long as the lattice structure of quantum mechanics was believed to be wedded to the modular identity, the problem of deriving the algebra of observables from the lattice structure was given up as intractable, even by von Neumann. We now know that the quantum lattice is isomorphic to a lattice of stochastically commuting σ-algebras. It may therefore be possible to derive the algebra of observables from the quantum lattice structure, in the spirit of Carathéodory and von Neumann. However, such a direct construction of the algebra of observables from the quantum lattice is likely to be a tour de force.

It is simpler to start with a direct definition of the algebra of observables, namely, the hyperfinite factor, and then to verify that the quantum lattice is indeed the algebra of "events" associated with such observables.

What is the hyperfinite factor? The hyperfinite factor is an algebra whose members are the quantum analogs of random variables. Since our point of view is not the usual one, it may be prudent to review the basic ideas.

There are several disparate ways of constructing the hyperfinite factor, and we will consider two of them.

The simplest construction is to be found in an obscure but enlightening paper of von Neumann, which he published in the journal *Portugaliae Mathematica* towards the end of his life. Since the algebra of observables is a noncommutative algebra, it is natural to construct it as a generalization of a matrix algebra with a trace. The trace will play the role of expectation. In this spirit, von Neumann considers the sequence of matrix algebras M_n with real entries, of size 2^n, in which the trace of the identity matrix is normalized to one. There is a natural isomorphism of M_n into M_{n+1} which preserves the trace, obtained by replacing the entries of a matrix in M_n by two-by-two matrices. Passing to the limit relative to a norm defined by the trace, one obtains a Banach algebra which intuitively consists of "continuous" matrices, and which is endowed with an involution and a trace. This Banach algebra is the hyperfinite factor. The self-adjoint elements of the hyperfinite factor are the observables, and their spectral resolutions give the analog of probability distributions. The idempotent self-adjoint elements can be shown to form a lattice which is isomorphic to the quantum lattice, with the trace playing the expected role of a generalized probability.

It is a common platitude to assert that there is in general no quantum analog of the notion of a joint distribution of a set of observables. We will see, however, that this statement is not entirely correct.

The hyperfinite factor has been used with striking success in quantum mechanics. However, quantum mechanics is not our present concern. Our present concern is that of defining stochastic processes that will have a chance of modeling some natural phenomena, in a spirit close to that of ordinary probability, aside from quantum mechanics. Judging from the pioneering work of Alain Connes and his collaborators, we surmise that stochastic processes that arise "naturally" happen to be also those that can be algebraically characterized.

Let me restate the program. It is known that any notion of ordinary probability that does not invoke joint distributions of random variables can be carried over to quantum probability. What is missing is a real world phenomenon, other than those of quantum mechanics, which will turn out to be better modeled by quantum probability than by ordinary probability. Irving Segal has suggested that any phenomenon where the act of observing alters the properties of the phenomenon observed might be a candidate for analysis by quantum probability, and he has recently tested his ideas on stock market fluctuations. Whether or not a phenomenon exists that can only be explained by quantum probability is at present a baffling open question.

10 Problem ten: the multivariate normal distribution and the Clifford distribution[1]

We hazard the guess that the quantum distribution that is most likely to work in quantum probability will be an analog of the multivariate normal distribution. Following the philosophy that demands that algebra should come first, we derive the quantum analog of the normal distribution by imitating the derivation of the multivariate normal distribution in the theory of Bosonic second quantization. In this theory, algebra

[1] Joint work with T.C. Wallstrom.

leads inexorably to the formula for the multivariate normal distribution of n random variables $X, X', \ldots, X''$, namely:

$$P((a < X < b) \cap (c < X' < d) \cap \cdots \cap (e < X'' < f))$$

$$\propto \int_a^b \int_c^d \cdots \int_e^f \exp\left(- \sum_{i,j} a_{ij} s_i s_j \right) ds_1\, ds_2 \cdots$$

where (a_{ij}) is a positive semidefinite symmetric matrix.

The algebra we are about to review differs from the algebraic formulation that was hinted at in Problem 1, as it does not rely on generators and relations on σ-algebras. It relies instead on an algebraic construction that goes by the name of second quantization. The advantage of the second quantization presentation of the normal distribution is that it leads one to guess the quantum analog of the normal distribution, which, as we will see, is a wholly new and quite simple multivariate distribution, called the Clifford distribution. We retrace the well-trodden path that leads to the formalism of Bosonic second quantization.

Consider an infinite dimensional real Hilbert space H whose vectors will be denoted by $a, b, \ldots, c$. We have in addition a symbol x to denote a variable vector. The inner product of two vectors is written $(a|b)$.

We define a polynomial over H in the variable x by recursion as follows:

1. $(a|x)$ is a polynomial for any element a of H.
2. The product of polynomials is a polynomial.
3. Any constant inner product $(a|b)$ is a constant polynomial (coefficient).
4. A linear combination with constant coefficients of polynomials is a polynomial.

Observe that one item in the preceding definition is redundant, since $(a|b)$ is a scalar. Nonetheless, we persist in our error, hoping to eventually give a rationale for it.

We denote a polynomial by $p(x)$. To every polynomial $p(x)$ one can associate a support, namely, a finite dimensional subspace H' of H such that $p(x) = 0$ for every vector x which is orthogonal to H'.

On the algebra of polynomials one defines polarization operators D_{ax} (read: polarize the letter x to the letter a) by setting $D_{ax}(b|x) = (b|a)$ and by requiring that the linear operator D_{ax} be a derivation.

If $p(x)$ is a polynomial, we denote by $p(D)$ the operator obtained by replacing every occurrence of $(a|x)$ in $p(x)$ by an occurrence of the polarization operator D_{ax}.

If $q(x)$ is another polynomial, we may define an inner product $\langle p(x)|q(x)\rangle$ by setting

$$\langle p(x)|q(x)\rangle = [q(D)p(x)]_{x=0}.$$

The bilinear form thus defined is symmetric. In particular:

$$\langle (a_1|x)(a_2|x)\cdots(a_n|x)|(b_1|x)(b_2|x)\cdots(b_n|x)\rangle$$

equals the permanent of the matrix $(a_i|b_j)$. It follows that the inner product $\langle p(x)|q(x)\rangle$ is orthogonally invariant:

$$\langle p(x)|q(x)\rangle = \langle p(Ux)|q(Ux)\rangle$$

for every orthogonal operator U in the Hilbert space H. If the supports H' and H'' of the polynomials $p(x)$ and $q(x)$ are orthogonal, then $\langle p(x)|q(x)\rangle = 0$.

We denote by H_{pol} the Hilbert space spanned by polynomials. The normal distribution arises from the following question: is it possible to find a linear map $p(x) \to: p(x):$ of H_{pol} into H_{pol}, and a linear functional E on H_{pol}, such that

$$\langle p(x)|q(x)\rangle = E(: p(x) :: q(x) :)?$$

Such a linear functional E will be positive and orthogonally invariant. Furthermore, if the supports of the polynomials $p(x)$ and $q(x)$ are orthogonal, then the random variables $: p(x) :$ and $: q(x) :$ are independent. The linear functional E will be given by integration relative to a probability on H which is orthogonally invariant. Since it is known since Mark Kac that the normal distribution is the only orthogonally invariant measure with the desired independence properties, it will follow that every finite set of polynomials $(a|x), (b|x), \ldots, (c|x)$ is mapped into a set of random variables having a multivariate normal distribution.

To obtain such a map $p(x) \to: p(x) :$, called the Wick map, we must make sure that every polynomial $(a|x)$ is mapped into some operator that comes close to being self-adjoint, so that the spectral resolution of such an operator will yield the probability distribution of a random variable. To this end, on the Hilbert space H_{pol} one defines the creation operator C_a as the operator mapping a polynomial $p(x)$ to the polynomial $(a|x)p(x)$. The adjoint of the creation operator C_a in the Hilbert space H_{pol} is the polarization operator D_{ax}.

Without going into details, it turns out that the operator $p(x) \to: p(x) :$ as well as the linear functional E exist and are unique. They are defined as follows.

If the support H' of the polynomial $p(x)$ is of dimension n, set

$$E(p(x)) = \frac{1}{(\sqrt{2\pi})^n} \int_{H'} p(x) e^{-(x|x)/2}\, dx,$$

where the integration runs over the space H'.

Second, one defines the Wick operator $: p(x) :$ by setting

$$: (a|x)(b|x) \cdots (c|x) := (C_a - D_{ax})(C_b - D_{bx}) \cdots (C_c - D_{cx})1.$$

In particular, one finds that

$$: (a|x)^n := (-1)^n e^{(x|x)/2} D_{ax}^n e^{-(x|x)/2},$$

and one sees that the right-hand side equals the n-th Hermite polynomial $H_n(x; a)$, as one finds in classical computations.

One finally verifies that

$$\langle p(x)|q(x)\rangle = E(: p(x) :: q(x) :).$$

Since a linear functional E with the above described invariance properties is unique, this argument may be taken as a definition of the normal distribution.

To conclude: the preceding argument shows that there is a unique way of implementing the inner product $\langle p(x)|q(x)\rangle$ by a linear functional, which has the properties of an expectation. In this implementation, every finite subset of vectors of H is represented by a set of random variables which have a multivariate normal distribution. I don't know a more natural way of introducing the multivariate normal distribution.

We will be led to the quantum analog of the normal distribution by performing the Fermionic analog of the preceding computation. Observe that the preceding construction could have been carried out without the aid of polynomials. It could have been expressed entirely in terms of the symmetric algebra over the Hilbert space H. There is a reason for our extravagant choice of the algebra of polynomials over the apparently simpler formalism of the symmetric algebra, which we are forced to pass over in silence because of time limitations. The reason, in brief, is that the formalism of second quantization comes into full bloom only when one employs the techniques of supersymmetric algebra.

In deriving the quantum analog of the normal distribution, we will deal with the exterior algebra $\Lambda(H)$ over the Hilbert space H. We denote the exterior product by $\wedge$ (there is no danger of confusion with meet in a lattice, since the two operations will never be used in the same context). The exterior algebra $\Lambda(H)$ is spanned by monomials $a \wedge b \wedge \cdots \wedge c$. We may define an inner product on $\Lambda(H)$ by setting

$$\langle a_1 \wedge a_2 \wedge \cdots \wedge a_n | b_1 \wedge b_2 \wedge \cdots \wedge b_n\rangle = \det(a_i|b_j).$$

We define the (Fermion) creation operator C_a by setting $C_a(b \wedge \cdots \wedge c) = a \wedge b \wedge \cdots \wedge c$, and we define the Annihilation operator A_a as the adjoint of the creation operator relative to the inner product $\langle .|.\rangle$.

We repeat in the present setup the question we asked in the classical case, namely: is there a Wick operator $: t :$ defined for every element $t \in \Lambda(H)$, such that $: t :$ is a linear operator from the exterior algebra $\Lambda(H)$ to linear operators over H, and a linear functional E on such operators, such that $\langle t|t'\rangle = E(: t^* :: t' :)$?

The answer to this question is affirmative; as a matter of fact it turns out to be simpler than in the Bosonic case. One simply defines $: a \wedge b \wedge \cdots \wedge c :=$ $(A_a + C_a)(A_b + C_b)\cdots(A_c + C_c)$, and one defines a linear functional E on the algebra generated by such operators to be the trace.

What, then, will the analog of the normal distribution be? In partial analogy with the Bosonic case, such a distribution will be obtained from the spectral resolution of the operators $A_a + C_a$. One checks by doing some algebra that the self-adjoint operator $A_a + C_a$ satisfies the identity $(A_a + C_a)^2 = (a|a)$, and thus its eigenvalues equal $\pm\sqrt{(a|a)}$. Thus, the quantum analog of the normal distribution is the Bernoulli distribution of a random variable taking values ± 1. This seems somewhat of a letdown, but let us not be discouraged. We are not going to end up with ordinary random walks after all this work. We are going to get something quite different and quite new. The algebra generated by the operators $A_a + C_a$, as a ranges over the Hilbert space H, generates (when closed in a suitable topology) an algebra which is isomorphic to the hyperfinite factor we saw in the preceding problem. In particular, the projection operators $E_a^{\pm}$ associated with each of the eigenvalues of each of the operators $A_a + C_a$ generate a lattice which is isomorphic to the quantum lattice of commuting σ-algebras of the preceding problem. The analog of the expectation is the trace, and the trace $\mathrm{Tr}(E_a^{\pm})$ is the same trace of the quantum lattice, that is, the quantum analog of probability.

What remains to be defined is the quantum analog of the multivariate normal distribution. Forget the old platitude that joint distributions do not exist in the quantum case. We present here an as yet unpublished proposal. The quantum analog of the joint distribution of observables $a, b, \ldots, c$ is the joint "probability" defined by the traces

$$\mathrm{Tr}(E_a^{\pm} E_b^{\pm} \cdots E_c^{\pm} E_b^{\pm} E_a^{\pm}).$$

Each of these traces is nonnegative, and the sum of these traces, as the signs vary in all possible ways, equals one. Thus, this is a multivariate probability distribution in the ordinary probabilistic sense, but its realization lives in the quantum lattice. We call this multivariate distribution the Clifford distribution.

Is there any hope that the multivariate Clifford distribution will find a practical application? We should like to propose a possible candidate: the analysis of sequences of ± 1's, or equivalently, of sequences whose entries belong to a finite alphabet. Statisticians working on such sequences, notably biologists working on DNA sequences, have often expressed the wish that there were a multivariate probability distribution that would enable a statistical analysis of such sequences in terms of means and variance alone, without invoking higher moments. But it is known that among all classical probability distributions, there is only one that is determined by mean and variance alone: the normal distribution. Nevertheless, if one is willing to admit quantum distributions as well, then one finds that the Clifford distribution is another other distribution that fits the bill.

We conjecture that the multivariate Clifford distribution will indeed prove useful in the statistical analysis of discrete time series. Observe that the joint Clifford distribution of observables $X_{a_1}, X_{a_2}, \cdots, X_{a_n}$ depends on the linear order of these observables. The joint Clifford distribution of a permutation of such a linearly ordered set of observables "almost" preserves the distribution: the difference between the joint Clifford distributions of two permutations of the same set of Clifford-distributed random variables can be shown to depend only on linear combinations of Clifford distributions of proper subsets of the sequence.

In closing, I should like to point out that the analogy between the Bosonic derivation of the multivariate normal distribution and the Fermionic derivation of quantum probability is deceptive. I will swallow my words: in this instance, the algebra has proved misleading. Let me explain.

Bosonic second quantization has been used in physics in connection with complex Hilbert spaces, not real Hilbert spaces. The reason for the choice of a complex rather than a real Hilbert space is ultimately the fact that one wishes to compute amplitudes, not probabilities (or rather, probabilities as the squares of amplitudes); besides, the Bosonic operators $C_a - D_{ax}$ are not self-adjoint in a real Hilbert space, and it is not clear how their spectral resolution might lead to the normal distribution, as I fraudulently gave us to understand. I brought into play Bosonic second quantization to justify a joint probability distribution, not an amplitude. This is not what Bosonic second quantization was meant for.

In the Fermionic case, however, the joint Clifford distribution is quadratic, and it squares with the ordinary notion of an amplitude. The joint Clifford distribution as defined above gives probabilities that are the squares of amplitudes.

I must leave to another occasion the discussion of the role of amplitude in quantum probability, and the fact that the emphasis on amplitudes, which in turn is patterned after the quantum mechanical analog of Brownian motion, has not helped the search for quantum stochastic processes divorced from quantum mechanics.

11 Problem eleven: cumulants

In the old days, the limit theorems of probability were derived by calculations with the moments $E(X^n)$ of a random variable X. It came as a shock, towards the middle of the last century, to discover that a random variable is not uniquely determined by its moments, and a huge literature, which shows no signs of abating, has developed on the moment problem, the problem of determining the probability density of a random variable in terms of its moments, as well as the uniqueness of such a density given by the moments.

In the first half of this century, the method of moments was replaced by Paul Lévy by a more pliable method relying upon the characteristic function $E(e^{itX})$, which is used to this day to derive the limit theorems of probability in their sharpest form. There is however one drawback to the characteristic function: it has no obvious probabilistic significance. My teacher William Feller was aware that the religious invocation of characteristic functions is extraneous to probabilistic reasoning. He managed to avoid characteristic functions in his treatise on probability. To be sure, characteristic functions made an occasional appearance in the second volume, when he just could not do without them. However, it was his intention to write a third volume, dealing with Brownian motion and diffusion processes, in which characteristic functions would be relegated to the dustbin of history. Unfortunately, he died before he could accomplish this task.

I have inherited from my teacher a dislike for characteristic functions, and I should like to make a suggestion, in this problem and the next, for a method that I believe should and will replace the use of characteristic functions, with one big proviso: a certain important problem will have to be solved first. Let us zero in on this problem.

Sometime in the last century, the Danish statistician Thiele observed that the variance of a random variable, namely, $\text{Var}(X) = E(X^2) - E(X)^2$, possesses notable properties, to wit:

1. It is invariant under translation: $\text{Var}(X + c) = \text{Var}(X)$ for any constant c.

2. If X and Y are independent random variables, then $\mathrm{Var}(X+Y) = \mathrm{Var}(X) + \mathrm{Var}(Y)$.
3. $\mathrm{Var}(X)$ is a polynomial in the moments of the random variable X.

He then proceeded to determine all nonlinear functionals of a random variable which have the same properties. The result of his investigations is striking. The identity of formal power series

$$\sum_{n=0}^{\infty} \frac{E(X^n)t^n}{n!} = \exp\left(\sum_{j=1}^{\infty} \frac{\kappa_j(X)t^j}{j!}\right)$$

determines uniquely the nonlinear functionals $\kappa_j(X)$ as polynomials in the moments, and it can be shown that every nonlinear functional of a random variable which enjoys the above three properties is a polynomial in the $\kappa_j(X)$. The nonlinear functionals $\kappa_j(X)$ are the cumulants of the random variable X. The second cumulant $\kappa_2(X)$ is the variance, the third is called the coefficient of skewness, and the fourth is called kurtosis. It is all very British.

Just as happened with the maximum entropy principle, it turns out that all the random variables occurring in the classical stochastic processes can be characterized by properties of their cumulants. For example, a normally distributed random variable is uniquely characterized by the fact that all its cumulants vanish beyond the second. Similarly, a random variable X which has the Poisson distribution with $E(X) = \alpha$ is the unique probability distribution for which all cumulants equal α.

It appears therefore that a random variable is better described by its cumulants than by its moments or its characteristic function. One can guess the central limit theorem in one minute by staring at cumulants of standardized averages of independent random variables. It is therefore natural to surmise that cumulants might be better functionals in dealing with random variables than characteristic functions. I personally believe that this dream will come true, and that cumulants will eventually displace characteristic functions and moments. However, there is one obstacle that stands in the way of dealing with cumulants. We do not know simple necessary and sufficient inequalities on cumulants that insure that a given sequence of real numbers is the sequence of cumulants of a random variable. For characteristic functions, such a necessary and sufficient condition is Bochner's theorem. To be sure, such inequalities exist in abundance if one expresses moments in terms of cumulants. The problem is finding the simplest such inequalities. One can prove that for every positive integer j there is one inequality involving κ_j and lower order cumulants, such that the sequence of all such inequalities characterizes the cumulants of a random variable. The simplest such inequality is Schwartz's inequality.

The explicit determination of the fundamental inequalities satisfied by the cumulants of a random variable is surely the most important open problem in probability theory.

Further insight into cumulants is achieved by linearizing them: one expands $\kappa_j(\lambda_1 X_1 + \lambda_2 X_2 + \cdots + \lambda_j X_j)$ as a polynomial in the λ's, and one takes the coefficient of $\lambda_1\lambda_2\cdots\lambda_j$. The resulting multilinear cumulant $\kappa(X_1, X_2, \cdots, X_j)$

has been axiomatized by Terry Speed. Its main property is the following. Suppose that ρ is a nontrivial partition of the index set $\{1, 2, \cdots, j\}$ such that, if i and j belong to the same block of the partition ρ, then the random variables X_i and X_j are uncorrelated, that is, $E(X_i X_j) = E(X_i)E(X_j)$. Then $\kappa(X_1, X_2, \cdots, X_j) = 0$.

Terry Speed has also obtained an expression for the multilinear cumulants in terms of the moments, by an application of the Möbius inversion formula on the lattice of partitions. Since we will shortly use this formula to motivate the next generalization of probability theory, it may be worth-while to review this formula now. I will therefore digress on a combinatorial topic that is dear to my heart.

One of the most useful computational devices in probability is the principle of inclusion-exclusion, which I will restate in terms that may not be the familiar ones. Consider a finite Boolean algebra $P(S)$, that is, the lattice of all subsets of a finite set S. If f and g are functions defined on all subsets of S such that

$$g(B) = \sum_{A \subseteq B} f(A),$$

then

$$f(B) = \sum_{A \subseteq B} (-1)^{|B|-|A|} g(A).$$

This inversion formula expresses the principle of inclusion-exclusion. The idea may be generalized to any finite partially ordered set; when this is done, the role of the factor $(-1)^{|B|-|A|}$ is taken by the Möbius function of the partially ordered set. One then obtains the Möbius inversion formula on a partially ordered set.

What interests us presently is the special case of the Möbius inversion formula one obtains when the partially ordered set is the lattice $\Pi(S)$ of all partitions of a finite set S. Partitions are ordered by refinement, like the Boolean algebras they generate: we say that a partition ρ is a refinement of a partition ρ' of the set S when every block of ρ is contained in a block of ρ'. The lattice $\Pi(S)$ of all partitions of S has a rich structure, much like the lattice of Boolean σ-subalgebras of a Boolean algebra.

The Möbius inversion formula on the lattice of partitions reads as follows. Suppose f and g are functions defined on partitions, and suppose that

$$g(\rho) = \sum_{\rho' \leq \rho} f(\rho').$$

Then

$$f(\rho) = \sum_{\rho' \leq \rho} \mu(\rho', \rho) g(\rho').$$

The Möbius function $\mu(\rho', \rho)$ has a simple formula which we need not recall here. For example, if $\hat{0}$ is the minimum partition and $\hat{1}$ is the maximum partition, then $\mu(\hat{0}, \hat{1}) = (-1)^n (n-1)!$.

Speed defines the function $g(\rho)$ to equal $\prod_{B \in \rho} E(\prod_{j \in B} X_j)$. By an application of the Möbius inversion formula on the lattice of partitions, he obtains for the cumulants the explicit expression

$$\kappa(X_1, X_2, \ldots, X_j) = \sum_{\rho \in \Pi(\{1,2,\ldots,j\})} \mu(\rho, \hat{1}) \prod_{B \in \rho} E(\prod_{j \in B} X_j).$$

For example,

$$\begin{aligned}\kappa(X_1, X_2, X_3) \\ = E(X_1X_2X_3) - E(X_1)E(X_2X_3) - E(X_2)E(X_1X_3) + 2E(X_3)E(X_1X_2).\end{aligned}$$

From this explicit formula, Speed was able to prove all known properties of cumulants.

For the inclusion-exclusion principle on a Boolean algebra of sets, a great deal of work has been done to obtain inequalities that provide approximations obtained by truncating the sum on the right-hand side of the inclusion-exclusion principle, as well as algorithms that in special cases do away with the awkward alternating sum expression of the inclusion-exclusion principle. Much of this work was motivated by the needs of reliability theory. No such similar work has been done in obtaining approximate estimations for Möbius inversions on the lattice of partitions of a set. The lack of this spadework is probably a stumbling block that stands in the way of solving the problem of the cumulants.

12 Problem twelve: free probability theory

It is a generally accepted platitude that two observables in quantum probability do not have a joint distribution; this is the very raison d'être of the quantum mechanical uncertainty principle. Shall we therefore throw up our hands and refuse to consider any form of correlation between random variables in quantum probability? Hardly. Even though a joint distribution of two observables X and Y may not exist, nevertheless we still have a powerful set of invariants: the joint moments. Given any finite sequence of observables $Z_1, Z_2, \cdots, Z_n$, where each of the Z_j equals either X or Y, we may define the mixed moments of the two observables X and Y as the trace

$$\mathrm{Tr}(Z_1, Z_2, \cdots, Z_n).$$

It is likely that these joint moments determine the two observables uniquely, although I have never seen a statement, much less a proof, of this fact. The analogous statement in a finite dimensional algebra of matrices has been proved in the affirmative by Procesi. Whether the answer is affirmative or not in the hyperfinite factor, recovering information on the observables X and Y from their mixed moments is likely to be a daunting problem.

This problem is in turn related to another sticky point in quantum probability: the definition of independence of two observables X and Y. When X and Y commute, then they have a joint probability density, and independence reduces to ordinary statistical independence. Statistical independence of two commuting observables, stated in algebraic terms, amounts to the statement that the meet of the informations provided by X and Y is isomorphic to the tensor product of the two Boolean σ-algebras.

But now, having shot the tensor product cannon to shoot the fly of statistical independence of commuting observables, how shall we define independence for a pair of noncommuting observables? Resorting again to tensor product is awkward, though not altogether meaningless. It does have the disadvantage of an ad hoc definition, which does not provide any of the bounties we expect of statistical independence, such as limit theorems for sums of independent random variables, or maybe even a central limit theorem.

The "right" definition of statistical independence in noncommuting algebras, that is, for observables, was discovered in 1985 by Voiculescu and developed by Voiculescu and his school, and it is at present the most promising branch of quantum probability. We will give a sketch of the main ideas.

Recall the construction of the hyperfinite factor as a limit of matrix algebras. Suppose that a sequence X_n of finite matrices is given, where X_n belongs to each of the matrix algebras M_n whose limit is the hyperfinite factor, and suppose in addition that the entries of the matrix X_n are independent normally distributed with mean equal to 0 and variance equal to $\frac{1}{2^n}$. These matrices converge in a suitable sense to an observable X in the hyperfinite factor, which may be imaginatively viewed as a continuous matrix with independent normal entries. As a matter of fact, Wigner proved that the probability distribution of this limit observable X is the Wigner circle distribution, given by

$$\text{dens}(X = t) = \frac{1}{2\pi}\sqrt{(4 - t^2)}$$

when $|t| \leq 2$, and equal to zero otherwise.

Wigner's discovery is the guide to the correct definition of independence of observables. Suppose that we have two such sequences of matrices, X_n and Y_n, each one with the properties stated above, and in addition suppose that each entry of X_n is independent of each entry of Y_n. In the limit, we obtain observables X and Y which will be, in a sense to be defined, independent. Two observables that are independent in this new sense are said to be free.

It is time to give a definition of freeness of two observables that is independent of the limiting construction. We will give two such definitions that appear to be unrelated, but that on closer inspections will reveal themselves to be equivalent. The first definition is due to Voiculescu, the second definition to Speicher. Both definitions confirm the guess that freeness is the right definition of independence for noncommuting observables.

The first definition takes its lead from the moments. Two ordinary random variables are independent if their joint moments factor in the obvious way. It turns out that there is a way of generalizing this idea of factorization of the moments for noncommuting observables. At first, the definition appears to be a sophisticated one, but we will see that it is not an arbitrary one. We will say that two observables X and Y are free when their moments satisfy the following conditions. Suppose we are given a finite sequence of polynomials of a real variable, say $p(t)$, $q(t)$, $r(t)$, $w(t), \ldots$, and suppose that whenever these polynomials are so chosen that $E(p(X)) = 0$, $E(q(Y)) = 0$, $E(r(X)) = 0$, $E(w(X)) = 0, \ldots$, then

$$E(p(X)q(Y)r(X)w(Y)\cdots) = 0.$$

When this happens for every choice of polynomials satisfying the above conditions, we say that the observables X and Y are free. For example, suppose that $p(t) = t - E(X)$ and $q(t) = t - E(Y)$. Freeness of X and Y implies that $E(p(X)q(Y)) = 0$. Expanding, we infer that $E(XY) = E(X)E(Y)$, just like in the old days. A similar argument shows that $E(XYXY) = E(X^2)E(Y)^2 + E(X)^2E(Y^2) + E(X)^2E(Y)^2$. The mixed moments expand in terms of products of moments of lower degree in a manner that at first appeared mysterious.

The second definition of independence takes its lead from the behavior of cumulants. As already mentioned, cumulants are the nonlinear functionals of random variables that are additive on independent random variables. If we could generalize the cumulants to observables, we would have another way of defining freeness. This is the program that was successfully carried out by Speicher as follows.

Speicher takes his lead from the definition of the classical multilinear cumulants in terms of the Möbius function. However, he changes the lattice over which the Möbius inversion formula is applied. Instead of using the lattice of all partitions of a finite set, he uses a smaller lattice that was first discovered by the French combinatorialist Germain Kreweras. This is the lattice of noncrossing partitions, a beautiful construct which we proceed to describe.

We consider partitions of the set of integers $\{1, 2, \cdots, n\}$, and we place this set in circular order as equidistant points on a circle in the plane. Thus, they are the vertices of a regular n-gon. We use the following graphical device to describe a partition ρ of the set $\{1, 2, \cdots, n\}$. If j and k are two integers belonging to the same block of ρ, and if no integers between j and k belong to the same block, then we draw a diagonal of the circle joining j to k. We repeat the same construction for all other occurrences of such pairs. Let us call the resulting set of diagonals the graph of the partition ρ.

We will say that a partition ρ is noncrossing when the graph of ρ contains no crossing diagonals. For example, the partition $\rho = \{\{1, 3, 5\}, \{2\}, \{4\}\}$ of the set $\{1, 2, 3, 4, 5\}$ is noncrossing, but the partition $\rho' = \{\{1, 3, 5\}, \{2, 4\}\}$ is crossing.

Noncrossing partitions form a lattice $NC\{1, 2, \cdots, n\}$ which remarkably turns out to be self-dual. The Möbius function of this lattice is computed in terms of the Catalan numbers; for example, one has $\mu_{NC}(\hat{0}, \hat{1}) = (-1)^{n-1}\frac{1}{n+1}\binom{2n}{n}$.

Speicher proceeds to define the multilinear cumulants of a sequence of observables $X_1, X_2, \cdots, X_n$ by analogy with Speed's definition in the lattice of noncrossing partitions. He sets

$$\kappa_{NC}(X_1, X_2, \ldots, X_n) = \sum_{\rho \in NC\{1,2,\ldots,n\}} \mu_{NC}(\rho, \hat{1}) \prod_{B \in \rho} E\left(\prod_{j \in B} X_j\right).$$

Here, the factors within each expectation are displayed in increasing order.

He then verifies that a sequence of observables is free in the sense defined above if and only if the noncrossing cumulants vanish in the following sense. For every sequence of distinct integers $i, j, \cdots, k$ (not necessarily monotonic) one has $\kappa_{NC}(X_i, X_j, \cdots, X_k) = 0$. By reversing the Möbius inversion, one obtains the expression of mixed moments of random variables in terms of products of moments of lower order, which formerly appeared mysterious.

It is natural to ask for a probability distribution all of whose noncrossing cumulants are zero except for the first two, in analogy with the normal distribution. Such a distribution exists and is unique, and it is (would you believe it?) the Wigner semicircle distribution. What is more amazing, all cumulants beyond the second (in the sense of free probability theory) vanish, and they characterize the Wigner distribution.

As a matter of fact, one can prove an analog of the central limit theorem for a sequence of free random variables. It goes as follows.

Let $X_1, X_2, \cdots$ be a sequence of free observables with zero expectation and with uniformly bounded moments $|E(X_j^n)| \leq 1$ for all j and all n. Assume that

$$\lim_{n\to\infty} \frac{\sum_{i=1}^n E(X_i^2)}{n} = \frac{r^2}{4}.$$

Then the sequence of observables

$$\frac{X_1 + X_2 + \cdots + X_n}{\sqrt{n}}$$

converges in distribution to the Wigner semicircle density.

Much work is presently being done on free probability theory, not only in the hyperfinite factor but in other algebras as well, and on this noncommutative version of probability we place our bets.

Recurrent Themes of Gian-Carlo Rota's Mathematical Thought

Resolution of Weyl modules: the Rota touch

D. A. Buchsbaum

1 Introduction

This article is offered as homage to Gian-Carlo Rota. In it, we'll indicate how the application of Rota's combinatorial methods to the resolution of Weyl modules changed the way the whole question of finding resolutions was addressed, and what kinds of additional information it led to. To do this, we first have to review briefly the methods that were being used prior to Rota's involvement. Then we look at how letter-place methods and polarizations provided information that had until his intervention not been forthcoming, such as homotopies and an explicit description of syzygies. Finally, we see how using (generalized) Capelli identities for polarizations leads to a better understanding of the known boundary maps for some resolutions, and provides a direction for finding the general boundary map. We end with a description of the terms of the general resolution of the Weyl module associated to the n-rowed *almost skew-shape* (see below for definition).

2 Pre-Rota period

In the early 1980's, Kaan Akin and I started working on the problem of resolving Schur modules in terms of direct sums of tensor products of exterior powers (or fundamental representations). For two-rowed Schur modules, by using the "fundamental exact sequence"

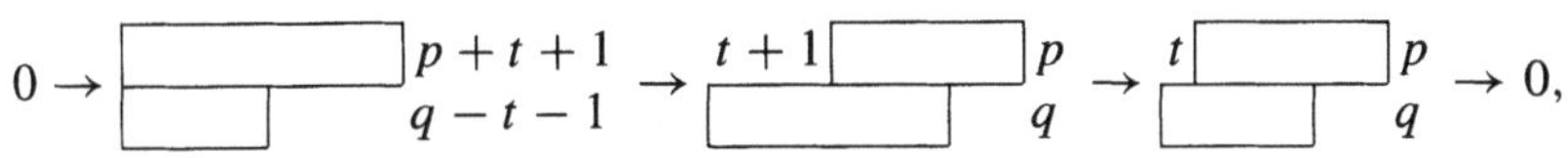

and a Koszul-like complex (which we called an "arithmetic Koszul complex", and which will be explained in the next section), an induction argument on the number of overlaps between the two rows led to a description of the desired type of resolution [1].

The three-rowed case presented quite another aspect. For one thing, the analogue of the exact sequence above led to a type of representation that had never been studied in the classical situation, what we tended to call "bad shapes" (simply because they were neither partitions nor skew-partitions). To be precise, the analogue of the above is the following:

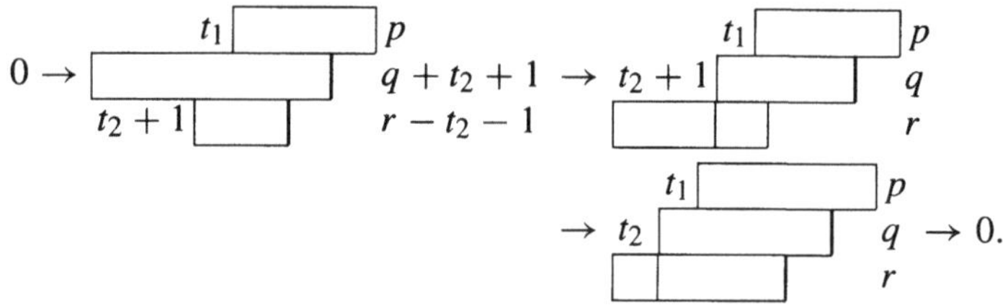

Thus one sees that the kernel of the epimorphism on the right is not any of the classical shapes, but what M. Klucznik calls an *"almost skew-shape"* (see [10]). If one starts with an epimorphism of the type above between two almost skew-shapes:

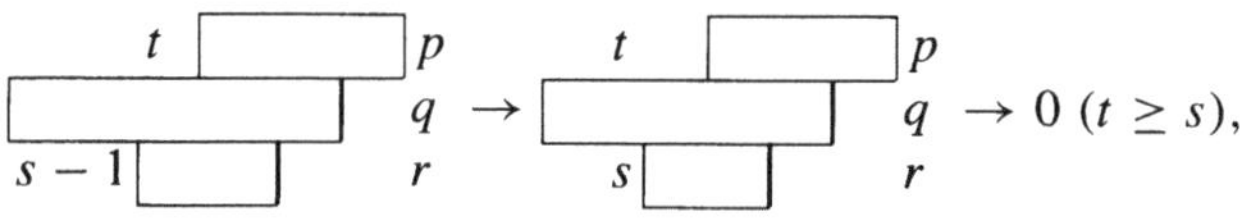

the kernel is again an almost skew-shape, and the fundamental exact sequence in this case is:

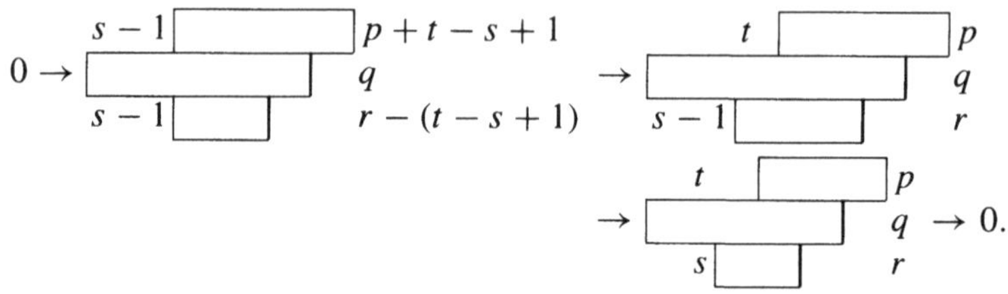

Of course, for three-rowed shapes, there is essentially only one type of almost skew-shape; in the case of n-rowed shapes, the number of distinct types of almost skew-shapes is $n - 2$.

Using the arithmetic Koszul complexes and induction on the number of triple overlaps, Akin and I had, in the summer of 1982, succeeded in constructing the resolutions of Schur modules of skew-shapes having no more than two triple overlaps. The computations were so *ad hoc*, and the way ahead so unclear, that we decided it was imprudent at that time to go ahead with the project of resolving skew-shapes in general; that it was almost impossible to think of resolving almost skew-shapes in general and, in fact, we realized we had to look for an existence proof for these types of resolutions (don't forget that the resolutions that we were looking for weren't projective resolutions). In [2], we turned our attention to Weyl modules and resolutions whose terms are direct sums of tensor products of divided powers. Terms of this kind are projective over the appropriate Schur algebra, so that, by using the same fundamental exact sequences for Weyl modules that we had been using for Schur modules, and a similar induction argument on the number of overlaps, an existence proof was obtained (see [2]). In fact, the existence of resolutions of the desired type was proved simultaneously for all the shapes that we were considering: partitions, skew-shapes and almost skew-shapes. By means of a duality between Schur and Weyl modules which carries exterior powers to divided powers and vice versa, the resolutions of Schur modules in terms of fundamental representations were also shown to exist.

Using these results, Akin and I were able to establish the finite global dimension of the Schur algebras over arbitrary fields and over the integers ([2], see also [9]), but we had pretty much given up on an explicit description of the resolutions of our modules, and we had no hope of ever obtaining resolutions of almost skew-shapes.

One interesting development along these lines is related to an attempt by Akin and myself to "resolve" almost skew-shapes in terms of "better" shapes. For example, we could show that a three-rowed almost skew-shape could be resolved by skew-shapes. As I pursued this problem a bit further, I succeeded in characteristic zero by using the fact that

$$\mathrm{Hom}(L_\lambda, L_\mu \otimes X) \approx \mathrm{Hom}(L_{\lambda/\mu}, X), \tag{Sk}$$

where the Hom is taken over the appropriate Schur algebras, and λ and μ are partitions. However, I discovered that this isomorphism isn't true in every characteristic (nor even over the integers). Using this as a basis for our construction, R. Sánchez and I showed that it was not possible to carry out the resolution of "bad" shapes by "better" shapes in a characteristic-free way (see [2]). This, however, made the question about the validity of the above isomorphism even more intriguing. A number of mathematicians thought about this problem without success until, in 1998, U. Kulkarni, in his thesis ([11]), described precisely in what situations a formula of type (Sk) above obtains.

3 Enter Rota

In the summer of 1990, Rota and I met in Rome. We had already been well aware of each other's work for a number of years, but in Rome it became painfully clear that, by not working together, we were both missing out on some very fruitful opportunities (and, in at least one case, duplicating each other's work – see [3, 5]). As a result we decided to get together regularly once we returned to the Boston area and see if we could learn each other's languages and methods.

With Rota's strong interest in invariants, it wasn't too surprising that one of the problems that he found most tantalizing was that of resolving Weyl modules. After all, with the resolution in hand, the next obvious step is to describe the syzygies at each stage, and thus to have generators of yet more invariants. When I showed him the presentation theorem for Schur and Weyl modules that Akin, Weyman and I had obtained in [4], he immediately said that the map in the presentation (what we called the "box map" and denoted by $\square$ in [4]) was a direct sum of "polarizations". Reading my perplexity correctly, Gian-Carlo began to educate me in "alphabets" (a notion I had always abhorred, feeling that they were too extrinsic, preferring instead to refer only to the free module the "alphabet" generated as a basis), polarizations, places, letter-place algebras and place polarizations. I have to confess to the expert reader of this article that I didn't *really* believe in the double standard tableaux letter-place basis until I had worked out for myself, going through various translations of the terms, that the double tableau

$$\left(\begin{array}{c|c} a^{(p)} & 1^{(p)} \\ b^{(q)} & 2^{(q)} \end{array}\right) = a^{(p)} \otimes b^{(q)},$$

with $a < b$, $1 < 2$ and $p < q$ can be straightened to

$$\sum_{k\geq 0} \binom{p-q}{k} \left(\begin{array}{c|c} a^{(p)}b^{(q-p+k)} & 1^{(p)}2^{(q-p+k)} \\ b^{(p-k)} & 2^{(p-k)} \end{array}\right).$$

However, by painstakingly going over the resolution of the two-rowed Weyl module that Akin and I had found, replacing the arithmetic Koszul complex by a Bar complex (something that Akin and I had already observed we could do, but could think of no good reason to do it), and reinterpreting the $\Box$ map as place polarizations, Rota and I not only had a complex defined in terms that were well understood, but we could also write down a splitting homotopy for the positive part of the resolution (see [6]). This enabled us to give an explicit description, in terms of double standard tableaux, of the syzygies of the two-rowed Weyl module. It was at this point in our collaboration that I became almost totally convinced that there was some merit to using alpahabets, places, etc. Moreover, when we reinterpreted the work that Akin and I had done in 1982 on the resolutions of skew-shapes with two triple overlaps, we found that the complicated identities and computations that were involved, and seemed so *ad hoc* at the time, could be explained as generalized Capelli identities on place polarizations. At this point, we both felt much more hopeful than Akin and I had felt years ago about the prospect of finding explicit resolutions of Weyl (and Schur) modules. Before going into further detail, let me clear up what is meant by "Koszul-like complex", and the "Bar complex"

The usual Koszul complex is a complex associated to a map, f, from an R-module, M, to the commutative ring, R, and is often denoted $\mathbb{K}(f)$. It is formed by taking, for each nonnegative integer, l, the lth exterior power of M, $\Lambda^l(M)$, and mapping it into $\Lambda^{l-1}(M)$ as follows:

$$\partial_f : \Lambda^l(M) \to \Lambda^{l-1}(M) \text{ sends } m_1 \wedge \cdots \wedge m_l \text{ to}$$

$$\sum_{j=1}^{l} (-1)^{j-1} f(m_j) m_1 \wedge \cdots \wedge \hat{m}_j \wedge \cdots \wedge m_l \text{ if } l > 0.$$

Here the circumflex over the m_j indicates that m_j is to be omitted. Note, that if $l = 1$, then $\Lambda^{l-1}(M) = \Lambda^0(M) = R$, and that, in this case, the map ∂_f is f itself. If M is a free R-module of rank n with basis $\{x_1, \ldots, x_n\}$, then a map f is completely determined by where the elements x_i are sent, and it's enough to define the boundary map on the elements of the form $x_{i_1} \wedge \cdots \wedge x_{i_l}$, with $i_1 < \cdots < i_l$ as above, namely:

$$x_{i_1} \wedge \cdots \wedge x_{i_l} \to \sum_{j=1}^{l} (-1)^{j-1} f(x_{i_j}) x_{i_1} \wedge \cdots \wedge \hat{x}_{i_j} \wedge \cdots \wedge x_{i_l}.$$

The arithmetic Koszul complex is one that was associated with two integers, n and k, in the following way. The integer n is the rank of the free $\mathbb{Z}$-module $\mathbb{Z}^n$ of rank n, and the integer k is what introduces the "arithmetic". The terms of the complex we are about to define are the exterior powers of $\mathbb{Z}^n$, but the boundary maps are a bit peculiar and are defined as follows:

$$\delta_f : \Lambda^l(\mathbb{Z}^n) \to \Lambda^{l-1}(\mathbb{Z}^n) \text{ sends } x_{i_1} \wedge \cdots \wedge x_{i_l} \text{ to}$$

$$\sum_{j=1}^{l}(-1)^{l-j}\binom{i_{j+1}-i_{j-1}}{i_j - i_{j-1}} x_{i_1} \wedge \cdots \wedge \hat{x}_{i_j} \wedge \cdots \wedge x_{i_l} \text{ if } l > 0.$$

Since the index of summation, j, is running from 1 to l, we have to agree to set $i_{l+1} = k$, and $i_0 = 0$. Notice that, when $l = 1$, the map $\delta_f : \mathbb{Z}^n \to \mathbb{Z}$ is the one that sends the (canonical) basis element x_i to the binomial coefficient $\binom{k}{i}$. Thus, the boundary map isn't really the one that we would obtain as ∂_f; it's a slight variation that depends on the arithmetic of the binomial coefficients. It was suggested, though, by the problem that we were trying to solve at the time.

The Bar complex is even more classical than the Koszul complex. It arises in the following way. If Λ is an algebra over the commutative ring R, and M a Λ-module, then one can form the following complex:

$$\cdots \to \underbrace{\Lambda \otimes \cdots \otimes \Lambda}_{l} \otimes M \to \underbrace{\Lambda \otimes \cdots \otimes \Lambda}_{l-1} \otimes M \to \cdots \to \Lambda \otimes M \to M$$

where the map $\Lambda \otimes M \to M$ is simply the action of Λ on M, and in general, the map $\underbrace{\Lambda \otimes \cdots \otimes \Lambda}_{l} \otimes M \to \underbrace{\Lambda \otimes \cdots \otimes \Lambda}_{l-1} \otimes M$ is defined by

$$\begin{aligned}\lambda_1 \otimes \cdots \otimes \lambda_l \otimes m \to & \sum_{j=1}^{l-1}(-1)^{j-1}\lambda_1 \otimes \cdots \lambda_j\lambda_{j+1} \otimes \cdots \otimes \lambda_l \otimes m \\ & + (-1)^{l-1}\lambda_1 \otimes \cdots \otimes \lambda_{l-1} \otimes \lambda_l m.\end{aligned}$$

In the case of a graded algebra, Λ, one usually restricts the degrees of the λ_j to be positive (it is usually also assumed that all the $\lambda's$ are homogeneous and that the module, M, is graded too). Observe, too, that in the graded case, we get a subcomplex by stipulating that the degree of λ_1 should always be greater than or equal to some fixed integer, s. Another useful observation is that we may always take as our module M the algebra Λ itself.

When this complex was originally introduced (in algebraic topology), the tensor product sign was not used; the symbol | ("bar") was used instead, so that the term $\lambda_1 \otimes \cdots \otimes \lambda_l \otimes m$ would have appeared as $\lambda_1 | \cdots |\lambda_l| m$. Hence the term "Bar complex".

To make the connection between arithmetic Koszul complexes and Bar complexes to which we alluded above, let's observe that a nice graded algebra that comes easily to hand is the divided power algebra, $\Lambda = D(\varepsilon)$, on one generator, ε, over the integers. This is the algebra which, in each degree, k, is free on one generator, denoted $\varepsilon^{(k)}$, and with multiplication given by $\varepsilon^{(k)}\varepsilon^{(l)} = \binom{k+l}{l}\varepsilon^{(k+l)}$. Then, for every pair of integers, t, s, with $s \leq t$, we have the sub-Bar complex given by

$$\cdots \to \sum_{t-|q|\geq s, q_i>0} D_{t-|q|}(\varepsilon) \otimes D_{q_1}(\varepsilon) \otimes D_{q_2}(\varepsilon) \to$$

$$\sum_{t-q\geq s, q>0} D_{t-q}(\varepsilon) \otimes D_q(\varepsilon) \to D_t(\varepsilon).$$

Since each $D_r(\varepsilon)$ is of rank one, we see that the rank of $\sum_{t-q\geq s,q>0} D_{t-q}(\varepsilon) \otimes D_q(\varepsilon)$ is $t - s$. This suggests that we might consider the arithmetic Koszul complex associated to the integers $t - s$ and k where k is equal to t (since the elements $\varepsilon^{(t-q)} \otimes \varepsilon^{(q)}$ map to $\binom{t}{q}\varepsilon^{(t)}$), and see if it is indeed isomorphic to this Bar complex. It's fairly straightforward to see that the correspondence

$$x_{i_1} \wedge \cdots \wedge x_{i_l} \to \varepsilon^{(t-i_l)} \otimes \varepsilon^{(i_l - i_{l-1})} \otimes \cdots \otimes \varepsilon^{(i_j - i_{j-1})} \otimes \cdots \otimes \varepsilon^{(i_1)}$$

does the job.

4 Some elementary results and ongoing methods

To see how letter-place methods and polarizations made it possible to define a homotopy for the resolution of the Weyl module associated to the skew-shape

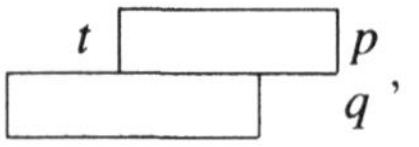

, (A)

we first observe that it is the image of $D_p \otimes D_q$ under the Weyl map. The 'box map' referred to above was described in [4] as the map $\sum_{k>t} D_{p+k} \otimes D_{q-k} \to D_p \otimes D_q$ which sends an element $x \otimes y$ of $D_{p+k} \otimes D_{q-k}$ to $\sum x_p \otimes x'_k y$, where $\sum x_p \otimes x'_k$ is the component of the diagonal of x in $D_p \otimes D_k$. To put this all in letter-place perspective, we see that, if we take a double standard tableau, say $\left(\begin{array}{c|c} w & 1^{(p)}2^{(k)} \\ w' & 2^{(q-k)} \end{array}\right)$, in $D_p \otimes D_q$, the Weyl map can be defined as the composition of place polarizations, from positive places $\{1, 2\}$ to negative places $\{1', \ldots, (p+t)'\}$, as follows: $\partial_{q'2} \cdots \partial_{1'2}\partial_{(p+t)'1} \cdots \partial_{(t+1)'1}$, where $\partial_{u'v}$ stands for the place polarization from (positive) place v to (negative) place u'. If we take an element of $D_{p+k} \otimes D_{q-k}$, the "box map" sending this to $D_p \otimes D_q$ is simply the kth divided power of the place polarization taking place 1 to place 2. It is now easy to see that a standard tableau $\left(\begin{array}{c|c} w & 1^{(p)}2^{(k)} \\ w' & 2^{(q-k)} \end{array}\right)$, with $k > t$, gets carried to zero by the Weyl map. Also, if we let $Z_{2,1}$ stand for the generator of a divided power algebra in one free generator, we see that $Z_{2,1}^{(k)}$ acts on $D_{p+k} \otimes D_{q-k}$ and carries it to $D_p \otimes D_q$. Thus, we may take homogeneous strands of the normalized Bar complex of this algebra acting on $\sum D_{p+k} \otimes D_{q-k}$ to get a complex over the Weyl module:

$$\cdots \to \sum_{k_i>0} Z_{2,1}^{(t+k_1)} x Z_{2,1}^{(k_2)} x \cdots x Z_{2,1}^{(k_{l+1})} x D_{t+p+|k|} \otimes D_{q-t-|k|} \to$$
$$\sum_{k_i>0} Z_{2,1}^{(t+k_1)} x Z_{2,1}^{(k_2)} x \cdots x Z_{2,1}^{(k_l)} x D_{t+p+|k|} \otimes D_{q-t-|k|} \to \cdots$$
$$\to \sum_{k>0} Z_{2,1}^{(t+k)} x D_{t+p+k} \otimes D_{q-t-k} \to D_p \otimes D_q,$$

where the symbol "x" is a separator variable (see [7]) used instead of the usual "bar" in the Bar complex, and $|k|$ stands for the sum of the indices k_i. In the usual Bar complex with bars, one defines the boundary operator as the map obtained by removing the bars one at a time and then multiplying in the divided power algebra or, in the case of the last place, acting on $D_{t+p+|k|} \otimes D_{q-t-|k|}$. Here, using the separator x, we say that the boundary operator is obtained by polarizing the variable x to the element 1. In this setup, the homotopy is trivial to define:

$$s_0 : D_p \otimes D_q \to \sum_{k>0} Z_{2,1}^{(t+k)} x D_{t+p+k} \otimes D_{q-t-k}$$

is defined by sending the double standard tableau $\left(\begin{array}{c|c} w & 1^{(p)}2^{(k)} \\ w' & 2^{(q-k)} \end{array} \right)$ to zero if $k \le t$, and to $Z_{2,1}^{(p+k)} x \left(\begin{array}{c|c} w & 1^{(p+k)} \\ w' & 2^{(q-k)} \end{array} \right)$ if $k > t$. For higher dimensions,

$$s_l : \sum_{k_i>0} Z_{2,1}^{(t+k_1)} x Z_{2,1}^{(k_2)} x \cdots x Z_{2,1}^{(k_l)} x D_{t+p+|k|} \otimes D_{q-t-|k|}$$
$$\to \sum_{k_i>0} Z_{2,1}^{(t+k_1)} x Z_{2,1}^{(k_2)} x \cdots x Z_{2,1}^{(k_{l+1})} x D_{t+p+|k|} \otimes D_{q-t-|k|},\ l > 0,$$

is defined by sending $Z_{2,1}^{(t+k_1)} x Z_{2,1}^{(k_2)} x \cdots x Z_{2,1}^{(k_l)} x \left(\begin{array}{c|c} w & 1^{(t+p+|k|)}2^{(m)} \\ w' & 2^{(q-t-|k|-m)} \end{array} \right)$ to zero if $m = 0$, and to $Z_{2,1}^{(t+k_1)} x Z_{2,1}^{(k_2)} x \cdots x Z_{2,1}^{(k_l)} x Z_{2,1}^{(m)} x \left(\begin{array}{c|c} w & 1^{(t+p+|k|+m)} \\ w' & 2^{(q-t-|k|-m)} \end{array} \right)$ if $m > 0$. It's a triviality to prove that this provides a splitting homotopy for the complex above. Since it's well-known that the Weyl module is the cokernel of the box map, the fact that we have a splitting homotopy for the rest of the complex gives us a quick proof of the fact that the above is a resolution of the Weyl module. Moreover, this homotopy provides us with an explicit description of the basis for the syzygies: we see that, in dimension 0, the basis for the syzygies can be taken to be the set

$$\left\{ Z_{2,1}^{(t+r)} x \left[\begin{array}{c|c} w & 1^{(p+t+r)} \\ w' & 2^{(q-t-r)} \end{array} \right] ; r > 0 \right\},$$

while, in positive dimension, $l > 0$, the basis can be taken to be the set

$$\left\{ Z_{2,1}^{(t+r_1)} x Z_{2,1}^{(r_2)} x \cdots x Z_{2,1}^{(r_l+1)} x \begin{bmatrix} w & 1^{(p+t+|r|)} \\ w' & 2^{(q-t-|r|)} \end{bmatrix} ; r_i > 0 \right\}.$$

Another observation about letter-place methods did totally convince me that one could do "intrinsic" module theory using them, namely, one could use the letter-place basis to define the equivariant filtration on, say, a two-rowed skew-shape that gives rise to the Pieri decomposition of the corresponding Weyl module. For suppose we have the skew-shape (A) above. As we've said, this is the image of $D_p \otimes D_q$ under the Weyl map, and the letter-place basis for $D_p \otimes D_q$ is the set of all double standard tableaux $\left\{ \begin{pmatrix} w & 1^{(p)} 2^{(l)} \\ w' & 2^{(q-l)} \end{pmatrix} \right\}$ with $q \leq p + l$, and where w and w' are words in the letter alphabet (the place alphabet consisting in this case merely of 1 and 2 in their usual order). Now the Pieri decomposition theorem for the above shape tells us that the Weyl module has an equivariant filtration such that the associated graded module is the direct sum of the Weyl modules corresponding to the partitions $(p + l, q - l)$ with $q - l \leq p$ and $l \leq t$. (The condition that $q - l \leq p$ arises from the fact that, when you take $t - l$ boxes away from the $p + t$ boxes in the filled-in top row, they can't overlap any boxes in the second row.) Except for the condition that $l \leq t$, these conditions are remarkably close to the standard basis for $D_p \otimes D_q$. However, if we use the fact, mentioned above, that the image of any basis element $\begin{pmatrix} w & 1^{(p)} 2^{(l)} \\ w' & 2^{(q-l)} \end{pmatrix}$ with $l > t$ is zero under the Weyl map, and that the straightening of a double tableau increases the shape of the tableau in the dominance order, we see that the equivariant Pieri filtration on our Weyl module is that obtained by taking, for each l, the image of the submodule of $D_p \otimes D_q$ generated by all standard tableaux $\left\{ \begin{pmatrix} w & 1^{(p)} 2^{(k)} \\ w' & 2^{(q-k)} \end{pmatrix} \right\}$ with $k \geq l$. For skew-shapes of Pieri type, this argument works for any number of rows; how to extend this to arbitrary (almost) skew-shapes in the form of a Littlewood-Richardson decomposition is still being explored.

5 Capelli operators and further results

Capelli identities don't enter the picture until we get to three-rowed shapes. In this case, if we have the skew-shape $(\lambda_1, \lambda_2, \lambda_3)/(\mu_1, \mu_2, \mu_3)$, with $p_i = \lambda_i - \mu_i$ for $i = 1, 2, 3$ and $t_i = \mu_i - \mu_{i+1}$ for $i = 1, 2$, then we know that the Weyl module is presented by the $\Box$ map ([4]):

$$\Box : \begin{array}{c} \sum_{k>0} D_{p_1+t_1+k} \otimes D_{p_2-t_1-k} \otimes D_{p_3} \\ \oplus \\ \sum_{l>0} D_{p_1} \otimes D_{p_2+t_2+l} \otimes D_{p_3-t_2-l} \end{array} \rightarrow D_{p_1} \otimes D_{p_2} \otimes D_{p_3}.$$

As in the two-rowed case, the maps $D_{p_1+t_1+k} \otimes D_{p_2-t_1-k} \otimes D_{p_3} \rightarrow D_{p_1} \otimes D_{p_2} \otimes D_{p_3}$ may be interpreted as the kth divided power of the place polarization from place 1 to place 2, and the maps $\sum_{l>0} D_{p_1} \otimes D_{p_2+t_2+l} \otimes D_{p_3-t_2-l} \to D_{p_1} \otimes D_{p_2} \otimes D_{p_3}$ may be interpreted as the lth divided power of the place polarization from place 2 to place 3. Again taking our cue from the two-rowed case, we now introduce two generators, $Z_{2,1}$ and $Z_{3,2}$ with their divided powers, and we write, in place of the above,

$$\begin{array}{ll} \sum_{k>0} Z_{2,1}^{(t_1+k)} x D_{p_1+t_1+k} \otimes D_{p_2-t_1-k} \otimes D_{p_3} & \\ \oplus & \to D_{p_1} \otimes D_{p_2} \otimes D_{p_3}, \\ \sum_{l>0} Z_{3,2}^{(t_2+l)} y D_{p_1} \otimes D_{p_2+t_2+l} \otimes D_{p_3-t_2-l} & \end{array}$$

where x and y stand for separator variables, and the boundary map is the sum of polarizing x and y to 1. In [7] we showed that the general term of the resolution of our Weyl module, assuming that the number of triple overlaps is no greater than one, is

$$\sum_{l>0} Z_{3,2}^{(t_2+l_1)} y Z_{3,2}^{(l_2)} y \cdots y Z_{3,2}^{(l_p)} y Z_{2,1}^{(t_1+t_2+|l|+k_1)} x Z_{2,1}^{(k_2)} x \cdots$$
$$\cdots x Z_{2,1}^{(k_q)} x D_{p_1+t_1+t_2+|l|+|k|} \otimes D_{p_2-t_1-|k|} \otimes D_{p_3-t_2-|l|},$$

where the boundary map is the sum of polarization of x and y to 1. Now, as we polarize the separators to 1, we have no problem recognizing the term that we get except when we eliminate the x between $Z_{2,1}^{(k_q)}$ and $D_{p_1+t_1+t_2+|l|+|k|} \otimes D_{p_2-t_1-|k|} \otimes D_{p_3-t_2-|l|}$, and when we eliminate the y between $Z_{3,2}^{(l_p)}$ and $Z_{2,1}^{(t_1+t_2+|l|+k_1)}$. In the first case, we simply replace the resulting term

$$Z_{2,1}^{(k_q)} D_{p_1+t_1+t_2+|l|+|k|} \otimes D_{p_2-t_1-|k|} \otimes D_{p_3-t_2-|l|}$$

by the term we get as a result of operating by the indicated divided power of the polarization map; in the second, we replace

$$Z_{3,2}^{(l_p)} Z_{2,1}^{(t_1+t_2+|l|+k_1)} x$$

by the term

$$\sum_{\alpha} Z_{2,1}^{(t_1+t_2+|l|+k_1-\alpha)} x Z_{3,2}^{(l_p-\alpha)} Z_{3,1}^{(\alpha)},$$

where $Z_{3,1}$ is the commutator of $Z_{3,2}$ and $Z_{2,1}$ and we postulate that $Z_{3,2}$ and $Z_{3,1}$ commute with x and that $Z_{3,1}$ also commutes with $Z_{2,1}$.

All of the above would seem like so much hocus pocus were it not for the fact that it comes directly from Capelli-like identities on the polarization operators $\partial_{2,1}$, $\partial_{3,2}$ and $\partial_{3,1}$, namely,

$$\partial_{3,2}^{(l)} \partial_{2,1}^{(k)} = \sum_{\alpha} \partial_{2,1}^{(k-\alpha)} \partial_{3,2}^{(l-\alpha)} \partial_{3,1}^{(\alpha)}.$$

We also have the "signed" Capelli identity

$$\partial_{2,1}^{(k)}\partial_{3,2}^{(l)} = \sum_{\alpha}(-1)^{\alpha}\partial_{3,2}^{(l-\alpha)}\partial_{2,1}^{(k-\alpha)}\partial_{3,1}^{(\alpha)};$$

this comes into play as we consider more general shapes. It was this kind of observation that explained the very complicated boundary operators Akin and I had obtained many years earlier.

With this kind of progress in hand, Rota and I were willing to undertake the general problem of finding resolutions of Weyl modules corresponding to almost skew-shapes. The presentation of such modules is a bit more complicated than that of skew-shapes; for example, in the three-rowed case, the polarization operator $\partial_{3,1}$ and its divided powers already come into play in dimension 1. As I've indicated, there are still many technical problems remaining here, and in a later article these will be detailed much more fully. Nevertheless, it is possible to describe what the terms of the resolutions of Weyl modules corresponding to these general shapes are. To do this we'll first have to introduce a little more notation. We're going to let $Z_{n,m}^{(l)}$ stand for polarizations from place m to place n, and we're going to suppress the various separators that come into play (as a rule, there will be a new separator introduced for each operator $Z_{n,m}$). For fixed n, m, we'll denote by

$$Z_{n,m}^{\tau} \circledcirc \underline{Z}_{n,m}^{(l)}$$

the homogeneous strand of the Bar complex of total degree $\tau + l$ with initial term of degree $\geq \tau$.

We will also denote an n-rowed almost skew-shape by $(p_1, \ldots, p_n; \tau_1, \ldots, \tau_{n-1})$, where the integer p_i is the number of boxes in row i, for $i = 1, \ldots, n$; τ_i is the protuberance of row $i+1$ beyond row i for $i = 1, \ldots n-1$; and τ_{n-1} is the protuberance of row n beyond row $n-1$ on the left. The important thing to note about this notation is that if our shape is a skew-shape, then $\tau_{n-1} \geq 0$. Otherwise, $\tau_{n-1} < 0$. We say that our shape is of *type* k if $\tau_{n-1}+\cdots+\tau_{n-k} < 0$ and $\tau_{n-1}+\cdots+\tau_{n-k-1} \geq 0$. Thus, it is of type 0 if it is a skew-shape; it is of type 1 if the bottom row doesn't extend as far to the left as the penultimate row, but does extend beyond (in the weak sense) row $n-2$, etc.

For our final bit of notation, we let $\operatorname{Res}(p_1, \ldots, p_m; \tau_1, \ldots, \tau_{m-1})$ stand for the resolution of the skew-shape $(p_1, \ldots, p_m; \tau_1, \ldots, \tau_{m-1})$. The term $\operatorname{Res}(p_1, \ldots, p_m; \tau_1, \ldots, \tau_{m-1}) \otimes (r)$ stands for the resolution tensored with D_r.

We'll now assume that we know $\operatorname{Res}(p_1, \ldots, p_m; \tau_1, \ldots, \tau_{m-1})$ for $m < n$, and we describe $\operatorname{Res}(p_1, \ldots, p_n; \tau_1, \ldots, \tau_{n-1})$ by induction on n.

For a skew-shape $(p_1, \ldots, p_n; \tau_1, \ldots, \tau_{n-1})$, the resolution looks like this:

$$\begin{aligned}\operatorname{Res}(p_1, \ldots, p_n; \tau_1, \ldots, \tau_{n-1}) &= \operatorname{Res}(p_1, \ldots, p_{n-1}; \tau_1, \ldots, \tau_{n-2}) \otimes (p_n)\\ \oplus \sum_{l_{n-1}\geq 0} Z_{n,n-1}^{\tau_{n-1}+1} &\circledcirc \underline{Z}_{n,n-1}^{(l_n-1)}\operatorname{Res}(p_1, \ldots, p_{n-2}, p_{n-1}+\overline{\tau}_{n-1}; \tau_1,\\ &\ldots, \tau_{n-3}, \tau_{n-2}+\overline{\tau}_{n-1}) \otimes (p_n - \overline{\tau}_{n-1})\end{aligned}$$

$$\oplus \sum_{l_{n-j}\geq 0} (Z_{n,n-1}^{\tau_{n-1}+1} \circledcirc \underline{Z}_{n,n-1}^{(l_{n-1})}) \otimes (Z_{n,n-2}^{\tau_{n-2}+1} \circledcirc \underline{Z}_{n,n-2}^{(l_{n-2})})$$

$$\otimes \mathrm{Res}(p_1, \dots p_{n-2} + \overline{\tau}_{n-2}, p_{n-1} + \overline{\tau}_{n-1}; \tau_1, \dots, \tau_{n-3} + \overline{\tau}_{n-2}, \tau_{n-2} + \overline{\tau}_{n-1} - \overline{\tau}_{n-2}) \otimes (p_n - \overline{\tau}_{n-1} - \overline{\tau}_{n-2})$$

$$\oplus \quad \vdots$$

$$\sum_{l_{n-j}\geq 0} (Z_{n,n-1}^{\tau_{n-1}+1} \circledcirc \underline{Z}_{n,n-1}^{(l_{n-1})}) \otimes (Z_{n,n-2}^{\tau_{n-2}+1} \circledcirc \underline{Z}_{n,n-2}^{(l_{n-2})}) \otimes \cdots \otimes (Z_{n,1}^{\tau_1+1} \circledcirc \underline{Z}_{n,1}^{(l_1)})$$

$$\otimes \mathrm{Res}(p_1 + \overline{\tau}_1, \dots, p_{n-1} + \overline{\tau}_{n-1}; \tau_1 + \overline{\tau}_2 - \overline{\tau}_1, \dots, \tau_{n-2} + \overline{\tau}_{n-1} - \overline{\tau}_{n-2}) \otimes (p_n - \sum_{1\leq j\leq n-1} \overline{\tau}_j),$$

where $\overline{\tau}_j = \tau_j + l_j + 1$ for $j = 1, \dots, n-1$.

For an almost skew-shape $(p_1, \dots, p_n; \tau_1, \dots, \tau_{n-1})$ of type k, let's denote by s the positive integer $-(\tau_{n-1} + \cdots + \tau_{n-k})$, and by t, the integer $(\tau_{n-1} + \cdots + \tau_{n-k-1})$. With this notation, the terms of the resolution are

$$\mathrm{Res}(p_1, \dots, p_n; \tau_1, \dots, \tau_{n-1}) = \mathrm{Res}(p_1, \dots, p_{n-1}; \tau_1, \dots, \tau_{n-2}) \otimes (p_n)$$

$$\oplus \sum_{l_{n-1}>0} \underline{Z}_{n,n-1}^{(l_{n-1})} \otimes \mathrm{Res}(p_1, \dots, p_{n-2}, p_{n-1} + l_{n-1}; \tau_1, \dots, \tau_{n-3}, \tau_{n-2} + l_{n-1})$$

$$\otimes (p_n - l_{n-1})$$

$$\oplus \sum_{l_{n-2}>0, l_{n-1}\geq 0} (\underline{Z}_{n,n-2}^{(l_{n-2})}) \otimes (\underline{Z}_{n,n-1}^{(l_{n-1})})$$

$$\otimes \mathrm{Res}(p_1, \dots p_{n-2} + l_{n-2}, p_{n-1} + l_{n-1}; \tau_1, \dots, \tau_{n-3} + l_{n-2}, \tau_{n-2} + l_{n-1} - l_{n-2}) \otimes (p_n - l_{n-1} - l_{n-2})$$

$$\oplus \quad \vdots$$

$$\sum_{l_{n-j}\geq 0} (Z_{n,n-k-1}^{t+1} \circledcirc \underline{Z}_{n,n-k-1}^{(l_{n-k-1})}) \otimes \underline{Z}_{n,n-k}^{(l_{n-k})} \otimes \cdots \otimes \underline{Z}_{n,n-1}^{(l_{n-1})})$$

$$\otimes \mathrm{Res}(p_1, p_2, \dots, p_{n-k-1} + t + 1 + l_{n-k-1}, p_{n-k} + l_{n-k}, \dots, p_{n-1} + l_{n-1}; \tau_1,$$

$$\dots, \tau_{n-k-2} + t + 1 + l_{n-k-1}, s - 1 + l_{n-k} - l_{n-k-1},$$

$$\dots, \tau_{n-2} + l_{n-1} - l_{n-2}) \otimes (p_n - \sum_{i=n-k-1}^{n-1} l_{n-i} - \sum_{j=n-k-1}^{n-1} \tau_j - 1)$$

$$\oplus \sum_{l_{n-j}\geq 0} (Z_{n,n-k-2}^{\tau_{n-k-2}+1} \underline{Z}_{n,n-k-2}^{(l_{n-k-2})}) \otimes (Z_{n,n-k-1}^{t+1}$$

$$\circledcirc \underline{Z}_{n,n-k-1}^{(l_{n-k-1})}) \otimes \underline{Z}_{n,n-k}^{(l_{n-k})} \otimes \cdots \otimes \underline{Z}_{n,n-1}^{(l_{n-1})})$$

$$\otimes \mathrm{Res}(p_1, p_2, \dots, p_{n-k-2} + \overline{\tau}_{n-k-2}, p_{n-k-1} + t + 1 + l_{n-k-1}, p_{n-k} + l_{n-k}, \dots,$$

$$p_{n-1}+l_{n-1};\tau_1,\dots\tau_{n-k-3}+\overline{\tau}_{n-k-2},\tau_{n-k-2}+t+1+l_{n-k-1}-\overline{\tau}_{n-k-2},$$
$$\dots,\tau_{n-2}+l_{n-1}-l_{n-2})\otimes(p_n-\sum_{i=n-k-2}^{n-1}l_{n-i}-\sum_{j=n-k-2}^{n-1}\tau_j-2)$$

$$\oplus \qquad \vdots$$
$$\sum_{l_{n-j}\geq 0}(Z_{n,1}^{\tau_1+1}\circledcirc \underline{Z}_{n,1}^{(l_1)})\otimes\cdots\otimes(Z_{n,n-k-2}^{\tau_{n-k-2}+1}\circledcirc \underline{Z}_{n,n-k-2}^{(l_{n-k-2})})$$
$$\otimes(Z_{n,n-k-1}^{t+1}\circledcirc \underline{Z}_{n,n-k-1}^{(l_{n-k-1})})\otimes \underline{Z}_{n,n-k}^{(l_{n-k})}\otimes\cdots\otimes \underline{Z}_{n,n-1}^{(l_{n-1})})$$
$$\otimes\mathrm{Res}(p_1+\overline{\tau}_1,\dots,p_{n-k-2}+\overline{\tau}_{n-k-2},p_{n-k-1}+t+1+l_{n-k-1},$$
$$p_{n-k}+l_{n-k},\dots,p_{n-1}+l_{n-1};$$
$$\tau_1+\overline{\tau}_2-\overline{\tau}_1,\dots,\tau_{n-k-3}+\overline{\tau}_{n-k-2}-\overline{\tau}_{n-k-3},$$
$$\tau_{n-k-2}+t+1+l_{n-k-1}-\overline{\tau}_{n-k-2},\dots,$$
$$\tau_{n-2}+l_{n-1}-l_{n-2})\otimes(p_n-\sum_{j=1}^{n-1}\overline{\tau}_j-k),$$

where $\overline{\tau}_j$ is defined as above.

It must be understood that, if the indicated shape isn't legitimate, then that term vanishes. This puts some constraints on the parameters over which we're summing.

If one takes a close look at the resolution of a skew-shape, one sees that the characteristic zero "skeleton" described by Lascoux ([16]) is in there. Recall that Lascoux has described the terms and their placement in the resolutions of Weyl modules (of partitions) in terms of the lengths of the permutations corresponding to the determinantal expansion of the Jacobi–Trudi matrix for that Weyl module. If we let S_n denote the symmetric group on n letters, and $\sigma\in S_n$ a permutation such that $\sigma(n)=i$, then σ can be written uniquely as a product:

$$\sigma=(n,n-1)(n,n-2)\cdots(n,i)\sigma',$$

where (n,j) stands for the transposition of n and j, and $\sigma'\in S_{n-1}$. (Notice that the length of σ is $n-i+\mathrm{length}(\sigma')$.) This is the way of recovering the terms of the Lascoux resolution within the resolution described above.

References

[1] Akin, K., Buchsbaum, D.A. (1985): Characteristic-free representation theory of the general linear group. Adv. Math. **58**, 149–200

[2] Akin, K., Buchsbaum, D.A. (1988): Characteristic-free representation theory of the general linear group. II. Homological considerations. Adv. Math. **72**, 171–210

[3] Akin, K., Buchsbaum, D.A. (1992): A note on the Poincaré resolution of the coordinate ring of the Grassmannian. J. Algebra **152**, 427–433

[4] Akin, K., Buchsbaum, D.A., Weyman, J. (1982): Schur functors and Schur complexes. Adv. Math. **44**, 207–278

[5] Anick, D., Rota, G.-C. (1991): Higher-order syzygies for the bracket algebra and for the ring of coordinates of the Grassmannian. Proc. Nat. Acad. Sci. U.S.A. **88**, 8087–8090

[6] Buchsbaum, D.A., Rota, G.-C. (1993): Projective resolutions of Weyl modules. Proc. Nat. Acad. Sci. U.S.A. **90**, 2448–2450

[7] Buchsbaum, D.A., Rota, G.-C. (1994): A new construction in homological algebra. Proc. Nat. Acad. Sci. U.S.A. **91**, 4115–4119

[8] Buchsbaum, D.A., Sánchez, R.(1994): On lifting maps between Weyl modules: can bad shapes be resolved by better shapes? Adv. Math. **105**, 59–75

[9] Donkin, S. (1986): Finite resolutions of modules for reductive algebraic groups. J. Algebra **101**, 473–488

[10] Klucznik, M.B. (1998): Exact sequences of Schur complexes. Ph.D. thesis, Brandeis University

[11] Kulkarni, U.B. (1997): Characteristic-free representation theory of $GL_n(\mathbb{Z})$: some homological aspects. Ph.D. thesis, Brandeis University

[12] Lascoux, A. (1977): Polynoms de Schur, fonctions de Schur, et grassmanniennes. Thèse, Université de Paris

Circulant recursive matrices

M. Barnabei, L. B. Montefusco

1 Introduction

Recursive matrices are bi-infinite matrices which can be recursively generated starting from two given Laurent series α and β, called the recurrence rule and the boundary value of the matrix, respectively. The ith row of a recursive matrix contains the coefficients of the series $\alpha^i \beta$. These matrices were introduced by Barnabei, Brini, and Nicoletti [1] to formulate a more general version of the umbral calculus developed in a series of fundamental papers by Rota and his school [7–9]. In [2] it was shown that the most important operators of the umbral calculus can be represented by recursive matrices. In particular, shift-invariant operators, which play a crucial role in Rota's theory, correspond to Toeplitz matrices, which can be characterized as recursive matrices with recurrence rule $\alpha(t) = t$. Recently, recursive matrices were found to be useful in studying algebraic aspects of signal processing, since they contain the classes of bi-infinite Toeplitz, Hankel and Hurwitz matrices as special cases [3, 4].

On the other hand, in a finite setting, a class of matrices sharing the most pleasant properties of bi-infinite Toeplitz matrices, such as closure with respect to product and transposition, is the class of circulant matrices. Such matrices have been studied since the 19th century, and are still of research interest because of their connections to problems in physics, probability and statistics, as well as applications to data smoothing, signal and image processing and matrix computations. A circulant matrix is a square $n \times n$ matrix whose first row is an arbitrary n-tuple, while the following rows are derived from it by successive cyclic shifts. It can be obtained from a bi-infinite banded Toeplitz matrix by folding it vertically and cutting its first n rows. The folding operation corresponds to listing in the ith row the coefficients of the polynomial $z^i p(z)$ reduced modulo $z^n - 1$, where $p(z)$ is the generating function of row 0.

Because of the many pleasant properties of circulant matrices, it is natural to look for an extension of the cutting-folding operation to general recursive matrices. To this aim, in the present paper we introduce and study a new class of finite matrices – circulant recursive matrices, as we propose to call them – which can be seen as finite analogs of recursive matrices and, at the same time, represent a generalization of circulant matrices.

The main notions concerning circulant recursive matrices are formally the same as those of bi-infinite recursive matrices. What is different is the algebraic framework: in the bi-infinite case it corresponds to the ring of Laurent series, while in the finite $n \times n$ case it is given by the quotient ring $\mathcal{P}_n$ of complex polynomials $p(z)$ modulo $(z^n - 1)$. In fact, a circulant recursive matrix is defined recursively, starting from two given complex polynomials $s(z)$ and $p(z)$ and performing computations in the quotient ring $\mathcal{P}_n$. More precisely, the entries of the ith row of the matrix are the coefficients of the polynomial $s^i(z)p(z)(\text{mod } z^n - 1)$. In order to make the definition consistent, the polynomial $s(z)$ must satisfy the requirement that $s^n(z) \equiv 1$ (mod $z^n - 1$). This condition is clearly satisfied by the polynomial $s(z) = z$; therefore circulant matrices are a particular case of circulant recursive matrices.

The study of recursive matrices is essentially performed using the technique, typical in combinatorics, of representing rows and columns of such matrices by means of their generating functions, which are Laurent series. This approach leads to a proof of the two fundamental properties of recursive matrices, the product rule and the double recurrence theorem. These results are the matrix form of two major results of the umbral calculus: umbral substitution and the transfer formula. In the same way, the fundamental features of circulant recursive matrices are proved by exploiting the properties of the quotient ring $\mathcal{P}_n$, and thus obtaining the finite analogs of the product rule and the double recurrence theorem.

In the finite case, the different algebraic setting leads to some interesting new results, which do not apply to the bi-infinite case. In fact, the fundamental role played by the polynomial $z^n - 1$ justifies the importance of the finite Fourier transform, which realises the bijection between the coefficients of polynomials in $\mathcal{P}_n$ and the corresponding values at the nth roots of unity. Hence, the Fourier transform becomes a fundamental tool for investigating spectral properties of circulant recursive matrices, whose eigenvalues can be explicitly evaluated.

This paper is a first attempt in the direction of studying circulant recursive matrices, which seem to be a promising tool both for theoretical and numerical purposes. Further investigations, and in particular a finite analog of the umbral calculus, will be the subject of a forthcoming paper.

2 Circulant matrices

In this section we recall the main definitions and results about circulant matrices. For a more detailed description, we refer the interested reader to the fundamental book by P. J. Davis [5].

A *circulant matrix* of order n is a square $n \times n$ matrix of the form

$$C = \begin{bmatrix} c_0 & c_1 & \cdots & c_{n-1} \\ c_{n-1} & c_0 & \cdots & c_{n-2} \\ \vdots & \vdots & & \vdots \\ c_1 & c_2 & \cdots & c_0 \end{bmatrix}$$

The sequence $(c_0, c_1, \ldots, c_{n-1})$ completely determines the whole matrix, since, for every $i, j = 0, 1, \ldots, n-1$, the $(i, j)^{\text{th}}$ element c_{ij} of C is given by

$$c_{ij} = c_k, \quad \text{where } k \equiv j - i \mod n.$$

Hence, we can associate to the circulant matrix C the generating function of the sequence $(c_0, c_1, \ldots, c_{n-1})$, namely, the polynomial of degree not exceeding $n - 1$,

$$p(z) := c_0 + c_1 z + \ldots + c_{n-1} z^{n-1},$$

which will be called the *generating function* of C.

In order to describe in a similar way the generating function of the other rows of the matrix C, we need to consider the quotient ring

$$\mathcal{P}_n := \frac{\mathbb{C}[z]}{(z^n - 1)}$$

of complex polynomials in the variable z modulo the principal ideal generated by the polynomial $z^n - 1$.

It is now easy to characterize circulant matrices as those square matrices $C := [c_{ij}]$, $i, j = 0, 1, \ldots, n-1$, such that, for every index $i = 1, 2, \ldots, n-1$, the generating function $C(i)$ of the ith row is given by

$$C(i) := c_{i0} + c_{i1} z + \ldots + c_{i,n-1} z^{n-1} \equiv z^i p(z) \qquad \text{mod } (z^n - 1),$$

where $p(z)$ denotes the generating function of the 0th row of C.

This characterization illuminates the analogy between circulant matrices and bi-infinite Toeplitz matrices, which can be defined as those bi-infinite matrices T such that, for every integer i, the generating function $T(i)$ of the ith row is given by

$$T(i) = t^i \alpha(t),$$

where $\alpha(t)$ is a given formal power series.

Moreover, as is the case for bi-infinite Toeplitz matrices, the product of two circulant matrices is again a circulant matrix.

Theorem 2.1. *Let C, D be circulant matrices of the same order, with generating functions $p(z), q(z)$, respectively. The product $C \times D$ is the circulant matrix with generating function $p(z)q(z) \mod (z^n - 1)$. As a consequence, circulant matrices commute.*

The simplest nontrivial example of circulant matrix is given by the "push" matrix

$$\Pi = \begin{bmatrix} 0 & 1 & 0 & \ldots & 0 \\ 0 & 0 & 1 & \ldots & 0 \\ \vdots & \vdots & & & \vdots \\ 1 & 0 & 0 & \ldots & 0 \end{bmatrix},$$

whose generating function is the polynomial $p(z) = z$.

Circulant matrices can be characterized by means of the push matrix.

Theorem 2.2. *Let C be a square matrix of order n. The following are equivalent:*

i) *C is a circulant matrix with generating function $p(z)$;*
ii) *$\Pi \times C = C \times \Pi$;*
iii) *$C = p(\Pi)$ for some complex polynomial $p(z)$ of degree at most $n - 1$.*

One of the most important properties of circulant matrices is that they are diagonalized by the Fourier matrix F, which is defined as

$$F = \frac{1}{\sqrt{n}} \begin{bmatrix} 1 & 1 & 1 & \dots & 1 \\ 1 & w^{n-1} & w^{n-2} & \dots & w \\ 1 & w^{n-2} & w^{n-4} & \dots & w^2 \\ \vdots & \vdots & & & \vdots \\ 1 & w & w^2 & \dots & w^{n-1} \end{bmatrix},$$

where $w := e^{\frac{2\pi i}{n}}$.

Note that

$$h, k \leq n-1, h \neq k \Longrightarrow w^h \neq w^k,$$

namely, the n complex numbers $1, w, w^2, \dots, w^{n-1}$ are precisely the nth roots of unity. Moreover:

$$\bar{w}^k = w^{n-k} = (w^k)^{-1}; \tag{1}$$

$$w^h = w^k \Longleftrightarrow h \equiv k \mod n; \tag{2}$$

$$\sum_{r=0}^{n-1} w^{kr} = \begin{cases} 0 & \text{if } k \neq 0 \\ n & \text{if } k = 0. \end{cases} \tag{3}$$

Hence, the conjugate transpose F^* of the Fourier matrix is

$$F^* = \frac{1}{\sqrt{n}} \begin{bmatrix} 1 & 1 & 1 & \dots & 1 \\ 1 & w & w^2 & \dots & w^{n-1} \\ 1 & w^2 & w^4 & \dots & w^{n-2} \\ \vdots & \vdots & & & \vdots \\ 1 & w^{n-1} & w^{n-2} & \dots & w \end{bmatrix}.$$

Theorem 2.3. *Let C be a circulant matrix of order n with generating function $p(z)$. Then*

$$FCF^* = \operatorname{Diag}(p(1), p(w), p(w^2), \dots, p(w^{n-1})). \tag{4}$$

3 The quotient ring $\mathcal{P}_n$

In this section we summarise some definitions and results concerning the ring $\mathcal{P}_n$, which will be widely used in what follows.

First of all, we recall that two complex polynomials $p(z), q(z)$ coincide in the quotient ring $\mathcal{P}_n$ whenever there exists a polynomial $g(z)$ such that

$$p(z) = q(z) + (z^n - 1)g(z). \tag{5}$$

Since the nth roots of unity $1, w, \ldots w^{n-1}$ are precisely the roots of the polynomial $z^n - 1$, condition (5) can be restated as follows.

Proposition 3.1. *Let $p(z), q(z)$ be complex polynomials. Then*

$$p(z) \equiv q(z) \mod (z^n - 1)$$

if and only if

$$p(w^k) = q(w^k) \qquad \textit{for } k = 0, 1, \ldots, n-1.$$

In other words, every equivalence class of the congruence relation $\mod (z^n - 1)$ can be represented by the ordered n-tuple $(p(1), p(w), \ldots, p(w^{n-1}))$ of the values at the nth roots of unity of any polynomial $p(z)$ in the class. Moreover, the canonical representative of the equivalence class, namely, the unique polynomial in the class whose degree does not exceed $n-1$, $p(z) = c_0 + c_1 z + \ldots + c_{n-1} z^{n-1}$, can be recovered from the above n-tuple by means of the Fourier matrix, as follows:

$$\begin{bmatrix} c_0 \\ c_1 \\ \vdots \\ c_{n-1} \end{bmatrix} = \frac{1}{\sqrt{n}} F \begin{bmatrix} p(1) \\ p(w) \\ \vdots \\ p(w^{n-1}) \end{bmatrix}, \tag{6}$$

and, hence,

$$\begin{bmatrix} p(1) \\ p(w) \\ \vdots \\ p(w^{n-1}) \end{bmatrix} = \sqrt{n}\ F^* \begin{bmatrix} c_0 \\ c_1 \\ \vdots \\ c_{n-1} \end{bmatrix}. \tag{7}$$

This implies, for example, that

$$p(0) = c_0 = \frac{1}{\sqrt{n}} \sum_{k=0}^{n-1} p(w^k). \tag{8}$$

Knowledge of the values taken by a polynomial at the nth roots of unity allow us to describe simply some important properties of the quotient ring $\mathcal{P}_n$. First of all, the invertible elements of the ring $\mathcal{P}_n$ can be characterized as follows.

Proposition 3.2. *A polynomial $p(z)$ is invertible in the ring $\mathcal{P}_n$ whenever*

$$p(w^k) \neq 0 \qquad \textit{for } k = 0, 1, \ldots, n-1. \tag{9}$$

Proof. Suppose that the polynomial $p(z)$ satisfies (9). Let $q(z)$ be the unique polynomial of degree $\leq n-1$ such that

$$q(w^k) = \frac{1}{p(w^k)} \qquad \text{for } k = 0, 1, \dots, n-1.$$

Then

$$(pq)(w^k) = 1 \qquad \text{for } k = 0, 1, \dots, n-1,$$

namely, by Proposition 3.1, $(pq)(z) \equiv 1 \mod (z^n - 1)$ for $z \in \mathcal{C}$. The converse is immediate. □

A particular class of polynomials which will play a crucial role in what follows is the class of permutation polynomials.

Definition. A *permutation polynomial* is a polynomial $s(z) \in \mathcal{P}_n$ such that

$$s^n(z) \equiv 1 \mod (z^n - 1).$$

Once again, these polynomials can be characterized by means of their values at the nth roots of unity, as follows.

Theorem 3.1. *A polynomial $s(z) \in \mathcal{P}_n$ is a permutation polynomial if and only if its values at the nth roots of unity are also n^{th} roots of unity. In other words, a polynomial $s(z) \in \mathcal{P}_n$ is a permutation polynomial if and only if there exists a generalized permutation of the set $\{0, 1, \dots, n-1\}$, namely, a function $\sigma : \{0, 1, \dots, n-1\} \longrightarrow \{0, 1, \dots, n-1\}$, such that*

$$s(w^k) = w^{\sigma(k)} \qquad \textit{for } k = 0, 1, \dots, n-1.$$

Proof. This follows immediately from Proposition 3.1, since:

$$s^n(z) \equiv 1 \pmod{z^n - 1} \iff s^n(w^k) = 1, \; k = 0, 1, \dots, n-1,$$

and the last condition is equivalent to requiring that $s(w^k) = w^h$ for some $h \in \{0, 1, \dots, n-1\}$. Setting $\sigma(k) := h$ we get the assertion. □

Example 3.1. In the quotient ring $\mathcal{P}_4$, the complex polynomial

$$s_1(z) = \frac{i}{2} - \frac{1}{2}z^2 + \frac{i+1}{2}z^3 \tag{10}$$

is the permutation polynomial associated with the generalized permutation

$$\sigma = \begin{pmatrix} 0 & 1 & 2 & 3 \\ 1 & 0 & 2 & 1 \end{pmatrix},$$

while the polynomial

$$s_2(z) = \frac{i}{2}z + \frac{i+1}{2}z^2 - \frac{1}{2}z^3 \tag{11}$$

is the permutation polynomial associated with the permutation

$$\tau = \begin{pmatrix} 0 & 1 & 2 & 3 \\ 1 & 2 & 0 & 3 \end{pmatrix}.$$

Now take $p(z), s(z) \in \mathcal{P}_n$, with $p(z) = c_0 + c_1 z + c_2 z^2 + \ldots + c_{n-1} z^{n-1}$ and $s(z)$ a permutation polynomial. We define the *composition* of p and s to be the polynomial

$$\begin{aligned}(p \circ s)(z) &:= p(s(z)) \\ &\equiv c_0 + c_1 s(z) + c_2 s^2(z) + \ldots + c_{n-1} s^{n-1}(z) \mod (z^n - 1).\end{aligned}$$

We point out that the condition $s^n(z) \equiv 1 \mod (z^n - 1)$ ensures that the above definition is consistent. Moreover, for every $q(z) \in \mathcal{P}_n$,

$$((pq) \circ s)(z) = (p \circ s)(z)(q \circ s)(z).$$

The set of permutation polynomials is closed under composition. In fact, we have the following result, whose proof is straightforward.

Proposition 3.3. *Let s_1, s_2 be two permutation polynomials, with associated generalized permutations σ_1, σ_2, respectively. Then, the composition $s_1 \circ s_2$ is the permutation polynomial associated with $\sigma_1 \circ \sigma_2$, the functional composition of σ_1, σ_2.*

As an immediate consequence, we have

Proposition 3.4. *A permutation polynomial $s(z)$ admits a two-sided inverse with respect to composition if and only if it is associated with a permutation σ, that is, with a bijective generalized permutation. If this is the case, the compositional inverse $\tilde{s}(z)$ of $s(z)$ is the permutation polynomial associated with σ^{-1}.*

A permutation polynomial which admits a compositional inverse will be called a *proper permutation polynomial.*

For example, the permutation polynomial $s_1(z)$ given in (10) is not proper, while $s_2(z)$, given in (11), is proper, and its compositional inverse is the polynomial

$$\tilde{s_2}(z) = -\frac{i}{2} z + \frac{i-1}{2} z^2 - \frac{1}{2} z^3,$$

whose associated permutation is

$$\tau^{-1} = \begin{pmatrix} 0 & 1 & 2 & 3 \\ 2 & 0 & 1 & 3 \end{pmatrix}.$$

We observe that any proper permutation polynomial has zero constant term. In fact:

Proposition 3.5. *Let $s(z)$ be a proper permutation polynomial. Then the polynomials $s^h(z)$, $h = 1, 2, \ldots, n-1$, have zero constant term.*

Proof. Let σ be the permutation associated with the polynomial $s(z)$. By (8) and (3) we see that

$$s(0) = \frac{1}{\sqrt{n}} \sum_{k=0}^{n-1} s(w^k) = \frac{1}{\sqrt{n}} \sum_{k=0}^{n-1} w^{\sigma(k)} = \frac{1}{\sqrt{n}} \sum_{h=0}^{n-1} w^h = 0.$$

A similar argument can be used for the powers of $s(z)$. □

A simple example of permutation polynomial is given by any monomial $s(z) = z^h$, $h = 1, \dots, n-1$. It is immediately seen that such a permutation polynomial is proper whenever h and n are relatively prime. This is true in the particular case $h = n-1$, which yields the permutation polynomial $m(z) := z^{n-1}$, whose associated permutation is the involution

$$\mu = \begin{pmatrix} 0 & 1 & 2 & \cdots & n-1 \\ 0 & n-1 & n-2 & \cdots & 1 \end{pmatrix}.$$

The polynomial $m(z)$ has many useful properties. In fact, $m(z)$ coincides both with its multiplicative inverse $m^{-1}(z)$ and with its compositional inverse $\tilde{m}(z)$. This last fact allows us to define the *star operator* $p(z) \longrightarrow p^\star(z)$ as the following involution: if $p(z) = c_0 + c_1 z + c_2 z^2 + \dots + c_{n-1} z^{n-1}$, then

$$p^\star(z) := (p \circ m)(z) = p(z^{n-1}) = c_0 + c_{n-1} z + c_{n-2} z^2 + \dots + c_1 z^{n-1}.$$

Also the multiplicative inverse $s^{-1}(z)$ of a permutation polynomial $s(z)$ can be represented as a composition involving $m(z)$, as follows:

$$s^{-1}(z) = (m \circ s)(z), \tag{12}$$

since $s^{-1}(z) = s^{n-1}(z) = (m \circ s)(z)$.

The preceding considerations immediately imply that, for every polynomial $p(z)$ in $\mathcal{P}_n$ and for every permutation polynomial $s(z)$,

$$(p^\star \circ s)(z) = (p \circ s^{-1})(z). \tag{13}$$

In fact,

$$(p^\star \circ s)(z) = (p \circ (m(z)) \circ s)(z) = (p \circ s^{-1})(z).$$

Moreover, the star operator allows us to describe the transpose of a circulant matrix, as follows.

Proposition 3.6. *Let C be the circulant matrix with generating function $p(z)$. Then the matrix C^T is the circulant matrix with generating function $p^\star(z)$.*

As an easy consequence of the above results, we get

Proposition 3.7. *Let $s(z)$ be a permutation polynomial. The circulant matrix with generating function $s(z)$ is orthogonal.*

Proposition 3.6 allows us to describe the convolution product in $\mathcal{P}_n$ as a matrix-vector product.

Corollary 3.1. *Let C be a circulant matrix with generating function $p(z)$, and let $q(z)$ be any polynomial of degree $\leq n-1$. Denoting by $[q(z)]$ the column vector of the coefficients of q, we have:*

$$[q(z)]^T \times C = [r(z)]^T,$$

and

$$C \times [q(z)] = [v(z)],$$

where

$$\begin{aligned} r(z) &\equiv q(z)p(z) \mod (z^n - 1), \\ v(z) &\equiv p(z)^\star q(z) \mod (z^n - 1). \end{aligned} \tag{14}$$

Proof. The row matrix $[q(z)]^T$ can be seen as the 0th row of the circulant matrix D with generating function $q(z)$. The first statement now follows from Theorem 2.1, while the second is a consequence of Proposition 3.6. □

4 Circulant recursive matrices

Let C be a circulant matrix, with generating function $s(z) \in \mathcal{P}_n$. Consider the problem of finding a square $n \times n$ complex matrix M such that

$$\Pi \times M = M \times C. \tag{15}$$

By Proposition 3.1, this problem can be restated equivalently in terms of generating functions as follows. Given a polynomial $s(z) \in \mathcal{P}_n$, we look for a square $n \times n$ complex matrix M such that the following conditions hold in the quotient ring $\mathcal{P}_n$:

$$\begin{aligned} M(i) &= M(i-1)\, s(z) \qquad \text{for } i = 1, 2, \ldots, n-1, \\ M(0) &= M(n-1)\, s(z). \end{aligned} \tag{16}$$

This last formulation shows that problem (15) admits a solution whenever $s(z)$ is a permutation polynomial. In fact, conditions (16) imply immediately that

$$M(0) \equiv M(0)\, s^n(z) \mod (z^n - 1),$$

which is possible if and only if $s^n(z) \equiv 1 \mod (z^n - 1)$. Moreover, if this is the case, the problem has infinitely many solutions, depending on the choice of the complex polynomial $M(0)$.

The above considerations can be summarized as follows.

Theorem 4.1. *Let $p(z), s(z)$ be polynomials in $\mathcal{P}_n$, and let C be the circulant matrix whose generating function is $s(z)$. There exists exactly one square $n \times n$ complex matrix M such that*

$$\begin{aligned} \Pi \times M &= M \times C, \\ M(0) &= p(z) \end{aligned} \tag{17}$$

if and only if $s(z)$ is a permutation polynomial.

The matrix M will be denoted by the symbol $CR(s, p)$, and called the (s, p)-*circulant recursive matrix,* with *recurrence rule* $s(z)$ and *generating function* $p(z)$. A circulant recursive matrix $CR(s, 1)$ with generating function $p(z) = 1$ will be called a *homogeneous circulant recursive matrix.*

We remark that circulant matrices can now be seen as circulant recursive matrices with recurrence rule $s(z) := z$. In particular, $\Pi = CR(z, z)$. Moreover, for every integer g, the choice $s(z) := z^g$ yields the so-called g-circulant matrices (see [5, Chap. 5].

The following characterization of circulant recursive matrices will be used often.

Theorem 4.2. *The square $n \times n$ complex matrix $M := [m_{i,j}]$ is a circulant recursive matrix if and only if there exists a permutation polynomial $s(z) := a_0 + a_1 z + \ldots + a_{n-1} z^{n-1}$ such that, for every index $i = 0, 1, \ldots, n-1$, the generating function $M(i)$ of the ith row of M satisfies*

$$M(i) \equiv M(0)\, s^i(z) \mod (z^n - 1),$$

or, equivalently, for $i = 0, 1, \ldots, n-2$, $j = 0, 1, \ldots, n-1$,

$$m_{i+1,j} = \sum_{\substack{h,k \\ h+k \equiv j \mod n}} m_{ih}\, a_k.$$

Example 4.1. Let $s(z) := s_2(z)$ be the permutation polynomial defined in (11), and $p(z) := 1 + z - 2z^2$. Then

$$CR(s, p) = \begin{bmatrix} 1 & 1 & -2 & 0 \\ -\frac{3}{2} - i & 1 + \frac{i}{2} & \frac{1}{2} + i & \frac{1}{2} i \\ -\frac{1}{2} + \frac{1}{2} i & \frac{1}{2} - \frac{3}{2} i & -\frac{1}{2} - \frac{1}{2} i & \frac{3}{2} i + \frac{1}{2} \\ \frac{1}{2} i - 1 & i - \frac{1}{2} & -\frac{1}{2} i & \frac{3}{2} - i \end{bmatrix}.$$

Circulant recursive matrices exhibit many typical features of circulant matrices. First of all, we show how the fundamental property of circulant matrices stated in Theorem 2.1 can be extended to circulant recursive matrices.

Theorem 4.3 (Product rule). *Let* $M := CR(s, p) = [m_{ij}]$, $N := CR(t, q) = [n_{ij}]$ *be two* $n \times n$ *circulant recursive matrices. Then, their product* $M \times N$ *is again a circulant recursive matrix. More precisely,*

$$CR(s, p) \times CR(t, q) = CR(s \circ t, (p \circ t)q), \tag{18}$$

where the operations are performed in the quotient ring $\mathcal{P}_n$.

Proof. Set $L := M \times N$. Then, by Theorem 4.2, for every index $i = 0, 1, \dots, n-1$, we have

$$\begin{aligned} L(i) &= \sum_{k=0}^{n-1} \sum_{j=0}^{n-1} m_{ij}\, n_{jk}\, z^k = \sum_{j=0}^{n-1} m_{ij}\, N(j) = \sum_{j=0}^{n-1} m_{ij}\, t^j(z)\, q(z) \\ &= (M(i) \circ t(z))\, q(z) = (s \circ t)^i(z)((p \circ t)(z)\, q(z)). \end{aligned}$$

Again by Theorem 4.2, this gives the assertion. □

As an immediate consequence of the product rule, we see that any circulant recursive matrix can be written as the product of the homogeneous matrix with the same recurrence rule and the circulant matrix with the same generating function.

Theorem 4.4 (Factorisation). *Let* $p(z)$ *be a polynomial in* $\mathcal{P}_n$, *and* $s(z)$ *a permutation polynomial. The circulant recursive matrix* $M := CR(s, p)$ *can be factored as*

$$M = H \times C,$$

where H *denotes the homogeneous matrix* $CR(s, 1)$, *and* C *the circulant matrix* $CR(z, p)$.

The product rule allows us to compute easily the product of a row vector and a circulant recursive matrix.

Proposition 4.1. *Let* $q(z)$ *be a polynomial in* $\mathcal{P}_n$, *and* $CR(s, p)$ *an* $n \times n$ *circulant recursive matrix. Then*

$$[q(z)]^T \times CR(s, p) = [(q \circ s)(z) p(z)]^T.$$

Proof. This follows by arguments similar to those used in the proof of Corollary 3.1. □

A second property of circulant matrices which can be extended to the present, more general setting concerns a characterization of invertible matrices. In fact, as is the case for circulant matrices, the inverse of a circulant recursive matrix, if it exists, is again circulant recursive.

Theorem 4.5. *The $n \times n$ circulant recursive matrix $CR(s, p)$ admits an inverse if and only if $s(z)$ is a proper permutation polynomial and the polynomial $p(z)$ is invertible in the ring $\mathcal{P}_n$. If this is the case, we have*

$$CR(s, p)^{-1} = CR(\tilde{s}, p^{-1} \circ \tilde{s}).$$

Proof. Suppose that the matrix $M = CR(s, p)$ is invertible, and denote by $p_0(z)$, $p_1(z), \ldots, p_{n-1}(z)$ the generating functions of the 0th, $1^{\text{th}}, \ldots, (n-1)^{\text{th}}$ rows of M^{-1}. By comparing the 0th rows of the product $M^{-1} \times M$ and of the identity matrix $I = CR(z, 1)$, and using Proposition 4.1, we get

$$(p_0 \circ s)(z)p(z) \equiv 1 \mod (z^n - 1).$$

This implies that the polynomial $p(z)$ is invertible in $\mathcal{P}_n$. A similar argument applied to the first row of $M^{-1} \times M$ gives

$$(p_1 \circ s)(p_0^{-1} \circ s) \equiv z \mod (z^n - 1),$$

which yields $(p_1 p_0^{-1}) \circ s \equiv z \mod (z^n - 1)$, and so $s(z)$ is a proper permutation polynomial. The converse is an immediate consequence of identity (18). □

Example 4.2. Consider the permutation polynomial in $\mathcal{P}_4$ $s(z) := \frac{i}{2}z - \frac{i+1}{2}z^2 - \frac{1}{2}z^3$, whose associated permutation is

$$\sigma = \begin{pmatrix} 0 & 1 & 2 & 3 \\ 2 & 1 & 3 & 0 \end{pmatrix}.$$

The homogeneous circulant recursive matrix with recurrence rule $s(z)$ is the following:

$$CR(s, 1) = \begin{bmatrix} 1 & 0 & 0 & 0 \\ 0 & \frac{i}{2} & -\frac{i+1}{2} & -\frac{1}{2} \\ 0 & \frac{i+1}{2} & 0 & \frac{1-i}{2} \\ 0 & -\frac{1}{2} & \frac{i-1}{2} & -\frac{i}{2} \end{bmatrix}.$$

Its inverse is the homogeneous circulant recursive matrix whose recurrence rule is the permutation polynomial associated with the inverse permutation of σ, namely,

$$\sigma^{-1} = \begin{pmatrix} 0 & 1 & 2 & 3 \\ 3 & 1 & 0 & 2 \end{pmatrix}.$$

This gives

$$\tilde{s}(z) = -\frac{i}{2}z + \frac{1-i}{2}z^2 - \frac{1}{2}z^3,$$

and

$$CR(s,1)^{-1} = CR(\tilde{s},1) = \begin{bmatrix} 1 & 0 & 0 & 0 \\ 0 & -\frac{i}{2} & \frac{1-i}{2} & -\frac{1}{2} \\ 0 & \frac{i-1}{2} & 0 & -\frac{i+1}{2} \\ 0 & -\frac{1}{2} & \frac{i+1}{2} & \frac{i}{2} \end{bmatrix}.$$

We remark that, in the particular case $s(z) = z$, Theorem 4.5 yields the usual invertibility condition for circulant matrices.

5 Double recursion

The present section is devoted to the generalization of another well-known result regarding circulant matrices, namely, the fact that the transpose of a circulant matrix is again circulant (see Proposition 3.6). We will show that a similar result holds for any circulant recursive matrix, provided that its recurrence rule is a proper permutation polynomial.

We begin with the homogeneous case.

Theorem 5.1. *Let $s(z)$ be a proper permutation polynomial in $\mathcal{P}_n$, and let $H := CR(s,1)$ be the homogeneous circulant recursive matrix with recurrence rule $s(z)$. Then the matrix H^T is again a homogeneous circulant recursive matrix. More precisely,*

$$H^T = CR(\tilde{s}^{\star^{-1}}, 1).$$

Proof. By Theorem 4.5, the matrix H is invertible, since $s(z)$ is a proper permutation polynomial, and H^{-1} is circulant recursive, with recurrence rule $\tilde{s}$. Hence, setting $C := CR(z, \tilde{s})$, we have

$$\Pi \times H^{-1} = H^{-1} \times C,$$

whence, as the matrix Π is orthogonal and C is invertible,

$$H \times \Pi^T = C^{-1} \times H,$$

which gives

$$\Pi \times H^T = H^T \times (C^{-1})^T.$$

Hence, H^T is a circulant recursive matrix whose recurrence rule equals the generating function of the circulant matrix $(C^{-1})^T$, namely, the proper permutation polynomial $\tilde{s}^{\star^{-1}}$. The homogeneity of H^T is an easy consequence of Proposition 3.5. □

The general case can now be easily derived by exploiting the factorization property of circulant recursive matrices and making use of relation (13).

Corollary 5.1 (Double recursion). *Let $s(z)$ be a proper permutation polynomial. For every polynomial $p(z) \in \mathcal{P}_n$ we have*

$$CR(s, p)^T = CR(\tilde{s}^{\star^{-1}}, (p \circ \tilde{s})^\star).$$

Example 5.1. The transpose of the circulant recursive matrix $CR(s, p)$ given in Example 4.1, namely,

$$CR(s, p)^T = \begin{bmatrix} 1 & -\frac{1}{2}(3+2i) & +\frac{1}{2}(i-1) & \frac{1}{2}(i-2) \\ 1 & \frac{1}{2}(2+i) & \frac{1}{2}(1-3i) & \frac{1}{2}(2i-1) \\ -2 & \frac{1}{2}(1+2i) & -\frac{1}{2}(1+i) & -\frac{1}{2}i \\ 0 & \frac{1}{2}i & \frac{1}{2}(3i+1) & \frac{1}{2}(3-2)i \end{bmatrix},$$

is the circulant recursive matrix $CR(\tilde{s}^{\star^{-1}}, (p \circ \tilde{s})^\star)$, with

$$\tilde{s}^{\star^{-1}} = -\frac{1}{2}z - \frac{i+1}{2}z^2 - \frac{1}{2}z^3,$$

$$(p \circ \tilde{s})^\star = 1 - (\frac{3}{2} - i)z - \frac{i-1}{2}z^2 + (\frac{1}{2}i - 1)z^3.$$

6 The Fourier transform of a circulant recursive matrix

As we have already remarked (see Theorem 2.3), circulant matrices are diagonalized by the Fourier transform. We will now show that a similar property also holds for circulant recursive matrices.

First of all, we associate to any given generalized permutation $\sigma : \{0, 1, \dots, n-1\} \longrightarrow \{0, 1, \dots, n-1\}$ the square $n \times n$ matrix $P(\sigma) := [p_{ij}]$ defined as follows:

$$p_{ij} = \begin{cases} 1 & \text{if } j = \sigma(i), \\ 0 & \text{otherwise.} \end{cases}$$

We get immediately that the rank of $P(\sigma)$ equals the cardinality of the image of σ, and $P(\sigma)$ is a permutation matrix if and only if σ is a bijection.

In any case, the matrix $P(\sigma)$ plays a central role in describing the Fourier transform of a circulant recursive matrix.

Theorem 6.1. *Let $s(z)$ be the permutation polynomial associated with the generalised permutation σ, and $p(z)$ any polynomial in $\mathcal{P}_n$. Then*

$$F \times CR(s, p) \times F^* = P^T \times D,$$

where $P = P(\sigma)$ and $D = \mathrm{Diag}(p(1), p(w), p(w^2), \dots, p(w^{n-1}))$.

Proof. Set $M := CR(s, p)$. By the factorization property of circulant recursive matrices, we can write $M = H \times C$, where $H := CR(s, 1)$ and $C := CR(z, p)$. We have that

$$\begin{aligned} F \times M \times F^* &= F \times H \times C \times F^* \\ &= (F \times H \times F^*) \times (F \times C \times F^*). \end{aligned}$$

Hence, we need only describe the Fourier transform of the homogeneous matrix H. Recalling that the rows of H contain the coefficients of the polynomials $1, s(z), s^2(z), \dots, s^{n-1}(z)$, by identity (7) we get

$$H \times F^* = \frac{1}{\sqrt{n}} \begin{bmatrix} 1 & 1 & \dots & 1 \\ s(1) & s(w) & \dots & s(w^{n-1}) \\ s^2(1) & s^2(w) & \dots & s^2(w^{n-1}) \\ \vdots & \vdots & & \vdots \\ s^{n-1}(1) & s^{n-1}(w) & \dots & s^{n-1}(w^{n-1}) \end{bmatrix}.$$

By Theorem 3.1, we obtain

$$H \times F^* = \frac{1}{\sqrt{n}} \begin{bmatrix} 1 & 1 & \dots & 1 \\ w^{\sigma(0)} & w^{\sigma(1)} & \dots & w^{\sigma(n-1)} \\ w^{2\sigma(0)} & w^{2\sigma(1)} & \dots & w^{2\sigma(n-1)} \\ \vdots & \vdots & & \vdots \\ w^{(n-1)\sigma(0)} & w^{(n-1)\sigma(1)} & \dots & w^{(n-1)\sigma(n-1)} \end{bmatrix} = F^* \times P^T,$$

whence

$$F \times H \times F^* = P^T.$$

The assertion now follows from Theorem 2.3. □

The preceding result closely follows the factorization property of circulant recursive matrices, and at the same time yields a simpler characterization of such matrices. Moreover, it shows that the problem of finding the eigenvalues of a circulant recursive matrix can be reduced to that of studying the eigenvalues of the matrix $P(\sigma)^T \times D$. This last problem has been extensively studied in the case when $P(\sigma)$ is a permutation matrix (see, e.g., [5, p. 166 ff.]). The case of generalized permutations can easily be treated in a similar way. In what follows we give two examples relevant to the above two situations.

Example 6.1. Consider the proper permutation polynomial defined in (11), that is,

$$s_2(z) = \frac{i}{2}z + \frac{i+1}{2}z^2 - \frac{1}{2}z^3.$$

The corresponding permutation matrix $P(\tau)$ is

$$P(\tau) = \begin{bmatrix} 0 & 1 & 0 & 0 \\ 0 & 0 & 1 & 0 \\ 1 & 0 & 0 & 0 \\ 0 & 0 & 0 & 1 \end{bmatrix}.$$

Now let $p(z)$ be the following polynomial in $\mathcal{P}_4$:

$$p(z) = \frac{1}{4}(i - 3 + (2 - 2i)z - (7 + i)z^2 + (4 + 2i)z^3).$$

The circulant recursive matrix with recurrence rule $s_2(z)$ and generating function $p(z)$ is the following:

$$CR(s_2, p) = \frac{1}{4}\begin{bmatrix} i-3 & 2(1-i) & -7-i & 2(2+i) \\ -5-i & 2(2+i) & -3-i & 4(1-i) \\ -1-i & 2(3-i) & i-5 & 2(2+i) \\ i-7 & 2(2+i) & i-1 & 4 \end{bmatrix}.$$

Since $p(1) = -1$, $p(i) = 2$, $p(-1) = -4$, $p(-i) = i$, Theorem 6.1 yields

$$F \times CR(s_2, p) \times F^* = \begin{bmatrix} 0 & 0 & -4 & 0 \\ -1 & 0 & 0 & 0 \\ 0 & 2 & 0 & 0 \\ 0 & 0 & 0 & i \end{bmatrix}.$$

By applying the results proved in [5, ibid.], the eigenvalues of $CR(s_2, p)$ are the complex number $p(-i) = i$ together with the three 3rd roots of $p(1)p(i)p(-1) = 8$.

Example 6.2. Let

$$s_1(z) = \frac{i}{2} - \frac{1}{2}z^2 + \frac{i+1}{2}z^3$$

be the permutation polynomial defined in (10), whose associated matrix $P(\sigma)$ is

$$P(\sigma) = \begin{bmatrix} 0 & 1 & 0 & 0 \\ 1 & 0 & 0 & 0 \\ 0 & 0 & 1 & 0 \\ 0 & 1 & 0 & 0 \end{bmatrix}.$$

Now let $q(z)$ be the polynomial in $\mathcal{P}_4$ such that $q(1) = 1, q(i) = \frac{1}{2}, q(-1) = -\frac{1}{2}, q(-i) = -1$, that is,

$$q(z) = \frac{1}{4}(\frac{3}{2}(1 - i)z + z^2 + \frac{3}{2}(1 + i)z^3).$$

We have

$$CR(s_1, q) = \frac{1}{4}\begin{bmatrix} 0 & \frac{3}{2}(1-i) & 1 & \frac{3}{2}(1+i) \\ 1 & \frac{i}{2}(1+i) & 2i & \frac{3}{2}(i-1) \\ 0 & \frac{1}{2}(i-1) & -3 & -\frac{1}{2}(i+1) \\ 1 & -\frac{3}{2}(i+1) & -2i & \frac{1}{2}(1-i) \end{bmatrix}.$$

By Theorem 6.1, the Fourier transform of the above matrix equals $P(\sigma)^T \times \text{Diag}(q(1), q(i), q(-1), q(-i))$, namely,

$$F \times CR(s_1, q) \times F^* = \begin{bmatrix} 0 & \frac{1}{2} & 0 & 0 \\ 1 & 0 & 0 & -1 \\ 0 & 0 & -\frac{1}{2} & 0 \\ 0 & 0 & 0 & 0 \end{bmatrix}.$$

It is easily checked that the nonzero eigenvalues of $CR(s_1, q)$ are the complex number $q(-1) = -\frac{1}{2}$ together with the two square roots of $q(1)q(i) = \frac{1}{2}$.

Acknowledgements. This work was supported by University of Bologna, ex 60% projects and by C.N.R. Project no. 98.01017.CT01.

References

[1] Barnabei, M., Brini, A., Nicoletti, G. (1982): Recursive matrices and umbral calculus. J. Algebra **75**, 546–573

[2] Barnabei, M., Brini, A., Nicoletti, G. (1986): A general umbral calculus. In: Science and computers. Advances in Math. Suppl. Studies, **10**. Academic Press, Orlando, FL, pp. 221–244

[3] Barnabei, M., Guerrini, C., Montefusco, L.B. (1998): Some algebraic aspects of signal processing. Linear Alg. Appl. **284**, 3–17

[4] Barnabei, M., Montefusco, L.B. (1998): Recursive properties of Toeplitz and Hurwitz matrices. Linear Alg. Appl. **274**, 367–388

[5] Davis, P.J. (1994): Circulant matrices. 2nd edition. Chelsea, New York

[6] Gantmacher, F.R. (1959): The theory of matrices, vols. 1, 2. Chelsea, New York

[7] Mullin, R., Rota, G.-C. (1970): On the foundations of combinatorial theory. III. Theory of binomial enumeration. In: Harris, B. (ed.), Graph theory and its applications. Academic Press, New York pp. 167–213

[8] Roman, S.M., Rota, G.-C. (1978): The umbral calculus. Adv. Math. **27**, 95–188

[9] Rota, G.-C., Kahaner, D., Odlyzko, A. (1973): On the foundations of combinatorial theory. VIII. Finite operator calculus. J. Math. Anal. Appl. **42**, 684–760

Remarks on invariant geometric calculus. Cayley–Grassmann algebras and geometric Clifford algebras

P. Bravi, A. Brini

1 Introduction

The invariant geometric calculus was founded by the German mathematician H.G. Grassmann in 1844 (*Ausdehnungslehre* [15, 16]). In this treatise, he introduced the modern notion of a vector in an abstract n-dimensional space and, in general, the notion of an extensor (decomposable antisymmetric tensor). Grassmann's plan was radically innovative; his aim was to found an intrinsic algebraic calculus for (projective, affine, euclidean) geometry, that was alternative to the cartesian idea of linking algebra and geometry by a system of coordinates. To this aim, in the *Ausdehnungslehre*, several algebraic operations on extensors are introduced, for example, the *progressive product*, the *regressive product* and the *Ergänzung operation* (in geometric language: projection, section and orthogonal duality).

Grassmann's approach has been the object of a deep and renewed interest during the last twenty years, in both pure and applied mathematics.

Two different algebraic formalisms for the invariant geometric calculus have been developed recently: one based on constructions that recover and develop the methods of the *Ausdehnungslehre* and the other based on geometric algebra in the sense of W.K. Clifford.

The first formalism is known as the theory of *Cayley–Grassmann algebras* and it is essentially due to Gian-Carlo Rota [2, 13]. Starting from an idea due to G. Peano, the vector space V (of finite dimension n) is endowed with a *bracket* $[\,,\ldots,\,]$ (non-degenerate skew-symmetric n-linear form); the pair $(V, [\,,\ldots,\,])$ is called *Peano space*. From a Peano space a natural ($SL(n)$-invariant) construction leads to an algebra with two bilinear operations, called *join* and *meet*, introduced by independent and closed forms, and a unary operation, the *Hodge $*$-operator*. Such operations are a strict and computationally useful formalization of the progressive product, the regressive product and the *Ergänzung* of Grassmann, respectively. They reflect the main operations on the subspaces of a vector (or projective) space, and therefore they constitute a natural algebraic language for the projective geometric calculus. The theory of Cayley–Grassmann algebras has recently been applied in other mathematical fields, for example, invariant theory, robotic and computational geometry [28].

The second formalism is based on the idea of *geometric Clifford algebra*, developed by D. Hestenes starting from a deep reflection on the fundamental work *Applications of Grassmann's Extensive Algebra* ([10, W.K. Clifford (1878)]). During this century the Clifford algebra has been largely studied from an algebraic point of view and has been especially applied in physics. Like Grassmann algebras the origin and geometric significance of Clifford algebras were almost completely forgotten. Hestenes' basic idea consists in fixing a special element, the *unit pseudoscalar* (*n-blade*), in the Clifford algebra of an n-vector space endowed with a non-degenerate symmetric bilinear form. Given the unit pseudoscalar (unique up to a multiplicative scalar), it is possible to construct products closely related to those introduced by Grassmann. The result is a structure with a deep geometric significance, of which Hestenes and his coworkers have shown several applications to both mathematics and physics [18–22].

In point of fact, the Cayley–Grassmann algebras and the geometric Clifford algebras are just two different formalizations of the invariant geometric calculus. Nevertheless, the substantial equivalence of these two formalisms has not yet been examined deeply nor made explicit.

In this work we analyse the equivalence between Cayley–Grassmann algebras and geometric Clifford algebras. Specifically, we show that such algebras can be regarded as algebras over the *same* vector space and we supply a *complete system* of identities that express the operations of Cayley–Grassmann type in terms of the operations of Clifford type, and vice versa. In fact there exists just one algebra, that we call the *Grassmann–Clifford algebra*, that provides a unified and richer language for the invariant geometric calculus.

2 A vector space isomorphism

Let $(V, (\,,\,))$ be a *metric space*, i.e., a vector space V over a fixed field of characteristic zero, endowed with a symmetric bilinear form $(\,,\,)$. Let $C(V)$ denote the Clifford algebra of $(V, (\,,\,))$, i.e., the quotient algebra of the tensor algebra $T(V)$ where $a \otimes a = (a, a)1$. Let ab denote the Clifford product of two vectors in the Clifford algebra. The algebra $C(V)$ is a filtered algebra. Specifically,

$$C(V) = \bigcup_{k \in \mathbb{N}} C^{(k)}(V),$$
$$C^{(k)}(V) \subseteq C^{(k+1)}(V),$$

where, if π denotes the projection from $T(V)$ onto $C(V)$,

$$C^{(k)}(V) = \pi\left(\bigoplus_{l=0}^{k} T^{l}(V)\right).$$

We next consider the *alternator*, namely, the linear operator A_k of $T^k(V)$ that assigns to a tensor $a_1 \otimes \ldots \otimes a_k$ of step k its full antisymmetric part

$$A_k(a_1 \otimes \ldots \otimes a_k) = \frac{1}{k!} \sum_{\sigma \in S_k} (-1)^{|\sigma|} a_{\sigma(1)} \otimes \ldots \otimes a_{\sigma(k)}. \tag{2.1}$$

Proposition 2.1. *The* k^{th} *filtered component* $C^{(k)}(V)$ *can be decomposed as the direct sum*

$$C^{(k)}(V) = \pi \circ A_k(T^k) \oplus C^{(k-1)}(V).$$

Proof. Let $\mathcal{E}$ be a (,)-orthogonal basis of V. The set

$$\{e_{i_1} \dots e_{i_l} | e_i \in \mathcal{E}, 0 \leq l \leq k-1, i_1 < \dots < i_l\}$$

is a basis of $C^{(k-1)}$ and the set

$$\{e_{i_1} \otimes \dots \otimes e_{i_k} | e_i \in \mathcal{E}\}$$

is a basis of T^k. Since $(e_i, e_j) = 0$, it follows that $e_i e_j = -e_j e_i$ for all $i, j, i \neq j$. Since the application $\pi \circ A_k$ is such that

$$\pi(A_k(e_{i_1} \otimes \dots \otimes e_{i_k})) = e_{i_1} \dots e_{i_k},$$

it follows that

$$\{e_{i_1} \dots e_{i_k} | e_i \in \mathcal{E}, i_1 < \dots < i_k\}$$

is a basis of $\pi \circ A_k(T^k)$. □

Corollary 2.2 (Direct sum decomposition).

$$C(V) = \bigoplus_k \pi \circ A_k(T^k).$$

Proposition 2.3. *Let* $G^k(V)$ *denote the* k^{th} *homogeneous component of the exterior algebra* $G(V)$ *of the vector space* V*. The application*

$$\alpha_V^k : G^k(V) \longrightarrow \pi \circ A_k(T^k) \subset C(V)$$
$$a_1 \vee \dots \vee a_k \mapsto \pi(A_k(a_1 \otimes \dots \otimes a_k)) = \frac{1}{k!} \sum (-1)^{|\sigma|} a_{\sigma(1)} \dots a_{\sigma(k)}$$

is a linear isomorphism from the vector space G^k *to the vector subspace* $\pi \circ A_k(T^k)$ *of* $C(V)$.

Proof. The assertion follows immediately by considering a (,)-orthogonal basis of V. □

Corollary 2.4. *Let* $a_1, \dots, a_k$ *denote (,)-orthogonal vectors of* V*. Then*

$$\alpha_V^k(a_1 \vee \dots \vee a_k) = a_1 \dots a_k.$$

Corollary 2.5. *Let α_V denote the linear application*

$$\alpha_V = \bigoplus_k \alpha_V^k : G(V) \longrightarrow C(V).$$

For any element x in $C^{(k)}(V)$, the decomposition

$$x = \sum_{l=0}^{k} \langle x \rangle_l,$$

where $\alpha_V^{-1}(\langle x \rangle_l) \in G^l$, is unique.

In plain words, the vector spaces of the exterior algebra $G(V)$ and the Clifford algebra $C(V)$ are not merely isomorphic, but their link is deeper.

Proposition 2.6. *The map $\alpha_V : G(V) \longrightarrow C(V)$ is a natural vector space isomorphism.*

Proof. For any pair of metric spaces $(V, (\ ,\ ))$ and $(W, (\ ,\ )')$ and for any linear application $f : V \longrightarrow W$ preserving the bilinear form (i.e., a morphism between two objects of the category of metric spaces), the following diagram is commutative.

$$\begin{array}{ccc} G(V) & \xrightarrow{G(f)} & G(W) \\ \downarrow \alpha_V & & \downarrow \alpha_W \\ C(V) & \xrightarrow{C(f)} & C(W) \end{array}$$

The morphisms $G(f)$ and $C(f)$ are the algebra homomorphisms induced by the morphism f (i.e., the images under the functors of exterior algebra and of Clifford algebra).

For $a_1, \dots, a_k$ in V, the properties of $G(f)$ and $C(f)$ lead to

$$\begin{aligned}(\alpha_W \circ G(f))(a_1 \vee \ldots \vee a_k) &= \alpha_W(f(a_1) \vee \ldots \vee f(a_k)) \\ &= \frac{1}{k!} \sum (-1)^{|\sigma|} f(a_{\sigma(1)}) \ldots f(a_{\sigma(k)}) \\ &= C(f) \left(\frac{1}{k!} \sum (-1)^{|\sigma|} a_{\sigma(1)} \ldots a_{\sigma(k)} \right) \\ &= (C(f) \circ \alpha_V)(a_1 \vee \ldots \vee a_k).\end{aligned}$$

Therefore $\alpha_W \circ G(f) = C(f) \circ \alpha_V$. □

Let $\widetilde{C}(V)$ denote the graded algebra associated to the filtered algebra $C(V)$, i.e.,

$$\widetilde{C}(V) = \bigoplus_k \frac{C^{(k)}}{C^{(k-1)}}.$$

Proposition 2.7. *Let α'_V be the natural isomorphism of graded algebras such that*

$$\begin{aligned} \alpha'_V &: G(V) \longrightarrow \widetilde{C}(V) \\ a_1 \vee \ldots \vee a_k &\mapsto [A_k(a_1 \ldots a_k)]_{C^{(k-1)}}. \end{aligned}$$

Let α''_V be the natural isomorphism of vector spaces such that

$$\begin{aligned} \alpha''_V &: \widetilde{C}(V) \longrightarrow C(V) \\ [A_k(a_1 \ldots a_k)]_{C^{(k-1)}} &\mapsto A_k(a_1 \ldots a_k). \end{aligned}$$

Then the map α_V satisfies

$$\alpha_V = \alpha''_V \circ \alpha'_V.$$

3 Blades, inner and outer products

In the Clifford algebra of an n-dimensional vector space V endowed with a non-degenerate symmetric bilinear form (,), the elements that correspond to extensors of $G(V)$ – under the isomorphism α_V (Proposition 2.3, Corollary 2.5) – are called *blades*. The blades are the main objects of the invariant geometric calculus in the sense of geometric Clifford algebra. A k-blade is the image under the isomorphism α_V of an extensor of step k and can be coherently associated to a k-subspace of the vector space V.

The set of blades is closed with respect to two different bilinear products, called, respectively, the outer and inner products [18]. These bilinear operations can be defined as follows using the notation of Corollary 2.5.

Definition 3.1. Let A be an r-blade and B an s-blade. Set

$$\begin{aligned} A \vee B &= \langle AB \rangle_{r+s} \\ A \cdot B &= \langle AB \rangle_{|r-s|}. \end{aligned}$$

Note that, if a and b are vectors, then $ab = \frac{1}{2}(ab - ba) + \frac{1}{2}(ab + ba)$; since $a \vee b = \frac{1}{2}(ab - ba)$ and $a \cdot b = \frac{1}{2}(ab + ba) = (a, b)$, it follows that

$$ab = a \vee b + a \cdot b. \tag{3.1}$$

In general, we have

Proposition 3.2. *Let a be a vector and A an r-blade, $r \geq 1$. Then*

$$aA = a \vee A + a \cdot A.$$

Proof. The identity can be demonstrated on the elements of a basis and then extended by linearity.

Let $\mathcal{E}$ be a $(\ ,\ )$-orthogonal basis of V. Let $A = \pi(A_r(e_1 \otimes \ldots \otimes e_r))$, where $e_1, \ldots, e_r$ belong to $\mathcal{E}$. Since $e_1, \ldots, e_r$ are $(\ ,\ )$-orthogonal vectors, it follows that $A = e_1 \ldots e_r$. Let a belong to $\mathcal{E}$. The Clifford product $a(e_1 \ldots e_r)$ is equal to

$$(-1)^{s-1}(a, e_s)\, e_1 \ldots e_{s-1} e_{s+1} \ldots e_r,$$

if $a \in \{e_1, \ldots, e_r\}$. Otherwise, it is equal to

$$\pi(A_{r+1}(a \otimes e_1 \otimes \ldots \otimes e_r)). \qquad \square$$

Proposition 3.3. *For A an r-blade and B an s-blade, it follows that*

$$AB = A \cdot B + \langle AB \rangle_{|r-s|+2} + \ldots + \langle AB \rangle_{r+s-2} + A \vee B.$$

Proposition 3.4. *The isomorphism α_V changes the join into the outer product, i.e.,*

$$\alpha_V(A \vee B) = \alpha_V(A) \vee \alpha_V(B).$$

Proof. Let $A = e_{i_1} \vee \ldots \vee e_{i_r}$ and $B = e_{j_1} \vee \ldots \vee e_{j_s}$ be two extensors, where $e_{i_1}, \ldots, e_{i_r}, e_{j_1}, \ldots, e_{j_s}$ are $(\ ,\ )$-orthogonal vectors of V. Since, from Corollary 2.4, the map α_V is such that $\alpha_V(A) = e_{i_1} \ldots e_{i_r}$ and $\alpha_V(B) = e_{j_1} \ldots e_{j_s}$, it follows that

$$\alpha_V(A) \vee \alpha_V(B) = \langle \alpha_V(A)\alpha_V(B) \rangle_{r+s} = \langle e_{i_1} \ldots e_{i_r} e_{j_1} \ldots e_{j_s} \rangle_{r+s}.$$

If $\{i_1, \ldots, i_r\} \cap \{j_1, \ldots, j_s\} \neq \emptyset$, both terms of the identity vanish. Otherwise, if $\{i_1, \ldots, i_r\} \cap \{j_1, \ldots, j_s\} = \emptyset$, then $e_{i_1} \ldots e_{i_r} e_{j_1} \ldots e_{j_s} = \alpha_V(e_{i_1} \vee \ldots \vee e_{i_r} \vee e_{j_1} \vee \ldots \vee e_{j_s})$ and hence $\alpha_V(A) \vee \alpha_V(B) = \alpha_V(A \vee B)$. $\square$

This proposition proves directly that the outer product of two blades is a blade and that the outer product is associative and anticommutative. In particular, by Proposition 3.4, the outer product in $C(V)$ of k given vectors $a_1, \ldots, a_k$ is a k-blade such that

$$a_1 \vee \ldots \vee a_k = \alpha_V(a_1 \vee \ldots \vee a_k). \tag{3.2}$$

In what follows, extensors and blades will be identified. Obviously, they *also* correspond to decomposable antisymmetric tensors in $T(V)$.

The main algebraic properties of inner product are demonstrated in Sect. 8 (Propositions 8.4, 8.5) by using the results obtained in the following sections. From the preceding notation and the proof of Proposition 3.2 we have the following useful result.

Proposition 3.5. *Let a be a vector and $A = a_1 \vee \ldots \vee a_k$ a k-blade. Then*

$$a \cdot A = \sum_{l=1}^{k} (-1)^{l-1}(a, a_l)\, a_1 \vee \ldots \widehat{a_l} \ldots \vee a_k.$$

4 The Caianiello identities

Let $(\ 1 \ \ \dots \ \ 2m\ )$ denote the pseudo-pfaffian of a given $2m \times 2m$ matrix (a_{pq}); explicitly,

$$(\ 1 \ \ \dots \ \ 2m\ ) = \sum_{\sigma} (-1)^{|\sigma|} \prod_{r=1}^{m} a_{\sigma(2r-1)\sigma(2r)},$$

for all permutations $\sigma \in S_{2m}$ such that, for every $r < m$, $\sigma(2r-1) < \sigma(2r+1)$ and $\sigma(2r-1) < \sigma(2r)$ [7]. The following decomposition rules are basic.

Lemma 4.1. *For a* $2m \times 2m$ *matrix* (a_{pq}), *it follows that:*

(i) $\forall h,\ 1 \le h \le 2m,$

$$\begin{aligned}(\ 1 \ \ \dots \ \ 2m\ ) = &\sum_{1 \le i < h} (-1)^{i+h+1} (\ i \ \ h\ )(\ 1 \dots \widehat{i} \dots \widehat{h} \dots 2m\ )\\ &+ \sum_{h < i \le 2m} (-1)^{i+h+1} (\ h \ \ i\ )(\ 1 \dots \widehat{h} \dots \widehat{i} \dots 2m\ );\end{aligned}$$

(ii) [1] $\forall l,\ 0 \le l \le m,$

$$\begin{aligned}\textstyle\sum_{i_1 < \dots < i_{2l}} \varepsilon_{i_1,\dots,i_{2l}} (\ i_1 \ \ \dots \ \ i_{2l}\ )(\ i'_1 \ \ \dots \ \ i'_{2(m-l)}\ )&\\ = \tbinom{m}{l}(\ 1 \ \ \dots \ \ 2m\ ).&\end{aligned}$$

Theorem 4.2. *Let* $a_1, \dots, a_k \in V$. *Then*

$$a_1 \dots a_k = \sum_{l=0}^{[k/2]} \sum_{i_1 < \dots < i_{2l}} \varepsilon_{i_1,\dots,i_{2l}} (\ i_1 \ \ \dots \ \ i_{2l}\ ) a_{i'_1} \vee \dots \vee a_{i'_{k-2l}},$$

where $(\ i_1 \ \ \dots \ \ i_{2l}\ )$ *is the pseudo-pfaffian of the symmetric matrix of the scalars* (a_{i_p}, a_{i_q}).

Proof. We proceed by induction on k. For $k = 1$ the identity is true. Assume that the identity is true for $k = p$ vectors. Let $a_0, a_1, \dots, a_p$ be vectors in V. From Proposition 3.2, by the decomposition of Corollary 2.5, we obtain

$$\langle a_0 a_1 \dots a_p \rangle_{p+1-2q} = a_0 \vee \langle a_1 \dots a_p \rangle_{p-2q} + a_0 \cdot \langle a_1 \dots a_p \rangle_{p-2(q-1)}.$$

The sum

$$\sum_{0 \le i_1 < \dots < i_{2q} \le p} \varepsilon_{i_1,\dots,i_{2q}} (\ i_1 \ \ \dots \ \ i_{2q}\ ) a_{i'_1} \vee \dots \vee a_{i'_{p-2q}}$$

[1] We denote by $(i'_1, \dots, i'_{k-2l})$ the ordered complementary set of the ordered set $(i_1, \dots, i_{2l})$ in $(1, \dots, k)$ and by $\varepsilon_{i_1,\dots,i_{2l}}$ the sign of the permutation $(i_1, \dots, i_{2l}, i'_1, \dots, i'_{k-2l})$.

can be subdivided into two parts, one consisting of the terms with $i_1 \geq 1$ and the other of the terms with $i_1 = 0$. Since by induction the first part is $a_0 \vee \langle a_1 \dots a_p \rangle_{p-2q}$, it is sufficient to prove that the second is $a_0 \cdot \langle a_1 \dots a_p \rangle_{p-2(q-1)}$.

Since

$$
\begin{aligned}
a_0 \cdot \langle a_1 \dots a_p \rangle_{p-2(q-1)} \\
&= \sum_{i_1 < \dots < i_{2(q-1)}} \varepsilon_{i_1,\dots,i_{2(q-1)}} (\, i_1 \quad \dots \quad i_{2(q-1)} \,) \\
&\quad \times a_0 \cdot (a_{i'_1} \vee \dots \vee a_{i'_{p-2(q-1)}}) \\
&= \sum_{i_1 < \dots < i_{2(q-1)}} \varepsilon_{i_1,\dots,i_{2(q-1)}} (\, i_1 \quad \dots \quad i_{2(q-1)} \,) \\
&\quad \times \sum_r (-1)^{r-1} (a_0, a_{i'_r})\, a_{i'_1} \vee \dots \widehat{a_{i'_r}} \dots \vee a_{i'_{p-2(q-1)}},
\end{aligned}
$$

by rearranging the sum and then applying (i) of Lemma 4.1, the last term is equal to

$$
\begin{aligned}
&\sum_{i_1 < \dots < i_{2q-1}} \varepsilon_{i_1,\dots,i_{2q-1}} \sum_r (-1)^{r-1} (a_0, a_{i_r}) \\
&\quad \times (\, i_1 \quad \dots \quad \widehat{i_r} \quad \dots \quad i_{2q-1} \,)\, a_{i'_1} \vee \dots \vee a_{i'_{p-2q+1}} \\
&= \sum_{i_1 < \dots < i_{2q-1}} \varepsilon_{i_1,\dots,i_{2q-1}} (\, 0 \quad i_1 \quad \dots \quad i_{2q-1} \,)\, a_{i'_1} \vee \dots \vee a_{i'_{p-2q+1}}. \qquad \square
\end{aligned}
$$

Corollary 4.3. *Let $a_1, \dots, a_k$ be linearly independent vectors of V. Then*

$$
a_1 \vee \dots \vee a_k = a_1 \dots a_k
$$

if and only if $a_1, \dots, a_k$ are $(\,,\,)$-orthogonal vectors.

Proposition 4.4. *Let $a_1, \dots, a_k \in V$. Then*

$$
a_1 \vee \dots \vee a_k = \sum_{l=0}^{[k/2]} (-1)^l \sum_{i_1 < \dots < i_{2l}} \varepsilon_{i_1,\dots,i_{2l}} (\, i_1 \quad \dots \quad i_{2l} \,)\, a_{i'_1} \dots a_{i'_{k-2l}}.
$$

Proof. From the preceding formula, it follows that

$$
\begin{aligned}
a_1 \vee \dots \vee a_k = a_1 \dots a_k - \sum_{l=1}^{[k/2]} \sum_{i_1 < \dots < i_{2l}} \varepsilon_{i_1,\dots,i_{2l}} (\, i_1 \quad \dots \quad i_{2l} \,) \\
\times a_{i'_1} \vee \dots \vee a_{i'_{k-2l}};
\end{aligned}
$$

by induction on k,

$$
\sum_{l=1}^{[k/2]} \sum_{i_1 < \dots < i_{2l}} \varepsilon_{i_1,\dots,i_{2l}} (\, i_1 \quad \dots \quad i_{2l} \,)\, a_{i'_1} \vee \dots \vee a_{i'_{k-2l}}
$$

$$= \sum_{l=1}^{[k/2]} \sum_{i_1<\ldots<i_{2l}} \varepsilon_{i_1,\ldots,i_{2l}} (\, i_1 \ \ldots \ i_{2l} \,)$$
$$\times \sum_{m=0}^{[(k-2l)/2]} (-1)^m \sum_{j_1<\ldots<j_{2m}} \varepsilon_{j_1,\ldots,j_{2m}} (\, j_1 \ \ldots \ j_{2m} \,) a_{j'_1} \ldots a_{j'_{k-2(l+m)}}$$

and, by (ii) of Lemma 4.1, the last term is equal to

$$\sum_l \sum_m (-1)^m \sum_{i_1<\ldots<i_{2(l+m)}} \varepsilon_{i_1,\ldots,i_{2(l+m)}} \binom{l+m}{m} (\, i_1 \ \ldots \ i_{2(l+m)} \,)$$
$$\times a_{i'_1} \ldots a_{i'_{k-2(l+m)}}$$
$$= \sum_{p=1}^{[k/2]} (-1)^p \sum_{i_1<\ldots<i_{2p}} \varepsilon_{i_1,\ldots,i_{2p}} (\, i_1 \ \ldots \ i_{2p} \,) a_{i'_1} \ldots a_{i'_{k-2p}}. \qquad \square$$

Lemma 4.5. *For any* $a_1, \ldots, a_k, b_1, \ldots, b_k \in V$,

$$(a_1 \vee \ldots \vee a_k) \cdot (b_k \vee \ldots \vee b_1) = \det\Big((a_i, b_j)\Big)_{1\le i,j\le k}.$$

Proof. Both terms are skew-symmetric k-linear forms with respect to each set of variables $a_1, \ldots, a_k$ and $b_1, \ldots, b_k$. Therefore, it is enough to choose a (,)-orthogonal basis e_i of V and prove the identity on $e_{i_1}, \ldots, e_{i_k}$ and $e_{j_1}, \ldots, e_{j_k}$ for any $i_1 < \ldots < i_k$ and $j_1 < \ldots < j_k$. From Theorem 4.2 we obtain

$$\begin{aligned}(e_{i_1} \vee \ldots \vee e_{i_k}) \cdot (e_{j_k} \vee \ldots \vee e_{j_1}) &= \langle e_{i_1} \ldots e_{i_k} e_{j_k} \ldots e_{j_1} \rangle_0 \\ &= (\, i_1 \ldots i_k \ \ j_k \ldots j_1 \,) \\ &= \prod_{l=1}^{k} (e_{i_l}, e_{j_l}) \\ &= \det\Big((e_{i_l}, e_{j_m})\Big)_{1\le l,m\le k}. \qquad \square\end{aligned}$$

Theorem 4.6. *For* $A = a_1 \vee \ldots \vee a_r$ *and* $B = b_1 \vee \ldots \vee b_s$, *it follows that*

$$AB = \sum_d \sum_{i_1<\ldots<i_{r-d}} \sum_{j_1<\ldots<j_d} \varepsilon_{i_1,\ldots,i_{r-d}} \varepsilon_{j_1,\ldots,j_d}$$
$$\times ((a_{i'_1} \vee \ldots \vee a_{i'_d}) \cdot (b_{j_1} \vee \ldots \vee b_{j_d}))$$
$$\times a_{i_1} \vee \ldots \vee a_{i_{r-d}} \vee b_{j'_1} \vee \ldots \vee b_{j'_{s-d}}.$$

Proof. We proceed by induction on r. The base step comes from Propositions 3.2, 3.5. Step r:

$$(a_0 \vee A)B = a_0(AB) - (a_0 \cdot A)B;$$

$$a_0(AB) = \sum_d \sum_{i_1<\ldots<i_{r-d}} \sum_{j_1<\ldots<j_d} \varepsilon_{i_1,\ldots,i_{r-d}} \varepsilon_{j_1,\ldots,j_d}$$
$$\times ((a_{i'_1} \vee \ldots \vee a_{i'_d}) \cdot (b_{j_1} \vee \ldots \vee b_{j_d}))$$
$$\times (a_0 \vee a_{i_1} \vee \ldots \vee a_{i_{r-d}} \vee b_{j'_1} \vee \ldots \vee b_{j'_{s-d}}$$
$$+ a_0 \cdot (a_{i_1} \vee \ldots \vee a_{i_{r-d}} \vee b_{j'_1} \vee \ldots \vee b_{j'_{s-d}}));$$

$$\sum_d \sum_{i_1<\ldots<i_{r-d}} \sum_{j_1<\ldots<j_d} \varepsilon_{i_1,\ldots,i_{r-d}} \varepsilon_{j_1,\ldots,j_d}$$
$$\times ((a_{i'_1} \vee \ldots \vee a_{i'_d}) \cdot (b_{j_1} \vee \ldots \vee b_{j_d}))$$
$$\times a_0 \cdot (a_{i_1} \vee \ldots \vee a_{i_{r-d}} \vee b_{j'_1} \vee \ldots \vee b_{j'_{s-d}})$$
$$= \sum_d \sum_{i_1<\ldots<i_{r-d}} \sum_{j_1<\ldots<j_d} \varepsilon_{i_1,\ldots,i_{r-d}} \varepsilon_{j_1,\ldots,j_d}$$
$$\times ((a_{i'_1} \vee \ldots \vee a_{i'_d}) \cdot (b_{j_1} \vee \ldots \vee b_{j_d}))$$
$$\times ((a_0 \cdot (a_{i_1} \vee \ldots \vee a_{i_{r-d}})) \vee (b_{j'_1} \vee \ldots \vee b_{j'_{s-d}})$$
$$+ (-1)^{r-d}(a_{i_1} \vee \ldots \vee a_{i_{r-d}}) \vee (a_0 \cdot (b_{j'_1} \vee \ldots \vee b_{j'_{s-d}})))$$
$$= \sum_d \sum_{i_1<\ldots<i_{r-d}} \sum_{j_1<\ldots<j_d} \varepsilon_{i_1,\ldots,i_{r-d}} \varepsilon_{j_1,\ldots,j_d}$$
$$\times ((a_{i'_1} \vee \ldots \vee a_{i'_d}) \cdot (b_{j_1} \vee \ldots \vee b_{j_d}))$$
$$\times (a_0 \cdot (a_{i_1} \vee \ldots \vee a_{i_{r-d}})) \vee (b_{j'_1} \vee \ldots \vee b_{j'_{s-d}})$$
$$+ \sum_d \sum_{i_1<\ldots<i_{r-d}} \sum_{j_1<\ldots<j_d} \varepsilon_{i_1,\ldots,i_{r-d}} \varepsilon_{j_1,\ldots,j_d}$$
$$\times ((a_{i'_1} \vee \ldots \vee a_{i'_d}) \cdot (b_{j_1} \vee \ldots \vee b_{j_d}))$$
$$\times (-1)^{r-d}(a_{i_1} \vee \ldots \vee a_{i_{r-d}}) \vee (a_0 \cdot (b_{j'_1} \vee \ldots \vee b_{j'_{s-d}}));$$

by rearranging the terms in the first sum and by applying Lemma 4.5 in the second one, from the last term we obtain

$$(a_0 \cdot A)B + \sum_d (-1)^{r-d} \sum_{i_1<\ldots<i_{r-d}} \sum_{j_1<\ldots<j_{d+1}} \varepsilon_{i_1,\ldots,i_{r-d}} \varepsilon_{j_1,\ldots,j_{d+1}}$$
$$\times ((a_0 \vee a_{i'_1} \vee \ldots \vee a_{i'_d}) \cdot (b_{j_1} \vee \ldots \vee b_{j_{d+1}}))$$
$$\times (a_{i_1} \vee \ldots \vee a_{i_{r-d}}) \vee (b_{j'_1} \vee \ldots \vee b_{j'_{s-d-1}}).$$

Moreover,

$$(a_0 \vee A)B = \sum_d \sum_{i_1<\ldots<i_{r-d}} \sum_{j_1<\ldots<j_d} \varepsilon_{i_1,\ldots,i_{r-d}} \varepsilon_{j_1,\ldots,j_d}$$
$$\times ((a_{i'_1} \vee \ldots \vee a_{i'_d}) \cdot (b_{j_1} \vee \ldots \vee b_{j_d}))$$

$$
\begin{aligned}
&\times a_0 \vee a_{i_1} \vee \ldots \vee a_{i_{r-d}} \vee b_{j'_1} \vee \ldots \vee b_{j'_{s-d}} \\
&+ \sum_d (-1)^{r-d} \sum_{i_1<\ldots<i_{r-d}} \sum_{j_1<\ldots<j_{d+1}} \varepsilon_{i_1,\ldots,i_{r-d}} \varepsilon_{j_1,\ldots,j_{d+1}} \\
&\times ((a_0 \vee a_{i'_1} \vee \ldots \vee a_{i'_d}) \cdot (b_{j_1} \vee \ldots \vee b_{j_{d+1}})) \\
&\times (a_{i_1} \vee \ldots \vee a_{i_{r-d}}) \vee (b_{j'_1} \vee \ldots \vee b_{j'_{s-d-1}}) \\
= &\sum_d \sum_{0 \le i_1<\ldots<i_{r+1-d}} \sum_{j_1<\ldots<j_d} \varepsilon_{i_1,\ldots,i_{r-d}} \varepsilon_{j_1,\ldots,j_d} \\
&\times ((a_{i'_1} \vee \ldots \vee a_{i'_d}) \cdot (b_{j_1} \vee \ldots \vee b_{j_{d+1}})) \\
&\times (a_{i_1} \vee \ldots \vee a_{i_{r+1-d}}) \vee (b_{j'_1} \vee \ldots \vee b_{j'_{s-d}}). \qquad \square
\end{aligned}
$$

Corollary 4.7 (Splitting rule for the inner product). *Let A be an extensor of step r and B an extensor of step s with $r \le s$. If $B = b_1 \vee \ldots \vee b_s$, then*

$$
A \cdot B = \sum_{i_1<\ldots<i_r} \varepsilon_{i_1,\ldots,i_r} (A \cdot (b_{i_1} \vee \ldots \vee b_{i_r}))\, b_{i'_1} \vee \ldots \vee b_{i'_{s-r}}.
$$

In Sect. 8 (Proposition 8.7) we will provide a demonstration of the preceding result in the setting of the Cayley–Grassmann algebra.

Proposition 4.8. *Let $a_1, \ldots, a_k, b_1, \ldots, b_l \in V$. Then*

$$
\begin{aligned}
(a_1 \ldots a_k) \vee (b_1 \ldots b_l) = &\sum_d (-1)^d \sum_{i_1<\ldots<i_{k-d}} \sum_{j_1<\ldots<j_d} \varepsilon_{i_1,\ldots,i_{k-d}} \varepsilon_{j_1,\ldots,j_d} \\
&\times ((a_{i'_1} \vee \ldots \vee a_{i'_d}) \cdot (b_{j_1} \vee \ldots \vee b_{j_d}) \\
&\times a_{i_1} \ldots a_{i_{k-d}} b_{j'_1} \ldots b_{j'_{l-d}}.
\end{aligned}
$$

Proof. By splitting $a_1 \ldots a_k$ and $b_1 \ldots b_l$ into blades and then $(a_1 \ldots a_k) \vee (b_1 \ldots b_l)$ into Clifford products of vectors, we get

$$
\begin{aligned}
&(a_1 \ldots a_k) \vee (b_1 \ldots b_l) \\
&= \sum_p \sum_{i_1<\ldots<i_{2p}} \varepsilon_{i_1,\ldots,i_{2p}} (\, i_1 \ \ldots \ i_{2p}\,)\, a_{i'_1} \vee \ldots \vee a_{i'_{k-2p}} \\
&\quad \vee \sum_q \sum_{j_1<\ldots<j_{2q}} \varepsilon_{j_1,\ldots,j_{2q}} (\, j_1 \ \ldots \ j_{2q}\,)\, b_{j'_1} \vee \ldots \vee b_{j'_{l-2q}} \\
&= \sum_p \sum_q \sum_{i_1<\ldots<i_{2p}} \sum_{j_1<\ldots<j_{2q}} \varepsilon_{i_1,\ldots,i_{2p}} (\, i_1 \ \ldots \ i_{2p}\,) \\
&\quad \times \varepsilon_{j_1,\ldots,j_{2q}} (\, j_1 \ \ldots \ j_{2q}\,) \sum_r (-1)^r \sum_{h_1<\ldots<h_{2r}} \varepsilon_{h_1,\ldots,h_{2r}} \\
&\quad \times (\, h_1 \ \ldots \ h_{2r}\,)\, c_{h'_1} \ldots c_{h'_{k+l-2(p+q+r)}}, \qquad (4.1)
\end{aligned}
$$

where $a_{i'_1} \vee \ldots \vee a_{i'_{k-2p}} \vee b_{j'_1} \vee \ldots \vee b_{j'_{l-2q}} = c_1 \vee \ldots \vee c_{k+l-2(p+q)}$.

Therefore, we can demonstrate the proposition by proving the following identity:

$$
\begin{aligned}
&\sum_p \sum_q \sum_{i_1<\ldots<i_{2p}} \sum_{j_1<\ldots<j_{2q}} \varepsilon_{i_1,\ldots,i_{2p}}(\, i_1 \;\ldots\; i_{2p}\,) \\
&\quad\times \varepsilon_{j_1,\ldots,j_{2q}}(\, j_1 \;\ldots\; j_{2q}\,) \sum_r (-1)^r \sum_{h_1<\ldots<h_{2r}} \varepsilon_{h_1,\ldots,h_{2r}} \\
&\quad\times (\, h_1 \;\ldots\; h_{2r}\,)\, c_{h'_1} \ldots c_{h'_{k+l-2(p+q+r)}} \\
&= \sum_d (-1)^d \sum_{i_1<\ldots<i_{k-d}} \sum_{j_1<\ldots<j_d} \varepsilon_{i_1,\ldots,i_{k-d}} \varepsilon_{j_1,\ldots,j_d} \\
&\quad\times ((a_{i'_1} \vee \ldots \vee a_{i'_d}) \cdot (b_{j_1} \vee \ldots \vee b_{j_d}) \\
&\quad\times a_{i_1} \ldots a_{i_{k-d}} b_{j'_1} \ldots b_{j'_{l-d}}.
\end{aligned} \tag{4.2}
$$

By an analogous procedure, we get

$$
\begin{aligned}
&(a_1 \vee \ldots \vee a_k)(b_1 \vee \ldots \vee b_l) \\
&\quad= \sum_p \sum_q (-1)^{p+q} \sum_{i_1<\ldots<i_{2p}} \sum_{j_1<\ldots<j_{2q}} \varepsilon(\, i_1 \;\ldots\; i_{2p}\,) \\
&\qquad\times \varepsilon(\, j_1 \;\ldots\; j_{2q}\,) \sum_r \sum_{h_1<\ldots<h_{2r}} \varepsilon(\, h_1 \;\ldots\; h_{2r}\,) \\
&\qquad\times c_{h'_1} \vee \ldots \vee c_{h'_{k+l-2(p+q+r)}}
\end{aligned} \tag{4.3}
$$

and, by Theorem 4.6, we obtain

$$
\begin{aligned}
&\sum_p \sum_q (-1)^{p+q} \sum_{i_1<\ldots<i_{2p}} \sum_{j_1<\ldots<j_{2q}} \varepsilon(\, i_1 \;\ldots\; i_{2p}\,) \varepsilon(\, j_1 \;\ldots\; j_{2q}\,) \\
&\quad\times \sum_r \sum_{h_1<\ldots<h_{2r}} \varepsilon(\, h_1 \;\ldots\; h_{2r}\,)\, c_{h'_1} \vee \ldots \vee c_{h'_{k+l-2(p+q+r)}} \\
&= \sum_d \sum_{i_1<\ldots<i_{k-d}} \sum_{j_1<\ldots<j_d} \varepsilon_{i_1,\ldots,i_{k-d}} \varepsilon_{j_1,\ldots,j_d} \\
&\quad\times ((a_{i'_1} \vee \ldots \vee a_{i'_d}) \cdot (b_{j_1} \vee \ldots \vee b_{j_d})) \\
&\quad\times a_{i_1} \vee \ldots \vee a_{i_{k-d}} \vee b_{j'_1} \vee \ldots \vee b_{j'_{l-d}}.
\end{aligned} \tag{4.4}
$$

To complete the proof it is sufficient to verify the equality between the corresponding coefficients in the last identity, since they are the same (up to a pairwise common change of sign) as those which occur in (4.2).

If $a_1, \ldots, a_k, b_1, \ldots, b_l$ are linearly independent, the statement follows directly from (4.4). In general we must fix $1 \le m_1 < \ldots < m_\alpha \le k$ and $1 \le n_1 < \ldots < n_\beta \le l$ ($\alpha + \beta$ even) and consider the coefficients of $a_{m'_1} \ldots a_{m'_{k-\alpha}} b_{n'_1} \ldots b_{n'_{l-\beta}}$ in the identity. Respectively they equal, up to a common change of sign, the scalar components of the identity obtained as in (4.3) and (4.4) by starting from $(a_{m_1} \ldots a_{m_\alpha}) \vee (b_{n_1} \ldots b_{n_\beta})$. □

5 Cayley–Grassmann algebras

Let $(V, [\,,\dots,\,])$ be a *Peano space*, i.e., an n-dimensional vector space V endowed with a non-degenerate skew-symmetric n-linear form $[\,,\dots,\,]$ called *bracket*. Let $\vee$ denote the *join* operation in the exterior algebra $G(V)$ of the vector space V.

Let Δ be the coproduct on the exterior algebra as Hopf algebra, i.e., the linear application

$$\Delta : G(V) \longrightarrow G(V) \otimes G(V)$$

such that, for an extensor $A = a_1 \vee \dots \vee a_r$,

$$\Delta(A) = \sum_{p=0}^{r} \sum_{i_1<\dots<i_p} \varepsilon_{i_1,\dots,i_p}(a_{i_1} \vee \dots \vee a_{i_p}) \otimes (a_{i'_1} \vee \dots \vee a_{i'_{r-p}})$$

$$= \sum_{(A)}^{\cdot} A_{(1)} \otimes A_{(2)}.$$

The last term is the Sweedler notation for $\Delta(A)$ [27].

In $(G(V), \vee)$ a second operation $\wedge$ [2], called *meet*, can be introduced. The meet depends on the bracket of the Peano space $(V, [\,,\dots,\,])$. Consider the following useful notation for $A = a_1 \vee \dots \vee a_r$ and $B = b_1 \vee \dots \vee b_s$:

$$[A, B] = [a_1, \dots, a_r, b_1, \dots, b_s]$$

if $r + s = n$ and $[A, B] = 0$ otherwise.

Definition 5.1. Given two extensors A, B, set:
if $\mathrm{step}(A) + \mathrm{step}(B) \geq n$,

$$A \wedge B = \sum_{(A)}^{\cdot} [A_{(1)}, B]\, A_{(2)} = \sum_{(B)}^{\cdot} [A, B_{(2)}]\, B_{(1)};$$

if $\mathrm{step}(A) + \mathrm{step}(B) < n$,

$$A \wedge B = 0.$$

The identity

$$\sum_{(A)}^{\cdot} [A_{(1)}, B]\, A_{(2)} = \sum_{(B)}^{\cdot} [A, B_{(2)}]\, B_{(1)} \tag{5.1}$$

in the preceding definition can be proved by linearity. It is fundamental and, in particular, it yields the associative and the anticommutative laws of the meet product.

For A, B extensors such that $\mathrm{step}(A) = r$ and $\mathrm{step}(B) = s$, it follows that

$$A \wedge B = (-1)^{(n-r)(n-s)} B \wedge A. \tag{5.2}$$

Let $(a_1, \ldots, a_n)$ be a unimodular basis of V (i.e., $[a_1, \ldots, a_n] = 1$). The extensor $I = a_1 \vee \ldots \vee a_n$ is the neutral element with respect to the meet. The element I is called the *integral*. Let $\langle A \rangle$ denote the subspace of V associated to the extensor A. For A and B extensors such that $\langle A \rangle \cap \langle B \rangle = \{0\}$, it follows that

$$\langle A \vee B \rangle = \langle A \rangle \oplus \langle B \rangle. \tag{5.3}$$

For A and B extensors, $A \wedge B$ is always an extensor and, if $\langle A \rangle + \langle B \rangle = V$, it follows that

$$\langle A \wedge B \rangle = \langle A \rangle \cap \langle B \rangle. \tag{5.4}$$

In the algebra $(G(V), \vee, \wedge)$ a further operator, called the *Hodge operator*, can be introduced. It implements the duality between join and meet, and translates completely this intrinsically geometric notion into an algebraic tool for the invariant calculus. The duality relations between join and meet are different and closely connected to metric properties. Analogously the Hodge operator depends on (actually is equivalent to) the choice of a non-degenerate symmetric bilinear form $(\ ,\ )$ on V.

Definition 5.2. Let $(\ ,\ )$ be a non-degenerate symmetric bilinear form on V, and let $(a_1, \ldots, a_n)$ be a $(\ ,\ )$-orthogonal basis of V. The *Hodge operator* $\varrho \in GL(G(V))$ is the following:

$$\begin{aligned} &\varrho(1) = I, \\ &\varrho(a_i) = (a_i, a_i)\, \alpha_i, \\ &\varrho(x \vee y) = \varrho(x) \wedge \varrho(y) \text{ for } x, y \in G(V), \end{aligned}$$

where I is the integral and $(\alpha_1, \ldots, \alpha_n)$ is the associated co-basis of the basis $(a_1, \ldots, a_n)$, i.e., the basis of $G^{n-1}(V)$ such that $[a_i, \alpha_j] = \delta_{ij}$ for all i, j, $1 \leq i, j \leq n$.

Proposition 5.3. *Let $(a_1, \ldots, a_n)$ be a $(\ ,\ )$-orthogonal basis of V. For any $a, b \in V$, it follows that*

$$a \wedge \varrho(b) = (a, b).$$

Proof. It is sufficient to verify the assertion on the $(\ ,\ )$-orthogonal basis $(a_1, \ldots, a_n)$:

$$a_i \wedge \varrho(a_j) = a_i \wedge \alpha_j (a_j, a_j) = (a_i, a_j). \qquad \square$$

Proposition 5.4. *The Hodge operator ϱ is independent of the choice of the $(\ ,\ )$-orthogonal basis $(a_1, \ldots, a_n)$.*

Proof. Let ϱ be the Hodge operator defined by the $(\ ,\ )$-orthogonal basis $(a_1, \ldots, a_n)$ and ϱ' the Hodge operator defined by $(a'_1, \ldots, a'_n)$. By applying Proposition 5.3 to ϱ and to ϱ', we obtain $a \wedge \varrho(b) = (a, b) = a \wedge \varrho'(b)$ for every $a, b \in V$. Since $a \wedge (\varrho - \varrho')(b) = 0$, it follows that $\varrho = \varrho'$. $\square$

Proposition 5.5. *For two r-blades $A = a_1 \vee \ldots \vee a_r$ and $B = b_1 \vee \ldots \vee b_r$, it follows that*

$$A \wedge \varrho(B) = \det\Big((a_i, b_j)\Big)_{1\le i,j\le r}.$$

Proof. From the definition of the Hodge operator ϱ, by the associativity of and the splitting rule for the meet, it follows that

$$\begin{aligned}
A \wedge \varrho(B) &= A \wedge (\varrho(b_1) \wedge \ldots \wedge \varrho(b_r)) \\
&= (A \wedge \varrho(b_1)) \wedge \varrho(b_2) \wedge \ldots \wedge \varrho(b_r) \\
&= \left(\sum_{(A)} [A_{(1)}, \varrho(b_1)]\, A_{(2)}\right) \wedge \varrho(b_2) \wedge \ldots \varrho(b_r) \\
&= \left(\sum_{(A)} (A_{(1)}, b_1)\, A_{(2)}\right) \wedge \varrho(b_2) \wedge \ldots \varrho(b_r) \\
&= \det\Big((a_i, b_j)\Big)_{1\le i,j\le r}. \qquad \square
\end{aligned}$$

Corollary 5.6. *Let $(a_1, \ldots, a_n)$ be a $(\,,\,)$-orthogonal basis of V. Then*

$$[a_{i_1} \vee \ldots \vee a_{i_r}, \varrho(a_{j_1} \vee \ldots \vee a_{j_r})] = \prod_{p=1}^{r} (a_{i_p}, a_{j_p})$$

($i_1 < \ldots < i_r$ and $j_1 < \ldots < j_r$).

The exterior algebra $(G(V), \vee)$ endowed with the meet and the Hodge operator is called the *Cayley–Grassmann algebra* (CG-algebra) of V.

6 Geometric Clifford algebras

In the Clifford algebra the n-blades form a one-dimensional subspace. They are called *pseudoscalars*. In the Clifford algebra we fix a distinguished pseudoscalar I. Following Hestenes [18], we call the element I the *unit pseudoscalar*.

The choice of a unit pseudoscalar is equivalent to the choice of a linear isomorphism between the field of the scalars of $C(V)$ and the one-dimensional subspace of the pseudoscalars, with $1 \mapsto I$. Such an isomorphism can be extended to a duality operator as follows.

Definition 6.1. Let I be the unit pseudoscalar of $C(V)$. The *Hestenes duality* is the operator of $GL(C(V))$ such that

$$\widetilde{A} = AI^{-1} = A \cdot I^{-1}$$

for every blade A.

The second identity in Definition 6.1 comes from the formula for the Clifford product of two blades of Theorem 4.6. The fact that the dual of a k-blade is an $(n-k)$-blade and the fact that the inner and outer products are in a duality relation is demonstrated in Sect. 8 (Corollary 8.6).

The Clifford algebra $C(V)$ of $(V, (\,,\,))$, endowed with the unit pseudoscalar, is called the *geometric Clifford algebra* of $(V, (\,,\,))$.

7 A natural equivalence

Here we consider the category of *metric Peano spaces*, that is, the category of vector spaces endowed with a bracket and a non-degenerate symmetric bilinear form. The Cayley–Grassmann algebra and the geometric Clifford algebra can be regarded as functors from the category of metric Peano spaces to the category of algebras with two binary operations and a unary operation and to the category of algebras with a binary operation and a constant, respectively. These functors map the triple $(V, [, \dots ,], (,))$ into the corresponding Cayley–Grassmann algebra and geometric Clifford algebra, respectively.

Lemma 7.1. *The choice of a Hodge operator in the Cayley–Grassmann algebra is equivalent to the choice of a non-degenerate symmetric bilinear form on V.*

Proof. For a non-degenerate symmetric bilinear form $(\,,\,)$ on V, a Hodge operator $\varrho_{(\,,)}$ can be defined as in Definition 5.2:

$$\begin{aligned}
&\varrho(1) = I,\\
&\varrho(a_i) = (a_i, a_i)\,\alpha_i,\\
&\varrho(x \vee y) = \varrho(x) \wedge \varrho(y) \text{ for } x, y \in G(V),
\end{aligned}$$

for any $(\,,\,)$-orthogonal basis $(a_1, \dots , a_n)$ of V.

For a Hodge operator ϱ on $G(V)$, the bilinear form such that

$$(a, b)_\varrho = a \wedge \varrho(b) \qquad \text{for all } a, b \in V$$

is symmetric and non-degenerate. By Proposition 5.3, we get the following identities:

$$\begin{aligned}
&(\,,\,)_{\varrho_{(\,,)}} = (\,,\,)\\
&\varrho_{(\,,)_\varrho} = \varrho. \qquad \square
\end{aligned}$$

Similarly, the choice of a non-degenarate symmetric bilinear form $(\,,\,)$ on V determines the Clifford product of $C(V)$ by the definition of the Clifford algebra of $(V, (\,,\,))$. Conversely, the bilinear form $(\,,\,)$ is the symmetric part of the restriction on V of the Clifford product of $C(V)$. For vectors a, b it follows that (Sect. 3)

$$(a, b) = \frac{1}{2}(ab + ba). \tag{7.1}$$

Lemma 7.2. *The choice of a unit pseudoscalar in the geometric Clifford algebra is equivalent to the choice of a bracket on V.*

Proof. For a bracket $[\,,\dots,\,]$ on V, the meet operation on $G(V)$ is well-defined and, hence, its neutral element I, i.e., the integral element. Since it is an extensor of step n, it is mapped to a pseudoscalar by the isomorphism α_V. From a pseudoscalar I, a non-degenerate skew-symmetric n-linear form can be defined as follows:

$$[a_1,\dots,a_n] = (a_1 \vee \dots \vee a_n)I^{-1}.$$

In particular, we have stated a bijective correspondence between pseudoscalars and non-degenerate skew-symmetric n-linear forms. □

Similarly, the choice of a bracket $[\,,\dots,\,]$ on V determines the meet operation on $(G(V), \vee)$ by Definition 5.1. Conversely, the bracket $[\,,\dots,\,]$ can be computed on n given vectors $a_1,\dots,a_n$ as follows:

$$[a_1,\dots,a_n] = a_1 \wedge (a_2 \vee \dots \vee a_n). \tag{7.2}$$

By combining the preceding results, we get the constructions inverse to the CG-algebra functor and to the geometric Clifford algebra functor. The combination of these functors and inverse constructions enables us to endow the isomorphism α_V with bijective maps which put in correspondence the choice of Hodge operator ϱ on $(G(V), \vee, \wedge)$ with the Clifford product of $C(V)$ and the choice of the unit pseudoscalar I in $C(V)$ with the meet operation $\wedge$ of the CG-algebra.

8 Grassmann–Clifford algebras

By using the isomorphism α_V (Proposition 2.3, Corollary 2.5) the algebraic structures of $G(V)$ and $C(V)$ can be superimposed and regarded as different algebra structures on the *same* vector space. The algebra consisting of all the operations of the Cayley–Grassmann algebra and the geometric Clifford algebra will be called the *Grassmann–Clifford algebra*. This construction, allowing us to assign to any metric Peano space $(V, [\,,\dots,\,], (\,,\,))$ its Grassmann–Clifford algebra, is functorial.

Since, by the isomorphism α_V, the Cayley–Grassmann algebra $(G(V)), \vee, \wedge, \varrho)$ and the geometric Clifford algebra $(C(V), I)$ can be regarded as algebras on the same vector space, the problem is now to establish a complete system of identities that express any operation of Cayley–Grassmann type in terms of operations of geometric Clifford type, and vice versa.

Proposition 3.4 states that the outer product and the join in the Grassmann–Clifford algebra are the same operation. The case of the inner product is quite different.

Theorem 8.1. *Let A be a k-blade and B an l-blade. Then* [2]

[2] By ${}^rA$ we denote the reverse of the extensor A, i.e., if $A = a_1 \vee \dots \vee a_k$, then ${}^rA = a_k \vee \dots \vee a_1 = (-1)^{\frac{k(k-1)}{2}} A$.

$$A \cdot B = B \wedge \varrho({}^r A) \text{ if } k \leq l;$$
$$A \cdot B = (-1)^{l(k-l)} A \wedge \varrho({}^r B) \text{ if } k \geq l.$$

Proof. Let $(e_1, \ldots, e_n)$ be a $(\ ,\ )$-orthogonal basis of V. For two blades $A = e_{i_1} \vee \ldots \vee e_{i_k} = e_{i_1} \ldots e_{i_k}$ and $B = e_{j_1} \ldots e_{j_l}$, it follows that the Clifford product $AB = e_{i_1} \ldots e_{i_k} e_{j_1} \ldots e_{j_l}$ is a blade.

If $k \leq l$, step$(AB) = l - k$ if and only if $\{i_1, \ldots, i_k\} \subseteq \{j_1, \ldots, j_l\}$. Since in this case we can write $B = e_{j_1} \ldots e_{j_l} = (-1)^d e_{i_k} \ldots e_{i_1} e_{h_1} \ldots e_{h_{l-k}}$, it follows that

$$A \cdot B = (-1)^d \prod_{p=1}^{k} (e_{i_p}, e_{i_p})\, e_{h_1} \ldots e_{h_{l-k}}.$$

Since $(e_1, \ldots, e_n)$ is a $(\ ,\ )$-orthogonal basis, the Hodge operator ϱ is such that $[e_{r_1} \ldots e_{r_k}, \varrho(e_{s_1} \ldots e_{s_k})]$ is equal to $\prod_{p=1}^{k} (e_{r_p}, e_{r_p})$ if $r_p = s_p$ for every p, while it is null if $\{r_1, \ldots, r_k\} \neq \{s_1, \ldots, s_k\}$. From the splitting rule for the meet it follows that, if $\{i_1, \ldots, i_k\} \subseteq \{j_1, \ldots, j_l\}$,

$$B \wedge \varrho({}^r A) = (-1)^d [e_{i_k} \ldots e_{i_1}, \varrho(e_{i_k} \ldots e_{i_1})]\, e_{h_1} \ldots e_{h_{l-k}};$$

otherwise, $B \wedge \varrho({}^r A) = 0$.

If $k \geq l$, the same argument leads in the non-null case to

$$\begin{aligned} A \cdot B &= ((-1)^d e_{h_1} \ldots e_{h_{k-l}} e_{j_l} \ldots e_{j_1})\,(e_{j_1} \ldots e_{j_l}) \\ &= (-1)^d \prod_{p=1}^{l} (e_{j_p}, e_{j_p})\, e_{h_1} \ldots e_{h_{k-l}} \\ &= (-1)^d [e_{j_l} \ldots e_{j_1}, \varrho(e_{j_l} \ldots e_{j_1})]\, e_{h_1} \ldots e_{h_{k-l}} \\ &= (-1)^{l(k-l)} A \wedge \varrho({}^r B). \qquad \square \end{aligned}$$

The inner product of two blades is hence a blade and, in particular, the Hestenes duality operator assigns to a k-blade an $(n-k)$-blade.

The Hodge operator and the Hestenes duality operator are closely related.

Theorem 8.2. *Let A be a blade. Then*

$$\widetilde{A} = (\varrho({}^r I))^{-1} \varrho({}^r A)$$

and

$$\varrho(A) = I^2 (\widetilde{{}^r A}).$$

Proof. Directly from the definitions, we obtain

$$\begin{aligned} \widetilde{A} &= I^{-1} \wedge \varrho({}^r A) = (I^2)^{-1} I \wedge \varrho({}^r A) \\ &= (I \cdot I)^{-1} \varrho({}^r A) = (I \wedge \varrho({}^r I))^{-1} \varrho({}^r A) \\ &= (\varrho({}^r I))^{-1} \varrho({}^r A) \end{aligned}$$

and

$$\varrho(A) = I \wedge \varrho(A) = I \wedge \varrho({}^r({}^r A)) = ({}^r A) \cdot I = ({}^r A)I = I^2 \widetilde{({}^r A)}. \qquad \square$$

By computing the meet using the inner product and Hestenes duality, the following result completes the translation system between Cayley–Grassmann algebra and geometric Clifford algebra.

Theorem 8.3. *Let A be a k-blade and B an l-blade. If $k + l \geq n$, then*

$$A \wedge B = \widetilde{B} \cdot A = (-1)^{k(n-k)} B \cdot \widetilde{A}.$$

Proof. Since the dual of B is an $(n-l)$-blade and $n - l \leq k$, it follows that

$$\widetilde{B} \cdot A = A \wedge \varrho({}^r(\widetilde{B})) = A \wedge (\widetilde{B} I) = A \wedge B.$$

Since the dual of A is an $(n-k)$-blade and $n - k \leq l$, it follows that

$$\begin{aligned} B \cdot \widetilde{A} &= (-1)^{(n-k)(k+l-n)} B \wedge \varrho({}^r(\widetilde{A})) \\ &= (-1)^{(n-k)(k+l-n)} B \wedge A = (-1)^{k(n-k)} A \wedge B. \qquad \square \end{aligned}$$

The above results are especially useful in proving geometric Clifford algebra identities in the Cayley–Grassmann algebra and vice versa. From Theorem 8.1 we can infer the main properties of the inner product.

Proposition 8.4 (Anticommutative law). *Let A be a k-blade and B an l-blade, with $k \leq l$. Then*

$$A \cdot B = (-1)^{k(l-k)} B \cdot A.$$

Proof. This follows directly from Theorem 8.1. $\square$

Proposition 8.5 (Rule of the middle factor). *Let A, B, C be blades of steps r, s, t, respectively. Then*

$$\begin{aligned} A \cdot (B \cdot C) &= (A \vee B) \cdot C \text{ if } r + s \leq t \\ A \cdot (B \cdot C) &= (A \cdot B) \cdot C \text{ if } r + t \leq s. \end{aligned}$$

Proof. Let $r + s \leq t$. Since $s \leq t$ and $r \leq t - s$, it follows that

$$\begin{aligned} A \cdot (B \cdot C) &= A \cdot (C \wedge \varrho({}^r B)) = (C \wedge \varrho({}^r B)) \wedge \varrho({}^r A) \\ &= C \wedge \varrho({}^r B \vee {}^r A) = C \wedge \varrho({}^r(A \vee B)) = (A \vee B) \cdot C. \end{aligned}$$

The case $r + t \leq s$ can be reduced to the preceding case by the anticommutative law:

$$\begin{aligned} A \cdot (B \cdot C) &= A \cdot ((-1)^{t(s-t)} C \cdot B) = (-1)^{t(s-t)} (A \vee C) \cdot B \\ &= (-1)^{t(s-t)} ((-1)^{rt} C \vee A) \cdot B = (-1)^{t(s-t)-rt} C \cdot (A \cdot B) \\ &= (A \cdot B) \cdot C. \qquad \square \end{aligned}$$

Corollary 8.6. *Let be A and B blades. Then*

$$\widetilde{(A \vee B)} = A \cdot \widetilde{B}.$$

Proof. In the non-degenerate case, from the rule of the middle factor it follows that

$$\widetilde{(A \vee B)} = (A \vee B) \cdot I^{-1} = A \cdot (B \cdot I^{-1}) = A \cdot \widetilde{B}. \qquad \square$$

The duality relation of the preceeding corollary between inner product and outer product can also be demonstrated in the Cayley–Grassmann algebra by using Theorem 8.2.

In the following we provide another demonstration of the splitting rule for the inner product of Corollary 4.7.

Proposition 8.7 (Splitting rule for the inner product). *Let A be a k-blade and B an l-blade, with $k \leq l$ and $B = b_1 \vee \ldots \vee b_l$. Then*

$$A \cdot B = \sum_{i_1 < \ldots < i_k} \varepsilon_{i_1, \ldots, i_k} (A \cdot (b_{i_1} \vee \ldots \vee b_{i_k}))\, b_{i'_1} \vee \ldots \vee b_{i'_{l-k}}.$$

Proof. The formula comes from the splitting rule for the meet in Definition 5.1, as follows:

$$\begin{aligned} A \cdot B = B \wedge \varrho({}^r A) &= \sum_{(B)}^{\cdot} [B_{(1)}, \varrho({}^r A)]\, B_{(2)} \\ &= \sum_{i_1 < \ldots < i_k} \varepsilon_{i_1, \ldots, i_k} [(b_{i_1} \vee \ldots \vee b_{i_k}), \varrho({}^r A)]\, b_{i'_1} \vee \ldots \vee b_{i'_{l-k}} \\ &= \sum_{i_1 < \ldots < i_k} \varepsilon_{i_1, \ldots, i_k} ((b_{i_1} \vee \ldots \vee b_{i_k}) \wedge \varrho({}^r A))\, b_{i'_1} \vee \ldots \vee b_{i'_{l-k}} \\ &= \sum_{i_1 < \ldots < i_k} \varepsilon_{i_1, \ldots, i_k} (A \cdot (b_{i_1} \vee \ldots \vee b_{i_k}))\, b_{i'_1} \vee \ldots \vee b_{i'_{l-k}}. \qquad \square \end{aligned}$$

The splitting rules for the inner product and the meet are equivalent. Indeed, the double splitting rule (5.1) can be demonstrated in the geometric Clifford algebra setting.

Proposition 8.8 (Double splitting for the meet product). *For two extensors A and B with step(A) + step$(B) \geq n$, we have*

$$A \wedge B = \sum_{(A)}^{\cdot} [A_{(1)}, B]\, A_{(2)} = \sum_{(B)}^{\cdot} [A, B_{(2)}]\, B_{(1)}.$$

Proof. From Theorem 8.3 and the splitting rule for the inner product it follows that:

$$\begin{aligned} A \wedge B &= \widetilde{B} \cdot A \\ &= \sum_{i_1 < \ldots < i_{n-l}} \varepsilon_{i_1,\ldots,i_{n-l}} (\widetilde{B} \cdot (a_{i_1} \vee \ldots \vee a_{i_{n-l}}))\, a_{i'_1} \vee \ldots \vee a_{i'_{k+l-n}} \\ &= \sum_{i_1 < \ldots < i_{n-l}} \varepsilon_{i_1,\ldots,i_{n-l}} ((a_{i_1} \vee \ldots \vee a_{i_{n-l}}) \wedge B)\, a_{i'_1} \vee \ldots \vee a_{i'_{k+l-n}} \\ &= \dot{\sum_{(A)}} [A_{(1)}, B]\, A_{(2)}; \end{aligned}$$

$$\begin{aligned} A \wedge B &= (-1)^{k(n-k)} B \cdot \widetilde{A} = (-1)^{(n-k)(n-l)} \widetilde{A} \cdot B \\ &= (-1)^{(n-k)(n-l)} \sum_{i_1 < \ldots < i_{n-k}} \varepsilon_{i_1,\ldots,i_{n-k}} (\widetilde{A} \cdot (b_{i_1} \vee \ldots \vee b_{i_{n-k}})) \\ &\quad \times b_{i'_1} \vee \ldots \vee b_{i'_{k+l-n}} \\ &= (-1)^{(n-k)(n-l)} \sum_{i_1 < \ldots < i_{n-k}} \varepsilon_{i_1,\ldots,i_{n-k}} ((b_{i_1} \vee \ldots \vee b_{i_{n-k}}) \wedge A) \\ &\quad \times b_{i'_1} \vee \ldots \vee b_{i'_{k+l-n}} \\ &= (-1)^{(n-k)(k+l-n)} \dot{\sum_{(B)}} (A \wedge B_{(1)})\, B_{(2)} \\ &= \dot{\sum_{(B)}} [A, B_{(2)}]\, B_{(1)}. \qquad \square \end{aligned}$$

References

[1] Abe, E. (1980): Hopf algebras. Cambridge University Press, Cambridge

[2] Barnabei, M., Brini, A., Rota, G.-C. (1985): On the exterior calculus of invariant theory. J. Algebra **96**, 120–160

[3] Bourbaki, N. (1970): Élements de mathématique. Algèbre. Chapitres 1 à 3. Hermann, Paris

[4] Brini, A., Huang, R.Q., Teolis, A.G.B. (1992): The umbral symbolic method for supersymmetric tensor. Adv. Math. **96**, 123–193

[5] Brini, A., Teolis, A.G.B. (1995): Capelli's method of *variabili ausiliarie*, superalgebras and geometric calculus. In: White, N. L. (ed.) Invariant Methods in Discrete and Computational Geometry. Kluwer, Dordrecht, pp. 59–75

[6] Brini, A., Teolis, A.G.B. (1996): Grassmann progressive and regressive products and CG-algebras. In: Schubring, G. (ed.) Hermann Günther Graßmann (1809–1877): Visionary Mathematician, Scientist and Neohumanist Scholar. Kluwer, Dordrecht, pp. 231–242

[7] Caianiello, E.R. (1973): Combinatorics and renormalization in quantum field theory. Benjamin, Reading, MA

[8] Cartan, E. (1952–1955): Oeuvres complètes. (6 vols.) Gauthier-Villars, Paris

[9] Chevalley, C. (1954): The algebraic theory of spinors. Columbia University Press, New York

[10] Clifford, W.K. (1878): Applications of Grassmann's extensive algebra. Amer. J. Math. **1**, 350–358

[11] De Concini, C., Procesi, C. (1976): A characteristic-free approach to invariant theory. Adv. Math. **21**, 330–354

[12] Dieudonné, J. (1979): The tragedy of Grassmann. Linear and Multilinear Algebra **8**, 1–14

[13] Doubilet, P., Rota, G.-C., Stein, J. (1974): On the foundations of combinatorial theory. IX. Combinatorial methods in invariant theory. Studies in Appl. Math. **53**, 185–216

[14] Forder, H.G. (1941): The calculus of extension. Cambridge University Press, Cambridge. Reprint: (1960 Chelsea), New York

[15] Grassmann, H.G. (1894–1911): Hermann Graßmanns gesammelte mathematische und physikalische Werke. (3 vols.) Engel. F. (ed.) Teubner, Leipzig

[16] Grassmann, H.G. (1995): A new branch of mathematics. The *Ausdehnungslehre* of 1844 and other works. Kannenberg, L.C. (translator). Open Court, Chicago

[17] Greub, W.H. (1967): Multilinear algebra. Springer, Berlin

[18] Hestenes, D., Sobczyk, G. (1984): Clifford algebra to geometric calculus. A unified language for mathematics and physics. Reidel, Dordrecht

[19] Hestenes, D. (1986): New foundations for classical mechanics. Reidel, Dordrecht

[20] Hestenes, D., Ziegler, R. (1991): Projective geometry with Clifford algebra. Acta Appl. Math. **23**, 25–63

[21] Hestenes, D. (1991): The design of linear algebra and geometry. Acta Appl. Math. **23**, 65–93

[22] Hestenes, D. (1996): Grassmann's vision. In: Schubring, G. (ed.) Hermann Günther Graßmann (1809–1877): Visionary Mathematician, Scientist and Neohumanist Scholar. Kluwer, Dordrecht, pp. 243–254

[23] Marcus, M. (1973–1975): Finite dimensional multilinear algebra. Parts I, II. Dekker, New York

[24] Peano, G. (1888): Calcolo geometrico secondo l'Ausdehnungslehre di H. Grassmann precedoto dalle operazioni della logica deduttiva. Bocca, Torino. Translation: (2000): Geometric calculus. Kannenberg, L.C. (translator), Birkhäuser Boston, Boston, MA

[25] Schubring, G. (ed.) (1996): Hermann Günther Graßmann (1809–1877): visionary mathematician, scientist and neohumanist scholar. Kluwer, Dordrecht

[26] Segre, B. (1951): Lezioni sulla teoria delle forme differenziali. Docet Edizioni Universitaria, Roma

[27] Sweedler, M.E. (1969): Hopf algebras. Benjamin, New York

[28] White, N.L. (ed.) (1995): Invariant methods in discrete and computational geometry. Kluwer, Dordrecht

[29] Whitehead, A.N. (1898): A treatise on universal algebra with applications. Cambridge University Press, Cambridge

[30] Zaddach, A. (1994): Graßmann Algebra in der Geometrie. Bibliographisches Institut, Mannheim

Grassmann geometric calculus, invariant theory and superalgebras

A. Brini, F. Regonati, A. G. B. Teolis

1 Introduction

The idea of exploring and developing the deep connections between the theory of Cayley-Grassmann algebras and the invariant theory of skew-symmetric tensors was a recurrent theme of Rota's mathematical work.

From an idea essentially due to G. Peano as point of departure, an early relevant step was to regard the exterior algebra as an algebraic object associated to a finite-dimensional vector space V endowed with a bracket (non-degenerate, n-multilinear, alternating form, $n = \dim(V)$), called a *Peano space* [16, 3, 22].

This point of view allows the exterior algebra to be endowed with two different associative products, called *join* and *meet*; the join is nothing but the usual wedge product, while the meet is a precise and intrinsic version of Grassmann's original regressive product [10].

The exterior algebra with meet and join is called a *Cayley–Grassmann algebra* and turns out to be a natural setting for stating and proving geometric results, since, under suitable non-degeneracy conditions, the meet and the join of two extensors are the algebraic counterparts of the geometric operations of taking the intersection and the sum of two subspaces in V.

A notable extension of the standard methods of the invariant theory of tensors was obtained in 1987 by Grosshans, Rota and Stein in the work *Invariant Theory and Superalgebras* [18], by using some ideas borrowed from the theory of Lie superalgebras.

Specifically, the authors introduced a new symbolic method (foreshadowed by Weitzenböck [23]) for the representation of invariants of skew-symmetric tensors.

The main feature of this method is that a skew-symmetric tensor may be symbolically represented by a divided power of a commuting variable, and the umbral operator turns out to be a mapping from an anticommutative letterplace algebra [16, 18, 6, 8, 5] to the commutative algebra of polynomial functions over tensors.

This new formalism allows several results in the theory of Cayley–Grassmann algebras to be proved by invariant theoretic methods.

A radically different approach to geometric calculus and invariant theory was outlined by Chan, Rota and Stein in their 1995 paper *The Power of Positive Thinking* [14].

Here, the attempt is to establish the existence of a universal supermodule for both geometric calculus and invariant theory.

In this module, called the White module, skew-symmetric tensors can be treated directly as divided powers of positive variables, and also covariants can be written as polynomials in divided powers of commuting variables; actual skew-symmetric tensors are obtained by a process of complete superpolarization from positive letters to negative letters and, in parallel, evaluations of covariants are obtained by the same process of complete superpolarization.

The purpose of the present paper is to explore and develop in a systematic way the deep connections between the three approaches mentioned above, namely, the theory of Cayley–Grassmann algebras, the superalgebraic invariant-theoretic method of Grosshans, Rota and Stein, and the direct positive method of Chan, Rota and Stein.

We begin by generalizing the construction of the Cayley–Grassmann algebra of a Peano space of dimension n.

Our first main result (see Sect. 9) establishes the existence of a $\mathbf{Z}_2$-graded double algebra (i.e., an algebra generated by negative variables, positive variables and their divided powers) endowed with two bilinear products (the meet and the join), whose negative part turns out to be epimorphic onto the Cayley–Grassmann algebra of a Peano space.

This double algebra, called the *White double algebra* of a Peano space, is obtained as the quotient of what we call the *symbolic algebra*, with respect to the kernel of a generalized Peano map taking values in a letterplace superalgebra.

The crucial point is that actions of the algebras generated by superpolarizations are defined on both the symbolic algebra and the letterplace superalgebra. The Peano map turns out to be an equivariant map with respect to these actions, and, therefore, the action of superpolarizations is well-defined on the White double algebra.

Superpolarizations are superderivations with respect to both the meet and the join products in the White double algebra.

The symbolic algebra is a Hopf algebra in the generalized sense of [7].

The fact that superpolarizations are coderivations in the symbolic algebra is perhaps the core of the symbolic method; in fact, this phenomenon allows us to translate every statement written in Hopf algebra language into a statement written in the language of superpolarizations of divided powers of positive letters [7].

Therefore, one may view divided powers of positive variables as symbolic representations of "generic" skew-symmetric tensors. Divided powers of positive variables take the role of Capelli's auxiliary variables in representation theory, along the lines of [6, 8, 10].

Let K be an infinite field. The White double algebra admits a significant invariant-theoretic interpretation, closely related to the previous one.

Specifically, by the process of complete superpolarization, an element in the positive part of the White double algebra may be regarded as a covariant polynomial function from a family of skew-symmetric tensors to a Cayley–Grassmann algebra.

Indeed, this construction induces an intrinsic and combinatorial description of an umbral operator over the White double algebra.

The main point (see Theorem 6) is that this umbral operator, viewed in the spirit of *The Power of Positive Thinking*, turns out to be the synthetic (i.e., coordinate-free) counterpart of the umbral operator defined in *Invariant Theory and Superalgebras*.

In the final part of the work, we focus our attention upon the covariant theory of just one tensor, since the generalization to the case of several tensors can be treated along the same lines.

The crucial concept is that of *generic vanishing* of a covariant at a tensor $\underline{t}$ (see Sect. 16); specifically, we say that a polynomial function F of an ordered $(p+1)$-tuple $(t; t_1, \dots, t_p)$ vanishes generically at a tensor $\underline{t}$ iff $F(\underline{t}; t_1, \dots, t_p) = 0$ for every choice of the tensors $(t_1, \dots, t_p)$. The variety of tensors at which a covariant F vanishes generically is a $GL(V)$-stable set.

We study in detail the generic vanishing of covariants that are umbrally represented by elements in the letterplace superalgebra or, equivalently, by elements in the White double algebra.

The meaning of what we call "Completeness Theorems" is that the generic vanishing at t of a covariant umbrally represented by an element C in the White double algebra is equivalent to the vanishing of C in the White double algebra under the action of a suitable superpolarization operator naturally associated to the tensor t (see Theorems 7 and 8).

These results take the notion of generic vanishing closer to its combinatorial and geometric meaning.

In arbitrary characteristic, it is not true that every covariant polynomial function may be represented by the umbral method. However, Gram's theorems (see Theorems 10 and 11) state that every $GL(V)$-stable algebraic set of tensors can be characterized by the generic vanishing of those special covariants that admit an umbral representation.

2 The supersymmetric Hopf algebra of a signed set

Let K be any fixed field of characteristic $\neq 2$; we use vector space, linear mapping, tensor product, algebra, ... as abbreviation for K-vector space, K-linear mapping, K-tensor product, K-algebra, ...

A *proper signed set* is a linearly ordered set L together with a composition $L = L^+ \cup L^-$ into a subset L^+, assumed to be countable, of positive elements and a subset L^- of negative elements. We define the *signature* $|x| \in \mathbf{Z}_2$ of an element $x \in L$ to be 0 if x is positive, and 1 if x is negative.

For a word in the free monoid $\mathrm{Mon}(L)$ over L we have the concepts of *content* and *length*; we define the *signature* $|w| \in \mathbf{Z}_2$ of a word $w = x_1 x_2 \dots x_p$ with $x_i \in L$ in $\mathrm{Mon}(L)$ by setting $|w| = |x_1| + |x_2| + \dots + |x_p|$.

The *extended signed set* L_d of a given proper signed set $L = L^+ \cup L^-$ is the proper signed set $L_d = (L^+ \times \mathbf{Z}^+)^+ \cup L^-$, where $(a, 1) = a$ for every $a \in L^+$. The pair (a, n), with $a \in L^+$ and $n \in \mathbf{N}$ is denoted by $a^{(n)}$ and called the nth divided power of a. Sending $a^{(n)} \mapsto a^n$ for $a \in L^+$ and $a \mapsto a$ for $a \in L^-$ induces a morphism, called *display* and denoted Disp, from the free monoid $\mathrm{Mon}(L_d)$ to the free monoid $\mathrm{Mon}(L)$. We define the *content*, the *length* and the *signature* of an element of $\mathrm{Mon}(L_d)$ as the content, the length and the signature of its display in $\mathrm{Mon}(L)$.

Definition 1. The *supersymmetric algebra* $S = \mathrm{Super}(L)$ *of the signed set* $L = L^+ \cup L^-$ is the associative unitary algebra generated by the extended signed set L_d, modulo the identities

$$\begin{aligned} uv &= (-1)^{|u||v|} vu, && u, v \in \mathrm{Mon}(L_d) \\ a^{(i)} a^{(j)} &= \tbinom{i+j}{i} a^{(i+j)}, && a \in L^+,\ i, j \in \mathbf{N}. \end{aligned}$$

The concepts of content, length and signature are unambiguously defined for nonzero monomials in $\mathrm{Super}(L)$.

We set $a^{(0)} = 1$, for every $a \in L^+$. By abuse of notation, we omit the symbol of signature, when it occurs in the exponent of -1; for example, we write the first identity as

$$uv = (-1)^{uv} vu.$$

The supersymmetric algebra $\mathrm{Super}(L)$ is a $\mathbf{Z}_2$-graded algebra

$$\mathrm{Super}(L) = \mathrm{Super}(L)_0 \oplus \mathrm{Super}(L)_1,$$

where $\mathrm{Super}(L)_0$ is the span of all the monomials having signature 0, and $\mathrm{Super}(L)_1$ is the span of all the monomials having signature 1.

The supersymmetric algebra $\mathrm{Super}(L)$ also admits finer gradings:

$$\mathrm{Super}(L) = \oplus_l \mathrm{Super}^l(L),$$

where $\mathrm{Super}^l(L)$ is the span of all the monomials having length l, and

$$\mathrm{Super}(L) = \oplus_\gamma \mathrm{Super}^\gamma(L),$$

where $\mathrm{Super}^\gamma(L)$ is the span of all the monomials having content γ.

Definition 2. The *supersymmetric Hopf* algebra of the signed set L is the structure $[\mathrm{Super}(L), \pi, \eta, \Delta, \epsilon, S]$, where $[\mathrm{Super}(L), \pi, \eta]$ is the supersymmetric algebra of L and

- $\begin{cases} \Delta(1) = 1 \otimes 1 \\ \Delta(a) = a \otimes 1 + 1 \otimes a, & a \in L^- \\ \Delta(a^{(n)}) = \sum_{i+j=n} a^{(i)} \otimes a^{(j)}, & a \in L^+ \end{cases}$
- $\epsilon(w) = \begin{cases} w & l(w) = 0 \\ 0 & l(w) > 0 \end{cases}$
- $S(w) = (-1)^{l(w)} w.$

3 The supersymmetric algebra of the braces of a signed set

Definition 3. A *brace of step n* is a variable $[u]$, where u ranges over the homogeneous elements of length n in $\text{Super}(L)$. The *supersymmetric algebra* of the braces of step n over the signed set L, denoted $Bc = Bc_n(L)$, is the unitary associative algebra generated by the braces of step n, modulo the identities

$$[au + bv] = a[u] + b[v], \qquad a, b \in K, u, v \in S$$
$$[u][v] = (-1)^{(u+n)(v+n)}[v][u], \qquad u, v \in S.$$

As a rule, we denote braces and brace monomials by upper case Roman letters. We define the *signature* of the brace monomial $A = [u_1][u_2]\dots[u_p]$ by setting $|A| = |u_1| + |u_2| + \dots + |u_p| + pn$. Thus in the algebra $Bc_n(L)$ we have the *supercommutation rule*

$$AB = (-1)^{AB} BA.$$

The supersymmetric algebra $Bc_n(L)$ is graded by setting $Bc_n(L)_0 =$ the span of all the brace monomials having even signature, and $Bc_n(L)_1 =$ the span of all the brace monomials having odd signature.

The conditions

$$u \mapsto \begin{cases} [u] & l(u) = n \\ 0 & l(u) \neq n \end{cases}$$

induce a linear mapping, homogeneous of grade the parity of n,

$$[\,] : \text{Super}(L) \to Bc_n(L).$$

4 The symbolic algebra of a signed set

The symbolic algebra is an algebra over the supersymmetric ring generated by the braces. This ring is, in general, not commutative. However, it is a supersymmetric ring in the sense of [7], and, thus, the main constructions and results of multilinear algebra still hold [7].

Let $S = \text{Super}(L)$ be the supersymmetric Hopf algebra, and let $Bc = Bc_n(L)$ be the supersymmetric algebra of the braces of step n. As a rule, we denote the elements of Bc by upper case Roman letters, and the elements of S by lower case Roman letters. Let A be the vector space $Bc \otimes S$, graded as usual by setting $|X \otimes y| = |X| + |y|$. At times we use lower case letters to denote the elements of A; when there is danger of misunderstanding, the range of each symbol is specified. We define the *length* $l(z)$ of a monomial $z = X \otimes y$ in A by setting $l(z) = l(y)$.

Definition 4. On the vector space A we define a product $\cdot$, called the *prejoin*, by setting

$$(X \otimes y) \cdot (Z \otimes t) = (-1)^{yZ}(XZ) \otimes (yt).$$

As a rule, we write X instead of $X \otimes 1$, y instead of $1 \otimes y$, and so we write Xy instead of $(X \otimes 1) \cdot (1 \otimes y) = X \otimes y$. The prejoin product is homogeneous of grade 0, associative and supercommutative:

$$x \cdot y = (-1)^{xy} y \cdot x, \qquad x, y \in A.$$

In particular, the vector space A may be regarded as a Bc-bimodule.

Definition 5. On the vector space A we define a product $*$, called the *premeet*, by setting

$$(Xy) * (Zt) = (-1)^{yZ} XZ \sum_{(t)} [yt_{(1)}] t_{(2)} (-1)^{nt_{(2)}}.$$

Note that

$$a^{(p)} * b^{(q)} = \begin{cases} [a^{(p)} b^{(n-p)}] b^{(p+q-n)} & p+q \geq n \\ 0 & p+q < n. \end{cases}$$

The premeet product is homogeneous of grade the parity of n and, in general, not associative. However, the premeet product is right Bc-bilinear, that is, the following identities hold:

$$(Xu) * v = X(u * v), \quad (uX) * v = u * (Xv),$$
$$u * (vX) = (u * v) X (-1)^{Xn},$$

where $X \in Bc$ and $u, v \in A$.

Definition 6. The *symbolic algebra* $A = A_n(L)$ is the structure

$$A = [Bc \otimes S; \ \cdot, \ *].$$

We define a coproduct $\Delta : Bc \otimes S \to (Bc \otimes S) \otimes_{Bc} (Bc \otimes S)$ by setting

$$\begin{aligned} \Delta(X \otimes y) &= X \Delta(1 \otimes y) \\ &= \sum_{(y)} (X \otimes y_{(1)}) \otimes_{Bc} (1 \otimes y_{(2)}) \\ &= \sum_{(y)} (-1)^{Xy_{(1)}} (1 \otimes y_{(1)}) \otimes_{Bc} (X \otimes y_{(2)}). \end{aligned}$$

The structure $[Bc \otimes S; \ \cdot, \ \Delta]$ is the Bc-bialgebra scalar extension of the supersymmetric bialgebra S to the supersymmetric algebra Bc of braces of step n. Indeed, this Bc-bialgebra has a natural structure of Hopf Bc-algebra [7].

We define a linear mapping $[\,] : A \to Bc$ by setting

$$[Xy] = X[y], \quad X \in Bc, \ y \in S.$$

This mapping is right Bc-linear, i.e., the following identities hold:

$$[Xy] = X[y], \quad [yX] = [y]X(-1)^{Xn}, \quad X \in Bc, \; y \in A.$$

The definition of the premeet product can now be given in the simpler form

$$x * y = \sum_{(y)} [xy_{(1)}]y_{(2)}(-1)^{ny_{(2)}}, \quad x, y \in A.$$

5 Polarizations

Definition 7. For every $x, y \in L$ we define the *polarization* $d_{x,y}$ *of* y *in* x as the linear operator of the supersymmetric Hopf algebra $S = \mathrm{Super}(L)$ into itself such that

- $d_{x,y}(z) = \begin{cases} \delta_{y,z}x & z \in L^- \\ \delta_{y,a}xa^{(p-1)} & z = a^{(p)}, \quad a \in L^+, \; p \geq 1 \end{cases}$
- $d_{x,y}$ is a left superderivation, homogeneous of grade $|x| + |y|$, that is,

$$d_{x,y}(zt) = (d_{x,y}z)t + (-1)^{(x+y)z}z(d_{x,y}t), \quad z, t \in S,$$

or, equivalently,

$$d_{x,y} \circ \pi = \pi \circ (d_{x,y} \otimes I + I \otimes d_{x,y}).$$

Proposition 1. *For every $x, y \in L$, the polarization $d_{x,y}$ of y in x is a left supercoderivation of the supersymmetric Hopf algebra $S = \mathrm{Super}(L)$, that is,*

$$\Delta(d_{x,y}(z)) = \sum_{(z)} \left(d_{x,y}z_{(1)} \otimes z_{(2)} + (-1)^{(x+y)z_{(1)}} z_{(1)} \otimes d_{x,y}z_{(2)} \right), \quad z \in S,$$

or, equivalently,

$$\Delta \circ d_{x,y} = (d_{x,y} \otimes I + I \otimes d_{x,y}) \circ \Delta.$$

Definition 8. For every $x, y \in L$ we define the *polarization* $\partial_{x,y}$ *of* y *in* x as the linear operator of the supersymmetric algebra $Bc_n(L)$ of braces into itself, such that

- $\partial_{x,y}[z] = [D_{x,y}z]$
- $\partial_{x,y}$ is a left superderivation, homogeneous of grade $|x| + |y|$, that is,

$$\partial_{x,y}(ZT) = (\partial_{x,y}Z)T + (-1)^{(x+y)Z}Z(\partial_{x,y}T), \quad Z, T \in Bc.$$

Definition 9. For every $x, y \in L$, we define the *polarization operator* D_{xy} on the symbolic algebra $A = Bc \otimes S$ by setting

$$D_{xy} = \partial_{xy} \otimes 1 + 1 \otimes d_{xy}.$$

Proposition 2. *The polarization operators D are superderivations with respect to the prejoin and premeet products.*

Definition 10. For every $x, y \in L^+$ we define the *divided powers* $d_{x,y}^{(p)}$, $p \in \mathbf{N}$, of the polarization $d_{x,y}$ of y in x as the linear operators of the supersymmetric Hopf algebra $S = \mathrm{Super}(L)$ into itself such that

- $d_{x,y}^{(p)}(z) = \begin{cases} x^{(p)} y^{(q-p)} & z = y^{(q)},\ q \geq p \\ 0 & \text{otherwise} \end{cases}$
- $d_{x,y}^{(0)} = I$
- the sequence $d_{x,y}^{(p)}$, $p \in \mathbf{N}$, satisfies the Leibniz rule, that is,

$$d_{x,y}^{(p)}(zt) = \sum_{q+r=p} (d_{x,y}^{(q)} z)(d_{x,y}^{(r)} t), \qquad z, t \in S,$$

or, equivalently,

$$d_{x,y}^{(p)} \circ \pi = \pi \circ \sum_{q+r=p} d_{x,y}^{(q)} \otimes d_{x,y}^{(r)}.$$

Proposition 3. *For every $x, y \in L^+$, the sequence of divided powers $d_{x,y}^{(p)}$ of the polarization $d_{x,y}$ of y in x satisfies the Leibniz rule with respect to the coproduct of the supersymmetric Hopf algebra $S = \mathrm{Super}(L)$, that is,*

$$\Delta(d_{x,y}^{(p)}(z)) = \sum_{(z);q+r=p} d_{x,y}^{(q)} z_{(1)} \otimes d_{x,y}^{(r)} z_{(2)}, \quad z \in S,$$

or, equivalently,

$$\Delta \circ d_{x,y}^{(p)} = (\sum_{q+r=p} d_{x,y}^{(q)} \otimes d_{x,y}^{(r)}) \circ \Delta.$$

Definition 11. For every $x, y \in L^+$ we define the *divided powers* $\partial_{x,y}^{(p)}$, $p \in \mathbf{N}$, of the polarization $\partial_{x,y}$ of y in x as the linear operators of the supersymmetric algebra $Bc_n(L)$ of braces into itself such that

- $\partial_{x,y}^{(p)}[z] = [D_{x,y}^{(p)} z]$
- the sequence $\partial_{x,y}^{(p)}$, $p \in \mathbf{N}$, satisfies the Leibniz rule, that is

$$\partial_{x,y}^{(p)}(ZT) = \sum_{q+r=p} (\partial_{x,y}^{(q)} Z)(\partial_{x,y}^{(r)} T), \quad Z, T \in Bc.$$

Definition 12. For every $x, y \in L^+$ and for every $p \in \mathbf{N}$, we define the *operators* $D_{xy}^{(p)}$ on the symbolic algebra $A = Bc \otimes S$ by setting

$$D_{xy}^{(p)} = \sum_{i+j=p} \partial_{xy}^{(i)} \otimes d_{xy}^{(j)}.$$

Equivalently, $D_{xy}^{(p)}$ can be defined as the pth divided power of the polarization operator D_{xy}.

Proposition 4. *The divided powers $D^{(p)}$ satisfy the Leibniz rule with respect to the prejoin and premeet products.*

We set

$$\begin{aligned}
d_{x^{(p)},y} &= d_{x,y}^{(p)}, &\quad x, y \in L^+, \ p \in \mathbf{N},\\
\partial_{x^{(p)},y} &= \partial_{x,y}^{(p)}, &\quad x, y \in L^+, \ p \in \mathbf{N},\\
D_{x^{(p)},y} &= D_{x,y}^{(p)}, &\quad x, y \in L^+, \ p \in \mathbf{N}.
\end{aligned}$$

Definition 13. For every element

$$m = \sum_i c_i x_{i1} x_{i2} \dots x_{ip_i}$$

of S, where $c_i \in K$ and $x_{ij} \in L_d$, homogeneous with respect to the length, and for every positive letter $a \in L^+$ which does not appear in m, we define the *polarization of a in m* on S as the operator

$$d_{m,a} = \sum_i c_i d_{x_{i1},a} d_{x_{i2},a} \dots d_{x_{ip_i},a}.$$

This definition is consistent because the operator $d_{m,a}$ does not depend on the given representation of m.

Proposition 5. *The polarization operators on S have the following properties:*

- $d_{m,a}(z) = \begin{cases} m a^{(q-l(m))} & z = a^{(q)},\ q \geq l(m) \\ 0 & \text{otherwise} \end{cases}$
- $d_{m_1 m_2,a} = d_{m_1,a} \circ d_{m_2,a}$
- $d_{m,a}(xy) = \sum_{(m)} (d_{m_{(1)},a} x)(d_{m_{(2)},a} y)(-1)^{m_{(2)} x}, \qquad x, y \in S.$

Definition 14. For every element

$$m = \sum_i c_i x_i$$

of S, where $c_i \in K$ and $x_i \in L$, and for every negative letter $a \in L^-$, we define the *polarization of a in m* on S as the operator

$$d_{m,a} = \sum_i c_i d_{x_i,a}.$$

Definition 15. For every element $m = \sum_i c_i x_{i1} x_{i2} \dots x_{ip_i}$ of S, where $c_i \in K$ and $x_{ij} \in L_d$, homogeneous with respect to the length, and for every positive letter $a \in L^+$ which does not appear in m, we define the *polarization of a in m* on Bc as the operator

$$\partial_{m,a} = \sum_i c_i \partial_{x_{i1},a} \partial_{x_{i2},a} \dots \partial_{x_{ip_i},a}.$$

Proposition 6. *The polarization operators on Bc have the following properties:*

- $\partial_{m,a}[z] = [D_{m,a} z]$
- $\partial_{xy,a} = \partial_{x,a} \circ \partial_{y,a}$
- $\partial_{m,a}(XY) = \sum_{(m)} (\partial_{m_{(1)},a} X)(\partial_{m_{(2)},a} Y)(-1)^{m_{(2)} X}, \qquad X, Y \in Bc.$

Definition 16. For every element $m = \sum_i c_i x_i$ of S, where $c_i \in K$ and $x_i \in L$, and for every negative letter $a \in L^-$, we define the *polarization of a in m* on Bc as the operator

$$\partial_{m,a} = \sum_i c_i \partial_{x_i,a}.$$

Proposition 7. *The mapping* [ ] : Super$(L) \to Bc_n(L)$ *is equivariant with respect to the polarization operators d and* ∂.

Definition 17. For every element

$$m = \sum_i C_i x_{i1} x_{i2} \dots x_{ip_i}$$

of A, where $C_i \in Bc$ and $x_{ij} \in L_d$, homogeneous with respect to the length, and for every positive letter $a \in L^+$ which does not appear in m, we define the *polarization of a in m* on Bc on A as the operator

$$D_{m,a} = \sum_i C_i D_{x_{i1},a} D_{x_{i2},a} \dots D_{x_{ip_i},a}.$$

This definition is consistent because the operator $D_{m,a}$ does not depend on the given representation of m.

Proposition 8. *The polarization operators on A have the following properties:*

- $D_{m,a}(z) = \begin{cases} m a^{(q-l(m))} & z = a^{(q)}, \quad q \geq l(m) \\ 0 & \text{otherwise} \end{cases}$
- $D_{m,a}[z] = [D_{m,a} z]$
- $D_{m_1 m_2,a} = D_{m_1,a} \circ D_{m_2,a}$
- $D_{m,a}(x \cdot y) = \sum_{(m)} (D_{m_{(1)},a} x) \cdot (D_{m_{(2)},a} y)(-1)^{m_{(2)} x}, \quad x, y \in A$
- $D_{m,a}(x * y) = \sum_{(m)} (D_{m_{(1)},a} x) * (D_{m_{(2)},a} y)(-1)^{m_{(2)} x}, \quad x, y \in A.$

Definition 18. For every element $m = \sum_i C_i x_i$ of A, where $C_i \in Bc$ and $x_i \in L$, and for every negative letter $a \in L^-$, we define the *polarization of* a *in* m on A as the operator

$$D_{m,a} = \sum_i C_i D_{x_i,a}.$$

Proposition 9. *The divided powers of the positive elements of* L^+ *span the symbolic algebra* $A_n(L)$ *under the action of the polarization operators and their divided powers.*

6 The letterplace superalgebra and the supersymmetric algebra of the brackets

In this section, we briefly summarize some basic definitions and results of [18].

A *signed set* is a linearly ordered set L together with a decomposition $L = L^+ \cup L^0 \cup L^-$ into a subset L^+ of positive elements, which we assume to be countable, a subset L^0 of neutral elements and a subset L^- of negative elements. If L is a signed set, and $\mathrm{Mon}(L)$ is the free monoid on L, then each element of $\mathrm{Mon}(L)$, i.e., each word $w = x_1 x_2 \ldots x_p$ with $x_i \in L$, has *signature* $|w| = 0$, resp., 1, if the number of occurrences of negative letters in w is even, resp., odd.

The *extended signed set* L_d corresponding to a given signed set $L = L^+ \cup L^0 \cup L^-$ is the set

$$L_d = (L^+ \times \mathbf{Z}^+)^+ \cup L^0 \cup L^-.$$

The pair (a, n), with $a \in L^+$ and $n \in \mathbf{N}$ is denoted by $a^{(n)}$ and is called the nth *divided power* of a. We set $a^{(1)} = a$. Sending $a^{(n)} \mapsto a^n$ for $a \in L^+$ and $a \mapsto a$ for $a \in L^-$ induces a morphism, called *display* and denoted Disp, from the free monoid $\mathrm{Mon}(L_d)$ to the free monoid $\mathrm{Mon}(L)$. We define the *length* and the *signature* of an element of $\mathrm{Mon}(L_d)$ as the length and the signature of its display in $\mathrm{Mon}(L)$.

We denote by $\mathrm{Tens}(L)$ the free associative algebra generated by the signed set L_d; in other words $\mathrm{Tens}(L)$ is the monoid algebra of the free monoid $\mathrm{Mon}(L_d)$. The *supersymmetric algebra* $\mathrm{Super}(L)$ is the quotient algebra of $\mathrm{Tens}(L)$ over the two-sided ideal I generated by the elements

$$uv = (-1)^{|v|\,|v|} vu, \qquad u, v \in \mathrm{Mon}(L_d)$$
$$a^{(i)} a^j = \binom{i+j}{i} a^{(i+j)}, \quad a \in L^+, i, j \in \mathbf{N}.$$

The concepts of length, and signature can be unambiguously defined for nonzero monomials in this quotient algebra. We set $\mathrm{Super}(L)_0 =$ the span of all the monomials containing an even number of negative letters and $\mathrm{Super}(L)_1 =$ the span of all the monomials containing an odd number of negative letters.

Let $L = L^+ \cup L^-$ and $P = P^+ \cup P^-$ be proper signed sets; the elements of L are called *letters*, the elements of P are called *places*, and the ordered pairs $(x|y)$, $x \in L$, $y \in P$, are called *letterplaces*. The *letterplace signed set* $(L|P)$ is defined by setting

$$(L|P)^+ = (L^+|P^+),\ (L|P)^0 = (L^-|P^-),\ (L|P)^- = (L^+|P^-) \cup (L^-|P^+).$$

Definition 19. The *letterplace superalgebra* $\mathrm{Super}(L|P)$ is the supersymmetric algebra over the signed set $(L|P)$.

For any two monomials $w \in \mathrm{Super}(L)$, $u \in \mathrm{Super}(P)$, the biproduct

$$(w|u) \in \mathrm{Super}(L|P)$$

may be defined by the following rules:

- $(w|u) = 0$ unless w and u have the same length
- $(1|1) = 1$
- the biproduct of a letter x and a place y is the corresponding letterplace $(x|y)$
- $(x^{(n)}|y^{(n)}) = (x|y)^{(n)}$ for $x \in L^+$, $y \in P^+$
- $(w|u'u'') = \sum_{(w)} (-1)^{|w_{(2)}||u'|} (w_{(1)}|u')(w_{(2)}|u'')$
- $(w'w''|u) = \sum_{(u)} (-1)^{|w''||u_{(1)}|} (w'|u_{(1)})(w''|u_{(2)})$.

We denote $(-1)^{\sum_{i>j}|w_i||u_j|}(w_1|u_1)(w_2|u_2)\dots(w_n|u_n)$ by

$$\begin{pmatrix} w_1 & | & u_1 \\ w_2 & | & u_2 \\ \dots & | & \dots \\ w_n & | & u_n \end{pmatrix},$$

and we call it a bi-tableau; we call the left part the letter-tableau, and the right part the place-tableau.

Proposition 10 (Straightening laws [18]). *For all monomials* $u, v, w \in \mathrm{Super}(L)$, $x, y \in \mathrm{Super}(P)$, *we have*

$$\sum_{(v)} \begin{pmatrix} u & v_{(1)} & | & x \\ v_{(2)} & w & | & y \end{pmatrix} = (-1)^{uv} \sum_{(u)\ (y)} (-1)^{l(u_{(2)})} \begin{pmatrix} v & u_{(1)} & | & x & y_{(1)} \\ u_{(2)} & w & | & y_{(2)} & \end{pmatrix}.$$

We say that a bi-tableau is *superstandard* if, in the displays of the letter-tableau and of the place-tableau, each row and each column are weakly increasing and, further, each row is strictly increasing in the negative elements and each column is strictly increasing in the positive elements.

Theorem 1 (Standard basis theorem [18]). *The superstandard bi-tableau form a linear basis of* $\mathrm{Super}(L|P)$.

Henceforth we will consider the letterplace superalgebra $\mathrm{Super}(L|P)$, where $P = P^- = \{n < \ldots < 2 < 1\}$.

Definition 20. For each monomial $w \in \mathrm{Super}(L)$, we define the *bracket* $[w]$ as the biproduct

$$(w|n, \ldots, 2, 1).$$

Therefore, a bracket is 0 unless the corresponding word has length n.

For any two brackets, we have the commutation rule

$$[w_1][w_2] = (-1)^{(w_1+n)(w_2+n)}[w_2][w_1].$$

Definition 21. The *bracket algebra* $B_n(L)$ *of step* n *over* L is the subalgebra of $\mathrm{Super}(L|P)$ generated by the brackets.

Proposition 11. *For all monomials* $u, v, w \in \mathrm{Super}(L)$,

$$\sum_{(v)}(-1)^{v_{(2)}n}[uv_{(1)}][v_{(2)}w] = (-1)^{uv+l(u_{(2)})}\sum_{(u)}(-1)^{u_{(2)}n}[vu_{(1)}][u_{(2)}w].$$

We define an action of the brace algebra $Bc_n(L)$ over the vector space $B_n(L)$ of brackets by setting $[w_1][w_2]\ldots[w_p]\cdot[w] = [w_1][w_2]\ldots[w_p][w]$, where $[w_i]$ denotes a brace on the left-hand side and denotes the corresponding bracket on the right-hand side. With respect to this action the bracket algebra is regarded as an algebra over the brace algebra. The linear operator $[\ ] : Bc_n(L) \to B_n(L)$ is an algebra homomorphism.

7 Polarizations

Definition 22. For every $x, y \in L$ we define the *polarization* $D_{x,y}$ *of* y *in* x as the linear operator of the letterplace superalgebra $\mathrm{Super}(L|P)$ into itself such that

- $$\begin{cases} D_{x,y}(z|t) = \delta_{y,z}(x|t) \\ D_{x,y}(a|b)^{(p)} = \delta_{y,a}(x|b)(a|b)^{(p-1)}, \quad a \in L^+,\ b \in P^+,\ p \geq 1 \end{cases}$$
- $D_{x,y}$ is a left superderivation, homogeneous of grade $|x| + |y|$, that is,

$$D_{x,y}(ZT) = (D_{x,y}Z)T + (-1)^{(x+y)Z}Z(D_{x,y}T), \quad Z, T \in \mathrm{Super}(L|P).$$

Definition 23. For every $x, y \in L^+$ we define the *divided powers* $D_{x,y}^{(p)}$, $p \in \mathbf{N}$, of the polarization $D_{x,y}$ of y in x as the linear operators of the letterplace superalgebra $\mathrm{Super}(L|P)$ into itself such that

- $D^{(p)}_{x,y} Z = \begin{cases} (x|b)^{(p)}(y|b)^{(q-p)} & Z = (y|b)^{(q)},\ q \geq p \\ 0 & \text{otherwise} \end{cases}$
- $D^{(0)}_{x,y} = I$
- the sequence $D^{(p)}_{x,y}$, $p \in \mathbf{N}$, satisfies the Leibniz rule, that is,

$$D^{(p)}_{x,y}(ZT) = \sum_{q+r=p} (D^{(q)}_{x,y} Z)(D^{(r)}_{x,y} T), \qquad Z, T \in \mathrm{Super}(L|P).$$

Proposition 12. *Given a biproduct $(w|u)$ in* Super$(L|P)$, *we have*

$$D_{x,y}(w|u) = (D_{x,y} w|u), \quad D^{(p)}_{x,y}(w|u) = (D^{(p)}_{x,y} w|u).$$

We set

$$D_{x^{(p)},y} = D^{(p)}_{x,y}, \quad x, y \in L^+, \ p \in \mathbf{N}.$$

Definition 24. For every element

$$m = \sum_i C_i x_{i1} x_{i2} \dots x_{ip_i}$$

of A, where $C_i \in Bc$ and $x_{ij} \in L_d$, homogeneous with respect to the length, and for every positive letter $a \in L^+$ which does not appear in m, we define the *polarization of a in m* as the operator

$$D_{m,a} = \sum_i C_i D_{x_{i1},a} D_{x_{i2},a} \dots D_{x_{ip_i},a}.$$

This definition is consistent because the operator $D_{m,a}$ does not depend on the given representation of m.

Proposition 13. *The polarization operators have the following properties:*

- $D_{m_1 m_2,a} = D_{m_1,a} \circ D_{m_2,a}$
- $D_{m,a}(XY) = \sum_{(m)} (D_{m_{(1)},a} X)(D_{m_{(2)},a} Y)(-1)^{m_{(2)} X}, \qquad X, Y \in \mathrm{Super}(L|P).$

Definition 25. For every element $m = \sum_i C_i x_i$ of A, where $C_i \in Bc$ and $x_i \in L$, and for every negative letter $a \in L^-$, we define the *polarization of a in m* as the operator

$$D_{m,a} = \sum_i C_i D_{x_i,a}.$$

8 The Peano map

In the sequel, we denote by $Bc = Bc_n(L)$ the supersymmetric algebra of the braces of step n over the signed set L, and by $B = B_n(L^*)$ the supersymmetric algebra of the brackets of step n over the signed set L^* obtained by adjoining to L a new positive letter ω. We denote braces and brackets by the same symbols; when there is danger of misunderstanding, the meaning of the symbol will be explicitly specified. At times we regard the algebra B as a Bc-module.

Definition 26. We define a linear map $\varphi : A = (Bc \otimes S) \to B$, which we call the *Peano map*, by setting

$$([u][v]\dots[w]x)\varphi = \begin{cases} [u][v]\dots[w][x\omega^{(n-p)}] & l(x) = p \leq n \\ 0 & l(x) > n \end{cases}$$

Here, $u, v, \dots, w, x$ are monomials in S; on the left-hand side $[u], \dots, [w]$ denote braces while, on the right-hand side, $[u], \dots, [x\omega^{(n-p)}]$ denote brackets.

The Peano map is homogeneous of grade equal to the parity of n and right Bc-linear, i.e.,

$$(Xy)\varphi = X(y)\varphi, \qquad (yX)\varphi = (y)\varphi X(-1)^{Xn}, \qquad X \in Bc,\ y \in A.$$

Thus the kernel of the Peano map is a graded subspace of the vector space A and a Bc-submodule of the Bc-module A. The elements of Ker φ can be written as linear combinations of elements of Ker φ of the form

$$\sum X_i y_i,$$

where the content and grade of $X_i y_i$ and the length of y_i are constant.

Proposition 14. *The Peano map is equivariant with respect to polarization operators, i.e., for every $x, y \in L$, we have*

$$(D_{xy}z)\varphi = D_{xy}((z)\varphi), \qquad z \in A.$$

The same statement holds for divided powers of polarization operators, and for polarization operators of the form $D_{m,a}$, with $m \in A$, $a \in L^+$.

Proposition 15. *The kernel of the Peano map is generated, as a Bc-submodule, by the elements of the following three kinds:*

$$u, \quad l(u) > n, \tag{1}$$

$$\sum_v (-1)^{v_2 n}[uv_{(1)}]v_{(2)}w - (-1)^{uv}\sum_u (-1)^{u_2 n + l(u_2)}[vu_{(1)}]u_{(2)}w, \tag{2}$$

$$\sum_v (-1)^{v_2 n}[uv_{(1)}][v_{(2)}w]y - (-1)^{uv}\sum_u (-1)^{u_2 n + l(u_2)}[vu_{(1)}][u_{(2)}w]y. \tag{3}$$

Here u, v, w, y are elements of S.

In particular, Ker φ contains the elements of the form

$$u * v - (-1)^{uv+l(u)+l(v)+n} v * u.$$

Proposition 16. *The kernel of the Peano map is generated, under the action of Bc and of the polarization operators, by the elements of the following three kinds:*

$$a^{(p)}, \quad p > n, \tag{4}$$

$$[a^{(p)}b^{(n-p)}]b^{(p+q-n)}c^{(r)} - (-1)^{p+q-n}[b^{(q)}a^{(n-q)}]a^{(p+q-n)}c^{(r)}, \tag{5}$$

$$\begin{aligned}&[a^{(p)}b^{(n-p)}][b^{(p+q-n)}c^{(2n-p-q)}]d^{(s)}\\ &\quad - (-1)^{p+q-n}[b^{(q)}a^{(n-q)}][a^{(p+q-n)}c^{(2n-p-q)}]d^{(s)}.\end{aligned} \tag{6}$$

Here a, b, c, d are elements of L^+.

Proposition 17. *The kernel of the Peano map is linearly generated by the elements obtained by polarization of monomials in A in which some positive letter appears with multiplicity greater than n.*

Proof (Sketch). We note that, if m is any monomial in A in which some positive letter appears with multiplicity greater than n, and D is any polarization operator, then

$$(D(m))\varphi = D((m)\varphi) = 0,$$

since $(m)\varphi = 0$ by the standard basis theorem (see Theorem 1).

We prove that the elements of the form (5), in which $r = 0$, can be obtained as linear combinations of polarizations of monomials in which some positive letter appears with multiplicity greater than n.

We set

$$E_{p,q} = [a^{(p)}b^{(n-p)}]b^{(p+q-n)} - (-1)^{(p+q-n)}[b^{(q)}a^{(n-q)}]a^{(p+q-n)},$$

$$F_s = [x^{(n)}]x^{(s-n)}, \quad x \neq a, b.$$

We show the result by induction on $p + q > n$.

If $p + q = n + 1$, then

$$E_{p,q} = [a^{(p)}b^{(n-p)}]b + [b^{(q)}a^{(n-q)}]a = D_{a^{(p)}b^{(q)},x} F_{n+1}.$$

We note that

$$\begin{aligned}\sum_{i=n-q}^{p} E_{i,q}a^{(p-i)} &= \sum_{i=n-q}^{p} [a^{(i)}b^{(n-i)}]b^{(q-n+i)}a^{(p-i)} \\ &\quad - \sum_{i=n-q}^{p} (-1)^{(q-n+i)}[b^{(q)}a^{(n-q)}]a^{(q-n+i)}a^{(p-i)} \\ &= D_{a^{(p)}b^{(q)},x}[x^{(n)}]x^{(p+q-n)} \\ &\quad - \sum_{i=n-q}^{p} (-1)^{(q-n+i)} \binom{p+q-n}{i+q-n} [b^{(q)}a^{(n-q)}]a^{(p+q-n)} \\ &= D_{a^{(p)}b^{(q)},x}F_{p+q} - 0.\end{aligned}$$

The claim follows by the induction hypothesis. □

Proposition 18. *The kernel of the Peano map is a two-sided ideal of the symbolic algebra, with respect to the prejoin product and with respect to the premeet product.*

The assertion follows from the properties of the polarization operators and the previous proposition.

We remark that the kernel of the Peano map is invariant with respect to the operator $[\,]: Bc \otimes S \to Bc \to Bc \otimes S$, i.e., we have

$$[\operatorname{Ker}\varphi] \subseteq \operatorname{Ker}\varphi.$$

9 The White double algebra of a signed set

Definition 27. The *White double algebra* $W = W_n(L)$ *of step* n over the signed set L is the structure

$$W = [W;\ \vee,\ \wedge],$$

where:

- W is the quotient of the symbolic algebra A over the kernel $\operatorname{Ker}\varphi$ of the Peano map
- the product $\vee$, called the *join*, is induced on W by the prejoin product $\cdot$ on A
- the product $\wedge$, called the *meet*, is induced on W by the premeet product $*$ on A.

Clearly, the Peano map $\varphi : A_n(L) \to B_n(L^*)$ induces a monomorphism

$$\overline{\varphi} : W_n(L) \to B_n(L^*).$$

The join product on W is associative. Under the canonical projection $A \to W$, the supersymmetric algebra Bc of the braces of step n is mapped onto a subalgebra of W which is isomorphic to the supersymmetric algebra B of the brackets of step n; by abuse of notation, we denote this subalgebra by B. Thus we can regard W as a B-bimodule.

Definition 28. We define the mapping [ ] : $W \to B$ as the mapping induced by the mapping [ ] : $A \to Bc$.

This mapping is homogeneous of grade the parity of n and right B-linear, i.e., we have the identities

$$[Xy] = X[y], \quad [yX] = [y]X(-1)^{Xn}, \qquad X \in B, \ y \in W.$$

Definition 29. For every $x, y \in L$ we define the *polarization operator* D_{xy} and, in the case of equally signed letters, the *divided powers* $D_{xy}^{(n)}$ *polarization operators*, on the White algebra W as the operator induced by the polarization operator D_{xy} and as the operators induced by the divided powers $D_{xy}^{(n)}$ polarization operators on the symbolic algebra A. In the same way we define on W the *polarization operators* of the form $D_{m,a}$, with $m \in W$, $a \in L^+$ not appearing in m.

We have to prove that this last definition is well-posed.

Proposition 19. *Let m, m' be elements of A whose images in the White module are the same. Then the operators $D_{m,a}$ and $D_{m',a}$ coincide, as endomorphisms of the White double algebra.*

Proof (Sketch). We consider "monomial" expressions X in the White module. If the occurrence of the symbol a in X exceeds n, then $X = 0$ in the White module. Otherwise, the monomial X can be rewritten (by applying the straightening laws) as a polynomial such that in every term the divided power $a^{(p)}$ (p equals the occurrence of the variable a) appears within a bracket. By the properties of the unary operation [ ] on W, the assertion follows. □

Proposition 20. *The polarization operators on W are superderivations with respect to the meet and join products. The divided powers of polarizations on W satisfy the Leibniz rule with respect to the meet and join products. Moreover, the polarization operators $D_{m,a}$ have the following properties:*

- $D_{m,a}(x \vee y) = \sum_{(m)} (D_{m_{(1)},a}x) \vee (D_{m_{(2)},a}y)(-1)^{m_{(2)}x}, \qquad x, y \in W$
- $D_{m,a}(x \wedge y) = \sum_{(m)} (D_{m_{(1)},a}x) \wedge (D_{m_{(2)},a}y)(-1)^{m_{(2)}x}, \qquad x, y \in W.$

Theorem 2. *Any polarization operator maps a monomial in meet and join into a linear combination of monomials of the same structure in meet and join.*

Proposition 21. *The meet product on W can be computed in two radically different ways:*

$$(-1)^{l(x)+l(y)-n} \sum_{(x)} x_{(1)}[x_{(2)}y] = x \wedge y = \sum_{(y)} [xy_{(1)}]y_{(2)}(-1)^{ny_{(2)}}.$$

The meet product on W is homogeneous of grade the parity of n and right B-bilinear, that is, the following identities hold:

$$(Zx) \wedge y = Z(x \wedge y), \quad (xZ) \wedge y = x \wedge (Zy),$$
$$x \wedge (yZ) = (x \wedge y)Z(-1)^{Zn} \quad x, y \in W, \ Z \in B.$$

Proposition 22. *The meet product on W is signed-commutative, i.e.,*

$$x \wedge y = (-1)^{xy+l(x)+l(y)-n} y \wedge x.$$

Proposition 23. *The meet product on W is signed-associative, i.e.,*

$$(x \wedge y) \wedge z = (-1)^{l(x)+nz} x \wedge (y \wedge z).$$

Proof. Indeed, for every $a, b, c \in L^+$ and every $0 < p, q, r \leq n$, on the one hand we have

$$\begin{aligned}(a^{(p)} \wedge b^{(q)}) \wedge c^{(r)} &= ([a^{(p)}b^{(n-p)}]b^{(p+q-n)}) \wedge c^{(r)} \\ &= [a^{(p)}b^{(n-p)}]b^{(p+q-n)} \wedge c^{(r)} \\ &= [a^{(p)}b^{(n-p)}][b^{(p+q-n)}c^{(2n-p-q)}]c^{(p+q+r-2n)};\end{aligned}$$

on the other hand we have

$$\begin{aligned}a^{(p)} \wedge (b^{(q)} \wedge c^{(r)}) &= a^{(p)} \wedge ([b^{(q)}c^{(n-q)}]c^{(q+r-n)}) \\ &= [b^{(q)}c^{(n-q)}]a^{(p)} \wedge c^{(q+r-n)} \\ &= [b^{(q)}c^{(n-q)}][a^{(p)}c^{(n-p)}]c^{(p+q+r-2n)}.\end{aligned}$$

Thus

$$(a^{(p)} \wedge b^{(q)}) \wedge c^{(r)} = (-1)^p a^{(p)} \wedge (b^{(q)} \wedge c^{(r)}).$$

Now

$$\begin{aligned}(x \wedge y) \wedge z &= (-1)^{nz} D_{x,a} D_{y,b} D_{z,c}((a^{(p)} \wedge b^{(q)}) \wedge c^{(r)}) \\ &= (-1)^{nz} D_{x,a} D_{y,b} D_{z,c}((-1)^p a^{(p)} \wedge (b^{(q)} \wedge c^{(r)})) \\ &= (-1)^{l(x)+nz} x \wedge (y \wedge z). \qquad \square\end{aligned}$$

We omit parentheses in each expression where all the possible parenthesizations lead to the same result.

10 Six straight lines

For every $a, b, c \in L^+$, in the White algebra $W_n(L)$ we have

$$\begin{aligned}[a^{(2)} \vee (b^{(n-1)} \wedge c^{(n-1)})] &= a^{(2)} \wedge (b^{(n-1)} \wedge c^{(n-1)}) \\ &= (a^{(2)} \wedge b^{(n-1)}) \wedge c^{(n-1)} \\ &= -a[ab^{(n-1)}] \wedge c^{(n-1)} \\ &= -[ab^{(n-1)}][ac^{(n-1)}].\end{aligned}$$

For every x_1, x_2, homogeneous and of length 1, and for every y_1, y_2, homogeneous and of length $n-1$, in the White algebra $W_n(L)$ we have

$$\begin{aligned}&[(x_1 \vee x_2) \vee (y_1 \wedge y_2)]\\&\quad = (-1)^{1+y_2 n + x_1(x_2+y_1+n)}[x_2 y_1][x_1 y_2] + (-1)^{1+y_2 n + x_2(n+y_1)}[x_1 y_1][x_2 y_2].\end{aligned}$$

Indeed,

$$\begin{aligned}&[(x_1 \vee x_2) \vee (y_1 \wedge y_2)]\\&\quad = D_{x_1 \vee x_2, a} D_{y_1, b} D_{y_2, c}[a^{(2)} \vee (b^{(n-1)} \wedge c^{(n-1)})]\\&\quad = -D_{x_1 \vee x_2, a} D_{y_1, b} D_{y_2, c}[ab^{(n-1)}][ac^{(n-1)}]\\&\quad = -D_{x_1, a} D_{x_2, a} D_{y_1, b} D_{y_2, c}[ab^{(n-1)}][ac^{(n-1)}]\\&\quad = -(-1)^{y_2 n} D_{x_1, a} D_{x_2, a} D_{y_1, b}[ab^{(n-1)}][a y_2]\\&\quad = -(-1)^{y_2 n} D_{x_1, a} D_{x_2, a}[a y_1][a y_2]\\&\quad = -(-1)^{y_2 n} D_{x_1, a}\left([x_2 y_1][a y_2] + (-1)^{x_2(n+y_1)}[a y_1][x_2 y_2]\right)\\&\quad = -(-1)^{y_2 n}\left((-1)^{x_1(x_2+y_1+n)}[x_2 y_1][x_1 y_2] + (-1)^{x_2(n+y_1)}[x_1 y_1][x_2 y_2]\right).\end{aligned}$$

For every x_1, x_2, negative and of length 1, and for every y_1, y_2, positive and of length 2, in the White algebra $W_3(L)$ we have

$$[(x_1 \vee x_2) \vee (y_1 \wedge y_2)] = -[x_2 y_1][x_1 y_2] + [x_1 y_1][x_2 y_2].$$

For every x_1, x_2, negative and of length 1, and for every y_1, y_2, positive and of length 2, in the White algebra $W_3(L)$ we have

$$[(y_1 \wedge y_2) \vee (x_1 \vee x_2)] = -[y_1 x_2][y_2 x_1] + [y_1 x_1][y_2 x_2].$$

Proposition 24. *In the White algebra $W_3(L)$, for any six divided powers $a^{(2)}$, $b^{(2)}$, $c^{(2)}$, $d^{(2)}$, $e^{(2)}$, $f^{(2)}$, the "positive Pappus" identity holds:*

$$\begin{aligned}&(a^{(2)} \wedge b^{(2)}) \vee (c^{(2)} \wedge d^{(2)}) \vee (e^{(2)} \wedge f^{(2)})\\&\quad -(b^{(2)} \wedge c^{(2)}) \vee (d^{(2)} \wedge e^{(2)}) \vee (f^{(2)} \wedge a^{(2)})\\&\quad +(a^{(2)} \wedge d^{(2)}) \vee (c^{(2)} \wedge f^{(2)}) \vee (e^{(2)} \wedge b^{(2)}) = 0.\end{aligned}$$

Proposition 24 is the virtual formulation in the positive part of the White double algebra of the following geometric statement: if the vertices of a hexagon lie alternatively on two straight lines, then the opposite sides intersect in collinear points.

Proof. Indeed,

$$\begin{aligned}
& [(a^{(2)} \wedge b^{(2)}) \vee (c^{(2)} \wedge d^{(2)}) \vee (e^{(2)} \wedge f^{(2)})] \\
& - [(b^{(2)} \wedge c^{(2)}) \vee (d^{(2)} \wedge e^{(2)}) \vee (f^{(2)} \wedge a^{(2)})] \\
& + [(a^{(2)} \wedge d^{(2)}) \vee (c^{(2)} \wedge f^{(2)}) \vee (e^{(2)} \wedge b^{(2)})] \\
= & [(a^{(2)} \wedge b^{(2)}) \vee e^{(2)}][(c^{(2)} \wedge d^{(2)}) \vee f^{(2)}] \\
& - [(c^{(2)} \wedge d^{(2)}) \vee e^{(2)}][(a^{(2)} \wedge b^{(2)}) \vee f^{(2)}] \\
& - [b^{(2)} \vee (d^{(2)} \wedge e^{(2)})][c^{(2)} \vee (f^{(2)} \wedge a^{(2)})] \\
& + [c^{(2)} \vee (d^{(2)} \wedge e^{(2)})][b^{(2)} \vee (f^{(2)} \wedge a^{(2)})] \\
& + [a^{(2)} \vee (c^{(2)} \wedge f^{(2)})][d^{(2)} \vee (e^{(2)} \wedge b^{(2)})] \\
& - [d^{(2)} \vee (c^{(2)} \wedge f^{(2)})][a^{(2)} \vee (e^{(2)} \wedge b^{(2)})] \\
= & (a^{(2)} \wedge b^{(2)} \wedge e^{(2)})(c^{(2)} \wedge d^{(2)} \wedge f^{(2)}) \\
& - (c^{(2)} \wedge d^{(2)} \wedge e^{(2)})(a^{(2)} \wedge b^{(2)} \wedge f^{(2)}) \\
& - (b^{(2)} \wedge d^{(2)} \wedge e^{(2)})(c^{(2)} \wedge f^{(2)} \wedge a^{(2)}) \\
& + (c^{(2)} \wedge d^{(2)} \wedge e^{(2)})(b^{(2)} \wedge f^{(2)} \wedge a^{(2)}) \\
& + (a^{(2)} \wedge c^{(2)} \wedge f^{(2)})(d^{(2)} \wedge e^{(2)} \wedge b^{(2)}) \\
& - (d^{(2)} \wedge c^{(2)} \wedge f^{(2)})(a^{(2)} \wedge e^{(2)} \wedge b^{(2)}) = 0. \qquad \square
\end{aligned}$$

Proposition 25. *In the White algebra $W_3(L)$, for any six positive letters a, b, c, d, e, f, the "positive dual Pappus" identity holds:*

$$ab \wedge cd \wedge ef + bc \wedge de \wedge fa + ad \wedge cf \wedge eb = 0.$$

Proof. Indeed, by direct computation we get

$$\begin{aligned}
& [abc][def] + [abd][cef] + [def][bca] \\
& \qquad + [dea][bcf] + [cfe][adb] + [cfb][ade] = 0,
\end{aligned}$$

since the brackets anticommute. □

Proposition 26. *In the White algebra $W_3(L^-)$, for any six negative letters r, s, t, u, v, w, the "negative dual Pappus" identity holds:*

$$rs \wedge tu \wedge vw - st \wedge uv \wedge wr + ru \wedge tw \wedge vs = 0.$$

Proof. Indeed, for any six positive letters a, b, c, d, e, f

$$\begin{aligned}
0 &= D_{ra} D_{sb} D_{tc} D_{ud} D_{ve} D_{wf}\, (ab \wedge cd \wedge ef + bc \wedge de \wedge fa + ad \wedge cf \wedge eb) \\
&= D_{ra} D_{sb} D_{tc} D_{ud} D_{ve} D_{wf}\, (ab \wedge cd \wedge ef) \\
&\quad - D_{sb} D_{tc} D_{ud} D_{ve} D_{wf} D_{ra}\, (bc \wedge de \wedge fa) \\
&\quad + D_{ra} D_{ud} D_{tc} D_{wf} D_{ve} D_{sb}\, (ad \wedge cf \wedge eb) \\
&= rs \wedge tu \wedge vw - st \wedge uv \wedge wr + ru \wedge tw \wedge vs. \qquad \square
\end{aligned}$$

Proposition 27. *In the White algebra $W_3(L^-)$, for any six monomials A, B, C, D, E, F of step 2, the "negative Pappus' identity holds:*

$$\begin{aligned}&(A \wedge B) \vee (C \wedge D) \vee (E \wedge F)\\ &\quad - (B \wedge C) \vee (D \wedge E) \vee (F \wedge A)\\ &\quad + (A \wedge D) \vee (C \wedge F) \vee (E \wedge B) = 0.\end{aligned}$$

Proposition 28. *In the White algebra $W_3(L^-)$, for any six negative letters p, p', q, q', r, r', the "Desargues" identity holds:*

$$[(pq \wedge p'q') \vee (q'r' \wedge qr) \vee (rp \wedge r'p')] = [p'q'r'][pqr](qq' \wedge rr' \wedge p'p).$$

Proof. Indeed, by the "negative Pappus" identity, with

$$A = pq,\ B = p'q',\ C = q'r',\ D = qr,\ E = rp,\ F = r'p',$$

we get

$$\begin{aligned}&[(pq \wedge p'q') \vee (q'r' \wedge qr) \vee (rp \wedge r'p')]\\ &= [(p'q' \wedge q'r') \vee (qr \wedge rp) \vee (r'p' \wedge pq)]\\ &\quad -[(pq \wedge qr) \vee (q'r' \wedge r'p') \vee (rp \wedge p'q')]\\ &= [p'q'r'][qrp][q'r \vee (r'p' \wedge pq)]\\ &\quad -[pqr][q'r'p'][qr' \vee (rp \wedge p'q')]\\ &= [p'q'r'][qrp](q'r \wedge r'p' \wedge pq)\\ &\quad -[pqr][q'r'p'](qr' \wedge rp \wedge p'q')\\ &= [p'q'r'][pqr](q'r \wedge r'p' \wedge pq - qr' \wedge rp \wedge p'q')\\ &= [p'q'r'][pqr](qq' \wedge rr' \wedge p'p).\end{aligned}$$

In the last step we used the "negative dual Pappus" identity. □

Proposition 29. *In the White algebra $W_3(L^-)$, for any six negative letters a, b, c, a', b', c', the "Bricard" identity holds:*

$$\begin{aligned}&[cab](a' \vee (bc \wedge b'c')) \wedge (b' \vee (ca \wedge c'a')) \wedge (c' \vee (ab \wedge a'b'))\\ &= [a'b'c']^2 \big((aa' \wedge bc) \vee (bb' \wedge ca)\big) \wedge (cc' \wedge ab).\end{aligned}$$

Proof. On the one hand, for any three positive letters $\delta, \epsilon, \varphi$ we have

$$\begin{aligned}&(a' \vee (bc \wedge b'c')) \wedge (b' \vee (ca \wedge c'a')) \wedge (c' \vee (ab \wedge a'b'))\\ &= D_{b'c',\delta} D_{c'a',\epsilon} D_{a'b',\varphi}\\ &\quad \Big((a' \vee (bc \wedge \delta^{(2)})) \wedge (b' \vee (ca \wedge \epsilon^{(2)})) \wedge (c' \vee (ab \wedge \varphi^{(2)}))\Big)\\ &= D_{b'c',\delta} D_{c'a',\epsilon} D_{a'b',\varphi} \big((a' \vee [bc\delta]\delta) \wedge (b' \vee [ca\epsilon]\epsilon) \wedge (c' \vee [ab\varphi]\varphi)\big)\\ &= D_{b'c',\delta} D_{c'a',\epsilon} D_{a'b',\varphi} \big([bc\delta][ca\epsilon][ab\varphi](a'\delta \wedge b'\epsilon \wedge c'\varphi)\big)\end{aligned}$$

$$= D_{b'c',\delta} D_{c'a',\epsilon} D_{a'b',\varphi} \left([bc\delta][ca\epsilon][ab\varphi](a'\delta \wedge [b'\epsilon c']\varphi - a'\delta \wedge [b'\epsilon\varphi]c')\right)$$
$$= D_{b'c',\delta} D_{c'a',\epsilon} D_{a'b',\varphi}$$
$$\left(-[bc\delta][ca\epsilon][ab\varphi][b'\epsilon c'][a'\delta\varphi] - [bc\delta][ca\epsilon][ab\varphi][b'\epsilon\varphi][a'\delta c']\right)$$
$$= [bcb'][cac'][aba'][b'a'c'][a'c'b'] + [bcc'][caa'][abb'][b'c'a'][a'b'c']$$
$$= [a'b'c']^2([bcb'][cac'][aba'] + [bcc'][caa'][abb']).$$

On the other hand, for any six positive letters α, β, γ, δ, ϵ, φ we have

$$\left((aa' \wedge bc) \vee (bb' \wedge ca)\right) \wedge (cc' \wedge ab)$$
$$= D_{aa',\alpha} D_{bc,\beta} D_{bb',\gamma} D_{ca,\delta} D_{cc',\epsilon} D_{ab,\varphi}$$
$$\left(((\alpha^{(2)} \wedge \beta^{(2)}) \vee (\gamma^{(2)} \wedge \delta^{(2)})) \wedge (\epsilon^{(2)} \wedge \varphi^{(2)})\right)$$
$$= D_{aa',\alpha} D_{bc,\beta} D_{bb',\gamma} D_{ca,\delta} D_{cc',\epsilon} D_{ab,\varphi}$$
$$\left((\alpha^{(2)} \wedge \beta^{(2)} \wedge \epsilon^{(2)})(\gamma^{(2)} \wedge \delta^{(2)} \wedge \varphi^{(2)})\right.$$
$$\left. -(\gamma^{(2)} \wedge \delta^{(2)} \wedge \epsilon^{(2)})(\alpha^{(2)} \wedge \beta^{(2)} \wedge \varphi^{2})\right)$$
$$= D_{aa',\alpha} D_{bc,\beta} D_{bb',\gamma} D_{ca,\delta} D_{cc',\epsilon} D_{ab,\varphi}$$
$$\left([\alpha^{(2)}\beta][\beta\epsilon^{(2)}][\gamma^{(2)}\delta][\delta\varphi^{(2)}] - [\gamma^{(2)}\delta][\delta\epsilon^{(2)}][\alpha^{(2)}\beta][\beta\varphi^{(2)}]\right)$$
$$= [aa'c][bcc'][bb'a][cab] - [bb'c][acc'][aa'b][cab]$$
$$= [cab]\left([aa'c][bcc'][bb'a] - [bb'c][acc'][aa'b]\right). \qquad \square$$

Proof (Alternative). On the one hand, by the "dual Pappus" identity we get

$$(a' \vee (bc \wedge b'c')) \wedge (b' \vee (ca \wedge c'a')) \wedge (c' \vee (ab \wedge a'b'))$$
$$= ((bc \wedge b'c') \vee b') \wedge ((ca \wedge c'a') \vee c') \wedge ((ab \wedge a'b') \vee a')$$
$$-(a' \vee (ca \wedge c'a')) \wedge (b' \vee (ab \wedge a'b')) \wedge (c' \vee (bc \wedge b'c'))$$
$$= -[bcb']c'b' \wedge [cac']a'c' \wedge [aba']b'a' - a'c'[caa'] \wedge b'a'[abb'] \wedge c'b'[bcc']$$
$$= -(c'b' \wedge a'c' \wedge b'a')([bcb'][cac'][aba'] + [caa'][abb'][bcc'])$$
$$= [c'b'a']^2([bcb'][cac'][aba'] + [caa'][abb'][bcc']).$$

On the other hand, by the "Pappus" identity we get

$$[(aa' \wedge bc) \vee (bb' \wedge ca) \vee (cc' \wedge ab)]$$
$$= [(bc \wedge bb') \vee (ca \wedge cc') \vee (ab \wedge aa')]$$
$$-[(aa' \wedge ca) \vee (bb' \wedge ab) \vee (cc' \wedge bc)]$$
$$= [[bcb']b \vee [cac']c \vee [aba']a] + [[aa'c]a \vee [bb'a]b \vee [cc'b]c]$$
$$= [bca]([bcb'][cac'][aba'] + [aa'c][bb'a][cc'b]). \qquad \square$$

11 Alternative laws

In this section, we consider the so-called "alternative laws" in the Cayley–Grassmann algebra [16, 3]. We exhibit a version of each alternative law in the positive setting, whose proof is essentially trivial. Under the complete superpolarization process, one gets all the forms of the alternative laws in the superalgebraic setting. Here, one can appreciate the replacement of Hopf algebra notation by the complete superpolarization process of divided powers of positive symbols. In the special case of superpolarization into negative variables, the alternative laws of [16, 3] are recovered.

Proposition 30. *For all $a_1, a_2, \dots a_r, b \in L^+$, in the White algebra we have*

$$a_r^{(p_r)} \wedge (\dots \wedge (a_2^{(p_2)} \wedge (a_1^{(p_1)} \wedge b^{(q)})) \dots)$$
$$= [a_1^{(p_1)} b^{(n-p_1)}][a_2^{(p_2)} b^{(n-p_2)}] \dots [a_r^{(p_r)} b^{(n-p_r)}] b^{(p_1+p_2+\dots+p_r+q-nr)}.$$

Proposition 31. *For all elements $A_1, A_2, \dots A_r, B$, homogeneous with respect to length and degree, in the White algebra, we have*

$$A_r \wedge (\dots \wedge (A_2 \wedge (A_1 \wedge B)) \dots)$$
$$= \sum_{(B)} (-1)^s [A_1 B_{(1)}][A_2 B_{(2)}] \dots [A_r B_{(r)}] B_{(r+1)},$$

where $s = n \sum_{i \le r+1} (i-1) B_{(i)} + \sum_{i \le r} A_i \sum_{j<i} (A_j + B_{(j)} + n)$.

Proposition 32. *For all elements $A_1, A_2, \dots, A_r$ of step $n-1$ and all letters $b_1, \dots, b_r$ in the White algebra $W_n(L^-)$, we have*

$$A_r \wedge (\dots \wedge (A_2 \wedge (A_1 \wedge (b_1 \vee \dots \vee b_r))) \dots)$$
$$= (-1)^{n\binom{r+1}{2}} \sum_{\sigma \in S_r} (-1)^{|\sigma|} [A_1 b_{\sigma 1}][A_2 b_{\sigma 2}] \dots [A_r b_{\sigma r}].$$

Proposition 33. *For every $a, x, b, c, y \in L^+$, in the White algebra we have*

$$\begin{aligned} &((a^{(k)} \vee x^{(n-k-h)}) \wedge b^{(h+p)}) \wedge (c^{(q)} \vee y^{(n-p-q)}) \\ = \ & [a^{(k)} x^{(n-k-h)} b^{(h)}][b^{(p)} c^{(q)} y^{(n-p-q)}] \\ = \ & ((a^{(k)} \vee b^{(h)}) \wedge x^{(n-k-h)}) \vee ((b^{(p)} \vee c^{(q)}) \wedge y^{(n-p-q)}). \end{aligned}$$

Proposition 34. *For any homogeneous elements A of length k, X of length $n-k-h$, B of length $h+p$, C of length q and Y of length $n-p-q$, in the White algebra we have*

$$((A \vee X) \wedge B) \wedge (C \vee Y)$$
$$= (-1)^{(p+n)(C+Y)} \sum_{(B)} (-1)^{nB_{(2)} + B_{(1)} X} ((A \vee B_{(1)}) \wedge X) \vee ((B_{(2)} \vee C) \wedge Y).$$

Proof. Indeed, for any $a, x, b, c, y \in L^+$

$$\begin{aligned} &((A \vee X) \wedge B) \wedge (C \vee Y) \\ = \ & ((D_{Aa} a^{(k)} \vee D_{Xx} x^{(n-k-h)}) \wedge D_{Bb} b^{(h+p)}) \wedge (D_{Cc} c^{(q)} \vee D_{Yy} y^{(n-p-q)}) \end{aligned}$$

$$= (-1)^{p(C+Y)} D_{Aa} D_{Xx} D_{Bb} D_{Cc} D_{Yy}((a^{(k)} \vee x^{(n-k-h)}) \\ \wedge b^{(h+p)}) \wedge (c^{(q)} \vee y^{(n-p-q)})$$

$$= (-1)^{p(C+Y)} D_{Aa} D_{Xx} D_{Bb} D_{Cc} D_{Yy}((a^{(k)} \vee b^{(h)}) \wedge x^{(n-k-h)}) \\ \vee ((b^{(p)} \vee c^{(q)}) \wedge y^{(n-p-q)})$$

$$= (-1)^{p(C+Y)} \sum_{(B)} (-1)^{n(Y+C+B_{(2)})+B_{(1)}X}((A \vee B_{(1)}) \\ \wedge X) \vee ((B_{(2)} \vee C) \wedge Y)$$

$$= (-1)^{(p+n)(C+Y)} \sum_{(B)} (-1)^{nB_{(2)}+B_{(1)}X}((A \vee B_{(1)}) \\ \wedge X) \vee ((B_{(2)} \vee C) \wedge Y). \qquad \square$$

Proposition 35. *For all $c_1, c_2, \dots c_r, b, a \in L^+$, in the White algebra we have*

$$c_r^{(p_r)} \wedge \left(\dots \wedge \left(c_1^{(p_1)} \wedge (b^{(k)} \vee a^{(s)})\right) \dots\right) \\ = \sum \left((c_r^{(p_r)} \vee b^{(k_1)}) \wedge \left(\dots \wedge \left((c_1^{(p_1)} \vee b^{(k_r)}) \wedge a^{(s)}\right) \dots\right)\right) \vee b^{(k_{r+1})},$$

where $(k_1, \dots, k_r, k_{r+1})$ ranges over all compositions of k subject to the conditions $p_i + k_{r-i+1} \leq n, \quad i = 1, 2, \dots, r$.

Proof. Indeed, we have

$$c_r^{(p_r)} \wedge \left(\dots \wedge \left(c_2^{(p_2)} \wedge \left(c_1^{(p_1)} \wedge b^{(k)} a^{(s)}\right)\right) \dots\right) \\ = \sum [c_1^{(p_1)} b^{(k_1)} a^{(s_1)}][c_2^{(p_2)} b^{(k_2)} a^{(s_2)}] \dots [c_r^{(p_r)} b^{(k_r)} a^{(s_r)}] b^{(k_{r+1})} a^{(s_{r+1})},$$

where $(k_1, \dots, k_r, k_{r+1})$ and $(s_1, \dots, s_r, s_{r+1})$ range over all compositions of k and s subject to the conditions $p_i + k_i + s_i = n$ for $i = 1, 2, \dots, r$,

$$= \sum \left(c_r^{(p_r)} b^{(k_r)} \wedge \left(\dots \wedge \left(c_2^{(p_2)} b^{(k_2)} \wedge \left(c_1^{(p_1)} b^{(k_1)} \wedge a^{(s)}\right)\right) \dots\right)\right) \vee b^{(k_{r+1})},$$

where $(k_1, \dots, k_r, k_{r+1})$ ranges over all compositions of k subject to the conditions $p_i + k_i \leq n$ for $i = 1, 2, \dots, r$. $\square$

Proposition 36. *For all elements $C_1, C_2, \dots C_r, B, A$, homogeneous with respect to length and degree, in the White algebra, we have*

$$C_r \wedge (\dots \wedge (C_1 \wedge (B \vee A)) \dots) \\ = \sum (-1)^s ((C_r \vee B_{(1)}) \wedge \left(\dots \wedge \left((C_1 \vee B_{(r)}) \wedge A\right) \dots\right)) \vee B_{(r+1)},$$

where the sum is over the r-fold diagonalization of B subject to the conditions $l(C_i)+l(B_{(r-i+1)}) \leq n$, $i = 1, 2, \dots, r$, and $s = (A+rn)B_{(r+1)} + \sum_{i+j \leq r} C_i B_{(j)}$.

12 The White double algebra of a Peano space

Let $(V, [\ ])$ be an n-dimensional Peano space, that is, $[\ ]$ is a non-degenerate alternating n-linear map from $V^{\oplus n}$ to the base field K. Let $E_n^- = (e_1, e_2, \dots, e_n)$ be a distinguished unimodular basis of V, that is, $[e_1, \dots, e_n] = 1$.

We consider the following signed linearly ordered sets:

- $L = L^+ \cup V^- \cup X^-$, where V^- is identified with the ground set of V
- $L^* = L^+ \cup \omega^+ \cup V^- \cup X^-$
- $\underline{L} = L^+ \cup E_n^- \cup X^-$
- $\underline{L}^* = L^+ \cup \omega^+ \cup E_n^- \cup X^-$.

On each of the vector spaces

$$\text{Super}(L),\ Bc_n(L),\ A_n(L),\ \text{Super}(L^*|P),\ B_n(L^*),\ W_n(L)$$

we impose linearity conditions on the symbols in V^-.

These conditions turn out to be congruences with respect to each operation on each algebra.

We denote by

$$\underline{\text{Super}}(L),\ \underline{Bc}_n(L),\ \underline{A}_n(L),\ \underline{\text{Super}}(L^*|P),\ \underline{B}_n(L^*),\ \underline{W}_n(L),$$

the corresponding quotient algebras.

They turn out to be isomorphic, respectively, to

$$\text{Super}(\underline{L}),\ Bc_n(\underline{L}),\ A_n(\underline{L}),\ \text{Super}(\underline{L}^*|P),\ B_n(\underline{L}^*),\ W_n(\underline{L}).$$

More precisely, the kernels of the linearity relations are:

- In $\text{Super}(L)$, the two-sided ideal I generated by the elements of the form
  $(ay +_V bz) - ay - bz$,
  where $y, z \in V^-, a, b \in K$.
- In $Bc_n(L)$, the two-sided ideal J generated by the elements of the form
  $[\dots(ay +_V bz)\dots] - a[\dots y \dots] - b[\dots z \dots]$,
  where $y, z \in V^-,\ a, b \in K$.
- In $A_n(L)$, the two-sided prejoin-ideal H generated by the elements of the form
  $[\dots(ax +_V by)\dots] - a[\dots x \dots] - b[\dots y \dots]$,
  $(ax +_V by) - ax - by$,
  where $x, y \in V^-,\ a, b \in K$. Note that $H = J * S + Bc * I$.
- In $\text{Super}(L^*|P^-)$, the two-sided ideal $(I|P)$ generated by the elements of the form
  $((ay +_V bz)|i) - a(y|i) - b(z|i)$,
  where $y, z \in V^-, a, b \in K, i = 1, \dots, n$.
- In $B_n(L^*)$, the two-sided ideal generated by the elements of the form $[\dots(ay +_V bz)\dots] - a[\dots y \dots] - b[\dots z \dots]$,
  where $y, z \in V^-,\ a, b \in K$.

- In $W_n(L)$, the two-sided join-ideal H generated by the elements of the form
  $[\dots(ax +_V by)\dots] - a[\dots x\dots] - b[\dots y\dots]$,
  $(ax +_V by) - ax - by$,
  where $x, y \in V^-$, $a, b \in K$. Note that $H = J \vee S + B \vee I$.

Definition 30. We call $[\underline{W}_n(L), \vee, \wedge]$ the *White double algebra* of the Peano space $(V, [\;])$.

Given two natural numbers r, l, with $l \leq n$, we denote by $\underline{W}_n^{rl}(L)$ the linear subspace of $\underline{W}_n(L)$ spanned by the linear combinations of elements of the form $X_1 \dots X_r y$, where each X_i is a bracket on L and y is an element of length l in $\underline{Super}(L)$. When there is no danger of misunderstanding, we drop the index n and write simply $\underline{W}^{rl}(L)$.

Each linear mapping f previously defined between the given algebras induces a linear mapping $\underline{f}$ on the corresponding quotient algebras, and every relation between the given algebras gives rise to an analogous relation on the corresponding quotient algebras.

There is only one exception: the various polarization operators of the form $D_{x,y}$, with $y \in L^-$, give rise to polarization operators on the linearized structures only if $y \in X^- \cup E_n^-$.

In particular, we have:

$$\underline{A}_n(L) \cong \underline{Bc}_n(L) \otimes \underline{\text{Super}}(L)$$

$$\underline{W}_n(L) \cong \frac{\underline{A}_n(L)}{\text{Ker}\,\underline{\varphi}}.$$

We have the following list of linear isomorphisms:

- $\underline{\text{Super}}(V^-) \cong \text{Super}(E_n^-) \cong \Lambda(V)$
- $\underline{Bc}_n(V^-) \cong Bc_n(E_n^-) \cong K[e]$
- $\underline{A}_n(V^-) \cong A_n(E_n^-) \cong K[e] \otimes \Lambda(V)$
- $\underline{B}_n(V^-) \cong B_n(E_n^-) \cong K[e]$
- $\underline{W}_n(V^-) \cong W_n(E_n^-) \cong K[e] \otimes \Lambda(V)$,

where $e = [e_1 \dots e_n]$ and $K[e]$ is the polynomial ring on the variable e.

Sending $[e]^p e_{i_1} \dots e_{i_q} \mapsto e_{i_1} \dots e_{i_q}$ induces a linear mapping $F : \underline{W}_n(V^-) \to \Lambda(V)$ such that

$$F(x \vee y) = Fx \vee Fy, \quad F(x \wedge y) = (-1)^{p(p+q-n)} Fx \wedge Fy,$$

for all elements $x, y \in \underline{W}_n(V^-)$, homogeneous of lengths p, q, respectively, where the operations on the right are the classical join and meet.

The vector space $\underline{W}_n^{rl}(V^-)$ is naturally isomorphic to the lth homogeneous component $\Lambda^l(V)$ of the Grassmann algebra of V.

13 Invariants of skew-symmetric tensors

From now on, K will denote an infinite field, with char$(K) \neq 2$. Let $(V, [\ ])$ be an n-dimensional Peano space over K. The general linear group $GL(V)$ of V acts in a natural way on the exterior algebra $\Lambda(V)$ of V, on each component $\Lambda^m(V)$, on any space

$$T = \oplus_{i=0}^{p} \Lambda^{m_i}(V) \oplus V^{\oplus q}$$

which is the direct sum of copies of such homogeneous components, and on the algebra

$$K[T] = K[\oplus_{i=0}^{p} \Lambda^{m_i}(V) \oplus V^{\oplus q}]$$

of polynomial functions on such a space.

These actions are given, for every $\rho \in GL(V)$, by

$$\rho \cdot (t_0, t_1, \ldots, t_p; v_1, \ldots, v_q) = (\rho \cdot t_0, \rho \cdot t_1, \ldots, \rho \cdot t_p; \rho \cdot v_1, \ldots, \rho \cdot v_q)$$
$$(\rho \cdot F)(t_0, t_1, \ldots, t_p; v_1, \ldots, v_q) = F(\rho^{-1} \cdot (t_0, t_1, \ldots, t_p; v_1, \ldots, v_q)).$$

We say that a polynomial function F in $K[T]$ is a *(relative) covariant of weight* g when, for every $\rho \in GL(V)$, we have

$$\rho \cdot F = (\det \rho)^g F.$$

Given a distinguished basis $(e_1, e_2, \ldots, e_n)$ of V, we denote by $(\eta_1, \eta_2 \ldots, \eta_n)$ the corresponding coordinate functions on V, and denote by $\tau_{j_1, j_2, \ldots, j_m}$, $1 \leq j_1 < j_2 < \ldots < j_m \leq n$, the corresponding Plücker coordinate functions on $\Lambda^m(V)$. So, for every $t \in \Lambda^m(V)$, we have

$$t = \sum_{1 \leq j_1 < j_2 < \ldots < j_m \leq n} \tau_{j_1, j_2, \ldots, j_m}(t) e_{j_1} e_{j_2} \ldots e_{j_m}.$$

We identify the general linear group $GL(V)$ of V with the group $GL_n(K)$ of all $n \times n$ invertible matrices over K. If the action of $GL(V)$ on the basis elements of V is described by setting

$$\rho^{-1} \cdot e_j = \sum_{i=1}^{n} e_i a_{ij},$$

then the action of $GL(V)$ on the coordinate functions of V is described by

$$\rho \cdot \eta_i = \sum_{j=1}^{n} a_{ij} \eta_j.$$

We define an action of $GL_n(K)$ over the letterplace superalgebra $\text{Super}(L|P)$ by setting

$$\rho \cdot (x|i) = \sum_{j=1}^{n} a_{ij}(x|j).$$

We say that a polynomial C in $\text{Super}(L|P)$ is a *(relative) covariant of weight* g when, for every $\rho \in GL_n(K)$, we have

$$\rho \cdot C = (\det \rho)^g C.$$

Theorem 3. *The algebra of invariants in* $\text{Super}(L|P)$ *is the bracket algebra* $B_n(L)$.

Henceforth, for the sake of readibility, we will denote a polarization operator by $D(a, d)$ instead of $D_{a,b}$. To every letter $\xi \in L$ and every tensor

$$t = \sum_{1 \leq j_1 < j_2 < \ldots < j_m \leq n} t_{j_1, j_2, \ldots, j_m} e_{j_1} e_{j_2} \ldots e_{j_m} \in \Lambda^m(V),$$

we associate a polarization operator (note the order)

$$D(t, \xi) = \sum_{1 \leq j_1 < j_2 < \ldots < j_m \leq n} t_{j_1, j_2, \ldots, j_m} D(e_{j_m}, \xi) \ldots D(e_{j_2}, \xi) D(e_{j_1}, \xi).$$

14 The umbral operator u

We consider the algebras:

- $K[T']$, with $T' = \Lambda^m(V)$, of polynomial functions in the Plücker coordinates $\tau_{j_1, j_2, \ldots, j_m}$
- $\text{Super}(L'|P)$, where L' is a countable linearly ordered set of positive letters $\alpha_1 < \alpha_2 < \ldots < \alpha_i < \ldots$, and $P^- = \{n < \ldots < 2 < 1\}$.

We say that all the letters α_i *belong to* $\Lambda^m(V)$ and have *arity* m.

Definition 31. The *umbral operator*

$$U' : \text{Super}(L'|P) \to K[T']$$

is the linear operator satisfying the following conditions

- $U'((\alpha_1|j_{11}) \ldots (\alpha_1|j_{1r_1})(\alpha_2|j_{21}) \ldots (\alpha_2|j_{2r_2}) \ldots) =$
  $U'((\alpha_1|j_{11}) \ldots (\alpha_1|j_{1r_1}))U'((\alpha_2|j_{21}) \ldots (\alpha_2|j_{2r_2})) \ldots,$
- $U'((\alpha_i|j_1) \ldots (\alpha_i|j_m)) = \tau_{j_1, j_2, \ldots, j_m},$
- $U'((\alpha_i|j_1) \ldots (\alpha_i|j_r)) = 0, \ r \neq m,$

where the places coupled with the umbral letters α_i are written in increasing order.

For all natural numbers $m_1, \ldots, m_p, q$, we consider the algebras:

- $K[T'']$, with $T'' = \oplus_1^p \Lambda^{m_i}(V) \oplus V^{\oplus q}$, of polynomial functions in the Plücker coordinates $\tau^i_{j_1,j_2,\dots,j_{m_i}}$, and τ^i_j,
- $\mathrm{Super}(L''|P)$, where $L'' = L_p^+ \cup X_q^-$, $L_p^+ = \{\omega_1 < \dots < \omega_p\}$, $X^- = \{x_1 < \dots < x_q\}$, with $\omega_p < x_1$, and $P^- = \{n < \dots < 1\}$.

We say that each letter ω_i *belongs to* the corresponding copy $\Lambda^{m_i}(V)$ and has *arity* m_i and each letter x_i *belongs to* the corresponding i^{th} copy of V.

Definition 32. The *umbral operator*

$$U'' : \mathrm{Super}(L''|P) \to K[T'']$$

is the linear operator satisfying the following conditions:

- $U''((\omega_1|h_{11})\dots(\omega_1|h_{1s_1})\dots(\omega_p|h_{p1})\dots(\omega_p|h_{ps_p})(x_{i_1}|k_1)\dots(x_{i_l}|k_l)) =$
$U''((\omega_1|h_{11})\dots(\omega_1|h_{1s_1}))\dots U''((\omega_p|h_{p1})\dots(\omega_p|h_{ps_p}))$
$U''((x_{i_1}|k_1))\dots U''((x_{i_l}|k_l))$,
- $U''\left((\omega_i|j_1)\dots(\omega_i|j_{m_i})\right) = \tau^i_{j_1,j_2,\dots,j_{m_i}}$,
- $U''\left((\omega_i|j_1)\dots(\omega_i|j_s)\right) = 0,\ s \neq m_i$,
- $U''((x_i|j)) = \tau^i_j$,

where the places coupled with the umbral letters ω_i are written in increasing order.

The operator U'' is injective and its image consists of the polynomials which are multilinear in the Plücker coordinates $\tau^i_{j_1,j_2,\dots,j_{m_i}}$.

Definition 33. We define the *umbral operator*

$$\begin{aligned} U : \mathrm{Super}(L' \cup L''|P) &\cong \mathrm{Super}(L'|P) \otimes \mathrm{Super}(L''|P) \to K[T'] \otimes K[T''] \\ &\cong K[T] \end{aligned}$$

by setting

$$U = U' \otimes U''.$$

Theorem 4. *The umbral operators U', U'' and U are equivariant with respect to the actions of $GL_n(K)$.*

Hence the umbral images of bracket polynomials are covariants. When $\mathrm{char}(K) \neq 0$, the converse does not hold.

We denote by val′, val″ and val the evaluation morphisms

$$\begin{aligned} &\mathrm{val}'(t) : K[T'] \to K, \\ &\mathrm{val}''(t_1,\dots,t_p; v_1,\dots,v_q) : K[T''] \to K, \\ &\mathrm{val}(t; \underline{t}_1,\dots,t_p; v_1,\dots,v_q) : K[T' \oplus T''] \to K. \end{aligned}$$

For every $t \in \Lambda^m(V)$ we now define a linear map

$$\mathcal{E}v'(t) : \mathrm{Super}(L'|P) \to \underline{\mathrm{Super}}(V^-|P) \cong \mathrm{Super}(E_n^-|P),$$

where $E_n^- = (e_1, e_2, \dots, e_n)$ is the distinguished basis of the vector space V.

Definition 34. Let C be a polynomial in $\mathrm{Super}(L'|P)$, homogeneous with respect to content, such that in each monomial every letter either does not occur or occurs with multiplicity m; let $\alpha_{i_1} < \cdots < \alpha_{i_r}$ be the letters occurring in C; then we set

$$\mathcal{E}v'(\underline{t})C = D(\underline{t}, \alpha_{i_r}) \ldots D(\underline{t}, \alpha_{i_1})C.$$

If C is a polynomial in $\mathrm{Super}(L'|P)$, homogeneous with respect to the content, such that in each monomial there is a letter that occours with multiplicity $\neq m$, then we set

$$\mathcal{E}v'(t)C = 0.$$

Then we define a family of maps

$$\mathcal{E}v''(\underline{t}_1, \ldots, \underline{t}_p; \underline{v}_1, \ldots, \underline{v}_q) : \mathrm{Super}(L''|P) \to \mathrm{Super}(E_n^-|P),$$

where $E_n^- = (e_1, e_2, \ldots, e_n)$ is the distinguished basis of the vector space V.

Definition 35. Let C be a polynomial in $\mathrm{Super}(L''|P)$, homogeneous with respect to content, such that in each monomial every letter either does not occur or occurs with multiplicity equal to its arity; let $\omega_{i_1} < \cdots < \omega_{i_r} < x_{j_1} < \cdots < x_{j_s}$ be the letters occurring in C; then we set

$$\mathcal{E}v''(t_1, \ldots, t_p; v_1, \ldots, v_q)C = \mathrm{val}(v_1, \ldots, v_q)D(t_{i_r}, \omega_{i_r}) \ldots D(t_{i_1}, \omega_{i_1})C,$$

where

$$\mathrm{val}(v_1, \ldots, v_q) : \mathrm{Super}(E_n^- \cup X_q^-|P) \to \mathrm{Super}(E_n^-|P)$$

is the algebra morphism induced by

$$\mathrm{val}(v_1, \ldots, v_q) : (x_i|j) \mapsto (v_i|j).$$

If C is a polynomial in $\mathrm{Super}(L''|P)$, homogeneous with respect to the content, such that in each monomial there is a letter that occours with multiplicity different by its arity, then we set

$$\mathcal{E}v''(t_1, \ldots, t_p; v_1, \ldots, v_q)C = 0.$$

Definition 36. Finally, we define a family of maps

$$\mathcal{E}v(\underline{t}; \underline{t}_1, \ldots, \underline{t}_p; \underline{v}_1, \ldots, \underline{v}_q) : \mathrm{Super}(L' \cup L''|P) \to \mathrm{Super}(E_n^-|P),$$

by setting

$$\mathcal{E}v(t; t_1, \ldots, t_p; v_1, \ldots, v_q) = \pi \circ \left(\mathcal{E}v'(t) \otimes \mathcal{E}v''(t_1, \ldots, t_p; v_1, \ldots, v_q)\right),$$

where π is the product on $\mathrm{Super}(E_n^-)$.

Note that the maps $\mathcal{E}v', \mathcal{E}v''$ and $\mathcal{E}v$ transform tableau in $\mathrm{Super}(L'|P)$, $\mathrm{Super}(L''|P)$ and $\mathrm{Super}(L' \cup L''|P)$, respectively, into tableau in $\mathrm{Super}(E_n^-|P)$.

We define an algebra morphism

$$\sigma : \mathrm{Super}(E_n^-|P) \to K$$

by setting

$$\sigma(e_i|j) = \delta_{ij},$$

or, equivalently

$$\sigma : \underline{\mathrm{Super}}(V^-|P) \to K$$

by setting

$$\sigma(v|j) = < v, \eta_j >,$$

where $(\eta_1, \dots, \eta_n)$ is the dual basis of $(e_1, \dots, e_n)$.

Theorem 5. *For every t in $\Lambda^m(V)$ and every $(t_1, \dots, t_p; v_1, \dots, v_q)$ in $\oplus_1^p \Lambda^{m_i}(V) \oplus V^{\oplus q}$, the following diagram commutes:*

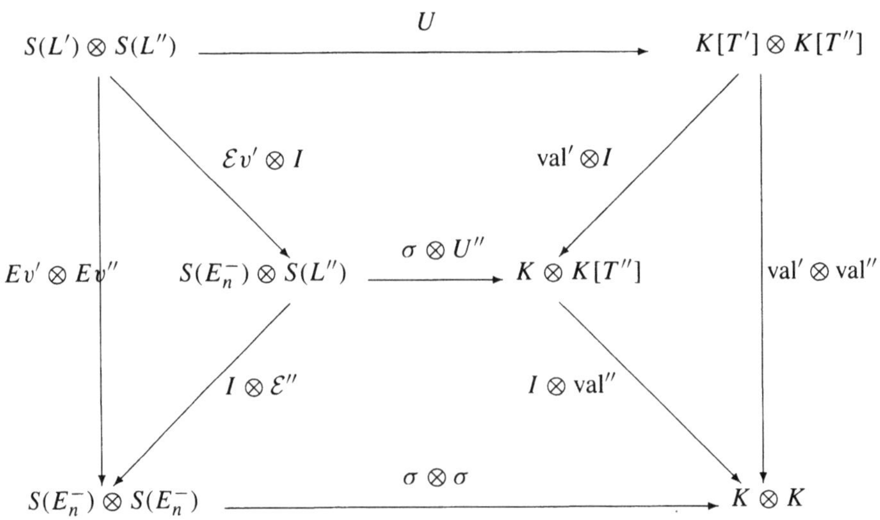

this implies,

$$\mathrm{val}(t; t_1, \dots, t_p; v_1, \dots, v_q) \circ U = \sigma \circ \mathcal{E}v(t; t_1, \dots, t_p; v_1, \dots, v_q).$$

Proof (sketch). Indeed, a straightforward computation shows that the linear operator on the left-hand side agrees with the one on the right-hand side on all the monomials in $\mathrm{Super}(L' \cup L'' P)$. One needs only to notice that

$$\mathrm{val}'(t)\,(U(\alpha_i|j_1)\dots(\alpha_i|j_m)) = \mathrm{val}'(t)\left(\tau_{j_1,\dots,j_m}\right) = t_{j_1\dots j_m},$$

and

$$\sigma\left(\mathcal{E}v'(t)(\alpha_i|j_1)\dots(\alpha_i|j_m)\right) = \sigma\left(D(t,\alpha_i)(\alpha_i^{(m)}|j_1\dots j_m)\right)$$
$$= \sigma\left(\sum t_{h_1\dots h_m} D(e_{h_m},\alpha_i)\dots D(e_{h_1},\alpha_i)(\alpha_i^{(m)}|j_1\dots j_m)\right)$$
$$= \sigma\left(\sum t_{h_1\dots h_m}(e_{h_m}\dots e_{h_1}|j_1\dots j_m)\right) = t_{j_1\dots j_m}. \qquad \square$$

15 The umbral operator u

We next consider the algebras

- $\Lambda^l[T' \oplus T'']$ of polynomial functions from the vector space $T = T' \oplus T''$ to the vector space $\Lambda^l(V)$, where $T' = \Lambda^m(V)$, $T'' = \oplus_1^p \Lambda^{m_i}(V) \oplus V^{\oplus q}$,
- $W^{rl}(L' \cup L'')$.

We denote by val$'$, val$''$ and val the evaluation morphisms

$$\mathrm{val}'(t) : \Lambda^l[T'] \to \Lambda^l,$$
$$\mathrm{val}''(t_1,\dots,t_p; v_1,\dots,\underline{v}_q) : \Lambda^l[T''] \to \Lambda^l,$$
$$\mathrm{val}(t; t_1,\dots,t_p; v_1,\dots,v_q) : \Lambda^l[T' \oplus T''] \to \Lambda^l.$$

For every $t \in \Lambda^m(V)$ we now define a linear map

$$\mathcal{E}v'(t) : W^{rl}(L' \cup L'') \to \underline{W}^{rl}(V^- \cup L'').$$

Definition 37. Let C be a polynomial in $W^{rl}(L' \cup L'')$, homogeneous with respect to content, such that in each monomial every letter of L' either does not occur or occurs with multiplicity m; let $\alpha_{i_1} < \dots < \alpha_{i_r}$ be the letters in L' occurring in C; then we set

$$\mathcal{E}v'(t)C = D(t,\alpha_{i_r})\dots D(t,\alpha_{i_1})C.$$

If C is a polynomial in $W^{rl}(L' \cup L'')$, homogeneous with respect to the content, such that in each monomial there is a letter of L' that occours with multiplicity $\neq m$, then we set

$$\mathcal{E}v'(t)C = 0.$$

Then we define a family of maps

$$\mathcal{E}v''(t_1,\dots,t_p; v_1,\dots,v_q) : \underline{W}^{rl}(V^- \cup L'') \to \underline{W}^{rl}(V^-).$$

Definition 38. Let C be a polynomial in $\underline{W}^{rl}(V^- \cup L'')$, homogeneous with respect to content, such that in each monomial every letter of L' either does not occur or occurs with multiplicity equal to its arity; let $\omega_{i_1} < \cdots < \omega_{i_r}, x_{j_1} < \cdots < x_{j_s}$ be the letters in L' occurring in C; then we set

$$\mathcal{E}v''(t_1, \dots, t_p; v_1, \dots, v_q)C = \mathrm{val}(v_1, \dots, v_q)D(t_{i_r}, \omega_{i_r}) \dots D(t_{i_1}, \omega_{i_1})C,$$

where

$$\mathrm{val}(v_1, \dots, v_q) : \underline{W}^{rl}(V^- \cup X_q^-) \to \underline{W}^{rl}(V^-)$$

is the algebra morphism induced by

$$\mathrm{val}(v_1, \dots, v_q) : x_i \mapsto v_i.$$

If C is a polynomial in $\underline{W}^{rl}(V^- \cup L'')$, homogeneous with respect to the content, such that in each monomial there is a letter of L'' that occours with multiplicity different by its arity, then we set

$$\mathcal{E}v''(t_1, \dots, t_p; v_1, \dots, v_q)C = 0.$$

Definition 39. Finally, we define a family of maps

$$\mathcal{E}v(t; t_1, \dots, t_p; v_1, \dots, v_q) : W^{rl}(L' \cup L'') \to \underline{W}^{rl}(V^-),$$

by setting

$$\mathcal{E}v(t; t_1, \dots, \underline{t}_p; \underline{v}_1, \dots, \underline{v}_q) = \mathcal{E}v''(t_1, \dots, t_p; v_1, \dots, v_q) \circ \mathcal{E}v'(t).$$

Definition 40. The *umbral operator*

$$u : W^{rl}(L' \cup L'') \to \Lambda^l[T' \oplus T'']$$

is the linear operator such that the following diagram commutes for every $(t; t_1, \dots, t_p; v_1, \dots, v_q) \in T' \oplus T''$:

$$\begin{array}{ccc}
W^{rl}(L' \cup L'') & \xrightarrow{\ u\ } & \Lambda^l[T' \oplus T''] \\
\mathcal{E}v \Big\downarrow & & \Big\downarrow \mathrm{val} \\
\underline{W}^{rl}(V^-) & \longrightarrow & \Lambda^l
\end{array}$$

In this diagram the unlabeled arrow denotes the natural isomorphism between $\underline{W}^{rl}(V^-)$ and Λ^l.

Definition 41. We define a function

$$\Phi : \Lambda^l \to K[\Lambda^{n-l}]$$

by setting, for every $t \in \Lambda^l$,

$$(\Phi t)(z) = [tz] \quad \forall z \in \Lambda^{n-l}.$$

We also define a function, which we denote, by abuse of notation, by the same symbol

$$\Phi : \Lambda^l[T' \oplus T''] \to K[T' \oplus T'' \oplus \Lambda^{n-l}]$$

by setting, for every $F \in \Lambda^l[T' \oplus T'']$,

$$(\Phi F)(t, z) = [F(t)z] \quad \forall t \in T' \oplus T'', \quad \forall z \in \Lambda^{n-l}.$$

Note the analogy between these maps and the map $\overline{\varphi} : W_n(L) \to B_n(L^*)$ induced by the Peano map (see Sect. 9).

Theorem 6. *The composite operators $\Phi \circ u$ and $U \circ \overline{\varphi}$ are equal as operators from $W^{rl}(L' \cup L'')$ to $K[T' \oplus T'' \oplus \Lambda^{n-l}]$.*

Furthermore, for every $C \in W^{rl}(L' \cup L'')$, the polynomial function $(\Phi \circ u)C$ is a covariant in $K[T' \oplus T'' \oplus \Lambda^{n-l}]$.

Proof. The first assertion follows from the fact that the following diagram commutes:

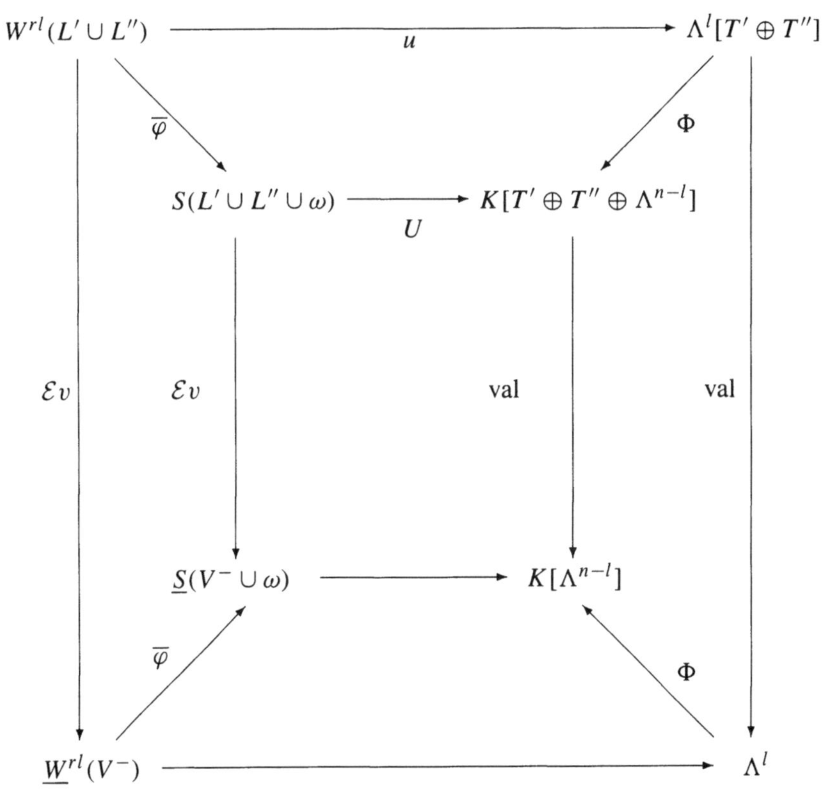

The second assertion follows from the fact that the umbral operator U is an equivariant operator and from the fact that the operator $\overline{\varphi}$ maps the elements of $W^{rl}(L' \cup L'')$ to bracket polynomials (i.e., invariants) in $S(L' \cup L'' \cup \omega)$. □

16 Generic vanishing and completeness theorems

Definition 42. We say that a polynomial function $F \in K[\Lambda^m(V) \oplus \oplus_1^p \Lambda^{m_i}(V) \oplus V^{\oplus q}]$ *vanishes "generically"* at $t \in \Lambda^m(V)$, and we write

$$F \sim_t 0,$$

when

$$F(t; t_1, \ldots, t_p; v_1, \ldots, v_q) = 0,$$

for every $(t_1, \ldots, t_p; v_1, \ldots, v_q)$ in $\oplus_1^p \Lambda^{m_i}(V) \oplus V^{\oplus q}$.

Note that the polynomial function F vanishes generically at t iff

$$(\mathrm{val}'(t) \otimes I)F = 0.$$

Proposition 37. *The following diagram commutes*

$$\begin{array}{ccccc}
S(L') \otimes S(L'') & \xrightarrow[U' \otimes I]{} & K[T'] \otimes S(L'') & \xrightarrow[I \otimes U'']{} & K[T'] \otimes K[T''] \\
\downarrow {\scriptstyle \mathcal{E}v' \otimes I} & & \downarrow {\scriptstyle \mathrm{val}' \otimes I} & & \downarrow {\scriptstyle \mathrm{val}' \otimes I} \\
S(E_n^-) \otimes S(L'') & \xrightarrow[\sigma \otimes I]{} & K \otimes S(L'') & \xrightarrow[I \otimes U'']{} & K \otimes K[T''] \\
\downarrow {\scriptstyle I \otimes \mathcal{E}v''} & & \downarrow {\scriptstyle I \otimes \mathcal{E}v''} & & \downarrow {\scriptstyle I \otimes \mathrm{val}''} \\
S(E_n^-) \otimes S(E_n^-) & \xrightarrow[\sigma \otimes I]{} & K \otimes S(E_n^-) & \xrightarrow[I \otimes \sigma]{} & K \otimes K
\end{array}$$

where $S(L') = \mathrm{Super}(L'|P)$, $S(L'') = \mathrm{Super}(L''|P)$ *and* $S(E_n^-) = \mathrm{Super}(E_n^-|P)$.

Definition 43. We say that $C \in \text{Super}(L' \cup L''|P)$ *vanishes generically* at $\underline{t} \in \Lambda^m(V)$, and we write

$$C \sim_{\underline{t}} 0,$$

when

$$(\mathcal{E}v'(\underline{t}) \otimes I)C = 0.$$

Theorem 7 (Completeness theorem). *For every bracket polynomial* $C \in \text{Super}(L' \cup L''|P)$*, homogeneous with respect to content, we have, for every* $t \in \Lambda^m(V)$,

$$U(C) \sim_t 0 \quad \Leftrightarrow \quad C \sim_t 0.$$

Proof. The implication $\Leftarrow$ is obvious; we prove the implication $\Rightarrow$. The hypothesis $U(C) \sim_t 0$ means:

$$(\text{val}' \otimes I) \circ (I \otimes U'') \circ (U' \otimes I)C = 0,$$

which is equivalent to

$$(I \otimes U'') \circ (\sigma \otimes I) \circ (\mathcal{E}v' \otimes I)C = 0.$$

Since U'' is a monomorphism, this happens iff

$$(\sigma \otimes I) \circ (\mathcal{E}v' \otimes I)C = 0.$$

Since C is a homogeneous bracket polynomial in the letters $\alpha_1, \ldots, \alpha_i, \ldots \in L'$, $\omega_1, \ldots, \omega_p, x_1, \ldots, x_q \in L''$, it is a linear combination of bi-tableau, each of p rows of length n.

Then $D = (\mathcal{E}v' \otimes I)C$ is a bracket polynomial in the letters $e_1, \ldots, e_n \in E_n^-$ and $\omega_1, \ldots, \omega_p, x_1, \ldots, x_q \in L''$, and can be written as a linear combination of standard bi-tableau of the same shape $n, \ldots, n$ (p times):

$$D = \sum_A c_A(A|Q) = \sum_A c_A \left(\begin{array}{c|cccc} & n & \ldots & 2 & 1 \\ A & \ldots & \ldots & \ldots & \ldots \\ & n & \ldots & 2 & 1 \end{array} \right).$$

Here we consider a linear order on $E_n^- \cup L''$ in which the last n letters are $e_1 < \ldots < e_n$.

By applying the operator $\sigma \otimes I$ to a standard bi-tableau $(A|Q)$ one gets a standard bi-tableau $(A'|Q_A)$, where A' is the standard subtableau obtained from A by deleting the e_i's, and in each row of Q_A a place appears iff it does not appear as an index of any of the e_i's in the corresponding row of A.[1]

Since all the bi-tableau in the expansion of D have the same number p of rows, any bi-tableau $(A|Q)$ can be uniquely reconstructed from the bi-tableau $(A'|Q_A)$.

Now, by the standard basis theorem, from

$$0 = (\sigma \otimes I)D = \sum_A c_A(A'|Q_A)$$

we get $c_A = 0$ for every A. Thus $(\mathcal{E}v' \otimes I)C = D = 0$, i.e.,

$$C \sim_t 0. \qquad \square$$

Definition 44. We say that $C \in W(L' \cup L''|P)$ *vanishes generically* at $t \in \Lambda^m(V)$, and we write

$$C \sim_t 0,$$

when

$$\mathcal{E}v'(t)C = 0.$$

---

[1] One can verify that any bi-tableau Q_A is standard by using the following argument. Let $I = \{i_p < i_{p+1} < \dots < i_n\}$ and $J = \{j_q < j_{q+1} < \dots < j_n\}$, with $q \leq p$, be two subsets of $\underline{n}$ such that

$$\begin{array}{ccccccccccc} & & & & i_p & < & i_{p+1} & < & \cdots & < & i_n \\ & & & & \wedge\!| & & \wedge\!| & & \cdots & & \wedge\!| \\ j_q & < & \cdots & < & j_p & < & j_{p+1} & < & \cdots & < & j_n \end{array}.$$

Then, denoting by $I' = \{i'_1 > i'_2 > \dots > i'_{n-p}\}$ the complement of I in $\underline{n}$ and by $J' = \{j'_1 > j'_2 > \dots > j'_{n-q}\}$ the complement of J in $\underline{n}$, we have

$$\begin{array}{ccccccccc} i'_1 & > & i'_2 & > & \cdots & > & i'_{n-q} & > & \cdots & > & i'_{n-p} \\ \vee\!| & & \vee\!| & & \cdots & & \vee\!| & & & & \\ j'_1 & > & j'_2 & > & \cdots & > & j'_{n-q} & & & & \end{array}.$$

Indeed, for every $h \in \underline{n}$, the number ξ_h of elements of I which are $\geq h$ does not exceed the number η_h of the elements of J which are $\geq h$. Thus

$$i'_h = n - h - \xi_h \geq n - h - \eta_h = j'_h.$$

Here is an example for $n = 4$:

$$(\sigma \otimes I)\left(\begin{array}{cccc|cccc} \dots & \dots & \dots & \dots & 4 & 3 & 2 & 1 \\ \dots & \dots & \dots & \dots & \dots & \dots & \dots & \dots \\ \dots & \dots & e_1 & e_3 & 4 & 3 & 2 & 1 \\ \dots & \dots & e_2 & e_4 & 4 & 3 & 2 & 1 \\ e_1 & e_2 & e_3 & e_4 & 4 & 3 & 2 & 1 \end{array}\right) = \left(\begin{array}{cccc|cccc} \dots & \dots & \dots & \dots & 4 & 3 & 2 & 1 \\ \dots & \dots & \dots & \dots & \dots & \dots & \dots & \dots \\ \dots & \dots & & & 4 & 2 & & \\ \dots & \dots & & & 3 & 1 & & \\ & & & & & & & \end{array}\right).$$

Proposition 38. *The following diagram commutes:*

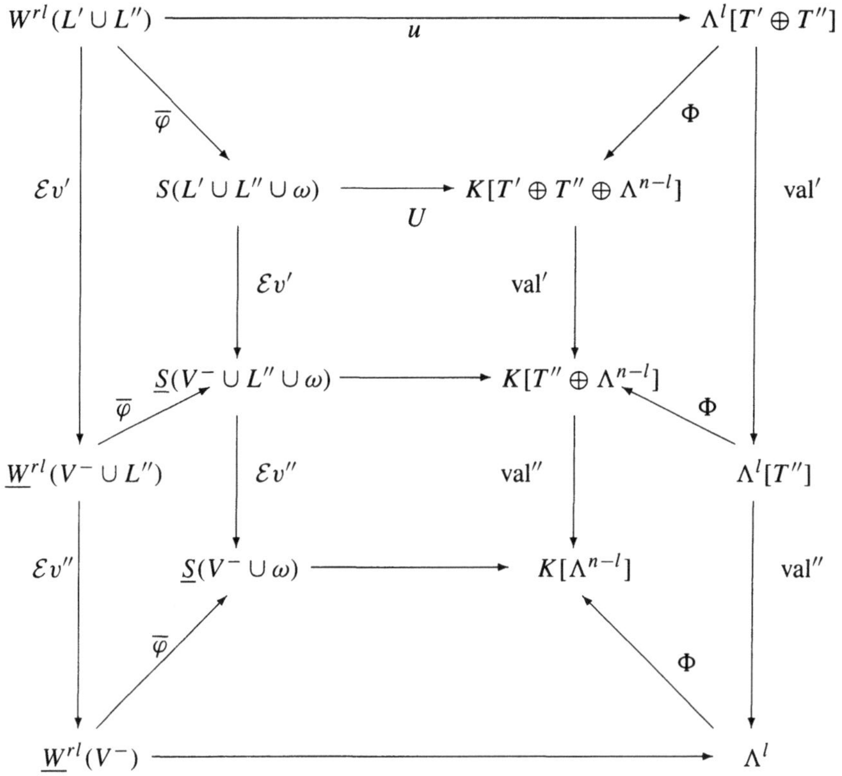

Theorem 8 (Completeness theorem)**.** *For every $C \in W^{rl}(L' \cup L'')$, we have, for every $t \in \Lambda^m(V)$,*

$$u(C) \sim_t 0 \quad \Leftrightarrow \quad C \sim_t 0.$$

17 Example: the space of the divisors of a skew-symmetric tensor

In the Cayley-Grassmann algebra $CG_n(V)$ of a Peano space $(V_n, [\ ])$, a vector v *divides* a tensor $t \in \Lambda^k$, in symbols $v|t$, iff there is a tensor $t' \in \Lambda^{k-1}$ such that

$$t = t' \vee v.$$

Proposition 39. *In the Cayley-Grassmann algebra $CG_n(V)$, a vector v divides a tensor t iff $t \vee v = 0$; in this case, we have*

$$t = \lambda(t \wedge v') \vee v, \quad \forall v' \in \Lambda^{n-1} : v \vee v' \neq 0, \quad \lambda \in K.$$

If the linearly independent vectors $v_1, \ldots, v_h$ divide the tensor t, then the extensor $v_1 \vee \ldots \vee v_h$ divides t; furthermore, we have

$$t = \lambda(t \wedge v') \vee v_1 \vee \ldots \vee v_h, \quad \forall v' \in \Lambda^{n-h} : v_1 \vee \ldots \vee v_h \vee v' \neq 0, \quad \lambda \in K.$$

Proof. In the White algebra $W_n(L \cup V)$ of the Peano space $(V_n, [\ ])$, for every two positive letters $\alpha, \gamma \in L^+$, we have

$$\begin{aligned}[v \vee \gamma^{(n-1)}]\alpha^{(k)} &= \pm[\alpha^{(k)} \vee v \vee \gamma^{(n-k-1)}]\gamma^{(k)} \pm [\alpha^{(k)} \vee \gamma^{(n-k)}]v \vee \gamma^{(k-1)} \\ &= \pm[\alpha^{(k)} \vee v \vee \gamma^{(n-k-1)}]\gamma^{(k)} \pm (\alpha^{(k)} \wedge \gamma^{(n-1)}) \vee v.\end{aligned}$$

By applying the operator $D_{t\alpha} D_{v'\gamma}$ to both sides, we get

$$[v \vee v']t = \pm(t \wedge v') \vee v.$$

The remaining part of the statement follows along the same lines. □

By the previous proposition, the set $\underline{V}(t)$ of the vectors which divide a given tensor t is a subspace of V.

Proposition 40. *In the Cayley-Grassmann algebra $CG_n(V)$, an extensor $x \in \Lambda^h$ divides a tensor $t \in \Lambda^k$ iff*

$$(t \vee w) \wedge x = 0 \qquad \forall w \in \Lambda^{n-k-1}.$$

Proof. First we prove the statement for a vector v: $v|t$ iff $t \vee v = 0$ iff

$$[t \vee v \vee w] = 0 \, \forall w \in \Lambda^{n-k-1} \quad \Leftrightarrow \quad (t \vee w) \wedge v = 0 \, \forall w \in \Lambda^{n-k-1}.$$

Then we prove the statement for an extensor $x \in \Lambda^h$:

$$\begin{aligned} x|t \;&\Leftrightarrow\; x \wedge y|t \;\forall y \in \Lambda^{n-h+1} \\ &\Leftrightarrow\; (t \vee w) \wedge (x \wedge y) = 0 \;\forall y \in \Lambda^{n-h+1} \;\forall w \in \Lambda^{n-k-1} \\ &\Leftrightarrow\; ((t \vee w) \wedge x) \wedge y = 0 \;\forall y \in \Lambda^{n-h+1} \;\forall w \in \Lambda^{n-k-1} \\ &\Leftrightarrow\; (t \vee w) \wedge x = 0 \;\forall w \in \Lambda^{n-k-1}. \end{aligned}$$ □

Proposition 41. *In the Cayley-Grassmann algebra $CG_n(V)$, if a vector v divides two tensors t_1, t_2, then v divides the tensor $t_1 \wedge t_2$.*

Definition 45. For $0 \leq q \leq k+1 \leq n$ we set

$$\begin{aligned} I_{k,q} &= \alpha_0^{(k)} \wedge (\alpha_1^{(k)} \vee \omega_1^{(n-k-1)}) \wedge \ldots \wedge (\alpha_q^{(k)} \vee \omega_q^{(n-k-1)}), \\ \mathcal{I}_{k,q} &= u(I_{k,q}). \end{aligned}$$

Note that, for every $0 \neq t \in \Lambda^k$,

$$I_{k,0} \not\sim_t 0, \qquad\qquad I_{k,k+1} = 0,$$
$$I_{k,q} \sim_t 0 \quad \Rightarrow \quad I_{k,q+1} \sim_t 0.$$

Thus there is exactly one index $\underline{q} = \underline{q}(t)$ such that

$$I_{k,\underline{q}} \not\sim_t 0, \quad I_{k,\underline{q}+1} \sim_t 0.$$

Note that

$$\mathcal{I}_{k,q} \in \Lambda^{k-q}[\Lambda^k \oplus (\Lambda^{n-k-1})^{\oplus q}],$$
$$\mathcal{I}_{k,q}(t; w_1, \ldots, w_q) = t \wedge (t \vee w_1) \wedge \ldots \wedge (t \vee w_q).$$

Theorem 9. *For a tensor $t \in \Lambda^k$ in the Cayley-Grassmann algebra $CG_n(V)$, we have*

$$d\underline{V}(t) = k - p \quad \Leftrightarrow \quad I_{k,p+1} \sim_t 0, \quad I_{k,p} \not\sim_t 0.$$

Furthermore, for every $w_1, \ldots, w_p \in \Lambda^{n-k-1}$ such that $\mathcal{I}_{k,p}(t; w_1, \ldots, w_p) \neq 0$, we have

$$\langle \mathcal{I}_{k,p}(t; w_1, \ldots, w_p) \rangle = \underline{V}(t).$$

In this statement $\langle z \rangle$ denotes the subspace corresponding to the extensor z.

Proof. We need only prove that, for all p, k, $0 \leq p \leq k \leq n$, the conditions

$$I_{k,p+1} \sim_t 0, \quad I_{k,p} \not\sim_t 0$$

imply that $d\underline{V}(t) = k - p$.

The first hypothesis implies that $\mathcal{I}_{k,p+1} \sim_t 0$, i.e.,

$$t \wedge (t \vee u_1) \wedge \ldots \wedge (t \vee u_{p+1}) = 0 \quad \forall u_1, \ldots, u_{p+1} \in \Lambda^{n-k-1}.$$

The second hypothesis implies, by the completeness theorem, that $\mathcal{I}_{k,p} \not\sim_t 0$, i.e.,

$$\exists w_1, \ldots, w_p \in \Lambda^{n-k-1} : \quad t \wedge (t \vee w_1) \wedge \ldots \wedge (t \vee w_p) \neq 0.$$

By definition,

$$0 \neq \mathcal{I}_{k,p}(t, w_1, \ldots, w_p) \in \Lambda^{k-p} \quad \Leftrightarrow$$
$$\exists x \text{ extensor } \in \Lambda^{n-k+p} : \mathcal{I}_{k,p}(t, w_1, \ldots, w_p) \wedge x \neq 0$$

We write x as the meet of $k - p$ covectors:

$$x = x_1 \wedge \ldots \wedge x_{k-p}, \; x_i \in \Lambda^{n-1};$$

set

$$y_i = x_1 \wedge \ldots \wedge \hat{x}_i \wedge \ldots \wedge x_{k-p}, \quad i = 1, \ldots, k-p,$$

and consider the $k - p$ vectors

$$v_i = \mathcal{I}_{k,p}(t, w_1, \ldots, w_p) \wedge y_i.$$

We note that

$$v_i \wedge x_i = \mathcal{I}_{k,p}(t, w_1, \ldots, w_p) \wedge y_i \wedge x_i = \pm\mathcal{I}_{k,p}(t, w_1, \ldots, w_p) \wedge x \neq 0,$$

and so

$$v_i \neq 0, \quad i = 1, \ldots, k-p.$$

Now

$$v_i | t \quad \forall i = 1, \ldots, k.$$

Indeed,

$$\begin{aligned}(t \vee w) \wedge v_i &= (t \vee w) \wedge (\mathcal{I}_{k,p}(t, w_1, \ldots, w_p) \wedge y_i) \\ &= ((t \vee w) \wedge \mathcal{I}_{k,p}(t, w_1, \ldots, w_p)) \wedge y_i \\ &= \pm\mathcal{I}_{k,p+1}(t, w_1, \ldots, w_p, w)) \wedge y_i = 0 \quad \forall w \in \Lambda^{n-k-1}.\end{aligned}$$

Finally,

$$v_1 \vee \ldots \vee v_k \neq 0.$$

Indeed,

$$\begin{aligned}&((\mathcal{I}_{k,p}(t, w_1, \ldots, w_p) \wedge y_1) \\ &\quad \vee \ldots \vee (\mathcal{I}_{k,p}(t, w_1, \ldots, w_p) \wedge y_{k-p})) \wedge (x_1 \wedge \ldots \wedge x_{k-p}) \\ &\quad = \pm \det(\mathcal{I}_{k,p}(t, w_1, \ldots, w_p) \wedge y_i \wedge x_j) \\ &\quad = \det \begin{pmatrix} \pm\mathcal{I}_{k,p}(t, w_1, \ldots, w_p) \wedge x & & \\ & \ddots & 0 \\ 0 & \ddots & \\ & & \pm\mathcal{I}_{k,p}(t, w_1, \ldots, w_p) \wedge x \end{pmatrix} \\ &\quad \neq 0.\end{aligned}$$

Thus we have proved that $d\underline{V}(t) \geq k - p$.
Set $\underline{V}(t) = \langle z \rangle$, with z an extensor, and observe that

$$z | \mathcal{I}_{k,p}(t, w_1, \ldots, w_p) \neq 0.$$

This implies that $d\underline{V}(t) \leq k - p$, so that we must have equality and

$$\underline{V}(t) = \left\langle \mathcal{I}_{k,p}(t, w_1, \ldots, w_p) \right\rangle. \qquad \square$$

Corollary 1 (Plücker relations). *A tensor $0 \neq t \in \Lambda^k$ in the Cayley-Grassmann algebra $CG_n(V)$ is decomposable iff*

$$I_{k,1} \sim_t 0, \quad I_{k,0} \not\sim_t 0.$$

18 GL(V)-stable algebraic sets and Gram's theorems

A subset Z of $\Lambda^m(V)$ is *$GL(V)$-stable* if $\rho \cdot Z \subseteq Z$ for every $\rho \in GL(V)$.

For a covariant $F \in K[\Lambda^m(V) \oplus V^{\oplus q}]$, the set Z of all $t \in \Lambda^m(V)$ such that $F \sim_t 0$ is $GL(V)$-stable.

For every increasing sequence $J : 1 \leq j_1 < \ldots < j_{n-m} \leq n$ we define a function

$$u_J \in K[\Lambda^m(V) \oplus V^{\oplus n}]$$

by setting

$$u_J(t, v_1, \ldots, v_n) = [t v_{j_1} \ldots v_{j_{n-m}}].$$

All these functions are covariants.

Note that

$$u_J(t, e_1, \ldots, e_n) = \pm t_{i_1,\ldots,i_m},$$

where the sign is the sign of the permutation $i_1 \ldots i_m j_1 \ldots j_{n-m}$, and $I : 1 \leq i_1 < \ldots < i_m \leq n$ is the complement of J in $\underline{n}$.

We denote by H the subalgebra of $K[\Lambda^m(V) \oplus V^{\oplus n}]$ generated by the polynomial functions u_J, where J ranges among the $(n-m)$-subsets of $\underline{n}$.

Theorem 10. *Let $G \in K[\Lambda^m(V)]$ be such that the set $Z = \{t \in \Lambda^m(V) : G(t) = 0\}$ is $GL(V)$-stable. Then there is a covariant $F \in H \subseteq K[\Lambda^m(V) \oplus V^{\oplus n}]$ such that $Z = \{t \in \Lambda^m(V) : F \sim_t 0\}$.*

The proof runs as follows. Define a mapping

$$\phi : K[\Lambda^m(V) \oplus V^{\oplus n}] \rightarrow K[\Lambda^m(V)]$$

by setting

$$(\phi F)(\underline{t}) = F(\underline{t}, \underline{e}_1, \ldots, \underline{e}_n).$$

This map is a valuation morphism and so it is an algebra morphism.

Note that

$$\phi u_J = \pm \tau_I.$$

Thus, the restriction of the algebra morphism ϕ to the subalgebra H generated by the covariant polynomial functions u_J is surjective. It follows that there is an $F \in H \subseteq K[\Lambda^m(V) \oplus V^{\oplus n}]$ such that $\phi F = G$.

Now, for every $t \in \Lambda^m(V)$, we have

$$G(t) = 0 \quad \Leftrightarrow F \sim_{\underline{t}} 0.$$

Indeed, if $G(t) = 0$, then, for every basis $(v_1, \ldots, v_n)$ of V, denoting by ρ the linear transformation carrying this basis to the distinguished basis $(e_1, \ldots, e_n)$, we have

$$\begin{aligned} F(t, v_1, \ldots, v_n) &= F(\rho^{-1}\rho t, \rho^{-1}\rho v_1, \ldots, \rho^{-1}\rho v_n) \\ &= F(\rho^{-1}\rho t, \rho^{-1}e_1, \ldots, \rho^{-1}e_n) \\ &= (\rho F)(\rho t, e_1, \ldots, e_n) = \lambda_\rho F(\rho t, e_1, \ldots, e_n) \\ &= \lambda_\rho G(\rho t) = 0. \end{aligned}$$

Since K is an infinite field, by Weyl's principle of irrelevance of algebraic inequalities, this argument implies that $F \sim_t 0$.

On the other hand, if $F \sim_t 0$, then, in particular, we get

$$G(t) = F(t, e_1, \ldots, e_n) = 0.$$

Let $U : Super(L' \cup X_q^- | P) \to K[\Lambda^m(V) \oplus V^{\oplus q}]$ be an umbral operator.

For every increasing sequence $J : 1 \leq j_1 < \ldots < j_{n-m} \leq n$ and every $\alpha \in L'$, we set

$$C_{J\alpha} = (\alpha^{(m)} x_{j_1} \ldots x_{j_{n-m}} | n \ldots 21).$$

Note that

$$\begin{aligned} \mathrm{val}(t, v_1, \ldots, v_n) \circ U(C_{J\alpha}) &= \sigma \circ \mathcal{E}v(t, v_1, \ldots, v_n)(C_{J\alpha}) \\ &= \sigma((t, v_{j_1}, \ldots, v_{j_{n-m}} | n \ldots 21)) \\ &= [t \, \underline{v}_{j_1} \ldots v_{j_{n-m}}], \end{aligned}$$

that is,

$$U(C_{J\alpha}) = u_J.$$

Thus, every $F \in H$ is the umbral image of a linear combination C of tableau of the form

$$\begin{pmatrix} \alpha_{i_1}^{(m)} & x_{j_{11}} & \ldots & x_{j_{1n-m}} \\ \alpha_{i_2}^{(m)} & x_{j_{21}} & \ldots & x_{j_{2n-m}} \\ \ldots & \ldots & \ldots & \ldots \\ \alpha_{i_p}^{(m)} & x_{j_{p1}} & \ldots & x_{j_{pn-m}} \end{pmatrix},$$

where $\alpha_{i_1}, \ldots, \alpha_{i_p}$ are distinct letters in L'.

Theorem 11 (Umbral version)**.** *Let $G \in K[\Lambda^m(V)]$ be such that the set $Z = \{t \in \Lambda^m(V) : G(t) = 0\}$ is $GL(V)$-stable. Then there is a bracket polynomial $C \in Super(L' \cup X_q^- | P)$ such that $Z = \{t \in \Lambda^m(V) : U(C) \sim_t 0\}$.*

References

[1] Abe, E. (1977): Hopf algebras. Cambridge University Press, Cambridge
[2] Barnabei, M.,Brini, A. (1988): The Littlewood–Richardson rule for co-Schur modules. Adv. Math. **67**, 143–173
[3] Barnabei, M., Brini, A., Rota, G.–C. (1985): On the exterior calculus of invariant theory. J. Algebra **96**, 120–160
[4] Bravi, P., Brini, A. (2000): Remarks on invariant geometric calculus. Cayley–Grassmann algebras and geometric Clifford algebras
[5] Brini, A., Huang, R.Q., Teolis, A. (1992): The umbral symbolic method for supersymmetric tensors. Adv. Math. **96**, 123–193
[6] Brini, A., Palareti, A., Teolis, A. (1988): Gordan–Capelli series in superalgebras. Proc. Nat. Acad. Sci. U.S.A. **85**, 1330–1333
[7] Brini, A., Regonati, F., Teolis, A. (1999): Multilinear algebra over supersymmetric rings. Adv. Math. **145**, 98–158
[8] Brini, A., Teolis, A. (1989): Young–Capelli symmetrizers in superalgebras. Proc. Nat. Acad. Sci. U.S.A. **86**, 775–778
[9] Brini, A., Teolis, A. (1995): Capelli's method of *variabili ausiliarie*, superalgebras and geometric calculus. In: White, N.L. (ed.) Invariant Methods in Discrete and Computational Geometry. Kluwer, Dordrecht, pp. 59–75
[10] Brini, A., Teolis, A. (1996): Grassmann progressive and regressive products and CG-Algebras. In: Schubring, G.(ed.) Hermann Günther Graßmann (1809–1877): Visionary Mathematician, Scientist and Neohumanist Scholar. Kluwer, Dordrecht, pp. 231–242
[11] Capelli, A. (1902): Lezioni sulla teoria delle forme algebriche: Pellerano, Napoli
[12] Chambadal, L., Ovaert, J.L. (1968): Algèbre linéaire et algèbre tensorielle. Dunod, Paris
[13] Chan, W. (1998): Classification of trivectors in 6-D space. In: Sagan, B.E. and Stanley, R.P. (eds.) Mathematical Essays in Honor of G.–C. Rota. Birkhäuser Boston, Boston, MA, pp. 63–110
[14] Chan, W., Rota, G.-C., Stein, J. (1995): The power of positive thinking. In: White, N.L. (ed.) Invariant Methods in Discrete and Computational Geometry. Kluwer, Dordrecht, pp. 1–36
[15] De Concini, C., Procesi, C. (1976): A characteristic-free approach to invariant theory. Adv. Math. **21**, 330–354
[16] Doubilet, P., Rota, G.-C., Stein, J. (1974): On the foundations of combinatorial theory: IX. Combinatorial methods in invariant theory. Studies in Appl. Math. **53**, 185–216
[17] Grassmann, H.G. (1894–1911): Hermann Graßmanns gesammelte mathematische und physikalische Werke. (3 vols.) Engel, F. (ed.) Teubner, Leipzig
[18] Grosshans, F.D., Rota, G.-C., Stein, J.A. (1987): Invariant theory and superalgebras. (CBMS Regional Conference Series in Mathematics, vol. 69, American Mathematical Society, Providence, RI
[19] Gurevich, G.B. (1964): Foundations of the theory of algebraic invariants. Noordhoff, Groningen
[20] Peano, G. (1888): Calcolo geometrico secondo l'Ausdehnungslehre di H.G. Grassmann preceduto dalle operazioni della logica deduttiva. Fratelli Bocca, Torino. Translation: (2000): Geometric calculus. Kannenberg, L.C. (translator). Birkhäuser Boston, Boston, MA
[21] Schubring, G. (ed.) (1996): Hermann Günther Graßmann (1809–1877): visionary mathematician, scientist and neohumanist scholar. Kluwer, Dordrecht
[22] Stewart, I. (1986): Herrmann Grassmann was right. Nature **321**, 17

[23] Weitzenböck, R. (1923): Invariantentheorie. Noordhoff, Groningen
[24] Weyl, H. (1946): The classical groups. Their invariants and representation, 2nd edition. Princeton University Press, Princeton, NJ
[25] White, N.L. (ed.) (1995): Invariant methods in discrete and computational geometry. Kluwer, Dordrecht

Rota–Metropolis cubic logic and Ulam–Rényi games

F. Cicalese*, D. Mundici**, U. Vaccaro

1 Cubic algebras, logic and games

1.1 The faces of the n-cube

In their paper [43] Rota and Metropolis considered the partially ordered set $\mathcal{F}_n$ of all nonempty faces of the n-cube $[0, 1]^n$ for each $n = 1, 2, \ldots$, equipped with the following operation:

($\sqcup$) taking the supremum $A \sqcup B$ of any two faces A and B of $\mathcal{F}_n$,
together with the following two *partially defined* operations:
($\sqcap$) taking the set-theoretic intersection $A \sqcap B$ of any two intersecting faces A and B of $\mathcal{F}_n$, and
(Δ) when a face A is contained in another face B, taking the antipode $\Delta(B, A)$ of A in B.

By a *finite cubic algebra* we understand a partial algebraic structure $(C, \sqcup, \sqcap, \Delta)$ which, for some (necessarily unique) integer $n \geq 1$, is isomorphic to $(\mathcal{F}_n, \sqcup, \sqcap, \Delta)$.

Among various realizations of finite cubic algebras, Rota and Metropolis considered the collection of all pairs $A = (A_0, A_1)$ of disjoint subsets of $\{1, 2, \ldots, n\}$. Here, A_0 (resp., A_1) is the set of coordinates where face A constantly takes value 0 (resp., value 1). The cubic operations are given by

$$(A_0, A_1) \sqcup (B_0, B_1) = (A_0 \cap B_0, A_1 \cap B_1);$$
$$(A_0, A_1) \sqcap (B_0, B_1) = (A_0 \cup B_0, A_1 \cup B_1),$$

provided $A_0 \cap B_1 = \emptyset = A_1 \cap B_0$.[1] Further,

$$\Delta((B_0, B_1), (A_0, A_1)) = (B_0 \cup (A_1 \setminus B_1), B_1 \cup (A_0 \setminus B_0)),$$

provided $A_0 \supseteq B_0$ and $A_1 \supseteq B_1$.

1.2 Algebras of three-valued logic

An equivalent realization of finite cubic algebras is given in this section, using MV_3 algebras, the Lindenbaum algebras of the three-valued Łukasiewicz calculus. This

* Partially supported by ENEA
** Partially supported by MURST Project on Logic
[1] To make the $\sqcap$ operation total, thus extending $\mathcal{F}_n$ to a lattice, Rota and Metropolis also considered the empty face, whose role is, however, immaterial for our purposes in this paper.

will be the main tool for our re-interpretation of the faces of the n-cubes. By an MV_3 *algebra* ([30, 17]) we mean a structure $M = (M, 1, \neg, \odot)$, with a distinguished element 1, a total unary operation $\neg$, and a total binary operation $\odot$ satisfying the following equations:

$$\begin{aligned} x \odot (y \odot z) &= (x \odot y) \odot z \\ x \odot y &= y \odot x \\ x \odot 1 &= x \\ x \odot \neg 1 &= \neg 1 \\ \neg\neg x &= x \\ \neg(y \odot \neg x) \odot y &= \neg(x \odot \neg y) \odot x \\ (x \odot x) \odot x &= x \odot x. \end{aligned}$$

Readers familiar with *MV* algebras will easily recognize in the above equations a reformulation of the usual definition of MV_3 algebras, based on Łukasiewicz conjunction $\odot$ rather than on disjunction (see, in particular, [30] and [17, 1.1 and 8.5]).

A basic example of an MV_3 algebra is given by the set $\{0, \frac{1}{2}, 1\}$ equipped with the operations $\neg x = 1 - x$ and $x \odot y = \max(0, x + y - 1)$.

In [30] it is shown that countable MV_3 algebras are in canonical one-to-one correspondence (via Grothendieck's functor K_0) with three-subhomogeneous AF C^*-algebras with Hausdorff structure space.

Following tradition, we shall consider the $\neg$ operation as more binding than all other operations. We shall also use the abbreviations

$$\begin{aligned} 0 &= \neg 1 \\ x^2 &= x \odot x \\ x \ominus y &= x \odot \neg y \\ x \oplus y &= \neg(\neg x \odot \neg y) \\ x \to y &= \neg x \oplus y \\ |x - y| &= (x \ominus y) \oplus (y \ominus x) \\ 2x &= x \oplus x \\ x \leftrightarrow y &= (x \to y) \odot (y \to x). \end{aligned}$$

Construction. Let M be a finite MV_3 algebra with a self-negated element $\epsilon = \neg\epsilon$. We enrich M with the following operation:

$$x \sqcup y = ((x \leftrightarrow y)^2 \odot y) \oplus ((\epsilon \to |x - y|)^2 \odot \epsilon),$$

together with the following two partially defined operations:

$$x \sqcap y = ((x \leftrightarrow \epsilon)^2 \odot y) \oplus ((x \leftrightarrow 2x)^2 \odot x),$$

for all $x, y \in M$ such that $|x - y| \to \epsilon = 1$, and

$$\Delta(x, y) = ((x \leftrightarrow y)^2 \odot y) \oplus ((x \leftrightarrow \epsilon)^2 \odot \neg y),$$

for all $x, y \in M$ such that $y \sqcup x = x$.

Theorem 1. *Let M be a finite MV_3 algebra with an element $\epsilon = \neg\epsilon$. Then ϵ is unique, and*

(i) *the structure $(M, \sqcup, \sqcap, \Delta)$ given by the above construction is an n-cubic algebra, where n is the number of homomorphisms of M into $\{0, \frac{1}{2}, 1\}$;*
(ii) *all finite cubic algebras arise from this construction;*
(iii) *nonisomorphic finite MV_3 algebras with self-negated elements determine nonisomorphic finite cubic algebras.*

Proofs will appear in [32].

1.3 Ulam–Rényi game, MV_3 algebras, cubic logic

Stanisław Ulam [48, p. 281] raised the following question.

> Someone thinks of a number between one and one million (which is just less than 2^{20}). Another person is allowed to ask up to twenty questions, to each of which the first person is supposed to answer only yes or no. Obviously the number can be guessed by asking first: Is the number in the first half-million? and then again reduce the reservoir of numbers in the next question by one-half, and so on. Finally the number is obtained in less than $\log_2(1{,}000{,}000)$. Now suppose one were allowed to lie once or twice, then how many questions would one need to get the right answer? One clearly needs more than n questions for guessing one of the 2^n objects because one does not know when the lie was told. This problem is not solved in general.

Since the same problem was also considered by Alfrèd Rényi in his book *A Diary on Information Theory* [41, p. 47], we shall henceforth refer to this as the *Ulam–Rényi problem.*

We first focus attention on the logical aspects of the game. In the classical game of Twenty Questions, in which every answer is reliable, a number x is a possible solution of the search problem over a preassigned *search space* S if, and only if, it satisfies the conjunction of answers. As soon as x falsifies an answer, it can be discarded by the questioner. Our state of knowledge is given by a function $\beta\colon S \to \{0, 1\}$, where $\beta^{-1}(1)$ is the set of elements of S satisfying all answers.

Assume now that one answer may be mendacious or erroneous. To fix ideas, suppose the question, "is the secret number x equal to 1?", has been answered "no". Then the number 1 is still a possible solution – because this answer might be false. If we now ask "is the number even?" and we get the answer "yes", then 1 must be discarded because it falsifies two answers.

To describe this state of affairs, since all our knowledge about the secret number is given by the conjunction ψ of received answers, skipping all syntactical details, one may conveniently regard ψ as a function assigning to each $y \in S$ one of *three* possible truth-values, as follows:

- $\psi(y) = 1$ iff y agrees with all answers
- $\psi(y) = 0$ iff y disagrees with two or more answers
- $\psi(y) = \frac{1}{2}$ iff y agrees with all answers, with exactly one exception.

Thus the questioner's state of knowledge can be equivalently represented by a triplet $\sigma = (A_1, A_{\frac{1}{2}}, A_0) = (\psi^{-1}(1), \psi^{-1}(1/2), \psi^{-1}(0))$ of pairwise disjoint subsets covering the search space S.

Rota and Metropolis gave the following logical interpretation of finite cubic algebras.

Each face $A = (A_0, A_1)$ of the n-cube is the result of sampling a population $S = \{1, \ldots, n\}$, with the view of testing the validity of a yes–no hypothesis. Here A_1 and A_0 are the subsets of S where the hypothesis does or does not hold, respectively. A third truth-value "not–yet–known" can be assigned to each element in the subset $A_{\frac{1}{2}} = S \setminus (A_0 \cup A_1)$. Two results A and B of this sampling are said to be incompatible if $A \sqcap B$ is undefined. There are 3^n faces.[2]

An alternative three-valued interpretation of finite cubic algebras is obtained by replacing the partial structure $(\mathcal{F}_n, \sqcup, \sqcap, \triangle)$ by its total counterpart $(M, 1, \epsilon, \neg, \odot)$ as given by Theorem 1. Owing to the existence of the self-negated element ϵ, up to isomorphism M consists of *all* functions $\psi : S \to \{0, 1/2, 1\}$ where S is the set of maximal ideals of M. There are 3^n such functions, where n is the cardinality of S (maximal ideals of S are in canonical one-to-one correspondence with homomorphisms of S into the three-element MV-algebra $\{0, 1/2, 1\}$).

By the above discussion, $M = \{0, \frac{1}{2}, 1\}^n$ is the set of states of knowledge in an Ulam-Rényi game of Twenty Questions with one lie. The element 1, i.e., the constant function 1 over S, is the *initial state*. The self-negated element ϵ is the constant function $\frac{1}{2}$ over $\{1, \ldots, n\}$. The zero element is the *incompatible*, or empty, state.

Suppose the questioner has received answers $R_1, \ldots, R_t$ to his questions $Q_1, \ldots, Q_t$. Then his *state* of knowledge is given by a function $\psi : \{1, \ldots, n\} \to \{0, \frac{1}{2}, 1\}$, or by its corresponding triplet $(A_1, A_{\frac{1}{2}}, A_0)$ as above.

Note that ψ is an element of M. Focusing attention on a single step $(Q_i \to R_i)$, we see that, if R_i is "yes" (resp., if R_i is "no"), then the resulting state of knowledge from this individual step is given by the function $\psi_i : S \to \{\frac{1}{2}, 1\}$ with $\psi_i^{-1}(1) = Q$ and $\psi_i^{-1}(\frac{1}{2}) = S \setminus Q$ (resp., $\psi_i^{-1}(1) = S \setminus Q$ and $\psi_i^{-1}(\frac{1}{2}) = Q$).

A moment's reflection shows that $\psi = \psi_1 \odot \cdots \odot \psi_t$, where $\odot$ is pointwise Łukasiewicz conjunction in M.[3] Starting from the initial state 1, the questioner's evolving state of knowledge is then represented by the decreasing set of functions $1 \geq \psi_1 \geq \psi_1 \odot \psi_2 \geq \psi_1 \odot \psi_2 \odot \psi_3 \geq \ldots$. Here the pointwise order $\psi \geq \gamma$ is naturally understood as "γ is more informative than ψ".

The negation operation $\neg$, too, has the following natural interpretation: $\neg\psi = 1 - \psi$ is the least informative state of knowledge ξ which is incompatible with ψ,

---

[2] See [10] for further results on infinite cubic algebras.

[3] In fact, the third truth-value was arithmetized as $\frac{1}{2}$ precisely for the purpose of having a simple conjunction operation, extending boolean conjunction.

in the sense that $\psi \odot \xi = 0$. Starting from negation and conjunction, one can now routinely define all other traditional connectives in Łukasiewicz propositional logic (see [31] and [17] for more details).

At the end of this introductory section we are left with two logical interpretations of the 3^n faces of the n-cube: the (A_0, A_1)-interpretation of Rota and Metropolis, and the $(A_1, A_{\frac{1}{2}}, A_0)$-interpretation given by the set of states of knowledge in an Ulam–Rényi game of Twenty Questions with one lie.

The latter game shall be our main concern in this paper: indeed, Ulam and Rényi not only described a searching game for x, but also asked for the *minimum* number of questions needed to find such an x. As we shall see, in order to solve this problem, one must attach numerical quantities to each state of knowledge, measuring its information content. One can then generalize such key notions as conservation and balancing, from the familiar game of Twenty Questions to its counterpart with lies, and then devise optimal searching strategies for the secret number x.

2 Fault-tolerant search à la Ulam–Rényi

2.1 Overview

Generalizing the above structure, we shall consider the following game between two players, Paul (the Questioner) and Carole (the Responder). Carole and Paul first fix a finite set $S = \{0, 1, \ldots, M-1\}$, called the *search space*, and an integer $e \geq 0$. Carole chooses a number x in S and Paul must guess x by asking questions of the form

"does x belong to T?"

where T is an arbitrary subset of S. Carole's only possible answers are "yes" or "no". Then what is the minimum number

$$N(M, e)$$

of questions to be asked by Paul in order to guess infallibly the number x, if we assume that Carole can lie at most e times? For the case $e = 0$, classical binary search yields $N(M, 0) = \lceil \log_2 M \rceil$, the smallest integer $\geq \log_2 M$.

In the last two decades, a rapidly increasing number of papers have been devoted to the Ulam–Rényi problem of computing $N(M, e)$. Earlier than both Ulam and Rényi, Berlekamp [6] had considered the same problem, in the context of block coding for the binary symmetric channel with noiseless feedback. We intend to survey the vast literature on the subject.

As a warm-up, we give the solution of the original game, that is, when $M =$ 1,000,000. Arbitrary cardinalities M and numbers of lies e shall be our main concern in later sections. To give further motivation, we also discuss the relationship between the Ulam–Rényi problem and Berlekamp's theory – with particular reference to the problem of reducing adaptiveness in the Ulam–Rényi game. In a final section we describe several variants of the game and their applications to error-correcting transmission.

2.2 Dynamics of states of knowledge

Suppose Carole answers "yes" to Paul's question T. If $e = 0$, then Paul discards all of $S \setminus T$ and reduces his search space from S to T. By contrast, if $e > 0$, then Paul's strategy must be more flexible. Paul considers the following two alternatives.

- Carole's answer is sincere. Then Paul's search continues not only over those elements of the set T which falsify precisely e answers, but also over those elements in the complementary set $S \setminus T$ which happened to falsify up to $e - 1$ answers before the last question T was asked.
- Carole's answer is mendacious. Then the search continues over those elements of the set $S \setminus T$ which falsify up to e answers, as well as over those elements of T which falsified up to $e - 1$ answers before the last question was asked.

Assume t questions have been answered. For each $j = 0, \ldots, e$, let L_j^t denote the set of elements of S falsifying exactly j answers. Thus, before the first question is asked, we can write $L_0^0 = \{0, 1, \ldots, M-1\}$, $L_1^0 = \cdots = L_e^0 = \emptyset$. Let T denote the $(t+1)^{\text{th}}$ question. Suppose the answer is "yes" (resp., suppose the answer is "no"). Then we can write, for each $j = 1, \ldots, e$,

$$\begin{cases} L_0^{t+1} = L_0^t \cap T & (\text{resp., } L_0^t \setminus T); \\ L_j^{t+1} = (L_j^t \cap T) \cup (L_{j-1}^t \setminus T) & (\text{resp., } (L_j^t \setminus T) \cup (L_{j-1}^t \cap T)) \end{cases} \tag{1}$$

At any stage t of the game, we say that $(L_0^t, L_1^t, \ldots, L_e^t)$ is Paul's *state* (of knowledge). Let $x_i = |L_i^t|$ for each $i = 0, 1, \ldots, e$. Then the state $\sigma = (L_0^t, L_1^t, \ldots, L_e^t)$ is said to be of *type* $(x_0, x_1, \ldots, x_e)$.

Our aim in this paper is to minimize the *number* of questions, rather than explicitly formulating these questions as subsets of S. Accordingly, we can safely restrict attention to the cardinalities x_i rather than to the actual sets L_i^t. Thus, by abuse of terminology, we say that the $(e+1)$-tuple of integers $(x_0, \ldots, x_e)$ is a *state*.

Definition 1. A *final state* is a state $(x_0, x_1, \ldots, x_e)$ such that $\sum_{j=0}^{e} x_j \leq 1$. Given states $\sigma = (x_0, \ldots, x_e)$ and $\sigma' = (y_0, \ldots, y_e)$, we say that σ' is a *substate* of σ if $y_j \leq x_j$ for each $j = 0, 1, \ldots, e$.

Again with sets replaced with by their cardinalities, a *question* T shall be denoted by $[t_0, \ldots, t_e]$, where $t_j = |T \cap L_j^t|$. Thus a question T is completely specified once we know the number t_j of elements in L_j^t quoted by T.

Suppose Paul's state is $(x_0, \ldots, x_e)$, and question $\delta = [a_0, \ldots, a_e]$ is asked. Suppose that Carole's answer is "yes" (resp., the answer is "no"). Then the resulting state is given by $(x_0', \ldots, x_e')$, where, for each $j = 1, \ldots, e$,

$$\begin{cases} x_0' = a_0 & (\text{resp., } x_0' = x_0 - a_0) \\ x_j' = a_j + (x_{j-1} - a_{j-1}) & (\text{resp., } x_j' = x_j - a_j + a_{j-1}). \end{cases} \tag{2}$$

For a state $\sigma = (x_0, \ldots, x_e)$ and a question δ, the two possible answers to δ determine two more informative states σ^{yes} and σ^{no}. Paul then adaptively asks a next question and, depending on Carole's answers, he is left in one of the four possible states $\sigma^{\text{yes,yes}}$, $\sigma^{\text{yes,no}}$, $\sigma^{\text{no,yes}}$, $\sigma^{\text{no,no}}$. Proceeding inductively, Paul can build a labelled binary tree $\mathcal{T}$, rooted at σ, as follows. The edges of $\mathcal{T}$ are labelled by the questions to be asked by Paul; nodes are labelled by the resulting states. We say that $\mathcal{T}$ is Paul's *strategy*. We say that the state σ has a *winning strategy of size t* if there exists a binary tree $\mathcal{T}$ of height t, rooted at σ, whose leaves are final states.

Definition 2. A *winning n-state* is a state $\sigma = (x_0, x_1, \dots, x_e)$ such that there exists a winning strategy of size n for it. We say that $(x_0, x_1, \dots, x_e)$ is a *borderline winning n-state* if it is a winning n-state but not a winning $(n-1)$-state.

If every component x_j in a state $\sigma = (x_0, \dots, x_e)$ is even, then, by an *even splitting question* for σ, we understand the question $\delta = [\frac{x_0}{2}, \frac{x_1}{2}, \dots, \frac{x_e}{2}]$.

When some component of σ is odd, by abuse of terminology, the question $\delta = [\frac{x_0}{2}, \frac{x_1}{2}, \dots, \frac{x_e}{2}]$ shall be called a *pseudo*-even splitting question.[4]

2.3 Basic inequalities

In this section we collect upper and lower bounds on the number of necessary questions for fault-tolerant search.

The following quantity shall find repeated use to derive lower bounds on the size of any winning strategy.

Definition 3 ([6]). The nth *volume*, $V_n(x_0, \dots, x_e)$, of a state $(x_0, \dots, x_e)$ is defined by

$$V_n(x_0, x_1, \dots, x_e) = \sum_{i=0}^{e} x_i \sum_{j=0}^{e-i} \binom{n}{j}.$$

Thus the nth volume of a state $\sigma = (x_0, \dots, x_e) = (|L_0|, \dots, |L_e|)$ is the number of answering strategies available to Carole to deceive Paul when he is in state σ and n questions are still to be asked.[5] As a matter of fact, for each one of the x_i elements of L_i, Carole can still lie up to $e - i$ times. If she decides to lie precisely j times $(j = 0, \dots, e - i)$, she can still choose where to lie, in $\binom{n}{j}$ ways. One then easily sees that, for the state $(x_0, \dots, x_e)$, the overall number of possibilities for Carole to lie coincides with $V_n(x_0, \dots, x_e)$.[6]

---

[4] Note that the two n-tuples of rationals resulting from Carole's answer to δ via (2) need no longer be states, because their components may be nonintegral. However, the dynamic laws given by (2) can still be applied without problems.

[5] Indeed, one can conveniently allow Carole to play a malicious answering strategy. Thus, for instance, x need not be chosen once and for all at the beginning of the game, but can be suitably changed, compatibly with the lie number e, so as to make Paul's task as hard as possible.

[6] Some authors [1, 19, 28] call $V_n(x_0, \dots, x_e)$ the nth weight (or more simply, the weight) of $(x_0, \dots, x_e)$.

The following theorem, first proved by Berlekamp [6], relates winning n-states and their n^{th} volumes. For the reader's convenience we include here a novel, induction-free proof. See [14] for a similar proof of a more general result.

Theorem 2 (Volume bound). *If $\sigma = (\hat{x}_0, \hat{x}_1, \dots, \hat{x}_e)$ is a winning n-state, then*

$$\sum_{i=0}^{e} \hat{x}_i \sum_{j=0}^{e-i} \binom{n}{j} \le 2^n.$$

Proof. By hypothesis there exists a strategy $\mathcal{T}$ of size n rooted in σ, whose leaves are states $(x'_0, \dots, x'_e)$ with $\sum_{j=0}^{e} x'_j \le 1$. For the moment we relax the requirement that states are n-tuples of integers. Replacing all questions in $\mathcal{T}$ by pseudo-even splitting questions, and formally applying the dynamical rules (2), we still get a tree rooted in σ, of height equal to n, whose leaves are vectors $(y'_0, \dots, y'_e)$ (possibly with rational, nonintegral components) and such that $\sum_{j=0}^{q} y'_j \le 1$. Thus, the question to be asked in state $(x_0, \dots, x_e)$, is given by $[\frac{x_0}{2}, \dots, \frac{x_e}{2}]$.

We use the notation $a = \frac{1}{2}$. By repeated application of (2), the ith component $x_i^{(j)}$ of the vector $(x_0^{(j)}, \dots, x_e^{(j)})$ obtained from $(\hat{x}_0, \hat{x}_1, \dots, \hat{x}_e)$ after j questions is given by

$$\begin{cases} x_0^{(j)} = a x_0^{(j-1)}, \quad x_0^{(0)} = \hat{x}_0 \\ x_i^{(j)} = a x_i^{(j-1)} + (1-a) x_{i-1}^{(j-1)}, \quad x_i^{(0)} = \hat{x}_i, \quad 1 \le i \le e. \end{cases} \tag{3}$$

All vectors $(x_0^{(n)}, \dots, x_e^{(n)})$ obtained after n pseudo-even splitting questions have the following property:

$$\sum_{i=0}^{e} x_i^{(n)} \le 1. \tag{4}$$

For all $i = 0, \dots, e$, let $F_i(t) = \sum_{j \ge 0} x_i^{(j)} t^j$ denote the *generating function* of the sequence $x_i^{(0)}, x_i^{(1)}, x_i^{(2)}, \dots$. Stated otherwise, $F_i(t) = \sum_{j \ge 0} x_i^{(j)} t^j$. From (3) we get:

$$\begin{aligned} F_0(t) &= at F_0(t) + \hat{x}_0 \\ F_i(t) &= at F_i(t) + (1-a) t F_{i-1}(t) + \hat{x}_i, \quad 1 \le i \le e, \end{aligned}$$

or, equivalently:

$$\begin{aligned} F_0(t) &= \frac{\hat{x}_0}{1-at} \\ F_i(t) &= (1-at)^{-1} \left((1-a) t F_{i-1}(t) + \hat{x}_i \right), \quad 1 \le i \le e. \end{aligned}$$

It follows that

$$F_0(t) = \frac{\hat{x}_0}{1-at}$$

$$
\begin{aligned}
F_1(t) &= (1-at)^{-1}\left((1-a)t\frac{\hat{x}_0}{1-at}+\hat{x}_1\right) = (1-a)\,t\frac{\hat{x}_0}{(1-at)^2}+\frac{\hat{x}_1}{1-at}\\
F_2(t) &= (1-at)^{-1}\left((1-a)tF_1(t)+\hat{x}_2\right)\\
&= (1-a)^2\,t^2\frac{\hat{x}_0}{(1-at)^3}+(1-a)\,t\frac{\hat{x}_1}{(1-at)^2}+\frac{\hat{x}_2}{1-at},
\end{aligned}
$$

and, more generally,

$$
F_i(t)=\sum_{j=0}^{i}(1-a)^j\,\frac{t^j\hat{x}_{i-j}}{(1-at)^{j+1}}=\sum_{j=0}^{i}\sum_{n\geq 0}\binom{n}{j}(1-a)^j a^{n-j}\hat{x}_{i-j}t^n, \tag{5}
$$

for all i, $0 \leq i \leq e$. Condition (4) now becomes

$$
[t^n]\left(\sum_{i=0}^{e}F_i(t)\right)\leq 1, \tag{6}
$$

where, as usual, $[t^n]f(t)$ denotes the coefficient of the nth power of t in the power series expansion of $f(t)$. Recalling that $a=\frac{1}{2}$, from (5) we see that inequality (6) can be reformulated as

$$
\sum_{i=0}^{e}\sum_{j=0}^{i}\binom{n}{j}\hat{x}_{i-j}\leq 2^n.
$$

To complete the proof it is sufficient to write

$$
\sum_{i=0}^{e}\sum_{j=0}^{i}\binom{n}{j}\hat{x}_{i-j}=\sum_{i=0}^{e}\hat{x}_i\sum_{j=0}^{e-i}\binom{n}{j}. \qquad \square
$$

Suppose a state $\sigma=(x_0,x_1,\ldots,x_e)$ fails to satisfy the condition

$$
V_n(x_0,x_1,\ldots,x_e)\leq 2^n. \tag{7}
$$

Then, as an immediate consequence of Theorem 2, σ cannot be a winning n-state. Instead of saying that a state σ satisfies condition (7) we shall henceforth say that σ *satisfies the volume bound for n questions*.

Definition 4. The *character* of a state $\sigma=(x_0,\ldots,x_e)$ is the least integer n such that σ satisfies the volume bound for n questions, in symbols,

$$
\mathrm{ch}(x_0,x_1,\ldots,x_e)=\min\{n \mid V_n(x_0,x_1,\ldots,x_e)\leq 2^n\}.
$$

A strategy $\mathcal{S}$ of size q for a state σ is said to be *perfect* if $\mathcal{S}$ is winning for σ and $q=\mathrm{ch}(\sigma)$.[7]

[7] There is no danger of confusion between the usual meaning of "perfect code" and the present generalization. Because a perfect strategy $\mathcal{S}$ uses the least possible number of questions, as given by Berlekamp's bound, $\mathcal{S}$ is *optimal*, in the sense that it cannot be superseded by a shorter strategy. On the other hand, this paper will exhibit several nonperfect optimal strategies.

For all integers $M \geq 0$ and $e \geq 0$, we define

$$N_{\min}(M, e) = \min \left\{ n \mid M \sum_{j=0}^{e} \binom{n}{j} \leq 2^n \right\}. \quad (8)$$

Then by Theorem 2 we immediately have the following lower bound on the size of the shortest winning strategy for the Ulam–Rényi game on e lies over a search space of cardinality M :

$$N(M, e) \geq N_{\min}(M, e).$$

Lemma 1. *Suppose that the states $\sigma = (x_0, \dots, x_e)$ and $\tau = (y_0, \dots, y_e)$ satisfy the condition $\sum_{j=0}^{k} x_j \leq \sum_{j=0}^{k} y_j$ for all $k = 0, \dots, e$. If τ is a winning n-state, then so is σ.*

Proof. By induction on $n \geq 0$. The basis is trivial. For the induction step, let τ be a winning n-state. Then there exists a question $\delta_\tau = [a_0, \dots, a_e]$, reducing τ to two winning $(n-1)$-states, in the sense that the two possible states, $\tau^{\text{yes}} = (y_0^{\text{yes}}, \dots, y_e^{\text{yes}})$ and $\tau^{\text{no}} = (y_0^{\text{no}}, \dots, y_e^{\text{no}})$, obtained from τ after question δ is answered, are winning $(n-1)$-states. Define the question $\delta_\sigma = [b_0, \dots, b_e]$ by

$$b_j = \min\{x_j, \ a_j\} \quad (j = 0, \dots, e). \quad (9)$$

For each $j = 0, 1, \dots, e$, we have the inequalities $b_j \leq a_j$ and $x_j - b_j \leq y_j - a_j$. Indeed, by (9) we have either $b_j = a_j$ or $b_j = x_j$. In the former case we can write $y_j - a_j \geq x_j - a_j = x_j - b_j$. In the latter case we get $x_j - b_j = 0 \leq y_j - a_j$. Let $\sigma^{\text{yes}} = (x_0^{\text{yes}}, \dots, x_e^{\text{yes}})$ and $\sigma^{\text{no}} = (x_0^{\text{no}}, \dots, x_e^{\text{no}})$ respectively denote the two states obtained from σ after Carole's answer "yes" or "no" to Paul's question δ_σ. Setting $x_{-1} = y_{-1} = a_{-1} = b_{-1} = 0$ for sake of definiteness, and recalling (2), for each $k = 0, 1, \dots, e$, we have

$$\sum_{j=0}^{k} x_j^{\text{yes}} = \sum_{j=0}^{k} b_j + \sum_{j=0}^{k-1} (x_j - b_j) \leq \sum_{j=0}^{k} a_j + \sum_{j=0}^{k-1} (y_j - a_j) = \sum_{j=0}^{k} y_j^{\text{yes}}$$

and

$$\sum_{j=0}^{k} x_j^{\text{no}} = \sum_{j=0}^{k} (x_j - b_j) + \sum_{j=0}^{k-1} b_j \leq \sum_{j=0}^{k} (y_j - a_j) + \sum_{j=0}^{k-1} a_j = \sum_{j=0}^{k} y_j^{\text{no}}.$$

Thus, by the induction hypothesis, both σ^{yes} and σ^{no} are winning $(n-1)$-states, whence σ is a winning n-state, as required to complete the proof. □

Lemma 2. *Let σ and σ' be two states, with σ a substate of σ'. Then the following conditions hold:*

(i) *for all integers* $i \geq 0$, $V_i(\sigma) \leq V_i(\sigma')$;
(ii) $\mathrm{ch}(\sigma) \leq \mathrm{ch}(\sigma')$;
(iii) *if* σ' *is a winning* k*-state, then so is* σ.

Proof. Conditions (i) and (ii) are immediate consequences of the definitions. Condition (iii) follows from Lemma 1. □

The proof of the following *conservation law* amounts to a straightforward verification.

Theorem 3. *For any state* $\sigma = (x_0, \dots, x_e)$ *and question* δ, *let us denote by* $(y_0^{\mathrm{yes}}, \dots, y_e^{\mathrm{yes}})$ *and* $(y_0^{\mathrm{no}}, \dots, y_e^{\mathrm{no}})$ *the two states resulting from* σ *after Carole's answer to* δ. *Then, for all integers* $n \geq 0$, *we have*

$$V_{n-1}(y_0^{\mathrm{yes}}, \dots, y_e^{\mathrm{yes}}) + V_{n-1}(y_0^{\mathrm{no}}, \dots, y_e^{\mathrm{no}}) = V_n(x_0, \dots, x_e).$$

Corollary 1. *Suppose that* $\sigma' = (y_0, \dots, y_e)$ *is the state resulting from* $\sigma = (x_0, \dots, x_e)$ *after an even splitting question. It follows that:*

(i) *if* σ *satisfies the volume bound for* n *questions, then* σ' *satisfies the volume bound for* $n-1$ *questions;*
(ii) $\mathrm{ch}(\sigma') = \mathrm{ch}(\sigma) - 1$.

Proof. From Theorem 3 we see that $2V_{n-1}(y_0, \dots, y_e) = V_n(x_0, \dots, x_e)$. By hypothesis, $V_n(x_0, \dots, x_e) \leq q^n$, whence $V_{n-1}(y_0, \dots, y_e) = \frac{1}{2}V_n(x_0, \dots, x_e) \leq 2^{n-1}$, which settles (i). Condition (ii) immediately follows from (i) by the definition of character. □

The following theorem, due to Berlekamp [6], yields a lower bound on the size of the smallest winning strategy in a game with $e+1$ lies, given the size of the smallest winning strategy in a game with e lies.

Theorem 4 (Translation bound). *If* $(x_0, \dots, x_e, x_{e+1})$ *is a winning* m*-state and* $(x_0, \dots, x_e)$ *is a borderline winning* n*-state, then* $m \geq n+3$.

2.4 Solving Ulam's original problem for $M = 10^6$

Recall that $N(M, e)$ denotes the size of the shortest winning strategy for the Ulam–Rényi game over a search space of cardinality M, when up to e answers may be mendacious.

In this section we present the solution of the problem for the case $M = 10^6$ (as originally proposed by Ulam). We first evaluate $N(2^{20}, e)$. Then we use the fact that $N(2^{20}, e) = N(10^6, e)$, for all $e \geq 0$, with the sole exception of $e = 4$. This latter case is taken care of separately.

For $e = 1, 2, 3$, the exact value of $N(2^{20}, e)$ was computed in the papers [37, 19, 35]. In these papers the reader can find the exact value of $N(2^m, e)$, for all integers $m \geq 0$, for the cases $e = 1$, $e = 2$, $e = 3$, respectively. The remaining values of $N(2^{20}, e)$ and $N(10^6, 4)$, were computed by Hill et al. in [24] (see also [23]). Here the authors make use of tables of winning states that were already known from [6]. Here is the main result of [24].

Table 1. The values of $N(2^{20}, e)$

e	0	1	2	3	4	5	6	7	8	9	...	e	...
$N(2^{20}, e)$	20	25	29	33	37	40	43	46	50	53	...	$3e + 26$	...

Theorem 5. *The values of $N(2^{20}, e)$ are given by Table 1. For all $e \geq 8$, we have $N(2^{20}, e) = 3e + 26$.*

Beginning of the proof. Starting from the initial state $\sigma_e^{(0)} = (2^{20}, \underbrace{0, \ldots, 0}_{e \text{ zeros}})$, and asking 20 even splitting questions, Paul is in a sequence of states $\sigma_e^{(1)}, \ldots, \sigma_e^{(20)}$, where

$$\sigma_e^{(i)} = \left(2^{20-i}, i2^{20-i}, \binom{i}{2}2^{20-i}, \ldots, \binom{i}{j}2^{20-i}, \ldots, \binom{i}{e}2^{20-i}\right),$$

for each $i = 1, \ldots, 20$. After these initial 20 questions are answered, and independently of Carole's answers, Paul's state is given by

$$\chi_e = \left(1, 20, \binom{20}{2}, \ldots, \binom{20}{j}, \ldots, \binom{20}{e}\right). \tag{10}$$

Being no longer able to split the first component of this state, Paul resorts to volume splitting questions. As we shall see, all he has to do is to reduce the state (10) to a substate of some state in Table 2 below. For any number of lies, this table displays an infinite sequence of states admitting a *perfect* winning strategy. Moreover, any such state σ is *maximal*, in the sense that $V_n(\sigma) = 2^n$, where $n = \mathrm{ch}(\sigma)$.

Table 2. States attaining the volume bound exactly

$\sigma_{i,j}$	$s_{\cdot,1}$	$s_{\cdot,2}$	$s_{\cdot,3}$	$s_{\cdot,4}$	$s_{\cdot,5}$	$s_{\cdot,6}$	$s_{\cdot,7}$	$s_{\cdot,8}$
$\sigma_{1,\cdot}$	$\overset{2}{4}$	$\overset{5}{8}$	$\overset{8}{36}$	$\overset{11}{152}$	$\overset{14}{644}$	$\overset{17}{2728}$	$\overset{20}{11556}$	...
$\sigma_{2,\cdot}$	$\overset{1}{2}$	$\overset{4}{6}$	$\overset{7}{22}$	$\overset{10}{94}$	$\overset{13}{398}$	$\overset{16}{1686}$	$\overset{19}{7142}$	...
$\sigma_{3,\cdot}$	$\overset{0}{1}$	$\overset{3}{4}$	$\overset{6}{14}$	$\overset{9}{58}$	$\overset{12}{246}$	$\overset{15}{1042}$	$\overset{18}{4414}$	...
$\sigma_{4,\cdot}$	$\overset{0}{1}$	$\overset{2}{1}$	$\overset{5}{10}$	$\overset{8}{36}$	$\overset{11}{152}$	$\overset{14}{644}$	$\overset{17}{2728}$	...
$\sigma_{5,\cdot}$	$\overset{0}{1}$	$\overset{0}{0}$	$\overset{4}{5}$	$\overset{7}{24}$	$\overset{10}{94}$	$\overset{13}{398}$	$\overset{16}{1686}$	...
$\vdots$	$\vdots$	$\vdots$	$\vdots$	$\vdots$	$\vdots$	$\vdots$	$\vdots$	$\ddots$

Let $s_{i,j}$ be the (i, j) entry of Table 2 occurring in row i and column j. Let $\sigma_{i,m}$ denote the state $(s_{i,1}, s_{i,2}, \ldots, s_{i,m})$. To signify that $\sigma_{i,m}$ is a winning n-state we place the integer n above the entry $s_{i,m}$ in the table.

Table 2 is constructed as follows. First of all, $s_{i,1} = 1$ and $s_{i,2} = 0$ for all $i = 5, 6, \ldots$. For each $i = 1, 2, 3, 4, \ldots$, the value $s_{i,1}$ is the largest possible cardinality n of a search space where Paul can successfully search by using $\lceil \log_2 n \rceil$ questions in the game with no lies. The values of $s_{i,2}$ are chosen so as to ensure that the state $(s_{i,1}, s_{i,2})$ is a winning n-state, with $n = \mathrm{ch}(s_{i,1}, s_{i,2})$, and there exists a question δ reducing $\sigma_{i,2}$ to $\sigma_{i+1,2}$, for all $i = 1, 2, 3, 4, \ldots$.

The remaining columns of Table 2 ($j \geq 3$) are given by the following recurrence:

$$\begin{cases} \text{for } i \geq 3, & s_{i,j} = s_{i-1,j-1} + s_{i-2,j-1} \\ \text{for } i = 2, & s_{2,j} = s_{3,j} + s_{1,j-1} \\ \text{for } i = 1, & s_{1,j} = s_{2,j} + s_{3,j} \end{cases} \tag{11}$$

Lemma 3. *With reference to Table 2 we have:*

(i) *for each $i = 1, 2$ and for all $j = 1, 2, \ldots$, there exists a question δ for the state $\sigma_{i,j}$ such that the two states resulting from $\sigma_{i,j}$ coincide with $\sigma_{i+1,j}$;*

(ii) *for all $i \geq 3$ and $j = 1, 2, \ldots$, there exists a question δ for the state $\sigma_{i,j}$ such that the two resulting states coincide with $\sigma_{i+1,j}$ and $\sigma_{i-2,j-1}$, respectively;*

(iii) *for all $i = 1, 2, \ldots$, and $j \geq i$, with the exceptions of $s_{1,1}$, $s_{1,2}$, $s_{2,2}$, the integer $s_{i,j}$ coincides with*

$$\left\lfloor 2(2+\sqrt{5})^j \left(\frac{\sqrt{5}-1}{2}\right)^{i+2} + \frac{1}{2} \right\rfloor.$$

(iv) *for all $i = 1, 2, \ldots$, and all integers $j \geq \max\{3, i\}$, the state $\sigma_{i,j}$ satisfies the recurrence law*

$$s_{ij} = \left\lfloor s_{i,j-1}\,(2+\sqrt{5}) + \frac{1}{2} \right\rfloor.$$

For a proof we refer the interested reader to [6]. The results in Lemma 3 were used in [23] to prove Theorem 5.

End of the proof of Theorem 5. We shall argue by cases.

Case 1. $e \geq 8$. Then, by direct inspection, $\mathrm{ch}(\sigma_8^{(0)}) = 50$. By the translation bound and Corollary 1, for all $e \geq 8$ no winning strategy exists for χ_e using $< 6 + 3e$ questions. Thus we have only to prove that a winning strategy of size $6 + 3e$ exists. As a matter of fact, the 64 states obtained by iterated applications of quasi-even splittings[8] in the next six questions, will be substates of

$$\sigma^{\sharp} = (1, 1, 5, 41, 233, 1028, 3597, 10278, 24411, 48821, \ldots).$$

[8] As a complement to Definition 2, for any state $\sigma = (x_0, x_1, \ldots, x_e)$ the question $\delta = [\lceil \frac{x_0}{2} \rceil, \lceil \frac{x_1}{2} \rceil, \ldots, \lceil \frac{x_e}{2} \rceil]$ is said to be *quasi-even* splitting.

We display the winning $3e$-state $\sigma_{3,e+1}$ in Table 2 as follows:

$$\sigma_{3,e+1} = (1, 4, 14, 58, 246, 1042, 4414, 18698, \ldots).$$

In the light of Lemma 2, it is sufficient to prove that $\sigma^\sharp$ is a substate of $\sigma_{3,e+1}$ for all $e \geq 8$. For $e = 8$ the claim follows by direct inspection. Proceeding by induction, and letting a_i be the ith component of $\sigma^\sharp$, we have $a_i \leq s_{3,i}$ for all $i = 1, 2, \ldots, 9$. It is not hard to see that, for all $i \geq 10$,

$$a_i < \left\lfloor a_{i-1}\,(2+\sqrt{5}) + \frac{1}{2} \right\rfloor .$$

Since each component of $\sigma^\sharp$ grows at a smaller rate than its corresponding component in $\sigma_{3,e+1}$, we conclude that

$$a_i < \left\lfloor a_{i-1}\,(2+\sqrt{5}) + \frac{1}{2} \right\rfloor \leq \left\lfloor s_{i-1}\,(2+\sqrt{5}) + \frac{1}{2} \right\rfloor = s_i,$$

thus settling the present case.

Case 2. $e \leq 7$. In the papers [37, 19, 35] one can find the proof that the state χ_e has a perfect strategy, for $e = 1, 2, 3$ respectively. The cases $e = 4$ and $e = 5$ are settled arguing as for Case 1 in the light of Berlekamp's Tables [6, Figs. 9, 11] (also see [23]). A detailed treatment of $e = 6, 7$ is also given in [23].

The proof is complete. □

Remark. The case $M = 10^6$ can now be settled without much effort. In the exceptional case $e = 4$, we have $\mathrm{ch}(10^6, 0, 0, 0, 0) = 36$ and $\mathrm{ch}(2^{20}, 0, 0, 0, 0) = 37$. Hill [23] provided a perfect strategy for the state $(10^6, 0, 0, 0, 0)$, whence $N(10^6, 4) = 36$. For the remaining values of e one has $N(10^6, e) = N(2^{20}, e)$ because $\mathrm{ch}(10^6, 0, \ldots, 0) = \mathrm{ch}(2^{20}, 0, \ldots, 0)$ and by Lemma 2 any winning strategy for $(2^{20}, 0, \ldots, 0)$ trivially yields a winning strategy for $(10^6, 0, \ldots, 0)$.

2.5 Asymptotic solutions

In [45] Spencer considered search spaces of arbitrary cardinality, and proved the following main result.

Theorem 6. *There exist constants K_e and Q_e (depending on e) having the following property: for all integers $n \geq Q_e$, if a state $(x_0, \ldots, x_e)$ satisfies $V_n(x_0, \ldots, x_e) \leq 2^n$ and $x_e \geq K_e n^e$, then the state is n-winning.*

Remark. There is an immediate corollary. Fix two integers $e, m \geq 0$. Let S be a search space of cardinality $M = 2^m$. Then, up to finitely many exceptional m's, there exists a perfect winning strategy for Paul. Stated otherwise, Paul can win the game over S with e lies using n questions, with n being the smallest integer satisfying the volume bound. Trivially, no such winning strategy can use less than n questions.

3 Ulam–Rényi games and error-correction

3.1 Coding in a channel with noiseless feedback

Led by Schalkwijk and Kailath, Berlekamp [6] proposed the following model for information transmission over a noisy channel equipped with a noiseless and delayless feedback channel of large capacity.

Suppose the source must send a message μ from a set $\mathcal{M}$ of M possible messages. To fix ideas, assume that the channel can only deliver bits, and that up to e of the bits can be distorted during transmission. In order to communicate the message to the receiver, the source sends a certain number n of bits over the noisy channel. By contrast with traditional e error-correcting codes, these n bits may depend adaptively on the information received by the source via the feedback channel.

As an equivalent reformulation, one may suppose that, after the source has chosen the message μ, the receiver chooses a subset Q_1 of $\mathcal{M}$, and then asks the following yes–no question over the noiseless feedback channel:

"is your message μ an element of Q_1?"

The source's answer (one bit) is then sent to the receiver via the noisy channel. The receiver gets a (possibly distorted) answer, and then adaptively asks the next question Q_2. *At each step t the question $Q_t \subseteq \mathcal{M}$ is adaptively asked knowing the answers to the preceding questions.*

Since the receiver's questions range over all possible subsets of a space of cardinality M, one might be led to think that any such question requires M bits to be sent over the noiseless channel. It turns out, however, that *only one feedback bit suffices, for each bit transmitted by the source.* This is so because we can safely assume a co-operative model, where the source knows the receiver's searching (question selecting) strategy. Thus, in particular, the first question Q_1 is known to the source: the first bit $A_1^* \in \{0, 1\} = \{\text{"no"},\text{"yes"}\}$ transmitted via the noisy channel is the answer to this question. This bit is received as A_1 where $A_1 = 1 - A_1^*$ or $A_1 = A_1^*$ according as distortion occurs during its transmission or not.

Proceeding inductively, let Q_{i+1} be the $(i + 1)^{\text{th}}$ question, A_{i+1}^* be the answer as transmitted from the source-end of the noisy channel, and A_{i+1} be the (possibly noisy) answer delivered at the receiver-end by the noisy channel, as obtained by the receiver. Since question Q_{t+1} depends only on the previous answers $A_1, \dots, A_t$, for the source to know question Q_{t+1} it is sufficient to know the bits $A_1, \dots, A_t$. By safely sending these bits over the feedback noiseless channel, the receiver enables the source to know question Q_{t+1}.

For any given M and e, one can naturally consider the problem of minimizing the number n of feedback bits – precisely as in the Ulam–Rényi game. As we shall see, any encoding scheme for Berlekamp's binary symmetric channel with noiseless feedback is essentially the same thing as a winning strategy for the Ulam–Rényi game. In particular, any winning strategy with a minimum number of questions amounts to a shortest error-correcting code for this sort of channel – and vice versa.

3.2 No feedback = error-correcting codes

Let us suppose that the feedback channel is not available. Then the problem is to send a message μ, using the minimum number of bits, in such a way that the receiver can recover μ, even if up to e of these bits may be distorted during transmission.

This is precisely the main issue of the theory of (binary) error-correcting codes. Here, source and receiver agree on fixing an injective map $c : \mathcal{M} \mapsto \{0, 1\}^n$. To transmit a message $\mu \in \mathcal{M}$, the source sends $\boldsymbol{x} = c(\mu)$, over the noisy channel. The original n-tuple of bits $\boldsymbol{x}$ is received as $\boldsymbol{x}'$ after transmission over the noisy channel, and in general $\boldsymbol{x}' \neq \boldsymbol{x}$, because of distortion. For the maximum number e of bits that can be distorted, a careful choice of the map c should allow the receiver to compute $\boldsymbol{x}$ from $\boldsymbol{x}'$ and then, by inverting c, to recover the original message μ.

Under this representation, one can safely regard the range of c as a set of possible sequences of answers to suitably chosen yes–no questions in an Ulam-Rényi game with e lies. More precisely, let

$$\mathcal{Q}_i = \{\mu \in \mathcal{M} \mid \text{the } i^{\text{th}} \text{ bit of } c(\mu) \text{ is } 1\}.$$

From the viewpoint of the source-end, computing $c(\mu)$ is the same as answering the questions "does the message μ belong to the set $\mathcal{Q}_i$?" for $i = 1, 2, \ldots, n$.

On the other hand, suppose $\mathcal{Q}_i \subseteq \mathcal{M}$ $(i = 1, 2, \ldots, n)$ is an n-tuple of (predetermined, nonadaptive) questions having the following property: even if up to e of the answers may be erroneous, from the n answers to the $\mathcal{Q}_i$ the secret number can always be guessed. Suppose that the map $c\colon \mathcal{M} \mapsto \{0, 1\}^n$ transforms each message $\mu \in \mathcal{M}$ into the n-tuple of bits $x_1, x_2, \ldots, x_n$, given by $x_i = 1$ or $x_i = 0$ according as $\mu \in \mathcal{Q}_i$ or $\mu \notin \mathcal{Q}_i$, respectively.

Then one can naturally say that the range $c(\mathcal{M})$ of c is the encoding of the set $\mathcal{M}$ of messages. Thus, a code is the same as the range of such a map c. Finding shortest codes amounts to finding shortest questioning strategies. By contrast with the original Ulam–Rényi game, here the receiver cannot adaptively ask his $(t + 1)^{\text{th}}$ question in the light of the previous t answers: all questions must be known by the source in advance, before any bit is sent.

3.3 More about error-correcting codes

Since shortest error-correcting codes are the same as solutions of the nonadaptive case of the Ulam–Rényi problem, for later use we collect here all necessary background material from the theory of error-correcting codes. We refer to [27] for more information. It is convenient to relax the assumption that the channel can deliver only binary digits (corresponding to the assumption that only yes–no answers are allowed).

For arbitrary integers $q \geq 2$ and $n > 0$, let $\boldsymbol{x}, \boldsymbol{y} \in \{0, 1, \ldots, q - 1\}^n$. Then the *Hamming distance* $d_H(\boldsymbol{x}, \boldsymbol{y})$ is defined by

$$d_H(\boldsymbol{x}, \boldsymbol{y}) = |\{i \in \{1, \ldots, n\} \mid x_i \neq y_i\}|,$$

where, as above, $|A|$ denotes the number of elements of A.

The *Hamming sphere* $\mathcal{B}_r(\boldsymbol{x})$ with *radius* r and *center* $\boldsymbol{x}$ is the set of elements of $\{0, 1, \ldots, q-1\}^n$ whose Hamming distance from $\boldsymbol{x}$ is $\leq r$, in symbols,

$$\mathcal{B}_r(\boldsymbol{x}) = \{\boldsymbol{y} \in \{0, 1, \ldots, q-1\}^n \mid d_H(\boldsymbol{x}, \boldsymbol{y}) \leq r\}.$$

For each $\boldsymbol{x} \in \{0, 1, \ldots, q-1\}^n$ we have

$$|\mathcal{B}_r(\boldsymbol{x})| = \sum_{i=0}^{r} \binom{n}{i} (q-1)^i. \tag{12}$$

The *Hamming weight* $w_H(\boldsymbol{x})$ is the number of nonzero digits of $\boldsymbol{x}$.

When q is clear from the context, by a *code* we mean a q-ary code in the following sense.

Definition 5. A *(q-ary) code C of length n* is a subset of $\{0, 1, \ldots, q-1\}^n$. When $q = 2$ we call C a *binary* code. Its elements are called *codewords*. The set $\{0, 1, \ldots, q-1\}$ is called the *alphabet* of the code C. The *minimum distance* of C is given by

$$\delta(C) = \min\{d_H(\boldsymbol{x}, \boldsymbol{y}) \mid \boldsymbol{x}, \boldsymbol{y} \in C, \boldsymbol{x} \neq \boldsymbol{y}\}.$$

We say that C is an (n, M, d)*-code*, if C has length n, $|C| = M$ and $\delta(C) = d$. The *minimum weight* of C is the minimum of the Hamming weights of its codewords, in symbols, $\mu(C) = \min\{w_H(\boldsymbol{x}) \mid \boldsymbol{x} \in C\}$.

Let C_1 and C_2 be two codes of length n. The *minimum distance* between C_1 and C_2 is defined by

$$\Delta(C_1, C_2) = \min\{d_H(\boldsymbol{x}, \boldsymbol{y}) \mid \boldsymbol{x} \in C_1, \boldsymbol{y} \in C_2\}.$$

By definition, the empty set $\emptyset$ is a q-ary $(n, 0, d)$ code for all integers $n, d \geq 0$ and $q \geq 2$. Further, for any code C and integer $d \geq 0$, we have the inequality $\Delta(\emptyset, C) \geq d$. Similarly, the code consisting of the single codeword $\underbrace{0 \cdots 0}_{n \text{ times}}$ is a q-ary $(n, 1, d)$-code for all integers $d \geq 0$ and $q \geq 2$.

Let C be a q-ary code and let $\boldsymbol{x}$ be a codeword of C. Suppose that we send $\boldsymbol{x}$ over a noisy channel. Under the assumption that at most e of the digits of $\boldsymbol{x}$ may be distorted, the received word $\boldsymbol{y} \in \{0, 1, \ldots, q-1\}^n$ trivially belongs to the Hamming sphere $\mathcal{B}_e(\boldsymbol{x})$.

If the Hamming spheres of radius e surrounding the codewords of C are pairwise disjoint, then for all received $\boldsymbol{y}$ there must exist exactly one $\boldsymbol{x}$ such that $\boldsymbol{y} \in \mathcal{B}_e(\boldsymbol{x})$. Thus at the receiving end it is safe to decode $\boldsymbol{y}$ as $\boldsymbol{x}$. Indeed, for any $\boldsymbol{z} \in C$ such that $\boldsymbol{z} \neq \boldsymbol{x}$, we have $d_H(\boldsymbol{z}, \boldsymbol{y}) \geq e + 1$.

Therefore any (n, m, d)-code C is capable of correcting e errors if and only if $d \geq 2e + 1$; when this is the case we say that C is an *e-error-correcting code*.

Thus, for an e-error-correcting code C, the Hamming spheres of radius e centered on the codewords of C must be pairwise disjoint. Since their union does not exceed the set $\{0, 1, \ldots, q-1\}^n$, we immediately have

Theorem 7 (Hamming, alias sphere packing, bound). *Let C be an e-error-correcting q-ary code of length n. Let $|C| = M$. Then*

$$M \sum_{j=0}^{e} \binom{n}{j} (q-1)^j \leq q^n. \tag{13}$$

Definition 6. A q-ary (n, m, d)-code C with minimum distance $d = 2e + 1$ is called *perfect* if for each $\boldsymbol{x} \in \{0, 1, \ldots, q-1\}^n$ there exists exactly one $\boldsymbol{y} \in C$ such that $d_H(\boldsymbol{x}, \boldsymbol{y}) \leq e$.

Perfect e-error-correcting codes are the most peculiar examples of perfect nonadaptive winning strategies for the Ulam–Rényi game. More generally, perfect strategies for the nonadaptive Ulam–Rényi game over a search space of cardinality M are the same as e-error-correcting codes that minimize n with respect to the bound in Theorem 7. These codes are our main concern in the rest of this section. As of today, however, aside from finitely many exceptional values of e and M, the best *known* codes are far from matching the lower bound on the length n provided in (13).

3.4 Linear and Hamming codes

The alphabet of every code C considered in this subsection is assumed to coincide with a finite field F_q, where $q = p^r$, the integer p is prime and $r \geq 1$. We let F_q^n denote the n-dimensional vector space over F_q. Thus every code C of length n satisfies $C \subseteq F_q^n$.

Adapting Definition 5, we give

Definition 7. A *linear code* C of length n is a linear subspace of F_q^n. If C is m-dimensional then C is called an $[n, m]$-code. We say for short that C is an $[n, m, d]$-*code* if C is an m-dimensional linear code of length n and minimum distance d.

By an $[n, m, d]$-*code* we understand an (n, q^m, d)-code whose codewords form a vector subspace of the vector space F_q^n.

Every linear code is specified by a pair of matrices whose effect is to simplify the encoding and decoding steps of the transmission protocol.

Definition 8. A *generator matrix* G for an $[n, m]$-code C is an $m \times n$ matrix whose rows yield a basis of C.

A *parity check matrix* H of an $[n, m]$-code C is an $(n - m) \times n$ matrix, such that, for all $\boldsymbol{x} \in C$, we have $H\,\boldsymbol{x}^T = \boldsymbol{0}$.

Let G be a generator matrix of an $[n, m]$-code C. If G has the form[9] $[I_m \mid A]$, where I_m is the identity $(m \times m)$-matrix, then the matrix $H = [-A^T \mid I_{n-m}]$ is a parity check matrix for C.

---

[9] For an $(m \times n_1)$-matrix A_1 and and an $(m \times n_2)$-matrix A_2, we denote by $[A_1 \mid A_2]$ the $(m \times (n_1 + n_2))$-matrix whose first n_1 columns are those of A_1 and the remaining columns are those of A_2, in the same order.

Remark. It is not hard to prove that, for any $[n, m, d]$-code C with parity check matrix H, the minimum distance d of C is equal to the minimum number of linearly dependent columns of H. One can now get a procedure to determine the minimum distance of a linear code, as well as its error-correcting capability.

Without loss of generality on identifying the source messages with vectors in $S = \{0, 1, \ldots, q-1\}^m$, it follows that, for each $\boldsymbol{u} \in S$, its corresponding codeword is given by $\boldsymbol{x} = \boldsymbol{u} \cdot G$. The map $\boldsymbol{u} \mapsto \boldsymbol{x}$ is easily computable. Hence, if a linear code is available, the encoding procedure becomes very efficient.

We next discuss the decoding procedure.

Definition 9. If C is a linear code with parity check matrix H, then for each $\boldsymbol{x} \in F_k^q$ we call $H\,\boldsymbol{x}^T$ the *syndrome* of $\boldsymbol{x}$.

By definition, codewords are characterized by having syndrome $\mathbf{0}$.

Let $\boldsymbol{x} \in C$ be the transmitted codeword and let $\boldsymbol{y} \in \{0, 1, \ldots, q-1\}^n$ be the received word. Then $\boldsymbol{e} = \boldsymbol{y} - \boldsymbol{x}$ is the error pattern. On the other hand, at the receiver end, the syndrome of the received word $\boldsymbol{y}$ is simply given by

$$H\,\boldsymbol{y}^T = H\,(\boldsymbol{x} + \boldsymbol{e})^T = H\,\boldsymbol{e}^T.$$

Since, by hypothesis, at most e errors have occurred, then $w_H(\boldsymbol{e}) \leq e$.

We now note that, for any two vectors $\boldsymbol{e}_1,\ \boldsymbol{e}_2 \in \{0, 1, \ldots, q-1\}^n$ of Hamming weight $\leq e$, we have $d_H(\boldsymbol{e}_1, \boldsymbol{e}_2) \leq 2e$, whence $H\,\boldsymbol{e}_1^T \neq H\,\boldsymbol{e}_2^T$. Thus, any syndrome corresponds to a unique error pattern.

Assuming that the number of errors is $\leq e$, the decoder must simply compute the syndrome of the received word $\boldsymbol{y}$. As we have seen, this amounts to recovering the error pattern $\boldsymbol{e}$, from which the decoder can easily obtain the original codeword $\boldsymbol{x} = \boldsymbol{y} - \boldsymbol{e}$. All that the decoder needs is a look-up table (syndrome/error-pattern) with q^{n-m} entries. This results in a significant improvement with respect to the general case of a code with no structure – where, using a table with q^n entries, one associates to every received vector $\boldsymbol{x} \in F_q^n$ its corresponding codeword $\boldsymbol{x}$.

Definition 10. For an integer $m \geq 1$, let $n = \frac{q^m - 1}{q - 1}$, where q is a prime power. An $[n, n-m]$-code over F_q is called a *Hamming code* if the columns of its parity check matrix are pairwise linearly independent vectors (over F_q). Stated otherwise, these columns yield a maximal set of pairwise linearly independent vectors.

Remark. The columns of the parity check matrix of a *binary* Hamming code ($q = 2$) precisely exhaust the set of nonzero vectors of F_2^m. Since the minimum number of linearly dependent columns in the parity matrix of a Hamming code, is 3, then the minimum distance of a Hamming code is equal to 3. Hence Hamming codes are one-error-correcting codes. Further, we have

Theorem 8. *Hamming codes are perfect one-error-correcting codes.*

Corollary 2. *For each* $m = 1, 2, 3, \ldots,$

$$N(2^m, 1) = N_{\min}(2^m, 1) = \min\{n = 0, 1, 2, \ldots \mid 2^n \geq 2^m(n+1)\}.$$

3.5 Bounds on codes

As we have seen above, the nonadaptive Ulam–Rényi problem with e lies over a search space of cardinality M amounts to finding a shortest *binary* e-error-correcting code with M codewords. Thus the nonadaptive Ulam–Rényi problem is an equivalent reformulation of the following main issue in combinatorial coding theory: *evaluate the maximum number $A(n, d)$ of codewords in a* binary *code of length n and minimum distance d*.

To make this paper more self-contained, we next recall some of the principal known bounds on $A(n, d)$.

Surprisingly enough, the best known lower bound on $A(n, d)$ is just the most trivial one:

Theorem 9 (Gilbert bound). *For all integers q, n, d, $q \geq 2$, $n \geq 1$, $1 \leq d \leq n$, there exists a q-ary (n, M, d)-code $\mathcal{C}$ with*

$$M \geq \frac{q^n}{\sum_{j=0}^{d-1} \binom{n}{j}(q-1)^j},$$

whence

$$A(n, d) \geq \frac{2^n}{\sum_{j=0}^{d-1} \binom{n}{j}}.$$

Proof. Starting with an arbitrary $x \in \{0, 1, \ldots, q-1\}^n$, one routinely keeps adding codewords lying at a distance at least d from all previously added codewords. The process stops, and the desired $\mathcal{C}$ is obtained, when, for each $z \in \{0, 1, \ldots, q-1\}^n$, there is at least one $x \in \mathcal{C}$ such that $y \in \mathcal{B}_{d-1}(x)$. Thus the Hamming spheres of radius $d-1$ surrounding the codewords of $\mathcal{C}$ cover the whole space $\{0, 1, \ldots, q-1\}^n$, whence

$$|\mathcal{C}| \sum_{j=0}^{d-1} \binom{n}{j}(q-1)^j \geq q^n,$$

as required to complete the proof. □

By definition, for all integers n and d, the *rate* of the largest binary code of length n and minimum distance d is given by

$$R(n, d) = \frac{1}{n} \log_2 A(n, d).$$

Most of the known upper bounds on $A(n, d)$ are expressed in terms of $R(n, d)$ as a function of d/n. The following theorem gives the best known upper bound on the largest size of a binary code of length n and minimum distance d, for all sufficiently large n. The theorem is due to McEliece–Rodemich–Rumsey–Welch [29].

Theorem 10 (McEliece, Rodemich, Rumsey, Welch). *For all sufficiently large n and for all d, $0 \leq d/n \leq 1/2$, we have*

$$R(n,d) \leq \min_{0 \leq u \leq 1-2d/n} \left\{ 1 + h(u^2) - h(u^2 + 2\frac{du}{n} + \frac{2d}{n}) \right\},$$

where $h(x) = \mathcal{H}(\frac{1}{2} - \frac{1}{2}\sqrt{1-x})$ *and* $\mathcal{H}(x)$ *denotes the binary entropy function* $\mathcal{H}(x) = -x \log_2 x - (1-x)\log_2(1-x)$.

In Table 3, we list, as a function of d/n, the rates of the largest binary codes allowed by the Gilbert, Hamming, and McEliece et al. bounds, respectively.

Table 3. Bounds on the rate of the best binary codes for large n

$\frac{d}{n}$	Gilbert-Varshamov lower-bound	Sphere Packing-Hamming upper-bound	McEliece et al. upper-bound
0	1	1	1
0.1	0.531	0.714	0.693
0.2	0.278	0.531	0.461
0.3	0.119	0.390	0.250
0.4	0.029	0.278	0.081
0.5	0	0.189	0

Evidently, there still exists a significant gap between the best known asymptotic lower and upper bounds on the size of the largest binary codes.

Moreover, the largest binary code of length n and minimum distance d as given by the McEliece et al. bound is far from matching the Hamming bound. Equivalently, in our terminology, the best known nonadaptive searching strategies for the Ulam–Rényi game over a search space of cardinality M and e lies, are far from being perfect.

In particular, when there exists an integer n such that $M \sum_{j=0}^{e} \binom{n}{j} = 2^n$, we have the following stronger negative result [47].

Theorem 11 (Tietäväinen's Theorem). *For each integer $e > 1$ no nontrivial perfect e-error-correcting code exists, except the* [23, 12, 7] *binary Golay code and the* [11, 6, 5] *ternary Golay code*[10].

As an interesting example, let $M = 2^{78}$ and $e = 2$. Then, for $n = 90$,

$$M\left(\binom{n}{2} + n + 1\right) = 2^n.$$

[10] Golay codes were introduced by M.J.E. Golay in [22] (see [27, Chap. 20] for more details).

By Tietäväinen's Theorem no nonadaptive perfect strategy exists for solving the Ulam–Rényi problem with the above parameters. On the other hand, as a particular case of the main result of [19, 33], fully adaptive perfect strategies do exist.

4 Fault-tolerant search: small feedback

4.1 The problem

We consider the case $M = 2^m$, and let the search space S now coincide with the set of m-bit integers, $S = \{0, 1, \ldots, 2^m - 1\}$. By Theorem 2, at least $N_{\min}(2^m, e)$ questions are *necessary* to find the secret number $x \in S$ in the adaptive and, a fortiori, in the nonadaptive Ulam–Rényi game with e lies.

In the fully adaptive case, the above mentioned result of Spencer [45] (Theorem 6 and the remark following it) shows that $N_{\min}(2^m, e)$ questions are always sufficient, up to finitely many exceptional m's. Optimal searching strategies are explicitly given in the papers [19, 35, 37] for the cases $e = 1$, $e = 2$ and $e = 3$ respectively. Altogether, fully adaptive fault-tolerant search can be performed in a very satisfactory manner.

In many practical situations, however, it is desirable to have searching strategies with a small degree of adaptiveness – that is, strategies in which most questions are predetermined, and can be asked in parallel. This is the case, e.g., when the Questioner and the Responder are far away from each other and can interact only on a slow channel.

Minimum feedback is also desirable in all situations in which the mere process of formulating the queries is so costly that the Questioner finds it more convenient to prepare them in advance. For instance, in certain applications of computational molecular biology [26], one preferably uses two-stage searching strategies, where the search is adapted only once. We refer to the monographs [2] and [21] for a discussion on the power of adaptive and nonadaptive searching strategies and their possible uses in different contexts.

As the reader will recall, in the fully nonadaptive case, finding a perfect strategy (i.e., a winning strategy of size $N_{\min}(2^m, e)$) amounts to finding an e-error-correcting code of length $N_{\min}(2^m, e)$ with 2^m codewords. By Theorem 8, Hamming codes yield perfect nonadaptive searching strategies (i.e., one round strategies) with the smallest possible number $N_{\min}(2^m, 1)$ of questions – indeed, Pelc [39] showed that adaptiveness in this case is irrelevant even under the stronger assumption that repetition of the same question is forbidden.

Despite the best efforts of several research groups[11], fully nonadaptive searching strategies with exactly $N_{\min}(2^m, e)$ questions – or equivalently, e-error-correcting codes with 2^m codewords of length $N_{\min}(2^m, e)$ – are very rare objects for $e \geq 2$.

[11] The situation is completely different in the case of no lies: here an optimal, fully nonadaptive searching strategy with $\lceil \log_2 |S| \rceil$ questions simply amounts to asking $\lceil \log_2 |S| \rceil$ queries about the occurrences of the bit 1 in the binary expansion of the unknown number $x \in S$.

A natural question to ask is then: *what happens if the Questioner is allowed a small amount of adaptiveness?*

In this section we show that, for each e and for all sufficiently large m, there exist searching strategies of shortest length (viz., using *exactly* the theoretical minimum number $N_{\min}(2^m, e)$ of questions) in which questions can be submitted to the Responder in *only two* rounds. Specifically, for the Questioner to guess infallibly the Responder's secret number $x \in S$ it is *sufficient* to ask a first batch of m nonadaptive questions, and then, depending only on the m-tuple of answers, ask a second mini-batch of s nonadaptive questions.

Our strategies are *perfect*, in that $m + s$ coincides with Berlekamp's minimum $N_{\min}(2^m, e)$, the number of questions that are a priori necessary to accommodate all possible answering strategies of the Responder – once he is allowed to lie up to e times.

Since the Questioner can adapt his strategy only once, we have indeed e-fault-tolerant search strategies with minimum (nonzero) adaptiveness and the least possible number of tests.

4.2 Strategies versus codes

Lemma 4. *Let e, n, m be integers > 0. Suppose that $\mathcal{C}$ is a binary (n, m, d)-code such that $\mu(\mathcal{C}) \geq e$ and $d \geq 3$. Then there exists a binary $(n + 2, 2m, d')$-code $\mathcal{D}$ such that $\mu(\mathcal{D}) \geq e$ and $d' \geq 3$.*

Proof. Given a (binary) code $\mathcal{G}$ of length n, n-tuples $\boldsymbol{x} = x_1 \cdots x_n \in \{0, 1\}^n$ and $\boldsymbol{a} = a_1 a_2 \cdots a_s \in \{0, 1\}^s$, we denote by $\{\mathcal{G} \oplus \boldsymbol{x}\} \otimes \boldsymbol{a}$ the code of length $n + s$ whose codewords are obtained by adding $\boldsymbol{x}$ (term-wise and modulo 2) to every codeword of $\mathcal{G}$, and then appending the suffix $\boldsymbol{a}$ to the resulting n-tuple. In symbols,

$$\begin{aligned}\{\mathcal{G} \oplus \boldsymbol{x}\} \otimes \boldsymbol{a} = \{&z_1 \cdots z_n a_1 \cdots a_s \mid z_i \equiv y_i + x_i \mod 2 \\ &\text{for some } y_1 \cdots y_n \in \mathcal{G}, \ \text{with } z_1, \ldots, z_n \in \{0, 1\}\}.\end{aligned}$$

The code $\mathcal{D}$ is now defined by

$$\mathcal{D} = \{\mathcal{C} \oplus 0\,\underbrace{00\cdots 0}_{n-1 \text{ times}}\} \otimes 00 \cup \{\mathcal{C} \oplus 1\,\underbrace{00\cdots 0}_{n-1 \text{ times}}\} \otimes 11. \tag{14}$$

We claim that $\mathcal{D}$ satisfies the requirements of the lemma. By definition, the length of $\mathcal{D}$ is $n + 2$. Since

$$\{\mathcal{C} \oplus 000\cdots 0\} \otimes 00 \cap \{\mathcal{C} \oplus 100\cdots 0\} \otimes 11 = \emptyset,$$

then $\mid \mathcal{D} \mid = 2 \mid \mathcal{C} \mid$.

We now show that $\delta(\mathcal{D}) \geq 3$. Indeed, any two distinct codewords $\boldsymbol{x}, \boldsymbol{y} \in \mathcal{D}$ have the form $\boldsymbol{x} = \{\boldsymbol{x}' \oplus i00\cdots 0\} \otimes ii$ and $\boldsymbol{y} = \{\boldsymbol{y}' \oplus j00\cdots 0\} \otimes jj$ for suitable codewords $\boldsymbol{x}', \boldsymbol{y}' \in \mathcal{C}$ and $i, j \in \{0, 1\}$.

We now argue by cases.

(i) If $i = j$, then $\mathbf{x}' \neq \mathbf{y}'$, whence

$$\begin{aligned} d_H(\mathbf{x}, \mathbf{y}) &= d_H(\{\mathbf{x}' \oplus i00\cdots0\} \otimes ii, \{\mathbf{y}' \oplus i00\cdots0\} \otimes ii) \\ &= d_H(\mathbf{x}', \mathbf{y}') \geq 3, \end{aligned}$$

by the hypothesis on $\delta(\mathcal{C})$.

(ii) If $i \neq j$, then

$$\begin{aligned} d_H(\mathbf{x}, \mathbf{y}) &= d_H(\{\mathbf{x}' \oplus i00\cdots0\} \otimes ii\,, \{\mathbf{y}' \oplus j00\cdots0\} \otimes jj) \\ &= d_H(\{\mathbf{x}' \oplus i00\cdots0\}\,, \{\mathbf{y}' \oplus j00\cdots0\}) + 2. \end{aligned}$$

If $\mathbf{x}' = \mathbf{y}'$, then $d_H(\{\mathbf{x}' \oplus i00\cdots0\}\,, \{\mathbf{y}' \oplus j00\cdots0\}) = 1$, and hence $d_H(\mathbf{x}, \mathbf{y}) = 3$.
If $\mathbf{x}' \neq \mathbf{y}'$, then $d_H(\{\mathbf{x}' \oplus i00\cdots0\}\,, \{\mathbf{y}' \oplus j00\cdots0\}) \geq d - 1 \geq 2$, whence $d_H(\mathbf{x}, \mathbf{y}) \geq 4$.

Finally, by definition, $\mu(\mathcal{D}) = \mu(\mathcal{C}) \geq e$. The proof is complete. □

We shall now introduce a correspondence between nonadaptive winning strategies and certain special binary codes. This will be our key tool in proving the existence of perfect searching strategies using the minimum possible amount of adaptiveness.

Unless otherwise specified, by a *code* we henceforth understand a *binary* code.

Lemma 5. *Let $(a_0, a_1, a_2, \ldots, a_e)$ be the type of $\sigma = (A_0, A_1, A_2, \ldots A_e)$. Let the integer n satisfy the condition $n \geq \mathrm{ch}(\sigma)$. Then a nonadaptive winning strategy for σ with n questions exists if and only if, for all $i = 0, 1, 2, \ldots, e-1$, there are integers $d_i \geq 2(e-i)+1$, together with an e-tuple of codes $\Gamma = \{\mathcal{C}_0, \mathcal{C}_1, \mathcal{C}_2, \ldots, \mathcal{C}_{e-1}\}$, such that each $\mathcal{C}_i$ is an (n, a_i, d_i)-code and $\Delta(\mathcal{C}_i, \mathcal{C}_j) \geq 2e - (i+j) + 1$, (for all $i, j, 0 \leq i < j \leq e-1$).*

Proof. We first prove the implication: *strategy* $\Rightarrow$ *codes*.

Assume $\sigma = (A_0, A_1, A_2, \ldots, A_e)$ to be a state of type $(a_0, a_1, a_2, \ldots, a_e)$ having a nonadaptive winning strategy $\mathcal{S}$ with n questions $T_1, \ldots, T_n$, $n \geq \mathrm{ch}(\sigma)$. Let the map

$$z \in A_0 \cup A_1 \cup A_2 \cup \cdots \cup A_e \mapsto \mathbf{z}^{\mathcal{S}} \in \{0, 1\}^n$$

send each $z \in A_0 \cup A_1 \cup A_2 \cup \cdots \cup A_e$ into the n-tuple of bits $\mathbf{z}^{\mathcal{S}} = z_1^{\mathcal{S}} \cdots z_n^{\mathcal{S}}$ arising from the sequence of "true" answers to the questions "does z belong to T_1 ?", "does z belong to T_2 ?", ..., "does z belong to T_n?", via the identifications 1 = "yes", 0 = "no". More precisely, for each $j = 1, \ldots, n$, $z_j^{\mathcal{S}} = 1$ iff $z \in T_j$. Let $\mathcal{C} \subseteq \{0, 1\}^n$ be the range of the map $z \mapsto \mathbf{z}^{\mathcal{S}}$. We first prove that, for every $i = 0, \ldots, e-1$, there exists an integer $d_i \geq 2(e-i)+1$ such that the set $\mathcal{C}_i = \{\mathbf{y}^{\mathcal{S}} \in \mathcal{C} \mid y \in A_i\}$ is an (n, a_i, d_i)-code.

Since $\mathcal{S}$ is winning, the map $z \mapsto \mathbf{z}^{\mathcal{S}}$ is one-to-one, whence in particular $|\mathcal{C}_i| = a_i$ for any $i = 0, 1, 2, \ldots, e-1$. Moreover, by definition, the $\mathcal{C}_i$'s are subsets of $\{0, 1\}^n$.

Claim. For any $i, j, 0 \leq i \leq j \leq e-1$, and for each $y \in A_i$ and $h \in A_j$, we have the inequality $d_H(\mathbf{y}^{\mathcal{S}}, \mathbf{h}^{\mathcal{S}}) \geq 2e - (i+j) + 1$.

For, otherwise (absurdum hypothesis), let $y \in A_i, h \in A_j$ be a counterexample, and $d_H(\boldsymbol{y}^{\mathcal{S}}, \boldsymbol{h}^{\mathcal{S}}) \leq 2e - (i + j)$. With $\boldsymbol{y}^{\mathcal{S}} = y_1^{\mathcal{S}} \dots y_n^{\mathcal{S}}$ and $\boldsymbol{h}^{\mathcal{S}} = h_1^{\mathcal{S}} \dots h_n^{\mathcal{S}}$, it is no loss of generality to assume that $h_k^{\mathcal{S}} = y_k^{\mathcal{S}}$ for all $k = 1, \dots, n - (2e - (i + j))$. Suppose that the answer to question T_k is "yes" or "no" according as $h_k^{\mathcal{S}} = 1$ or $h_k^{\mathcal{S}} = 0$, respectively. Then the state resulting from these answers has the form $\sigma'' = (A_0'', A_1'', A_2'', \dots, A_e'')$, where $y \in A_i''$ and $h \in A_j''$. Since, by [6, Lemma 2.5], $\mathrm{ch}(\sigma'') \geq \mathrm{ch}(0, \dots, 0, 1, 0, \dots, 0, 1, 0, \dots, 0) = 2e - (i + j) + 1$[12], Theorem 2 shows that $2e - (i + j)$ additional questions will not suffice to find the unknown number. This contradicts the assumption that $\mathcal{S}$ is a winning strategy.

By the above claim, when $i = j$ we immediately have $\delta(C_i) \geq 2(e - i) + 1$. In conclusion, for all $i = 0, 1, \dots, e-1$, C_i is an (n, a_i, d_i)-code with $d_i \geq 2(e-i)+1$. Similarly we have the desired inequality $\Delta(C_i, C_j) \geq 2e - (i + j) + 1i$ for all i, j, $0 \leq i < j \leq e - 1$.

Now we prove the converse implication: *strategy* $\Leftarrow$ *codes*.

Let $\Gamma = \{C_0, C_1, C_2, \dots, C_{e-1}\}$ be a family of codes satisfying the hypotheses. Let

$$\mathcal{H} = \bigcup_{i=0}^{e-1} \bigcup_{\boldsymbol{x} \in C_i} \mathcal{B}_{e-i}(\boldsymbol{x}).$$

For all $i, j \in \{0, 1, \dots, e - 1\}$ and $\boldsymbol{x} \in C_i$, $\boldsymbol{y} \in C_j$, we have $d_H(\boldsymbol{x}, \boldsymbol{y}) \geq 2e - (i + j) + 1$. Thus the Hamming spheres $\mathcal{B}_{e-i}(\boldsymbol{x})$, $\mathcal{B}_{e-j}(\boldsymbol{y})$ are pairwise disjoint and hence

$$|\mathcal{H}| = \sum_{i=0}^{e-1} a_i \sum_{j=0}^{e-i} \binom{n}{j}. \tag{15}$$

Let $\mathcal{D} = \{0, 1\}^n \setminus \mathcal{H}$. From the inequality $n \geq \mathrm{ch}(a_0, a_1, a_2, \dots, a_e)$ we get $2^n \geq \sum_{i=0}^{e} a_i \sum_{j=0}^{e-i} \binom{n}{j}$.

From the above identity it follows that

$$|\mathcal{D}| = 2^n - \sum_{i=0}^{e-1} a_i \sum_{j=0}^{e-i} \binom{n}{j} \geq a_e. \tag{16}$$

Let $\sigma = (A_0, A_1, A_2, \dots, A_e)$ be an arbitrary state of type $(a_0, a_1, a_2, \dots, a_e)$. We now fix, once and for all, $e+1$ one-to-one maps $f_i \colon A_i \to C_i$, for $i = 0, 1, \dots, e-1$, and $f_e \colon A_e \to \mathcal{D}$. The existence of the map f_i, for each $i = 0, 1, \dots, e$, is ensured by our assumptions about Γ, together with (16).

[12] Note that, in the particular case $i = j$, again by [6, Lemma 2.5], we have $\mathrm{ch}(\sigma'') \geq \mathrm{ch}(0, \dots, 0, 2, 0, \dots, 0) = 2(e - i) + 1 = 2e - (i + j) + 1$.

Define the map $f\colon A_0 \cup A_1 \cup A_2 \cup \cdots \cup A_e \to \{0,1\}^n$ by cases as follows:

$$f(y) = \begin{cases} f_0(y), & y \in A_0 \\ f_1(y), & y \in A_1 \\ \vdots \\ f_e(y), & y \in A_e \end{cases} \tag{17}$$

Note that f is one-to-one. For each $y \in A_0 \cup A_1 \cup A_2 \cup \cdots \cup A_e$ and $j = 1, \ldots, n$, let $f(y)_j$ be the jth bit of the binary vector corresponding to y via f. We can now explicitly formulate the questions T_j of our searching strategies.

For each $j = 1, \ldots, n$, let the set $T_j \subseteq S$ be defined by $T_j = \{z \in \bigcup_{i=0}^{e} A_i \mid f(z)_j = 1\}$. Intuitively, if x_* denotes the unknown number, T_j asks "is the jth bit of $f(x_*)$ equal to one ?".

Again with "yes" = 1 and "no" = 0, the answers to questions $T_1, \ldots, T_n$ determine an n-tuple of bits $\boldsymbol{b} = b_1 \cdots b_n$. We shall show that the sequence $T_1, \ldots, T_n$ yields an optimal nonadaptive winning strategy for σ. Let $\sigma_1 = \sigma^{b_1}$, $\sigma_2 = \sigma_1^{b_2}, \ldots, \sigma_n = \sigma_{n-1}^{b_n}$. Arguing by cases we shall show that $\sigma_n = (A_0^*, A_1^*, \ldots, A_e^*)$ is a final state.

By (1)–(2), for all $i = 0, 1, \ldots, e$, any $z \in A_{e-i}$ that falsifies $> i$ answers does not survive in σ_n – in the sense that $z \notin A_0^* \cup A_1^* \cup \cdots \cup A_e^*$.

Case 1. $\boldsymbol{b} \notin \bigcup_{i=0}^{e} \bigcup_{y \in A_i} \mathcal{B}_{e-i}(f(y))$. For all $i = 0, 1, \ldots, e$ and for each $y \in A_i$ we must have $y \notin A_0^* \cup A_1^* \cup \cdots \cup A_e^*$. Indeed, the assumption $\boldsymbol{b} \notin \mathcal{B}_{e-i}(f(y))$ implies that $d_H(f(y), \boldsymbol{b}) > e - i$, whence y falsifies $> e - i$ of the answers to $T_1, \ldots, T_n$, and y does not survive in σ_n. We have proved that $A_0^* \cup A_1^* \cup \cdots \cup A_e^*$ is empty, and σ_n is a final state.

Case 2. $\boldsymbol{b} \in \mathcal{B}_{e-i}(f(y))$ *for some* $i \in \{0, 1, \ldots, e\}$ *and* $y \in A_i$. Then $y \in A_0^* \cup A_1^* \cup \cdots \cup A_e^*$ because $d_H(f(y), \boldsymbol{b}) \le e - i$, whence y falsifies $\le e - i$ answers. Our assumptions about Γ ensure that, for all $j = 0, 1, \ldots, e$ and for all $y' \in A_j$ and $y \ne y'$, we have $\boldsymbol{b} \notin \mathcal{B}_{e-j}(f(y'))$. Thus, $d_H(f(y'), \boldsymbol{b}) > e - j$ and y' falsifies $> e - j$ of the answers to $T_1, \ldots, T_n$, whence y' does not survive in σ_n. This shows that, for any $y' \ne y$, $y' \notin A_0^* \cup A_1^* \cup \cdots \cup A_e^*$. Therefore, $A_0^* \cup A_1^* \cup \cdots \cup A_e^*$ only contains the element y, and σ_n is a final state. □

4.3 Perfect fault-tolerant search with minimum adaptiveness

Recall that $N_{\min}(2^m, e) = \mathrm{ch}(2^m, 0, \ldots, 0)$ is the smallest integer $n \ge 0$ such that $2^n \ge 2^m(\binom{n}{e} + \binom{n}{e-1} + \cdots + \binom{n}{2} + n + 1)$. By Lemma 2, at least $N_{\min}(2^m, e)$ questions are *necessary* to guess the unknown number $x_{\text{Carole}} \in S = \{0, 1, \ldots, 2^m - 1\}$, allowing up to e answers to be erroneous.

The aim of this section is to prove that, conversely, up to finitely many exceptional m, $N_{\min}(2^m, e)$ questions are *sufficient* by using the following procedure: first we use a predetermined nonadaptive batch of m questions $D_1, \ldots, D_m$, and then, depending only on the answers, we ask the remaining $N_{\min}(2^m, e) - m$ questions in a second nonadaptive batch.

The first batch of questions is easily described as follows.

> For each $i = 1, 2, \ldots, m$, let $D_i \subseteq S$ denote the question "is the ith binary digit of x_{Carole} equal to 1?". Thus a number $y \in S$ belongs to D_i iff the ith bit y_i of its binary expansion $\boldsymbol{y} = y_1 \cdots y_m$ is equal to 1.

Writing 1 = "yes" and 0 = "no", let $b_i \in \{0, 1\}$ be the answer to question D_i. Let $\boldsymbol{b} = b_1 \cdots b_m$. Repeated applications of (1)–(2), beginning with the initial state $\sigma = (S, \emptyset, \ldots, \emptyset)$, yield the state $\sigma^{\boldsymbol{b}} = (A_0, A_1, \ldots, A_e)$, where

$$A_i = \{y \in S \mid d_H(\boldsymbol{y}, \boldsymbol{b}) = i\} \quad \text{for all } i = 0, 1, \ldots, e.$$

Direct verification shows that

$$|A_0| = 1, \quad |A_1| = m, \ldots, |A_e| = \binom{m}{e}.$$

Thus $\sigma^{\boldsymbol{b}}$ has type $(1, m, \binom{m}{2}, \ldots, \binom{m}{e})$. Let σ_i be the state determined by the first i answers, beginning with $\sigma_0 = \sigma$. Since each question D_i is balanced for σ_{i-1}, an easy induction using Corollary 1 yields $\mathrm{ch}(\sigma^{\boldsymbol{b}}) = N_{\min}(2^m, e) - m$.

For each m-tuple $\boldsymbol{b} \in \{0, 1\}^m$ of possible answers, we shall construct a nonadaptive strategy $\mathcal{S}_{\boldsymbol{b}}$ with $\mathrm{ch}(1, m, \binom{m}{2}, \ldots, \binom{m}{e})$ questions, which turns out to be winning for the state $\sigma^{\boldsymbol{b}}$.

The resulting perfect strategy is "canonical" in the following sense.

Definition 11. A strategy $\mathcal{S}$ for a state σ of type $(2^m, 0, \ldots, 0)$ is said to be *canonical* if $\mathcal{S}$ is winning for σ and consists of two batches of nonadaptive questions, where the questions in the first batch ask for the binary digits of x_{Carole}, and the second batch depends only on the m-tuple of Carole's answers to these questions.

Next we focus attention on the values of $\mathrm{ch}(1, m, \binom{m}{2}, \ldots, \binom{m}{e})$ for $m \geq 1$.

Definition 12. Let $e \geq 0$ and $n \geq 2e$ be arbitrary integers. The *critical index* $m_{n,e}$ is the largest integer $m \geq 0$ such that $\mathrm{ch}(1, m, \binom{m}{2}, \ldots, \binom{m}{e}) = n$.

A moment's reflection shows that, for our present purposes in this paper, it is sufficient to prove that, for all $n \geq 1$ and $e \geq 2$, there exists a winning strategy of size n for the state $(1, m_{n,e}, \binom{m_{n,e}}{2}, \ldots, \binom{m_{n,e}}{e})$.

4.4 Canonical search: $e = 2$

We start our analysis of perfect least-adaptive e-fault-tolerant searching strategies by considering the case $e = 2$. By Corollary 2, this is the first significant case where the dichotomy between adaptive and nonadaptive search makes its appearance.

We present an inductive algorithm that effectively computes the second batch of nonadaptive questions for all $m \neq 2$. This includes all cases of practical interest, and should be contrasted with the asymptotic results presented in other sections of this survey. We rely on existing tables of error-correcting binary codes (notably [8] and [9]).

For each m-tuple $\boldsymbol{b} \in \{0, 1\}^m$ of Carole's answers, we construct a nonadaptive strategy with $\mathrm{ch}(1, m, \binom{m}{2})$ questions, and we prove that this strategy is winning for the above state $\sigma^{\boldsymbol{b}}$. For this purpose, we consider the values of $\mathrm{ch}(1, m, \binom{m}{2})$ for $m \geq 1$. Direct computation yields $\mathrm{ch}(1, 1, 0) = 4$, $\mathrm{ch}(1, 2, 1) = 5$, $\mathrm{ch}(1, 3, 3) = \mathrm{ch}(1, 4, 6) = 6$, $\mathrm{ch}(1, 5, 10) = \cdots = \mathrm{ch}(1, 8, 28) = 7$, $\mathrm{ch}(1, 9, 36) = \cdots = \mathrm{ch}(1, 14, 91) = 8, \ldots$

With reference to Definition 12 we have

Lemma 6. *Let $n \geq 4$ be an arbitrary integer.*

(i) *If n is odd, then $m_{n,2} = 2^{\frac{n+1}{2}} - n - 1$.*
(ii) *If n is even, then, letting $m^* = \lfloor 2^{\frac{n+1}{2}} \rfloor - n - 1$, we have either $m_{n,2} = m^*$ or $m_{n,2} = m^* + 1$.*

Proof. The case $n = 4$ is settled by direct verification (recall that $m_{4,2} = 1$). For the case $n \geq 5$ see [33, Lemma 4.2], where our present $m_{n,2}$ is denoted n_χ and is called the first critical index. □

As a key tool for the construction of the second batch of questions we prepare the following

Corollary 3. *Let $m = 1, 2, \ldots$ and $n = \mathrm{ch}(1, m, \binom{m}{2})$. Let $\sigma = (A_0, A_1, A_2)$ be a state of type $(1, m, \binom{m}{2})$. Then there exists a nonadaptive winning strategy for σ with n questions if and only if, for some integer $d \geq 3$, there exists an (n, m, d)-code with minimum Hamming weight ≥ 4.*

Proof. If there exists a nonadaptive winning strategy for σ with n questions, then, by Lemma 5, there exist an $(n, 1, d_0)$-code $\mathcal{C}_0$ and an (n, m, d_1)-code $\mathcal{C}_1$ with $d_0 \geq 5$, $d_1 \geq 3$, and $\Delta(\mathcal{C}_0, \mathcal{C}_1) \geq 4$. Let $\mathcal{C}_0 = \{\boldsymbol{h}\}$. Define

$$\mathcal{C} = \{\boldsymbol{y} \oplus \boldsymbol{h} \mid \boldsymbol{y} \in \mathcal{C}_1\},$$

where $\oplus$ stands for bit-wise sum modulo 2. For any two distinct codewords $\boldsymbol{a}, \boldsymbol{b} \in \mathcal{C}$ we have $\boldsymbol{a} = \boldsymbol{c} \oplus \boldsymbol{h}$ and $\boldsymbol{b} = \boldsymbol{d} \oplus \boldsymbol{h}$ for uniquely determined elements $\boldsymbol{c}, \boldsymbol{d} \in \mathcal{C}_1$. Thus we get $d_H(\boldsymbol{a}, \boldsymbol{b}) = d_H(\boldsymbol{c} \oplus \boldsymbol{h}, \boldsymbol{d} \oplus \boldsymbol{h}) = d_H(\boldsymbol{c}, \boldsymbol{d}) \geq d_1 \geq 3$, whence $\delta(\mathcal{C}) \geq 3$. Using the abbreviation

$$\mathbf{0} = \underbrace{0 \cdots 0}_{n \text{ times}},$$

we have $w_H(\boldsymbol{a}) = d_H(\boldsymbol{a}, \mathbf{0}) = d_H(\boldsymbol{c} \oplus \boldsymbol{h}, \boldsymbol{h} \oplus \boldsymbol{h}) = d_H(\boldsymbol{c}, \boldsymbol{h}) \geq 4$, whence $\mu(\mathcal{C}) \geq 4$. In conclusion, $\mathcal{C}$ is an (n, m, d)-code with $d \geq 3$ and $\mu(\mathcal{C}) \geq 4$, as required.

Conversely, let $\mathcal{C}$ be an (n, m, d)-code with $d \geq 3$ and $\mu(\mathcal{C}) \geq 4$. Let $\mathcal{D} = \{\mathbf{0}\}$. Then $\mathcal{D}$ is an $(n, 1, d')$-code for every $d' \geq 1$. Furthermore, we have $\Delta(\mathcal{C}, \mathcal{D}) \geq 4$. Thus by Lemma 5 there exists a nonadaptive winning strategy for σ with n questions. The proof is complete. □

Lemma 7. *For each integer $n \geq 7$ there exists an integer $d \geq 3$ and an $(n, m_{n,2}, d)$-code $\mathcal{C}_n$ such that $\mu(\mathcal{C}_n) \geq 4$, where $m_{n,2}$ is as in Definition* 12.

Proof. We argue by cases.

Case 1. $7 \leq n < 11$.

We have $m_{7,2} = 8$, $m_{8,2} = 14$, $m_{9,2} = 22$ and $m_{10,2} = 34$. By direct inspection in [9, Table I-A], for suitable integers $e_1, e_2 > 0$ with $e_1 + e_2 \geq m_{n,2}$ there exist an $(n, e_1, 4)$-code $\mathcal{A}_1$ and an $(n, e_2, 4)$-code $\mathcal{A}_2$. Moreover, for every $\boldsymbol{x} \in \mathcal{A}_1$, $w_H(\boldsymbol{x}) = 4$ and, for all $\boldsymbol{y} \in \mathcal{A}_2$, $w_H(\boldsymbol{y}) = 7$; hence $d_H(\boldsymbol{x}, \boldsymbol{y}) \geq 3$. It follows that every set $\mathcal{C}_n \subseteq \mathcal{A}_1 \cup \mathcal{A}_2$ such that $|\mathcal{C}_n| = m_{n,2}$ is an $(n, m_{n,2}, 3)$-code with the additional property $\mu(\mathcal{C}_n) = 4$, as required.

Case 2. $n \geq 11$.

Claim. There exists an (n, e, d)-code $\mathcal{D}_n$ such that $e \geq 2^{\frac{n+1}{2}}$, $d \geq 3$ and $\mu(\mathcal{D}_n) \geq 4$.

We argue by induction on n.

Base. $(n = 11, 12)$.

Then direct inspection in [9, Table I-A] yields two codes $\mathcal{A}_1, \mathcal{A}_2$, such that

- $\mathcal{A}_1$ is an $(11, 66, 4)$-code and, for every $\boldsymbol{x} \in \mathcal{A}_1$, $w_H(\boldsymbol{x}) = 6$
- $\mathcal{A}_2$ is a $(12, 132, 4)$-code and, for every $\boldsymbol{x} \in \mathcal{A}_2$, $w_H(\boldsymbol{x}) = 6$.

Let $\mathcal{D}_{11} = \mathcal{A}_1$ and $\mathcal{D}_{12} = \mathcal{A}_2$. The inequalities $132 > 2^{\frac{13}{2}}$ and $66 > 2^6$ now settle our claim for $n \in \{11, 12\}$.

Induction Step. Assuming the claim to hold for $n \geq 11$, we see by Lemma 4 that the claim also holds for $n + 2$, as required.

From Lemma 6 we get $2^{(n+1)/2} > m_{n,2}$, whence the desired conclusion now follows from our claim by picking arbitrarily a subcode $\mathcal{C}_n \subseteq \mathcal{D}_n$ with $|\mathcal{C}_n| = m_{n,2}$.
□

Corollary 4. *For $m = 5, 6, 7, \ldots$, let σ be a state of type $(1, m, \binom{m}{2})$. Then there exists a perfect strategy $\mathcal{S}$ for σ. In other words, $\mathcal{S}$ is winning for σ and the number of questions in $\mathcal{S}$ coincides with Berlekamp's lower bound* $\mathrm{ch}(\sigma) = \mathrm{ch}(2^m, 0, 0) - m$.

Proof. Let $n = \mathrm{ch}(\sigma)$. From the assumption $m \geq 5$ we get $n \geq 7$. By Definition 12, $m \leq m_{n,2}$. By Lemma 7 there exists an $(n, m_{n,2}, d)$-code $\mathcal{C}_n$ with $d \geq 3$ and $\mu(\mathcal{C}_n) \geq 4$. Now picking a subcode $\mathcal{C}'_n \subseteq \mathcal{C}_n$ such that $|\mathcal{C}'_n| = m$ and applying Corollary 3 we obtain the desired conclusion. □

Turning our attention to the remaining cases, we prove that Corollary 4 also holds when $m = 1$ and $m = 3$. For $m = 2$ and $m = 4$ we prove that the shortest nonadaptive winning strategy for a state of type $(1, m, \binom{m}{2})$ requires $\mathrm{ch}(1, m, \binom{m}{2}) + 1$ questions.

Lemma 8. *Let $\mathcal{C}$ be a largest code of length* 6 *such that $\mu(\mathcal{C}) = 4$ and $\delta(\mathcal{C}) \geq 3$. Then $|\mathcal{C}| = 3$.*

Proof. The code $\mathcal{D} = \{111100, 110011, 001111\}$ satisfies $\delta(\mathcal{D}) \geq 3$, $\mu(\mathcal{D}) = 4$ and $|\mathcal{D}| = 3$. It follows that a largest code $\mathcal{C}$ satisfying the requirements of the lemma necessarily has ≥ 3 elements.

Conversely, we prove that any such $\mathcal{C}$ has ≤ 3 (and, hence, exactly 3) elements. We write $\mathcal{C} = \mathcal{C}^{(4)} \cup \mathcal{C}^{(5)} \cup \mathcal{C}^{(6)}$, where $\mathcal{C}^{(i)} = \{\boldsymbol{x} \in \mathcal{C} \mid w_H(\boldsymbol{x}) = i\}$. We prove the following easy facts:

(a) $|\mathcal{C}^{(5)} \cup \mathcal{C}^{(6)}| \leq 1$,
(b) $|\mathcal{C}^{(4)}| \leq 3$;
(c) if $|\mathcal{C}^{(4)}| = 3$, then $|\mathcal{C}^{(5)} \cup \mathcal{C}^{(6)}| = 0$.

There cannot exist two distinct codewords $\boldsymbol{y}_1, \boldsymbol{y}_2 \in \mathcal{C}^{(5)} \cup \mathcal{C}^{(6)}$, for, otherwise, $d_H(\boldsymbol{y}_1, \boldsymbol{y}_2) \leq 2$, contrary to the hypothesis $\delta(\mathcal{C}) \geq 3$. This settles (a).

To prove (b), let $\boldsymbol{x}_1, \dots, \boldsymbol{x}_n$ be the list of codewords of $\mathcal{C}^{(4)}$. For each $i = 1, \dots, n$, let $N_i = \{\boldsymbol{y} \in \{0,1\}^6 \mid d_H(\boldsymbol{x}_i, \boldsymbol{y}) \leq 1 \text{ and } w_H(\boldsymbol{y}) = 5\}$. Each N_i has exactly two elements, and whenever $i \neq j$ we have $N_i \cap N_j = \emptyset$. It follows that $|\bigcup_{i=1}^n N_i| = \sum_{i=1}^n |N_i| = 2n$. Therefore,

$$2n = |\bigcup_{i=1}^{n} N_i| \leq |\{\boldsymbol{x} \in \{0,1\}^6 \mid w_H(\boldsymbol{x}) = 5\}| = 6,$$

and $n \leq 3$, as desired.

Finally, to prove (c), assume that $|\mathcal{C}^{(4)}| = 3$. Since, by the above proof of (b), $\bigcup_{i=1}^n N_i$ exhausts the set of 6-tuples of bits having Hamming weight 5, every 6-tuple of bits having Hamming weight 5 is contained in the Hamming sphere of radius 1 centered at some codeword in $\mathcal{C}^{(4)}$. From the assumption $\delta(\mathcal{C}) \geq 3$ it follows that $|\mathcal{C}^{(5)}| = 0$. Finally, $\mathcal{C}^{(6)}$ must be empty, because its only element 111111 has Hamming distance 2 from every element of $\mathcal{C}^{(4)}$. □

Lemma 9. *For each* $m = 1, 2, 3, 4$, *let* $\lambda(m)$ *be the length of the shortest nonadaptive winning strategy for some (equivalently, for every) state of type* $(1, m, \binom{m}{2})$. *Then*

$$\lambda(1) = 4, \quad \lambda(2) = 6, \quad \lambda(3) = 6, \quad \lambda(4) = 7.$$

For $m \in \{1, 3\}$, *and only for these values of* m, *the number* $\lambda(m)$ *satisfies the condition* $\lambda(m) = \mathrm{ch}(1, m, \binom{m}{2})$.

Proof. For $m = 1$ we have $\lambda(1) \geq \mathrm{ch}(1, 1, 0) = 4$. Conversely, by Corollary 3, using the singleton code $\{1111\}$, we also get $\lambda(1) \leq 4$.

For $m = 2$, by [19, pp. 75–76], any winning strategy for a state of type $(1, 2, 1)$ necessarily uses ≥ 6 questions – even in the fully interactive model, where each question is adaptively asked after receiving the answer to the previous questions. A fortiori, in our present nonadaptive case, $\lambda(2) \geq 6$. On the other hand, taking the code $\mathcal{C} = \{111100, 001111\}$ and using Corollary 3, one obtains a nonadaptive strategy with 6 questions which is winning for every state of type $(1, 2, 1)$. Thus $\lambda(2) \leq 6$.

For $m = 3$ we have $\lambda(3) \geq \mathrm{ch}(1, 3, 3) = 6$. Conversely, combining Corollary 3 and Lemma 8, we get $\lambda(3) \leq 6$.

Finally, we consider the case $m = 4$. On the one hand, Corollary 3 and Lemma 8 imply that $\lambda(4) \geq 7$. On the other hand, taking the $(7, 4, 3)$-code $C = \{1111000, 0001111, 0110011, 1111111\}$ and again using Corollary 3, we obtain a nonadaptive winning strategy with 7 questions for any state of type $(1, 4, 6)$. Therefore, $\lambda(4) \leq 7$, and the proof is complete. □

Combining Corollary 4 and Lemma 9 we have

Theorem 12. *For each integer $m = 1, 3, 5, 6, 7, 8, \ldots$, there is a strategy $\mathcal{S}$ to guess a number $x \in \{0, \ldots, 2^m - 1\}$ with up to two lies in the answers, which is perfect and canonical. Thus, $\mathcal{S}$ uses a first batch of m nonadaptive questions asking for the bits of the binary expansion of x and then, depending only on the answers to these questions, a second batch of* $\mathrm{ch}(2^m, 0, 0) - m$ *nonadaptive questions. In case $m \in \{2, 4\}$, let $\mathcal{S}$ be the shortest canonical strategy to guess a number $x \in \{0, \ldots, 2^m - 1\}$ with up to two lies in the answers. Then $\mathcal{S}$ requires precisely* $\mathrm{ch}(2^m, 0, 0) + 1$ *questions.*

The above theorem leaves open the possibility that there exist perfect noncanonical strategies for the exceptional cases $m = 2$ and $m = 4$. The following lemma shows that this is indeed the case for $m = 4$. A final remark in this section takes care of the case $m = 2$ (negatively).

Lemma 10. *There exists a perfect strategy to guess a number $x \in S = \{0, \ldots, 15\}$ with up to two lies in the answers, using a first batch of 5 nonadaptive questions and then, depending only on the answers to these questions, a second batch of 5 nonadaptive questions.*

Proof. Let x_{Carole} denote Carole's secret number. We can safely identify each $x \in S$ with the four bit string $x_1x_2x_3x_4$ yielding the binary expansion of x. Paul's first batch $[Q_1, \ldots, Q_5]$ of nonadaptive questions is as follows. For each $i = 1, \ldots, 4$, question Q_i asks

"is the ith bit of (the binary expansion of) x_{Carole} equal to 1?".

Question Q_5 asks

"is the sum modulo 2 of the first three bits of x_{Carole} equal to 1?".

Let $\sigma = (A_0, A_1, A_2)$ denote Paul's state resulting from Carole's answers to questions $Q_1, \ldots, Q_4$. There is precisely one element $h = h_1h_2h_3h_4 \in S$ such that $A_0 = \{h\}$. Specifically, with the identifications "yes" $= 1$ and "no" $= 0$, the ith bit h_i of the only element of A_0 coincides with Carole's answer to the ith question. Each element $x = x_1x_2x_3x_4 \in A_1$ has precisely one *discrepancy* from h (in the sense that $x_j = h_j$ for all $j = 1, 2, 3, 4$ except one.) Each element $y = y_1y_2y_3y_4 \in A_2$ has exactly two discrepancies from h. Direct inspection shows that the type of σ is $(1, 4, 6)$. Let $\sigma^{\surd} = (A_0^{\surd}, A_1^{\surd}, A_2^{\surd})$ be the state resulting from the first five answers. Then, if we denote by $b \in \{0, 1\}$ Carole's answer to the fifth question, the state $\sigma^{\surd}$ arises from σ in accordance with the formation rules (1)–(2).

Claim 1. The type of $\sigma^{\surd}$ is either $(1, 1, 6)$ or $(0, 4, 4)$.

We shall argue by cases as follows.

Case 1. h satisfies Carole's fifth answer, i.e., $h_1 + h_2 + h_3 \equiv b \mod 2$.

Then $A_0^{\surd} = A_0 = \{h\}$. An element $x = x_1x_2x_3x_4 \in A_1$ satisfies Carole's fifth answer iff its unique discrepancy from h occurs in the *fourth* position – so that the sum modulo 2 of its first three bits is the same as for h. Only one element $x^* \in A_1$ satisfies this condition, and $A_1^{\surd}$ consists only of this element. The three remaining elements of A_1 will survive as elements of $A_2^{\surd}$, because they falsify exactly two of Carole's answers to $Q_1, \ldots, Q_5$. An element $y = y_1y_2y_3y_4 \in A_2$ satisfies Carole's fifth answer iff its two discrepancies from h both occur in the first three positions – so that the sum modulo 2 of the first three bits of y is the same as for h. The three elements in A_2 satisfying this condition will survive in $A_2^{\surd}$, together with the three elements of A_1 other than x^*. Since the remaining elements of A_2 falsify three answers, they do not survive in $\sigma^{\surd}$. For the case under consideration, we have proved that $\sigma^{\surd}$ is of type $(1, 1, 6)$.

Case 2. h falsifies Carole's fifth answer, i.e., $h_1 + h_2 + h_3 \equiv 1 - b \mod 2$.

Then no element of S satisfies all five answers, and $A_0^{\surd} = \emptyset$. Since h only falsifies Carole's answer to Q_5, we have $h \in A_1^{\surd}$. An element $x = x_1x_2x_3x_4 \in A_1$ belongs to $A_1^{\surd}$ iff it satisfies the fifth answer, iff its unique discrepancy from h occurs among its first three bits – so that the sum of these three bits modulo 2 coincides with Carole's answer. The three elements in A_1 satisfying this condition will be members of $A_1^{\surd}$, together with h. The remaining element $x^{\dagger} \in A_1$ will survive in $A_2^{\surd}$. An element $y = y_1y_2y_3y_4 \in A_2$ belongs to $A_2^{\surd}$ iff it satisfies Carole's fifth answer, iff its two discrepancies from h do not both occur in the first three positions: this latter condition is necessary and sufficient for $y_1 + y_2 + y_3 \equiv b \equiv 1 - (h_1 + h_2 + h_3) \mod 2$. The three elements in A_2 satisfying this condition belong to $A_2^{\surd}$, together with $x^{\dagger}$. The remaining members of A_2 do not survive in $\sigma^{\surd}$. We have proved that, in the present case, $\sigma^{\surd}$ is of type $(0, 4, 4)$, and the claim is proved.

Claim 2. For any state $\sigma^{\surd}$ of either type $(1, 1, 6)$ or $(0, 4, 4)$, Paul has a nonadaptive winning strategy with 5 questions.

Indeed, if $\sigma^{\surd}$ is of type $(1, 1, 6)$, then $\mathrm{ch}(\sigma^{\surd}) = 5$. Let $\mathcal{C}_0 = \{00000\}$ and $\mathcal{C}_1 = \{11111\}$. Then $\Delta(\mathcal{C}_0, \mathcal{C}_1) = 5$. Further, for all integers $d_0 \geq 5$ and $d_1 \geq 3$, $\mathcal{C}_i$ is a $(5, 1, d_i)$-code (for each $i \in \{0, 1\}$). By Lemma 5 there exists a nonadaptive winning strategy for $\sigma^{\surd}$ with 5 questions.

If, on the other hand, $\sigma^{\surd}$ is of type $(0, 4, 4)$ then, again, $\mathrm{ch}(\sigma^{\surd}) = 5$. Let $\mathcal{C}_0 = \emptyset$ and $\mathcal{C}_1 = \{11100, 10111, 01011, 00000\}$. According to Definition 5, we can write $\Delta(\mathcal{C}_0, \mathcal{C}_1) \geq 4$. In the same way, $\mathcal{C}_0$ is a $(5, 0, d)$-code for every integer $d \geq 5$, and $\mathcal{C}_1$ is a $(5, 4, 3)$-code. By Lemma 5 there exists a nonadaptive winning strategy for $\sigma^{\surd}$ with 5 questions. Our second claim is settled.

From the above claims we obtain the desired perfect strategy, upon noting that $\mathrm{ch}(2^4, 0, 0) = 10$. □

Remark. As proved in [19], in the fully adaptive game with two lies, a perfect strategy exists to find an m-bit number iff $m \neq 2$. Therefore, combining Theorem 12 and the above Lemma 10, we now have the stronger result that: *even if Paul is allowed to adapt his strategy only once (as is the case in the game with one-shot feedback), a perfect strategy exists to find an m-bit number with two lies if and only if* $m \neq 2$.

4.5 Asymptotics for arbitrary numbers of errors

Recall that $m_{n,e} = \max\left\{m \mid V_n\left(1, m, \binom{m}{2}, \ldots, \binom{m}{e}\right) \leq 2^n\right\}$.

Lemma 11. *Let* $e \geq 0$ *and* $n \geq 2e$ *be arbitrary integers. Then* $m_{n,e} < \sqrt[e]{e!}\,2^{\frac{n}{e}} + e$.

Proof. Set $m^* = \sqrt[e]{e!}\,2^{\frac{n}{e}} + e$. Then the desired result follows directly from the inequality $V_n\left(1, m^*, \binom{m^*}{2}, \ldots, \binom{m^*}{e}\right) > 2^n$. Indeed, we have

$$V_n\left(1, m^*, \binom{m^*}{2}, \ldots, \binom{m^*}{e}\right) > V_n\left(0, 0, \ldots, 0, \binom{m^*}{e}\right)$$
$$= \binom{m^*}{e} = \frac{m^*(m^* - 1)\cdots(m^* - e + 1)}{e!}$$
$$\geq \frac{\left(\sqrt[e]{e!}\,2^{n/e}\right)^e}{e!} = 2^n. \qquad \square$$

We now prove that, for all sufficiently large m, there exists a second batch of

$$\operatorname{ch}\left(1, m, \binom{m}{2}, \ldots, \binom{m}{e}\right)$$

nonadaptive questions allowing the Questioner to guess infallibly the Responder's secret number. We first need the following lemma.

Lemma 12. *For any fixed e and for all sufficiently large n, there exists an e-tuple of codes* $\Gamma = (C_0, C_1, \ldots, C_{e-1})$ *together with integers* $d_i \geq 2(e - i) + 1$ $(i = 0, 1, \ldots, e - 1)$ *such that*

(i) *each* C_i *is an* $(n, \binom{m_{n,e}}{i}, d_i)$*-code;*
(ii) $\Delta(C_i, C_j) \geq 2e - (i + j) + 1$ *whenever* $0 \leq i < j \leq e - 1$.

Proof. Let $n' = n - e^2$. First we prove the existence of an $(n', \binom{m_{n,e}}{e-1}, d')$-code, with $d' = 2e + 1$. From Lemma 11 together with the well-known inequality $e! \leq \frac{(e+1)^e}{2^e}$, it follows that, for all sufficiently large n,

$$
\begin{aligned}
\binom{m_{n,e}}{e-1} &< (m_{n,e})^{e-1} < (\sqrt[e]{e!}\,2^{\frac{n}{e}} + e)^{e-1} \\
&\leq (e\,2^{\frac{n}{e}})^{e-1} = e^{e-1}\,2^{n\frac{e-1}{e}} \\
&\leq \frac{2^{n-e^2}}{\sum_{j=0}^{2e}\binom{n-e^2}{j}}.
\end{aligned}
$$

The existence of the desired $(n', \binom{m_{n,e}}{e-1}, d')$-code now follows from the Gilbert bound (see Theorem 9).

In summary, for all sufficiently large n, there exists an $(n-e^2, \binom{m_{n,e}}{e-1}, d')$-code $\mathcal{C}'$ with $d' \geq 2e+1$. For any $i = 0, 1, \ldots, e-1$, define the e^2-tuple $\boldsymbol{a}_i$ by

$$
\boldsymbol{a}_i = \underbrace{00\ldots0}_{(i-1)e}\underbrace{11\ldots1}_{e}\underbrace{0\ldots0}_{e^2-ie}.
$$

Furthermore, let $\mathcal{C}_i''$ be the code obtained by appending the suffix $\boldsymbol{a}_i$ to the codewords of $\mathcal{C}'$, in symbols,

$$
\mathcal{C}_i'' = \mathcal{C}' \otimes \boldsymbol{a}_i.
$$

Trivially, $\mathcal{C}_i''$ is an $(n, \binom{m_{n,e}}{e-1}, 2e+1)$-code for all $i = 0, 1, \ldots, e$. Furthermore, we have $\Delta(\mathcal{C}_i'', \mathcal{C}_j'') = 2e \geq 2e - (i+j) + 1$ whenever $0 \leq i < j \leq e-1$. For each $i = 0, 1, \ldots, e-1$, pick a subcode $\mathcal{C}_i \subseteq \mathcal{C}_i''$ with $|\mathcal{C}_i| = \binom{m_{n,e}}{i}$. Then the new e-tuple of codes $\Gamma = (\mathcal{C}_0, \mathcal{C}_1, \ldots, \mathcal{C}_{e-1})$ satisfies both conditions (i) and (ii) and the proof is complete. □

The following corollary yields perfect searching strategies with minimum adaptiveness.

Corollary 5. *Fix an integer $e \geq 0$. Then, for all sufficiently large integers m and for every state σ of type $(1, m, \binom{m}{2}, \ldots, \binom{m}{e})$, there exists a nonadaptive winning strategy $\mathcal{S}$ such that the number of questions in $\mathcal{S}$ for σ coincides with* $\mathrm{ch}(\sigma)$.

Proof. Let $n = \mathrm{ch}(\sigma)$. By definition, $n \to \infty$ as $m \to \infty$. Lemmas 12 and 5 yield a nonadaptive winning strategy with n questions for any state of type $(1, m_{n,e}, \binom{m_{n,e}}{2}, \ldots, \binom{m_{n,e}}{e})$. By Definition 12, $m \leq m_{n,e}$ and, a fortiori, for all sufficiently large m, a nonadaptive winning strategy with n questions exists for any state of type $(1, m, \binom{m}{2}, \ldots, \binom{m}{e})$. □

Remark. In terms of Berlekamp's theory of error-correcting transmission with feedback, the results of this section show that the minimum amount of redundancy which, by Berlekamp's bound, is *necessary* for e-error-correction of an m-tuple of bits, turns out to be *sufficient* – as soon as the receiver is allowed to send just one feedback message to the satellite. This is achieved via the following protocol:

(i) the original m-tuple $\boldsymbol{x}$ is sent by the satellite, and is received as $\boldsymbol{x}'$;
(ii) the receiver feeds $\boldsymbol{x}'$ back to the satellite via a noiseless feedback channel;

(iii) depending only on $\boldsymbol{x}'$, the satellite sends a final tip $\boldsymbol{r}$ of $q_e(m) - m$ bits, which are received as $\boldsymbol{r}'$, in such a way that

(iv) from $\boldsymbol{x}'\boldsymbol{r}'$ the receiver is able to recover the original m-tuple $\boldsymbol{x}$ (as well as $\boldsymbol{r}$), even if distortion has corrupted at most e of the bits of $\boldsymbol{x}\boldsymbol{r}$ (causing $\boldsymbol{x}'\boldsymbol{r}'$ to be received instead of $\boldsymbol{x}\boldsymbol{r}$).

5 Fault-tolerant search: variations

5.1 More powerful questions: q-ary search

A natural generalization of the Ulam–Rényi game is obtained by assuming that Paul asks questions allowing Carole to choose among q possible alternative answers. One is then concerned with *fault-tolerant q-ary search.* In particular, the classical Ulam–Rényi problem pertains to q-ary search with lies and $q = 2$.

Typically, a q-ary *question* **T** has the form

"to which one of $T_0, T_1, \ldots, T_{q-1}$ does the secret number x_{Carole} belong ?",

where $\mathbf{T} = (T_0, T_1, \ldots, T_{q-1})$ is a q-tuple of of (possibly empty) pairwise disjoint subsets of S whose union is S.

Carole's answer is an integer $i \in \{0, 1, \ldots, q-1\}$, telling Paul that x_{Carole} belongs to T_i.

Thus a winning strategy for q-ary search with e lies is essentially the same thing as an e-error-correcting code for the q-ary symmetric channel with noiseless delayless feedback channel.

We denote by $N^{[q]}(M, e)$ the size of a shortest winning strategy for finding an unknown number in a set of cardinality M by using only q-ary questions, when up to e of the answers may be mendacious. Also, let $N^{[q]}_{\min}(M, e)$ denote the minimum integer n such that n, M, e satisfy the Hamming bound for q-ary codes (13). Generalizing Theorem 2, we can easily obtain

$$N^{[q]}(M, e) \geq N^{[q]}_{\min}(M, e).$$

The problems of q-ary search with $e > 0$ lies were first considered by Malinowski [28] and Aigner [1],[13] who independently evaluated the minimum number of questions needed by Paul to win the game when at most one of Carole's answers are mendacious.

The authors of [14] evaluated $N^{[q]}(M, 2)$ exactly, which counts the shortest strategies to solve the q-ary search with lies, when $e = 2$ and the search space has cardinality $M = q^m$ for all $m \geq 0$. The result of [14] states that

$$N^{[q]}(M, 2) = \min\{n \mid n \geq N^{[q]}_{\min}(M, 2) \text{ and } n - 2 \geq N^{[q]}_{\min}(m, 1)\}.$$

[13] Trivially, when $e = 0$, $\lceil \log_q M \rceil$ questions are necessary and sufficient to find an unknown number $x \in \{0, 1, \ldots, M - 1\}$.

In the same paper it is also proved that, for any search space of arbitrary cardinality n, there exists a search strategy using at most one additional question, with respect to the theoretical minimum number of questions needed in q-ary search with two lies, i.e., for all $M \geq 1$,

$$N_{\min}^{[q]}(M, 2) \leq N^{[q]}(M, 2) \leq N_{\min}^{[q]}(M, 2) + 1.$$

These results were strengthened in [12], where the authors proved that winning strategies with $N_{\min}^{[q]}(M, 2)$ q-ary questions (indeed, perfect strategies) for the Ulam–Rényi game with $e = 2$ lies, still exist under the stronger hypothesis that adaptiveness is used only once. More precisely, in [12] it is proved that, if there exists a fully adaptive strategy of size n finding an unknown number $x \in \{0, 1, \ldots q^m - 1\}$, then there exists a winning strategy of size n using adaptiveness only once.

A general asymptotic solution of the Ulam–Rényi game with q-ary questions was given by Muthukrishnan [34], who generalized Spencer's results for the binary case. More precisely, Muthukrishnan [34] proved that, for each $e \geq 1$ and for all sufficiently large m, Paul can infallibly guess an unknown number in a search space of cardinality $M = q^m$, using q-ary questions with e lies – by asking the theoretical minimum number of questions. Stated otherwise, we have that, for all integers $q \geq 2$, $e \geq 1$ and for all sufficiently large m,

$$N^{[q]}(q^m, e) = N_{\min}^{[q]}(q^m, e).$$

For all sufficiently large cardinalities other than q^m, a similar result holds: in the worst case one additional question needs to be asked, with respect to the theoretical minimum.

5.2 Less powerful questions: comparison search

We now turn to the variant of the Ulam–Rényi game in which only *comparison* questions are allowed, that is, questions in the form "is the secret number $x \leq a$?" for some $a \in S$. Let

$$N^{(\mathrm{cmp})}(M, e)$$

denote the least number of questions sufficient to find an unknown number $x \in \{0, 1, \ldots, M - 1\}$ by using comparison questions only, when up to e of the answers may be mendacious.

Obviously, $N^{(\mathrm{cmp})}(M, e) \geq N(M, e)$ for all $e \geq 0$ and $M \geq 1$. Further, for $e = 0$, a moment's reflection shows that $N^{(\mathrm{cmp})}(M, 0) = N(M, 0)$ for all $M \geq 1$.

The Ulam–Rényi game with comparison questions was considered for the first time in [42]. In that paper the Ulam–Rényi problem with comparison search over $S = \{0, 1, 2, \ldots, M - 1\}$ is solved by embedding S into the half-open real interval $]0, 1]$, and then looking for a subset A of $]0, 1]$ such that A contains the secret number x and $|A| \leq \epsilon$ for an arbitrarily chosen ϵ, $0 < \epsilon < 1$.

From [42] one has the inequalities

$$N_{\min}(M, e) \leq N^{(\mathrm{cmp})}(M, e) \leq \min \left\{ n | 2^{n-e} \geq M \sum_{j=0}^{e} \binom{n-e}{j} \right\},$$

which are still valid when e is a function of M. From the above result we can easily obtain

$$N^{(\mathrm{cmp})}(M, e) = \log_2 M + e \log_2 \log_2 M + O(e \log_2 e). \tag{1}$$

The analysis of [42] does not provide an exact answer to the question raised by Ulam – nor is it sufficient to determine whether 25 or 26 is the least number of comparison questions to solve the problem for the case $M = 10^6$, when at most one of the answers is mendacious. In [44], the author focused on $e = 1$, and $S = \{0, 1, 2, \ldots, M - 1\}$, and proved the following:

- whenever n satisfies the inequality $\frac{2^n}{n+1} < M$, no winning strategy exists using n comparison questions
- if $n \leq \frac{5\,2^n}{8\,(n+1)}$, then there exists a winning strategy for Paul using n comparison questions.

Thus, for all $M \geq 1$, we have the estimate $N^{(\mathrm{cmp})}(M, 1) \leq N(M, 1) + 1$. Again, this result does not decide whether 25 or 26 comparison questions are the solution of the original Ulam problem ($M = 10^6,\ e = 1$). Finally, Duncan Innes [25] proved that $N^{(\mathrm{cmp})}(10^6, 1) = N(10^6, 1) = 25$. As a matter of fact, Innes [25] provided a tighter bound to the effect that, whenever

$$M \leq \frac{27}{128} \frac{2^n}{n+1},$$

there exists a winning strategy for Paul using exactly $n - 2$ comparison questions to search for an unknown number x in the set $\{0, 1, \ldots, M - 1\}$ when at most one answer is a lie.

5.3 Less powerful questions: bicomparison search ($e = 2$)

In [33] the authors considered *bicomparison* questions, in which queries have the form "does x satisfy either condition $a \leq x \leq b$ or $c \leq x \leq d$?".

Focusing on the case $e = 2$, in [33] they prove that, for all $m \geq 1$ and $m \neq 2$, an unknown m-bit number can always be found by asking $N_{\min}(2^m, 2)$ bicomparison questions. Therefore, despite the limitations in Paul's expressive power imposed by restriction to bicomparison questions, there exist optimal searching strategies having precisely the same number of questions as in the general unrestricted case (arbitrary yes–no questions).

In [33] it was also proved that, if all questions are assumed to be of the form "does x satisfy the condition $a \leq x \leq b$?", then no search strategy can exist to find an unknown m-bit number using $N(2^m, 2)$ such questions.

As a final by-product of the analysis in [33] we find that bicomparison questions not only provide a much simpler formal representation of questions themselves but, surprisingly enough, they also yield a radical simplification of the representation of states and, ultimately, of the actual computation of coding and decoding procedures. This is because each state arising from bicomparison questions turns out to have a simple geometric shape that can be completely specified by eleven numbers in S.

5.4 Linearly bounded numbers of lies

Up to this point we have been assuming that a *fixed* maximum number e of bits can be distorted by noise during each transmission. On the other hand, it is quite natural to assume that the number e of expected noisy bits grows proportionally to the number of bits actually sent. Since bits are the same as yes–no answers, we are led to consider the Ulam–Rényi problem with linearly bounded numbers of errors.

Here one assumes the existence of a fixed real value r in the open interval $]0, 1[$, known to both Paul and Carole, such that, if Paul asks n questions, then Carole is allowed to tell at most $\lfloor r\, n \rfloor$ lies. Thus, after agreeing upon the size M of the search space, and also agreeing on the lie-fraction r, Paul announces that he is going to ask at most n questions, whence allowing Carole to fix the maximum number of lies that she is allowed to tell.

Again, we are interested in the minimum number n of questions sufficient for Paul to guess Carole's secret number infallibly, for all possible values of the parameters M and r.

The linearly bounded lie model bears an interesting relationship with the *probabilistic* model, in which the responder lies at random, and an answer is mendacious with a constant probability $p < 1/2$. Here, lies are no longer malicious, and p arises as the result of a probabilistic analysis of the noisy environment. See [38] and [36] for further discussion.

In [46] Spencer and Winkler investigated the linearly bounded lie model, both for the fully adaptive and for the fully nonadaptive model. Their main result is as follows.

Theorem 13. 1. *For nonadaptive search over* $S = \{0, 1, \ldots, M-1\}$, *we have:*
  (i) *if* $r < 1/4$, *then Paul has a winning strategy with* $\Theta(\log_2 M)$ *questions;*
  (ii) *if* $r = 1/4$, *then Paul has a winning strategy with* $\Theta(M)$ *questions;*
  (iii) *if* $r > 1/4$, *then no winning strategy exists for Paul for any* $M \geq 9r/(r-1/4)$, *no matter the number of questions.*

2. *For fully adaptive search over* $S = \{0, 1, \ldots, M-1\}$, *we have:*
  (iv) *If* $r < 1/3$, *then Paul has a winning strategy with* $\Theta(\log_2 M)$ *questions;*
  (v) *if* $r \geq 1/3$, *then no winning strategy exists for Paul for all* $M \geq 5$, *no matter the number of questions.*

In [20], Dhagat et al. investigated the linearly bounded lie model in which Paul is allowed to use more restrictive types of questions. In [20] a *comparison question* is taken to mean a question of the form "is $x < a$?". Further, they call *bit question* a query asking for the value of a bit in the binary expansion of Carole's secret number, that is, a question of the form "is the ith bit of x equal to 1?".

The main result of [20] is summarized in the following

Theorem 14. 1. *For nonadaptive search over $S = \{0, 1, \ldots, M-1\}$, the following hold:*

(i) *Paul has a winning strategy with n bit questions iff* $M \leq 2^{\left\lfloor \frac{n}{2\lfloor r n\rfloor+1} \right\rfloor}$;

(ii) *Paul has a winning strategy with n comparison questions iff* $M \leq \frac{n}{2\lfloor r n\rfloor+1}+1$.

2. *For fully adaptive search over $S = \{0, 1, \ldots, M-1\}$, we have:*

(iii) *Paul has a winning strategy with n bit questions iff* $M \leq 2^{\left\lfloor \frac{n-\lfloor r n\rfloor}{\lfloor r n\rfloor+1} \right\rfloor}$;

(iv) *Paul has a winning strategy using only comparison questions iff $r < 1/3$; moreover, for all $r < 1/3$ there exists a winning strategy with n comparison questions, where*

$$n = \left\lceil \frac{8 \log_2 M}{(1-3r)^2} \right\rceil. \tag{19}$$

The bounds in (19) were subsequently strengthened by Pedrotti [36] who proved that [14]

$$n = \left\lceil \frac{8 \ln 2}{3} \frac{\log_2 M}{(1-3r)^2(1+3r)} \right\rceil. \tag{20}$$

As pointed out by Yossi Azar, a winning strategy with $O(\log_2 M)$ comparison questions in the linearly bounded lie model can be obtained using the main result of [42], despite the fact that this paper mainly dealt with fixed numbers of lies. In particular, the above identity (20) is obtained by a refined analysis of the searching algorithm of [42]. Remarkably enough, (20) improves (19) by a factor of $\frac{3(1-3r)}{\ln 2}$.

5.5 Prefix bounded numbers of lies

A particular case of the linearly bounded lie model is obtained when the fractional bound on the number of lies is assumed to hold at any point of the game. The resulting model is known as the *prefix bounded lie model.* Here one assumes that, for a given real value r in the open interval $]0, 1[$, at any point of the game, when i answers have been given, no more than $\lfloor r\, i \rfloor$ lies have been told.

It turns out that in this case the bounds on r for the existence of a winning strategy for Paul become weaker. Indeed, Spencer and Winkler [46] proved that, for $S = \{0, 1, \ldots, M-1\}$, the following hold.

(a) For $r \geq 1/2$, no winning strategy exists for Paul ($M \geq 3$), no matter how many questions he asks.

(b) For $r < 1/2$, Paul has a winning strategy with $\Theta(\log_2 M)$ arbitrary yes–no questions.

---

[14] Here ln denotes the natural logarithm.

The following result is due to Aslam and Dhagat [4] and concerns the case in which only comparison questions are allowed.

Theorem 15. *For any $r < 1/2$, Paul has a winning strategy in the prefix bounded lie model over $S = \{0, 1, \ldots, M-1\}$, with $O(M^{\log_2(\frac{1}{1-r})})$ comparison questions.*

In [20] the authors proved that the same result also holds when only bit questions are used.

The bound of Theorem 15 was improved by Borgstrom and Kosaraju [7], whose results give added evidence to the conjecture that $O(\log_2 M)$ comparison questions are sufficient for successful search in the set $\{0, 1, \ldots, M-1\}$ in the prefix bounded lie model, for any value of the error fraction $r < 1/2$. Moreover in [7] an algorithm whose overall running time is $O(\log_2 M)$ is designed.[15]

5.6 Less powerful lies: half-lies ($e = 1$)

In [42] Rivest et al. considered a variant of the Ulam–Rényi game, called the *half-lie game*, in which only negative answers may be mendacious. Naturally enough, the half-lie game also has an interesting counterpart in the framework of error-correcting codes with feedback.

We limit ourselves to a couple of examples. In optical communication systems, photons may fail to be detected ($1 \to 0$), but the creation of spurious photons ($0 \to 1$) is impossible [40]. Similarly, in most LSI memory protection schemes one can safely assume that only one of the two possible error patterns ($1 \to 0$) and ($0 \to 1$) can occur [18].

Optimal error-correcting codes for these asymmetric channels with feedback are the same as shortest binary strategies to find an m-bit number, when only negative tests can be faulty, as required in the half-lie Ulam–Rényi game.

In [42] the authors proved an asymptotic result to the effect that the half-lie game has the same complexity as the original Ulam problem – henceforth referred to as the "full-lie" game. Focusing on the case $e = 1$, in the paper [13] the authors strengthen this result, by giving very tight and nonasymptotic lower and upper bounds for the half-lie game. This is our main concern in the rest of this section.

In the half-lie game, Carole (the Responder) can lie at most once, but only by answering "no" to a question whose correct (sincere) answer is "yes". Thus, *only Carole's negative answers are doubtful for Paul.* Accordingly, for any state $\sigma = (x_0, x_1)$ and question $[t_0, t_1]$, the two possible states resulting from Carole's answer are given by

$$\sigma^{\text{yes}} = (t_0, t_1) \qquad \text{and} \qquad \sigma^{\text{no}} = (x_0 - t_0,\ \ t_0 + x_1 - t_1). \tag{21}$$

[15] This is a stronger result than simply proving that $O(\log_2 M)$ questions are sufficient, in that we are claiming here that $O(\log_2 M)$ steps are also sufficient when all the necessary computation supporting the search is considered, e.g., operations like state-update and question-computation.

Definition 13. For any two integers $m = 1, 2, 3, \ldots$ and $q > m$, the integer $\kappa(m, q)$ is the smallest k such that $\sum_{i=0}^{k} \binom{q}{i} > 2^m$. The integer $\rho = \rho(m, q)$ is defined by $\rho = 2^m - \sum_{i=0}^{\kappa(m,q)-1} \binom{q}{i}$.

Trivially, $1 \leq \kappa(m, q) \leq q$ and $0 \leq \rho < \binom{q}{\kappa(m,q)}$.

Theorem 16. *Fix an integer $m = 1, 2, 3, \ldots$. Let $q^{\ddagger}$ be the smallest integer q such that*

$$2^q \geq 2^m + \sum_{i=0}^{\kappa(m,q)-1} i\binom{q}{i} + \kappa(m, q)\rho(m, q).$$

Assume that there exists a winning strategy with q^* questions for finding an unknown integer in the set $S = \{0, 1, \ldots 2^m - 1\}$ in the half-lie game, i.e., when at most one of the "no" answers may be mendacious and all "yes" answers are true. Then $q^* \geq q^{\ddagger}$.

Proof. Let $\mathcal{S}$ be such a strategy, where we can safely assume that the number q^* of questions in $\mathcal{S}$ is the smallest possible. By a path in $\mathcal{S}$ we understand a path from the root to a leaf of $\mathcal{S}$. For each $x \in S$ there exists precisely one path μ_x in $\mathcal{S}$ leading to the final state $(\{x\}, \emptyset)$. This final state is obtained if Carole chooses x as the secret number and then decides to give true answers to all q^* questions. Let $southwest_x$ be the number of left branches (i.e., branches whose label is "yes") in this path. The $southwest_x$ branches of μ_x are a record of Carole's "yes" answers, once she decides to choose x as the secret number and always to tell the truth. If, on the other hand, Carole decides to give one mendacious answer, she can deviate from this path in $southwest_x$ ways, replacing a true "yes" answer by a mendacious "no" answer. [16] Since $\mathcal{S}$ is a winning strategy, there exist in $\mathcal{S}$ precisely $southwest_x$ paths leading to the final state $(\emptyset, \{x\})$. Furthermore, whenever $x \neq y \in S$ the state $(\emptyset, \{y\})$ will be obtained by a different path. Now, each path in $\mathcal{S}$ has q^* branches, and paths are in one-to-one correspondence with subsets of $\{1, \ldots, q^*\}$ (giving the depths of their left branches). To obtain a set of 2^m paths having the smallest possible total number of left branches, with reference to Definition 13, one must proceed as follows: first take the only path in $\mathcal{S}$ with no left branch, then the q^* paths with one left branch, the $\binom{q^*}{2}$ paths with two left branches, $\ldots$, the $\binom{q^*}{\kappa(m,q^*)-1}$ paths with $\kappa(m, q^*) - 1$ left branches, and finally, the $\rho(m, q^*)$ paths with $\kappa(m, q^*)$ left branches. Then the total number N_{once} of possibilities for Carole to lie precisely once in $\mathcal{S}$ satisfies the inequality

$$N_{\text{once}} \geq \sum_{i=0}^{\kappa(m,q^*)-1} i\binom{q^*}{i} + \kappa(m, q^*)\rho(m, q^*).$$

On the other hand, if Carole decides not to lie, she can still choose $x \in S$, and the corresponding path leading to the leaf $(\{x\}, \emptyset)$, in 2^m ways. Summing up, we see

[16] Note the effect of the assumed asymmetry in Carole's lies: in the classical Ulam-Rényi game Carole can choose to lie in q^* ways.

that the 2^{q^*} paths of the winning strategy $\mathcal{S}$ cannot be less than the total number of Carole's answering strategies, whence

$$2^{q^*} \geq 2^m + N_{\text{once}} \geq \sum_{i=0}^{\kappa(m,q^*)-1} i\binom{q^*}{i} + \kappa(m,q^*)\rho(m,q^*).$$

This completes the proof. □

Recall that, by Corollary 2, $N(2^m, 1) = N_{\min}(2^m, 1)$.

Theorem 17. *For each $m \geq 1$, at least $N(2^m, 1) - 2$ questions are necessary to win the half-lie game (with one lie) over the search space $\{0, 1, \ldots 2^m - 1\}$.*

Recall that the critical index $m_{n,1}$ is a particular case of Definition 12. The following result is proved in [13].

Theorem 18. *Fix an integer $s \geq 1$, and let $m = m_{s,1}$. Then at least $N(2^m, 1) - 1$ questions are necessary to solve the half-lie game with one lie, over any search space of cardinality 2^m.*

For all $m \geq 1$ we now construct a winning strategy for solving the half-lie game over the search space $\{0, 1, \ldots, 2^m - 1\}$ with one lie, using no more than $N(2^m, 1) - 1$ questions. Moreover, we show that, for infinitely many m, this strategy uses no more than $N(2^m, 1) - 2$ questions.

Theorem 19. *For every integer $m \geq 1$, let $\sigma = (2^m, 0)$. Then there exists a winning strategy for σ in the half-lie game, using exactly $m - 1 + \lceil \log_2(m + 3) \rceil$ questions.*

Proof. The proof proceeds through several claims.

Claim 1. For each integer $r \geq 1$, the state $\sigma = (1, 2^r - 1)$ has a winning r-strategy.

By induction on r. For $r = 1$ we have $\sigma = (1, 1)$. Upon asking the question $[0, 1]$, we obtain the two final states $(0, 1)$ and $(1, 0)$. For the induction step, suppose that the claim is true for $r \geq 1$, with the object of proving it for the state $(1, 2^{r+1} - 1)$. Let $\sigma = (1, 2^{r+1} - 1)$ and ask the question $[0, 2^r]$. We then obtain the two states $\sigma^{\text{yes}} = (0, 2^r)$ and $\sigma^{\text{no}} = (1, 2^r - 1)$. It is easy to see that σ^{yes} is a winning r-state. Moreover, σ^{no} is a winning r-state by the induction hypothesis. Therefore, σ is a winning $(r + 1)$-state, and the claim is settled.

Claim 2. For each integer $r \geq 1$, the state $\sigma = (2, 2^r - 3)$ is a winning r-state.

Upon asking question $[1, 2^{r-1} - 1]$, the two resulting states are $\sigma^{\text{yes}} = \sigma^{\text{no}} = (1, 2^{r-1} - 1)$. By Claim 1, both states have a winning strategy with $r - 1$ questions. A fortiori, there exists a winning strategy with r questions for the state $(2, 2^r - 3)$.

Claim 3. For every integer $m \geq 1$, the state $\sigma = (2, m)$ has a winning strategy with $\lceil \log_2(m + 3) \rceil$ questions.

Let $r = \lceil \log_2(m + 3) \rceil$. Then $r = \min\{i \mid 2^{i-1} - 3 < m \leq 2^i - 3\}$. Therefore, σ is a substate of $(2, 2^r - 3)$ and the desired result follows directly from Claim 2 together with Lemma 2.

Claim 4. For any two integers $m \geq 1$ and $k = 1, 2, \ldots, m$, there exists a winning strategy with $k-1+\lceil \log_2(m+3) \rceil$ questions for the state $\sigma = \left(2^k, (m-k+1)2^{k-1}\right)$.

We proceed by inductionon k. The base case is a direct consequence of Claim 3. For the induction step, let $k = i + 1$. Then $\sigma = \left(2^{i+1}, (m-i)2^i\right)$. By asking the question $\left[2^i, (m-i+1)2^{i-1}\right]$, the two resulting states coincide – specifically, $\sigma^{\text{yes}} = \sigma^{\text{no}} = \left(2^i, (m-i+1)2^{i-1}\right)$. By induction one immediately sees that $\sigma^{\text{yes}} = \sigma^{\text{no}}$ is a winning $\left(i - 1 + \lceil \log_2(m+3) \rceil\right)$-state. Therefore, σ is a winning $\left(k - 1 + \lceil \log_2(m+3) \rceil\right)$-state, as required to settle the claim.

End of the Proof of Theorem 19. Let $\sigma = (2^m, 0)$. Then σ is a substate of $(2^m, 2^{m-1})$ and we obtain the desired result by Claim 4 together with Lemma 2. □

Lemma 13. *For every integer $m \geq 1$, let $S = \{0, 1, \ldots, 2^m - 1\}$. Then there exists a winning strategy to solve the half-lie game over S, using $\leq N(2^m, 1) - 1$ questions.*

Proof. By Theorem 19 there exists a winning strategy using $m - 1 + \lceil \log_2(m+3) \rceil$ questions. Then we need only show that, for all $m \geq 1$,

$$m + \lceil \log_2(m+3) \rceil \leq N(2^m, 1).$$

Letting $k = m - 1 + \lceil \log_2(m+3) \rceil$, we must only settle the inequality $k+1 > 2^{k-m}$, that is,

$$m + \lceil \log_2(m+3) \rceil > 2^{\lceil \log_2(m+3) \rceil - 1}. \tag{22}$$

To this purpose, let $s \geq 0$ be the smallest integer such that $m = 2^s - t$ for some $0 \leq t < 2^{s-1}$. Then we argue by cases.

Case 1. $t \geq 3$. Then (22) becomes $2^s - t + s > 2^{s-1}$. This latter inequality holds, because $2^{s-1} > t \geq 3$ implies $s > 2$, whence $2^{s-1} + s > t$.

Case 2. $0 \leq t \leq 2$. Then (22) becomes $2^s - t + s + 1 > 2^s$, or, equivalently,

$$s + 1 > t. \tag{23}$$

Since $2^{s-1} > t$, the only possible cases to be considered are:

$t = 0$ and $s \geq 0$,
$t = 1$ and $s \geq 2$,
$t = 2$ and $s \geq 3$.

In every case, inequality (23) is satisfied and the proof is complete. □

Lemma 14. *Fix integers n, t, $n \geq 1$ and $3 \leq t \leq n$. Let $m = 2^n - t$. Then there exists a winning strategy for the state $\sigma = (2^m, 0)$, using $\leq N(2^m, 1) - 2$ questions.*

Proof. Let $k = m - 1 + \lceil \log_2(m+3) \rceil$. By Theorem 19 there exists a winning strategy for σ using k questions. Then we need only prove that $k < N(2^m, 1) - 1$. It remains to prove the inequality $k + 2 > 2^{k+1-m}$, that is, $m + 1 + \lceil \log_2(m+3) \rceil > 2^{\lceil \log_2(m+3) \rceil}$. From our hypotheses $m = 2^n - t \leq 2^n - 3$ and $t \leq n \leq 2^{n-1}$, we conclude that $\lceil \log_2(m+3) \rceil = n$. It follows that $m + 1 + \lceil \log_2(m+3) \rceil = 2^n - t + 1 + n > 2^n = 2^{\lceil \log_2(m+3) \rceil}$, as required. □

The following main result of [13] shows that *the sphere packing bound is ineffective in games with half-lies.*

Theorem 20. *For each* $m = 1, 2, \ldots$, *let* $N^{\text{half}}(2^m, 1)$ *denote the number of questions of the shortest winning strategy for the half-lie game over the search space* $S = \{0, 1, \ldots, 2^m - 1\}$. *Let* $\mathcal{C} \subseteq \mathbf{Z}$ *be defined by*

$$\mathcal{C} = \{m \geq 0 \mid N(2^m, 1) = 2^{N(2^m,1)-m} - 1\}.$$

For each $s = 1, 2, 3, \ldots$, *let* $\mathcal{E}_s \subseteq \mathbf{Z}$ *be defined by*

$$\mathcal{E}_s = \{m \geq 0 \mid m_{s,1} + 1 \leq m \leq m_{s,1} + s - 2\}.$$

It follows that:

(i) $N(2^m, 1) - 2 \leq N^{\text{half}}(2^m, 1) \leq N(2^m, 1) - 1$ *for each* $m = 1, 2, \ldots;$
(ii) $N^{\text{half}}(2^m, 1) = N(2^m, 1) - 1$ *for all* $m \in \mathcal{C}$;
(iii) $N^{\text{half}}(2^m, 1) = N(2^m, 1) - 2$ *for each* $s = 1, 2, \ldots$ *and* $m \in \mathcal{E}_s$.

Proof. Condition (i) follows from Lemma 13 and Theorem 17 and condition (ii) from Lemma 13 and Theorem 18. Condition (iii) is a consequence of Theorem 17 and Lemma 14. □

5.7 Time-outs instead of errors

The last variant of the Ulam–Rényi problem considered in this survey was investigated in [15]. Here an error is no longer understood as a mendacious answer, but rather as a lost answer.

In terms of Berlekamp's error-correcting transmission with a feedback channel, the setting of [15] corresponds to the case in which the feedback channel is only used to acknowledge the receipt (or the nonreceipt) of the source-bits. Any bit delivered to the receiver is taken to be correct, but some of the bits may be lost during transmission. Moreover, it is assumed that the noisy channel is much slower in delivering bits than the source is to produce them. Thus the (positive or negative) acknowledgement of the ith bit reaches the source only after the $(i + d)^{\text{th}}$ bit has been sent.

Erasure of an answer can also be the effect of (software or hardware) devices whose task is to clear spurious bits from the channel. Altogether, we may assume that the quantity d represents the maximum time the receiver is to wait before the sent bit reaches him. After the deadline of d time units has expired, an undelivered bit is automatically destroyed (as a *time-out* bit) in order to prevent de-synchronization of the communication channel.

Among other topics, in [15] the authors discuss in which sense "delayed search strategies" [3], and broadcast of messages in networks with link latency [5], can be considered as isomorphic problems.

In the framework of the Ulam–Rényi game, the problem of searching with delays and cancellations is formally stated as follows. Paul and Carole now agree on three integers $d, c \geq 0$ and $M \geq 1$, the latter denoting as usual the cardinality of the search space S. Then Carole chooses a number x in S, and Paul has to determine x by using only comparison questions, e.g., "is $x \leq a$?", for some $a \in S$. For each $i = 1, 2, \ldots$, Paul must ask a question precisely at time i. This is his i^{th} question. Paul knows Carole's answer during the open interval $]i + d, i + d + 1[$. Thus, except for trivial extremal situations,[17] in general Paul asks his i^{th} question when the answers to his previous d questions are still pending.

It is also assumed that Carole can choose not to answer up to c questions – or, equivalently, up to c answers may be lost.

The problem is then to find the (minimum) number n of questions sufficient for Paul to guess the secret number x. We then say that Paul *(optimally) wins* the (M, d, c)-game with n questions.

By definition, if the total number of questions asked by Paul is t, then the game actually ends at time $t + d + 1$.

Remark. The problem of coping with lost answers in the classical setting (i.e., when there is no delay between questions and answers) is trivial – for, there is no better strategy than repeating all unanswered questions. Analogously, if *arbitrary* yes–no questions are permitted, then, even if we assume delayed answers, an optimal strategy is given by asking for the binary encoding of the secret number, and repeating all unanswered questions.

The results of [15] are given for a dual counterpart of the above problem, limited to the case $c \in \{0, 1\}$.[18] Let $A_d^{(c)}(t)$ be the largest integer M such that Paul wins the (M, d, c)-game with t questions.

Theorem 21 ([15, 3]). *For all $t \geq 0$, $d \geq 0$ and $c \in \{0, 1\}$, the following holds:*

$$A_d^{(c)}(t) = \begin{cases} \left\lfloor \frac{t}{c+1} \right\rfloor + 1 & \text{if } t \leq d + 1 \\ A_d^{(c)}(t-1) + A_d^{(c)}(t-d-1) & \text{if } t \geq d + 2. \end{cases} \tag{24}$$

Corollary 6. *Let ϕ_d be the largest (positive) real root of $x^{d+1} = x^d + 1$. Then, for all $t \geq 0$, $d \geq 0$, and $c \in \{0, 1\}$, we have $A_d^{(c)}(t) \in \Theta(\phi_d^t)$, and*

$$\log_{\phi_d}(n + 1) + O(1)$$

questions are necessary and sufficient to win the (n, d, c)-game.

In [15] an alternative formulation of the recurrence (24) is given. According to the authors, this should be more suitable for generalizations to the case of an arbitrary (fixed) number of cancellations ($c \geq 2$).

---

[17] Trivially, if n is the overall number of questions to be asked by Paul, then, for each $i = 0, 1, \ldots, d - 1$, when Carole answers the $(n - i)^{\text{th}}$ question, only i questions are pending. A dual remark applies to the first d questions.

[18] The case $c = 0$ was settled in [3].

Definition 14. For all integers $d, t, c \geq 0$, let

$$\tilde{B}_d^{(c)}(t) = \begin{cases} 1 & \text{for } t \leq 0 \\ \sum_{i=1}^{t+1} G_i^{(c)}(t, d) & \text{otherwise,} \end{cases} \tag{25}$$

where

- for $i = 0, 1, \ldots, c+1$,

$$G_i^{(c)}(t, d) = \min_{0 \leq j \leq i-1} \left\{ \tilde{B}_d^{(c-j)}(t - d - i + j) - \sum_{k=1}^{j} G_{i-k}^{(c)}(t, d) \right\},$$

- for $i \geq c+2$,

$$G_i^{(c)}(t, d) = \min_{0 \leq j \leq c} \left\{ \tilde{B}_d^{(c-j)}(t - d - i + j) - \sum_{k=1}^{j} G_{i-k}^{(c)}(t, d) \right\}.$$

It turns out that, for $c = 0, 1$ and for all $d, t \geq 0$, $A_d^{(c)}(t) = \tilde{B}_d^{(c)}(t)$. In [15] the authors conjectured that, under the hypothesis $c \leq d$, $\tilde{B}_d^{(c)}(t) = A_d^{(c)}(t)$, but they proved only the inequality $\tilde{B}_d^{(c)}(t) \leq A_d^{(c)}(t)$, by providing a winning strategy for the $(\tilde{B}_d^{(c)}(t), d, c)$-game with t questions. Whether the converse inequality holds is an open question.

As we have seen, while the Ulam-Rényi game yields one more three-valued logic interpretation of cubic algebras, its scope is well beyond three-valued logic, and also includes optimal error-correcting codes, binary search with faulty tests and Berlekamp's theory of communication with feedback. In surveying the rapidly growing literature on the subject we have naturally come across a significant portion of mathematics, and in particular, of combinatorial mathematics, to which Gian-Carlo Rota has made fundamental contributions.

Ultimately, our review is intended as a tribute to Gian-Carlo, for his 360° view of mathematics as a unified body of knowledge.

References

[1] Aigner, M. (1996): Searching with lies. J. Combin. Theory Ser. A **74**, 43–56

[2] Aigner, M. (1988): Combinatorial search. Wiley–Teubner, Chichester–Stuttgart

[3] Ambainis, A., Bloch, S.A., Schweizer, D.L. (1999): Delayed binary search, or playing twenty questions with a procrastinator. In: Proceedings of Tenth Annual ACM SIAM Symposium on Discrete Algorithms. Association for Computing Machinery, New York, pp. 844–845

[4] Aslam, J., Dhagat, A. (1991): Searching in the presence of linearly bounded errors. In: Proceedings of the Twenty-Third ACM Symposium on the Theory of Computing. Association for Computing Machinery, New York, pp. 486–493

[5] Bar-Noy, A., Kipnis, S. (1994): Designing broadcast algorithms in the postal model for message-passing systems. Math. Systems Theory **27**, 431–452
[6] Berlekamp, E.R. (1969): Block coding for the binary symmetric channel with noiseless, delayless feedback. In: Mann, H.B. (ed.) Error Correcting Codes, Wiley, New York, pp. 61–88
[7] Borgstrom, R.S., Kosaraju, S. Rao (1993): Comparison-based search in the presence of errors. In: Proceedings of the Twenty-Fifth ACM Symposium on the Theory of Computing. Association for Computing Machinery, New York, pp. 130–136
[8] Bose, R.C., Shrikhande, S.S., Parker, E.T. (1960): Further results on the construction of mutually orthogonal Latin squares and the falsity of a Euler's conjecture. Canad. J. Math. **12**, 189–203
[9] Brouwer, A.E., Shearer, J.B., Sloane, N.J.A., Smith, W.D. (1990): A new table of constant weight codes. IEEE Trans. Inform. Theory **36**, 1334–1380
[10] Chen, W.Y.C., Oliveira, J.S. (1995): Implication algebras and the Metropolis-Rota axioms for cubic lattices. J. Algebra **171**, 383–396
[11] Cicalese, F., Mundici, D. (1999): Optimal binary search with two unreliable tests and minimum adaptiveness. In: Nešetřil, J. (ed.) Algorithms ESA 99. (Lecture Notes in Computer Science, vol. 1643) Springer, Berlin, pp. 257–266
[12] Cicalese, F., Mundici, D. (2000): Perfect two-fault tolerant search with minimum adaptiveness. Adv. Appl. Math. **25**, 65–101
[13] Cicalese, F., Mundici, D. (2000): Optimal coding with one asymmetric error: below the sphere packing bound. In: Du, D.-Z. et. al. (eds.) Computing and Combinatorics. (Lecture Notes in Computer Science, vol. 1858). Springer, Berlin, pp. 159–169
[14] Cicalese, F., Vaccaro, U. (2000): Optimal strategies against a liar. Theoret. Comput. Sci. **230**, 167–193
[15] Cicalese, F., Vaccaro, U. (2000): Coping with delays and time-outs in binary search procedures. Preprint
[16] Cicalese, F., Mundici, D., Vaccaro, U. (2000): Least adaptive optimal search with unreliable tests. In: Halldorsson, M.M. (ed.) Algorithm Theory SWAT 2000. (Lecture Notes in Computer Science, vol. 1851). Springer, Berlin
[17] Cignoli, R., D'Ottaviano, I.M.L., Mundici, D. (2000): Algebraic foundations of many-valued reasoning. (Trends in Logic, vol. 7). Kluwer, Dordrecht
[18] Constantin, S.D., Rao, T.R.N. (1979): On the theory of binary asymmetric error correcting codes. Inform. and Control **40**, 20–26
[19] Czyzowicz, J., Mundici, D., Pelc, A. (1989): Ulam's searching game with lies. J. Combin. Theory Ser. A **52**, 62–76
[20] Dhagat, A., Gács, P., Winkler, P. (1992): On playing "Twenty Questions" with a liar. In: Proceedings of the Third Annual ACM-SIAM Symposium on Discrete Algorithms. Association for Computing Machinery, New York, pp. 16–22
[21] Du, D.-Z., Hwang, F.K. (1993): Combinatorial group testing and its applications. World Scientific, Singapore
[22] Golay, M.J.E. (1949): Notes on digital coding. Proc. IRE **37**, 657
[23] Hill, R. (1995): Searching with lies. In: Rowlinson, P. (ed.) Surveys in Combinatorics. Cambridge University Press, Cambridge, pp. 41–70
[24] Hill, R., Karim, J., Berlekamp, E.R. (1998): The solution of a problem of Ulam on searching with lies. In: Proceedings 1998 IEEE International Symposium on Information Theory. IEEE, Piscatawy, NJ, p. 244
[25] Innes, D. (2000): Searching with a lie using only comparison questions. Preprint

[26] Knill, E. (1995): Lower bounds for identifying subset members with subset queries. In: Proceedings of the Sixth Annual ACM-SIAM Symposium on Discrete Algorithms. Association for Computing Machinery, New York, pp. 369–377
[27] MacWilliams, F.J., Sloane, N.J.A. (1977): The theory of error-correcting codes. North-Holland, Amsterdam
[28] Malinowski, A. (1994): K-ary searching with a lie. Ars Combin. **37**, 301–308
[29] McEliece, R.J., Rodemich, E.R., Rumsey, H.C., Welch, L.R. (1977): New upper bounds on the rate of a code via the Delsarte–MacWilliams inequalities. IEEE Trans. Inform. Theory **23**, 157–166
[30] Mundici, D. (1989): The C^*-algebras of three-valued logic. In: Ferro, R. et al. (eds.), Logic Colloquium '88. (Studies in Logic and the Foundations of Mathematics vol 127). North-Holland, Amsterdam, pp. 61–77
[31] Mundici, D. (1992): The logic of Ulam's game with lies. In: Bicchieri, C., Dalla Chiara, M. L. (eds.) Knowledge, Belief, and Strategic Interaction. (Cambridge Studies in Probability, Induction, and Decision Theory) Cambridge Universtity Press, Cambridge, pp. 275–284
[32] Mundici, D. (2001): Fault-tolerance and Rota-Metropolis cubic logic. In: Proceedings of the Second World Congress on Paraconsistency. Dekker, New York
[33] Mundici, D., Trombetta, A. (1997): Optimal comparison strategies in Ulam's searching game with two errors. Theoret. Comput. Sci. **182**, 217–232
[34] Muthukrishnan, S. (1994): On optimal strategies for searching in presence of errors. In: Proceedings of the Fifth Annual ACM-SIAM Symposium on Discrete Algorithms. Association for Computing Machinery, New York, pp. 680–689
[35] Negro, A., Sereno, M. (1992): Ulam's searching game with three lies. Adv. Appl. Math. **13**, 404–428
[36] Pedrotti, A. (1998): Reliable RAM computation in the presence of noise. PH.D. thesis, Scuola Normale Superiore, Pisa, Italy
[37] Pelc, A. (1987): Solution of Ulam's problem on searching with a lie. J. Combin. Theory. Ser. A **44**, 129–142
[38] Pelc, A. (1989): Searching with known error probability. Theoret. Comput. Sci. **63**, 185–202
[39] Pelc, A. (1994): Searching with permanently faulty tests. Ars Combin. **38**, 65–76
[40] Pierce, J.R. (1978): Optical channels: practical limits with photon counting. IEEE Trans. Comm. **26**, 1819–1821
[41] Rényi, A. (1976): Napló az információelméletről. Gondolat, Budapest (Hungarian) [English transl.: A diary on information theory. Wiley, New York (1984)]
[42] Rivest, R.L., Meyer, A.R., Kleitman, D.J., Winklmann,K., Spencer, J. (1980): Coping with errors in binary search procedures. J. Comput. System Sci. **20**, 396–404
[43] Rota, G.-C., Metropolis, N. (1978): Combinatorial structure of the faces of the n-cube. SIAM J. Appl. Math. **35**, 689–694
[44] Spencer, J. (1984): Guess a number–with lying. Math. Mag. **57**, 105–108
[45] Spencer, J. (1992): Ulam's searching game with a fixed number of lies. Theoret. Comput. Sci. **95**, 307–321
[46] Spencer, J., Winkler, P. (1992): Three thresholds for a liar. Combin. Probab. Comput. **1**, 81–93
[47] Tietäväinen, A. (1973): On the nonexistence of perfect codes over finite fields. SIAM J. Appl. Math. **24**, 88–96
[48] Ulam, S.M. (1976): Adventures of a mathematician. Scribner's, New York

Umbral nature of the Poisson random variables

E. Di Nardo, D. Senato

1 Introduction

The symbolic method, nowadays known as *umbral calculus*, has been extensively used since the nineteenth century although the mathematical community was sceptical of it, perhaps because of its lack of foundation. This method was fully developed by Rev. John Blissard in a series of papers beginning in 1861 [6–16]; nevertheless it is impossible to attribute the credit for the original idea to him alone since Blissard's calculus has a mathematical source in symbolic differentiation. In [22] Lucas even claimed that the umbral calculus has its historical roots in the writing of Leibniz for the successive derivatives of a product with two or more factors; moreover Lucas held that this symbolic method had been developed subsequently by Laplace, by Vandermonde, by Herschel, and augmented by the works of Cayley and of Sylvester in the theory of forms. Lucas' papers attracted considerable attention and the predominant contribution of Blissard to this method was kept in the background. Bell reviewed the whole subject in several papers, restoring the purport of the Blissard's idea [4], and in 1940 he tried to give a rigorous foundation for the mystery at the bottom of the umbral calculus [5] but his attempt did not gain a hold. Indeed, in the first modern textbook of combinatorics [24], Riordan largely employed this symbolic method without giving any formal justification. It was Gian-Carlo Rota [26] who six years later disclosed the "umbral magic art" consisting in lowering and raising exponents bringing to the light the underlying linear functional. The ideas from [26] led Rota and his collaborators to conceive a beautiful theory which gave rise to a large variety of applications ([23, 27]). Some years later, Roman and Rota [25] gave rigorous form to the umbral tricks in the setting of Hopf algebras. On the other hand, as Rota himself has written in [28]: "...Although the notation of Hopf algebra satisfied the most ardent advocate of spic-and-span rigor, the translation of "classical", umbral calculus into the newly found rigorous language made the method altogether unwieldy and unmanageable. Not only was the eerie feeling of witchcraft lost in the translation, but, after such a translation, the use of calculus to simplify computation and sharpen our intuition was lost by the wayside..." Thus in 1994 Rota and Taylor [28] started a rigorous and simple presentation of the umbral calculus in the spirit of the founders. The present article takes this last point of view.

As sometimes happens in the practice of mathematical investigation, the subject we deal with does not develop the original idea from which our research began in the spring of 1997, but this paper is closely related to it. In that period, Gian-Carlo Rota was visiting professor at the University of Basilicata and, during one of our last conversations before his departure, he shared with us his close interest in a certain research project: a combinatorial random variable theory. The delicate question arising from the underlying foundational problems and lack of time led us to protract the discussion via e-mail, intertwining it with different activities for several months. The following year, Gian-Carlo Rota held his last course in Cortona and we did not miss the opportunity to spend some time with him. We resumed the thread of our conversations and presented him with the doubts that had gradually taken hold of us. As usual, his contribution disclosed new horizons that have led us to write these pages.

Our starting point is the umbral notion of the Bell numbers. Many classical identities relating to these numbers are expressed in umbral notation, making use of a new umbra, the partition umbra, connected with the so-called "partition polynomials" generated by expanding the exponential function $\exp(f(x))$ into an exponential power series. The theory of the Bell umbrae developed here is not only an example of the computational power of the umbral calculus but also offers, we would like to believe, a natural way to interpret the functional composition of exponential power series as tested by a new proof of the Lagrange inversion formula. Here the point operations, extended by a new operation, play a central role. From a probabilistic point of view, functional composition of exponential power series is closely related to the family of Poisson random variables so that these random variables have found a natural umbral interpretation through the Bell umbrae. In particular the probabilistic counterpart of the partition umbra is the compound Poisson random variable. Also the less familiar randomized Poisson random variable unexpectedly finds an umbral correspondent and in turn the umbral composition gives a way of generalizing this Poisson random variable.

What is more, we believe that the probabilistic interpretation of the partition umbra and of the umbral composition could give a probabilistic meaning to Joyal species theory [21], namely, a combinatorial random variable theory that we hope to deal in a future publication.

2 The classical umbral calculus

We take a step forward in the program of building a rigorous foundation for the classical umbral calculus initiated by Rota and Taylor [28] (see also [29] and [32]).

In what follows we denote by R a commutative integral domain whose quotient field is of characteristic zero and by $A = \{\alpha, \beta, \dots\}$ a set whose elements are called *umbrae*. An umbral calculus is given when a linear functional $E : R[A, x, y] \rightarrow R[x, y]$ is assigned such that:

i) $E[1] = 1$;
ii) $E[\alpha^i \beta^j \cdots \gamma^k x^n y^m] = x^n y^m E[\alpha^i] E[\beta^j] \cdots E[\gamma^k]$ for any set of distinct umbrae in A and for $i, j, \dots, k, n, m$ nonnegative integers (*uncorrelation property*);

iii) there exists an element $\epsilon \in A$ such that $E[\epsilon^n] = \delta_{0,n}$ for any nonnegative integer n, where

$$\delta_{i,j} = \begin{cases} 1 & \text{if } i = j \\ 0 & \text{if } i \neq j \end{cases} \qquad i, j \in N;$$

iv) there exists an element $u \in A$ such that $E[u^n] = 1$ for any nonnegative integer n.

The umbra ϵ is named *augmentation*, as Roman and Rota first called it [25]. We will call the umbra u the *unity* umbra.

A sequence $a_0, a_1, a_2, \ldots$ in $R[x, y]$ is said to be umbrally represented by an umbra α when

$$E[\alpha^i] = a_i, \quad i = 0, 1, 2, \ldots$$

so that the linear functional E plays the role of an evaluation map.

As Rota suggested, there is an analogy between umbrae and random variables (r.v.) (see also [32]), so we will refer to the elements a_i in $R[x, y]$ as *moments* of the umbra α. The umbra ϵ can be viewed as the r.v. which takes the value 0 with probability 1 and the umbra u as the r.v. which takes the value 1 with probability 1.

An umbra is said to be a *scalar* umbra if the moments are elements of R while it is said to be a *polynomial* umbra if the moments are polynomials. Note that, if the sequence $a_0, a_1, a_2, \ldots$ is umbrally represented by a scalar umbra α, then $a_0 = 1$. In the same way, for polynomial umbrae, a sequence of polynomials $p_0(x), p_1(x), p_2(x), \ldots$ will always denote a sequence of polynomials with coefficients in R such that $p_0(x) = 1$ and $p_n(x)$ is of degree n for every positive integer n.

A polynomial $p \in R[A]$ is called an *umbral polynomial*. The *support* of p is defined to be the set of all umbrae of A which occur. Two umbral polynomials are said to be *uncorrelated* when their supports are disjoint.

If α and β either scalar or polynomial umbrae, we will say that α and β are *umbrally equivalent* when

$$E[\alpha] = E[\beta],$$

in symbols, $\alpha \simeq \beta$. Two scalar (or polynomial) umbrae are said to be *similar* when $E[\alpha^k] = E[\beta^k]$, $k = 0, 1, 2, \ldots$, or

$$\alpha^k \simeq \beta^k, \; k = 0, 1, 2, \ldots,$$

in symbols, $\alpha \equiv \beta$[1]. The notion of equivalence and similarity for umbral polynomials is obvious.

[1] In [28], [29] and [32] Rota and his collaborators call *exchangeable* the umbrae that we call similar. Our choice is motivated by the analogy with random variables; on the other hand Rota himself agreed with it during one of our last conversations in the summer of 1998.

The formal power series

$$e^{\alpha t} = u + \sum_{n \geq 1} \alpha^n \frac{t^n}{n!}$$

is said to be the *generating function* of the umbra α. Moreover, if the sequence $a_0, a_1, a_2, \ldots$ has (exponential) generating function $f(t)$ and is umbrally represented by an umbra α, then $E[e^{\alpha t}] = f(t)$, in symbols, $e^{\alpha t} \simeq f(t)$. When α is regarded as a r.v., $f(t)$ is its *moment* generating function. The notion of equivalence and similarity are extended coefficientwise to the generating functions of umbrae so that $\alpha \equiv \beta$ if and only if $e^{\alpha t} \simeq e^{\beta t}$. Note that $e^{\epsilon t} \simeq 1$ and $e^{ut} \simeq e^t$.

2.1 The point operations

The notion of similarity among umbrae is useful in expressing sequences such as

$$\sum_{i=0}^{n} \binom{n}{i} a_i a_{n-i}, \quad n = 0, 1, 2, \ldots \tag{1}$$

as moments of umbrae. The sequence (1) cannot be represented by using only the umbra α with moments $a_0, a_1, a_2, \ldots$ because $a_i a_{n-i}$ could not be written as $E[\alpha^i \alpha^{n-i}]$, α being correlated to itself. If we assume that the umbral calculus we deal is saturated [28], then the sequence $a_0, a_1, a_2, \ldots$ in $R[x, y]$ is represented by infinitely many distinct (and thus similar) umbrae. Therefore, if we choose two similar umbrae α, α', they are uncorrelated and

$$\sum_{i=0}^{n} \binom{n}{i} a_i a_{n-i} = E\left[\sum_{i=0}^{n} \binom{n}{i} \alpha^i \alpha'^{n-i}\right] = E[(\alpha + \alpha')^n].$$

Then the sequence (1) represents the moments of the umbra $(\alpha + \alpha')$. This matter was first explicitly pointed out by E.T. Bell [5] who was not able to provide an effective notation[2]:

[2] The quotation needs more details. It is

$$\alpha x \dotplus \cdots \dotplus \varepsilon x \equiv (\alpha x_0 + \cdots + \varepsilon x_0, \ldots, \alpha x_N + \cdots + \varepsilon x_N, \ldots),$$

where x is an umbra. Formula (1.22) is

$$(\alpha a \dotplus \cdots \dotplus \varepsilon x)^N = \sum M_{S_1, \ldots, S_T} \alpha^{S_1} \cdots \varepsilon^{S_T} a^{S_1} \cdots x^{S_T}$$

with $M_{S_1, \ldots, S_T}$ the coefficient of $x_1^{S_1} \cdots x_T^{S_T}$ in the expansion of $(x_1 + \cdots + x_T)^N$ through the multinomial theorem. Formula (1.20) is

$$(\alpha a \dotplus \cdots \dotplus \varepsilon x)^N \equiv \sum M_{S_1, \ldots, S_T} \alpha^{S_1} \cdots \varepsilon^{S_T} a_{S_1} \cdots x_{S_T}.$$

"...If in $\alpha x\dot{+}\cdots\dot{+}\xi x$ there are precisely T summands $\alpha x, \ldots, \xi x$ each of which is a scalar product of a scalar and x, we replace ($\rightarrow$) the T x's by T distinct umbrae, say $a, \ldots, x$ in any order, and indicate this replacement by writing

$$\alpha x\dot{+}\cdots\dot{+}\xi x \rightarrow \alpha a\dot{+}\cdots\dot{+}\xi x.$$

Then $(\alpha a\dot{+}\cdots\dot{+}\xi x)^N$ is to be calculated by (1.22) and the exponents are degraded as in (1.20). In the result, each of $a, \ldots, x$ is replaced ($\leftarrow$) by x; the resulting polynomial is defined to be the Nth power $(\alpha x\dot{+}\cdots\dot{+}\xi x)^N$ of the sum $(\alpha x\dot{+}\cdots\dot{+}\xi x)$. For example

$$\begin{aligned}
(\alpha x+\beta x)^3 &\rightarrow (\alpha a+\beta x)^3;\\
(\alpha a+\beta x)^3 &= \alpha^3 a_3 x_0 + 3\alpha^2\beta a_2 x_1 + 3\alpha\beta^2 a_1 x_2 + \beta^3 a_0 x_3,\\
&\leftarrow \alpha^3 x_3 x_0 + 3\alpha^2\beta x_2 x_1 + 3\alpha\beta^2 x_1 x_2 + \beta^3 x_0 x_3;\\
(\alpha x+\beta x)^3 &= (\alpha^3+\beta^3)x_0x_3 + 3\alpha\beta(\alpha+\beta)x_1x_2 \ldots\text{''}
\end{aligned}$$

The last identity makes sense when the left-hand side is replaced by $(\alpha x + \beta x')^3$ with $x \equiv x'$, but Bell did not have the notion of similar umbrae. However, the need for handling sequences like (1) leads us to introduce some new operations between umbrae, as shown in the subsequent sections.

2.2 The point product

We denote by the symbol $n.\alpha$ an auxiliary umbra similar to the sum $\alpha'+\alpha''+\ldots+\alpha'''$ where $\alpha', \alpha'', \ldots, \alpha'''$ are a set of n distinct umbrae each of which is similar to the umbra α. A similar notation $n.p$ is introduced for any umbral polynomial. We assume that $0.\alpha$ is an umbra similar to the augmentation ϵ. The following statements are easy to prove.

Proposition 1. (i) *If $n.\alpha \equiv n.\beta$ for some integer $n \neq 0$, then $\alpha \equiv \beta$.*
(ii) *If $c \in R$, then $n.(c\alpha) \equiv c(n.\alpha)$ for any nonnegative integer n.*
(iii) *$n.(m.\alpha) \equiv (nm).\alpha \equiv m.(n.\alpha)$ for any two nonnegative integers n, m.*
(iv) *$(n+m).\alpha \equiv n.\alpha + m.\alpha'$ for any two nonnegative integers n, m and any two distinct umbrae $\alpha \equiv \alpha'$.*
(v) *$(n.\alpha + n.\beta) \equiv n.(\alpha+\beta)$ for any nonnegative integer n and any two distinct umbrae α and β.*

Proposition 2. *If α is an umbra with generating function $e^{\alpha t} \simeq f(t)$, then the umbra $n.\alpha$ has generating function $e^{(n.\alpha)t} \simeq [f(t)]^n$.*

Proof. This follows from the definition of the auxiliary umbra $n.\alpha$. □

The moments of the umbra $n.\alpha$ are the following polynomials in the variable n:

$$E[(n.\alpha)^k] = q_k(n) = \sum_{i=0}^{k} (n)_i B_{k,i}, \quad k = 0, 1, 2, \ldots \tag{2}$$

where $B_{k,i} = B_{k,i}(a_1, a_2, \ldots, a_{k-i+1})$ for $i \leq k$ are the (partial) Bell exponential polynomials [2], $(n)_i$ is the lower factorial and a_i are the moments of the umbra α. Recalling that

$$\sum_{k=i}^{\infty} B_{k,i} \frac{t^k}{k!} = \frac{1}{i!}[f(t) - 1]^i, \tag{3}$$

identity (2) follows from

$$[f(t)]^n = \sum_{i=0}^{\infty} (n)_i \frac{[f(t) - 1]^i}{i!} = \sum_{k=0}^{\infty} \left(\sum_{i=0}^{k} (n)_i B_{k,i} \right) \frac{t^k}{k!}. \tag{4}$$

If we set $\alpha = u$ in (2), then $q_k(n) = n^k$. Note that $q_0(n) = 1$, $q_k(0) = 0$ and the polynomial sequence $\{q_k(n)\}$ is of binomial type as follows by using statement (iv) of Proposition 1:

$$[(n+m).\alpha]^k \simeq [n.\alpha + m.\alpha']^k \simeq \sum_{i=0}^{k} \binom{k}{i} (n.\alpha)^i (m.\alpha')^{k-i}.$$

Moreover a variety of combinatorial identities can be interpreted umbrally. As instance in point, the classical Abel identity becomes

$$(\alpha + \beta)^n \simeq \sum_{k=0}^{n} \binom{n}{k} \alpha(\alpha + k.\gamma)^{k-1} (\beta + k.\gamma)^{n-k}, \quad n = 0, 1, 2, \ldots. \tag{5}$$

The expression of the polynomial sequence $\{q_k(n)\}$ in (2) suggests a way to define the auxiliary umbra $x.\alpha$ when $x \in R$; however it is impossibile to give an intrinsic definition. Up to similarity, the umbra $x.\alpha$ is the polynomial umbra whose moments are

$$E[(x.\alpha)^k] = q_k(x) = \sum_{i=0}^{k} (x)_i B_{k,i}, \quad k = 0, 1, 2, \ldots. \tag{6}$$

Note that $q_k(x) = x^k$ when $\alpha = u$.

Proposition 3. *If α is an umbra with generating function $e^{\alpha t} \simeq f(t)$, then the umbra $x.\alpha$ has generating function $e^{(x.\alpha)t} \simeq [f(t)]^x$.*

Proof. This follows from (4) and (3) with n replaced by x. □

Corollary 1. (i) *If $x.\alpha \equiv x.\beta$ for $x \in R - \{0\}$, then $\alpha \equiv \beta$.*
(ii) *If $c \in R$, then $x.(c\alpha) \equiv c(x.\alpha)$ for any $x \in R$.*

(iii) $x.(y.\alpha) \equiv (xy).\alpha \equiv y.(x.\alpha)$ *for any* $x, y \in R$.
(iv) $(x+y).\alpha \equiv x.\alpha + y.\alpha'$ *for any* $x, y \in R$ *and any two distinct umbrae* $\alpha \equiv \alpha'$.
(v) $(x.\alpha + x.\beta) \equiv x.(\alpha+\beta)$ *for any* $x \in R$ *and any two distinct umbrae* α *and* β.

Theorem 1. *Up to similarity, each polynomial sequence of binomial type is umbrally represented by an auxiliary umbra* $x.\alpha$, *and conversely.*

Proof. From statement (iv) of Corollary 1, it follows that the polynomial sequence $\{q_k(x)\}$ is of binomial type. For the converse, first observe that from (6) we have

$$D_x[q_k(x)]_{x=0} = a_k + F(a_1, a_2, \dots, a_{k-1}) \tag{7}$$

where F is a function of the moments $a_1, a_2, \dots, a_{k-1}$. Let $\{p_k(x)\}$ be a polynomial sequence of binomial type. From (7), the moments of the umbra α are uniquely determined by knowing the first derivative with respect to x of $p_k(x)$ evaluated at 0. Moreover, the sequence of first derivatives with respect to x of $p_k(x)$ evaluated at 0 uniquely determines a sequence of binomial type. □

Similarly to what was done for the auxiliary umbra $x.\alpha$, we define a point product among umbrae. Up to similarity, the umbra $\beta.\alpha$ is an auxiliary umbra whose moments are umbrally expressed through the umbral polynomials $q_{\alpha,k}(\beta)$:

$$(\beta.\alpha)^k \simeq q_{\alpha,k}(\beta) = \sum_{i=0}^{k} (\beta)_i B_{k,i}, \quad k = 0, 1, 2, \dots . \tag{8}$$

If α is an umbra with generating function $e^{\alpha t} \simeq f(t)$, then identity (4) can be rewritten as

$$[f(t)]^{\beta} \simeq \sum_{i=0}^{\infty} (\beta)_i \frac{[f(t)-1]^i}{i!} \simeq \sum_{k=0}^{\infty} \left(\sum_{i=0}^{k} (\beta)_i B_{k,i} \right) \frac{t^k}{k!} \tag{9}$$

so that $e^{(\beta.\alpha)t} \simeq [f(t)]^{\beta}$. Moreover, if β is an umbra with generating function $e^{\beta t} \simeq g(t)$, then

$$[f(t)]^{\beta} \simeq e^{\beta \log f(t)} \simeq g\left[\log f(t)\right].$$

This proves the following proposition.

Proposition 4. *If* α *is an umbra with generating function* $e^{\alpha t} \simeq f(t)$ *and* β *is an umbra with generating function* $e^{\beta t} \simeq g(t)$, *then the umbra* $\beta.\alpha$ *has generating function*

$$e^{(\beta.\alpha)t} \simeq [f(t)]^{\beta} \simeq g\left[\log f(t)\right]. \tag{10}$$

Corollary 2. *If* $\gamma \equiv \gamma'$, *then*

$$(\alpha+\beta).\gamma \equiv \alpha.\gamma + \beta.\gamma'.$$

Proof. Let $e^{\gamma t} \simeq h(t)$ be the generating function of the umbra γ. We then have

$$e^{[(\alpha+\beta).\gamma]t} \simeq [h(t)]^{\alpha+\beta} \simeq [h(t)]^{\alpha}[h(t)]^{\beta} \simeq e^{(\alpha.\gamma)t} e^{(\beta.\gamma')t}$$

from which the result follows. □

Remark 1. As Taylor suggests in [32], the auxiliary umbra $\beta.\alpha$ provides an umbral interpretation of the random sum since the moment generating function $g[\log f(t)]$ corresponds to the r.v. $S_N = X_1 + X_2 + \cdots + X_N$ with X_i independent identically distributed (i.i.d.) r.v. having moment generating function $f(t)$ and with N a discrete r.v. having moment generating function $g(t)$. The probabilistic interpretation of Corollary 2 states that the random sum S_{N+M} is similar to $S_N + S_M$, where N and M are two independent discrete r.v.

The left distributive property of the point product with respect to the sum does not hold since

$$e^{[\alpha.(\beta+\gamma)]t} \simeq [g(t)]^{\alpha}[h(t)]^{\alpha} \not\simeq f[\log g(t)] f[\log h(t)]$$

where $g(t) \simeq e^{\beta t}$, $h(t) \simeq e^{\gamma t}$ and $f(t) \simeq e^{\alpha t}$.

Again this result runs in parallel with probability theory. In fact, let $Z = X + Y$ be an r.v. with X and Y two independent r.v. The random sum $S_N = Z_1 + Z_2 + \cdots + Z_N$, with Z_i i.i.d. r.v. similar to Z, is not similar to the r.v. $S_N^X + S_N^Y$ where $S_N^X = X_1 + X_2 + \cdots + X_N$ with X_i i.i.d. r.v. similar to X, and where $S_N^Y = Y_1 + Y_2 + \cdots + Y_N$ with Y_i i.i.d. r.v. similar to Y.

Corollary 3. (i) *If $\beta.\alpha \equiv \beta.\gamma$, then $\alpha \equiv \gamma$.*
(ii) *If $c \in R$, then $\beta.(c\alpha) \equiv c(\beta.\alpha)$ for any two distinct umbrae α and β.*
(iii) $\beta.(\gamma.\alpha) \equiv (\beta.\gamma).\alpha$.

Proof. Via generating functions. □

To end this section, we consider the notion of the inverse of an umbra. Two umbrae α and β are said to be *inverse* to each other when $\alpha + \beta \equiv \varepsilon$. Recall that, in dealing with a saturated umbral calculus, the inverse of an umbra is not unique, but any two inverse umbrae of the umbra α are similar.

Proposition 5. *If α is an umbra with generating function $e^{\alpha t} \simeq f(t)$, then its inverse β has generating function $e^{\beta t} \simeq [f(t)]^{-1}$.*

Proof. The result follows by observing that $e^{(\alpha+\beta)t} \simeq 1$. □

Similarly, for every positive integer n and for every umbra $\alpha \in A$, the inverse of the auxiliary umbra $n.\alpha$, written as $-n.\alpha'$ with $\alpha \equiv \alpha'$, is similar to $\beta' + \beta'' + \cdots + \beta'''$ where $\beta', \beta'', \ldots, \beta'''$ is any set of n distinct umbrae similar to β, β being the inverse of α. The notation $-n.\alpha'$ is justified by noting that

$$n.\alpha - n.\alpha' \equiv (n-n).\alpha \equiv 0.\alpha \equiv \varepsilon.$$

Proposition 6. *If α is an umbra with generating function $e^{\alpha t} \simeq f(t)$, then the inverse of $n.\alpha$ has generating function $e^{(-n.\alpha')t} \simeq [f(t)]^{-n}$.*

Proof. The result follows by observing that $e^{(n.\alpha - n.\alpha')t} \simeq 1$. □

The inverse of the umbra $x.\alpha$ is the umbra $-x.\alpha'$ with $\alpha \equiv \alpha'$ defined by

$$x.\alpha - x.\alpha' \equiv (x - x).\alpha \equiv 0.\alpha \equiv \varepsilon.$$

2.3 The point power

As one might expect, the definition of the power of moments requires the use of similar umbrae and so of a point operation. This notion comes into this picture in a natural way, providing as well a useful tool for umbral manipulation of generating functions.

We denote by the symbol $\alpha^{.n}$ an auxiliary umbra similar to the product $\alpha'\alpha'' \cdots \alpha'''$ where $\alpha', \alpha'', \ldots, \alpha'''$ are a set of n distinct umbrae each of which is similar to the umbra α. A similar notation is introduced for any umbral polynomial p. We assume that $\alpha^{.0}$ is an umbra similar to the unity umbra u. The following statements are easy to prove.

Proposition 7. (i) *If $c \in R$, then $(c\alpha)^{.n} \equiv c^n \alpha^{.n}$ for any nonnegative integer $n \neq 0$.*
(iii) *$(\alpha^{.n})^{.m} \equiv \alpha^{.nm} \equiv (\alpha^{.m})^{.n}$ for any two nonnegative integers n, m.*
(iii) *$\alpha^{.(n+m)} \equiv \alpha^{.n} (\alpha')^{.m}$ for any two nonnegative integers n, m and any two distinct umbrae $\alpha \equiv \alpha'$.*
(iv) *$(\alpha^{.n})^k \equiv (\alpha^k)^{.n}$ for any two nonnegative integers n, k.*

By the last statement, the moments of $\alpha^{.n}$ for any integer n are:

$$E[(\alpha^{.n})^k] = E[(\alpha^k)^{.n}] = a_k^n, \quad k = 0, 1, 2, \ldots \tag{11}$$

so that the moments of the umbra $\alpha^{.n}$ are the nth power of the moments of the umbra α. What is more, we have

$$(\alpha + \beta)^{.n} \equiv \sum_{i=0}^{n} \binom{n}{i} \alpha^{.i} \beta^{.(n-i)}.$$

Proposition 8. *The generating function of the nth point power of the umbra α is the nth power of the generating function of the umbra α.*

Note that the generating function of the point product between umbrae is umbrally equivalent to the series

$$e^{(\beta.\alpha)t} \simeq \sum_{i=0}^{\infty} (\beta)_i \frac{[e^{\alpha t} - u]^{.i}}{i!} \tag{12}$$

by relation (9) and Proposition 8. Moreover, by virtue of Propositions 2 and 8, we have

$$e^{(n.\alpha)t} \simeq (e^{\alpha t})^{.n}. \tag{13}$$

Relation (13) restores the natural umbral interpretation of $[f(t)]^n$.

The point power operation leads us to define the point exponential of an umbra. We denote by the symbol $e.^{\alpha}$ the auxiliary umbra

$$e.^{\alpha} = u + \sum_{n=1}^{\infty} \frac{\alpha^{.n}}{n!}. \tag{14}$$

We immediately have $e.^{\epsilon} \equiv u$.

Proposition 9. *For any umbra α, we have*

$$e.^{(n.\alpha)} \simeq (e.^{\alpha})^{.n}. \tag{15}$$

Proof. We have

$$E[(e.^{\alpha})^{.n}] = E[e.^{\alpha}]^n = e^{nE[\alpha]} = \sum_{k\geq 0} \frac{n^k E[\alpha]^k}{k!}$$

and also

$$E[e.^{(n.\alpha)}] = \sum_{k\geq 0} \frac{E[(n.\alpha)^{.k}]}{k!} = \sum_{k\geq 0} \frac{n^k E[\alpha.^k]}{k!},$$

from which (15) follows. □

Up to similarity, the expression of the moments given in (11) justifies the definition of the auxiliary umbra $\alpha^{.x}$ as the umbra whose moments are

$$E[(\alpha^{.x})^k] = a_k^x, \quad k = 0, 1, 2, \dots .$$

Proposition 10. *Let α be an umbra and $e^{\alpha t} \simeq f(t)$ its generating function. Then*

$$e^{(x.\alpha)t} \simeq (e^{\alpha t})^{.x} \simeq [f(t)]^x.$$

Via moments, it is possible to prove the analogue of Proposition 7 where n and m are replaced by x and y with $x, y \in R$.

Once again, we define the auxiliary umbra $\alpha^{.\beta}$ as the umbra whose moments are umbrally equivalent to

$$(\alpha^{.\beta})^k \simeq a_k^{\beta}, \quad k = 0, 1, 2, \dots$$

and we set $\epsilon^{.\alpha} \equiv \epsilon$.

Proposition 11. (i) $(\alpha^{.\beta})^{.\gamma} \equiv \alpha^{.(\gamma.\beta)}$.
(ii) $\alpha^{.(\beta+\gamma)} \equiv \alpha^{.\beta}(\alpha')^{.\gamma}$ *for any two distinct umbrae* $\alpha \equiv \alpha'$.

Proof. This follows via moments. □

Finally, the next proposition provides an umbral expression for the generating function of the umbra $\beta.\alpha$.

Proposition 12. *Let* $e^{\alpha t} \simeq f(t)$ *be the generating function of the umbra* α. *Then*

$$(e^{\alpha t})^{.\beta} \simeq [f(t)]^{\beta}.$$

3 Bell umbrae

The Bell numbers B_n have a long history and their origin is unknown. Bell ascribes them to Euler although without a specific reference [3]. Usually they are referred as the number of partitions of a finite nonempty set with n elements or as the coefficients of the Taylor series expansion of the function $\exp(e^t - 1)$. It is precisely in writing about the Bell numbers that Gian-Carlo Rota [26] gives the first glimmering of the effectiveness of the umbral calculus in manipulating number sequences; indeed, his proof of Dobinski's formula is implicitly of umbral nature.

In this section, the umbral definition of the Bell numbers allows the proofs of several classical identities (cf. [33]) to be given by elementary arguments and smooths the way to the umbral interpretation of Poisson random variables.

Definition 1. An umbra β is said to be a *Bell scalar umbra* if

$$(\beta)_n \simeq 1, \qquad n = 0, 1, 2, \dots,$$

where $(\beta)_0 = 1$ and $(\beta)_n = \beta(\beta - 1) \cdots (\beta - n + 1)$ is the lower factorial.

Up to similarity, the Bell number sequence is umbrally represented by the Bell scalar umbra. Indeed, as

$$\beta^n = \sum_{k=0}^{n} S(n,k)(\beta)_k$$

where $S(n,k)$ are the Stirling numbers of second kind,

$$E(\beta^n) = \sum_{k=0}^{n} S(n,k) E[(\beta)_k] = \sum_{k=0}^{n} S(n,k) = B_n$$

where B_n are the Bell numbers.

The following theorem provides a characterization of the Bell umbra.

Theorem 2. *A scalar umbra* β *is a Bell umbra iff*

$$\beta^{n+1} \simeq (\beta + u)^n, \quad n = 0, 1, 2, \dots. \tag{16}$$

Proof. If β is the Bell scalar umbra, then, as $\beta(\beta - u)_n \simeq (\beta)_{n+1}$, we have $E[\beta(\beta - u)_n] = 1 = E[(\beta)_n]$. By linearity it follows that

$$E[\beta p(\beta - u)] = E[p(\beta)]$$

for every polynomial p in β. So, formula (16) follows on setting $p(\beta) = (\beta + u)^n$. Conversely, relation (16) gives

$$E[\beta^{n+1}] = E[(\beta + u)^n] = \sum_{k=0}^{n} \binom{n}{k} E[\beta^k]$$

or, setting $E[\beta^n] = B_n$,

$$B_{n+1} = \sum_{k=0}^{n} \binom{n}{k} B_k$$

that is, the recursion formula for the Bell numbers. □

Corollary 4. *If β is the Bell scalar umbra, then*

$$D_t[e^{\beta t}] \simeq e^{(\beta+u)t}. \tag{17}$$

Proposition 13. *If β is the Bell scalar umbra, then, for any integer $k > 0$ and for any polynomial $p(x)$, the following relation holds:*

$$p(\beta + k.u) \simeq (\beta)_k p(\beta) \simeq p(\beta).$$

Proof. For $n \geq k$, by Definition 1 it follows that

$$(\beta)_n \simeq (\beta)_{n+k} \simeq (\beta)_k (\beta - k.u)_n.$$

Thus, for any polynomial q, we have

$$q(\beta) \simeq (\beta)_k q(\beta - k.u)$$

from which it follows that

$$(\beta + k.u)^n \simeq (\beta)_k \beta^n, \quad n = 0, 1, 2, \ldots$$

by setting $q(\beta) = (\beta + k.u)^n$ and the result follows by linearity. □

Proposition 14. *The generating function of the Bell umbra is*

$$e^{\beta t} \simeq e_{.}^{\,e^{ut} - u}. \tag{18}$$

Proof. By Definition 1 and relation (12) we have

$$e^{\beta t} \simeq e^{(\beta . u)t} \simeq \sum_{i=0}^{\infty} \frac{[e^{ut} - u]^{.i}}{i!}.$$

Thus relation (18) follows from (14). □

Remark 2. We continue with our probabilistic counterpoint by noting that the Bell umbra can be view as a Poisson r.v. with parameter $\lambda = 1$. Indeed, the moment generating function of the Bell umbra is equal to the moment generating function of the Poisson random variable with parameter 1, i.e., $\exp(e^t - 1)$ (see (18)). It follows that the moments of such a Poisson r.v. are the Bell numbers and its factorial moments are equal to 1.

The following theorem makes clear how the proof of Dobinski's formula becomes natural through the umbral expression of the Bell umbra generating function.

Theorem 3 (Umbral Dobinski's formula). *The Bell umbra β satisfies the formula:*

$$\beta^n \simeq e.^{-u} \sum_{k=0}^{\infty} \frac{(k.u)^n}{k!}.$$

Proof. As $e^{\beta t} \simeq e.^{-u} e.^{e^{ut}}$, we have

$$e^{\beta t} \simeq e.^{-u} \sum_{k=0}^{\infty} \frac{e^{(k.u)t}}{k!} \simeq e.^{-u} \sum_{k=0}^{\infty} \frac{1}{k!} \left\{ \sum_{n=0}^{\infty} \frac{(k.u)^n t^n}{n!} \right\}$$

from which the result follows. □

3.1 The Bell polynomial umbra

Definition 2. An umbra ϕ is said to be a *Bell polynomial umbra* if

$$(\phi)_n \simeq x^n, \qquad n = 0, 1, 2, \ldots .$$

Note that $\phi \equiv \beta$ for $x = 1$. Moreover, as

$$\phi^n = \sum_{k=0}^{n} S(n,k)(\phi)_k,$$

by Definition 2 it follows that

$$E(\phi^n) = \sum_{k=0}^{n} S(n,k) E[(\phi)_k] = \sum_{k=0}^{n} S(n,k) x^k = \Phi_n(x). \tag{19}$$

The polynomials $\Phi_n(x)$ have a statistical origin and are known in the literature as *exponential polynomials*. Indeed, they were first introduced by Steffensen [31] and studied further by Touchard [33] and others. Rota, Kahaner and Odlyzko [27] state their basic properties via umbral operators.

Proposition 15. *The generating function of the Bell polynomial umbra is*

$$e^{\Phi t} \simeq e.^{x.(e^{ut} - u)} \tag{20}$$

Proof. By Definition 2 and relation (12) we have

$$e^{\phi t} \simeq e^{(\phi.u)t} \simeq \sum_{i=0}^{\infty} x^i \frac{[e^{ut} - u]^{.i}}{i!}.$$

Thus relation (20) follows from (14). □

The following theorem provides a characterization of the Bell polynomial umbra.

Theorem 4. *An umbra ϕ is the Bell polynomial umbra iff*

$$\phi \equiv x.\beta$$

where β is the Bell scalar umbra.

Proof. The result follows from (20). □

Remark 3. The Bell polynomial umbra can be view as a Poisson r.v. with parameter $\lambda = x$. Indeed, the moment generating function of the Bell polynomial umbra is equal to the moment generating function of the Poisson random variable with parameter x, i.e., $\exp[x(e^t - 1)]$ (see (20)). It follows that the moments of such a Poisson r.v. are the exponential polynomials and its factorial moments are equal to x^n.

When $x = n$, the Bell polynomial umbra $n.\beta$ is the sum of n similar uncorrelated Bell scalar umbrae, just as in probability theory where a Poisson r.v. of parameter n can be viewed as the sum of n i.i.d. (eventually uncorrelated) Poisson r.v. with parameter 1. More generally, the closure under convolution of the Poisson probability distributions, i.e., $F_s \star F_t = F_{s+t}$, where F_t is a Poisson probability distribution depending on the parameter t, is umbrally translated by $x.\beta + y.\beta' \equiv (x + y).\beta$ (cf. statement (iv) of Corollary 1).

The next theorem is the polynomial analogue of Theorem 2.

Theorem 5. *An auxiliary umbra $x.\beta$ is a Bell polynomial umbra iff*

$$(x.\beta)^{n+1} \simeq x(x.\beta + u)^n, \quad n = 0, 1, 2, \ldots, \tag{21}$$

Proof. Observe that $D_t[e^{(x.\beta)t}] \simeq D_t[(e^{\beta t})^{.x}] \simeq x e^{[(x-1).\beta]t} D_t[e^{\beta' t}]$, where $\beta' \equiv \beta$. From (17) we have $D_t[e^{(x.\beta)t}] \simeq x e^{(x.\beta+u)t}$ from which the result follows immediately. Conversely, relation (21) gives (16) for $x = 1$, from which it follows that β is the Bell scalar umbra. □

Relation (21) represents the umbral equivalent of the well-known recursive formula for the exponential polynomials:

$$\Phi_{n+1}(x) = x \sum_{k=0}^{n} \binom{n}{k} \Phi_k(x).$$

Similarly, the next proposition gives an umbral analogue of Rodrigues' formula for the exponential polynomials (cf. [27]).

Proposition 16. *The Bell polynomial umbra $x.\beta$ has the following property:*

$$D_x[(x.\beta)^n] \simeq (x.\beta + u)^n - (x.\beta)^n.$$

Proof. From (18) we have

$$D_x[e^{(x.\beta)t}] \simeq e^{(x.\beta)t}(e^{ut} - u) \simeq e^{(x.\beta+u)t} - e^{(x.\beta)t}$$

from which the result follows immediately. □

In closing we state the polynomial version of the umbral Dobinski's formula.

Proposition 17. *The Bell polynomial umbra $x.\beta$ satisfies the relation*

$$(x.\beta)^n \simeq e^{-x.u} \sum_{k=0}^{\infty} \frac{(k.u)^n x^k}{k!}.$$

Proof. As $e^{(x.\beta)t} \simeq e.^{-x.u} e.^{(x.e^{ut})}$ we have

$$e^{(x.\beta)t} \simeq e.^{-x.u} \sum_{k=0}^{\infty} \frac{(x.e^{ut})^{.k}}{k!} \simeq e.^{-x.u} \sum_{k=0}^{\infty} \frac{x^k e^{(k.u)t}}{k!}$$

from which the result follows. □

3.2 The exponential umbral polynomials

We now introduce a new family of umbral polynomials that will turn out to be a useful tool in umbral composition, and disclosing an unexpected probabilistic interpretation.

Set

$$\Phi_n(\alpha) = \sum_{k=0}^{n} S(n,k)\alpha^k, \quad n = 0, 1, 2, \ldots ; \tag{22}$$

we will call $\Phi_n(\alpha)$ *exponential umbral polynomials*. By identity (19), as $(x.\beta)^n \simeq \Phi_n(x)$, where β is the Bell scalar umbra, it is obvious that

$$\Phi_n(\alpha) \simeq (\alpha.\beta)^n, \quad n = 0, 1, 2, \ldots ,$$

and

$$(\alpha.\beta)_n \simeq \alpha^n, \quad n = 0, 1, 2, \ldots , \tag{23}$$

a formal proof being available by arguments similar to those already produced for the umbra $x.\beta$.

Proposition 18. *Let β be the Bell scalar umbra. If $e^{\alpha t} \simeq f(t)$ is the generating function of the umbra α, then*

$$e^{(\alpha.\beta)t} \simeq f[e^t - 1]. \tag{24}$$

Proof. The result follows from relation (10) on observing that $e^{\beta t} \simeq e^{e^t - 1}$. □

When $f(t)$ is considered as the moment generating function of an r.v. X, a probabilistic interpretation of (24) suggests that the umbra $\alpha.\beta$ represents a Poisson r.v. N_X with random parameter X. Indeed, the probability generating function of N_X is

$$\begin{aligned} P(s) &= \sum_{k=0}^{\infty} \mathbf{P}(N_X = k) s^k = \sum_{k=0}^{\infty} s^k \int_0^{\infty} \mathbf{P}(N_X = k | X = x) d F_X(x) \\ &= \sum_{k=0}^{\infty} \frac{s^k}{k!} \int_0^{\infty} x^k e^{-x} d F_X(x) = \sum_{k=0}^{\infty} \frac{(s-1)^k}{k!} E[X^k] = f(s-1); \end{aligned}$$

hence the moment generating function of N_X is $f(e^t - 1)$. To the best of our knowledge, this r.v. was first introduced in [20] as the randomized Poisson r.v. Once more, closure under convolution of the Poisson probability distributions leads us to claim that the point product $\alpha.\beta$ is the umbral correspondent of the random sum of independent Poisson r.v. with parameter 1 indexed by an integer r.v. X.

4 The partition umbra

As suggested in [27], there is a connection between polynomials of binomial type and compound Poisson processes. Two different approaches can be found in [30] and in [17]. In this section, we suggest a way, that we believe to be natural, of making clear this connection.

Definition 3. An umbra ψ is said to be an *α-partition umbra* if

$$\psi \equiv \beta.\alpha$$

with β the Bell scalar umbra.

Note that the u-partition umbra is the Bell scalar umbra.

Proposition 19. *The generating function of the α-partition umbra ψ is*

$$e^{\psi t} \simeq e.^{(e^{\alpha t} - u)}. \tag{25}$$

Proof. From (12) and Definition 1 we have

$$e^{(\beta.\alpha)t} \simeq \sum_{n=0}^{\infty} \frac{(e^{\alpha t} - u)^{.n}}{n!}.$$

Thus formula (25) follows from (14). □

The generating function (25) leads us to interpret a partition umbra as a compound Poisson r.v. with parameter 1. As is well-known (cf. [19]), a compound Poisson r.v. with parameter x is introduced as a random sum $S_N = X_1 + X_2 + \cdots + X_N$ where N

has a Poisson distribution with parameter x. The point product of Definition 3 fits this probabilistic notion perfectly by taking into consideration the fact that the Bell scalar umbra β plays the role of a Poisson r.v. with parameter 1. What is more, since the Poisson r.v. with parameter x is umbrally represented by the Bell polynomial umbra $x.\beta$, a compound Poisson r.v. with parameter x is represented by the *polynomial α-partition umbra* $x.\psi \equiv x.\beta.\alpha$ with generating function

$$e^{(x.\psi)t} \simeq e.^{[x.(e^{\alpha t}-u)]}. \tag{26}$$

The name "partition umbra" also has a probabilistic ground. Indeed, the parameter of a Poisson r.v. is usually denoted by $x = \lambda t$, with t representing a time interval, so that, when this interval is partitioned into non-overlapping intervals, their contributions are stochastically independent and sum to S_N. The last condition is umbrally expressed by the relation

$$(x+y).\beta.\alpha \equiv x.\beta.\alpha + y.\beta.\alpha \tag{27}$$

which also ensures the binomial property for the polynomial sequence defined by $x.\beta.\alpha$. In terms of generating functions, formula (27) means that

$$h_{x+y}(t) = h_x(t)h_y(t) \tag{28}$$

where $h_x(t)$ is the generating function of $x.\beta.\alpha$. Conversely, every generating function $h_x(t)$ satisfying the equality (28) is the generating function of a polynomial $\alpha-$partition umbra, namely, $h_x(t)$ has an umbral expression of the form (26).

Going back to the moments of a partition umbra, according to the definition of the Bell scalar umbra and from (8), we have

$$E[(\beta.\alpha)^n] = \sum_{k=1}^{n} B_{n,k}(a_1, a_2, \ldots, a_{n-k+1}) = Y_n(a_1, a_2, \ldots, a_n) \tag{29}$$

where $Y_n = Y_n(a_1, a_2, \ldots, a_n)$ are the *partition polynomials* (or complete Bell exponential polynomials) and a_i are the moments of the umbra α. Despite the complexity of the partition polynomial expressions, their umbral interpretation (29) allows an easy proof of the fact that they are of binomial type, simply by observing that $\beta.\alpha + \beta.\gamma \equiv \beta.(\alpha + \gamma)$.

Partition polynomials were first introduced by Bell [1] who gave a pioneering umbral version of them in [2]. Because of their generality, they include a variety of other polynomials such as the cycle indicator of the symmetric group and others of interest in number theory.

As already done for the Bell scalar umbra, the next theorem characterizes the partition umbrae and also provides the following recursive formula for the partition polynomials:

$$Y_{n+1}(a_1, a_2, \cdots, a_{n+1}) = \sum_{k=0}^{n} \binom{n}{k} a_{n-k+1} Y_k(a_1, a_2, \cdots, a_k).$$

Theorem 6. *Every α-partition umbra satisfies the relation*

$$(\beta.\alpha)^{n+1} \simeq \alpha'(\beta.\alpha + \alpha')^n, \quad \alpha' \equiv \alpha, \; n = 0, 1, 2, \dots, \tag{30}$$

and conversely.

Proof. Let ψ be an α-partition umbra. Then, from (25), we have $D_t[e^{(\beta.\alpha)t}] \simeq e^{(\beta.\alpha)t} D_t[e^{\alpha' t}]$, where $\alpha' \equiv \alpha$. Formula (30) follows by observing that $D_t[e^{\alpha' t}] \simeq \alpha' e^{\alpha' t}$. Going back to the previous steps, from (30) we see that $\beta.\alpha$ has generating function (25) and so it is an α-partition umbra. □

The moments of the polynomial α-partition umbra are

$$\begin{aligned} E[(x.\beta.\alpha)^n] &= \sum_{k=1}^{n} (x.\beta)_k B_{n,k}(a_1, a_2, \dots, a_{n-k+1}) \\ &= \sum_{k=1}^{n} x^k B_{n,k}(a_1, a_2, \dots, a_{n-k+1}) \end{aligned} \tag{31}$$

according to Definition 2. The arguments given in the proof of Theorem 6 also allow us to assert that every polynomial partition umbra satisfies the formula

$$(x.\beta.\alpha)^{n+1} \simeq x\alpha'(x.\beta.\alpha + \alpha')^n, \quad \alpha' \equiv \alpha, \; n = 0, 1, 2, \dots,$$

and conversely.

4.1 Umbral expression of functional composition

An umbral expression for the functional composition of exponential formal power series is a thorny matter. It was broached by Rota, Shen and Taylor in [29] by passing through the sequence of Abel polynomials. In this last section, we give an intrinsic umbral expression of this operation via the notion of partition umbra.

Definition 4. A *composition umbra* of the umbrae α and γ is the umbra

$$\chi \equiv \gamma.\beta.\alpha,$$

where β is the Bell scalar umbra.

In other words, the composition umbra χ is the point product of the umbra γ and the α-partition umbra $\beta.\alpha$.

Remark 4. As already emphasized in Sect. 3.2, the umbra $\gamma.\beta$ represents a randomized Poisson r.v. Hence it is natural to look at the composition umbra as a new r.v. which we call the *compound randomized Poisson r.v.* Moreover, as $(\gamma.\beta).\alpha \equiv \gamma.(\beta.\alpha)$ (cf. statement (iii) of Corollary 3), the previous relation allows us to view this new r.v. from another side: the umbra $\gamma.(\beta.\alpha)$ generalizes the concept of a random sum of i.i.d. compound Poisson r.v. with parameter 1 indexed by an integer r.v. X, i.e., a randomized compound Poisson r.v. with random parameter X.

Proposition 20. *The generating function of the composition umbra $\gamma.\beta.\alpha$ is the functional composition of the generating functions $e^{\alpha t} \simeq f(t)$ and $e^{\gamma t} \simeq g(t)$.*

Proof. Via (25) we have $e^{(\beta.\alpha)t} \simeq e^{f(t)-1}$. The result follows from (10) by observing that $e^{[\gamma.(\beta.\alpha)]t} \simeq g\left\{\log[e^{f(t)-1}]\right\}$. □

The moments of the composition umbra are

$$(\gamma.\beta.\alpha)^n \simeq \sum_{k=0}^{n} \gamma^k B_{n,k}(a_1, a_2, \dots, a_{n-k+1}), \tag{32}$$

where a_i are the moments of the umbra α. Indeed, by (8) we have

$$(\gamma.\beta.\alpha)^n \simeq \sum_{k=0}^{n} (\gamma.\beta)_k B_{n,k}(a_1, a_2, \dots, a_{n-k+1})$$

and (32) follows from (23).

Once more, we give a characterization of the composition umbra in the next theorem.

Theorem 7. *Every composition umbra satisfies the relation*

$$(\gamma.\beta.\alpha)^{n+1} \simeq \gamma\alpha'(\gamma.\beta.\alpha + \alpha')^n, \quad \alpha \equiv \alpha', \ n = 0, 1, 2, \dots, \tag{33}$$

and conversely.

Proof. Let χ be a composition umbra of α and γ. Then, from Proposition 20, we have $D_t[e^{\chi t}] \simeq g'[f(t) - 1]f'(t)$. Formula (33) follows since $f'(t) \simeq \alpha' e^{\alpha' t}$ with $\alpha' \equiv \alpha$ and $g'[f(t) - 1] \simeq \gamma e^{\chi t}$. Going back to the previous steps, from (33) we see that $\gamma.\beta.\alpha$ has generating function $g[f(t) - 1]$ and so it is a composition umbra of α and γ. □

At this point, as usual, we put Definition 4 of composition umbra to the test, by giving a proof of the Lagrange inversion formula. In the literature (cf. [18] for plentiful references) different forms of the Lagrange inversion formula are derived using umbral calculus. The main tool of our proof is the umbral expression of the (partial) Bell exponential polynomials that we state in the next lemma.

Lemma 15. *We have*

$$B_{n,k}(a_1, a_2, \dots, a_{n-k+1}) \simeq \binom{n}{k} \alpha^{.k} (k.\overline{\alpha})^{n-k}, \tag{34}$$

where $\overline{\alpha}$ is the umbra with moments $E\left[\overline{\alpha}^n\right] = \dfrac{a_{n+1}}{a_1(n+1)}$, $n = 1, 2, \dots$.

Proof. By identity (3) it follows that

$$B_{n,k}(a_1, a_2, \ldots, a_{n-k+1}) = \frac{1}{k!} D_t^{(n)}, [(f(t) - 1)^k]_{t=0}$$

where $D_t^{(n)}[\cdot]_{t=0}$ is the nth derivative with respect to t evaluated at $t = 0$ and $f(t) \simeq e^{\alpha t}$, so that

$$B_{n,k}(a_1, a_2, \ldots, a_{n-k+1}) \simeq \frac{1}{k!} D_t^{(n)}[(e^{\alpha t} - u)^{.k}]_{t=0}.$$

On the other hand, by the moment expression of umbra $\overline{\alpha}$, it follows that $e^{\alpha t} - u \simeq \alpha\, t\, e^{\overline{\alpha} t}$. Therefore we obtain

$$\begin{aligned} D_t^{(n)}[(e^{\alpha t} - u)^{.k}] &\simeq \alpha^{.k} D_t^{(n)}[t^k e^{k.\overline{\alpha}}] \\ &\simeq \alpha^{.k} \sum_{j=0}^{k} \binom{n}{j} D_t^{(j)}[t^k] D_t^{(n-j)}[e^{k.\overline{\alpha}}], \end{aligned}$$

by using the binomial property of the derivative operator. Finally, the result follows by evaluating the right-hand side of the previous formula at $t = 0$ and observing that $D_t^{(n-k)}[e^{k.\overline{\alpha}}]_{t=0} \simeq (k.\overline{\alpha})^{n-k}$. □

Remark 5. Set $\alpha \equiv u$. Then $\alpha^{.k} \simeq 1$ for $k = 0, 1, 2, \ldots$ and $B_{n,k}(1, 1, \ldots, 1) = S(n, k)$, the Stirling number of the second kind. Moreover we have $\overline{\alpha} \equiv (-1.\delta)$ where δ is the Bernoulli umbra that is the umbra whose moments are the Bernoulli numbers (cf. [28]). From Lemma 1 it follows that

$$S(n, k) \simeq \binom{n}{k} (-k.\delta)^{n-k}$$

as already obtained by Rota and Taylor using a different approach (cf. Proposition 9.1 [28])

Theorem 8 (Lagrange inversion formula). *Let $e^{\alpha t} \simeq f(t)$ and $e^{\gamma t} \simeq g(t)$. If $g[f(t) - 1] = f[g(t) - 1] = 1 + t$, then*

$$\alpha^{.k} \gamma^k \simeq (-k.\overline{\alpha})^{k-1}, k = 1, 2, \ldots. \tag{35}$$

Proof. By formulas (32) and (34), we have

$$\chi^n \simeq \sum_{k=0}^{n} \binom{n}{k} \alpha^{.k} \gamma^k (-k.\overline{\alpha})^{n-k}. \tag{36}$$

On the other hand, the Abel identity (5) gives

$$\chi^n \simeq \sum_{k=0}^{n} \binom{n}{k} \chi(\chi - k.\overline{\alpha})^{k-1} (k.\overline{\alpha})^{n-k}. \tag{37}$$

Comparing (36) with (37) we obtain

$$\alpha^{.k}\gamma^k \simeq \chi(\chi - k.\overline{\alpha})^{k-1}$$

by which the result follows on expanding the right-hand side of the previous formula by the binomial theorem and observing that, from $g[f(t) - 1] = 1 + t$, we have $\chi \simeq 1$ and $\chi^j \simeq 0$, $j = 2, 3, \ldots$. □

More explicitly, formula (35) says that the kth coefficient of the generating function $g(t)$ is equal to the $(k-1)^{\text{th}}$ coefficient of the generating function $[(f(t) - 1)/t]^{-k}$, when $g[f(t) - 1] = 1 + t$. Note that, if $f(t) - 1 = te^{-t}$, then $\alpha^{.k} \simeq 1$, $\overline{\alpha} \equiv -1.u$, and, from (35), we have $\gamma^k \simeq (k.u)^{k-1}$.

In closing, we observe that, if $a_1 = 1$, then $f(t) - 1 \simeq te^{\overline{\alpha}t}$ and the Lagrange inversion formula (35) becomes

$$\gamma^k \equiv (-k.\overline{\alpha})^{k-1}.$$

On the other hand, if the generating function $g(t)$ is written as $g(t) - 1 \simeq te^{\overline{\gamma}t}$, then the Lagrange inversion formula (35) becomes

$$k\overline{\gamma}^{k-1} \simeq (-k.\overline{\alpha})^{k-1},$$

which is equivalent to the version given in [29] by using the Abel polynomial sequence and its delta operator.

References

[1] Bell, E.T. (1927): Partition polynomials. Ann. Math. (2) **29**, 38–46
[2] Bell, E.T. (1934): Exponential polynomials. Ann. Math. (2) **35**, 258–277
[3] Bell, E.T. (1938): The iterated exponential integers. Ann. Math. (2) **39**, 539–557
[4] Bell, E.T. (1938): The history of Blissard's symbolic method with a sketch of its inventor's life. Amer. Math. Monthly **45**, 414–421
[5] Bell, E.T. (1940): Postulational bases for the umbral calculus. Amer. J. Math. **62**, 717–724
[6] Blissard, J. (1861): Theory of generic equations. Quart. J. Pure Appl. Math. **4**, 279–305
[7] Blissard, J. (1862): Theory of generic equations. Quart. J. Pure Appl. Math. **5**, 58–75, 185–208
[8] Blissard, J. (1862): Note on certain remarkable properties of numbers. Quart. J. Pure Appl. Math. **5**, 184
[9] Blissard, J. (1862): On the discovery and properties of a peculiar class of algebraic formulae. Quart. J. Pure Appl. Math. **5**, 325-335
[10] Blissard, J. (1864): Examples of the use and application of representative notation. Quart. J. Pure Appl. Math. **6**, 49–64
[11] Blissard, J. (1864): On the generalization of certain formulae investigated by Mr. Walton. Quart. J. Pure Appl. Math. **6**, 167–179
[12] Blissard, J. (1864): Researches in analysis. Quart. J. Pure Appl. Math. **6**, 142–257
[13] Blissard, J. (1867): On the properties of the $\Delta^m 0^n$ class of numbers and of others analogous to them, as investigated by means of representative notation. Quart. J. Pure Appl. Math. **8**, 85–110

[14] Blissard, J. (1868): On the properties of the $\Delta^m 0^n$ class of numbers. (continued from Vol. VIII - p.110). Quart. J. Pure Appl. Math. **9**, 82–94, 154–171
[15] Blissard, J. (1868): Note on a certain formula. Quart. J. Pure Appl. Math. **9**, 71–76
[16] Blissard, J. (1868): On certain properties of the gamma function. Quart. J. Pure Appl. Math. **9**, 280–296
[17] Cerasoli, M. (1979): Enumerazione binomiale e processi stocastici di Poisson composti. Boll. Un. Mat. Ital. (A 5) **16**, 310–315
[18] Di Bucchianico, A., Loeb, D. (1995): A selected survey of umbral calculus. Electron. J. Combin. **2**, Dynamic Survey 3, 28 pp. (electronic; updated (2000))
[19] Feller, W. (1950): An introduction to probability theory and its applications. Vol. I. Wiley, New York
[20] Feller, W. (1966): An introduction to probability theory and its applications. Vol. II. Wiley, New York
[21] Joyal, A. (1981): Une théorie combinatoire des séries formelles. Adv. Math. **42**, 1–82
[22] Lucas, E. (1891): Théorie des nombres. Gauthier-Villars, Paris
[23] Mullin, R., Rota, G.-C. (1970): On the foundations of combinatorial theory. III. Theory of binomial enumeration. In: Harris, B. (ed.) Graph Theory and its Applications. Academic Press, New York, pp. 167–213
[24] Riordan, J. (1958): An introduction to combinatorial analysis. Wiley, New York
[25] Roman, S.M., Rota, G.-C. (1978): The umbral calculus. Adv. Math. **27**, 95–188
[26] Rota, G.-C. (1964): The number of partitions of a set. Amer. Math. Monthly **71**, 498-504
[27] Rota, G.-C., Kahaner D., Odlyzko, A. (1973): On the foundations of combinatorial theory. VIII. Finite operator calculus. J. Math. Anal. Appl. **42**, 684–760
[28] Rota, G.-C., Taylor, B.D. (1994): The classical umbral calculus. SIAM J. Math. Anal. **25**, 694–711
[29] Rota, G.-C., Shen, J., Taylor, B.D. (1997): All polynomials of binomial type are represented by Abel polynomials. Ann. Scuola Norm. Sup. Pisa Cl. Sci. (4) **25**, 731–738
[30] Stam, A. J. (1988): Polynomials of binomial type and compound Poisson processes. J. Math. Anal. Appl. **130**, 493–508
[31] Steffensen, J.F. (1950): Interpolation. 2nd edition. Chelsea, New York
[32] Taylor, B.D. (2000): Umbral presentations for polynomial sequences. Comput. Math. Appl., to appear
[33] Touchard, J. (1956): Nombres exponentiels et nombres de Bernoulli. Canad. J. Math. **8**, 305–320

A formal theory of resultants (I): an algorithm in invariant theory

G.-C. Rota, J. A. Stein

1 Background in letterplace algebras

Let $A = \{a < b < \dots\}$ be an infinite alphabet of negative *letters*, and let $P = \{u_1 < u_2 < \cdots < u_n\}$ be an alphabet containing n negative *places*. The elements of the *letterplace* alphabet

$$[A|P] = \{[a|u_1] < [a|u_2] < \cdots < [a|u_n] < [b|u_1] < \dots\}$$

will be denoted by $[a|u_i]$, rather than by $(a|u_i)$ as in our previous work. The letterplaces $[a|u_i]$ of the alphabet $[A|P]$ are neutral elements, and thus the supersymmetric algebra Super$[A|P]$ is an ordinary polynomial algebra generated by the "variables" $[a|u_i]$, $a \in A$, $i = 1, 2, \dots, n$. For $a_1, a_2, \dots, a_k \in A$ we denote by $[a_1 a_2 \dots a_k | u_1 u_2 \dots u_k]$ the ordinary biproduct, that is, the determinant

$$[a_1 a_2 \dots a_k | u_1 u_2 \dots u_k] = \sum_{\theta} (-1)^{\theta} [a_1|u_{\theta 1}][a_2|u_{\theta 2}] \dots [a_k|u_{\theta k}].$$

We shall take as ring of coefficients of the polynomial algebra Super$[A|P]$ the field Q of rational numbers, rather than the ring Z of integers.

An element w of Super$[A|P]$ is said to be *homogeneous of degree p* in the letter a if it is a linear combination with coefficients in Q of products

$$\sum w' w'',$$

where w' is a homogeneous polynomial with coefficients in Q of degree p in the variables $[a|u_1], [a|u_2], \dots, [a|u_n]$ alone, and w'' is an arbitrary element in Super$[A|P]$ in which the letter a does not occur.

We denote by K the algebraic closure of the quotient field of the integral domain Super$[A|P]$. If S is a subset of A, an element w of K will be said to be *S-free* when it lies in the algebraic closure of a subfield of K obtained by adjoining a (necessarily finite) subset T of $[A|P]$ such that the letter content of T is disjoint from S.

2 Background in plethystic algebras

The algebra Pleth$(A|P)$ will be taken with coefficients in the field K; more precisely, as

$$\text{Pleth}(A|P) = \text{Super}[A|P] \otimes \text{Pleth}(\text{Super}[A|P]).$$

In other terms: the elements of $\mathrm{Pleth}(A|P)$ are linear combinations of monomials in $\mathrm{Pleth}(\mathrm{Super}[A|P])$ with coefficients in K.

A *monomial element* W of $\mathrm{Pleth}(A|P)$ is an element of the form

$$W = (w_1)(w_2)\dots(w_r)$$

where $w_1, w_2, \dots$ are monomials belonging to $\mathrm{Super}[A|P]$. We say W is *homogeneous of inner degree d* if

$$d = \mathrm{degree}[w_1] = \cdots = \mathrm{degree}[w_r].$$

The *inner degree* of W is defined to be the integer

$$\mathrm{degree}[w_1] = \max\{\mathrm{degree}[w_1], \dots, \mathrm{degree}[w_r]\};$$

we may assume that in all cases

$$\mathrm{degree}[w_1] \geq \cdots \geq \mathrm{degree}[w_r].$$

The *shape* of W is then defined to be the partition

$$(\mathrm{degree}[w_1], \dots, \mathrm{degree}[w_r]).$$

The *outer degree* of $W = (w_1)(w_2)\dots(w_r)$ is defined to be the number of juxtaposition factors of W and is equal to the integer r. We also say that W has *r parts.* The *degree* of W is defined to be the integer

$$\mathrm{degree}[W] = \mathrm{degree}[w_1] + \cdots + \mathrm{degree}[w_r]$$

so that the shape of W is a partition of the degree of W. Shapes that are partitions of the same integer are partially ordered by the *dominance* order. An *elementary monomial* W in $\mathrm{Pleth}(A|P)$, of letter content $p, q, \cdots, r$ in the letters $a, b, \cdots, c$ respectively, is a special homogeneous monomial element of the form

$$W = \prod_{i=1}^{n}([a|u_i])^{(p_i)} \prod_{i=1}^{n}([b|u_i])^{(q_i)} \cdots \prod_{i=1}^{n}([c|u_i])^{(r_i)}.$$

The elementary monomial W is of inner degree 1 and of outer degree $p+q+r$. Recall that the circle product in $\mathrm{Pleth}(\mathrm{Super}[A|P])$ is defined for W and W' in $\mathrm{Pleth}(\mathrm{Super}[A|P])$ by

$$W \circ W' = \sum W_{(1)}(W_{(2)}|W'_{(1)})W'_{(2)},$$

where, for monomial elements

$$\begin{aligned} W &= (w_1)^{(p_1)} \dots (w_r)^{(p_r)} \\ W' &= (w'_1)^{(q_1)} \dots (w'_s)^{(q_s)}, \end{aligned}$$

we have

$$\sum W_{(1)} \otimes W_{(2)} = \sum (w_1)^{(p_{11})} \dots (w_r)^{(p_{r1})} \otimes (w_1)^{(p_{12})} \dots (w_r)^{(p_{r2})}$$
$$\sum W'_{(1)} \otimes W'_{(2)} = \sum (w'_1)^{(q_{11})} \dots (w'_s)^{(q_{s1})} \otimes (w'_1)^{(q_{12})} \dots (w'_s)^{(q_{s2})},$$

where the sums are over all choices of (p_{ij}) and (q_{ij}) satisfying

$$p_{i1} + p_{i2} = p_i \ (1 \le i \le r)$$
$$q_{j1} + q_{j2} = q_j \ (1 \le j \le s),$$

and the bar product is defined by

$$(W|W') = \sum \prod_{i=1}^{r} \prod_{j=1}^{s} (w_i w'_j)^{(m_{ij})}$$

where the sum is over all matrices of nonnegative integers (m_{ij}) with row sums $(p_1, \cdots, p_r)$ and column sums $(q_1, \cdots, q_s)$. The bar product is associative, $((W|W')|W'') = (W|(W'|W''))$.

On monomial elements the circle product thus takes the form

$$(w_1)^{(p_1)} \dots (w_r)^{(p_r)} \circ (w'_1)^{(q_1)} \dots (w'_s)^{(q_s)}$$
$$= \sum \left[\prod_{i=1}^{r} (w_i)^{(c_i)} \right] \left[\prod_{i=1}^{r} \prod_{j=1}^{s} (w_i w'_j)^{(m_{ij})} \right] \left[\prod_{j=1}^{s} (w'_j)^{(d_j)} \right]$$

where the sum is over all posssible choices of c_i, m_{ij}, d_j which satisfy

$$c_i + \sum_{j=1}^{s} m_{ij} = p_i \ (1 \le i \le r)$$
$$d_j + \sum_{i=1}^{r} m_{ij} = q_j \ (1 \le j \le s).$$

3 Forms

A *homogeneous form of degree d* in the variables $x_1, x_2, \dots, x_n$ is a homogeneous polynomial

$$\Phi(x) = \Phi(x_1, \dots, x_n) = \sum_{i_1 + \cdots + i_n = d} \phi_{i_1, \dots, i_n} x_1^{i_1} \dots x_n^{i_n}.$$

The coefficients $\phi_{i_1, \dots, i_n}$ are either elements of the field K or else transcendentals not contained in K. When all coefficients of a form Φ are independent transcendentals, we say that the form Φ is *generic*.

A distinguished subset $S = \{a_1, \ldots, a_n\}$ of A is chosen once and for all, and forms $\Phi_1, \ldots, \Phi_n$ will be understood to be S-free forms of degrees $d_1, \ldots, d_n$ in the n variables $x_1, \ldots, x_n$, whose coefficients are denoted by

$$\Phi_r(x_1, \ldots, x_n) = \sum_{i_1+\cdots+i_n=d_r} \phi^r_{i_1,\ldots,i_n} x_1^{i_1} \ldots x_n^{i_n}.$$

4 Cayley forms

For each letter a in A we write $a(x) = [a|x]$ to denote the linear form

$$a(x) = [a|x] = [a|u_1]x_1 + \cdots + [a|u_n]x_n.$$

A form

$$\begin{aligned} f(x_1, x_2, \ldots, x_n) &= \prod_{i=1}^{d} f_i(x) \\ &= \prod_{i=1}^{d} f_{i1}x_1 + \cdots + f_{in}x_n \\ &= \sum_{i_1+\cdots+i_n=d} f_{i_1,\ldots,i_n} x_1^{i_1} \ldots x_n^{i_n}, \end{aligned}$$

where the coefficients f_{ij} of the linear forms f_i are either members of the field K or independent transcendentals disjoint from K, is said to be a *Cayley form*.

Proposition 1. *Every binary form is a Cayley form.*

By extension, an element Φ of Super$[A|P]$ which can be expressed as a product of linear forms

$$\Phi = \prod_{i=1}^{r} [\alpha_i|\beta_i] = \prod_{i=1}^{r} \left[\sum_j \alpha_{ij} a_j | \sum_{k=1}^{n} \beta_{ik} u_k \right]$$

will be called a *supersymmetric Cayley form*.

5 The Brill conditions

We recall the classical *Brill conditions* which give necessary and sufficient conditions for $f(x)$ to be a Cayley form. We follow the modern treatment given in [3].

(a) Define *polarization operators*

$$D_{y,x} = \sum_{i=1}^{n} y_i \frac{\partial}{\partial x_i},$$

(b) We then have for all scalar variables t that

$$f(tx+y) = \sum_{k=1}^{d} \binom{d}{k} f_{x^k}(x, y)t^k,$$

where

$$\binom{d}{k} f_{x^k}(x, y) = D_{y,x}^{(k)} f(x)$$

and $D_{y,x}^{(k)}$ denotes the kth divided power of $D_{y,x}$.

(c) Define for all forms $f(x)$ and $g(x)$ of degree d the form

$$(f \odot g)(x, y) = \frac{1}{d+1} \sum_{k=0}^{d} (-1)^k \binom{d}{k} f_{y^k}(y, x) g_{x^k}(x, y)$$

called the *apolar covariant* of $f(x)$ and $g(x)$.

(d) Note that when $g(x)$ is a power of a linear form $[a|x]^d$ we have

$$(f \odot [a|x]^d)(x, y) = \frac{1}{d+1} \sum_{k=0}^{d} (-1)^k \binom{d}{k} f_{y^k}(y, x)[a|x]^k [a|y]^{d-k}$$

so that

$$(f \odot [a|x]^d)(x, y) = \frac{1}{d+1} \left(f(x)[a|y]^d + \Delta \right),$$

where Δ is divisible by $[a|x]$. We conclude that

$$(f \odot [a|x]^d)(x, y) = 0 \Rightarrow f(x) \text{ is divisible by } [a|x].$$

(e) We have easily that

$$f(z + tf(z)x) = f(z)E(t),$$

where

$$E(t) = \sum_{k=0}^{d} e_k t^k$$

and

$$e_k = \binom{d}{k} f_{x^k}(x, z) f(z)^{k-1}.$$

(f) Note that when

$$f(x) = [f_1|x] \cdots [f_d|x]$$

we have

$$E(t) = \prod_{i=1}^{d} \left(1 + t\frac{f(z)}{[f_i|z]}[f_i|x]\right).$$

(g) Define the power sum polynomials

$$p_k(E) = (-1)^k k \sum_{i_1+2i_2+\cdots+di_d=k} \binom{i_1+i_2+\cdots+i_d}{i_1, i_2, \cdots, i_d} \frac{e_1^{i_1} \cdots e_d^{i_d}}{i_1+i_2+\cdots+i_d}$$

so that when $f(x) = [f_1|x] \cdots [f_d|x]$ we have

$$p_k(E) = \sum_{i=1}^{d} \frac{f(z)^k}{[f_i|z]^k}[f_i|x]^k.$$

(h) Let

$$P_{f,z}(x) = p_d(E)$$

and define the *Brill polynomials*

$$B_f(x, y, z) = (f \odot P_{f,z})(x, y)$$

which are homogeneous forms in (x, y, z) of multidegree $(d, d, d(d-1))$ whose coefficients are forms of degree $d+1$ in the coefficients of $f(x)$.

(i) Brill's theorem is as follows.

Theorem 1 (Brill conditions). *The form $f(x)$ is a Cayley form if and only if $B_f(x, y, z) = 0$.*

6 The algebra of coefficients of a Cayley form

Let $f(x)$ be the Cayley form

$$f(x) = f(x_1, \cdots, x_n) = \sum_{i_1+\cdots+i_n=r} f_{i_1\cdots i_n} x_1^{i_1} \cdots x_n^{i_n} = [f_1|x] \cdots [f_r|x].$$

We do not know the vectors $f_1, \cdots, f_r$ explicitly, only that their homogeneous elementary monomials of shape (1^r) are the coefficients $f_{i_1\cdots i_n}$ of $f(x)$.

Which symmetric functions of the vectors $f_1, \cdots, f_r$ can be expressed in terms of the coefficients of $f(x)$? To answer this question we introduce a variant of the plethystic algebra.

7 The plethystic algebra $\mathrm{Pleth}_r(P_a)$

For every $a \in A$ define the alphabet

$$P_a = \{[a|u_1] < \cdots < [a|u_n]\}.$$

Let $\mathrm{Pleth}_r[P_a]$ denote the subalgebra of $\mathrm{Pleth}(P_a)$ generated by the homogeneous monomial elements of shapes (d^r) for all $d \geq 0$. Let

$$\mathrm{Pleth}_r(P_a) = \mathrm{Pleth}_r[P_a]/\Delta_r,$$

where

$$\Delta_r = \mathrm{Pleth}_r[P_a] \cap \Delta'_r$$

and where Δ'_r denotes the ideal generated by all monomial elements in $\mathrm{Pleth}(P_a)$ with more than r parts.

Note that for all monomial elements ϕ and ψ of $\mathrm{Pleth}_r[P_a]$ we have

$$\phi \circ \psi = (\phi|\psi) + \Delta$$

where Δ is in Δ_r. Thus $\mathrm{Pleth}_r(P_a)$ is a ring with circle multiplication given by the bar product.

Suppose Φ is a monomial element in $\mathrm{Pleth}_r(P_a)$ which is homogeneous of shape (i^r) for some value of $i > 1$. We may write

$$\Phi = ([a|u_1]^{p_{11}} \cdots [a|u_n]^{p_{n1}})^{(p_1)} \cdots ([a|u_1]^{p_{1d}} \cdots [a|u_n]^{p_{nd}})^{(p_d)}$$

where $p_1 + \cdots + p_d = r$ and for $1 \leq j \leq d$ we have $p_{1j} + \cdots + p_{nj} = i$. Either Φ is *irreducible* or for some irreducible form Φ' we have

$$\Phi = U_1^u \circ U_2^v \circ \cdots \circ U_n^w \circ \Phi'$$

where $u + v + \cdots + w \leq i$ and $U_j = ([a|u_j])^{(r)}$ so that

$$U_j^k = ([a|u_j])^{(r)} \circ \cdots \circ ([a|u_j])^{(r)} = ([a|u_j]^k)^{(r)}.$$

We consider a system α of vectors, $\alpha_1, \cdots, \alpha_r$, given by

	u_1	$\cdots$	u_n
α_1	α_{11}	$\cdots$	α_{1n}
$\vdots$			
α_r	α_{r1}	$\cdots$	α_{rn}

where the coordinate functions of the vectors $\alpha_1, \cdots, \alpha_r$ are $u_1, \cdots, u_n$ and the elements $\alpha_{ij} = (\alpha_i|u_j)$ are either elements of K or variables. We say that the system α is *invertible* when the elements α_{ij} are all nonzero, satisfying $\alpha_{ij} \neq 0$ for all i and j.

Proposition 2. *A monomial symmetric function $\Phi(\alpha)$ of the system α of vectors $\alpha_1, \cdots, \alpha_r$ is determined by each monomial element Φ of the algebra* $\text{Pleth}_r(P_a)$. *The correspondence is an algebra map. For any two monomial elements Φ and Ψ of the algebra* $\text{Pleth}_r(P_a)$ *we have*

$$[\Phi \circ \Psi](\alpha) = \Phi(\alpha)\ \Psi(\alpha).$$

Example. Suppose that Φ is the monomial element

$$\Phi = ([a|u_1]^i)([a|u_2]^i) \cdots ([a|u_r]^i).$$

Then we have

$$\Phi(\alpha) = \sum_{\theta} (\alpha_{i_1}|u_1)^i (\alpha_{j_1}|u_2)^i \cdots (\alpha_{h_1}|u_r)^i,$$

where the sum is over all permutations

$$\theta(1) \cdots \theta(r) = i_1 j_1 \cdots h_1.$$

Let $Q(f)$ denote the ring of rational functions of the coefficients of the Cayley form

$$f(x) = f(x_1, \cdots, x_n) = \sum_{i_1+\cdots+i_n=r} f_{i_1\cdots i_n} x_1^{i_1} \cdots x_n^{i_n} = [f_1|x] \cdots [f_r|x].$$

Which symmetric functions of the vectors $f_1, \cdots, f_r$ can be expressed in terms of the coefficients of $f(x)$? The algebra $\text{Pleth}_r(P_a)$ would appear to be a natural candidate for constructing an evaluation map

$$E_{f,a} : \text{Pleth}_r(P_a) \to Q(f)$$

which will be a map of algebras. A rational extension $\text{Pleth}_r\{P_a\}$ of $\text{Pleth}_r(P_a)$ which will prove to be the proper candidate for our purposes will be introduced next.

8 The rational plethystic algebra

We begin by defining the rational analogue of $\text{Super}[A|P]$. First define the *rational monomials*

$$w = \frac{w'}{w''},$$

where w' and w'' are monomials in $\text{Super}[A|P]$ and suppose that w' and w'' have degrees g and h respectively. Suppose further that w' and w'' have no common factors (we say w is in *reduced form).* The rational monomial w is said to be *proper* if $g \geq h$. It is said to be *strict* if $g > h$ and *neutral* if $g = h$. The *length* of w is defined to be the integer

$$\text{length}[w] = \text{degree}[w'] = g.$$

The *degree* of w is defined to be the integer

$$\text{degree}[w] = \text{degree}[w'] - \text{degree}[w''] = g - h.$$

The *rational supersymmetric algebra* Super$\{A|P\}$ is then the ring generated by all proper rational monomials. We shall take the ring of coefficients of the rational supersymmetric algebra Super$\{A|P\}$ to be the field Q of rational numbers.

The *rational plethystic algebra* Pleth$\{A|P\}$ is defined to be

$$\text{Pleth}\{A|P\} = \text{Super}\{A|P\} \otimes \text{Pleth}(\text{Super}\{A|P\}).$$

Let W be a rational monomial element in Pleth$\{A|P\}$ expressed in reduced form as

$$W = (w_1) \cdots (w_r) = \left(\frac{w_1'}{w_1''}\right) \cdots \left(\frac{w_r'}{w_r''}\right).$$

The *shape* of W is defined to be the partition

$$\text{shape}[W] = (\text{length}[w_1], \cdots, \text{length}[w_r]) .$$

The *inner degree of* W is the integer

$$\max\{\text{degree}[w_1], \cdots, \text{degree}[w_r]\}.$$

W is said to be *homogeneous of inner degree* d if

$$d = \text{degree}[w_1] = \cdots = \text{degree}[w_r].$$

The *inner length* of W is defined to be the integer

$$\max\{\text{length}[w_1], \cdots, \text{length}[w_r]\}.$$

The *outer degree* of W is defined to be the number of juxtaposition factors, counting any divided powers. The outer degree of $(w_1)^{(r)}$ is thus equal to r, as is the outer degree of $(w_1) \cdots (w_r)$. We also say that W has r *parts*.

Proposition 3. *Each rational monomial element W of the algebra* Pleth$\{P_a\}$ *determines a rational symmetric function of each invertible sequence α of vectors $\alpha_1, \cdots, \alpha_r$. If W has more than r parts then $W(\alpha) = 0$. The correspondence is an algebra map. For any two rational monomial elements Φ and Ψ of the algebra* Pleth$\{P_a\}$ *we have*

$$[\Phi \circ \Psi](\alpha) = \Phi(\alpha)\, \Psi(\alpha).$$

Example. Consider the neutral rational monomial element W in the algebra Pleth$\{P_a\}$ given by

$$W = \left(\frac{[a|u_1]^i}{[a|u_2]^i}\right)\left(\frac{[a|u_2]^j}{[a|u_3]^j}\right)\left(\frac{[a|u_3]^k}{[a|u_4]^k}\right)\cdots\left(\frac{[a|u_c]^h}{[a|u_1]^h}\right).$$

Then for all $r \geq c$ and for all invertible systems α of vectors $\alpha_1, \cdots, \alpha_r$ we have

$$W(\alpha) = \sum_{\theta} \frac{(\alpha_{i_1}|u_1)^i}{(\alpha_{i_1}|u_2)^i} \frac{(\alpha_{j_1}|u_2)^j}{(\alpha_{j_1}|u_3)^j} \frac{(\alpha_{k_1}|u_3)^k}{(\alpha_{k_1}|u_4)^k} \cdots \frac{(\alpha_{h_1}|u_c)^h}{(\alpha_{h_1}|u_1)^h},$$

where the sum is over all permutations

$$\theta(1) \cdots \theta(r) = i_1 j_1 k_1 \cdots h_1 \theta(c+1) \cdots \theta(r)$$

which are increasing in the last $r - c$ positions.

9 The plethystic algebra $\mathrm{Pleth}_r\{P_a\}$

Let $\mathrm{Pleth}_r[[P_a]]$ be the subalgebra of $\mathrm{Pleth}\{P_a\}$ generated by the *neutral* homogeneous proper rational monomial elements with *at most* r parts and the *strict* homogeneous proper rational monomial elements with *exactly* r parts.

Let Δ_r'' denote the ideal generated by all monomial elements in $\mathrm{Pleth}\{P_a\}$ with more than r parts.

Define $\mathrm{Pleth}_r\{P_a\}$ by

$$\mathrm{Pleth}_r\{P_a\} = \mathrm{Pleth}_r[[P_a]]/\Delta_r''',$$

where

$$\Delta_r''' = \mathrm{Pleth}_r[[P_a]] \cap \Delta_r''.$$

Note that we have a natural inclusion of algebras

$$\mathrm{Pleth}_r(P_a) \subset \mathrm{Pleth}_r\{P_a\}$$

and we will identify $\mathrm{Pleth}_r(P_a)$ with its image under the canonical embedding.

$\mathrm{Pleth}_r\{P_a\}$ is a ring with circle multiplication given by the ordinary circle product in $\mathrm{Pleth}\{P_a\}$ modulo any terms with more than r parts. On strict proper homogeneous rational monomial elements, circle multiplication reduces to the bar product.

Proposition 4. *Each homogeneous proper rational monomial element Φ of the algebra* $\mathrm{Pleth}_r\{P_a\}$ *determines a rational symmetric function $\Phi(\alpha)$ of each invertible sequence α of vectors $\alpha_1, \cdots, \alpha_r$. The correspondence is an algebra map. For any two rational monomial elements Φ and Ψ of the algebra* $\mathrm{Pleth}_r\{P_a\}$ *we have that*

$$[\Phi \circ \Psi](\alpha) = \Phi(\alpha)\, \Psi(\alpha).$$

Example. In the algebra $\mathrm{Pleth}_r\{P_a\}$ consider the strict proper homogeneous rational monomial element Φ given by

$$\Phi = \left(\frac{[a|u_1]^i}{[a|u_2]^j}\right)\left(\frac{[a|u_2]^i}{[a|u_3]^j}\right) \cdots \left(\frac{[a|u_r]^i}{[a|u_1]^j}\right),$$

where $i > j$. Then for all invertible systems α of vectors $\alpha,_1 \cdots, \alpha_r$

$$\Phi(\alpha) = \sum_\theta \frac{(\alpha_{i_1}|u_1)^i}{(\alpha_{i_1}|u_2)^j} \frac{(\alpha_{j_1}|u_2)^i}{(\alpha_{j_1}|u_3)^j} \cdots \frac{(\alpha_{h_1}|u_r)^i}{(\alpha_{h_1}|u_1)^j},$$

where the sum is over all permutations

$$\theta(1) \cdots \theta(r) = i_1 j_1 \cdots h_1.$$

We will now show in $\text{Pleth}_r\{P_a\}$ that, for all homogeneous proper rational monomial elements W, there are homogeneous monomial elements U and V with exactly r parts satisfying

$$W \circ U = V$$

and we may write

$$W = \frac{V}{U}.$$

Proposition 5. *Let W be a strict homogeneous proper rational monomial element in* $\text{Pleth}_r\{P_a\}$ *of inner degree $p > 0$, inner length i and outer degree r, expressed in reduced form as*

$$W = (w_1) \cdots (w_r) = \left(\frac{w_1'}{w_1''}\right) \cdots \left(\frac{w_r'}{w_r''}\right)$$

so that

$$\begin{aligned} p &= \text{degree}[w_1] = \cdots = \text{degree}[w_r], \\ i &= \text{length}[w_1] \geq \cdots \geq \text{length}[w_r]. \end{aligned}$$

Let u be the least common multiple of the denominators,

$$u = lcm\{w_1'', \cdots, w_r''\} = [a|u_1]^{m_1} \cdots [a|u_n]^{m_n}.$$

Note that

$$\text{degree}[u] \geq \text{degree}[w_1''] = \text{length}[w_1] - \text{degree}[w_1]$$

and define

$$U = (u)^{(r)}.$$

Then for some $h \geq 0$ the homogeneous monomial element U is of inner degree $d' = i - p + h$ and of outer degree r, and

$$W \circ U = V,$$

where V is the homogeneous monomial element of inner degree $d'' = i + h$ and of outer degree r given by

$$V = (uw_1) \cdots (uw_r) = (v_1) \cdots (v_r),$$

where $v_1, \cdots, v_r$ are monomials of degree $i + h$ in $\text{Super}[A|P]$.

Proof. We have easily that $d' \geq i - p$ since

$$\begin{aligned} d' &= \text{inner degree}\left[(u)^{(r)}\right] = \text{degree}[u] \\ &\geq \text{deg}[w_1''] = \text{length}[w_1] - \text{degree}[w_1] = i - p. \end{aligned}$$

Letting $h = \text{degree}[u] - \text{degree}[w_1'']$ we have $h \geq 0$ and

$$d' = i - p + h.$$

Similarly,

$$\begin{aligned} d'' &= \text{inner degree}\,[V] \\ &= \text{degree}[uw_1] \\ &= \text{degree}[u] + \text{degree}[w_1] \\ &= h + \text{degree}[w_1''] + \text{degree}[w_1] \\ &= h + i - p + p \\ &= h + i. \end{aligned}$$

This concludes the proof. □

Proposition 6. *Let W be a neutral homogeneous proper rational monomial element in* $\text{Pleth}_r\{P_a\}$ *of inner degree $p = 0$, inner length i and outer degree at most r, expressed in reduced form as*

$$W = (w_1) \cdots (w_c) = \left(\frac{w_1'}{w_1''}\right) \cdots \left(\frac{w_c'}{w_c''}\right),$$

where $c \leq r$ and

$$\begin{aligned} 0 &= \text{degree}[w_1] = \cdots = \text{degree}[w_c], \\ i &= \text{length}[w_1] \geq \cdots \geq \text{length}[w_c]. \end{aligned}$$

Let u be the least common multiple of the denominators,

$$u = lcm\{w_1'', \cdots, w_c''\} = [a|u_1]^{m_1} \cdots [a|u_n]^{m_n}$$

and let

$$U = (u)^{(r)}.$$

Then for some $h \geq 0$ the homogeneous monomial element U is of inner degree

$$d' = i + h$$

and of outer degree r, and

$$W \circ U = V$$

where V is the homogeneous monomial element of inner degree $d'' = i + h$ and of outer degree r, given by

$$\begin{aligned} V &= (uw_1)\cdots(uw_c)(u)^{(r-c)} \\ &= (v_1)\cdots(v_c)(u)^{(r-c)}, \end{aligned}$$

where $v_1, \cdots, v_r$ are monomials of degree $i + h$ in Super$[A|P]$.

Proof. We have easily that $d' \geq i$ since

$$\begin{aligned} d' &= \text{inner degree}\left[(u)^{(r)}\right] = \text{degree}[u] \\ &\geq \text{degree}[w_1''] = \text{length}[w_1] - \text{degree}[w_1] = i - 0. \end{aligned}$$

Letting $h = \text{degree}[u] - \text{degree}[w_1'']$ we have $h \geq 0$ and

$$d' = h + i - 0 = h + i.$$

Similarly,

$$\begin{aligned} d'' &= \text{inner degree}\,[V] \\ &= \text{degree}[uw_1] \\ &= \text{degree}[u] + \text{degree}[w_1] \\ &= h + \text{degree}[w_1''] + \text{degree}[w_1] \\ &= h + i - 0 + 0 \\ &= h + i. \qquad \square \end{aligned}$$

Example. Consider the monomial element W in the algebra $\text{Pleth}_r(P_a)$ given by

$$W = \left([a|u_1]^i\right)^{(j_1)}\left([a|u_2]^i\right)^{(j_2)}\cdots\left([a|u_n]^i\right)^{(j_n)}$$

with $i > 1$ and

$$j_1 + \cdots + j_n = r.$$

Identifying $\text{Pleth}_r(P_a)$ with its image in $\text{Pleth}_r\{P_a\}$ we may write in $\text{Pleth}_r\{P_a\}$ that

$$W = ([a|u_1])^{(r)} \circ W',$$

where the strict proper homogeneous rational monomial element W' is given by

$$W' = \left([a|u_1]^{i-1}\right)^{(j_1)}\left(\frac{[a|u_2]^i}{[a|u_1]}\right)^{(j_2)}\cdots\left(\frac{[a|u_n]^i}{[a|u_1]}\right)^{(j_n)}$$

and similarly

$$W = \left([a|u_1]^i\right)^{(r)} \circ W_1,$$

where the neutral homogeneous proper rational monomial element W_1 is of outer degree $r - j_1$ and is given by

$$W_1 = \left(\frac{[a|u_2]^i}{[a|u_1]^i}\right)^{(j_2)} \cdots \left(\frac{[a|u_n]^i}{[a|u_1]^i}\right)^{(j_n)}.$$

For all invertible systems α of vectors $\alpha_1, \cdots, \alpha_r$ we then have

$$\begin{aligned} W'(\alpha) &= \sum_\theta \left[(\alpha_{i_{11}}|u_1)^{i-1} \cdots (\alpha_{i_{1j_1}}|u_1)^{i-1}\right] \left[\frac{(\alpha_{i_{21}}|u_2)^i}{(\alpha_{i_{21}}|u_1)} \cdots \frac{(\alpha_{i_{2j_2}}|u_2)^i}{(\alpha_{i_{2j_2}}|u_1)}\right] \\ &\quad \cdots \left[\frac{(\alpha_{i_{n1}}|u_n)^i}{(\alpha_{i_{n1}}|u_1)} \cdots \frac{(\alpha_{i_{nj_n}}|u_n)^i}{(\alpha_{i_{nj_n}}|u_1)}\right] \end{aligned}$$

and

$$W_1(\alpha) = \sum_\theta \left[\frac{(\alpha_{i_{21}}|u_2)^i}{(\alpha_{i_{21}}|u_1)^i} \cdots \frac{(\alpha_{i_{2j_2}}|u_2)^i}{(\alpha_{i_{2j_2}}|u_1)^i}\right] \cdots \left[\frac{(\alpha_{i_{n1}}|u_n)^i}{(\alpha_{i_{n1}}|u_1)^i} \cdots \frac{(\alpha_{i_{nj_n}}|u_n)^i}{(\alpha_{i_{nj_n}}|u_1)^i}\right]$$

where both sums are over all permutations

$$\theta(1)\cdots\theta(r) = i_{11}\cdots i_{1j_1}\, i_{21}\cdots i_{2j_2}\, \cdots\, i_{n1}\cdots i_{nj_n}$$

which are increasing in the first j_1 positions, the next j_2 positions, $\cdots$, the last j_n positions. We will then find that

$$W'(\alpha) = \frac{\Phi(\alpha)}{(\alpha_1|u_1)\cdots(\alpha_r|u_1)}$$

and that

$$W_1(\alpha) = \frac{\Phi(\alpha)}{(\alpha_1|u_1)^i\cdots(\alpha_r|u_1)^i}.$$

10 The fundamental theorem of symmetric functions for $\text{Pleth}_r\{P_a\}$

In this section we will show that the elementary homogeneous proper rational monomial elements generate $\text{Pleth}_r\{P_a\}$ rationally under the circle product. Suppose first that Φ is an irreducible monomial element in $\text{Pleth}_r(P_a) \subset \text{Pleth}_r\{P_a\}$ which is homogeneous of shape (i^r) for some value of $i > 1$. We may write

$$\begin{aligned} \Phi &= ([a|u_1]^{p_{11}} \cdots [a|u_n]^{p_{n1}})^{(q_1)} \cdots ([a|u_1]^{p_{1b}} \cdots [a|u_n]^{p_{nb}})^{(q_b)} \\ &= (\phi_1)^{(q_1)}(\phi_2)^{(q_2)} \cdots (\phi_b)^{(q_b)}, \end{aligned}$$

where

$$q_1 + \cdots + q_b = r$$

and for $1 \leq j \leq b$

$$p_{1j} + \cdots + p_{nj} = i$$

and we may assume that the inner terms $\phi_1, \phi_2, \cdots, \phi_b$ are all distinct. Since Φ is irreducible we must have

$$p_{s1} p_{s2} \cdots p_{sb} = 0$$

for each s.

We wish to expand Φ as a rational expression in *elementary* monomials. The fundamental theorem of symmetric functions for $\mathrm{Pleth}_r\{P_a\}$ asserts that this is possible.

Express Φ in the form

$$\Phi = [([a|u_1])^{(r)}]^i \circ \Phi_1,$$

where

$$\begin{aligned}\Phi_1 &= \left(\frac{\phi_1}{[a|u_1]^i}\right)^{(q_1)} \cdots \left(\frac{\phi_b}{[a|u_1]^i}\right)^{(q_b)} \\ &= \left(\frac{[a|u_2]^{p_{21}}}{[a|u_1]^{p_{21}}} \cdots \frac{[a|u_n]^{p_{n1}}}{[a|u_1]^{p_{n1}}}\right)^{(q_1)} \cdots \left(\frac{[a|u_2]^{p_{2b}}}{[a|u_1]^{p_{2b}}} \cdots \frac{[a|u_n]^{p_{nb}}}{[a|u_1]^{p_{nb}}}\right)^{(q_b)}.\end{aligned}$$

Φ_1 is thus a neutral homogeneous rational monomial element in $\mathrm{Pleth}_r\{P_a\}$. Since the inner terms $\phi_1, \phi_2, \cdots, \phi_b$ are all distinct, at most one of the terms can satisfy

$$\frac{\phi_t}{[a|u_1]^i} = \frac{[a|u_2]^{p_{2t}}}{[a|u_1]^{p_{2t}}} \cdots \frac{[a|u_n]^{p_{nt}}}{[a|u_1]^{p_{nt}}} = 1.$$

If there is such a term, set

$$\left(\frac{\phi_t}{[a|u_1]^i}\right)^{(q_t)} = \left(\frac{[a|u_2]^{p_{2t}}}{[a|u_1]^{p_{2t}}} \cdots \frac{[a|u_n]^{p_{nt}}}{[a|u_1]^{p_{nt}}}\right)^{(q_t)} = 1.$$

The resulting expression for Φ_1 will be written

$$\Phi_1 = \left(\frac{\phi_1'}{\phi_1''}\right)^{(h_1)} \left(\frac{\phi_2'}{\phi_2''}\right)^{(h_2)} \cdots \left(\frac{\phi_c'}{\phi_c''}\right)^{(h_c)}$$

where c equals b or $b-1$ and for each $s < t$

$$\left(\frac{\phi_s'}{\phi_s''}\right)^{(h_s)} = \left(\frac{\phi_s}{[a|u_1]^i}\right)^{(q_s)}$$

while for $s \geq t$

$$\left(\frac{\phi_s'}{\phi_s''}\right)^{(h_s)} = \left(\frac{\phi_{s+1}}{[a|u_1]^i}\right)^{(q_{s+1})}.$$

Note that Φ_1 is of inner length i as follows from the identity $p_{s1}p_{s2}\cdots p_{sb}=0$ for all s, which implies that for some s we have

$$\text{length}\left[\frac{\phi'_s}{\phi''_s}\right]=i.$$

Expand Φ_1 as a linear combination of circle products of neutral elementary rational monomials in $\text{Pleth}_r\{P_a\}$. Note that the circle product degree $i+d$, for some $d\geq 0$ of this expansion, is at least the inner length i of Φ_1. The expansion will be of the form

$$\begin{aligned}\Phi_1 &= \Psi'(\cdots w''_{j_2\cdots j_n}\cdots)\\ &= \sum_{k=1}^{i+d}\sum_j c_{jk}w''_{j_{12}\cdots j_{1n}}\circ\cdots\circ w''_{j_{k2}\cdots j_{kn}},\end{aligned}$$

where the inner sum is over all matrices of nonnegative integers (j_{mq}) with $1\leq m\leq k$ and $1\leq q\leq n$ satisfying

$$j_{m1}+\cdots+j_{mn}=r$$

and where for $1\leq m\leq k$ we have

$$w''_{j_{m2}\cdots j_{mn}}=\left(\frac{[a|u_2]}{[a|u_1]}\right)^{(j_{m2})}\cdots\left(\frac{[a|u_n]}{[a|u_1]}\right)^{(j_{mn})}.$$

The homogeneous rational monomial element $w''_{j_{m2}\cdots j_{mn}}$ is an elementary neutral homogeneous rational monomial of outer degree $r-j_{m1}$ in $\text{Pleth}_r\{P_a\}$. It satisfies, for all invertible sequences α of vectors $\alpha_1,\cdots,\alpha_r$,

$$\begin{aligned}w''_{j_{m2}\cdots j_{mn}}(\alpha) = \sum_\theta &\left[\frac{(\alpha_{i_{21}}|u_2)}{(\alpha_{i_{21}}|u_1)}\cdots\frac{(\alpha_{i_{2j_{m2}}}|u_2)}{(\alpha_{i_{2j_{m2}}}|u_1)}\right]\\ &\cdots\left[\frac{(\alpha_{i_{n1}}|u_n)}{(\alpha_{i_{n1}}|u_1)}\cdots\frac{(\alpha_{i_{nj_{mn}}}|u_n)}{(\alpha_{i_{nj_{mn}}}|u_1)}\right],\end{aligned}$$

where the sum is over all permutations

$$\theta(1)\cdots\theta(r)=i_{11}\cdots i_{1j_{m1}}i_{21}\cdots i_{2j_{m2}}\cdots i_{n1}\cdots i_{nj_{mn}}$$

which are increasing in the first j_{m1} positions, the next j_{m2} positions, $\cdots$, the last j_{mn} positions. Note that in $\text{Pleth}_r\{P_a\}$ we have

$$\begin{aligned}[([a|u_1])^{(r)}]^d\circ\Phi &= [([a|u_1])^{(r)}]^{i+d}\circ\Phi_1\\ &= [([a|u_1])^{(r)}]^{i+d}\circ\Psi'(\cdots w''_{j_2\cdots j_n}\cdots)\\ &= [([a|u_1])^{(r)}]^{i+d}\circ\sum_{k=1}^{i+d}\sum_j c_{jk}w''_{j_{12}\cdots j_{1n}}\circ\cdots\circ w''_{j_{k2}\cdots j_{kn}}.\end{aligned}$$

Expanding this further we obtain

$$\begin{aligned}
[([a|u_1])^{(r)}]^d \circ \Phi = \; & [([a|u_1])^{(r)}]^{i+d} \circ \sum_{k=1}^{i+d} \sum_j c_{jk} \\
& \left[\left(\frac{[a|u_2]}{[a|u_1]} \right)^{(j_{12})} \cdots \left(\frac{[a|u_n]}{[a|u_1]} \right)^{(j_{1n})} \right] \\
& \circ \cdots \circ \left[\left(\frac{[a|u_2]}{[a|u_1]} \right)^{(j_{k2})} \cdots \left(\frac{[a|u_n]}{[a|u_1]} \right)^{(j_{kn})} \right] \\
= \; & \sum_{k=1}^{i+d} [([a|u_1])^{(r)}]^{i+d-k} \circ \sum_j c_{jk} [([a|u_1])^{(r)}]^k \circ \\
& \left[\left(\frac{[a|u_2]}{[a|u_1]} \right)^{(j_{12})} \cdots \left(\frac{[a|u_n]}{[a|u_1]} \right)^{(j_{1n})} \right] \\
& \circ \cdots \circ \left[\left(\frac{[a|u_2]}{[a|u_1]} \right)^{(j_{k2})} \cdots \left(\frac{[a|u_n]}{[a|u_1]} \right)^{(j_{kn})} \right] \\
= \; & \sum_{k=1}^{i+d} [([a|u_1])^{(r)}]^{i+d-k} \circ \sum_j c_{jk} \\
& \left[([a|u_1])^{(r)} \circ \left(\frac{[a|u_2]}{[a|u_1]} \right)^{(j_{12})} \cdots \left(\frac{[a|u_n]}{[a|u_1]} \right)^{(j_{1n})} \right] \\
& \circ \cdots \circ \left[([a|u_1])^{(r)} \circ \left(\frac{[a|u_2]}{[a|u_1]} \right)^{(j_{k2})} \cdots \left(\frac{[a|u_n]}{[a|u_1]} \right)^{(j_{kn})} \right].
\end{aligned}$$

Note that

$$\begin{aligned}
& ([a|u_1])^{(r)} \circ \left(\frac{[a|u_2]}{[a|u_1]} \right)^{(j_{m2})} \cdots \left(\frac{[a|u_n]}{[a|u_1]} \right)^{(j_{mn})} \\
= \; & ([a|u_1])^{(j_{m1})} ([a|u_2])^{(j_{m2})} \cdots ([a|u_n])^{(j_{mn})}
\end{aligned}$$

for $1 \leq m \leq k$ and therefore

$$\begin{aligned}
[([a|u_1])^{(r)}]^d \circ \Phi = \; & [([a|u_1])^{(r)}]^{i+d} \circ \Psi'(\cdots w''_{j_2 \cdots j_n} \cdots) \\
= \; & \sum_{k=1}^{i+d} [([a|u_1])^{(r)}]^{i+d-k} \circ \sum_j c_{jk} \\
& \left[([a|u_1])^{(j_{11})} ([a|u_2])^{(j_{12})} \cdots ([a|u_n])^{(j_{1n})} \right] \\
& \circ \cdots \circ \left[([a|u_1])^{(j_{k1})} ([a|u_2])^{(j_{k2})} \cdots ([a|u_n])^{(j_{kn})} \right]
\end{aligned}$$

$$= \sum_{k=1}^{i+d} [w_{r0\cdots 0}]^{i+d-k} \circ \sum_j c_{jk} \, w_{j_{11}\cdots j_{1n}} \circ \cdots \circ w_{j_{k1}\cdots j_{kn}}$$
$$= \Psi(\cdots w_{i_1\cdots i_n} \cdots)$$

so that

$$[([a|u_1])^{(r)}]^{i+d} \circ \Psi'(\cdots w''_{j_2\cdots j_n} \cdots) = \Psi(\cdots w_{i_1\cdots i_n} \cdots).$$

Here

$$w_{j_{m1} j_{m2}\cdots j_{mn}} = ([a|u_1])^{(j_{m1})}([a|u_2])^{(j_{m2})} \cdots ([a|u_n])^{(j_{mn})}$$

and for each m we have

$$j_{m1} + j_{m2} + \cdots + j_{mn} = r.$$

We therefore have

$$[([a|u_1])^{(r)}]^d \circ \Phi = \sum_{k=1}^{i+d} [w_{r0\cdots 0}]^{i+d-k} \circ \sum_j c_{jk} w_{j_{11}\cdots j_{1n}} \circ \cdots \circ w_{j_{k1}\cdots j_{kn}}$$

so that

$$[w_{r0\cdots 0}]^d \circ \Phi = \sum_{k=1}^{i+d} [w_{r0\cdots 0}]^{i+d-k} \circ \sum_j c_{jk} w_{j_{11}\cdots j_{1n}} \circ \cdots \circ w_{j_{k1}\cdots j_{kn}}$$

or

$$[w_{r0\cdots 0}]^d \circ \Phi = \Psi(\cdots w_{i_1\cdots i_n} \cdots)$$

and therefore

$$\Phi = \frac{\Psi(\cdots w_{i_1\cdots i_n} \cdots)}{[w_{r0\cdots 0}]^d}.$$

We have now proved the polynomial version of the fundamental theorem of symmetric functions for $\text{Pleth}_r\{P_a\}$.

Theorem 2 (Fundamental theorem). *Every irreducible monomial element Φ in* $\text{Pleth}_r(P_a)$ *of inner degree i can be expanded in the form*

$$[([a|u_1])^{(r)}]^d \circ \Phi = \Psi(w)$$

where $\Psi(w) = \Psi(\cdots w_{i_1\cdots i_n} \cdots)$ is a polynomial function of circle product degree $i + d$ for some $d \geq 0$ in the elementary homogeneous monomials

$$w_{i_1\cdots i_n} = ([a|u_1])^{(i_1)} \cdots ([a|u_n])^{(i_n)}$$

of outer degree r in $\mathrm{Pleth}_r(P_a)$*. The degree of* $\Psi(w)$ *is the circle product degree* $i+d$ *of the expansion of the neutral homogeneous rational monomial element* Φ_1 *of inner length* i *satisfying*

$$\Phi = [([a|u_1])^{(r)}]^i \circ \Phi_1$$

as a linear combination of circle products of elementary neutral homogeneous rational monomials

$$\Phi_1 = \Psi'(w''),$$

where $\Psi'(w'') = \Psi'(\cdots w''_{j_2\cdots j_n}\cdots)$ *is a polynomial function in the elementary neutral homogeneous rational monomials of circle product degree* $i+d$ *for some* $d \geq 0$ *given by*

$$\Psi'(\cdots w''_{j_2\cdots j_n}\cdots) = \sum_{k=1}^{i+d}\sum_{j} c_{jk} w''_{j_{12}\cdots j_{1n}} \circ \cdots \circ w''_{j_{k2}\cdots j_{kn}},$$

where the inner sum is over all matrices of nonnegative integers (j_{mq}) *with* $1 \leq m \leq k$ *and* $1 \leq q \leq n$ *satisfying*

$$j_{m1} + \cdots + j_{mn} = r$$

and where for $1 \leq m \leq k$ *the elementary neutral homogeneous rational monomials*

$$w''_{j_{m2}\cdots j_{mn}} = \left(\frac{[a|u_2]}{[a|u_1]}\right)^{(j_{m2})} \cdots \left(\frac{[a|u_n]}{[a|u_1]}\right)^{(j_{mn})}$$

are of outer degree $r - j_{m1}$ *in* $\mathrm{Pleth}_r\{P_a\}$*. The polynomial function* $\Psi(w)$ *satisfies*

$$\Psi(w) = [([a|u_1])^{(r)}]^{i+d} \circ \Psi'(w'').$$

Hence the homogeneous monomial element Φ *can be expressed rationally in terms of the homogeneous elementary monomials* $w_{i_1\cdots i_n}$ *of shape* (1^r)*.*

We also have the following generalization of the fundamental theorem for neutral proper rational monomial elements.

Theorem 3 (Fundamental theorem). *Let* Φ *be an irreducible neutral homogeneous proper rational monomial element of the algebra* $\mathrm{Pleth}_r\{P_a\}$ *of inner degree* 0*, inner length* i *and outer degree* $c \leq r$*. Then for some* $h \geq 0$ *and for some homogeneous monomial elements* U *and* V *in* $\mathrm{Pleth}_r(P_a)$ *with* U *and* V *both of inner degree* $i+h$ *we have*

$$U \circ \Phi = V$$

and we have that Φ *can be expanded in the form*

$$[([a|u_1])^{(r)}]^d \circ U \circ \Phi = [([a|u_1])^{(r)}]^d \circ V = \Psi(w),$$

where $\Psi(w) = \Psi(\cdots w_{i_1 \cdots i_n} \cdots)$ is a polynomial function of degree $i + h + d$ for some $d \geq 0$ in the elementary homogeneous monomials

$$w_{i_1 \cdots i_n} = ([a|u_1])^{(i_1)} \cdots ([a|u_n])^{(i_n)}$$

in $\mathrm{Pleth}_r(P_a)$ *of outer degree r. The degree of $\Psi(w)$ is the circle product degree $i + h + d$ for some $d \geq 0$ of the expansion of the neutral homogeneous rational monomial element V_1 of inner length $i + h$ satisfying*

$$V = [([a|u_1])^{(r)}]^{i+h} \circ V_1$$

as a linear combination of circle products of elementary neutral homogeneous rational monomials

$$V_1 = \Psi'(w''),$$

where $\Psi'(w'') = \Psi'(\cdots w''_{j_2 \cdots j_n} \cdots)$ is given by

$$\Psi'(\cdots w''_{j_2 \cdots j_n} \cdots) = \sum_{k=1}^{i+d} \sum_{j} c_{jk} w''_{j_{12} \cdots j_{1n}} \circ \cdots \circ w''_{j_{k2} \cdots j_{kn}},$$

where the inner sum is over all matrices of nonnegative integers (j_{mq}) with $1 \leq m \leq k$ and $1 \leq q \leq n$ satisfying

$$j_{m1} + \cdots + j_{mn} = r$$

and where for $1 \leq m \leq k$ the elementary neutral homogeneous rational monomials

$$w''_{j_{m2} \cdots j_{mn}} = \left(\frac{[a|u_2]}{[a|u_1]}\right)^{(j_{m2})} \cdots \left(\frac{[a|u_n]}{[a|u_1]}\right)^{(j_{mn})}$$

are of outer degree $r - j_{m1}$ in $\mathrm{Pleth}_r\{P_a\}$. *The polynomial function $\Psi(w)$ satisfies*

$$\Psi(w) = [([a|u_1])^{(r)}]^{i+h+d} \circ \Psi'(w'').$$

Hence the neutral homogeneous monomial element Φ can be expressed rationally in terms of the homogeneous elementary monomials $w_{i_1 \cdots i_n}$ in $\mathrm{Pleth}_r(P_a)$ *of shape* (1^r).

Proof. We have seen previously that for some $h \geq 0$ the neutral homogeneous rational monomial element Φ satisfies

$$\Phi \circ U = V,$$

where U and V are homogeneous monomial elements in $\mathrm{Pleth}_r(P_a)$ of inner degree $i+h$ and of outer degree r. Expand U and V by the previous theorems and the result follows. □

We have the analogous result for all strict homogeneous proper rational monomial elements.

Theorem 4 (Fundamental theorem). *Let Φ be an irreducible strict homogeneous proper rational monomial element of the algebra* $\mathrm{Pleth}_r\{P_a\}$ *of inner degree $p > 0$, inner length i and outer degree r. Then, for some $h \geq 0$ and for some homogeneous monomial elements U and V in* $\mathrm{Pleth}_r(P_a)$ *with U of inner degree $i - p + h$ and V of inner degree $i + h$, we have*

$$U \circ \Phi = V$$

and for some $d \geq 0$ we have that Φ can be expanded in the form

$$[([a|u_1])^{(r)}]^d \circ U \circ \Phi = \Psi(w)$$

where $\Psi(w) = \Psi(\cdots w_{i_1 \cdots i_n} \cdots)$ is a polynomial function of degree $i + h + d$ for some $d \geq 0$ in the elementary homogeneous monomials

$$w_{i_1 \cdots i_n} = ([a|u_1])^{(i_1)} \cdots ([a|u_n])^{(i_n)}$$

in $\mathrm{Pleth}_r(P_a)$ *of outer degree r. The degree of $\Psi(w)$ is the circle product degree $i + h + d$ for some $d \geq 0$ of the expansion of the neutral homogeneous rational monomial element V_1 of inner length $i + h$ satisfying*

$$V = [([a|u_1])^{(r)}]^{i+h} \circ V_1$$

as a linear combination of circle products of elementary neutral homogeneous rational monomials

$$V_1 = \Psi'(w''),$$

where $\Psi'(w'') = \Psi'(\cdots w''_{j_2 \cdots j_n} \cdots)$ is given by

$$\Psi'(\cdots w''_{j_2 \cdots j_n} \cdots) = \sum_{k=1}^{i+d} \sum_{j} c_{jk} w''_{j_{12} \cdots j_{1n}} \circ \cdots \circ w''_{j_{k2} \cdots j_{kn}},$$

where the inner sum is over all matrices of nonnegative integers (j_{mq}) with $1 \leq m \leq k$ and $1 \leq q \leq n$ satisfying

$$j_{m1} + \cdots + j_{mn} = r$$

and where for $1 \leq m \leq k$ the elementary neutral homogeneous rational monomials

$$w''_{j_{m2} \cdots j_{mn}} = \left(\frac{[a|u_2]}{[a|u_1]}\right)^{(j_{m2})} \cdots \left(\frac{[a|u_n]}{[a|u_1]}\right)^{(j_{mn})}$$

are of outer degree $r - j_{m1}$ in $\mathrm{Pleth}_r\{P_a\}$*. The polynomial function $\Psi(w)$ satisfies*

$$\Psi(w) = [([a|u_1])^{(r)}]^{i+h+d} \circ \Psi'(w'').$$

Hence the strict homogeneous rational monomial element Φ can be expressed rationally in terms of the homogeneous elementary monomials $w_{i_1 \cdots i_n}$ in $\mathrm{Pleth}_r(P_a)$*. of shape (1^r).*

Proof. We have seen previously that for some $h \geq 0$ the strict homogeneous proper rational monomial element Φ satisfies

$$\Phi \circ U = V,$$

where U and V are homogeneous monomial elements in $\mathrm{Pleth}_r(P_a)$ with U of inner degree $d' = i - p + h$ and V of inner degree $d'' = i + h$. Expand U and V by the previous theorems and the result follows. □

Corollary 1 (Fundamental theorem of symmetric functions). *The elementary homogeneous monomial elements generate* $\mathrm{Pleth}_r\{P_a\}$ *rationally under the circle product.*

11 An algorithm in invariant theory

Let $f(x)$ be the Cayley form

$$f(x) = f(x_1, \cdots, x_n) = \sum_{i_1+\cdots+i_n=r} f_{i_1\cdots i_n} x_1^{i_1} \cdots x_n^{i_n} = [f_1|x] \cdots [f_r|x].$$

We do not know the vectors $f_1, \cdots, f_r$ explicitly, only that their homogeneous elementary monomials of shape (1^r) are the coefficients $f_{i_1\cdots i_n}$ of $f(x)$.

Let $Q(f)$ denote the ring of rational funcions of the coefficients of the Cayley form $f(x)$. The evaluation operator

$$E_{f,a} : \mathrm{Pleth}_r\{P_a\} \to Q(f)$$

is defined next.

(1) We first define for all homogeneous monomial elements ϕ and ψ of $\mathrm{Pleth}_r(P_a)$ that

$$E_{f,a}[\phi \circ \psi] = E_{f,a}[(\phi|\psi)] = E_{f,a}[\phi]\, E_{f,a}[\psi].$$

(2) For the homogeneous elementary monomials

$$w_{i_1 i_2 \cdots i_n} = ([a|u_1])^{(i_1)} \cdots ([a|u_n])^{(i_n)}$$

in $\mathrm{Pleth}_r(P_a)$ of shape (1^r) we define

$$E_{f,a}\left[w_{i_1 i_2 \cdots i_n}\right] = f_{i_1 i_2 \cdots i_n}.$$

(3) Suppose Φ is a homogeneous monomial element of $\mathrm{Pleth}_r(P_a)$ of shape (i^r) for some value of $i > 1$. We may write

$$\Phi = ([a|u_1]^{p_{11}} \cdots [a|u_n]^{p_{n1}})^{(q_1)} \cdots ([a|u_1]^{p_{1b}} \cdots [a|u_n]^{p_{nb}})^{(q_b)}.$$

Either Φ is irreducible or for some irreducible element Φ' we have

$$\Phi = U_1^u \circ U_2^v \circ \cdots \circ U_n^w \circ \Phi',$$

where $U_j = ([a|u_j])^{(r)}$ and

$$U_j^k = ([a|u_j])^{(r)} \circ \cdots \circ ([a|u_j])^{(r)} = ([a|u_j]^k)^{(r)}.$$

In such a case define

$$E_{f,a}[\Phi] = E_{f,a}[U_1]^u E_{f,a}[U_2]^v \cdots E_{f,a}[U_n]^w E_{f,a}[\Phi'].$$

(4) Suppose Φ is an irreducible monomial element in $\text{Pleth}_r(P_a)$ which is homogeneous of shape (i^r) for some value of $i > 1$. We may write

$$\begin{aligned} \Phi &= ([a|u_1]^{p_{11}} \cdots [a|u_n]^{p_{n1}})^{(q_1)} \cdots ([a|u_1]^{p_{1b}} \cdots [a|u_n]^{p_{nb}})^{(q_b)} \\ &= (\phi_1)^{(q_1)} (\phi_2)^{(q_2)} \cdots (\phi_b)^{(q_b)}, \end{aligned}$$

where $q_1 + \cdots + q_b = r$ and for $1 \le j \le b$

$$p_{1j} + \cdots + p_{nj} = i.$$

We may assume the inner terms $\phi_1, \phi_2, \cdots, \phi_b$ are all distinct. Since Φ is irreducible we must have

$$p_{s1} p_{s2} \cdots p_{sb} = 0$$

for each s. Express Φ in the form

$$\Phi = [([a|u_1])^{(r)}]^i \circ \Phi_1,$$

where

$$\begin{aligned} \Phi_1 &= \left(\frac{\phi_1}{[a|u_1]^i}\right)^{(q_1)} \cdots \left(\frac{\phi_b}{[a|u_1]^i}\right)^{(q_b)} \\ &= \left(\frac{[a|u_2]^{p_{21}}}{[a|u_1]^{p_{21}}} \cdots \frac{[a|u_n]^{p_{n1}}}{[a|u_1]^{p_{n1}}}\right)^{(q_1)} \cdots \left(\frac{[a|u_2]^{p_{2b}}}{[a|u_1]^{p_{2b}}} \cdots \frac{[a|u_n]^{p_{nb}}}{[a|u_1]^{p_{nb}}}\right)^{(q_b)}. \end{aligned}$$

Φ_1 is thus a neutral homogeneous rational monomial element in $\text{Pleth}_r\{P_a\}$. Since the inner terms $\phi_1, \phi_2, \cdots, \phi_b$ are all distinct, at most one of the terms can satisfy

$$\frac{\phi_t}{[a|u_1]^i} = \frac{[a|u_2]^{p_{2t}}}{[a|u_1]^{p_{2t}}} \cdots \frac{[a|u_n]^{p_{nt}}}{[a|u_1]^{p_{nt}}} = 1.$$

If there is such a term, set

$$\left(\frac{\phi_t}{[a|u_1]^i}\right)^{(q_t)} = \left(\frac{[a|u_2]^{p_{2t}}}{[a|u_1]^{p_{2t}}} \cdots \frac{[a|u_n]^{p_{nt}}}{[a|u_1]^{p_{nt}}}\right)^{(q_t)} = 1.$$

The resulting expression for Φ_1 will be written

$$\Phi_1 = \left(\frac{\phi_1'}{\phi_1''}\right)^{(h_1)} \left(\frac{\phi_2'}{\phi_2''}\right)^{(h_2)} \cdots \left(\frac{\phi_c'}{\phi_c''}\right)^{(h_c)},$$

where c equals b or $b-1$ and for each $s < t$

$$\left(\frac{\phi_s'}{\phi_s''}\right)^{(h_s)} = \left(\frac{\phi_s}{[a|u_1]^i}\right)^{(q_s)}$$

while for $s \geq t$

$$\left(\frac{\phi_s'}{\phi_s''}\right)^{(h_s)} = \left(\frac{\phi_{s+1}}{[a|u_1]^i}\right)^{(q_{s+1})}.$$

Note that Φ_1 is of inner length i as follows from the identity $p_{s1}p_{s2}\cdots p_{sb} = 0$ for all s, which implies that for some s we have

$$\text{length}\left[\frac{\phi_s'}{\phi_s''}\right] = i.$$

We have seen previously that the neutral homogeneous rational monomial element Φ_1 satisfies

$$\Phi_1 = \Psi'(\cdots w''_{j_2\cdots j_n}\cdots) = \sum_{k=1}^{i+d}\sum_j c_{jk} w''_{j_{12}\cdots j_{1n}} \circ \cdots \circ w''_{j_{k2}\cdots j_{kn}},$$

where the inner sum is over all matrices of nonnegative integers (j_{mq}) with $1 \leq m \leq k$ and $1 \leq q \leq n$ satisfying

$$j_{m1} + \cdots + j_{mn} = r$$

and where, for $1 \leq m \leq k$, the elementary neutral homogeneous rational monomials

$$w''_{j_{m2}\cdots j_{mn}} = \left(\frac{[a|u_2]}{[a|u_1]}\right)^{(j_{m2})} \cdots \left(\frac{[a|u_n]}{[a|u_1]}\right)^{(j_{mn})}$$

are of outer degree $r - j_{m1}$ in $\text{Pleth}_r\{P_a\}$, while

$$\begin{aligned} [([a|u_1])^{(r)}]^d \circ \Phi &= [([a|u_1])^{(r)}]^{i+d} \circ \Phi_1 \\ &= [([a|u_1])^{(r)}]^{i+d} \circ \Psi'(w'') \\ &= \Psi(w), \end{aligned}$$

where $\Psi(w) = \Psi(\cdots w_{i_1\cdots i_n}\cdots)$ is a polynomial function of circle product degree $i+h+d$ for some $d \geq 0$ in the elementary homogeneous monomials

$$w_{i_1\cdots i_n} = ([a|u_1])^{(i_1)} \cdots ([a|u_n])^{(i_n)}$$

in $\mathrm{Pleth}_r(P_a)$ of outer degree r. Furthermore, we have obtained the explicit expansion

$$\begin{aligned}
[([a|u_1])^{(r)}]^d \circ \Phi &= [([a|u_1])^{(r)}]^{i+d} \circ \Phi_1 \\
&= [([a|u_1])^{(r)}]^{i+d} \circ \Psi(\cdots w''_{j_2\cdots j_n}\cdots) \\
&= [([a|u_1])^{(r)}]^{i+d} \circ \sum_{k=1}^{i+d}\sum_j c_{jk}\, w''_{j_{12}\cdots j_{1n}} \circ \cdots \circ w''_{j_{k2}\cdots j_{kn}} \\
&= \sum_{k=1}^{i+d} [w_{r0\cdots 0}]^{i+d-k} \circ \sum_j c_{jk}\, w_{j_{11}\cdots j_{1n}} \circ \cdots \circ w_{j_{k1}\cdots j_{kn}} \\
&= \Psi(\cdots w_{i_1\cdots i_n}\cdots).
\end{aligned}$$

We can therefore define

$$\begin{aligned}
&E_{f,a}\left[[w_{r0\cdots 0}]^d \circ \Phi\right] \\
&\quad = E_{f,a}\left[\sum_{k=1}^{i+d}[w_{r0\cdots 0}]^{i+d-k} \circ \sum_j c_{jk} w_{j_{11}\cdots j_{1n}} \circ \cdots \circ w_{j_{k1}\cdots j_{kn}}\right]
\end{aligned}$$

and note that

$$\begin{aligned}
E_{f,a}\left[[w_{r0\cdots 0}]^d \circ \Phi\right] &= E_{f,a}\left[[w_{r0\cdots 0}]^d\right] E_{f,a}[\Phi] \\
&= [f_{r0\cdots 0}]^d\, E_{f,a}[\Phi]
\end{aligned}$$

while

$$\begin{aligned}
&E_{f,a}\left[\sum_{k=1}^{i+d}[w_{r0\cdots 0}]^{i+d-k} \circ \sum_j c_{jk} w_{j_{11}\cdots j_{1n}} \circ \cdots \circ w_{j_{k1}\cdots j_{kn}}\right] \\
&\quad = \sum_{k=1}^{i+d} E_{f,a}\left[[w_{r0\cdots 0}]^{i+d-k}\right] \sum_j c_{jk} E_{f,a}\left[w_{j_{11}\cdots j_{1n}}\right] \cdots E_{f,a}\left[w_{j_{k1}\cdots j_{kn}}\right] \\
&\quad = \sum_{k=1}^{i+d} [f_{r0\cdots 0}]^{i+d-k} \sum_j c_{jk}\, f_{j_{11}\cdots j_{1n}} \cdots f_{j_{k1}\cdots j_{kn}}
\end{aligned}$$

so that

$$[f_{r0\cdots 0}]^d\, E_{f,a}[\Phi] = \sum_{k=1}^{i+d} [f_{r0\cdots 0}]^{i+d-k} \sum_j c_{jk}\, f_{j_{11}\cdots j_{1n}} \cdots f_{j_{k1}\cdots j_{kn}}.$$

We may therefore define

$$
\begin{aligned}
E_{f,a}(\Phi) &= \frac{\sum_{k=1}^{i+d} [f_{r0\cdots 0}]^{i+d-k} \sum_j c_{jk}\, f_{j_{11}\cdots j_{1n}} \cdots f_{j_{k1}\cdots j_{kn}}}{[f_{r0\cdots 0}]^d} \\
&= \frac{\Psi(\cdots E_{f,a}[w_{i_1\cdots i_n}]\cdots)}{\left[E_{f,a}[w_{r0\cdots 0}]\right]^d} \\
&= \frac{\Psi(\cdots f_{i_1\cdots i_n}\cdots)}{[f_{r0\cdots 0}]^d}.
\end{aligned}
$$

(5) Let Φ be an irreducible homogeneous proper rational monomial element of the algebra $\mathrm{Pleth}_r\{P_a\}$ of inner degree $p \geq 0$, inner length i and outer degree $c \leq r$ with $c = r$ if Φ is strict. Then, for some $h \geq 0$ and for some homogeneous monomial elements U and V in $\mathrm{Pleth}_r(P_a)$ with U of inner degree $i - p + h$ and V of inner degree $i + h$, we have seen that

$$U \circ \Phi = V$$

and for some $d \geq 0$ we have seen that Φ can be expanded in the form

$$[([a|u_1])^{(r)}]^d \circ U \circ \Phi = \Psi(w),$$

where $\Psi(w) = \Psi(\cdots w_{i_1\cdots i_n}\cdots)$ is a polynomial function of degree $i + h + d$ in the homogeneous elementary monomials

$$w_{i_1\cdots i_n} = ([a|u_1])^{(i_1)} \cdots ([a|u_n])^{(i_n)}$$

in $\mathrm{Pleth}_r(P_a)$ of outer degree r. The polynomial function $\Psi(w)$ satisfies

$$\Psi(w) = [([a|u_1])^{(r)}]^{i+h+d} \circ \Psi'(w''),$$

where $\Psi'(w'') = \Psi'(\cdots w''_{j_2\cdots j_n}\cdots)$ is the polynomial function of degree $i + h + d$ in the elementary neutral homogeneous rational monomials

$$w''_{j_2\cdots j_n} = \left(\frac{[a|u_2]}{[a|u_1]}\right)^{(j_2)} \cdots \left(\frac{[a|u_n]}{[a|u_1]}\right)^{(j_n)}$$

of outer degree $r - j_1$ in $\mathrm{Pleth}_r\{P_a\}$ expressing the neutral homogeneous rational monomial element V_1 of inner length $i + h$ satisfying

$$V = [([a|u_1])^{(r)}]^{i+h} \circ V_1$$

as a linear combination of circle products of elementary neutral homogeneous rational monomials $V_1 = \Psi'(w'')$. The polynomial functions $\Psi'(w'')$ and $\Psi(w)$ have the explicit form

$$\begin{aligned}\Psi'(w'') &= \sum_{k=1}^{i+h+d} \sum_j c_{jk} w''_{j_{12}\cdots j_{1n}} \circ \cdots \circ w''_{j_{k2}\cdots j_{kn}} \\ \Psi(w) &= \sum_{k=1}^{i+h+d} [w_{r0\cdots 0}]^{i+h+d-k} \circ \sum_j c_{jk} w_{j_{11}\cdots j_{1n}} \circ \cdots \circ w_{j_{k1}\cdots j_{kn}}\end{aligned}$$

where the inner sums are over all matrices of nonnegative integers (j_{mq}) with $1 \leq m \leq k$ and $1 \leq q \leq n$ satisfying $j_{m1} + \cdots + j_{mn} = r$. Consider the identity

$$[([a|u_1])^{(r)}]^d \circ U \circ \Phi = \Psi(w)$$

and recall that $U = (u)^{(r)}$ is of the form

$$\begin{aligned}U &= \left([a|u_1]^{m_1}[a|u_2]^{m_2} \cdots [a|u_n]^{m_n}\right)^{(r)} \\ &= \left[([a|u_1])^{(r)}\right]^{m_1} \circ \left[([a|u_2])^{(r)}\right]^{m_2} \circ \cdots \circ \left[([a|u_n])^{(r)}\right]^{m_n} \\ &= [w_{r0\cdots 0}]^{m_1} \circ [w_{0r0\cdots 0}]^{m_2} \circ \cdots \circ [w_{0\cdots 0r}]^{m_n} .\end{aligned}$$

We therefore have

$$\begin{aligned}&E_{f,a}\left[[([a|u_1])^{(r)}]^d \circ U \circ \Phi\right] \\ &\quad = E_{f,a}\left[[([a|u_1])^{(r)}]^d\right] E_{f,a}[U]\, E_{f,a}[\Phi] \\ &\quad = [f_{r0\cdots 0}]^{d+m_1} [f_{0r0\cdots 0}]^{m_2} \cdots [f_{0\cdots 0r}]^{m_r} E_{f,a}[\Phi] .\end{aligned}$$

Since we also have

$$\begin{aligned}&E_{f,a}\left[[([a|u_1])^{(r)}]^d \circ U \circ \Phi\right] \\ &\quad = E_{f,a}\left[[([a|u_1])^{(r)}]^d \circ V\right] \\ &\quad = E_{f,a}\left[[w_{r0\cdots 0}]^{i+h+d} \circ V_1\right] \\ &\quad = E_{f,a}\left[[w_{r0\cdots 0}]^{i+h+d} \circ \sum_{k=1}^{i+h+d} \sum_j c_{jk} w''_{j_{12}\cdots j_{1n}} \circ \cdots \circ w''_{j_{k1}\cdots j_{kn}}\right] \\ &\quad = E_{f,a}\left[\sum_{k=1}^{i+h+d} [w_{r0\cdots 0}]^{i+h+d-k} \circ \sum_j c_{jk} w_{j_{11}\cdots j_{1n}} \circ \cdots \circ w_{j_{k1}\cdots j_{kn}}\right] \\ &\quad = \sum_{k=1}^{i+h+d} [f_{r0\cdots 0}]^{i+h+d-k} \sum_j c_{jk}\, f_{j_{11}\cdots j_{1n}} \cdots f_{j_{k1}\cdots j_{kn}},\end{aligned}$$

we may define

$$\begin{aligned} E_{f,a}(\Phi) &= \frac{\sum_{k=1}^{i+h+d} [f_{r0\cdots0}]^{i+h+d-k} \sum_j c_{jk}\, f_{j_{11}\cdots j_{1n}} \cdots f_{j_{k1}\cdots j_{kn}}}{[f_{r0\cdots0}]^{d+m_1} [f_{0r0\cdots0}]^{m_2} \cdots [f_{0\cdots0r}]^{m_r}} \\ &= \frac{\Psi(\cdots f_{i_1\cdots i_n} \cdots)}{[f_{r0\cdots0}]^{d+m_1} [f_{0r0\cdots0}]^{m_2} \cdots [f_{0\cdots0r}]^{m_r}}. \end{aligned}$$

The evaluation map $E_{f.a}$ has now been defined on all of $\mathrm{Pleth}_r\{P_a\}$.

12 An example: ternary cubic Cayley forms

We use the notation

$$[a|u_1] = U, \quad [a|u_2] = V, \quad [a|u_3] = W.$$

We consider the evaluation in $\mathrm{Pleth}_r(A|P)$ of the monomial function Φ of outer degree $r = 3$ and inner degree $i = 2$ given by

$$\Phi = (U^2)(V^2)(W^2)$$

for the ternary cubic Cayley form

$$f(X) = \sum_{i+j+k=3} f_{ijk} x_1^i x_2^j x_3^k = (f_1|X)(f_2|X)(f_3|X).$$

In this example we will use Φ_3 instead of Φ_1. Let

$$\frac{U}{W} = u, \quad \frac{V}{W} = v.$$

We begin by considering the neutral homogeneous rational monomial element Φ_3 satisfying

$$\Phi = \left[(W)^{(3)}\right]^2 \circ \Phi_3 = \left[(W)^{(3)}\right]^2 \circ (u^2)(v^2).$$

Expanding Φ_3 as a linear combination of circle products of elementary neutral homogeneous rational monomials gives

$$\begin{aligned} \Phi_3 = {} & (u^2)(v^2) \\ = {} & \frac{2}{3}(u)^{(2)}(v)^{(2)} - \frac{2}{3}(u)^{(2)}(v) \circ (v) - \frac{2}{3}(u)(v)^{(2)} \circ (u) \\ & + \frac{10}{3}(u)^{(2)} \circ (v)^{(2)} - \frac{1}{3}(u)(v) \circ (u)(v) \\ & - \frac{4}{3}(u)^{(2)} \circ (v) \circ (v) - \frac{4}{3}(v)^{(2)} \circ (u) \circ (u) \\ & + \frac{4}{3}(u)(v) \circ (u) \circ (v). \end{aligned}$$

The term $\frac{2}{3}(u)^{(2)}(v)^{(2)}$ is zero for $f(X) = (f_1|X)(f_2|X)(f_3|X)$ and may be dropped, since Φ is in $\mathrm{Pleth}_3(P_a)$. The monomial function Φ_3 is of inner degree $i = 2$, while the expansion of Φ_3 as a linear combination of circle products of elementary neutral homogeneous rational monomials is of degree $i+d = 2+1 = 3$ with respect to the circle product. We conclude that in $\mathrm{Pleth}_3(P_a)$ where $P_a = \{U < V < W\}$ we have

$$\begin{aligned}(W)^{(3)} \circ \Phi &= (W)^{(3)} \circ (U^2)(V^2)(W^2)\\
&= -\frac{2}{3}(U)^{(2)}(V) \circ (V)(W)^{(2)} \circ (W)^{(3)}\\
&\quad -\frac{2}{3}(U)(V)^{(2)} \circ (U)(W)^{(2)} \circ (W)^{(3)}\\
&\quad +\frac{10}{3}(U)^{(2)}(W) \circ (V)^{(2)}(W) \circ (W)^{(3)}\\
&\quad -\frac{1}{3}(U)(V)(W) \circ (U)(V)(W) \circ (W)^{(3)}\\
&\quad -\frac{4}{3}(U)^{(2)}(W) \circ (V)(W)^{(2)} \circ (V)(W)^{(2)}\\
&\quad -\frac{4}{3}(V)^{(2)}(W) \circ (U)(W)^{(2)} \circ (U)(W)^{(2)}\\
&\quad +\frac{4}{3}(U)(V)(W) \circ (U)(W)^{(2)} \circ (V)(W)^{(2)}.\end{aligned}$$

We can justify this as follows. Expand the linear factors of $f(X)$ by

$$\begin{aligned}(f_1|X) &= \sum_{j=1}^{3} f_{1j}x_j = f_{13}\sum_{j=1}^{3}\frac{f_{1j}}{f_{13}}x_j = f_{13}\sum_{j=1}^{3} f''_{1j}x_j = f_{13}(f''|X)\\
(f_2|X) &= \sum_{j=1}^{3} f_{2j}x_j = f_{23}\sum_{j=1}^{3}\frac{f_{2j}}{f_{23}}x_j = f_{23}\sum_{j=1}^{3} f''_{1j}x_j = f_{23}(f''|X)\\
(f_3|X) &= \sum_{j=1}^{3} f_{3j}x_j = f_{33}\sum_{j=1}^{3}\frac{f_{3j}}{f_{31}}x_j = f_{33}\sum_{j=1j}^{3} f''_{1j}x_j = f_{33}(f''|X).\end{aligned}$$

We have

$$f(X) = f_{003}(f_1''|X)(f_2''|X)(f_3''|X),$$

where

$$f_{003} = f_{13}f_{23}f_{33}.$$

The expansion of

$$\Phi_3 = (u^2)(v^2)$$

given by

$$\begin{aligned}\Phi_3 = {} & (u^2)(v^2)\\ = {} & -\frac{2}{3}(u)^{(2)}(v)\circ(v) - \frac{2}{3}(u)(v)^{(2)}\circ(u) + \frac{10}{3}(u)^{(2)}\circ(v)^{(2)}\\ & -\frac{1}{3}(u)(v)\circ(u)(v) - \frac{4}{3}(u)^{(2)}\circ(v)\circ(v)\\ & -\frac{4}{3}(v)^{(2)}\circ(u)\circ(u) + \frac{4}{3}(u)(v)\circ(u)\circ(v)\end{aligned}$$

applied to the vectors (f_1, f_2, f_3) leads to an identity in the components of these vectors, which may be obtained as follows. The monomial expression for Φ_3 applied to the vectors (f_1, f_2, f_3) gives

$$\begin{aligned}\Phi_3(f_1, f_2, f_3) = {} & [f''_{11}]^2[f''_{22}]^2 + [f''_{12}]^2[f''_{21}]^2\\ & + [f''_{11}]^2[f''_{32}]^2 + [f''_{12}]^2[f''_{31}]^2\\ & + [f''_{21}]^2[f''_{32}]^2 + [f''_{22}]^2[f''_{31}]^2.\end{aligned}$$

Multiply this expression for Φ_3 by $[f_{003}]^{i+d} = [f_{13}f_{23}f_{33}]^{i+d}$ where $i + d = 2+1 = 3$ is the circle product degree of the expansion of Φ_3 as a linear combination of circle products of elementary homogeneous neutral rational monomials. We obtain $[f_{13}f_{23}f_{33}]^3\Phi_3(f_1, f_2, f_3)$ which equals

$$\begin{aligned}& [f_{003}]^{i+d}\Phi_3(f_1, f_2, f_3)\\ & = [f_{003}]^3\Phi_3(f_1, f_2, f_3)\\ & = [f_{13}f_{23}f_{33}]^3\Phi_3(f_1, f_2, f_3)\\ & = [f_{13}f_{23}f_{33}]^3\begin{bmatrix} [f''_{11}]^2[f''_{22}]^2 + [f''_{12}]^2[f''_{21}]^2\\ +[f''_{11}]^2[f''_{32}]^2 + [f''_{12}]^2[f''_{31}]^2\\ +[f''_{21}]^2[f''_{32}]^2 + [f''_{22}]^2[f''_{31}]^2\end{bmatrix}\\ & = \begin{bmatrix} [f_{11}]^2[f_{22}]^2[f_{33}]^2 + [f_{12}]^2[f_{21}]^2[f_{33}]^2\\ +[f_{11}]^2[f_{32}]^2[f_{23}]^2 + [f_{12}]^2[f_{31}]^2[f_{23}]^2\\ +[f_{21}]^2[f_{32}]^2[f_{13}]^2 + [f_{22}]^2[f_{31}]^2[f_{13}]^2\end{bmatrix}[f_{13}f_{23}f_{33}]\\ & = [f_{003}]^1\Phi(f_1, f_2, f_3)\\ & = [f_{003}]^d\Phi(f_1, f_2, f_3).\end{aligned}$$

This illustrates the identity

$$[f_{003}]^{i+d}\Phi_3(f_1, f_2, f_3) = [f_{003}]^d\Phi(f_1, f_2, f_3).$$

The expression for Φ_3 as a sum of circle products of elementary elements given by

$$\begin{aligned}\Phi_3 = (u^2)(v^2) = {} & \frac{2}{3}(u)^{(2)}(v)^{(2)} \\ & - \frac{2}{3}(u)^{(2)}(v) \circ (v) - \frac{2}{3}(u)(v)^{(2)} \circ (u) \\ & + \frac{10}{3}(u)^{(2)} \circ (v)^{(2)} - \frac{1}{3}(u)(v) \circ (u)(v) \\ & - \frac{4}{3}(u)^{(2)} \circ (v) \circ (v) - \frac{4}{3}(v)^{(2)} \circ (u) \circ (u) \\ & + \frac{4}{3}(u)(v) \circ (u) \circ (v)\end{aligned}$$

applied to the vectors (f_1, f_2, f_3) gives

$$\begin{aligned}& \Phi_3(f_1, f_2, f_3) \\ & = -\frac{2}{3}\begin{bmatrix} [f''_{11}][f''_{21}][f''_{32}] \\ +[f''_{11}][f''_{22}][f''_{12}] \\ +[f''_{12}][f''_{21}][f''_{31}] \end{bmatrix}\begin{bmatrix} [f''_{12}] \\ +[f''_{22}] \\ +[f''_{32}] \end{bmatrix} \\ & \quad - \frac{2}{3}\begin{bmatrix} [f''_{11}][f''_{22}][f''_{32}] \\ +[f''_{12}][f''_{21}][f''_{32}] \\ +[f''_{12}][f''_{22}][f''_{31}] \end{bmatrix}\begin{bmatrix} [f''_{11}] \\ +[f''_{21}] \\ +[f''_{31}] \end{bmatrix} \\ & \quad + \frac{10}{3}\begin{bmatrix} [f''_{11}][f''_{21}] \\ +[f''_{11}][f''_{31}] \\ +[f''_{21}][f''_{31}] \end{bmatrix}\begin{bmatrix} [f''_{12}][f''_{22}] \\ +[f''_{22}][f''_{21}] \\ +[f''_{12}][f''_{32}] \end{bmatrix} - \frac{1}{3}\begin{bmatrix} [f''_{11}][f''_{22}] + [f''_{12}][f''_{21}] \\ +[f''_{11}][f''_{32}] + [f''_{12}][f''_{31}] \\ +[f''_{21}][f''_{32}] + [f''_{22}][f''_{31}] \end{bmatrix}^2 \\ & \quad - \frac{4}{3}\begin{bmatrix} [f''_{11}][f''_{21}] \\ +[f''_{11}][f''_{31}] \\ +[f''_{21}][f''_{31}] \end{bmatrix}\begin{bmatrix} [f''_{12}] \\ +[f''_{22}] \\ +[f''_{32}] \end{bmatrix}^2 - \frac{4}{3}\begin{bmatrix} [f''_{12}][f''_{22}] \\ +[f''_{22}][f''_{32}] \\ +[f''_{12}][f''_{32}] \end{bmatrix}\begin{bmatrix} [f''_{11}] \\ +[f''_{21}] \\ +[f''_{31}] \end{bmatrix}^2 \\ & \quad + \frac{4}{3}\begin{bmatrix} [f''_{11}][f''_{22}] + [f''_{12}][f''_{21}] \\ +[f''_{11}][f''_{32}] + [f''_{12}][f''_{31}] \\ +[f''_{21}][f''_{32}] + [f''_{22}][f''_{31}] \end{bmatrix}\begin{bmatrix} [f''_{11}] \\ +[f'_{21}] \\ +[f''_{31}] \end{bmatrix}\begin{bmatrix} [f''_{12}] \\ +[f''_{22}] \\ +[f''_{32}] \end{bmatrix}\end{aligned}$$

so that, with $f_{003} = f_{13}f_{23}f_{33}$, we have

$$[f_{003}]^1\Phi_3(f_1, f_2, f_3) = [f_{003}]^d\Phi(f_1, f_2, f_3) = [f_{003}]^{i+d}\Phi_3(f_1, f_2, f_3)$$

$$= [f_{003}]^3\Phi_3(f_1, f_2, f_3)$$

$$= -\frac{2}{3}\left[\left[\begin{matrix}[f''_{11}][f''_{21}][f''_{32}]\\ +[f''_{11}][f''_{22}][f''_{12}]\\ +[f''_{12}][f''_{21}][f''_{31}]\end{matrix}\right][f_{003}]\right]\left[\left[\begin{matrix}[f''_{12}]\\ +[f''_{22}]\\ +[f''_{32}]\end{matrix}\right][f_{003}]\right][f_{003}]$$

$$-\frac{2}{3}\left[\left[\begin{matrix}[f''_{11}][f''_{22}][f''_{32}]\\ +[f''_{12}][f''_{21}][f''_{32}]\\ +[f''_{12}][f''_{22}][f''_{31}]\end{matrix}\right][f_{003}]\right]\left[\left[\begin{matrix}[f''_{11}]\\ +[f''_{21}]\\ +[f''_{31}]\end{matrix}\right][f_{003}]\right][f_{003}]$$

$$+\frac{10}{3}\left[\left[\begin{matrix}[f''_{11}][f''_{21}]\\ +[f''_{11}][f''_{31}]\\ +[f''_{21}][f''_{31}]\end{matrix}\right][f_{003}]\right]\left[\left[\begin{matrix}[f''_{12}][f''_{22}]\\ +[f''_{22}][f''_{21}]\\ +[f''_{12}][f''_{32}]\end{matrix}\right][f_{003}]\right][f_{003}]$$

$$-\frac{1}{3}\left[\begin{matrix}[f''_{11}][f''_{22}]+[f''_{12}][f''_{21}]\\ +[f''_{11}][f''_{32}]+[f''_{12}][f''_{31}]\\ +[f''_{21}][f''_{32}]+[f''_{22}]][f''_{31}][f_{003}]\end{matrix}\right]^2[f_{003}]$$

$$-\frac{4}{3}\left[\left[\begin{matrix}[f''_{11}][f''_{21}]\\ +[f''_{11}][f''_{31}]\\ +[f''_{21}][f''_{31}]\end{matrix}\right][f_{003}]\right]\left[\left[\begin{matrix}[f''_{12}]\\ +[f''_{22}]\\ +[f''_{32}]\end{matrix}\right][f_{003}]\right]^2$$

$$-\frac{4}{3}\left[\left[\begin{matrix}[f''_{12}][f''_{22}]\\ +[f''_{22}][f''_{32}]\\ +[f''_{12}][f''_{32}]\end{matrix}\right][f_{003}]\right]\left[\left[\begin{matrix}[f''_{11}]\\ +[f''_{21}]\\ +[f''_{31}]\end{matrix}\right][f_{003}]\right]^2$$

$$+\frac{4}{3}\left[\begin{matrix}\left[\left[\begin{matrix}[f''_{11}][f''_{22}]+[f''_{12}][f''_{21}]\\ +[f''_{11}][f''_{32}]+[f''_{12}][f''_{31}]\\ +[f''_{21}][f'_{32}]+[f''_{22}][f''_{31}]\end{matrix}\right][f_{003}]\right]\\ \left[\left[\begin{matrix}[f''_{11}]\\ +[f'_{21}]\\ +[f''_{31}]\end{matrix}\right][f_{003}]\right]\left[\left[\begin{matrix}[f''_{12}]\\ +[f''_{22}]\\ +[f''_{32}]\end{matrix}\right][f_{003}]\right]\end{matrix}\right].$$

Substituting $f_{003} = f_{13}f_{23}f_{33}$ we have the symmetric function identities

$$E_{f,a}\left[(u)^{(2)}(v)\circ(W)^{(3)}\right] = E_{f,a}\left[(U)^{(2)}(V)\right] = f_{210}$$

$$\left[\begin{array}{c} [f''_{11}][f''_{21}][f''_{32}] \\ +[f''_{11}][f''_{22}][f''_{12}] \\ +[f''_{12}][f''_{21}][f''_{31}] \end{array}\right][f_{003}] = \begin{array}{c} [f_{11}][f_{21}][f_{32}] \\ +[f_{11}][f_{22}][f_{31}] \\ +[f_{12}][f_{21}][f_{31}]. \end{array}$$

$$E_{f,a}\left[(v)\circ(W)^{(3)}\right] = E_{f,a}\left[(V)(W)^{(2)}\right] = f_{012}$$

$$\left[\begin{array}{c} [f''_{12}] \\ +[f''_{22}] \\ +[f''_{32}] \end{array}\right][f_{003}] = \begin{array}{c} [f_{12}][f_{23}][f_{33}] \\ +[f_{22}][f_{13}][f_{33}] \\ +[f_{32}][f_{13}][f_{23}]. \end{array}$$

$$E_{f,a}\left[(u)(v)^{(2)}\circ(W)^{(3)}\right] = E_{f,a}\left[(U)^{(2)}(V)\right] = f_{120}$$

$$\left[\begin{array}{c} [f''_{11}][f''_{22}][f''_{32}] \\ +[f''_{12}][f''_{21}][f''_{32}] \\ +[f''_{12}][f''_{22}][f''_{31}] \end{array}\right][f_{003}] = \begin{array}{c} [f_{11}][f_{22}][f_{32}] \\ +[f_{12}][f_{21}][f_{32}] \\ +[f_{12}][f_{22}][f_{31}]. \end{array}$$

$$E_{f,a}\left[(u)\circ(W)^{(3)}\right] = E_{f,a}\left[(U)(W)^{(2)}\right] = f_{102}$$

$$\left[\begin{array}{c} [f''_{11}] \\ +[f''_{21}] \\ +[f''_{31}] \end{array}\right][f_{003}] = \begin{array}{c} [f_{11}][f_{23}][f_{33}] \\ +[f_{21}][f_{13}][f_{33}] \\ +[f_{31}][f_{13}][f_{23}]. \end{array}$$

$$E_{f,a}\left[(u)^{(2)}\circ(W)^{(3)}\right] = E_{f,a}\left[(U)^{(2)}(W)\right] = f_{201}$$

$$\left[\begin{array}{c} [f''_{11}][f''_{21}] \\ +[f''_{11}][f''_{31}] \\ +[f''_{21}][f''_{31}] \end{array}\right][f_{003}] = \begin{array}{c} [f_{11}][f_{21}][f_{33}] \\ +[f_{11}][f_{31}][f_{23}] \\ +[f_{21}][f_{31}][f_{13}]. \end{array}$$

$$E_{f,a}\left[(v)^{(2)}\circ(W)^{(3)}\right] = E_{f,a}\left[(V)^{(2)}(W)\right] = f_{021}$$

$$\left[\begin{array}{c} [f''_{12}][f''_{22}] \\ +[f''_{22}][f''_{21}] \\ +[f''_{12}][f''_{32}] \end{array}\right][f_{003}] = \begin{array}{c} [f_{12}][f_{22}][f_{33}] \\ +[f_{22}][f_{32}][f_{13}] \\ +[f_{12}][f_{32}][f_{23}]. \end{array}$$

$$E_{f,a}\left[(u)(v)\circ(W)^{(3)}\right] = E_{f,a}\left[(U)(V)(W)\right] = f_{111}$$

$$\left[\begin{array}{c} [f''_{11}][f''_{21}][f''_{32}] \\ +[f''_{11}][f''_{22}][f''_{12}] \\ +[f''_{12}][f''_{21}][f''_{31}] \end{array}\right][f_{003}] = \begin{array}{c} [f_{11}][f_{22}][f_{33}]+[f_{12}][f_{21}][f_{33}] \\ +[f_{11}][f_{32}][f_{23}]+[f_{12}][f_{31}][f_{23}] \\ +[f_{21}][f_{32}][f_{13}]+[f_{22}][f_{31}][f_{13}]. \end{array}$$

We therefore have

$$
\begin{aligned}
&[f_{003}]^1\Phi[f_1, f_2, f_3] \\
&= -\frac{2}{3}\begin{bmatrix} [f_{11}][f_{21}][f_{32}] \\ +[f_{11}][f_{22}][f_{31}] \\ +[f_{12}][f_{21}][f_{31}] \end{bmatrix}\begin{bmatrix} [f_{12}][f_{23}][f_{33}] \\ +[f_{22}][f_{13}][f_{33}] \\ +[f_{32}][f_{13}][f_{23}] \end{bmatrix}[f_{13}f_{23}f_{33}] \\
&\quad -\frac{2}{3}\begin{bmatrix} [f_{11}][f_{22}][f_{32}] \\ +[f_{12}][f_{21}][f_{32}] \\ +[f_{12}][f_{22}][f_{31}] \end{bmatrix}\begin{bmatrix} [f_{11}][f_{23}][f_{33}] \\ +[f_{21}][f_{13}][f_{33}] \\ +[f_{31}][f_{13}][f_{23}] \end{bmatrix}[f_{13}f_{23}f_{33}] \\
&\quad +\frac{10}{3}\begin{bmatrix} [f_{11}][f_{21}][f_{33}] \\ +[f_{11}][f_{31}][f_{23}] \\ +[f_{21}][f_{31}][f_{13}] \end{bmatrix}\begin{bmatrix} [f_{12}][f_{22}][f_{33}] \\ +[f_{22}][f_{32}][f_{13}] \\ +[f_{12}][f_{32}][f_{23}] \end{bmatrix}[f_{13}f_{23}f_{33}] \\
&\quad -\frac{1}{3}\begin{bmatrix} [f_{11}][f_{22}][f_{33}] + [f_{12}][f_{21}][f_{33}] \\ +[f_{11}][f_{32}][f_{23}] + [f_{12}][f_{31}][f_{23}] \\ +[f_{21}][f_{32}][f_{13}] + [f_{22}][f_{31}][f_{13}] \end{bmatrix}^2 [f_{13}f_{23}f_{33}] \\
&\quad -\frac{4}{3}\begin{bmatrix} \begin{bmatrix} [f_{11}][f_{21}][f_{32}] \\ +[f_{11}][f_{22}][f_{31}] \\ +[f_{12}][f_{21}][f_{31}] \end{bmatrix}\begin{bmatrix} [f_{12}][f_{23}][f_{33}] \\ +[f_{22}][f_{13}][f_{33}] \\ +[f_{32}][f_{13}][f_{23}] \end{bmatrix} \end{bmatrix}^2 \\
&\quad -\frac{4}{3}\begin{bmatrix} \begin{bmatrix} [f_{11}][f_{22}][f_{32}] \\ +[f_{12}][f_{21}][f_{32}] \\ +[f_{12}][f_{22}][f_{31}] \end{bmatrix}\begin{bmatrix} [f_{11}][f_{23}][f_{33}] \\ +[f_{21}][f_{13}][f_{33}] \\ +[f_{31}][f_{13}][f_{23}] \end{bmatrix} \end{bmatrix}^2 \\
&\quad +\frac{4}{3}\begin{bmatrix} \begin{bmatrix} [f_{11}][f_{22}][f_{33}] + [f_{12}][f_{21}][f_{33}] \\ +[f_{11}][f_{32}][f_{33}] + [f_{12}][f_{31}][f_{23}] \\ +[f_{21}][f_{32}][f_{33}] + [f_{22}][f_{31}][f_{13}] \end{bmatrix} \\ \begin{bmatrix} [f_{11}][f_{23}][f_{33}] \\ +[f_{21}][f_{13}][f_{33}] \\ +[f_{31}][f_{13}][f_{23}] \end{bmatrix}\begin{bmatrix} [f_{12}][f_{23}][f_{33}] \\ +[f_{22}][f_{13}][f_{33}] \\ +[f_{32}][f_{13}][f_{23}] \end{bmatrix} \end{bmatrix}.
\end{aligned}
$$

Hence we have that

$$[f_{003}]^1\Phi(f_1, f_2, f_3)$$

equals

$$-\frac{2}{3}[f_{210}][f_{012}][f_{003}]-\frac{2}{3}[f_{120}][f_{102}][f_{003}]$$
$$+\frac{10}{3}[f_{201}][f_{021}][f_{003}]-\frac{1}{3}[f_{111}]^2[f_{003}]$$
$$-\frac{4}{3}[f_{210}][f_{012}]^2-\frac{4}{3}[f_{201}][f_{102}]^2$$
$$+\frac{4}{3}[f_{111}][f_{102}][f_{012}]$$

and this equals the value at

$$f(X)=\sum_{i+j+k=3} f_{ijk}x_1^i x_2^j x_3^k$$
$$=(f_1|X)(f_2|X)(f_3|X)$$

of the plethystic expression for $(W)^{(3)}\circ\Phi$ given by

$$-\frac{2}{3}(U)^{(2)}(V)\circ(V)(W)^{(2)}\circ(W)^{(3)}-\frac{2}{3}(U)(V)^{(2)}\circ(U)(W)^{(2)}\circ(W)^{(3)}$$
$$+\frac{10}{3}(U)^{(2)}(W)\circ(V)^{(2)}(W)\circ(W)^{(3)}-\frac{1}{3}(U)(V)(W)\circ(U)(V)(W)\circ(W)^{(3)}$$
$$-\frac{4}{3}(U)^{(2)}(W)\circ(V)(W)^{(2)}\circ(V)(W)^{(2)}$$
$$-\frac{4}{3}(V)^{(2)}(W)\circ(U)(W)^{(2)}\circ(U)(W)^{(2)}$$
$$+\frac{4}{3}(U)(V)(W)\circ(U)(W)^{(2)}\circ(V)(W)^{(2)}.$$

We have now explicitly justified the previous expansion for

$$(W)^{(3)}\circ\Phi=(W)^{(3)}\circ(U^2)(V^2)(W^2).$$

13 Extensionals

The expansion

$$\begin{aligned}\Phi_3 &= (u^2)(v^2)\\ &= \frac{2}{3}(u)^{(2)}(v)^{(2)}-\frac{2}{3}(u)^{(2)}(v)\circ(v)-\frac{2}{3}(u)(v)^{(2)}\circ(u)\\ &\quad+\frac{10}{3}(u)^{(2)}\circ(v)^{(2)}-\frac{1}{3}(u)(v)\circ(u)(v)-\frac{4}{3}(u)^{(2)}\circ(v)\circ(v)\\ &\quad-\frac{4}{3}(v)^{(2)}\circ(u)\circ(u)+\frac{4}{3}(u)(v)\circ(u)\circ(v)\end{aligned}$$

applied to the *quartic* Cayley form

$$\begin{aligned} q(X) &= \sum_{i+j+k=4} q_{ijk} x_1^i x_2^j x_3^k \\ &= (q_1|X)(q_2|X)(q_3|X)(q_4|X) \end{aligned}$$

yields additional identities. In this case consider the evaluation for $q(X)$ of the monomial function

$$\Phi = (U^2)(V^2)(W^2)^{(2)}$$

and write

$$\Phi = \left[(W)^{(4)}\right]^2 \circ \Phi_3$$

for which we again have $\Phi_3 = (u^2)(v^2)$.

The monomial function Φ_3 is of inner degree $i = 2$, while the expansion by circle products of elementary monomials is of degree $i + d = 2 + 1 = 3$ with respect to the circle product. We conclude that in $\mathrm{Pleth}_4(P_a)$ we have

$$\begin{aligned} (W)^{(4)} \circ \Phi = \ & (W)^{(4)} \circ (U^2)(V^2)(W^2)^{(2)} \\ = \ & \frac{2}{3}(U)^{(2)}(V)^{(2)} \circ (W)^{(4)} \circ (W)^{(4)} \\ & - \frac{2}{3}(U)^{(2)}(V)(W) \circ (V)(W)^{(3)} \circ (W)^{(4)} \\ & - \frac{2}{3}(U)(V)^{(2)}(W) \circ (U)(W)^{(3)} \circ (W)^{(4)} \\ & + \frac{10}{3}(U)^{(2)}(W)^{(2)} \circ (V)^{(2)}(W)^{(2)} \circ (W)^{(4)} \\ & - \frac{1}{3}(U)(V)(W)^{(2)} \circ (U)(V)(W)^{(2)} \circ (W)^{(4)} \\ & - \frac{4}{3}(U)^{(2)}(W)^{(2)} \circ (V)(W)^{(3)} \circ (V)(W)^{(3)} \\ & - \frac{4}{3}(V)^{(2)}(W)^{(2)} \circ (U)(W)^{(3)} \circ (U)(W)^{(3)} \\ & + \frac{4}{3}(U)(V)(W)^{(2)} \circ (U)(W)^{(3)} \circ (V)(W)^{(3)} \end{aligned}$$

from which the evaluation may be concluded. Since all terms of the expansion of $(u^2)(v^2)$ by elementary monomials have been employed in this expansion, we may obtain without further work analogous expansions, called *extensionals*, by adding higher powers of (W^2) to the basic function Φ. Thus, we may consider the expansion in $\mathrm{Pleth}_{2+h}(P_a)$ for $h \geq 2$ of the monomial function

$$\Phi = (U^2)(V^2)(W^2)^{(h)}$$

and we will find that

$$
\begin{aligned}
(W)^{(2+h)} \circ \Phi = {} & (W)^{(2+h)} \circ (U^2)(V^2)(W^2)^{(h)} \\
= {} & \frac{2}{3}(U)^{(2)}(V)^{(2)}(W)^{(h-2)} \circ \left[(W)^{(2+h)}\right]^2 \\
& - \frac{2}{3}(U)^{(2)}(V)(W)^{(h-1)} \circ (V)(W)^{(h+1)} \circ (W)^{(2+h)} \\
& - \frac{2}{3}(U)(V)^{(2)}(W)^{(h-1)} \circ (U)(W)^{(h+1)} \circ (W)^{(2+h)} \\
& + \frac{10}{3}(U)^{(2)}(W)^{(h)} \circ (V)^{(2)}(W)^{(h)} \circ (W)^{(2+h)} \\
& - \frac{1}{3}(U)(V)(W)^{(h)} \circ (U)(V)(W)^{(h)} \circ (W)^{(2+h)} \\
& - \frac{4}{3}(U)^{(2)}(W)^{(h)} \circ (V)(W)^{(h+1)} \circ (V)(W)^{(h+1)} \\
& - \frac{4}{3}(V)^{(2)}(W)^{(h)} \circ (U)(W)^{(h+1)} \circ (U)(W)^{(h+1)} \\
& + \frac{4}{3}(U)(V)(W)^{(h)} \circ (U)(W)^{(h+1)} \circ (V)(W)^{(h+1)}.
\end{aligned}
$$

14 Symmetric functions of the system of solutions to $n-1$ homogeneous equations in n variables

Suppose now that

$$f(x) = f(x_1, \cdots, x_n) = \sum_{i_1+\cdots+i_n=r} f_{i_1\cdots i_n} x_1^{i_1} \cdots x_n^{i_n} = (f_1|x)(f_2|x)\cdots(f_r|x)$$

where the vectors $f_1, f_2, \cdots, f_r$ are the simultaneous solutions to a set of $n-1$ homogeneous equations $\Phi^j(x) = 0$ in n variables, where $1 \le j \le n-1$. We do not know the vectors $f_1, f_2, \cdots, f_r$ explicitly, only that their homogeneous elementary symmetric functions of shape (1^r) are the coefficients of $f(x)$. These coefficients may be computed explicitly by the techniques introduced in Part II of this paper. The function $f(x)$ is known as the *eliminant* of the given equations. The coefficients $f_{i_1\cdots i_n}$ are polynomials in the coefficients $\Phi^j_{i_1\cdots i_n}$ of the underlying equations. Our next goal is to prove the following theorem.

Theorem 5. *The evaluation operator $E_{f,a}$ for an eliminant $f(x)$ has the additional property that the value $E_{f,a}[\Phi]$ at any homogeneous monomial element Φ of shape (i^r) in* Pleth$_r(P_a)$ *can be expressed as a polynomial in the coefficients of the underlying equations.*

Proof. Recall the identities

$$\left(E_{f,a}\left[w_{r0\cdots0}\right]\right)^d E_{f,a}\left[\Phi\right] = E_{f,a}\left[\Psi(\cdots w_{i_1\cdots i_n}\cdots)\right]$$

which we have written in the form

$$\begin{aligned}[f_{r0\cdots0}]^d\, E_{f,a}(\Phi) &= [f_{r0\cdots0}]^{i+d}\, E_{f,a}(\Phi_1) \\ &= [f_{r0\cdots0}]^{i+d}\, \Psi'(\cdots f'_{j_2\cdots j_n}\cdots) \\ &= \Psi(\cdots f_{i_1\cdots i_n}\cdots),\end{aligned}$$

where, as previously,

$$E_{f,a}[w_{i_1\cdots i_n}] = f_{i_1\cdots i_n}.$$

We shall write instead

$$[f_{r0\cdots0}]^{d_1}\, E_{f,a}(\Phi) = \Psi_1(\cdots f_{i_1\cdots i_n}\cdots),$$

where we write Ψ_1 for Ψ and d_1 for d to emphasize the special role of $[a|u_1]$. Using $[a|u_n]$ instead of $[a|u_1]$ in the special role, we would arrive at a similar identity

$$[f_{0\cdots0r}]^{d_n}\, E_{f,a}(\Phi) = \Psi_n(\cdots f_{i_1\cdots i_n}\cdots)$$

and we may conclude that

$$[f_{r0\cdots0}]^{d_1}\, \Psi_n(\cdots f_{i_1\cdots i_n}\cdots) = [f_{0\cdots0r}]^{d_n}\, \Psi_1(\cdots f_{i_1\cdots i_n}\cdots).$$

The coefficient $f_{0\cdots0r}$ is the resultant of the $n-1$ homogeneous equations obtained from the original equations by setting $x_n = 0$, while the coefficient $f_{r0\cdots0}$ is the resultant of the $n-1$ homogeneous equations obtained from the original equations by setting $x_1 = 0$. When the underlying forms $\Phi^1, \cdots, \Phi^{n-1}$ are generic, the coefficients $f_{0\cdots0r}$ and $f_{r0\cdots0}$ are irreducible polynomials in the coefficients of these equations. Hence we may conclude that after the coefficients $f_{i_1\cdots i_n}$ are replaced by their values as polynomials in the coefficients $\Phi^j_{i_1\cdots i_n}$ of the underlying system of $n-1$ homogeneous equations the polynomial $\Psi_1(\cdots f_{i_1\cdots i_n}\cdots)$ must be divisible by $[f_{r0\cdots0}]^{d_1}$ and similarly the polynomial $\Psi_n(\cdots f_{i_1\cdots i_n}\cdots)$ must be divisible by $[f_{0\cdots0r}]^{d_n}$. Thus the evaluation operator, when expressed in terms of the coefficients of the underlying equations, yields a polynomial expression for any homogeneous monomial element of shape (i^r). □

15 An example: ternary cubic eliminants

We have obtained for $\Phi = (U^2)(V^2)(W^2)$ the expansion in $\mathrm{Pleth}_3(P_a)$

$$\begin{aligned}
(W)^{(3)} \circ \Phi = & \ (W)^{(3)} \circ (U^2)(V^2)(W^2) \\
= & -\frac{2}{3}(U)^{(2)}(V) \circ (V)(W)^{(2)} \circ (W)^{(3)} \\
& -\frac{2}{3}(U)(V)^{(2)} \circ (U)(W)^{(2)} \circ (W)^{(3)} \\
& +\frac{10}{3}(U)^{(2)}(W) \circ (V)^{(2)}(W) \circ (W)^{(3)} \\
& -\frac{1}{3}(U)(V)(W) \circ (U)(V)(W) \circ (W)^{(3)} \\
& -\frac{4}{3}(U)^{(2)}(W) \circ (V)(W)^{(2)} \circ (V)(W)^{(2)} \\
& -\frac{4}{3}(V)^{(2)}(W) \circ (U)(W)^{(2)} \circ (U)(W)^{(2)} \\
& +\frac{4}{3}(U)(V)(W) \circ (U)(W)^{(2)} \circ (V)(W)^{(2)}.
\end{aligned}$$

We may similarly conclude that

$$\begin{aligned}
(V)^{(3)} \circ \Phi = & \ (V)^{(3)} \circ (U^2)(V^2)(W^2) \\
= & -\frac{2}{3}(U)^{(2)}(W) \circ (W)(V)^{(2)} \circ (V)^{(3)} \\
& -\frac{2}{3}(U)(W)^{(2)} \circ (U)(V)^{(2)} \circ (V)^{(3)} \\
& +\frac{10}{3}(U)^{(2)}(V) \circ (W)^{(2)}(V) \circ (V)^{(3)} \\
& -\frac{1}{3}(U)(W)(V) \circ (U)(W)(V) \circ (V)^{(3)} \\
& -\frac{4}{3}(U)^{(2)}(V) \circ (W)(V)^{(2)} \circ (W)(V)^{(2)} \\
& -\frac{4}{3}(W)^{(2)}(V) \circ (U)(V)^{(2)} \circ (U)(V)^{(2)} \\
& +\frac{4}{3}(U)(W)(V) \circ (U)(V)^{(2)} \circ (W)(V)^{(2)}.
\end{aligned}$$

We therefore consider the expansion of

$$(V)^{(3)} \circ (U^2)(V^2)(W^2) \circ (W)^{(3)}$$

and by associativity we have

$$\begin{aligned}
\big[(V)^{(3)} \circ & \quad (U^2)(V^2)(W^2)\big] \circ (W)^{(3)} \\
= & \quad (V)^{(3)} \circ \Big[(U^2)(V^2)(W^2) \circ (W)^{(3)}\Big]
\end{aligned}$$

and therefore

$$\begin{bmatrix} -\frac{2}{3}(U)^{(2)}(W)\circ(W)(V)^{(2)}\circ(V)^{(3)} \\ -\frac{2}{3}(U)(W)^{(2)}\circ(U)(V)^{(2)}\circ(V)^{(3)} \\ +\frac{10}{3}(U)^{(2)}(V)\circ(W)^{(2)}(V)\circ(V)^{(3)} \\ -\frac{1}{3}(U)(W)(V)\circ(U)(W)(V)\circ(V)^{(3)} \\ -\frac{4}{3}(U)^{(2)}(V)\circ(W)(V)^{(2)}\circ(W)(V)^{(2)} \\ -\frac{4}{3}(W)^{(2)}(V)\circ(U)(V)^{(2)}\circ(U)(V)^{(2)} \\ +\frac{4}{3}(U)(W)(V)\circ(U)(V)^{(2)}\circ(W)(V)^{(2)} \end{bmatrix} \circ (W)^{(3)}$$

equals

$$(V)^{(3)}\circ \begin{bmatrix} -\frac{2}{3}(U)^{(2)}(V)\circ(V)(W)^{(2)}\circ(W)^{(3)} \\ -\frac{2}{3}(U)(V)^{(2)}\circ(U)(W)^{(2)}\circ(W)^{(3)} \\ +\frac{10}{3}(U)^{(2)}(W)\circ(V)^{(2)}(W)\circ(W)^{(3)} \\ -\frac{1}{3}(U)(V)(W)\circ(U)(V)(W)\circ(W)^{(3)} \\ -\frac{4}{3}(U)^{(2)}(W)\circ(V)(W)^{(2)}\circ(V)(W)^{(2)} \\ -\frac{4}{3}(V)^{(2)}(W)\circ(U)(W)^{(2)}\circ(U)(W)^{(2)} \\ +\frac{4}{3}(U)(W)(V)\circ(U)(W)^{(2)}\circ(V)(W)^{(2)} \end{bmatrix}.$$

Evaluating this identity with $E_{f,a}$ for the ternary cubic Cayley form

$$f(X) = \sum_{i+j+k=3} f_{ijk}x_1^i x_2^j x_3^k = (f_1|X)(f_2|X)(f_3|X)$$

gives

$$\begin{bmatrix} -\frac{2}{3}f_{201}f_{021}f_{030} \\ -\frac{2}{3}f_{102}f_{120}f_{030} \\ +\frac{10}{3}f_{201}f_{012}f_{012} \\ -\frac{1}{3}f_{111}f_{111}f_{030} \\ -\frac{4}{3}f_{201}f_{102}f_{102} \\ -\frac{4}{3}f_{021}f_{120}f_{120} \\ +\frac{4}{3}f_{111}f_{102}f_{012} \end{bmatrix} f_{003} = f_{030} \begin{bmatrix} -\frac{2}{3}f_{210}f_{012}f_{003} \\ -\frac{2}{3}f_{120}f_{102}f_{003} \\ +\frac{10}{3}f_{201}f_{021}f_{003} \\ -\frac{1}{3}f_{111}f_{111}f_{003} \\ -\frac{4}{3}f_{201}f_{012}f_{012} \\ -\frac{4}{3}f_{021}f_{102}f_{102} \\ +\frac{4}{3}f_{111}f_{102}f_{012} \end{bmatrix}.$$

Suppose now that

$$\begin{aligned} f(X) &= \sum_{i_1+i_2+i_3=3} f_{i_1 i_2 i_3} x_1^{i_1} x_2^{i_2} x_3^{i_3} \\ &= (f_1|X)(f_2|X)(f_3|X) \end{aligned}$$

is the eliminant of two underlying ternary equations. Since the degree of an eliminant is the product of the degrees of the underlying equations, these equations are necessarily of degree 3 and degree 1. If the underlying equations are

$$\begin{aligned} \Phi(X) &= \sum_{i+j+k=3} \phi_{ijk} x_1^i x_2^j x_3^k = 0, \\ b(X) &= \sum_{m=1}^{n} [b|u_m] x_m = 0, \end{aligned}$$

the eliminant is given explicitly by

$$\begin{aligned} f(a) &= \sum_{i_1+i_2+i_3=3} f_{i_1 i_2 i_3} [a|u_1]^{i_1} [a|u_2]^{i_2} [au_3]^{i_3} \\ &= \Phi([ab|u_2 u_3], [ab|u_3 u_1], [ab|u_1 u_2]) \\ &= f_1(a) f_2(a) f_3(a). \end{aligned}$$

Recalling that the polynomial expressions for f_{003} and f_{030} are then irreducible polynomials in the coefficients ϕ_{ijk} and $[b|u_m]$ of the underlying equations, it follows that, after substitution of the polynomial expressions in the ϕ_{ijk} and $[b|u_m]$ for the coefficients $f_{i_1 i_2 i_3}$, we must have that

$$\begin{bmatrix} -\frac{2}{3} f_{210} f_{012} f_{003} - \frac{2}{3} f_{120} f_{102} f_{003} \\ +\frac{10}{3} f_{201} f_{021} f_{003} - \frac{1}{3} f_{111} f_{111} f_{003} \\ -\frac{4}{3} f_{201} f_{012} f_{012} - \frac{4}{3} f_{021} f_{102} f_{102} \\ +\frac{4}{3} f_{111} f_{102} f_{012} \end{bmatrix}$$

is divisible by f_{003} and that

$$\begin{bmatrix} -\frac{2}{3} f_{201} f_{021} f_{030} - \frac{2}{3} f_{102} f_{120} f_{030} \\ +\frac{10}{3} f_{201} f_{012} f_{012} - \frac{1}{3} f_{111} f_{111} f_{030} \\ -\frac{4}{3} f_{210} f_{021} f_{021} - \frac{4}{3} f_{012} f_{120} f_{120} \\ +\frac{4}{3} f_{111} f_{120} f_{021} \end{bmatrix}$$

is divisible by f_{030}. We shall see below that the common value of these expressions, first obtained by Cayley, is

$$-\frac{2}{3} f_{120} f_{102} - \frac{2}{3} f_{012} f_{210} - \frac{2}{3} f_{201} f_{021} + \frac{2}{3} f_{111} f_{111}.$$

In other words, the coefficients $f_{i_1 i_2 i_3}$ of the eliminant satisfy the additional relations, not deducible from the fact that $f(X)$ is a Cayley form,

$$\begin{bmatrix} -\frac{2}{3} f_{210} f_{012} f_{003} - \frac{2}{3} f_{120} f_{102} f_{003} \\ +\frac{10}{3} f_{201} f_{021} f_{003} - \frac{1}{3} f_{111} f_{111} f_{003} \\ -\frac{4}{3} f_{201} f_{012} f_{012} - \frac{4}{3} f_{021} f_{102} f_{102} \\ +\frac{4}{3} f_{111} f_{102} f_{012} \end{bmatrix}$$

$$= \begin{bmatrix} -\frac{2}{3} f_{120} f_{102} - \frac{2}{3} f_{012} f_{210} \\ -\frac{2}{3} f_{201} f_{021} + \frac{2}{3} f_{111} f_{111} \end{bmatrix} f_{003}$$

and

$$\begin{bmatrix} -\frac{2}{3} f_{201} f_{021} f_{030} - \frac{2}{3} f_{102} f_{120} f_{030} \\ +\frac{10}{3} f_{201} f_{012} f_{012} - \frac{1}{3} f_{111} f_{111} f_{030} \\ -\frac{4}{3} f_{210} f_{021} f_{021} - \frac{4}{3} f_{012} f_{120} f_{120} \\ +\frac{4}{3} f_{111} f_{120} f_{021} \end{bmatrix}$$

$$= \begin{bmatrix} -\frac{2}{3} f_{120} f_{102} - \frac{2}{3} f_{012} f_{210} \\ -\frac{2}{3} f_{201} f_{021} + \frac{2}{3} f_{111} f_{111} \end{bmatrix} f_{030},$$

and these relations are a consequence of the additional fact that the vectors f_1, f_2, f_3 satisfy a common linear relation

$$b(f_1) = 0, b(f_2) = 0, b(f_3) = 0.$$

In this case, as there are sufficiently strong relations among the coefficients $f_{i_1 i_2 i_3}$ of the eliminant themselves to effect the reduction, we do not require any equations relating the coefficients $f_{i_1 i_2 i_3}$ of the eliminant to the coefficients ϕ_{ijk} and $[b|u_m]$ of the underlying equations.

16 Quadratic Cayley forms

We will show next that in the case of an arbitrary n-ary Cayley form $f(X)$ of degree *two*, not necessarily an eliminant, all the monomial symmetric functions of the linear factors can be expressed as *polynomials* in the coefficients of the form.

Theorem 6. *In the case of a quadratic Cayley form,*

$$f(x) = \sum_{i_1+\cdots+i_n=2} f_{i_1 \cdots i_n} x_1^{i_1} \cdots x_n^{i_n} = (a|x)(b|x),$$

the value of the evaluation operator

$$E_{f.a} : \mathrm{Pleth}_2(P_a) \to Q(f)$$

at any monomial symmetic function can be expressed as a polynomial in the coefficients $f_{i_1 \cdots i_n}$.

Proof. The theorem follows from the identity in $\mathrm{Pleth}_2(P_a)$

$$\begin{aligned}([a|u_1]^2)([a|u_2]^2) &= ([a|u_1])([a|u_2]) \circ ([a|u_1])([a|u_2]) \\ &\quad -2([a|u_1])^{(2)} \circ ([a|u_2])^{(2)}\end{aligned}$$

and the recursion ($d \geq 2$) in $\mathrm{Pleth}_2(P_a)$

$$\begin{aligned}&([a|u_1]^{d+1})([a|u_2]^{d+1}) \\ &= ([a|u_1]^{d})([a|u_2]^{d}) \circ ([a|u_1])([a|u_2]) \\ &\quad - ([a|u_1]^{d-1})([a|u_2]^{d-1}) \circ ([a|u_1])^{(2)} \circ ([a|u_2])^{(2)}\end{aligned}$$

from which explicit formulas can be obtained for all monomial functions by polarization. □

17 Cubic eliminants

It is often the case that the symmetric functions of the linear factors of an eliminant $f(x)$ can be expressed as *polynomials* in the coefficients of the eliminant, without recourse to the expressions for these coefficients as polynomials in the coefficients of the underlying equations.

By way of example, we consider the case of an n-ary cubic eliminant, first studied by Cayley in the ternary case,

$$f(x) = \sum_{i_1+\cdots+i_n=3} f_{i_1\cdots i_n} x_1^{i_1} \cdots x_n^{i_n} = (f_1|x)(f_2|x)(f_3|x).$$

Observe again that $f(x)$ must be the eliminant of one cubic equation and $n-2$ linear equations. Since the vectors f_1, f_2, f_3 satisfy $n-2$ common linear equations, all determinants of order three $[f_1 f_2 f_3 | u_i u_j u_k]$ must vanish. Although the determinants $[f_1 f_2 f_3 | u_i u_j u_k]$ are antisymmetric functions of the vectors f_1, f_2, f_3, we may obtain from them other functions which are symmetric in the vectors f_1, f_2, f_3 such as $[f_1 f_2 f_3 | u_i u_j u_k]^2$ and $[f_1 f_2 f_3 | u_{i_1} u_{j_1} u_{k_1}][f_1 f_2 f_3 | u_{i_2} u_{j_2} u_{k_2}]$. We find that

$$\begin{aligned}&[f_1 f_2 f_3 | u_i u_j u_k]^2 \\ &= \sum_\theta (-1)^\theta (f_1|u_{\theta i})(f_2|u_{\theta j})(f_3|u_{\theta k}) \sum_\tau (-1)^\tau (f_1|u_{\tau i})(f_2|u_{\tau j})(f_3|u_{\tau k}) \\ &= \sum_\tau (-1)^\tau \sum_\theta (-1)^\theta (f_1|u_{\theta i})(f_1|u_{\tau i})(f_2|u_{\theta j})(f_2|u_{\tau j})(f_3|u_{\theta k})(f_3|u_{\tau k})\end{aligned}$$

$$= \sum_{\tau}(-1)^{\tau}\sum_{\theta}(-1)^{\theta}(f_1^{(2)}|u_{\theta i}u_{\tau i})(f_2^{(2)}|u_{\theta j}u_{\tau j})(f_3^{(2)}|u_{\theta k}u_{\tau k})$$
$$= \sum_{\tau}\sum_{\theta}(-1)^{\theta}(f_{\tau 1}^{(2)}|u_{\theta i}u_i)(f_{\tau 2}^{(2)}|u_{\theta j}u_j)(f_{\tau 3}^{(2)}|u_{\theta k}u_k)$$
$$= \Big[[a|u_i][a|u_j][a|u_k]\Big|[a|u_i][a|u_j][a|u_k]\Big](f_1, f_2, f_3).$$

Here

$$\Big[[a|u_i][a|u_j][a|u_k]\Big|[a|u_i][a|u_j][a|u_k]\Big]$$

denotes the monomial determinant

$$\sum_{\theta}(-1)^{\theta}([a|u_i][a|u_{\theta i}])([a|u_j][a|u_{\theta j}])([a|u_k][a|u_{\theta k}]).$$

We would similarly find that

$$[f_1f_2f_3|u_{i_1}u_{j_1}u_{k_1}][f_1f_2f_3|u_{i_2}u_{j_2}u_{k_2}]$$
$$= \Big[[a|u_{i_1}][a|u_{j_1}][a|u_{k_1}]\Big|[a|u_{i_2}][a|u_{j_2}][a|u_{k_2}]\Big](f_1, f_2, f_3).$$

Observe that for n-ary forms in $\mathrm{Pleth}_3(P_a)$ with

$$P_a = \{[a|u_1] < \cdots < [a|u_n]\}$$

we have

$$\begin{pmatrix} ([a|u_1])([a|u_2])^{(2)} \circ ([a|u_1])([a|u_3])^{(2)} \\ ([a|u_2])([a|u_3])^{(2)} \circ ([a|u_2])([a|u_1])^{(2)} \\ ([a|u_3])([a|u_1])^{(2)} \circ ([a|u_3])([a|u_2])^{(2)} \\ ([a|u_1])([a|u_2])([a|u_3]) \circ ([a|u_1])([a|u_2])([a|u_3]) \\ [[a|u_1][a|u_2][a|u_3]|[a|u_1][a|u_2][a|u_3]] \end{pmatrix}$$
$$= \begin{pmatrix} 1 & 0 & 0 & 0 & 1 \\ 0 & 1 & 0 & 0 & 1 \\ 0 & 0 & 1 & 0 & 1 \\ 2 & 2 & 2 & 1 & 2 \\ -2 & -2 & -2 & 1 & 2 \end{pmatrix}$$
$$\begin{pmatrix} ([a|u_1]^2)([a|u_2][a|u_3])^{(2)} \\ ([a|u_2]^2)([a|u_3][a|u_1])^{(2)} \\ ([a|u_3][a|u_1])([a|u_1][a|u_2])^{(2)} \\ ([a|u_1]^2)([a|u_2]^2)([a|u_3]^2) \\ ([a|u_1][a|u_2])([a|u_2][a|u_3])([a|u_3][a|u_1]) \end{pmatrix}$$

which can be inverted to give

$$\begin{pmatrix} ([a|u_1]^2)([a|u_2][a|u_3])^{(2)} \\ ([a|u_2]^2)([a|u_3][a|u_1])^{(2)} \\ ([a|u_3][a|u_1])([a|u_1][a|u_2])^{(2)} \\ ([a|u_1]^2)([a|u_2]^2)([a|u_3]^2) \\ ([a|u_1][a|u_2])([a|u_2][a|u_3])([a|u_3][a|u_1]) \end{pmatrix}$$

$$= \frac{1}{12}\begin{pmatrix} 8 & -4 & -4 & 1 & -1 \\ -4 & 8 & -4 & 1 & -1 \\ -4 & -4 & 8 & 1 & -1 \\ -8 & -8 & -8 & 8 & 4 \\ 4 & 4 & 4 & -1 & 1 \end{pmatrix}$$

$$\begin{pmatrix} ([a|u_1])([a|u_2])^{(2)} \circ ([a|u_1])([a|u_3])^{(2)} \\ ([a|u_2])([a|u_3])^{(2)} \circ ([a|u_2])([a|u_1])^{(2)} \\ ([a|u_3])([a|u_1])^{(2)} \circ ([a|u_3])([a|u_2])^{(2)} \\ ([a|u_1])([a|u_2])([a|u_3]) \circ ([a|u_1])([a|u_2])([a|u_3]) \\ [[a|u_1][a|u_2][a|u_3] | [a|u_1][a|u_2][a|u_3]] \end{pmatrix}$$

where the monomial determinant satisfies

$$\Big[[a|u_1][a|u_2][a|u_3]\Big|[a|u_1][a|u_2][a|u_3]\Big](f_1, f_2, f_3) = 0$$

since the vectors f_1, f_2, f_3 satisfy $n-2$ common linear relations. In particular, the equation for

$$([a|u_1]^2)([a|u_2]^2)([a|u_3]^2)$$

when evaluated by $E_{f,a}$ gives Cayley's formula

$$\begin{aligned} &E_{f,a}([a|u_1]^2)([a|u_2]^2)([a|u_3]^2) \\ &\quad = -\frac{2}{3} f_{120} f_{102} - \frac{2}{3} f_{012} f_{210} - \frac{2}{3} f_{201} f_{021} + \frac{2}{3} f_{111} f_{111}. \end{aligned}$$

The plethystic identity

$$\begin{aligned} &([a|u_1]^2)([a|u_2]^2)([a|u_3]^2) \\ = \quad & -\frac{2}{3}([a|u_1])([a|u_2])^{(2)} \circ ([a|u_1])([a|u_3])^{(2)} \\ & -\frac{2}{3}([a|u_2]^2)([a|u_3][a|u_1])^{(2)} \circ ([a|u_2])([a|u_1])^{(2)} \end{aligned}$$

$$-\frac{2}{3}([a|u_3])([a|u_1])^{(2)} \circ ([a|u_3])([a|u_2])^{(2)}$$
$$+\frac{2}{3}([a|u_1])([a|u_2])([a|u_3]) \circ ([a|u_1])([a|u_2])([a|u_3])$$
$$+\frac{1}{3}\Big[[a|u_1][a|u_2][a|u_3]\Big|[a|u_1][a|u_2][a|u_3]\Big]$$

when polarized gives the expression for an arbitrary monomial element of shape (2^3) as a linear combination of circle products of elementary monomials, plus a sum of terms obtained from polarizing

$$\Big[[a|u_1][a|u_2][a|u_3]\Big|[a|u_1][a|u_2][a|u_3]\Big]$$

which evaluates to zero for the system of vectors f_1, f_2, f_3.

Can these computations be extended to give expressions for the monomial elements $([a|u_1]^i)([a|u_2]^i)([a|u_3]^i)$ as polynomials in the coefficients of the cubic eliminant? We believe so.

Conjecture. We are led to conjecture that the value

$$E_{f,a}\,[(W_1)(W_2)\cdots(W_r)]$$

of the evaluation operator $E_{f,a}$ for the eliminant

$$\begin{aligned} f(a) &= \sum_{i_1+\cdots+i_n=r} f_{i_1\cdots i_n}[a|u_1]^{i_1}\cdots[a|u_n]^{i_n} \\ &= \Phi((-1)^0\,[aa_2\cdots a_{n-1}|u_2\cdots u_n], \\ &\quad \cdots,(-1)^{n-1}\,[aa_2\cdots a_{n-1}|u_1\cdots u_{n-1}]) \\ &= f_1(a)\cdots f_r(a). \end{aligned}$$

of one arbitrary n-ary form $\Phi(x)$ of degree r and $n-2$ linear forms $a_2(x),\cdots,$ $a_{n-1}(x)$ at an arbitrary monomial element

$$W = (W_1)(W_2)\cdots(W_r)$$

of $\mathrm{Pleth}_r(P_a)$ can be expressed as a polynomial in the coefficients $f_{i_1\cdots i_n}$ of the eliminant.

18 Polarization techniques

We wish to illustrate how the expansion as a linear combination of circle products of elementary monomials of a typical monomial element in $\mathrm{Pleth}_r(P_a)$ can be obtained from the expansion as a linear combination of circle products of elementary monomials of the special monomial elements

$$([a|u_1]^i)\cdots([a|u_r]^i)$$

in $\mathrm{Pleth}_r(P''_a)$ where, for $n \geq r$, the alphabet P''_a is given by

$$\{u_1 < u_{11} < \cdots < u_{1n} < \cdots < u_n < u_{n1} < \cdots < u_{nn}\} \ (n \geq r)$$

and, for $n < r$ the alphabet P''_a is given by

$$\{u_1 < u_{11} < \cdots < u_{1n} < \cdots < u_r < u_{r1} < \cdots < u_{rn}\} \ (n < r).$$

The plethystic algebra $\mathrm{Pleth}_r(P''_a)$ now contains the special monomial element of shape (i^r)

$$\Phi^0 = ([a|u_1]^i)([a|u_2]^i)\cdots([a|u_r]^i)$$

with

$$\Phi^0_1 = (\frac{[a|u_2]^i}{[a|u_1]^i})\cdots(\frac{[a|u_r]^i}{[a|u_1]^i}).$$

Suppose the special monomial element Φ^0 given is replaced by an arbitrary irreducible monomial element Φ of shape (i^r) in $\mathrm{Pleth}_r(P_a)$, say

$$\Phi = \left(\prod_{k=1}^{n}[a|u_k]^{i_k}\right)\left(\prod_{k=1}^{n}[a|u_k]^{j_k}\right)\cdots\left(\prod_{k=1}^{n}[a|u_k]^{h_k}\right)$$

with

$$\Phi_1 = \left(\prod_{k=2}^{n}\frac{[a|u_k]^{i_k}}{[a|u_1]^{i_k}}\right)\left(\prod_{k=2}^{n}\frac{[a|u_k]^{j_k}}{[a|u_1]^{j_k}}\right)\cdots\left(\prod_{k=2}^{n}\frac{[a|u_k]^{h_k}}{[a|u_1]^{h_k}}\right).$$

Consider the product of polarization operators

$$\mathbf{P} = \left[\prod_{k=2}^{n} D^{(j_k)}_{\frac{[a|u_2]}{[a|u_1]},\frac{[a|u_{2k}]}{[a|u_{21}]}}\cdots\prod_{k=2}^{n} D^{(h_k)}_{\frac{[a|u_r]}{[a|u_1]},\frac{[a|u_{rk}]}{[a|u_{r1}]}}\right]$$
$$\cdot\left[\prod_{k=2}^{n} D^{(j_k)}_{\frac{[a|u_{2k}]}{[a|u_{21}]},\frac{[a|u_2]}{[a|u_1]}}\cdots\prod_{k=2}^{n} D^{(h_k)}_{\frac{[a|u_{rk}]}{[a|u_{r1}]},\frac{[a|u_r]}{[a|u_1]}}\right].$$

We have easily that

$$\mathbf{P}\left[\Phi^0_1\right] = \binom{i}{j_1,\cdots,j_n}\cdots\binom{i}{h_1,\cdots,h_n}\Phi_1.$$

We can now obtain the expansion for Φ_1 in $\mathrm{Pleth}_r(P_a)$ as a linear combination of circle products of rational elementary monomials by applying $\mathbf{P}$ to the corresponding expansion for Φ^0_1 in $\mathrm{Pleth}_r(P''_a)$.

References

[1] Cayley, A. (1889–1898): The collected mathematical papers of Arthur Cayley, vols. 1–14. Cambridge University Press, Cambridge
[2] Cullis, C. E. (1913–1925): Matrices and determinoids. Cambridge University Press, Cambridge
[3] Gel'fand, I.M., Kapranov, M.M., Zelevinsky, A.V. (1994): Discriminants, resultants and multidimensional determinants. Birkhäuser Boston, Boston, MA
[4] Grosshans, F., Rota, G.-C., Stein, J. A. (1987): Invariant theory and superalgebras. (CBMS Regional Conference Series in Mathematics, vol. 69). American Mathematical Society, Providence, RI
[5] Mac Mahon, P.A. (1978, 1986): Collected papers I, II. MIT Press, Cambridge, MA
[6] Netto, E. (1896–1900): Vorlesungen über Algebra. Teubner, Leipzig
[7] Rota, G.-C., Stein, J. A. (1994): Plethystic Hopf algebras. Proc. Nat. Acad. Sci. U.S.A. **91**, 13057–13061
[8] Rota, G.-C., Stein, J. A. (1994): Plethystic algebras and vector symmetric function. Proc. Nat. Acad. Sci. U.S.A. **91**, 13062–13066
[9] Schinzel, A. (1982): Selected topics on polynomials. University of Michigan Press, Ann Arbor, MI

A formal theory of resultants (II): a constructive definition of the resultant

G.-C. Rota, J. A. Stein

1 Parenthesis operators

Let A and P denote negative alphabets so that the supersymmetric algebra $\text{Super}[A|P]$ is an ordinary polynomial algebra in the neutral elements $[a|u]$ for a in A and u in P.

We define *parenthesis operators* $\mathbb{P}_a$ for every $a \in A$. The operator $\mathbb{P}_a$ is a linear operator from $\text{Super}[A|P]$ to $\text{Pleth}(A|P)$, defined as follows. Let $w = w_1 w_2$, with $w_1 = [a|u_1]^{k_1}[a|u_2]^{k_2}\ldots$ and let w_2 be an $\{a\}$-free monomial. Set

$$\mathbb{P}_a w = (w_1)w_2$$

and extend by linearity over Q. For every $w \in \text{Super}[A|P]$, the element $\mathbb{P}_a w$ is a primitive element of $\text{Pleth}(A|P)$. More specifically, the element $\mathbb{P}_a w$ is a linear combination with coefficients in Q of elements of the form $\mathbb{P}_a w = \sum_i (w_{1i})w_{2i}$, where w_{1i} is a monomial with coefficients in Q containing only the letter a, and where w_2 is an $\{a\}$-free monomial.

Proposition 1. *The parenthesis operator $\mathbb{P}_a$ satisfies $\mathbb{P}_a[ww'] = (\mathbb{P}_a w|\mathbb{P}_a w')$.*

Proof. Indeed, $\mathbb{P}_a w = \sum_i (w_{1i})w_{2i}$, and $\mathbb{P}_a w' = \sum_j (w'_{1j})w'_{2j}$, so that

$$\begin{aligned}(\mathbb{P}_a w|\mathbb{P}_a w') &= \sum_{i,j}(w_{1i}w'_{1j})w_{2i}w'_{2j}\\ &= \mathbb{P}_a[\sum_{i,j} w_{1i}w'_{1j}w_{2i}w'_{2j}] = \mathbb{P}_a[ww'].\end{aligned}$$ □

Proposition 2. *Let $w, w' \in$ $\text{Super}[A|P]$ be homogeneous of degrees p and q in the letter a. Then*

$$[\mathbb{P}_a w]^{(d)} \circ \left[\mathbb{P}_a w'\right]^{(d)} = (\mathbb{P}_a w|\mathbb{P}_a w')^{(d)} + \Delta.$$

The element $(\mathbb{P}_a w|\mathbb{P}_a w')^{(d)}$ is a linear combination of monomial elements of $\text{Pleth}(A|P)$*, all of them of shape $(p+q)^d$, while Δ is a linear combination of monomial elements whose shapes have more than d parts.*

Proof. Let $W = [\mathbb{P}_a w]^{(d)}$ and let $W' = \left[\mathbb{P}_a w'\right]^{(d)}$. Then

$$W \circ W' = \sum W_{(1)}(W_{(2)}|W'_{(1)})W'_{(2)}.$$

Expanding the right side, we find that

$$\sum W_{(1)}(W_{(2)}|W'_{(1)})W'_{(2)} = \sum_{i+j=d} ([\mathbb{P}_a w]^{(i)}|[\mathbb{P}_a w']^{(i)})[\mathbb{P}_a w]^{(j)}[\mathbb{P}_a w']^{(j)}$$
$$= \sum_{i+j=d} (\mathbb{P}_a w|\mathbb{P}_a w')^{(i)}[\mathbb{P}_a w]^{(j)}[\mathbb{P}_a w']^{(j)}.$$

In the last sum, every term is a linear combination of monomial elements of shapes strictly smaller than $((p+q)^d)$, and containing more than d parts, except for the single term

$$(\mathbb{P}_a w|\mathbb{P}_a w')^{(d)}$$

which is a linear combination of monomial elements of shape equal to $(p+q)^d$. □

2 Resultant operators

Given a form $g(x)$ of degree d,

$$g(x) = g(x_1, \cdots, x_n) = \sum_{i_1+\cdots+i_n=d} g_{i_1\cdots i_n} x_1^{i_1} \cdots x_n^{i_n},$$

we define the form

$$g(a) = \sum_{i_1+\cdots+i_n=d} g_{i_1\cdots i_n} [a|u_1]^{i_1} \cdots [a|u_n]^{i_n} .$$

The resultant operator of the form $g(x)$ will be an operator on Cayley forms. Suppose $f(x)$ is a Cayley form of degree r,

$$f(x) = \prod_{j=1}^{r} f_j(x) = \sum_{j_1+\cdots+j_n=r} f_{j_1\cdots j_n} x_1^{j_1} \cdots x_n^{j_n}.$$

We then define

$$R_{g,a} f(a) = E_{f,a}\left[[\mathbb{P}_a\,[g(a)]]^{(r)}\right]$$
$$= E_{f,a}\left[\left[\sum_{i_1+\cdots+i_n=d} g_{i_1\cdots i_n}\left([a|u_1]^{i_1}\cdots[a|u_n]^{i_n}\right)\right]^{(r)}\right]$$
$$= g(f_1)\cdots g(f_r).$$

Observe that, by the definition of the evaluation operator of a Cayley form, the expression $[\mathbb{P}_a\,[g(a)]]^{(r)}$ is to be treated above as an element of the algebra $\mathrm{Pleth}_r\{P_a\}$.

Proposition 3. *Suppose that* $g(x)$ *is also a Cayley form*

$$g(x) = \prod_{i=1}^{d} g_i(x).$$

Then we have

$$R_{g,a} f(a) = R_{f,a} g(a).$$

Proof. Indeed,

$$\begin{aligned} R_{g,a} f(a) &= \prod_{j=1}^{r} g(f_j) = \prod_{i=1}^{d} \prod_{j=1}^{r} g_i(f_j) \\ &= \prod_{i=1}^{d} \prod_{j=1}^{r} f_j(g_i) = R_{f,a} g(a). \qquad \square \end{aligned}$$

Proposition 4.

$$E_{[b|x],a}\left([a|u_1]^{i_1} \cdots [a|u_n]^{i_n}\right) = b_1^{i_1} \cdots b_n^{i_n}.$$

Proof. Indeed,

$$\begin{aligned} &E_{[b|x],a}\left([a|u_1]^{i_1} \cdots [a|u_n]^{i_n}\right) \\ &= E_{[b|u_1]x_1+\cdots+[b|u_n]x_n,a}\left([a|u_1]^{i_1} \cdots [a|u_n]^{i_n}\right) \\ &= b_1^{i_1} \cdots b_n^{i_n}. \qquad \square \end{aligned}$$

Proposition 5.

$$R_{g,a}[b|x] = g(b).$$

Proof. Indeed,

$$\begin{aligned} R_{g,a}[b|x] &= E_{[b|x],a}\left[[\mathbb{P}_a\,[g(a)]]^{(1)}\right] \\ &= E_{[b|x],a}\left[\sum_{i_1+\cdots+i_n=d} g_{i_1\cdots i_n}\left([a|u_1]^{i_1} \cdots [a|u_n]^{i_n}\right)\right] \\ &= \sum_{i_1+\cdots+i_n=d} g_{i_1\cdots i_n} E_{[b|x],a}\left[\left([a|u_1]^{i_1} \cdots [a|u_n]^{i_n}\right)\right] \\ &= \sum_{i_1+\cdots+i_n=d} g_{i_1\cdots i_n} b_1^{i_1} \cdots b_n^{i_n} \\ &= g(b) \end{aligned}$$

as desired. $\square$

Proposition 6.

$$R_{g,a_1}[a_1 \cdots a_n] = g\left((-1)^0[a_2 \cdots a_n|u_2 \cdots u_n], \cdots, (-1)^{n-1}[a_2 \cdots a_n|u_1 \cdots u_{n-1}]\right),$$

Proof. Indeed,

$$\begin{aligned}
R_{g,a_1}[a_1 \cdots a_n] &= E_{[a_1 \cdots a_n],a_1}\left[[\mathbb{P}_a[g(a_1)]]^{(1)}\right] \\
&= E_{\left[(-1)^0\left[a_2 \cdots a_n \middle| u_2 \cdots u_n\right][a_1|u_1]+\cdots+(-1)^{n-1}\left[a_2 \cdots a_n|u_1 \cdots u_{n-1}\right][a_1|u_n]\right],a_1} \\
&\quad \cdot \left[\sum_{i_1+\cdots+i_n=d} g_{i_1 \cdots i_n}\left([a_1|u_1]^{i_1} \cdots [a_1|u_n]^{i_n}\right)\right] \\
&= \sum_{i_1+\cdots+i_n=d} g_{i_1 \cdots i_n} \\
&\quad \cdot E_{\left[(-1)^0\left[a_2 \cdots a_n \middle| u_2 \cdots u_n\right][a_1|u_1]+\cdots+(-1)^{n-1}\left[a_2 \cdots a_n|u_1 \cdots u_{n-1}\right][a_1|u_n]\right],a_1} \\
&\quad \cdot \left[\left([a_1|u_1]^{i_1} \cdots [a_1|u_n]^{i_n}\right)\right] \\
&= \sum_{i_1+\cdots+i_n=d} g_{i_1 \cdots i_n} \\
&\quad \cdot \left[(-1)^0[a_2 \cdots a_n|u_2 \cdots u_n]\right]^{i_1} \cdots \left[(-1)^{n-1}[a_2 \cdots a_n|u_1 \cdots u_{n-1}]\right]^{i_n} \\
&= g\left((-1)^0[a_2 \cdots a_n|u_2 \cdots u_n], \cdots, (-1)^{n-1}[a_2 \cdots a_n|u_1 \cdots u_{n-1}]\right). \qquad \square
\end{aligned}$$

Similarly we have

Proposition 7.

$$R_{g,a_n}[a_1 \cdots a_n] = g\left((-1)^{n-1}[a_1 \cdots a_{n-1}|u_2 \cdots u_n], \cdots, (-1)^0[a_1 \cdots a_{n-1}|u_1 \cdots u_{n-1}]\right).$$

Observe that the resultant operators are not linear. Nonetheless, they enjoy the following remarkable multiplicative property.

Theorem 1. *Let Φ, Φ' be forms and let $w \epsilon$ Super$[A|P]$ be a supersymmetric Cayley form of degree d. Then*

$$R_{\Phi\Phi',a}w = R_{\Phi,a}w R_{\Phi',a}w.$$

Proof. Indeed,

$$R_{\Phi,a}w R_{\Phi',a}w = E_{w,a}[\mathbb{P}_a\Phi(a)]^{(d)} E_{w,a}[\mathbb{P}_a\Phi'(a)]^{(d)}.$$

Since the evaluation operator $E_{w,a}$ is multiplicative relative to the circle product in $\text{Pleth}_d\{A|P\}$, the last expression equals

$$E_{w,a}[[\mathbb{P}_a\Phi]^{(d)} \circ [\mathbb{P}_a\Phi']^{(d)}] = E_{w,a}[([\mathbb{P}_a\Phi]|[\mathbb{P}_a\Phi'])^{(d)}] = E_{w,a}[[\mathbb{P}_a\Phi\Phi']^{(d)}].$$

Thus,

$$R_{\Phi,a}w R_{\Phi',a}w = E_{w,a}[\mathbb{P}_a\Phi\Phi']^{(d)} = R_{\Phi\Phi',a}w. \qquad \square$$

When $\Phi(x)$ is also a Cayley form we have an additional multiplicative property.

Theorem 2. *Let $\Phi(x)$ be a Cayley form of degree r and let $w, w' \in$ $\text{Super}[A|P]$ be supersymmetric Cayley forms. Then*

$$R_{\Phi,a}ww' = \left[R_{\Phi,a}w\right]\left[R_{\Phi,a}w'\right].$$

Proof. Indeed,

$$\begin{aligned}\left[R_{\Phi,a}w\right]\left[R_{\Phi,a}w'\right] &= \left[R_{w,a}\Phi\right]\left[R_{w',a}\Phi\right]\\ &= E_{\Phi,a}[[\mathbb{P}_a w]^{(r)} E_{\Phi,a}[\mathbb{P}_a w']^{(r)}].\end{aligned}$$

Since the evaluation operator $E_{\Phi,a}$ is multiplicative relative to the circle product in $\text{Pleth}_r\{A|P\}$, the last expression satisfies

$$\begin{aligned}E_{\Phi,a}[[\mathbb{P}_a w]^{(r)} E_{\Phi,a}[\mathbb{P}_a w']^{(r)}] &= E_{\Phi,a}[[\mathbb{P}_a w]^{(r)} \circ [\mathbb{P}_a w']^{(r)}]\\ &= E_{\Phi,a}[\left([\mathbb{P}_a w]|[\mathbb{P}_a w']\right)^{(r)}]\\ &= E_{\Phi,a}[[\mathbb{P}_a ww']^{(r)}].\end{aligned}$$

Thus,

$$R_{\Phi,a}wR_{\Phi,a}w' = E_{\Phi,a}[\mathbb{P}_a ww']^{(r)} = R_{ww',a}\Phi = R_{\Phi,a}ww'. \qquad \square$$

Proposition 8. *For t a variable, we have*

$$R_{\Phi+t\Psi,a}\prod_{i=1}^{d}[a|\alpha^{(i)}] = \sum_{i+j=d} t^j \sum_{\substack{\theta(1)<\cdots<\theta(i)\\ \theta(i+1)<\cdots<\theta(i+j)}} \prod_{u=1}^{i}\Phi(\alpha^{(\theta u)})\prod_{v=1}^{j}\Phi(\alpha^{(\theta v)}).$$

3 A constructive definition of the resultant

Let $\Phi_1, \Phi_2, \ldots, \Phi_n$ be $\{a_1, \ldots, a_n\}$-free forms in the variables $x_1, x_2, \ldots, x_n$ of degrees $d_1, d_2, \ldots, d_n$. We define the *resultant* of $\Phi_1, \Phi_2, \ldots, \Phi_n$ to be the element of the field K given by

$$\text{Res}(\Phi_1, \Phi_2, \ldots, \Phi_n) = R_{\Phi_1,a_1}R_{\Phi_2,a_2}\ldots R_{\Phi_n,a_n}[a_1a_2\ldots a_n|u_1u_2\ldots u_n].$$

Proposition 9. *If each of* $\Lambda_1, \Lambda_2, \cdots \Lambda_n$ *is a linear form, say*

$$\Lambda_i(x_1, x_2, \cdots, x_n) = \sum_{j=1}^{n} \lambda_{ij} x_j,$$

then

$$\mathrm{Res}(\Lambda_1, \Lambda_2, \cdots \Lambda_n) = \det(\lambda_{ij}).$$

The definition given above

$$\mathrm{Res}(\Phi_1, \Phi_2, \ldots, \Phi_n) = R_{\Phi_1,a_1} R_{\Phi_2,a_2} \ldots R_{\Phi_n,a_n}[a_1 a_2 \ldots a_n | u_1 u_2 \ldots u_n]$$

gives an explicit expression for the resultant, which can be computed by evaluating the resultant operators in turn, starting with the innermost. We call attention to the fact that such evaluation, while in principle trivial, is computationally far from routine. Consider, by way of example, the first step in the evaluation, to wit:

$$R_{\Phi_{n-1},a_{n-1}} R_{\Phi_n,a_n}[a_1 a_2 \ldots a_n | u_1 u_2 \ldots u_n].$$

This is computed in the following steps.

Step 1. Compute first $R_{\Phi_n,a_n}[a_1 a_2 \ldots a_n | u_1 u_2 \ldots u_n]$ obtaining as above

$$\Phi_n\Big((-1)^{n-1}[a_1 \cdots a_{n-1} | u_2 \cdots u_n], \cdots, (-1)^0[a_1 \cdots a_{n-1} | u_1 \cdots u_{n-1}]\Big)$$

and expand the bracket arguments with respect to a_{n-1} obtaining

$$\Phi_n \left[\begin{array}{c} (-1)^{n-1} \sum_{\theta:\theta(1)<\cdots<\theta(n-2)} (-1)^{\theta} [a_1 \ldots a_{n-2} | u_{\theta 1} \ldots u_{\theta(n-2)}][a_{n-1} | u_{\theta(n-1)}], \\ \cdots, \\ (-1)^0 \sum_{\theta:\theta(2)<\cdots<\theta(n-1)} (-1)^{\theta} [a_1 \ldots a_{n-2} | u_{\theta 2} \ldots u_{\theta(n-1)}][a_{n-1} | u_{\theta(n)}] \end{array} \right].$$

Step 2. This polynomial is now expanded as a polynomial in the variables $[a_{n-1}|u_1]$, $\ldots$, $[a_{n-1}|u_n]$, say

$$\begin{aligned} &\Phi_n((-1)^{n-1}[a_1 \ldots a_{n-1} | u_1 \ldots u_{n-1}], \ldots, (-1)^0[a_1 \ldots a_{n-1} | u_2 \ldots u_n]) \\ &\quad = \sum_{m_1+\cdots+m_n=d_n} \gamma_{m_1 \ldots m_n} [a_{n-1}|u_1]^{m_1} \ldots [a_{n-1}|u_n]^{m_n}, \end{aligned}$$

where the coefficients $\gamma_{m_1 \ldots m_n}$ are $\{a_{n-1}\}$-free. In order to apply the next resultant operator, it is necessary to assume that the result obtained thus far is a supersymmetric Cayley form. This crucial fact will be verified later in the paper.

Step 3. Apply the resultant operator $R_{\Phi_{n-1},a_{n-1}}$ to this polynomial, thereby obtaining

$$R_{\Phi_{n-1},a_{n-1}}\left[\sum_{m_1+\cdots+m_n=d_n} \gamma_{m_1\ldots m_n}[a_{n-1}|u_1]^{m_1}\ldots[a_{n-1}|u_n]^{m_n}\right]$$
$$= E_{\sum_{m_1+\cdots+m_n=d_n}\gamma_{m_1\ldots m_n}[a_{n-1}|u_1]^{m_1}\ldots[a_{n-1}|u_n]^{m_n},a_{n-1}}$$
$$\left[\left[\mathbb{P}_{a_{n-1}}\left[\Phi_{n-1}(a_{n-1})\right]\right]^{(d_n)}\right].$$

Step 4. Now consider the inner term

$$\mathbb{P}_{a_{n-1}}\left[\Phi_{n-1}(a_{n-1})\right].$$

Applying the parenthesis operator $\mathbb{P}_a$, we obtain

$$\mathbb{P}_{a_{n-1}}\left[\Phi_{n-1}(a_{n-1})\right] = \sum_{i_1+\cdots+i_n=d_{n-1}} \Phi^{n-1}_{i_1\ldots i_n}\left([a_{n-1}|u_1]^{i_1}\ldots[a_{n-1}|u_n]^{i_n}\right).$$

Observe that the terms

$$([a_{n-1}|u_1]^{i_1}\ldots[a_{n-1}|u_n]^{i_n})$$

are now primitive elements of $\text{Pleth}(\text{Super}[A|P])$.

Step 5. Raising the last displayed expression in Step 4 to the divided power d_n we obtain

$$R_{\Phi_{n-1},a_{n-1}}R_{\Phi_n,a_n}[a_1a_2\ldots a_n|u_1u_2\ldots u_n]$$
$$= E_{\sum_{m_1+\cdots+m_n=d_n}\gamma_{m_1\ldots m_n}[a_{n-1}|u_1]^{m_1}\ldots[a_{n-1}|u_n]^{m_n},a_{n-1}}$$
$$\left[\left[\sum_{i_1+\cdots+i_n=d_{n-1}} \Phi^{n-1}_{i_1\ldots i_n}\left([a_{n-1}|u_1]^{i_1}\ldots[a_{n-1}|u_n]^{i_n}\right)\right]^{(d_n)}\right]$$

where for the purposes of evaluation the expression

$$\left[\sum_{i_1+\cdots+i_n=d_{n-1}} \Phi^{n-1}_{i_1\ldots i_n}\left([a_{n-1}|u_1]^{i_1}\ldots[a_{n-1}|u_n]^{i_n}\right)\right]^{(d_n)}$$

is to be treated as an element of $\text{Pleth}_{d_n}(\text{Super}[A|P])$.

Expanding this expression, we obtain a linear combination of monomial elements with $\{a_{n-1}\}$-free coefficients in $\text{Pleth}_{d_n}(\text{Super}[A|P])$. Each such monomial element is of outer degree d_n and inner degree d_{n-1} and is of the form

$$([a_{n-1}|u_1]^{i_{11}} \ldots [a_{n-1}|u_n]^{i_{n1}})^{(j_1)} \ldots ([a_{n-1}|u_1]^{i_{1m}} \ldots [a_{n-1}|u_n]^{i_{nm}})^{(j_m)},$$

so that $j_1 + \cdots + j_m = d_n$ and $i_{1s} + \cdots + i_{ns} = d_{n-1}$ for $1 \leq s \leq m$.

Step 6. Thus, we have

$$\left[\sum_{i_1+\cdots+i_n=d_{n-1}} \Phi^{n-1}_{i_1\ldots i_n}\left([a_{n-1}|u_1]^{i_1} \ldots [a_{n-1}|u_n]^{i_n}\right)\right]^{(d_n)}$$
$$= \sum \delta_{j,j',\ldots,j''}$$
$$([a_{n-1}|u_1]^{i_1} \ldots [a_{n-1}|u_n]^{i_n})^{(j)} \ldots ([a_{n-1}|u_1]^{i''_1} \ldots [a_{n-1}|u_n]^{i''_n})^{(j'')},$$

where the coefficients $\delta_{j,j',\ldots,j''}$ are $\{a_{n-1}\}$-free.

Step 7. We now apply the evaluation operator

$$E_{\sum_{m_1+\cdots+m_n=d_n} \gamma_{m_1\ldots m_n}[a_{n-1}|u_1]^{m_1}\ldots[a_{n-1}|u_n]^{m_n},a_{n-1}}.$$

This gives

$$E_{\sum_{m_1+\cdots+m_n=d_n} \gamma_{m_1\ldots m_n}[a_{n-1}|u_1]^{m_1}\ldots[a_{n-1}|u_n]^{m_n},a_{n-1}}$$
$$\cdot\left[\left[\sum_{i_1+\cdots+i_n=d_{n-1}} \Phi^{n-1}_{i_1\ldots i_n}\left([a_{n-1}|u_1]^{i_1} \ldots [a_{n-1}|u_n]^{i_n}\right)\right]^{(d_n)}\right]$$
$$= E_{\sum_{m_1+\cdots+m_n=d_n} \gamma_{m_1\ldots m_n}[a_{n-1}|u_1]^{m_1}\ldots[a_{n-1}|u_n]^{m_n},a_{n-1}}$$
$$\cdot\left[\begin{matrix}\sum \delta_{j,j',\ldots,j''} \\ ([a_{n-1}|u_1]^{i_1} \ldots [a_{n-1}|u_n]^{i_n})^{(j)} \ldots ([a_{n-1}|u_1]^{i''_1} \ldots [a_{n-1}|u_n]^{i''_n})^{(j'')}\end{matrix}\right]$$
$$= \sum \delta_{j,j',\ldots,j''} E_{\sum_{m_1+\cdots+m_n=d_n} \gamma_{m_1\ldots m_n}[a_{n-1}|u_1]^{m_1}\ldots[a_{n-1}|u_n]^{m_n},a_{n-1}}$$
$$\cdot\left[([a_{n-1}|u_1]^{i_1} \ldots [a_{n-1}|u_n]^{i_n})^{(j)} \ldots ([a_{n-1}|u_1]^{i''_1} \ldots [a_{n-1}|u_n]^{i''_n})^{(j'')}\right].$$

Step 8. We must now evaluate each of the expressions

$$E_{\sum_{m_1+\cdots+m_n=d_n} \gamma_{m_1\ldots m_n}[a_{n-1}|u_1]^{m_1}\ldots[a_{n-1}|u_n]^{m_n},a_{n-1}}$$
$$\left[([a_{n-1}|u_1]^{i_1} \ldots [a_{n-1}|u_n]^{i_n})^{(j)} \ldots ([a_{n-1}|u_1]^{i''_1} \ldots [a_{n-1}|u_n]^{i''_n})^{(j'')}\right]$$

where $i_1 + \cdots + i_n = d_{n-1}, \cdots, i''_1 + \cdots + i''_n = d_{n-1}$ and where $j + \cdots + j'' = d_n$. We arrive at the crucial point of the computation. Using the algorithm developed in Part I of this paper, each of the monomial elements of inner degree d_{n-1}

$$([a_{n-1}|u_1]^{i_1} \ldots [a_{n-1}|u_n]^{i_n})^{(j)} \ldots ([a_{n-1}|u_1]^{i''_1} \ldots [a_{n-1}|u_n]^{i''_n})^{(j'')}$$

is now expressed as a linear combination of rational expressions in elementary monomials with $\{a_{n-1}\}$-free coefficients c_{jk} in $\mathrm{Pleth}_{d_n}(\mathrm{Super}[A|P])$, say

$$
\begin{aligned}
&([a_{n-1}|u_1]^{i_1}\dots[a_{n-1}|u_n]^{i_n})^{(j)}\dots([a_{n-1}|u_1]^{i''_1}\dots[a_{n-1}|u_n]^{i''_n})^{(j'')}\\
&=\sum_{k=1}^{d_{n-1}+d}\sum_j c_{jk}\left[([a_{n-1}|u_1])^{(d_n)}\right]^{d_{n-1}+d-k}\circ\\
&\left[\frac{\left[([a|u_1])^{(j_{11})}\cdots([a|u_n])^{(j_{1n})}\right]\circ\cdots\circ\left[([a|u_1])^{(j_{k1})}\cdots([a|u_n])^{(j_{kn})}\right]}{\left[([a_{n-1}|u_1])^{(d_n)}\right]^d}\right]
\end{aligned}
$$

for some $d\geq 0$, where for each m we have

$$j_{m1}+\cdots+j_{mn}=d_n.$$

Step 9. We are now ready to apply the evaluation operator

$$E_{\sum_{m_1+\cdots+m_n=d_n}\gamma_{m_1\dots m_n}[a_{n-1}|u_1]^{m_1}\dots[a_{n-1}|u_n]^{m_n},a_{n-1}}$$

in Step 8. We thereby obtain

$$
\begin{aligned}
&E_{\sum_{m_1+\cdots+m_n=d_n}\gamma_{m_1\dots m_n}[a_{n-1}|u_1]^{m_1}\dots[a_{n-1}|u_n]^{m_n},a_{n-1}}\\
&\quad\cdot[([a_{n-1}|u_1]^{i_1}\dots[a_{n-1}|u_n]^{i_n})^{(j)}\dots([a_{n-1}|u_1]^{i''_1}\dots[a_{n-1}|u_n]^{i''_n})^{(j'')}]\\
&=E_{\sum_{m_1+\cdots+m_n=d_n}\gamma_{m_1\dots m_n}[a_{n-1}|u_1]^{m_1}\dots[a_{n-1}|u_n]^{m_n},a_{n-1}}\\
&\quad\cdot\sum_{k=1}^{d_{n-1}+d}\sum_j c_{jk}\left[([a_{n-1}|u_1])^{(d_n)}\right]^{d_{n-1}+d-k}\circ\\
&\quad\cdot\left[\frac{\left[([a|u_1])^{(j_{11})}\cdots([a|u_n])^{(j_{1n})}\right]\circ\cdots\circ\left[([a|u_1])^{(j_{k1})}\cdots([a|u_n])^{(j_{kn})}\right]}{\left[([a_{n-1}|u_1])^{(d_n)}\right]^d}\right]\\
&=\sum_{k=1}^{d_{n-1}+d}\sum_j c_{jk}E_{\sum_{m_1+\cdots+m_n=d_n}\gamma_{m_1\dots m_n}[a_{n-1}|u_1]^{m_1}\dots[a_{n-1}|u_n]^{m_n},a_{n-1}}\\
&\quad\cdot\left[([a_{n-1}|u_1])^{(d_n)}\right]^{d_{n-1}+d-k}\circ\\
&\quad\cdot\left[\frac{\left[([a|u_1])^{(j_{11})}\cdots([a|u_n])^{(j_{1n})}\right]\circ\cdots\circ\left[([a|u_1])^{(j_{k1})}\cdots([a|u_n])^{(j_{kn})}\right]}{\left[([a_{n-1}|u_1])^{(d_n)}\right]^d}\right]
\end{aligned}
$$

where

$$
\begin{aligned}
\left[([a_{n-1}|u_1])^{(d_n)}\right]^d &= \left[([a_{n-1}|u_1])^{(d_n)}\right]\circ\cdots\circ\left[([a_{n-1}|u_1])^{(d_n)}\right]\\
&=\left([a_{n-1}|u_1]^d\right)^{(d_n)}.
\end{aligned}
$$

Step 10. Recalling that the evaluation operators are multiplicative over circle products, we see that the evaluation equals

$$\sum_{k=1}^{d_{n-1}+d} \sum_{j} c_{jk}$$
$$\left[E_{\sum \gamma_{m_1\dots m_n}[a_{n-1}|u_1]^{m_1}\dots[a_{n-1}|u_n]^{m_n},a_{n-1}} ([a_{n-1}|u_1])^{(d_n)}\right]^{d_{n-1}+d-k}$$
$$\left[\frac{\prod_{s=1}^{k} E_{\sum \gamma_{m_1\dots m_n}[a_{n-1}|u_1]^{m_1}\dots[a_{n-1}|u_n]^{m_n},a_{n-1}} \left[([a|u_1])^{(j_{s1})}\cdots([a|u_n])^{(j_{sn})}\right]}{\left[E_{\sum \gamma_{m_1\dots m_n}[a_{n-1}|u_1]^{m_1}\dots[a_{n-1}|u_n]^{m_n},a_{n-1}} \left[([a_{n-1}|u_1])^{(d_{n-1})}\right]\right]^{d}}\right].$$

Step 11. Each application of the evaluation operator

$$\begin{aligned}
&E_{\sum \gamma_{m_1\dots m_n}[a_{n-1}|u_1]^{m_1}\dots[a_{n-1}|u_n]^{m_n},a_{n-1}}\\
&\quad = E_{\sum_{m_1+\dots+m_n=d_n} \gamma_{m_1\dots m_n}[a_{n-1}|u_1]^{m_1}\dots[a_{n-1}|u_n]^{m_n},a_{n-1}}\\
&\quad = E_{\left[\Phi_n\left((-1)^{n-1}[a_1\dots a_{n-1}|u_1\dots u_{n-1}],\dots,(-1)^0[a_1\dots a_{n-1}|u_2\dots u_n]\right)\right],a_{n-1}}
\end{aligned}$$

to an elementary monomial gives a coefficient of the form

$$\begin{aligned}
&\Phi_n\Big((-1)^{n-1}[a_1\dots a_{n-1}|u_1\dots u_{n-1}],\dots,(-1)^0[a_1\dots a_{n-1}|u_2\dots u_n]\Big)\\
&\quad = \sum_{m_1+\dots+m_n=d_n} \gamma_{m_1\dots m_n}[a_{n-1}|u_1]^{m_1}\dots[a_{n-1}|u_n]^{m_n},
\end{aligned}$$

namely,

$$E_{\sum \gamma_{m_1\dots m_n}[a_{n-1}|u_1]^{m_1}\dots[a_{n-1}|u_n]^{m_n},a_{n-1}} \left[\prod_{i=1}^{n}([a_{n-1}|u_i])^{(r_i)}\right] = \gamma_{r_1\dots r_n}$$

so that

$$\begin{aligned}
&E_{\sum \gamma_{m_1\dots m_n}[a_{n-1}|u_1]^{m_1}\dots[a_{n-1}|u_n]^{m_n},a_{n-1}}\\
&\quad\cdot\left[([a_{n-1}|u_1]^{i_1}\dots[a_{n-1}|u_n]^{i_n})^{(j)}\dots([a_{n-1}|u_1]^{i''_1}\dots[a_{n-1}|u_n]^{i''_n})^{(j'')}\right]\\
&\quad = \sum_{k=1}^{d_{n-1}+d} \left[\gamma_{d_n 0\dots 0}\right]^{d_{n-1}+d-k} \sum_{j} c_{jk}\left[\frac{\prod_{h=1}^{k}\gamma_{j_1\dots j_n}}{\left[\gamma_{d_n 0\dots 0}\right]^{d}}\right].
\end{aligned}$$

Step 12. The result of evaluating

$$R_{\Phi_{n-1},a_{n-1}} R_{\Phi_n,a_n}[a_1a_2\dots a_n|u_1u_2\dots u_n]$$

is therefore a rational function of the coefficients of $\Phi_{n-1}(x)$ and the coefficients $\gamma_{m_1\ldots m_n}$ of the form

$$\Phi_n\Big((-1)^{n-1}[a_1\ldots a_{n-1}|u_1\ldots u_{n-1}],\ldots,(-1)^0[a_1\ldots a_{n-1}|u_2\ldots u_n]\Big)$$
$$= \sum_{m_1+\cdots+m_n=d_n} \gamma_{m_1\ldots m_n}[a_{n-1}|u_1]^{m_1}\ldots[a_{n-1}|u_n]^{m_n},$$

which by the analysis given above is known to be a polynomial in the coefficients of the forms $\Phi_n(x)$, $\Phi_{n-1}(x)$ and the biproducts $[a_1\ldots a_{n-2}|u_{i_1}\ldots u_{i_{n-2}}]$ with rational numbers for coefficients. It will later be shown that the numerical coefficients are actually integers.

Step 13. We are now ready to apply the next resultant operator, after verifying that the result obtained thus far is again a supersymmetric Cayley form. Again, this crucial fact will be demonstrated later in the paper.

4 Reduced forms and eliminants

Given an integer k such that $0 \le k \le n$, given $n-k$ letters $a_{k+1}, a_{k+2}, \ldots, a_n$ in A, given a form $\Phi(x)$ and assuming that the form $\Phi(x)$ is $\{a, a_{k+1}, \ldots, a_n\}$-free, we define the *reduced form of step* k of Φ to be the form

$$\begin{aligned}\Phi^{[k]}(x_1, x_2, \ldots, x_k) &= \Phi(x_1, x_2, \ldots, x_k; a_{k+1}, a_{k+2}, \ldots, a_n)\\ &= R_{\Phi,a}[aa_{k+1}a_{k+2}\ldots a_n|xu_{k+1}u_{k+2}\ldots u_n]\\ &= R_{\Phi,a}\left[\sum_{i=1}^{k}[aa_{k+1}a_{k+2}\ldots a_n|u_iu_{k+1}u_{k+2}\ldots u_n]x_i\right].\end{aligned}$$

Occasionally, we shall write $\Phi^{[k]}$ in place of $\Phi^{[k]}(x_1, x_2, \ldots, x_k)$ or $\Phi(x_1, x_2, \ldots, x_k; a_{k+1}, a_{k+2}, \ldots, a_n)$.

Proposition 10. *Let* Φ, Ψ *be forms. Then*

$$(\Phi+\Psi)^{[k]} = \Phi^{[k]} + \Psi^{[k]}$$

and

$$(\Phi\Psi)^{[k]} = \Phi^{[k]}\Psi^{[k]}.$$

Proof. Observe that

$$\begin{aligned}\Phi(x_1, x_2, \ldots, x_k; a_{k+1}, a_{k+2}, \ldots, a_n)\\ &= R_{\Phi,a}\left[\sum_{i=1}^{k}[aa_{k+1}a_{k+2}\ldots a_n|u_i u_{k+1}u_{k+2}\ldots u_n]x_i\right]\\ &= R_{\Phi,a}\left[\sum_{i=1}^{n}[a|u_i]t_{ij}x_j\right]\end{aligned}$$

for suitable $\{a\}$-free elements $t_{ij} \in \text{Super}[A|P]$. Furthermore,

$$R_{\Phi,a}\left[\sum_{i=1}^{n}[a|u_i]t_{ij}x_j\right] = \Phi(\sum_{j=1}^{n} t_{1j}x_j, \ldots, \sum_{j=1}^{n} t_{nj}x_j),$$

whence the conclusion. □

The *eliminant of step k* of k forms $\Phi_1, \Phi_2, \ldots, \Phi_k$ is defined next, by the formula

$$\begin{aligned}\text{Elim}(\Phi_1, \Phi_2, \ldots, \Phi_k; a_{k+1}, a_{k+2}, \ldots, a_n)\\ &= R_{\Phi_1,a_1}R_{\Phi_2,a_2}\ldots R_{\Phi_k,a_k}[a_1a_2\ldots a_n|u_1u_2\ldots u_n]\end{aligned}$$

The eliminant of step n is the resultant. The eliminant of step $n-1$ will be referred to as "the eliminant".

Proposition 11. *The eliminant of step one is given by*

$$\begin{aligned}\text{Elim}(\Phi; a_2, \cdots, a_n)\\ &= \Phi\Big([a_2\cdots a_n|u_2\cdots u_n](-1)^0, \cdots, [a_2\cdots a_n|u_1\cdots u_{n-1}](-1)^{n-1}\Big).\end{aligned}$$

Proposition 12 (Reduction formula).

$$\begin{aligned}[a_{k+1}\cdots a_n|u_{k+1}\cdots u_n]^{(k-1)d_1\cdots d_k}\text{Elim}(\Phi_1, \cdots, \Phi_k; a_{k+1}, \cdots a_n)\\ &= \text{Res}\left(\Phi_1^{[k]}, \cdots \Phi_k^{[k]}\right)\\ &= \text{Res}\Big(\Phi_1(x_1, \cdots, x_k; a_{k+1}, \cdots, a_n), \cdots, \Phi_k(x_1, \cdots, x_k; a_{k+1}, \cdots, a_n)\Big).\end{aligned}$$

Proof.

$$\text{Res}(\Phi_1^{[k]}, \Phi_2^{[k]}, \ldots, \Phi_k^{[k]})$$

$$= R_{\Phi_1,a_1}R_{\Phi_2,a_2}\ldots R_{\Phi_k,a_k}\sum_{\theta}(-1)^{\theta}\left[\begin{array}{c|c} a_1a_{k+1}\ldots a_n & u_{\theta 1}u_{k+1}\ldots u_n\\ a_2a_{k+1}\ldots a_n & u_{\theta 2}u_{k+1}\ldots u_n\\ \cdots & \cdots\\ a_ka_{k+1}\ldots a_n & u_{\theta k}u_{k+1}\ldots u_n\end{array}\right]$$

$$= R_{\Phi_1,a_1} R_{\Phi_2,a_2} \dots R_{\Phi_k,a_k}$$
$$\Big[[a_1 \dots a_n | u_1 \dots u_n][a_{k+1} \dots a_n | u_{k+1} \dots u_n]^{k-1}\Big]$$
$$= [a_{k+1} \dots a_n | u_{k+1} \dots u_n]^{(k-1)d_1 \dots d_k} R_{\Phi_1,a_1} R_{\Phi_2,a_2}$$
$$\dots R_{\Phi_k,a_k} [a_1 \dots a_n | u_1 \dots u_n]. \qquad \square$$

Proposition 13 (Second reduction formula). *Let the linear forms* $\Lambda_j(y)$ *for* $y = (y_1, \cdots y_{n-1})$ *and for* $j = 1, \cdots, n$ *be given by*

$$\Lambda_j(y) = [b_1 | u_j] y_1 + \cdots + [b_{n-1} | u_j] y_{n-1},$$

and let $B(x_1, \cdots, x_n)$ *be the linear form*

$$B(x_1, \cdots, x_n) = \sum_{\theta(1) < \cdots < \theta(n-1)} (-1)^\theta [b_1 \cdots b_{n-1} | u_{\theta 1} \cdots u_{\theta(n-1)}] x_{\theta n}.$$

Then

$$\operatorname{Res}(\Phi_1, \cdots, \Phi_{n-1}, B)$$
$$= \operatorname{Res}\Big(\Phi_1(\Lambda_1(y), \cdots, \Lambda_n(y)), \cdots, \Phi_{n-1}(\Lambda_1(y), \cdots, \Lambda_n(y))\Big).$$

Proof. We have

$$\Phi_i(x_1, \dots, x_n) = R_{\Phi_i,a_i}[a_i | x].$$

Therefore

$$\Phi_i(\Lambda_1(y), \dots, \Lambda_n(y)) = R_{\Phi_i,a_i}\left[\sum_{j=1}^{n-1} \sum_{k=1}^{n} [a_i | u_k][b_j | u_k] y_j\right].$$

Hence we have

$$\operatorname{Res}(\Phi_1, \dots, \Phi_{n-1}, B)$$
$$= R_{\Phi_1,a_1} \dots R_{\Phi_{n-1},a_{n-1}} R_{B,a_n}[a_1 \dots a_n]$$
$$= R_{\Phi_1,a_1} \dots R_{\Phi_{n-1},a_{n-1}} \Bigg[\sum_{\theta : \theta(1) < \theta(2) < \cdots < \theta(n-1)}$$
$$\cdot \left[\begin{array}{c|c} a_1 \dots a_{n-1} & u_{\theta 1} \dots u_{\theta(n-1)} \\ b_1 \dots b_{n-1} & u_{\theta 1} \dots u_{\theta(n-1)} \end{array} \right] \Bigg]$$
$$= R_{\Phi_1,a_1} \dots R_{\Phi_{n-1},a_{n-1}} \left[\sum_\theta (-1)^\theta \prod_{m=1}^{n-1} \left[\sum_{k=1}^{n} [a_m | u_k][b_{\theta m} | u_k] \right] \right],$$

as desired. $\square$

Observe that, whereas the first reduction formula is essentially an elimination algorithm, the second reduction formula is of a different nature in that, rather than eliminate one of the x_i and retain the others, the entire set of n variables $x_1, \cdots, x_n$ is replaced by an unrelated set of $n-1$ new variables $y_1, \cdots, y_{n-1}$, an interesting twist on the usual reduction process.

Proposition 14 (Third reduction formula). *Let* $\Lambda(x) = [a_n|x]$. *Then*

$$[a_1 \ldots a_n]^{(n-2)d_1 \cdots d_{n-1}} Res(\Phi_1, \ldots, \Phi_{n-1}, \Lambda)$$

$$= \mathrm{Res}\begin{bmatrix} R_{\Phi_1,a}\left[\sum_{\theta:\theta(2)<\cdots<\theta(n-1)}(-1)^\theta [a a_{\theta 2} \ldots a_{\theta(n-1)} a_n] x_{\theta 1}\right] \\ ,\ldots, \\ R_{\Phi_{n-1},a}\left[\sum_{\theta:\theta(2)<\cdots<\theta(n-1)}(-1)^\theta [a a_{\theta 2} \ldots a_{\theta(n-1)} a_n] x_{\theta 1}\right] \end{bmatrix}.$$

Proof. Observe that

$$[a_1 \ldots a_n]^{(n-1)d_1 \ldots d_{n-1}} \mathrm{Res}(\Phi_1, \ldots, \Phi_{n-1}, \Lambda)$$

$$= \mathrm{Res}\begin{bmatrix} R_{\Phi_1,a}\left[\sum_{\theta:\theta 2<\cdots<\theta n}(-1)^\theta [a a_{\theta 2} \ldots a_{\theta(n-1)} a_{\theta n}] x_{\theta 1}\right], \\ \ldots, \\ R_{\Phi_{n-1},a}\left[\sum_{\theta:\theta 2<\cdots<\theta n}(-1)^\theta [a a_{\theta 2} \ldots a_{\theta(n-1)} a_{\theta n}] x_{\theta 1}\right], \\ R_{\Lambda,a}\left[\sum_{\theta:\theta 2<\cdots<\theta n}(-1)^\theta [a a_{\theta 2} \ldots a_{\theta(n-1)} a_{\theta n}] x_{\theta 1}\right] \end{bmatrix}$$

$$= \mathrm{Res}\begin{bmatrix} R_{\Phi_1,a}\left[\sum_{\theta:\theta 2<\cdots<\theta n}(-1)^\theta [a a_{\theta 2} \ldots a_{\theta(n-1)} a_{\theta n}] x_{\theta 1}\right], \\ \ldots, \\ R_{\Phi_{n-1},a}\left[\sum_{\theta:\theta 2<\cdots<\theta n}(-1)^\theta [a a_{\theta 2} \ldots a_{\theta(n-1)} a_{\theta n}] x_{\theta 1}\right], \\ [a_1 \ldots a_n] x_n \end{bmatrix}$$

$$= [a_1 \ldots a_n]^{d_1 \ldots d_{n-1}}$$

$$\mathrm{Res}\begin{bmatrix} R_{\Phi_1,a}\left[\sum_{\theta:\theta(1)<\cdots<\theta(n-1)}(-1)^\theta [a a_{\theta 2} \ldots a_{\theta(n-1)} a_n] x_{\theta 1}\right], \\ \ldots, \\ R_{\Phi_{n-1},a}\left[\sum_{\theta:\theta(1)<\cdots<\theta(n-1)}(-1)^\theta [a a_{\theta 2} \ldots a_{\theta(n-1)} a_n] x_{\theta 1}\right] \end{bmatrix}$$

whence the conclusion follows. □

5 Resultants of binary forms

In this section, let $\Phi(x_1, x_2)$ and $\Psi(x_1, x_2)$ be $\{a, b\}$-free binary forms of degrees p and q, respectively, and let

$$\Phi(x_1, x_2) = \prod_{i=1}^{p}(\alpha_1^{(i)} x_1 + \alpha_2^{(i)} x_2)$$

and

$$\Psi(x_1, x_2) = \prod_{i=1}^{q} (\beta_1^{(i)} x_1 + \beta_2^{(i)} x_2)$$

be their factorizations into linear factors, which we may assume to be "distinct".

Proposition 15.

$$\mathrm{Res}(\Phi, \Psi) = \prod_{i=1}^{p} \prod_{j=1}^{q} [\alpha^{(i)} \beta^{(j)}].$$

Proof. We have

$$\begin{aligned}
\mathrm{Res}(\Phi, \Psi) &= R_{\Phi,a} R_{\Psi,b} [[a|u_1][b|u_2] - [a|u_2][b|u_1]] \\
&= R_{\Phi,a} \Psi(-[a|u_2], [a|u_1]) \\
&= R_{\Phi,a} \prod_{j=1}^{q} (-[a|u_2]\beta_1^{(i)} + [a|u_1]\beta_2^{(i)}) \\
&= \prod_{j=1}^{q} R_{\Phi,a} (-[a|u_2]\beta_1^{(i)} + [a|u_1]\beta_2^{(i)}) \\
&= \prod_{j=1}^{q} \Phi(\beta_2^{(i)}, -\beta_1^{(i)}) = \prod_{i=1}^{p} \prod_{j=1}^{q} [\alpha^{(i)} \beta^{(j)}]. \qquad \square
\end{aligned}$$

Corollary 1.

$$\mathrm{Res}(\Phi, \Psi) = (-1)^{pq} \mathrm{Res}(\Psi, \Phi).$$

Corollary 2. *Suppose in addition that $p = q$. Then*

$$\mathrm{Res}([c|u_1]\Phi + [c|u_2]\Psi, [d|u_1]\Phi + [d|u_2]\Psi) = [cd]^p \mathrm{Res}(\Phi, \Psi).$$

Corollary 3. *Again when $p = q$, the resultant* $\mathrm{Res}(\Phi, \Psi)$ *is a polynomial in the elements*

$$\Phi_{i,j}\Psi_{r,s} - \Phi_{r,s}\Psi_{i,,j}.$$

Corollary 4. *The resultant of two binary forms vanishes if and only if the forms have a common zero in K.*

Proof. By the preceding proposition, we have

$$\mathrm{Res}(\Phi, \Psi) = \prod_{i=1}^{p} \prod_{j=1}^{q} [\alpha^{(i)} \beta^{(j)}].$$

Thus, the resultant vanishes if and only if

$$[\alpha^{(1)} \beta^{(1)}] = 0,$$

say. Therefore, there exists $x^{(0)}$ such that $[\alpha^{(1)}|x^{(0)}] = 0 = [\beta^{(1)}|x^{(0)}]$. We conclude that $\Phi(x^{(0)}) = 0 = \Psi(x^{(0)})$. $\square$

6 Properties of resultants

Proposition 16. *Let $\Lambda_1, \Lambda_2, \cdots, \Lambda_n$ be linear forms with*

$$\Lambda_i(x) = \sum_{j=1}^{n} \lambda_{ij} x_j,$$

and let

$$\Phi_i \circ \Lambda(x) = \Phi_i(\Lambda_1(x), \cdots, \Lambda_n(x)).$$

Then

$$\mathrm{Res}(\Phi_1 \circ \Lambda, \ldots, \Phi_n \circ \Lambda) = \det(\lambda_{ij})^{d_1 \ldots d_n} \mathrm{Res}(\Phi_1, \ldots, \Phi_n).$$

Proof. Indeed,

$$\begin{aligned}
&\mathrm{Res}(\Phi_1 \circ \Lambda, \ldots, \Phi_n \circ \Lambda) \\
&\quad = R_{\Phi_1,a_1} \ldots R_{\Phi_n,a_n} \left[\textstyle\sum_{i=1}^n a_i \lambda_{i1}, \ldots, \sum_{i=1}^n a_i \lambda_{in} \;\middle|\; u_1 \ldots u_n \right] \\
&\quad = R_{\Phi_1,a_1} \ldots R_{\Phi_n,a_n} [det(\lambda_{ij})[a_1 \ldots a_n | u_1 \ldots u_n]] \\
&\quad = \det(\lambda_{ij})^{d_1 \ldots d_n} R_{\Phi_1,a_1} \ldots R_{\Phi_n,a_n} [a_1 a_2 \ldots a_n | u_1 u_2 \ldots u_n]. \qquad \square
\end{aligned}$$

Theorem 3 (The product formula). *Suppose that the form Φ_n factors as the product of two forms:*

$$\Phi_n = \Phi\Psi.$$

Then

$$\mathrm{Res}(\Phi_1, \ldots, \Phi_n) = \mathrm{Res}(\Phi_1, \ldots, \Phi_{n-1}, \Phi)\mathrm{Res}(\Phi_1, \ldots, \Phi_{n-1}, \Psi).$$

Proof. Observe that

$$\begin{aligned}
&\mathrm{Res}(\Phi_1, \ldots, \Phi_{n-1}, \Phi_n) \\
&\quad = R_{\Phi_1,a_1} \ldots R_{\Phi_{n-1},a_{n-1}} R_{\Phi_n,a_n} [a_1 \ldots a_{n-1} a_n | u_1 \ldots u_{n-1} u_n] \\
&\quad = R_{\Phi_1,a_1} \ldots R_{\Phi_{n-1},a_{n-1}} \\
&\qquad \Phi_n\Big((-1)^{n-1}[a_1 \ldots a_{n-1} | u_1 \ldots u_{n-1}], \ldots, (-1)^0 [a_1 \ldots a_{n-1} | u_2 \ldots u_n]\Big) \\
&\quad = R_{\Phi_1,a_1} \ldots R_{\Phi_{n-1},a_{n-1}} \\
&\qquad \Phi\Psi\Big((-1)^{n-1}[a_1 \ldots a_{n-1} | u_1 \ldots u_{n-1}], \ldots, (-1)^0 [a_1 \ldots a_{n-1} | u_2 \ldots u_n]\Big) \\
&\quad = R_{\Phi_1,a_1} \ldots R_{\Phi_{n-1},a_{n-1}} \\
&\qquad \Big[\Phi\Big((-1)^{n-1}[a_1 \ldots a_{n-1} | u_1 \ldots u_{n-1}], \ldots, (-1)^0 [a_1 \ldots a_{n-1} | u_2 \ldots u_n]\Big)\Big] \\
&\qquad R_{\Phi_1,a_1} \ldots R_{\Phi_{n-1},a_{n-1}} \\
&\qquad \Big[\Psi\Big((-1)^{n-1}[a_1 \ldots a_{n-1} | u_1 \ldots u_{n-1}], \ldots, (-1)^0 [a_1 \ldots a_{n-1} | u_2 \ldots u_n]\Big)\Big] \\
&\quad = \mathrm{Res}(\Phi_1, \Phi_2, \ldots, \Phi)\mathrm{Res}(\Phi_1, \Phi_2, \ldots, \Psi). \qquad \square
\end{aligned}$$

Proposition 17. *If $\Phi_1, \cdots, \Phi_n$ are generic forms, then* $\mathrm{Res}(\Phi_1, \cdots, \Phi_n)$ *is an irreducible polynomial in the coefficients of the forms.*

Proof. This follows immediately from the preceding theorem and from the fact that the resultant of linear forms is the determinant. □

Proposition 18. *Suppose that $\Phi_n = \Phi_{n-1}\Psi_{n-1} + \Psi$. Then*

$$\mathrm{Res}(\Phi_1, \Phi_2, \ldots, \Phi_{n-1}, \Phi_n) = \mathrm{Res}(\Phi_1, \Phi_2, \ldots, \Phi_{n-1}, \Psi).$$

Proof. It suffices to show that

$$\begin{aligned} &R_{\Phi_{n-1},a_{n-1}} R_{\Phi_n,a_n}[a_1a_2\ldots a_n|u_1u_2\ldots u_n] \\ &\quad = R_{\Phi_{n-1},a_{n-1}} R_{\Psi,a_n}[a_1a_2\ldots a_n|u_1u_2\ldots u_n]. \end{aligned}$$

By the reduction formula,

$$\begin{aligned} &[a_1\ldots a_{n-2}|u_1\ldots u_{n-2}]^{d_{n-1}d_n} R_{\Phi_{n-1},a_{n-1}} R_{\Phi_n,a_n}[a_1a_2\ldots a_n|u_1u_2\ldots u_n] \\ &\quad = \mathrm{Res}(\Phi_{n-1}^{[2]}, \Phi_n^{[2]}) \\ &\quad = \mathrm{Res}(\Phi_{n-1}^{[2]}, \Phi_{n-1}^{[2]}\Psi_{n-1}^{[2]} + \Psi^{[2]}) \\ &\quad = \mathrm{Res}(\Phi_{n-1}^{[2]}, \Psi^{[2]}) \\ &\quad = [a_1\ldots a_{n-2}|u_1\ldots u_{n-2}]^{d_{n-1}d_n} \\ &\qquad \cdot R_{\Phi_{n-1},a_{n-1}} R_{\Psi,a_n}[a_1a_2\ldots a_n|u_1u_2\ldots u_n]. \qquad \square \end{aligned}$$

7 The eliminant of ternary forms

Recall that the definition given for the resultant assumes that the successive eliminants are supersymmetric Cayley forms. In this section we verify this fact for ternary forms.

Let $\Phi_1(x_1, x_2, x_3)$ and $\Phi_2(x_1, x_2, x_3)$ be ternary forms of degrees d_1 and d_2, and let $[a|x]$ be a linear form. Our first objective is to prove the following

Theorem 4. *The eliminant*

$$E(a) = \mathrm{Res}(\Phi_1, \Phi_2, [a|x])$$

is a Cayley form in the variables $[a|u_1], [a|u_2], [a|u_3]$. Furthermore,

$$\begin{aligned} E(a) &= \prod_{i=1}^{d_1d_2} [a|x^{(i)}] \\ &= \prod_{i=1}^{d_1d_2} [a|u_1]x_1^{(i)} + [a|u_2]x_2^{(i)} + [a|u_3]x_3^{(i)}, \end{aligned}$$

where the $x^{(i)}$ are the common zeros of Φ_1 and Φ_2.

The theorem will be a consequence of the following propositions.

Proposition 19. *Let* $x^{(0)} = (x_1^{(0)}, x_2^{(0)}, x_3^{(0)})$, *where* $x_1^{(0)}, x_2^{(0)}, x_3^{(0)} \in K$, *and let* $x^{(0)}$ *be a common zero of the three forms* $\Phi_1(x)$, $\Phi_2(x)$ *and* $[a|x]$, *that is,*

$$0 = \Phi_1(x^{(0)}) = \Phi_2(x^{(0)}) = [a|x^{(0)}].$$

Then

$$E(a) = 0.$$

Proof. Since Φ_1 and Φ_1 are homogeneous we must have

$$0 = \Phi_1([a|u_3]x^{(0)}) = \Phi_2([a|u_3]x^{(0)}).$$

Since $[a|x^{(0)}] = 0$, we have

$$[a|u_3]x_3^{(0)} = -[a|u_1]x_1^{(0)} - [a|u_2]x_2^{(0)}$$

so that

$$\begin{aligned} 0 &= \Phi_1([a|u_3]x^{(0)}) \\ &= \Phi_1([a|u_3]x_1^{(0)}, [a|u_3]x_2^{(0)}, [a|u_3]x_3^{(0)}) \\ &= \Phi_1([a|u_3]x_1^{(0)}, [a|u_3]x_2^{(0)}, -[a|u_1]x_1^{(0)} - [a|u_2]x_2^{(0)}). \end{aligned}$$

The binary form

$$A(x_1, x_2) = \Phi_1([a|u_3]x_1, [a|u_3]x_2, -[a|u_1]x_1 - [a|u_2]x_2)$$

factors into a product of linear factors in K:

$$A(x_1, x_2) = \prod_{i=1}^{d_1} (\alpha_1^{(i)} x_1 + \alpha_2^{(i)} x_2).$$

Since $A(x_1^{(0)}, x_2^{(0)}) = 0$ we may assume that

$$\alpha_1^{(1)} x_1^{(0)} + \alpha_2^{(1)} x_2^{(0)} = 0.$$

Similarly, defining $B(x_1, x_2)$ in terms of Φ_2, we have

$$B(x_1, x_2) = \prod_{i=1}^{d_2} (\beta_1^{(i)} x_1 + \beta_2^{(i)} x_2),$$

and since $B(x_1^{(0)}, x_2^{(0)}) = 0$ we may assume that

$$\beta_1^{(1)} x_1^{(0)} + \beta_2^{(1)} x_2^{(0)} = 0,$$

say. By the reduction formula we have

$$[a|u_3]^{d_1 d_2} E(a) = \prod_{i=1}^{d_1} \prod_{j=1}^{d_2} [\alpha^{(i)} \beta^{(j)}].$$

But

$$[\alpha^{(1)} \beta^{(1)}] = 0$$

since $\alpha^{(1)}$ and $\beta^{(1)}$ satisfy the same linear equation. □

Proposition 20. *Let* $0 = \Phi_1(x^{(0)}) = \Phi_2(x^{(0)})$. *Then*

$$E(a) = [a|x^{(0)}] E_1(a).$$

Proof. We will show that every $a^{(0)}$ such that $[a^{(0)}|x^{(0)}] = 0$ also satisfies $E(a^{(0)}) = 0$.

Indeed, if $[a^{(0)}|x^{(0)}] = 0$, then

$$0 = \Phi_1([a^{(0)}|u_3]x_1^{(0)}, [a^{(0)}|u_3]x_2^{(0)}, -[a^{(0)}|u_1]x_1^{(0)} - [a^{(0)}|u_2]x_2^{(0)})$$

and

$$0 = \Phi_2([a^{(0)}|u_3]x_1^{(0)}, [a^{(0)}|u_3]x_2^{(0)}, -[a^{(0)}|u_1]x_1^{(0)} - [a^{(0)}|u_2]x_2^{(0)}).$$

By the previous proposition it follows that $E(a^{(0)}) = 0$, as desired. □

Proposition 21. *Suppose that* $E(a^{(1)}) = 0$. *Then there exists an* $x^{(0)}$ *such that*

$$E(a) = [a|x^{(0)}] E_1(a)$$

and

$$\Phi_1(x^{(0)}) = \Phi_2(x^{(0)}) = [a^{(1)}|x^{(0)}] = 0.$$

Proof. Since $E(a^{(0)}) = 0$, we must have

$$0 = \mathrm{Res} \begin{bmatrix} \Phi_1([a^{(0)}|u_3]x_1, [a^{(0)}|u_3]x_2, -[a^{(0)}|u_1]x_1 - [a^{(0)}|u_2]x_2), \\ \Phi_2([a^{(0)}|u_3]x_1, [a^{(0)}|u_3]x_2, -[a^{(0)}|u_1]x_1 - [a^{(0)}|u_2]x_2) \end{bmatrix}.$$

But

$$\begin{aligned} \Phi_1([a^{(0)}|u_3]x_1, [a^{(0)}|u_3]x_2, -[a^{(0)}|u_1]x_1 - [a^{(0)}|u_2]x_2) \\ = A(x_1, x_2) \\ = \prod_{i=1}^{d_1} (\alpha_1^{(i)} x_1 + \alpha_2^{(i)} x_2), \end{aligned}$$

and similarly

$$\begin{aligned}\Phi_2([a^{(0)}|u_3]x_1, [a^{(0)}|u_3]x_2, -[a^{(0)}|u_1]x_1 - [a^{(0)}|u_2]x_2) \\ = B(x_1, x_2) \\ = \prod_{i=1}^{d_2}(\beta_1^{(i)}x_1 + \beta_2^{(i)}x_2).\end{aligned}$$

By the reduction formula we have

$$[a|u_3]^{d_1 d_2} E(a) = \prod_{i=1}^{d_1}\prod_{j-1}^{d_2}[\alpha^{(i)}\beta^{(j)}] = 0.$$

Thus, $[\alpha^{(1)}\beta^{(1)}] = 0$, say, so both $\alpha^{(1)}$ and $\beta^{(1)}$ satisfy the same linear equation, to wit:

$$[\alpha^{(1)}|u_1]x_1^{(0)} + [\alpha^{(1)}|u_2]x_2^{(0)} = 0,$$

and

$$[\beta^{(1)}|u_1]x_1^{(0)} + [\beta^{(1)}|u_2]x_2^{(0)} = 0.$$

Hence,

$$\Phi_1([a^{(0)}|u_3], [a^{(0)}|u_3]x_2^{(0)}, -[a^{(0)}|u_1]x_1^{(0)} - [a^{(0)}|u_2]x_2^{(0)}) = 0,$$

and

$$\Phi_2([a^{(0)}|u_3]x_1^{(0)}, [a^{(0)}|u_3]x_2^{(0)}, -[a^{(0)}|u_1]x_1^{(0)} - [a^{(0)}|u_2]x_2^{(0)}) = 0.$$

Define $x_3^{(0)}$ by the equation

$$[a^{(0)}|u_1]x_1^{(0)} + [a^{(0)}|u_2]x_2^{(0)} + [a^{(0)}|u_3]x_3^{(0)} = 0 = [a^{(0)}|x^{(0)}].$$

Note that

$$\Phi_1([a^{(0)}|u_3]x_1^{(0)}, [a^{(0)}|u_3]x_2^{(0)}, [a^{(0)}|u_3]x_3^{(0)}) = 0,$$

and

$$\Phi_2([a^{(0)}|u_3]x_1^{(0)}, [a^{(0)}|u_3]x_2^{(0)}, [a^{(0)}|u_3]x_3^{(0)}) = 0.$$

In other words, $x^{(0)}$ is a common zero of Φ_1 and Φ_2.

We will show that

$$E(a) = [a|x^{(0)}]E_1(a).$$

This will be done by showing that every zero of $[a|x^{(0)}]$ is also a zero of $E(a)$. Suppose that there is an $a^{(1)}$ such that $[a^{(1)}|x^{(0)}] = 0$. Then $x^{(0)}$ is a common zero of $\Phi_1(x)$, $\Phi_2(x)$ and $[a^{(1)}|x]$. By the previous propositions $E(a^{(1)}) = 0$. □

We are now ready to prove the theorem presented at the beginning of this section, stating that the eliminant

$$E(a) = \mathrm{Res}(\Phi_1, \Phi_2, [a|x])$$

is a Cayley form in the variables $[a|u_1], [a|u_2], [a|u_3]$.

Proof of Theorem 4. We have shown that every zero of $E(a)$ is a zero of some linear factor of $E(a)$, and that the coefficients of each linear factor are common zeros of Φ_1, Φ_2. The proof is therefore complete. □

8 Eliminants

Recall again that the definition given for the resultant assumes that the successive eliminants are Cayley forms. Our objective in this section is to prove this crucial result in the general case.

Theorem 5. *The eliminant*

$$E(a) = \mathrm{Res}(\Phi_1, \cdots, \Phi_{n-1}, [a|x])$$

is a Cayley form in the variables $[a|u_1], \cdots [a|u_n]$*. That is, for suitable elements* $x_j^{(i)}$ *of the field* K*, the vectors*

$$x^{(i)} = (x_1^{(i)}, x_2^{(i)}, \cdots, x_n^{(i)})$$

satisfy

$$\begin{aligned} E(a) &= \prod_{i=1}^{d_1 d_2 \cdots d_{n-1}} [a|x^{(i)}] \\ &= \prod_{i=1}^{d_1 d_2 \cdots d_{n-1}} [a|u_1]x_1^{(i)} + [a|u_2]x_2^{(i)} + \cdots [a|u_n]x_n^{(i)} \end{aligned}$$

where the $x^{(i)}$ *are the common zeros of* $\Phi_1, \cdots, \Phi_{n-1}$.

This result will be a consequence of the following propositions.

Proposition 22. *Let* $x^{(0)} = (x_1^{(0)}, x_2^{(0)}, \cdots, x_n^{(0)})$*, where* $x_1^{(0)}, x_2^{(0)}, \cdots, x_n^{(0)} \in K$*, and let* $x^{(0)}$ *be a common zero of the forms* $\Phi_1(x), \cdots, \Phi_{n-1}(x)$ *and* $[a|x]$*, that is,*

$$0 = \Phi_1(x^{(0)}) = \cdots = \Phi_{n-1}(x^{(0)}) = [a|x^{(0)}].$$

Then

$$E(a) = 0.$$

Proof. Reasoning as in the ternary case we have for $i = 1, \ldots, n-1$,

$$\begin{aligned} 0 &= \Phi_i([a|u_n]x_1^{(0)}, \ldots, [a|u_n]x_n^{(0)}) \\ &= \Phi_i([a|u_n]x_1^{(0)}, \ldots, [a|u_n]x_{n-1}^{(0)}, -\sum_{j=1}^{n-1}[a|u_j]x_j^{(0)}). \end{aligned}$$

By the reduction formula we have

$$[a|u_n]^{(n-2)d_1 d_2 \ldots d_{n-1}} E(a)$$

$$= \text{Res}\begin{bmatrix} \Phi_1([a|u_n]x_1, \ldots, [a|u_n]x_{n-1}, -\sum_{j=1}^{n-1}[a|u_j]x_j), \\ \ldots, \\ \Phi_{n-1}([a|u_n]x_1, \ldots, [a|u_n]x_{n-1}, -\sum_{j=1}^{n-1}[a|u_j]x_j) \end{bmatrix}.$$

Since $x^{(0)}$ is a common zero of each of these forms, it follows by induction from the ternary case that the above resultant equals zero, and hence $E(a) = 0$. □

Proposition 23. *Let* $0 = \Phi_1(x^{(0)}) = \cdots = \Phi_{n-1}(x^{(0)})$. *Then*

$$E(a) = [a|x^{(0)}]E_1(a).$$

Proof. We will show that every $a^{(0)}$ such that $[a^{(0)}|x^{(0)}] = 0$ also satisfies $E(a^{(0)}) = 0$. Indeed, if $[a^{(0)}|x^{(0)}] = 0$, then for $i = 1, 2, \ldots, n-1$ we have

$$0 = \Phi_i([a^{(0)}|u_n]x_1^{(0)}, [a^{(0)}|u_n]x_2^{(0)}, \ldots, [a|u_n]x_{n-1}, -\sum_{j=1}^{n-1}[a^{(0)}|u_j]x_j^{(0)}).$$

By the previous proposition it follows that $E(a^{(0)}) = 0$, as desired. □

Proposition 24. *Suppose that $a^{(0)}$ is a zero of $E(a)$, that is, $E(a^{(0)}) = 0$. Then there exists an $x^{(0)}$ such that*

$$E(a) = [a|x^{(0)}]E_1(a)$$

and

$$0 = \Phi_1(x^{(0)}) = \cdots = \Phi_{n-1}(x^{(0)}) = [a^{(0)}|x^{(0)}].$$

Proof. Since $E(a^{(0)}) = 0$, we must have

$$0 = \text{Res}\begin{bmatrix} \Phi_1([a^{(0)}|u_n]x_1, \ldots, [a^{(0)}|u_n]x_{n-1}, -\sum_{i-1}^{n-1}[a^{(0)}|u_i]x_i), \\ \ldots, \\ \Phi_2([a^{(0)}|u_n]x_1, \ldots, [a^{(0)}|u_n]x_{n-1}, -\sum_{i-1}^{n-1}[a^{(0)}|u_i]x_i) \end{bmatrix}.$$

It follows by induction that the forms

$$\Phi_i([a^{(0)}|u_n]x_1, \ldots, [a^{(0)}|u_n]x_{n-1}, -\sum_{i-1}^{n-1}[a^{(0)}|u_i]x_i)$$

have a common zero, say $(x_1^{(0)}, \ldots, x_{n-1}^{(0)})$.

Define $x_n^{(0)}$ by the equation

$$[a^{(0)}|u_1]x_1^{(0)} + \cdots + [a^{(0)}|u_{n-1}]x_{n-1}^{(0)} + [a^{(0)}|u_n]x_n^{(0)} = 0 = [a^{(0)}|x^{(0)}].$$

Note that

$$\Phi_i([a^{(0)}|u_n]x_1^{(0)}, \ldots, [a^{(0)}|u_n]x_n^{(0)}) = 0.$$

In other words, $x^{(0)}$ is a common zero of $\Phi_1, \ldots, \Phi_{n-1}$.

We will show that $E(a) = [a|x^{(0)}]E_1(a)$. This will be done by showing that every zero of $[a|x^{(0)}]$ is also a zero of $E(a)$.

Suppose that there is an $a^{(1)}$ such that $[a^{(1)}|x^{(0)}] = 0$. Then $x^{(0)}$ is a common zero of $\Phi_1(x), \ldots, \Phi_{n-1}(x)$ and $[a^{(1)}|x]$. By the previous propositions it follows that

$$E(a^{(1)}) = 0. \qquad \square$$

We are now ready to prove the theorem stated at the beginning of this section, that the eliminant

$$E(a) = \text{Res}(\Phi_1, \cdots, \Phi_{n-1}, [a|x])$$

is a Cayley form in the variables $[a|u_1], \cdots, [a|u_n]$.

Proof of Theorem 5. We have shown that every zero of $E(a)$ is a zero of some linear-factor of $E(a)$, and that these linear factors give the common zeros of $\Phi_1, \ldots, \Phi_{n-1}$. Therefore, the proof is complete. $\square$

Corollary 5. *The eliminant $E(a)$ of $\Phi_2, \cdots, \Phi_n$ satisfies*

$$E(a) = \sum_{i=1}^{n}[a|u_i]^{d_2\cdots d_n}\text{Res}(\Phi_2, \cdots, \Phi_n, \Lambda_i) + \Delta,$$

where $\Lambda_i(x) = x_i$ and Δ is a sum of terms, each of which contains a product of at least two distinct letterplaces among $[a|u_1], \cdots, [a|u_n]$.

Corollary 6. *The resultant satisfies*

$$\begin{aligned}\text{Res}(\Phi, \Phi_2, \cdots, \Phi_n) = \; & \left[\Phi_{d_1,0,\cdots,0}\right]^{d_2\cdots d_n} \text{Res}\Lambda_1, \Phi_2, \cdots, \Phi_n) + \cdots \\ & + \left[\Phi_{0,\cdots,0,d_n}\right]^{d_1\cdots d_{n-1}} \text{Res}\Lambda_n, \Phi_2, \cdots, \Phi_n) + \Delta\end{aligned}$$

where Δ is a sum of terms, each of strictly lower degree in each of the variables $\Phi_{d_1,0,\cdots,0}, \cdots, \Phi_{0,\cdots,0,d_n}$

At this point we believe that the integrality of the resultant $\text{Res}(\Phi_1, \ldots, \Phi_n)$ as a polynomial in the coefficients $\phi^r_{i_1,\ldots,i_n}$ may be inferred.

9 Commutativity of the resultant operators

Proposition 25 (Commutativity of resultant operators).

$$R_{\Phi_1,a_1}R_{\Phi_2,a_2}[a_1a_2a_3\dots a_n] = R_{\Phi_2,a_1}R_{\Phi_1,a_2}[a_1a_2a_3\dots a_n]$$

Proof. By the reduction formula we have

$$\begin{aligned}
&[a_3\dots a_n|u_3\dots u_n]^{d_1d_2}R_{\Phi_1,a_1}R_{\Phi_2,a_2}[a_1a_2a_3\dots a_n]\\
&\quad = \mathrm{Res}(\Phi_1^{[2]}(x_1,x_2),\Phi_2^{[2]}(x_1,x_2))\\
&\quad = (-1)^{d_1d_2}\mathrm{Res}(\Phi_2^{[2]}(x_1,x_2),\Phi_1^{[2]}(x_1,x_2))\\
&\quad = [a_3\dots a_n|u_3\dots u_n]^{d_1d_2}R_{\Phi_2,a_1}R_{\Phi_1,a_2}[a_1a_2a_3\dots a_n].\square
\end{aligned}$$

Theorem 6. *Let $\Phi_1,\Phi_2,\cdots,\Phi_n$ be generic forms. Then*

$$\mathrm{Res}(\Phi_1,\Phi_2,\cdots,\Phi_n) = (-1)^{d_1d_2}\mathrm{Res}\Phi_2,\Phi_1,\cdots,\Phi_n).$$

Proof. We have shown that the eliminant $E(a_1)$ of $\Phi_2,\dots,\Phi_n$ factors as

$$E(a_1) = \prod_{i=1}^{d_2\dots d_n}\left[[a_1|u_1]x_1^{(i)}+\dots+[a_1|u_n]x_n^{(i)}\right],$$

where the $x^{(i)}$ are the common zeros of the forms $\Phi_2,\dots,\Phi_n$. From the definition of the resultant, it follows that

$$\begin{aligned}
\mathrm{Res}(\Phi_1,\Phi_2,\Phi_3,\dots,\Phi_n) &= R_{\Phi_1,a_1}E(a_1)\\
&= \prod_{i=1}^{d_2\dots d_n}\Phi_1(x^{(i)})
\end{aligned}$$

where $\Phi_2(x^{(i)})=0$ for $1\le i\le d_2\dots d_n$.

Suppose now that

$$\mathrm{Res}(\Phi_1,\Phi_2,\Phi_3,\dots,\Phi_n)=0.$$

From the last displayed equation it follows that $\Phi_1(x^{(1)})=0$, say. Now consider $\mathrm{Res}(\Phi_2,\Phi_1,\Phi_3,\dots,\Phi_n)$, and let $E_1(a_1)$ be the eliminant of $\Phi_1,\Phi_3,\dots,\Phi_n$ so that

$$E_1(a_1) = \prod_{i=1}^{d_1d_3\dots d_n}\left[[a_1|u_1]y_1^{(i)}+\dots+[a_1|u_n]y_n^{(i)}\right],$$

since $y^{(1)},\dots,y^{(d_1d_3\dots d_n)}$ are the common zeros of $\Phi_1,\Phi_3,\dots,\Phi_n$. We may assume that $y^{(1)}=x^{(1)}$. Now

$$\mathrm{Res}(\Phi_2,\Phi_1,\Phi_3,\dots,\Phi_n) = R_{\Phi_2,a_1}E_1(a_1).$$

Again we find that

$$\operatorname{Res}(\Phi_2, \Phi_1, \Phi_3, \ldots, \Phi_n) = \prod_{i=1}^{d_1 d_3 \ldots d_n} \Phi_2(y^{(i)}).$$

Since $y^{(1)} = x^{(1)}$, we infer that $\Phi_2(y^{(1)}) = 0$.

Thus, we have shown that, whenever the coefficients $\phi^r_{i_1 i_2 \ldots i_n}$ are assigned values in K, if the resultant satisfies

$$\operatorname{Res}(\Phi_1, \Phi_2, \Phi_3, \ldots, \Phi_n) = 0,$$

then the resultant satisfies

$$\operatorname{Res}(\Phi_2, \Phi_1, \Phi_3, \ldots, \Phi_n) = 0.$$

Since we have shown that the resultant is an irreducible polynomial, the two expressions can only differ in sign. The conclusion now follows easily. □

Corollary 7. *Let θ be a permutation of $1, \cdots, n$. Then*

$$\operatorname{Res}(\Phi_1, \Phi_2, \Phi_3, \ldots, \Phi_n) = \left[(-1)^{\theta}\right]^{d_1 \cdots d_n} \operatorname{Res}(\Phi_{\theta 1}, \Phi_{\theta 2}, \Phi_{\theta 3}, \ldots, \Phi_{\theta n}).$$

Corollary 8 (Commutativity of resultant operators). *Let θ be a permutation of $1, \cdots, n$. Then*

$$\begin{aligned} &R_{\Phi_1, a_1} R_{\Phi_2, a_2} \ldots R_{\Phi_n, a_n} [a_1 a_2 \ldots a_n] \\ &\quad = R_{\Phi_{\theta 1}, a_1} R_{\Phi_{\theta 2}, a_2} \ldots R_{\Phi_{\theta n}, a_n} [a_1 a_2 \ldots a_n]. \end{aligned}$$

10 Jouanolou's formula

Let $\Psi_1, \ldots, \Psi_n$ be forms all of degree d in the variables $x_1, \ldots, x_n$. We assume that $\Phi_1, \ldots, \Phi_n$ as well as $\Psi_1, \ldots, \Psi_n$ are S-free, where S is the set of letters $S = \{a_1, \ldots, a_n, b_1, \ldots, b_n\}$.

We set

$$\Phi_i \circ \Psi(x_1, \ldots, x_n) = \Phi_1(\Psi_1(x_1, \ldots, x_n), \ldots, \Psi_n(x_1, \ldots, x_n))$$

for $i = 1, \ldots, n$, and we define the form $[a_i | \Psi]$ to be

$$[a_i | \Psi(x)] = [a_i | u_1] \Psi_1(x) + \cdots + [a_i | u_n] \Psi_n(x).$$

Proposition 26.

$$\operatorname{Res}([a_1 | \Psi], \ldots, [a_n | \Psi]) = [a_1 \ldots a_n]^{d^{n-1}} \operatorname{Res}(\Psi_1, \ldots, \Psi_n).$$

Proof. We have seen that if $\Phi_n = \Phi_{n-1}\Psi_{n-1} + \Psi$ then

$$\mathrm{Res}(\Phi_1, \Phi_2, \ldots, \Phi_{n-1}, \Phi_n) = \mathrm{Res}(\Phi_1, \Phi_2, \ldots, \Phi_{n-1}, \Psi).$$

By the commutativity of resultant operators it follows that

$$\mathrm{Res}([a_1|\Psi], \ldots, [a_n|\Psi]) = \mathrm{Res}([a_1 \ldots a_n]\Psi_1, \ldots, \Psi_n)$$

from which the result follows. One may also reason that

$$[a_1 \ldots a_n]^d R_{[a_1|\Psi],b_1} \ldots R_{[a_n|\Psi],b_n}[b_1 \ldots b_n]$$

equals

$$[a_1 \ldots a_n]^{d^n} R_{\Psi_1,b_1} \ldots R_{\Psi_n,b_n}[b_1 \ldots b_n].$$

We may alternatively proceed as follows. Define a symmetric bilinear forms on the alphabet A with values in Super$[A|P]$ by

$$[b|a] = [b|u_1][a|u_1] + \cdots + [b|u_n][a|u_n] \in \mathrm{Super}[A|P].$$

Define compound inner products by

$$[a_1 \ldots a_n|b_1 \ldots b_n] = \sum_\theta (-1)^\theta \prod_{i=1}^n \left[\sum_{j=1}^n [a_i|u_j][b_{\theta i}|u_j] \right].$$

The proposition then follows by expanding in two ways

$$R_{\Psi_1,b_1} \ldots R_{\Psi_n,b_n}[a_1 \ldots a_n|b_1 \ldots b_n]. \qquad \square$$

Corollary 9. *The resultant* $\mathrm{Res}(\Psi^{(1)}, \ldots, \Psi^{(n)})$ *is a polynomial in the elements*

$$\sum_\theta (-1)^\theta \Psi^{(\theta 1)}_{i_{11}\cdots i_{1n}} \cdots \Psi^{(\theta n)}_{i_{n1}\cdots i_{nn}}.$$

Proposition 27.

$$R_{\Phi_i \circ \Psi, b}[b|x] = R_{\Phi_i, a_i} R_{[a_i|\Psi],b}[b|x].$$

Proof. Indeed,

$$\begin{aligned} R_{\Phi_i \circ \Psi, b_i}[b_i|x] &= \Phi_i(\Psi_1(x_1, \ldots, x_n), \ldots, \Psi_n(x_1, \ldots, x_n)) \\ &= R_{\Phi_i, a_i}([a_i|u_1]\Psi_1(x) + \cdots + [a_i|u_n]\Psi_n(x)) \\ &= R_{\Phi_i, a_i}[a_i|\Psi(x)] = R_{\Phi_i, a_i} R_{[a_i|\Psi],b_i}[b_i|x]. \qquad \square \end{aligned}$$

Proposition 28.

$$\prod_j \Phi_i \circ \Psi(y^{(j)}) = R_{\Phi_i, a_i} R_{[a_i|\Psi],b_i} \prod_j [b_i|y^{(j)}].$$

Proof. Immediate from the product formula. □

Proposition 29.

$$R_{\Phi_1,a_1} R_{\Phi_2,a_2} \dots R_{\Phi_n,a_n} \mathrm{Res}([a_1|\Psi], \cdots, [a_n|\Psi])$$
$$= \mathrm{Res}(\Phi_1 \circ \Psi, \cdots, \Phi_n \circ \Psi).$$

Proof.

$$R_{\Phi_1,a_1} \dots R_{\Phi_n,a_n} \mathrm{Res}[a_1|\Psi], \dots, [a_n|\Psi])$$
$$= R_{\Phi_1,a_1} \dots R_{\Phi_n,a_n} R_{[a_1|\Psi],b_1} \dots R_{[a_n|\Psi],b_n} [b_1 \dots b_n].$$

By the definition of the eliminant, we have that

$$R_{\Phi_1,a_1} \dots R_{\Phi_n,a_n} R_{[a_n|\Psi],b_n} R_{[a_{n-1}|\Psi],b_{n-1}} \dots R_{[a_1|\Psi],b_1} [b_1 \dots b_n]$$

equals

$$R_{\Phi_1,a_1} \dots R_{\Phi_{n-1},a_{n-1}} R_{\Phi_n,a_n} R_{[a_n|\Psi],b_n} E(b_n),$$

where $E(b_n)$ is the eliminant of the forms $[a_1|\Psi], \dots, [a_{n-1}|\Psi]$,

$$E(b_n) = \mathrm{Res}([a_1|\Psi], \dots, [a_{n-1}|\Psi], [b_n|x])$$

so that, since the eliminant is a Cayley form, $E(b_n) = \prod_j [b_n|x^{(j)}]$. Hence,

$$R_{\Phi_n,a_n} R_{[a_n|\Psi],b_n} E(b_n) = R_{\Phi_n \circ \Psi,b_n} E(b_n),$$

so that

$$R_{\Phi_1,a_1} \dots R_{\Phi_n,a_n} R_{[a_n|\Psi],b_n} R_{[a_{n-1}|\Psi],b_{n-1}} \dots R_{[a_1|\Psi],b_1} [b_1 \dots b_n],$$
$$= R_{\Phi_1,a_1} \dots R_{\Phi_{n-1},a_{n-1}} R_{\Phi_n \circ \Psi,b_n} R_{[a_{n-1}|\Psi],b_{n-1}} \dots R_{[a_1|\Psi],b_1} [b_1 \dots b_n]$$
$$= R_{\Phi_1,a_1} \dots R_{\Phi_{n-1},a_{n-1}} R_{[a_{n-1}|\Psi],b_{n-1}} \dots R_{[a_1|\Psi],b_1} R_{\Phi_n \circ \Psi,b_n} [b_1 \dots b_n]$$
$$= R_{\Phi_1,a_1} \dots R_{\Phi_{n-1},a_{n-1}} R_{[a_{n-1}|\Psi],b_{n-1}} E_1(b_{n-1}),$$

where $E_1(b_{n-1})$ is the eliminant of the forms $[a_1|\Psi], \dots, [a_{n-2}|\Psi], \Phi_n \circ \Psi$.

By parity of reasoning, the conclusion follows. □

Theorem 7 (Jouanolou's formula).

$$\mathrm{Res}(\Phi_1 \circ \Psi, \dots, \Phi_n \circ \Psi)$$
$$= \mathrm{Res}(\Psi_1, \dots, \Psi_n)^{d_1 \dots d_n} \mathrm{Res}(\Phi_1, \dots \Phi_n)^{d^{n-1}}.$$

Proof. By the previous proposition,

$$\begin{aligned}
&\mathrm{Res}(\Phi_1 \circ \Psi, \ldots, \Phi_n \circ \Psi) \\
&\quad = R_{\Phi_1,a_1} \ldots R_{\Phi_n,a_n} Res([a_1|\Psi], \ldots, [a_n|\Psi]) \\
&\quad = R_{\Phi_1,a_1} \ldots R_{\Phi_n,a_n} [a_1 \ldots a_n]^{d^{n-1}} \mathrm{Res}(\Psi_1, \ldots, \Psi_n),
\end{aligned}$$

as follows from the identity

$$\mathrm{Res}([a_1|\Psi], \ldots, [a_n|\Psi]) = [a_1 \ldots a_n]^{d^{n-1}} \mathrm{Res}(\Psi_1, \ldots, \Psi_n).$$

But the last expression equals

$$\mathrm{Res}(\Psi_1, \ldots, \Psi_n)^{d_1 \ldots d_n} R_{\Phi_1,a_1} \ldots R_{\Phi_n,a_n} [a_1 \ldots a_n]^{d^{n-1}},$$

and by the product formula this equals

$$\mathrm{Res}(\Psi_1, \ldots, \Psi_n)^{d_1 \ldots d_n} \left(R_{\Phi_1,a_1} \ldots R_{\Phi_n,a_n} [a_1 \ldots a_n] \right)^{d^{n-1}},$$

which is what was to be shown. □

Corollary 10. *Let $\Phi_1, \cdots, \Phi_n$ be forms of degree d_Φ, and let $\Psi_1, \ldots, \Psi_n$ be forms of degree d_Ψ. Then*

$$\begin{aligned}
&\mathrm{Res}(\Phi_1, \cdots, \Phi_n)^{d_\Psi} \mathrm{Res}(\Phi_1 \circ \Psi, \ldots, \Phi_n \circ \Psi) \\
&\quad = \mathrm{Res}(\Psi_1, \ldots, \Psi_n)^{d_\Phi} \mathrm{Res}(\Psi_1 \circ \Phi, \ldots, \Psi_n \circ \Phi).
\end{aligned}$$

References

[1] Cayley, A. (1989–1898): The collected mathematical papers of Arthur Cayley, vols. 1–14. Cambridge University Press, Cambridge

[2] Cullis, C. E. (1913–1925): Matrices and determinoids. Cambridge University Press, Cambridge

[3] Gel'fand, I.M., Kapranov, M.M., Zelevinsky, A.V. (1994): Discriminants, resultants and multdimensional determinants. Birkhäuser Boston, Boston, MA

[4] Grosshans, F., Rota, G.-C., Stein, J. A. (1987): Invariant theory and superalgebras. (CBMS Regional Conference Series in Mathematics, vol. 69). American Mathematical Society, Providence, RI

[5] Jouanolou, J.-P. (1980): Ideaux résultants. Adv. Math. **37**, 212–238

[6] Jouanolou, J.-P. (1991): Le formalisme du résultant. Adv. Math. **90**, 117–263

[7] Mac Mahon, P.A. (1978–1986): Collected papers I, II. MIT Press, Cambridge, MA

[8] Netto, E. (1896–1900): Vorlesungen über Algebra. Teubner, Leipzig

[9] Poisson, J.J. (1802): J. Ecole Polytech. **4**, 199–203

[10] Rota, G.-C., Stein, J. A. (1994): Plethystic Hopf algebras. Proc. Nat. Acad. Sci. U.S.A. **91**, 13057–13061

[11] Rota, G.-C., Stein, J. A. (1994): Plethystic algebras and vector symmetric function. Proc. Nat. Acad. Sci. U.S.A **91**, 13062–13066

[12] Schinzel, A. (1982): Selected topics on polynomials. University of Michigan Press, Ann Arbor, MI

Focus on Catalan Numbers and Combinatorics on Words

Foreword to the surveys by Aigner and Perrin

M. Barnabei, F. Bonetti

Gian-Carlo Rota always considered the academic role of great importance and spent as much time and effort in teaching as in doing research. In light of this we mention his very noteworthy quote (Los Alamos Science 12, 1985):

"A good teacher brings out the best in his students. What young people need the most is encouragement. Left to themselves, students may not know how to decide what is worthwhile. They may drop an original idea because they think someone else must have thought of it already."

For the same reason, going against the stream as he commonly did, he gave much consideration to survey articles, which he described in "Ten rules for the survival of a mathematics department" as follows:

"Not only is it good for you to write an expository paper once in a while, but such writing is essential for the survival of mathematics. Look at the most influential writings in mathematics of the last hundred years. At least half of them, from Hilbert's Zahlbericht on down, would have to be classified as expository."

Many, maybe all, of his Italian students became familiar with combinatorics by attending one of the many summer courses that Rota taught in Italy, thanks to his fascinating style of exposing ideas and problems. His lectures were at the same time a source of knowledge, a guide to research and a precious lesson in the methods of teaching.

Therefore, it will be no surprise that in a volume dedicated to him there will be space for two elegant survey articles, which introduce and develop two subjects both connected, although in different ways, to Rota's ideas. As was customary for Gian-Carlo himself, these subjects will be treated in depth but with a gentle exposition.

In the first of these surveys, Martin Aigner analyzes a subject of classic combinatorics, introducing variations on the theme of the Catalan numbers, which in his own words "...are next to the binomial coefficients the best studied of all combinatorial counting numbers".

In the second work, Dominique Perrin presents some combinatorial aspects of the theory of formal languages, thus describing an important application of typical combinatorial techniques to theoretical computer science.

These two articles efficiently show two different, but in a certain way complementary, aspects of research in the field of combinatorics.

Catalan and other numbers: a recurrent theme

M. Aigner

1 Six ways to look at Catalan numbers

The Catalan numbers C_n are, next to the binomial coefficients, the best studied of all combinatorial counting numbers. The wonderful book by Stanley [28] lists more than 70 instances of enumeration problems which are counted by the sequence (C_n). We will see several of them as we go along.

To start our survey we present six ways to introduce these numbers. Every approach has its own flavor and will be treated in subsequent chapters.

(A) Recursion. The most natural way, which we call the *combinatorial definition*, is the convolution recursion:

$$C_{n+1} = \sum_{k=0}^{n} C_k C_{n-k} \ (n \geq 0), \quad C_0 = 1. \tag{1.1}$$

We obtain as initial values:

n	0	1	2	3	4	5	6	7
C_n	1	1	2	5	14	42	132	429

As an example, we immediately find that C_n is the number of rooted binary plane trees with n vertices of degree 2 (by classifying the subtrees at the root):

$C_0 = 1$

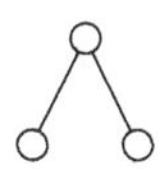

$C_1 = 1$

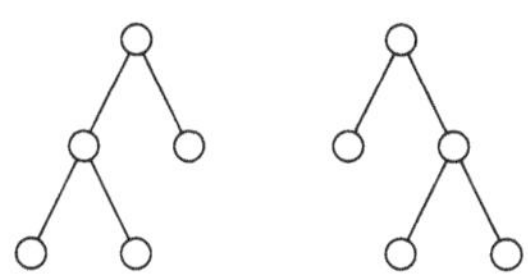

$C_2 = 2$

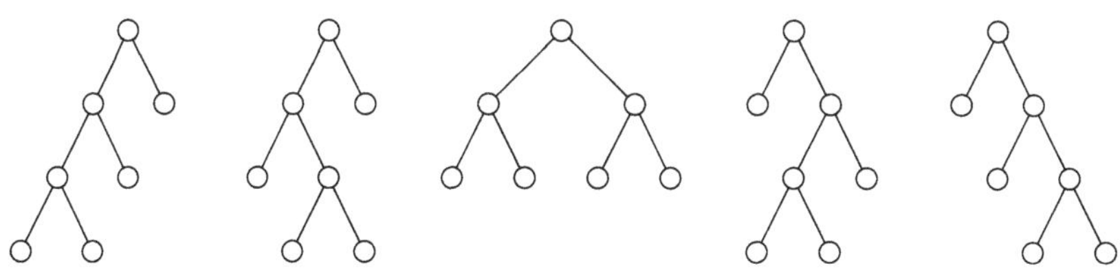

$C_3 = 5$

(B) Representation as an integral. We have

$$C_n = \frac{2^{2n+1}}{\pi} \int_{-1}^{1} x^{2n}(1-x^2)^{1/2} dx. \tag{1.2}$$

How this representation comes about will be explained in the next section. The integral is easily evaluated, resulting in the usual expression

$$C_n = \frac{1}{n+1}\binom{2n}{n}. \tag{1.3}$$

(C) Continued fractions. The ordinary generating function of $(C_0, 0, C_1, 0, C_2, 0, \dots)$ may be expressed as the following so–called J-fraction:

$$\sum_{n\geq 0} C_n z^{2n} = \cfrac{1}{1-\cfrac{z^2}{1-\cfrac{z^2}{1-\cfrac{z^2}{\cdots}}}} \tag{1.4}$$

We take a brief look at continued fractions in Sect. 3.

(D) Determinants of Hankel matrices. To the sequence (C_n) we associate the following two matrices for $n \geq 0$:

$$H_n = \begin{pmatrix} C_0 & C_1 & \dots & C_n \\ C_1 & C_2 & \dots & C_{n+1} \\ \vdots & & \dots & \\ C_n & C_{n+1} & \dots & C_{2n} \end{pmatrix}, \quad \widetilde{H}_n = \begin{pmatrix} C_1 & C_2 & \dots & C_{n+1} \\ C_2 & C_3 & \dots & C_{n+2} \\ \vdots & & \dots & \\ C_{n+1} & C_{n+2} & \dots & C_{2n+1} \end{pmatrix}.$$

H_n and $\widetilde{H}_n$ are called the *Hankel matrices* (of orders 0 and 1) of the sequence (C_n). We find

$$\det H_n = \det \widetilde{H}_n = 1 \text{ for all } n. \tag{1.5}$$

Now we may, of course, associate Hankel matrices H_n, $\widetilde{H}_n$ to any given sequence (B_n) of real numbers. It is clear that the sequence

$$\det H_0, \det \widetilde{H}_0, \det H_1, \det \widetilde{H}_1, \dots$$

uniquely determines the original number sequence provided that $\det H_n$ and $\det \widetilde{H}_n$ are $\neq 0$ for all n. Hence, the Catalan numbers (C_n) are the *unique* sequence with $\det H_n = \det \widetilde{H}_n = 1$. We will return to this beautiful characterization in Sect. 8.

(E) Ballot recursion Consider the infinite lower triangular matrix $B = (b_{n,k})$ defined as follows:

$$\begin{aligned} &b_{0,0} = 1, \quad b_{0,k} = 0 \quad (k > 0) \\ &b_{n,k} = b_{n-1,k-1} + b_{n-1,k} + \ldots + b_{n-1,n-1} \quad (n \geq 1). \end{aligned} \tag{1.6}$$

We obtain as initial rows and columns (leaving out the 0's):

$n \backslash k$	0	1	2	3	4	5
0	1					
1	1	1				
2	2	2	1			
3	5	5	3	1		
4	14	14	9	4	1	
5	42	42	28	14	5	1

We find that, for all n,

$$\begin{aligned} C_n &= b_{n,0}, \\ C_{n+1} &= \sum_{k=0}^{n} b_{n,k}. \end{aligned} \tag{1.7}$$

The numbers $b_{n,k}$ are usually called the *ballot numbers* (see, e.g., [24]) and are easily computed:

$$b_{n,k} = \frac{k+1}{n+1}\binom{2n-k}{n}$$

We will study generalized ballot numbers in some detail in Sect. 9.

(F) 3-term recursion. We come to our last and, for our purposes, most important way to introduce the Catalan numbers. Consider the following 3-term recursion which again gives rise to an infinite lower triangular matrix $A = (a_{n,k})$:

$$\begin{aligned} &a_{0,0} = 1, \quad a_{0,k} = 0 \quad (k > 0), \\ &a_{n,k} = a_{n-1,k-1} + a_{n-1,k+1} \quad (n \geq 1). \end{aligned} \tag{1.8}$$

The initial rows and columns are as follows:

$n \backslash k$	0	1	2	3	4	5	6
0	$\underline{1}$						
1	0	1					
2	$\underline{1}$	0	1				
3	0	2	0	1			
4	$\underline{2}$	0	3	0	1		
5	0	5	0	4	0	1	
6	$\underline{5}$	0	9	0	5	0	1

Then $C_n = a_{2n,0}$.

We have presented six rather different ways to introduce the Catalan numbers, but we still have to show that all approaches indeed yield the same sequence (C_n). The fascinating connections between the topics outlined in (A) to (F) will be precisely the content of this paper. For general background see any of the standard texts such as [16, 24, 27].

2 Recursive matrices and orthogonal systems of polynomials

Our starting point is a general form of (1.8). Given two sequences

$$\sigma = (s_0, s_1, s_2, \dots), \quad \tau = (t_1, t_2, t_3, \dots), \quad t_i \neq 0 \text{ for all } i,$$

we define the matrix $A = A^{\sigma,\tau} = (a_{n,k})$ recursively:

$$\begin{aligned} &a_{0,0} = 1, \quad a_{0,k} = 0 \quad (k > 0) \\ &a_{n,k} = a_{n-1,k-1} + s_k a_{n-1,k} + t_{k+1} a_{n-1,k+1} \quad (n \geq 1). \end{aligned} \tag{2.1}$$

We call $A = A^{\sigma,\tau}$ the σ, τ-matrix. A matrix A is called *recursive* if $A = A^{\sigma,\tau}$ for some pair of sequences σ and τ. The term "recursive matrix" appears in a different context in the literature (see, e.g., [5, 6, 25, 26]), but in this paper we always mean $A = A^{\sigma,\tau}$. With $A = A^{\sigma,\tau} = (a_{n,k})$, the numbers $B_n = B_n^{\sigma,\tau} = a_{n,0}$ are called the *Catalan-like numbers* corresponding to σ, τ.

Looking at (1.8) we see that the sequence $(C_0, 0, C_1, 0, C_2, 0, \dots)$ corresponds to $\sigma \equiv 0$, $\tau \equiv 1$.

The matrix $A = A^{\sigma,\tau} = (a_{n,k})$ is lower triangular with diagonal equal to 1. Furthermore, we find from (2.1) by induction that

$$a_{n+1,n} = s_0 + s_1 + \dots + s_n. \tag{2.2}$$

Given $\tau = (t_1, t_2, t_3, \dots)$, in the sequel we set

$$T_n = t_1 t_2 \dots t_n \quad (n \geq 1), \qquad T_0 = 1. \tag{2.3}$$

The following theorem (which is easily proved by induction) is basic to all that is to come.

Fundamental Theorem. *If $A = A^{\sigma,\tau} = (a_{n,k})$, then, for all m and n,*

$$\sum_k a_{m,k} a_{n,k} T_k = a_{m+n,0} = B_{m+n}. \tag{2.4}$$

We can write (2.4) in a compact matrix form. Define the diagonal matrix T and the infinite Hankel matrix of the sequence (B_n) as follows:

$$T = \begin{pmatrix} T_0 & & \\ & T_1 & & 0 \\ & 0 & T_2 & \\ & & & \ddots \end{pmatrix}, \qquad H = \begin{pmatrix} B_0 & B_1 & B_2 & \dots \\ B_1 & B_2 & B_3 & \dots \\ B_2 & B_3 & B_4 & \dots \\ & & \dots & \end{pmatrix}.$$

Then (2.4) can be written as

$$ATA^T = H. \tag{2.5}$$

Thus $A = A^{\sigma,\tau}$ satisfies (2.2) and (2.5). Conversely, it is readily seen that recursion (2.5) together with the sequence $(a_{n+1,n})$ *uniquely* determines A. Hence we have the following characterization.

Corollary 1. *$A = (a_{n,k})$ is a recursive matrix if and only if $ATA^T = H$ with $T_n \neq 0$ for all n, $T_0 = 1$. The sequences σ and τ are then given by $s_k = a_{k+1,k} - a_{k,k-1}$, $t_k = T_k / T_{k-1}$.*

If we look at (2.5) and consider the submatrices consisting of rows and columns 0 to n, that is,

$$A_n \begin{pmatrix} T_0 & & \\ & T_1 & & 0 \\ & & \ddots & \\ & 0 & & \\ & & & T_n \end{pmatrix} A_n^T = H_n,$$

we find that

$$\det H_n = T_0 T_1 \dots T_n \neq 0 \tag{2.6}$$

where H_n is the nth Hankel matrix of the sequence $(B_0, B_1, B_2, \dots)$. Again the converse holds, and we obtain the following characterization.

Corollary 2. *A sequence* (B_n) *is Catalan-like if and only if* $\det H_n \neq 0$ *for all* n.

For $A = A^{\sigma,\tau}$ we may consider the inverse matrix $U = A^{-1} = (u_{n,k})$. U is again lower triangular with diagonal equal to 1, and satisfies the recursion

$$\begin{aligned} &u_{0,0} = 1, \quad u_{0,k} = 0 \quad (k > 0) \\ &u_{n+1,k} = u_{n,k-1} - s_n u_{n,k} - t_n u_{n-1,k} \ \ (n \geq 0). \end{aligned} \tag{2.7}$$

Define the polynomial sequence $(p_n(x))$ by

$$p_n(x) = \sum_{k=0}^{n} u_{n,k} x^k; \tag{2.8}$$

then (2.7) translates into the recursive system

$$\begin{aligned} &p_0(x) = 1 \\ &p_{n+1}(x) = (x - s_n) p_n(x) - t_n p_{n-1}(x) \quad (n \geq 0). \end{aligned} \tag{2.9}$$

Hence taking $A = A^{\sigma,\tau}$ and $U = A^{-1}$ together we find that

$$\begin{array}{ccc} A = A^{\sigma,\tau} & \Longleftrightarrow & ATA^T = H \\ \Updownarrow & & \Updownarrow \\ p_{n+1}(x) = (x - s_n) p_n(x) - t_n p_{n-1}(x) & & UHU^T = T \end{array}$$

and have thus obtained the equivalence

$$(p_n(x)) \text{ satisfies (2.9)} \Longleftrightarrow UHU^T = T \tag{2.10}$$

for some Hankel matrix H.

It remains to interpret the right-hand side of (2.10) in terms of the sequence $(p_n(x))$, and this leads us to orthogonal systems of polynomials.

A sequence of real polynomials $(p_n(x))$, where $\deg p_n(x) = n$ for all n, is said to form an *orthogonal system* (OPS) if there exists a linear operator $L : \mathbb{R}[x] \longrightarrow \mathbb{R}$ and numbers $T_n \neq 0$ $(n \geq 1)$, $T_0 = 1$, such that

$$L(p_m(x) p_n(x)) = \delta_{m,n} T_n,$$

where $\delta_{m,n}$ is the Kronecker symbol.

Suppose $UHU^T = T$, where H is the Hankel matrix of the sequence (B_n). Define the linear operator L by

$$L : x^n \longrightarrow B_n \text{ for all } n.$$

Then

$$\begin{aligned} UHU^T = T &\iff \sum_{i,k} u_{m,i} u_{n,k} B_{i+k} = \delta_{m,n} T_n \\ &\iff L((\sum_i u_{m,i} x^i)(\sum_k u_{n,k} x^k)) = \delta_{m,n} T_n \\ &\iff L(p_m(x) p_n(x)) = \delta_{m,n} T_n \text{ for all } m, n, \end{aligned}$$

that is, $(p_n(x))$ forms an OPS. The converse is just as easily established, whence, in view of (2.10), we obtain the famous theorem of Favard.

Corollary 3. *A sequence $(p_n(x))$ forms an orthogonal system if and only if $p_{n+1}(x) = (x - s_n)p_n - t_n p_{n-1}(x)$, $p_0(x) = 1$, for some pair of sequences $\sigma = (s_k)$, $\tau = (t_k)$.*

The analytical study of orthogonal systems goes back to Chebyshev and Stieltjes and led to the invention of the Stieltjes integral (see [9]). In this context, the sequence $B_n = Lx^n$ is called the *moment sequence* of L. The classical results in this area concern the structure of the operator L for the OPS $(p_n(x))$: given $B_n = Lx^n$, as before let H_n be the nth Hankel matrix. Then

$$\begin{aligned} \det H_n > 0 \text{ for all } n &\implies Lq = \int_{-\infty}^{\infty} q\,d\psi \\ \det H_n > 0, \det \widetilde{H}_n > 0 \text{ for all } n &\implies Lq = \int_{0}^{\infty} q\,d\psi \end{aligned}$$

where $\psi(x)$ is bounded and non-decreasing.

Example. Here we analyze our initial sequence (B_n) arising from $\sigma \equiv 0$, $\tau \equiv 1$. We have $T_n = 1$ for all n, and therefore $\det H_n = 1$ by (2.6). According to (2.9) the associated OPS satisfies the recursion

$$p_{n+1}(x) = xp_n(x) - p_{n-1}(x), \quad p_0(x) = 1.$$

Now it is a classical result [21] that the *Chebyshev polynomials* $U_n(x) = \frac{\sin(n+1)\vartheta}{\sin\vartheta}$, $x = \cos\vartheta$, satisfy the recursion

$$U_{n+1}(x) = 2xU_n(x) - U_{n-1}(x), \quad U_0(x) = 1.$$

Hence we obtain $p_n(x) = U_n(\frac{x}{2})$. Another classical result says that the orthogonality relation for $(U_n(x))$ is given by

$$\int_{-1}^{1} U_m(x)U_n(x)(1-x^2)^{1/2}dx = 0 \quad \text{for } m \neq n.$$

Making the sustitution $y = \frac{x}{2}$ and normalizing, we arrive at the operator L:

$$Lq(x) = \frac{2}{\pi}\int_{-1}^{1} q(2x)(1-x^2)^{1/2}dx.$$

Hence the moment sequence (for even index) $B_{2n} = Lx^{2n}$ is precisely as given in (1.2).

3 Recursive matrices and continued fractions

Again this is a time-honored subject and intimately connected with the name Jacobi [9]. In a seminal paper Flajolet [14] studied many combinatorial enumeration problems from this point of view.

Consider a recursive matrix $A = A^{\sigma,\tau}$ associated with the sequences $\sigma = (s_k)$, $\tau = (t_k)$. We recall the recursion

$$\begin{aligned} a_{0,0} &= 1, \quad a_{0,k} = 0 \quad (k > 0) \\ a_{n,0} &= \phantom{a_{n-1,k-1}+} s_0 a_{n-1,0} + t_1 a_{n-1,1} \\ &\qquad\qquad\qquad\qquad\qquad\qquad (n \geq 1) \\ a_{n,k} &= a_{n-1,k-1} + s_k a_{n-1,k} + t_{k+1} a_{n-1,k+1}. \end{aligned} \tag{3.1}$$

Consider the ordinary generating function of the Catalan-like numbers $B_n = a_{n,0}$:

$$G(z) = \sum_{n \geq 0} B_n z^n.$$

We make the following crucial observation: the first $m+1$ columns $0, 1, \ldots, m$ of A determine the numbers $B_0, B_1, \ldots, B_{2m+1}$.

Now consider the system (3.1) only for the columns $0, 1, \ldots, m$; in other words we set $t_{m+1} = 0$. Let $\widehat{C}_k(z)$ be the generating function of the kth column of this truncated system.

For an arbitrary generating function $F(z) = \sum_{n \geq 0} F_n z^n$ we define the shift operator Δ by

$$\Delta F(z) = \sum_{n \geq 0} F_{n+1} z^n.$$

Clearly,

$$\Delta F(z) = \frac{F(z) - F(0)}{z}. \tag{3.2}$$

Hence the truncated system (3.1) translates into

$$\begin{aligned} \Delta\widehat{C}_0(z) &= \phantom{\widehat{C}_{k-1}(z)+} s_0\widehat{C}_0(z) + t_1\widehat{C}_1(z) \\ \Delta\widehat{C}_k(z) &= \widehat{C}_{k-1}(z) + s_k\widehat{C}_k(z) + t_{k+1}\widehat{C}_{k+1}(z) \quad (k \geq 1). \end{aligned}$$

According to (3.2) we find that

$$\Delta\widehat{C}_0(z) = \frac{\widehat{C}_0(z) - 1}{z}, \quad \Delta\widehat{C}_k = \frac{\widehat{C}_k(z)}{z} \quad (k \geq 1),$$

and thus

$$\begin{aligned}\widehat{C}_0(z) - 1 &= \qquad\qquad s_0 z\widehat{C}_0(z) + t_1 z\widehat{C}_1(z)\\ \widehat{C}_k(z) &= z\widehat{C}_{k-1}(z) + s_k z\widehat{C}_k(z) + t_{k+1} z\widehat{C}_{k+1}(z).\end{aligned}$$

Writing this out in detail, we arrive at the following system of linear equations:

$$\begin{array}{rcccccl} (1-s_0z)\widehat{C}_0 & - & t_1z\widehat{C}_1 & & & = & 1\\ -z\widehat{C}_0 & + & (1-s_1z)\widehat{C}_1 & - & t_2z\widehat{C}_2 & = & 0\\ & & & \cdots & & & \cdots\\ & & & - & z\widehat{C}_{m-1} + (1-s_m)\widehat{C}_m & = & 0 \end{array} \tag{3.3}$$

Applying Cramer's rule to (3.3) we find that

$$\widehat{C}_0(z) = \frac{\det\begin{pmatrix} 1-s_1z & -t_2z & & \\ -z & 1-s_2z & & \\ & & \cdots & \\ & & -z & 1-s_mz \end{pmatrix}}{\det\begin{pmatrix} 1-s_0z & -t_1z & & \\ -z & 1-s_1z & & \\ & & \cdots & \\ & & -z & 1-s_mz \end{pmatrix}} =: \frac{D_1}{D_0}.$$

Developing the determinant D_0 according to the first column gives

$$D_0 = (1-s_0z)D_1 - t_1z^2D_2,$$

and thus

$$\widehat{C}_0(z) = \frac{D_1}{(1-s_0z)D_1 - t_1z^2D_2} = \frac{1}{(1-s_0z) - t_1z^2\frac{D_2}{D_1}}. \tag{3.4}$$

Iteration of (3.4) leads therefore to the continued fraction

$$\widehat{C}_0(z) = \cfrac{1}{1-s_0z-\cfrac{t_1z^2}{1-s_1z-\cfrac{\cdots}{\cdots-\cfrac{t_mz^2}{1-s_mz}}}}$$

and we *know* from our observation that $\widehat{C}_0(z)$ agrees with $G(z) = \sum B_n z^n$ up to index $2m+1$. Going to infinity with m yields therefore the so-called infinite J-fraction expansion of $G(z)$:

$$G(z) = \cfrac{1}{1 - s_0 z - \cfrac{t_1 z^2}{1 - s_1 z - \cfrac{t_2 z^2}{1 - s_2 z - \quad \dots}}}$$

Looking once more at our initial example $\sigma \equiv 0$, $\tau \equiv 1$, we find precisely the expansion (1.4).

4 Recursive matrices and lattice paths

In his fundamental work [29] Viennot studied orthogonal systems of polynomials – and hence recursive matrices – from the point of view of *weighted lattice paths*. We introduce his basic ideas in this section.

Consider once more a recursive matrix $A = A^{\sigma,\tau} = (a_{n,k})$ given by

$$a_{n,k} = a_{n-1,k-1} + s_k a_{n-1,k} + t_{k+1} a_{n-1,k+1}.$$

Turning the matrix by 90° we associate $a_{n,k}$ with the point (n, k). Hence our recursion looks as follows:

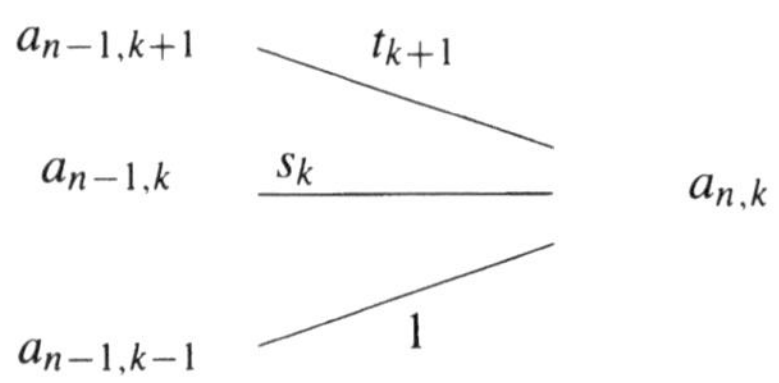

A *Motzkin path* of length n is a lattice path with the following properties:

- it starts at $(0, 0)$ and ends at $(n, 0)$;
- it has steps $(1, 1)$, $(1, 0)$, $(1, -1)$ (diagonally up, horizontal, diagonally down);
- it never falls below $y = 0$.

To the steps we assign weights w as follows:

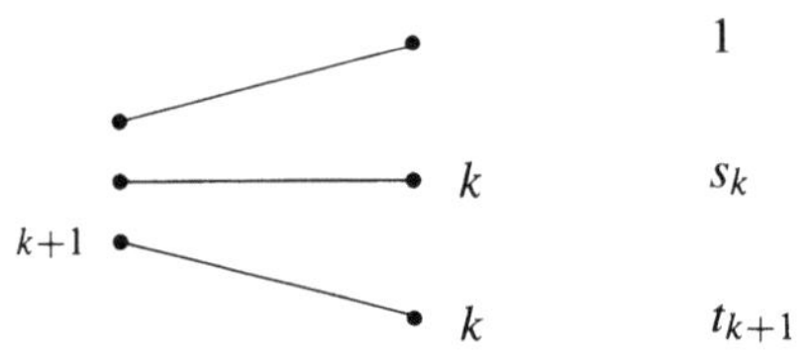

Let $\mathcal{P}_n$ be the set of Motzkin paths of length n. For $P \in \mathcal{P}_n$ we define the *weight* $w(P)$ as

$$w(P) = \prod w\text{ (steps)}.$$

A moment's thought should convince the reader that

$$B_n = \sum_{P \in \mathcal{P}_n} w(P).$$

Example. For our initial example $\sigma \equiv 0$, $\tau \equiv 1$, this interpretation just means that $B_{2n} = C_n$ counts the number of Motzkin paths of length $2n$ without horizontal steps ($\sigma \equiv 0$). These paths are usually called Dyck paths, and their enumeration by means of the Catalan numbers is one of the best-known of all counting results. As an example, for $B_6 = C_3 = 5$ we obtain the 5 paths:

5 Recursive matrices: umbral calculus

In the last section we discussed some interesting connections of recursive matrices $A = A^{\sigma,\tau}$ to other fields, but how do we compute the Catalan-like numbers $B_n = B_n^{\sigma,\tau}$? Or rather, what do these numbers count? One way (certainly not the only one) to tackle this question is the umbral calculus, so dear to Rota's work. There is a great body of literature on this subject under various names, e.g., [5, 22, 23, 25, 26].

We start once more with the defining recursion

$$a_{n,k} = a_{n-1,k-1} + s_k a_{n-1,k} + t_{k+1} a_{n-1,k+1}. \tag{5.1}$$

We set

$$t_n = q_n u_n \quad (n \geq 1), \qquad Q_n = q_1 \ldots q_n, \quad Q_0 = 1.$$

Now we consider Q-generating functions

$$G(z) = \sum_{n \geq 0} \frac{G_n}{Q_n} z^n.$$

The (linear) shift operator Δ is defined by

$$\Delta : \begin{array}{l} z^0 \longrightarrow 0 \\ z^n \longrightarrow q_n z^{n-1} \quad (n \geq 1); \end{array}$$

hence

$$\Delta G(z) = \sum_{n\geq 0} \frac{G_{n+1}}{Q_n} z^n.$$

For $A = A^{\sigma,\tau}$ let $C_k(z)$ be the Q-generating function of the kth column, that is,

$$C_k(z) = \sum_{n\geq 0} \frac{a_{n,k}}{Q_n} z^n \quad (k \geq 0).$$

The recursion (5.1) therefore translates into

$$\begin{aligned} \Delta C_k &= C_{k-1} + s_k C_k + u_{k+1} q_{k+1} C_{k+1} \quad (k \geq 1) \\ \Delta C_0 &= aC_0 + bq_1 C_1, \end{aligned} \tag{5.2}$$

where we set $a = s_0$, $b = u_1$. (The numbers a and b will play a special role.)

Here comes the umbral calculus idea. We consider only solutions of (5.2) of the form

$$C_k(z) = C_0(z)\frac{F(z)^k}{Q_k}, \quad F(0) = 0. \tag{5.3}$$

Hence, with (5.3), (5.2) transforms into

$$\begin{aligned} \Delta(C_0 F^k) &= q_k C_0 F^{k-1} + s_k C_0 F^k + u_{k+1} C_0 F^{k+1} \quad (k \geq 1) \\ \Delta C_0 &= aC_0 + bC_0 F, \end{aligned} \tag{5.4}$$

and the goal is to compute

$$C_0(z) = \sum_{n\geq 0} \frac{B_n}{Q_n} z^n.$$

According to (5.4) we need a multiplication rule for $\Delta(GH)$. To find this we assume the following special case. For any Q-generating functions $G(z)$, $H(z)$,

$$\Delta(GH) = (\Delta G)H(qz) + (\Delta H)G \text{ for some } q \in \mathbb{R}. \tag{5.5}$$

There are three cases:

(A) $q = 0$: Normalizing $q_1 = 1$, we obtain $q_n = 1$ for all n, resulting in *ordinary generating functions* $G(z) = \sum\limits_{n\geq 0} G_n z^n$.

(B) $q = 1$: Normalizing $q_1 = 1$ we obtain $q_n = n$ for all n, and hence $Q_n = n!$, resulting in *exponential generating functions* $G(z) = \sum\limits_{n\geq 0} \frac{G_n}{n!} z^n$.

(C) $q \neq 0, 1$: Normalizing $q_1 = q - 1$, we obtain $q_n = q^n - 1$, and hence $Q_n = \prod\limits_{i=1}^{n} (q^i - 1)$, resulting in *$q$-generating functions* $G(z) = \sum\limits_{n\geq 0} \frac{G_n}{\prod_{i=1}^{n} (q^i - 1)} z^n$.

Applying (5.5) to F^k we find that

$$\Delta F^k = (\Delta F)(F^{k-1}(z) + F^{k-2}(z)F(qz) + \ldots + F^{k-1}(qz)),$$

which yields for our three cases:

$$\begin{array}{lll} \text{(A)} & q = 0: & \Delta F^k = (\Delta F)F^{k-1}, \ \ \Delta F = \frac{F(z)-F(0)}{z}, \\ \text{(B)} & q = 1: & \Delta F^k = (\Delta F)kF^{k-1}, \ \ \Delta F = F', \\ \text{(C)} & q \neq 0, 1: & \Delta F^k = (\Delta F)(F^{k-1}(z) + \ldots + F^{k-1}(qz)), \\ & & \Delta F = \frac{F(qz)-F(z)}{z}. \end{array} \tag{5.6}$$

Now we substitute (5.6) into the system (5.4) and try to compute $F(z)$ and $C_0(z)$.

We will treat the solutions for the three cases separately.

(A) Ordinary generating functions (divided difference calculus). We obtain

$$\Delta(C_0F^k) = C_0(\Delta F)F^{k-1},$$

and hence, by using (5.4) and cancelling,

$$\begin{aligned} \Delta F &= 1 + s_k F + u_{k+1}F^2 \quad (k \geq 1) \\ \Delta C_0 &= aC_0 + bC_0F. \end{aligned} \tag{5.7}$$

We see that (5.7) has a solution of the form (5.3) if and only if $s_k = s$ $(k \geq 1)$, $u_k = u$ $(k \geq 2)$. Hence the possible sequences σ and τ are of the form

$$\sigma = (a, s, s, s, \ldots), \quad \tau = (b, u, u, u, \ldots), \tag{5.8}$$

and (5.7) becomes

$$\begin{aligned} \Delta F &= 1 + sF + uF^2 \\ \Delta C_0 &= aC_0 + bC_0F. \end{aligned} \tag{5.9}$$

In terms of the coefficients of $F(z)$ and $C_0(z)$, (5.9) translates into

$$\begin{aligned} F_{n+1} &= sF_n + u\sum_{k=1}^{n-1} F_kF_{n-k} \quad (n \geq 1), \qquad F_0 = 0, F_1 = 1 \\ B_{n+1} &= aB_n + b\sum_{k=0}^{n-1} B_kF_{n-k} \quad (n \geq 0), \qquad B_0 = 1. \end{aligned} \tag{5.10}$$

This latter system now leads to the desired combinatorial interpretations which we will discuss in detail in the next section.

(B) Exponential generating functions (classical calculus). A similar analysis leads to the result that the solutions are precisely $s_k = a + ks$, $u_{k+1} = b + ku$ for fixed a, s, b, u. Hence the sequences σ and τ are of the form

$$\sigma = (a + ks), \quad \tau = (k(b + (k-1)u)), \tag{5.11}$$

and the system (5.4) becomes

$$\begin{aligned} F' &= 1 + sF + uF^2 \\ C_0' &= aC_0 + bC_0F. \end{aligned} \tag{5.12}$$

In terms of the coefficients $F(z)$ and $C_0(z)$, (5.12) translates into

$$\begin{aligned} F_{n+1} &= sF_n + u\sum_{k=1}^{n-1}\binom{n}{k}F_kF_{n-k} \quad (n \ge 1), \ F_0 = 0, F_1 = 1 \\ B_{n+1} &= aB_n + b\sum_{k=0}^{n-1}\binom{n}{k}B_kF_{n-k} \quad (n \ge 0), \ B_0 = 1. \end{aligned} \tag{5.13}$$

The system (5.13) again leads to manifold combinatorial interpretations as will be studied in the next section.

(C) q-generating functions (q-calculus), $q \neq 0, 1$. Here the solution is incomplete. If the quadratic coefficient of $F(z)$ is $F_2 = 0$, then it can be shown that $F(z) = z$ is the only solution. The case $F_2 \neq 0$ is open. For $F(z) = z$, it is readily seen that $s_k = aq^k$, $u_{k+1} = bq^k$ are the only possibilities. In other words, the sequences σ and τ are of the form

$$\sigma = (aq^k), \quad \tau = (bq^{k-1}(q^k - 1)). \tag{5.14}$$

In terms of the Catalan-like numbers B_n, the system (5.4) translates into

$$B_{n+1} = aB_n + b(q^n - 1)B_{n-1} \quad (n \ge 0), \quad B_0 = 1. \tag{5.15}$$

Again we will discuss combinatorial interpretations of (5.15) in the next section.

Before going on to these combinatorial interpretations we take another quick look at orthogonal systems of polynomials via the umbral calculus.

As before let $A = (a_{n,k})$, $C_k(z) = \sum_{n \ge 0} \frac{a_{n,k}}{Q_n} z^n$. To state that (5.3) holds, we use the abbreviated notation

$$A \longleftrightarrow (C_0, F).$$

The following result is well-known (see [25]):

$$A \longleftrightarrow (C_0, F) \Longleftrightarrow A^{-1} \longleftrightarrow \left(\frac{1}{C_0(\overline{F})}, \overline{F}\right), \tag{5.16}$$

where $\overline{F}$ is the *composition inverse* of F, that is,

$$\overline{F}(F(z)) = F(\overline{F}(z)) = z.$$

As in Sect. 3 set $A^{-1} = (u_{n,k})$ and

$$p_n(x) = \sum_{k=0}^{n} u_{n,k} x^k. \tag{5.17}$$

In umbral calculus, the sequence $(p_n(x))$ is called the *Sheffer sequence* corresponding to $A \longleftrightarrow (C_0, F)$.

Define the infinite (column) vectors $Z = (\frac{z^n}{Q_n})$, $X = (x^n)$. Then by (5.16) and (5.17) we obtain

$$\sum_{n\geq 0} \frac{p_n(x)}{Q_n} z^n = Z^T A^{-1} X = \frac{1}{C_0(\overline{F}(z))} \sum_{k\geq 0} \frac{(x\overline{F}(z))^k}{Q_k}. \tag{5.18}$$

Question. Given a Q-umbral calculus, which OPS $(p_n(x))$ are Sheffer sequences?

The answer is given by our analysis of the cases (A) to (C).

(A) Ordinary generating functions. Here (5.8) and (5.18) yield the following result. An OPS $(p_n(x))$ is a Sheffer sequence with

$$\sum_{n\geq 0} p_n(x) z^n = g(z) \frac{1}{1 - xf(z)}$$

if and only if $s_k = s \ (k \geq 1)$, $t_k = u \ \ (k \geq 2)$.

(B) Exponential generating functions. Here (5.11) and (5.18) give a classical result of Meixner [9]. An OPS $(p_n(x))$ is a Sheffer sequence with

$$\sum_{n\geq 0} \frac{p_n(x)}{n!} z^n = g(z) e^{xf(z)}$$

if and only if $s_k = a + ks$, $t_k = k(b + (k-1)u)$.

(C) q-generating functions. With the assumption $F_2 = 0$, we obtain the following result from (5.14) and (5.18). An OPS $(p_n(x))$ is a Sheffer sequence with

$$\begin{aligned}\sum_{n\geq 0} \frac{p_n(x)}{(q^n-1)\dots(q-1)} z^n &= g(z) \sum_{k\geq 0} \frac{(xz)^k}{(q^k-1)\dots(q-1)} \\ &= g(z) \prod_{k\geq 0} \frac{1}{1 + q^k xz}\end{aligned}$$

if and only if $s_k = aq^k$, $t_k = bq^{k-1}(q^k - 1)$.

Example. We look at the particularly simple case when $F(z) = z$, thus $A \longleftrightarrow (C_0(z), z)$. This is called the *Appel case* in the literature [22].

(A) Ordinary generating functions. According to (5.9) we obtain

$$\Delta F = 1 = 1 + sz + uz^2.$$

Hence there is no solution since $u \neq 0$.

(B) Exponential generating functions. By (5.12) we obtain

$$F' = 1 = 1 + sz + uz^2,$$

and thus $s = u = 0$, that is, $s_k = a$, $t_k = bk$ by (5.11). Looking again at (5.12) we see that

$$C_0' = aC_0 + bC_0 z,$$

and therefore as solution

$$C_0 = e^{az + \frac{bz^2}{2}}.$$

The corresponding OPS $(p_n(x))$ is given by (5.18):

$$\begin{aligned}\sum_{n \geq 0} \frac{p_n(x)}{n!} z^n &= e^{-az - \frac{bz^2}{2}} \sum_{k \geq 0} \frac{(xz)^k}{k!} \\ &= e^{(-a+x)z - \frac{bz^2}{2}},\end{aligned}$$

and by an easy computation we obtain for $p_n(x)$ the classical *Hermite polynomials* $H_n^{a,b}(x)$:

$$H_n^{a,b}(x) = \sum_j \Big(\sum_k \frac{(-1)^{j-k} a^{j-2k} b^k}{(j-2k)! 2^k k!}\Big) n^{\underline{j}} x^{n-j},$$

where $n^{\underline{j}} = n(n-1)\dots(n-j+1)$ are the falling factorials. In particular, for $a = 0$ and hence $s_k = 0$, $t_k = bk$, we find that

$$H_n^{0,b}(x) = \sum_{k \geq 0} \Big(-\frac{b}{2}\Big)^k \frac{n^{\underline{2k}}}{k!} x^{n-2k}.$$

Example. We look again at our initial example $\sigma \equiv 0$, $\tau \equiv 1$. According to (5.7) we have

$$\Delta F = \frac{F}{z} = 1 + F^2,$$

and thus $F = z + zF^2$. A simple computation yields $\overline{F} = \frac{z}{1+z^2}$, and $C_0(\overline{F}) = 1 + z^2$. Hence we infer from (5.18) that

$$\sum_{n \geq 0} p_n(x) z^n = \frac{1}{1+z^2} \frac{1}{1 - \frac{zx}{1+z^2}} = \frac{1}{1 - xz + z^2}.$$

Comparing coefficients this gives the usual expression for the Chebyshev polynomials $U_n(x)$:

$$p_n(x) = \sum_i (-1)^i \binom{n-i}{i} x^{n-2i},$$

and hence

$$U_n(x) = p_n(2x) = \sum_i (-1)^i \binom{n-i}{i} 2^{n-2i} x^{n-2i}.$$

We close this section with a natural problem.

Question. Which recursive matrices A have an inverse $U = A^{-1}$ which is also recursive? Suppose $A = A^{\sigma,\tau}$ and $U = U^{\sigma',\tau'}$. Then it is easily seen that

$$\sigma' = -\sigma.$$

Furthermore, in the Sheffer case $A \longleftrightarrow (C_0, F)$, there are two possibilities:

(a) $\tau' = -\tau$: This implies $\sigma = (s, s, s, \dots)$, $\tau = (t, 2t, 3t, \dots)$, and we obtain the *Hermite polynomials* $H_n^{s,t}(x)$ from above.
(b) $\tau' = \tau$: Here we get $\sigma = (s_k)$, $s_k = a + 2mk$, $\tau = (t_k)$, $t_k = k(am + m^2(k-1))$ for some m. The corresponding OPS are the classical *Laguerre polynomials* $L_n^{a,m}(x)$.

Nothing substantial is known in the non-Sheffer case.

6 Catalan-like numbers: combinatorial interpretations

Starting with the recursion formulae (5.10), (5.13), (5.15) we give a combinatorial interpretation in each case, then derive convolution-type formulae, and give a host of examples, including some of the most basic combinatorial counting numbers.

(A) Ordinary generating functions. Given the sequences $\sigma = (a, s, s, \dots)$, $\tau = (b, u, u, \dots)$, the Catalan-like numbers $B_n = B_n^{\sigma,\tau}$ are computed by the system (5.10):

$$F_{n+1} = sF_n + u \sum_{k=1}^{n-1} F_k F_{n-k} \quad (n \geq 1), \quad F_0 = 0, F_1 = 1 \tag{6.1}$$

$$B_{n+1} = aB_n + b \sum_{k=0}^{n-1} B_k F_{n-k} \quad (n \geq 0), \quad B_0 = 1. \tag{6.2}$$

Our combinatorial objects are *Motzkin trees*, that is, rooted plane trees in which every vertex has out-degree ≤ 2. The figure shows the trees with $n \leq 4$ vertices.

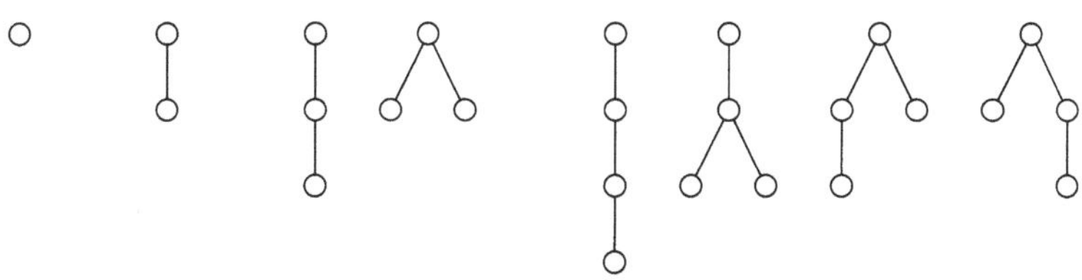

Let $\mathcal{T}_n$ be the set of Motzkin trees with n vertices. For $T \in \mathcal{T}_n$ define the weight

$$w(T) = s^{\ell} u^{m} \text{ where } \begin{array}{rcl} \ell & = & \#\text{ vertices of out-degree } 1, \\ m & = & \#\text{ vertices of out-degree } 2. \end{array}$$

Then, for F_n in (6.1), we have for $n \geq 2$

$$F_n = \sum_{T \in \mathcal{T}_n} w(T) \quad (F_0 = 0,\, F_1 = 1).$$

The figure below shows the proof:

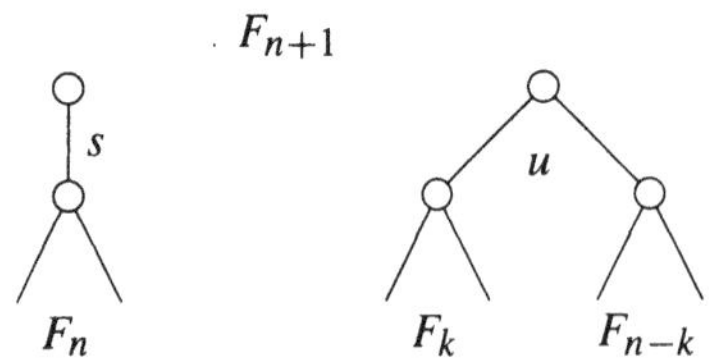

Now define the refined weight

$$W(T) = a^{i} b^{j} s^{h} u^{k}$$

where i = # vertices of out-degree 1 on the left–most branch,
j = # vertices of out-degree 2 on the left–most branch,
h = # vertices of out-degree 1 not on the left–most branch,
k = # vertices of out-degree 2 not on the left–most branch.

Then

$$B_n = \sum_{T \in \mathcal{T}_{n+1}} W(T) \quad (B_0 = 1).$$

The proof is again sketched in a picture:

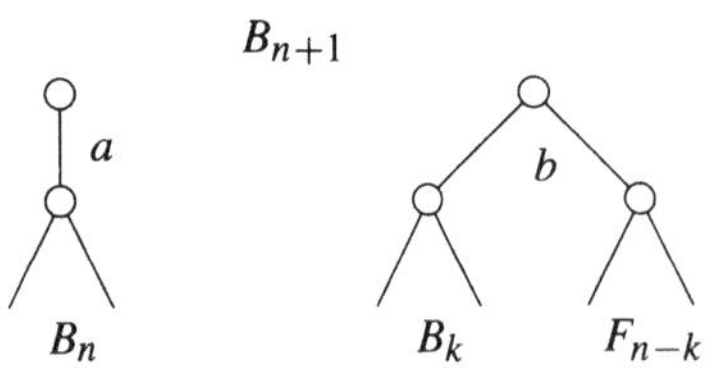

Example. For $a = b = s = u = 1$ we obtain $B_n = |\mathcal{T}_{n+1}|$ and hence $B_n = M_n$, the Motzkin number, which counts all n-edge Motzkin trees [13, 28]. Our initial example $\sigma \equiv 0$, $\tau \equiv 1$ is the case $a = s = 0$, $b = u = 1$. Here B_n counts the n-edge Motzkin trees with all out-degrees equal to 2, and we obtain precisely the example in Sect. 1.

Now we look at convolution recursions as in (1.1). From (6.1) and (6.2) the following formula can be derived:

$$\begin{aligned} bB_{n+1} = \; & abB_n + (u-b)\sum_{k=0}^{n-1} B_{k+1}B_{n-k} + (bs - 2au + ab)\sum_{k=0}^{n-1} B_k B_{n-k} \\ & +(a^2u - abs + b^2)\sum_{k=0}^{n-1} B_k B_{n-1-k}. \end{aligned}$$

The following cases are especially interesting:

$$b = u, a = s: \qquad B_{n+1} = sB_n + b\sum_{k=0}^{n-1} B_k B_{n-1-k},$$

$$b = u, a = s-1: \quad B_{n+1} = aB_n + \sum_{k=0}^{n-1} B_k B_{n-k} + (b-a)\sum_{k=0}^{n-1} B_k B_{n-1-k},$$

$$b = 2u, a = s: \qquad B_n \doteq \frac{1}{2^n}\sum_{k\geq 0}\binom{2k}{k}\binom{k}{n-k}s^{2k-n}(4u-s^2)^{n-k}.$$

From these formulae or from the combinatorial interpretation we can now derive a number of interesting examples:

$\tau \equiv 1$ $(b = u = 1)$:

a	s		
0	0	$(C_0, 0, C_1, 0, C_2, 0, C_3, \ldots)$	Catalan
0	1	$(R_0, R_1, R_2, R_3, R_4, \ldots)$	Riordan [7]
		$= (1, 0, 1, 1, 3, 6, 15, \ldots)$	
0	2	$(F_0, F_1, F_2, F_3, \ldots)$	Fine [12]
		$= (1, 0, 1, 2, 6, 18, 57, \ldots)$	
1	0	$\binom{n}{\lfloor n/2 \rfloor}$	middle binomials
		$= (1, 1, 2, 3, 6, 10, 20, \ldots)$	
1	1	(M_n)	Motzkin
		$= (1, 1, 2, 4, 9, 21, 51, \ldots)$	
1	2	(C_n)	Catalan
		$= (1, 1, 2, 5, 14, 42, 132, \ldots)$	

a	s		
2	2	(C_{n+1})	Catalan
3	2	$\binom{2n+1}{n}$	central binomials
3	3	(H_n)	restricted hexagonals [18]

$b = 2, u = 1$:

$a = s = 1$:	$(1, 1, 3, 7, 19, 51, \ldots)$	trinomials
$a = s = 2$:	$\binom{2n}{n}$	binomials

$b = 2, u = 2$:

$a = 2, s = 3$:	$(1, 2, 6, 22, 90, 394, \ldots)$	Schröder

Of course, there are a host of other relations between the various numbers that can be deduced, such as Schröder$_n = \sum_k \binom{2n-k}{k} C_{n-k}$, $M_n = R_n + R_{n+1}$, or $C_n = 2F_n + F_{n-1}$.

(B) Exponential generating functions. Given $\sigma = (a+ks)$, $\tau = (k(b+(k-1)u))$, the Catalan-like numbers $B_n = B_n^{\sigma,\tau}$ are given by the system (5.13):

$$F_{n+1} = sF_n + u\sum_{k=1}^{n-1}\binom{n}{k}F_kF_{n-k} \ (n \geq 1),\ F_0 = 0,\ F_1 = 1 \tag{6.3}$$

$$B_{n+1} = aB_n + b\sum_{k=0}^{n-1}\binom{n}{k}B_kF_{n-k} \ (n \geq 0),\ B_0 = 1. \tag{6.10}$$

We distinguish two cases.

(B1) $u = 0$**:** Here $s_k = a + ks$, $t_k = bk$, and (6.3) yields $F_n = s^{n-1} (n \geq 1)$, $F_0 = 0$. In this case the combinatorial object constitutes *partitions* of an n-set. The figure shows the partitions up to $n = 4$.

1	12	123	1234	12 − 34	2 − 3 − 14
	1 − 2	12 − 3	123 − 4	13 − 24	1 − 23 − 4
		2 − 13	3 − 124	23 − 14	1 − 3 − 24
		1 − 23	2 − 134	12 − 3 − 4	1 − 2 − 34
		1 − 2 − 3	1 − 234	2 − 13 − 4	1 − 2 − 3 − 4

The *Bell number* Bell_n counts the total number of partitions. Let $\mathcal{P}_n$ be the partitions of an n-set. For $P \in \mathcal{P}_n$ define the weight

$$W(P) = a^{n_1} b^{n_2} s^{n-n_1-2n_2},$$

where

$$\begin{aligned} n_1 &= \#\text{ singleton blocks}, \\ n_2 &= \#\text{ blocks of size } \geq 2. \end{aligned}$$

Then it is readily verified that

$$B_n = \sum_{P \in \mathcal{P}_n} W(P).$$

Hence we obtain the following examples:

$a = b = s = 1:$	$B_n = \text{Bell}_n$	Bell
$\sigma = (1, 2, 3, \dots)$	$= (1, 1, 2, 5, 15, 52, 203, \dots)$	
$\tau = (1, 2, 3, \dots)$		
$a = b, s = 1:$	$B_n = \sum_{k=0}^{n} S_{n,k} a^k$	Stirling polynomial
$\sigma = (a, a+1, a+2, \dots)$	where $S_{n,k}$ are the Stirling	
$\tau = (a, 2a, 3a, \dots)$	numbers of the second kind	
$a = b = 1, s = 0:$	$B_n = \#$ involutions	involutions
$\sigma = (1, 1, 1, \dots)$	$= (1, 1, 2, 4, 10, 26, 76, \dots)$	
$\tau = (1, 2, 3, \dots)$		
$a = 0, b = 1, s = 0:$	$B_n = 1 \cdot 3 \cdot 5 \cdot \ldots \cdot (2n-1)$	fixed-point-free
$\sigma = (0, 0, 0, \dots)$	or 0	involutions
$\tau = (1, 2, 3, \dots)$	$= (1, 0, 1, 0, 3, 0, 15, \dots)$	

(B2) $u \neq 0$**:** The combinatorial objects are *permutations* of $\{1, 2, \dots, n\}$. Consider a permutation $\pi = \pi_1 \pi_2 \dots \pi_n$, and set $\pi_0 = \pi_{n+1} = n+1$. The weight $w(\pi)$ is defined as

$$w(\pi) = s^{\ell} u^m,$$

where

$$\ell = \#\text{ double rises, i.e., } \pi_{i-1} < \pi_i < \pi_{i+1},$$

$$m = \#\text{ local maxima.}$$

Let $\mathcal{S}_n$ be the set of permutations of $\{1, 2, \ldots, n\}$ *without double falls* (including $\pi_0 > \pi_1 > \pi_2$). Then

$$F_n = \sum_{\pi \in \mathcal{S}_n} w(\pi).$$

The proof is suggested by the following figure:

$$\pi_{n+1} = n + 1 \quad \Longrightarrow \quad sF_n \text{ (recall } \pi_{n+2} = n + 2)$$

$$\pi_{n+1} \neq n + 1 \quad \Longrightarrow \quad u \sum_{k=1}^{n-1} F_k F_{n-k}$$

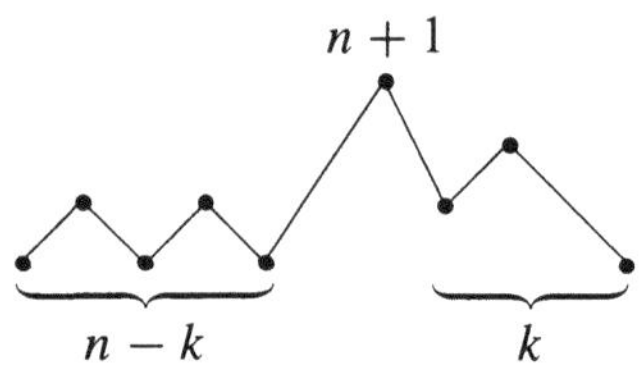

Now we consider the refined weight for $\pi \in \mathcal{S}_n$:

$$W(\pi) = a^i b^j s^h u^k,$$

where $\pi_0 = 0$, $\pi_{n+1} = n + 1$.

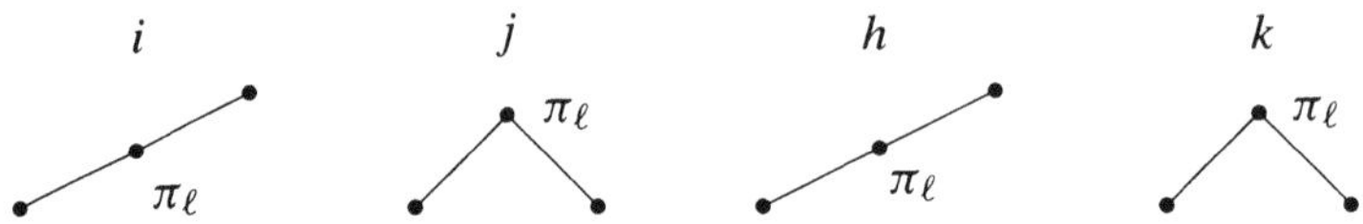

$\pi_i < \pi_\ell (i < \ell)$ $\quad$ $\pi_i < \pi_\ell (i < \ell)$ $\quad$ $\exists i < \ell, \pi_i > \pi_\ell$ $\quad$ $\exists i < \ell : \pi_i > \pi_\ell$

Then

$$B_n = \sum_{\pi \in \mathcal{S}_n} W(\pi).$$

The proof is similar to the previous, by considering the cases $\pi_{n+1} = n + 1$ or $\pi_{n+1} \neq n + 1$. Again we obtain some very interesting examples.

$a = s = 0, b = u = 1$: $\quad$ $B_n =$ secant numbers $\quad$ alternating

$\sigma \equiv 0$ $\quad$ $= (1, 0, 1, 0, 5, 0, 61, \ldots)$ $\quad$ permutations

$\tau = (k^2)$

$a = s = b = u = 1$: $\quad$ $B_n = |\mathcal{S}_n|$ $\quad$ without double falls

$\sigma = (k + 1)$ $\quad$ $= (1, 1, 2, 5, 17, 70, 349, \ldots)$

$\tau = (k^2)$

For $s = 2$, $u = 1$, we readily obtain $F_n = n!$ $(n \geq 1)$ in (6.3), and hence

$$B_{n+1} = aB_n + b \sum_{k=0}^{n-1} \binom{n}{k} B_k (n - k)!.$$

If we define a new weight $\overline{W}(\pi)$ of any permutation π of $\{1, 2, \ldots, n\}$ by

$$\overline{W}(\pi) = a^v b^w,$$

where

$$v = \#\text{ fixed points},$$

$$w = \#\text{ cycles of length } \geq 2,$$

then it follows that

$$B_n = \sum_{\pi} \overline{W}(\pi).$$

From this we deduce the following examples:

$a = b = 1, s = 2, u = 1$:	$B_n = n!$	all permutations
$\sigma = (2k+1)$	$= (1, 1, 2, 6, 24, 120, 720, \ldots)$	
$\tau = (k^2)$		
$a = 0, b = 1, s = 2, u = 1$:	$B_n = D_n = \#$ derangements	fixed-point-free
$\sigma = (2k)$	$= (1, 0, 1, 2, 9, 44, 265, \ldots)$	
$\tau = (k^2)$		
$a = b, s = 2, u = 1$:	$B_n = \sum_k s_{n,k} a^k$	Stirling polynomial
$\sigma = (a + 2k)$	where $s_{n,k}$ are the	
$\tau = (k(a + k - 1))$	Stirling numbers of the first kind	

When $a = s$ and $b = 2u$, then (6.3), (6.4) yield $B_n = F_{n+1}$. From this, we deduce as examples:

$a = s = 0, b = 2, u = 1$:	$B_n =$ tangent numbers	alternating
$\sigma \equiv 0$	$= (1, 0, 2, 0, 16, 0, 272, \ldots)$	permutations
$\tau = (k(k+1))$		

Another case which is easily analyzed is $s = 2m$, $u = m^2$. We obtain from (6.3), $F_n = m^{n-1} n!$, and from (6.4),

$$B_{n+1} = (mn + a)B_n + (b - am)nB_{n-1}.$$

In particular, this gives again for $m = 1$

$$a = b, \quad s = 2, \quad u = 1: \qquad B_n = a^{\overline{n}} = a(a+1)\ldots(a+n-1)$$
$$= \sum_k s_{n,k} a^k$$

and for

$$\begin{array}{lll} a = b = 2,\ s = 2, & u = 1: & B_n = (n+1)! \\ \sigma = (2+2k), & \tau = (k(k+1)). & \end{array}$$

Finally, for $a = b = u$, $s = u + 1$, we find that $B_n = uF_n$, and from this

$$B_n = \sum_k A_{n,k} u^k \quad \text{Eulerian polynomial,}$$

where $A_{n,k}$ is the number of permutations of $\{1, \ldots, n\}$ with $k - 1$ descents.

(C) q-generating functions. Given the sequences $\sigma = (aq^k)$, $\tau = (bq^{k-1}(q^k - 1))$, we have the recursion (5.15):

$$B_{n+1} = aB_n + b(q^n - 1)B_{n-1}. \tag{6.5}$$

What do these Catalan-like numbers count? The nicest result is obtained in the case $a = b + 1$. We claim that, for $a = b + 1$,

$$B_n = \sum_k \binom{n}{k}_q b^k \quad \text{Galois polynomial,} \tag{6.6}$$

where $\binom{n}{k}_q = \frac{(q^n-1)\ldots(q^{n-k+1}-1)}{(q^k-1)\ldots(q-1)}$ are the *Gaussian coefficients*. In particular, when q is a prime power, then $\binom{n}{k}_q$ counts the number of k-dimensional subspaces of an n-dimensional vector space over $GF(q)$.

Here is a proof of (6.6) à la Rota. Let

$$\begin{aligned} g_k(x) &= (x-1)(x-q)\cdots(x-q^{k-1}) \ (k \geq 1), \\ g_0(x) &= 1 \end{aligned} \tag{6.7}$$

be the *Gaussian polynomials*.

As before denote by Δ the shift operator of the q-calculus, that is,

$$\Delta x^n = \begin{cases} (q^n - 1)x^{n-1} & (n \geq 1) \\ 0 & (n = 0). \end{cases}$$

It is immediately verified that

$$\Delta g_k(x) = (q^k - 1)g_{k-1}(x) \ (k \geq 1). \tag{6.8}$$

Now we define the linear operator $L : \mathbb{R}[x] \longrightarrow \mathbb{R}$ by

$$L : g_k(x) \longrightarrow b^k. \tag{6.9}$$

From the well-known relation $x^n = \sum_k \binom{n}{k}_q g_k(x)$ it follows that

$$Lx^n = \sum_k \binom{n}{k}_q b^k.$$

So we have to show that

$$Lx^n = B_n$$

or, in other words, that Lx^n satisfies the recursion (6.5), that is,

$$Lx^{n+1} = (b+1)Lx^n + b(q^n - 1)Lx^{n-1}.$$

From (6.7) we have

$$g_{n+1}(x) = xg_n(x) - q^n g_n(x);$$

hence by (6.9)

$$\begin{aligned} Lxg_n(x) &= Lg_{n+1}(x) + q^n Lg_n(x) \\ &= b^{n+1} + q^n b^n \\ &= b^n(b+1) + (q^n - 1)b^n. \end{aligned} \tag{6.10}$$

Furthermore by (6.8),

$$L\Delta g_n(x) = L(q^n - 1)g_{n-1}(x) = (q^n - 1)b^{n-1},$$

and thus by (6.10)

$$Lxg_n(x) = (b+1)Lg_n(x) + bL\Delta g_{n(x)}. \tag{6.11}$$

Since the Gaussian polynomials form a basis, (6.11) holds for *all* polynomials $p(x)$:

$$Lxp(x) = (b+1)Lp(x) + bL\Delta p(x).$$

Setting $p(x) = x^n$, we thus infer that

$$Lx^{n+1} = (b+1)Lx^n + b(q^n - 1)Lx^{n-1},$$

as claimed.

As example for $a = 2, b = 1$, that is, the sequences $\sigma = (2q^k), \tau = (q^{k-1}(q^k - 1))$ we obtain:

$$B_n = \sum_{k=0}^{n} \binom{n}{k}_q \quad \text{Galois number,}$$

which, for q a prime power, is the total number of subspaces of an n-dimensional vector space over $GF(q)$.

To end this section we return to the general case and derive some binomial relations between Catalan-like numbers which unify many known instances.

Let $A = A^{\sigma,\tau} = (a_{n,k})$ be a recursive matrix given as in (5.1):

$$a_{n,k} = a_{n-1,k-1} + s_k a_{n-1,k} + t_{k+1} a_{n-1,k+1}. \tag{6.12}$$

Denote by $P = \left(\binom{n}{k}\right)$ the infinite Pascal matrix, having $\binom{n}{k}$ as its (n,k)-entry. By checking (6.12), the following two results are readily seen:

$$\begin{aligned} &\textit{Let } A = A^{\sigma,\tau},\ A' = PA, \\ &\quad \textit{then } A' = A^{\sigma+1,\tau}, \textit{ where } \sigma + 1 = (s_0+1, s_1+1, s_2+1, \dots). \end{aligned} \tag{6.13}$$

$$\begin{aligned} &\textit{Let } A = A^{\sigma,\tau} = (a_{n,k}),\ B = (b_{n,k}) \textit{ with } b_{n,k} = (-1)^{n-k} a_{n,k}, \\ &\quad \textit{then } B = A^{-\sigma,\tau},\ \textit{ where } -\sigma = (-s_0, -s_1, -s_2, \dots). \end{aligned} \tag{6.14}$$

Iterating (6.13) we have the following relation between Catalan-like numbers B_n for σ, τ and a constant λ:

$$B_n^{\sigma+\lambda,\tau} = \sum_{k=0}^{n} \binom{n}{k} \lambda^{n-k} B_k^{\sigma,\tau} \tag{6.15}$$

$$B_n^{-\sigma,\tau} = (-1)^n B_n^{\sigma,\tau}. \tag{6.16}$$

We can now consult our list of examples and apply these formulae. Here are a few examples:

$$\sigma \equiv 0, \quad \tau \equiv 1, \quad \lambda = 1: \quad M_n = \sum_k \binom{n}{2k} C_k$$

$$\sigma \equiv 1, \quad \tau \equiv 1, \quad \lambda = 1: \quad C_{n+1} = \sum_k \binom{n}{k} M_k.$$

Comparing $\sigma \equiv 2$, $\tau = (2, 1, 1, \dots)$ and $-\sigma$, τ, we obtain

$$\binom{2n}{n} = \sum_k (-1)^k \binom{n}{k} 4^{n-k} \binom{2k}{k}.$$

Comparing $\sigma = (3, 2, 2, \dots)$, $\tau \equiv 1$, $\sigma' = (-1, -2, -2, \dots)$, $\tau' \equiv 1$, yields

$$\binom{2n+1}{n} = \sum_k (-1)^k \binom{n}{k} 4^{n-k} C_k.$$

Finally, comparing $\sigma = (1, 2, 2, \dots)$, $\tau \equiv 1$, and $\sigma' = (-1, 0, 0, \dots)$, $\tau' \equiv 1$, we find that

$$C_n = \sum_k (-1)^k \binom{n}{k} 2^{n-k} \binom{k}{\lfloor k/2 \rfloor},$$

and comparing $\sigma = (1, 1, 1, \dots)$, $\tau \equiv 1$, and $\sigma' = (-2, -2, -2, \dots)$, $\tau' \equiv 1$, we have

$$M_n = \sum_k (-1)^k \binom{n}{k} 3^{n-k} C_{k+1}.$$

For more on this topic the reader is referred to [2].

7 Non-Sheffer sequences

In the case when the basic recursion (5.1) does not satisfy the Sheffer-relation (5.3), only a few results are known. These results are of two types. Either new sequences $\widehat{\sigma}, \widehat{\tau}$ are constructed out of a given pair σ, τ (the binomial formulae (6.15) are examples), or the Catalan-like numbers B_n satisfy certain simple recursions.

A. As an illustration of the first type, we have the following result. Suppose we are given $\sigma = (0, 0, 0, \dots)$, $\tau = (t_k)$ arbitrary. Then, for the Catalan-like numbers $B_n = B_n^{\sigma,\tau}$, we clearly have

$$B_{2n+1} = 0 \quad (n \geq 0).$$

In the literature this is called the *symmetric* case.

Define the sequences $\widehat{\sigma} = (\widehat{\sigma_k})$, $\widehat{\tau} = (\widehat{\tau_k})$ by

$$\begin{aligned} \widehat{\tau_0} &= t_1, \quad \widehat{s_1} = t_2 + t_3, \dots, \quad \widehat{s_k} = t_{2k} + t_{2k+1}, \dots \\ \widehat{t_1} &= t_1 t_2, \quad \widehat{t_2} = t_3 t_4, \dots, \quad \widehat{t_k} = t_{2k-1} t_{2k}, \dots. \end{aligned} \tag{7.1}$$

Then it is easily checked that, for $\widehat{B}_n = B_n^{\widehat{\sigma},\widehat{\tau}}$, we have

$$\widehat{B}_n = B_{2n}. \tag{7.2}$$

Examples. i) $\sigma \equiv 0$, $\tau \equiv 1$ gives $B_{2n} = C_n$ (Catalan), and once more we obtain by (7.1) that $\widehat{\sigma} = (1, 2, 2, \dots)$, $\widehat{\tau} \equiv 1$, yields $\widehat{B}_n = C_n$.

ii) $\sigma \equiv 0$, $\tau = (a, b, a, b, a, b, \dots)$. Here we have $\widehat{\sigma} = (a, a+b, a+b, \dots)$, $\widehat{\tau} = (ab, ab, ab, \dots)$. It is readily seen that

$$\widehat{B}_n = B_n^{\widehat{\sigma},\widehat{\tau}} = \sum_{k=0}^{n-1} N(n,k) a^{n-k} b^k,$$

where

$$N(n,k) = \frac{1}{n} \binom{n}{k} \binom{n}{k+1}$$

are the *Narayana numbers*. In particular, we see that the case $a = 1$, that is, $\widehat{\sigma} = (1, b+1, b+1, \dots)$, $\widehat{\tau} = (b, b, b, \dots)$, yields the Narayana polynomial

$$\widehat{B}_n = \sum_{k=0}^{n-1} N(n,k)b^k.$$

Since $a = b = 1$ gives $\widehat{\sigma} = (1, 2, 2, \dots)$, $\widehat{\tau} \equiv 1$, we obtain the well-known relation $C_n = \sum_{k=0}^{n-1} N(n,k)$.

This approach can also be used to derive some binomial identities. As an example, consider $\sigma' = (1, 1, 1, \dots)$, $\tau' = (2, 1, 1, \dots)$. The Catalan-like number B'_n is the trinomial number, and we know from Sect. 6 (A) that

$$B'_n = \frac{1}{2^n} \sum_k \binom{2k}{k} \binom{k}{n-k} 3^{n-k}.$$

Now we want to show that, in addition,

$$B'_n = \sum_k \binom{n}{2k} \binom{2k}{k}.$$

To prove this, consider $\sigma \equiv 0$, $\tau = (2, 1, 1, \dots)$. The associated sequences $\widehat{\sigma}, \widehat{\tau}$ are then, by (7.1),

$$\widehat{\sigma} = (2, 2, 2, \dots), \quad \widehat{\tau} = (2, 1, 1, \dots),$$

which according to our list in Sect. 6 gives

$$B_{2n} = \widehat{B}_n = \binom{2n}{n}, \quad B_{2n+1} = 0.$$

Applying the binomial formula (6.15) to σ, τ we obtain for $\sigma' = \sigma + 1$, $\tau' = \tau$, precisely

$$B'_n = \sum_k \binom{n}{2k} \binom{2k}{k}.$$

B. Here is a simple but important illustration of the second type of result. Suppose the sequence (B_n) is given by the following recursion:

$$\begin{aligned} B_{2n+2} &= \frac{\alpha n + \beta}{\gamma n + \delta} B_{2n}, \quad B_0 = 1, \\ B_{2n+1} &= 0 \quad \text{for all } n. \end{aligned} \tag{7.3}$$

Then it can be shown that (B_n) is Catalan-like if and only if, for all k,

$$\alpha k + \beta \neq 0, \gamma k + \delta \neq 0, \alpha\delta - \beta\gamma + k\alpha\gamma \neq 0.$$

The associated sequences are then $\sigma \equiv 0$, $\tau = (t_k)$ with

$$\begin{aligned} t_{2k} &= k\frac{(\alpha\delta - \beta\gamma) + (k-1)\alpha\gamma}{((2k-2)\gamma + \delta)((2k-1)\gamma + \delta)} \ (k \geq 1) \\ t_{2k+1} &= \frac{(k\alpha + \beta)((k-1)\gamma + \delta)}{((2k-1)\gamma + \delta)(2k\gamma + \delta)} \ (k \geq 1), \ t_1 = \frac{\beta}{\delta}. \end{aligned} \tag{7.4}$$

When $\alpha = 2\beta$, we may combine (7.4) into a single expression

$$t_k = \frac{k\beta(\delta + (k-3)/2\gamma)}{((k-2)\gamma + \delta)((k-1)\gamma + \delta)} \quad (k \geq 2), \quad t_1 = \frac{\beta}{\delta}. \tag{7.5}$$

There are two cases of (7.5) which have been studied intensively [9].

Consider $\alpha = 2$, $\beta = 1$, $\gamma = 2$, $\delta = 2\lambda + 2$ with $\lambda \notin \{0, -1, -2, \ldots, -\frac{1}{2}, -\frac{3}{2}, -\frac{5}{2}, \ldots\}$. Here

$$t_k = \frac{k(k + 2\lambda - 1)}{(2k + 2\lambda - 2)(2k + 2\lambda)} \quad (k \geq 1).$$

i) For $\lambda \in \mathbb{N}$ this gives by (7.3)

$$B_{2n} = \frac{1}{4^n}\frac{\binom{2n}{n}}{\binom{n+\lambda}{n}},$$

and, in particular, for $\lambda = 1$,

$$B_{2n} = \frac{1}{4^n}C_n.$$

ii) For $\lambda = \frac{2m-1}{2}$, $m \in \mathbb{N}$, this gives

$$B_{2n} = \frac{\binom{m+n}{m}}{\binom{2m+2n}{2m}},$$

and, in particular, for $\lambda = \frac{1}{2}$,

$$B_{2n} = \frac{1}{2n+1}.$$

The associated orthogonal polynomial systems are for arbitrary λ the *Gegenbauer polynomials*, and for $\lambda = \frac{1}{2}$ the *Legendre polynomials*.

Another classical case is $\alpha = 2$, $\beta = 2b + 2$, $\gamma = 1$, $\delta = a + b + 2$ which leads to the *Jacobi polynomials.*

C. Another simple recursion is

$$\begin{aligned} B_{2n+2} &= c^n B_{2n}, \ B_0 = 1, \\ B_{2n+1} &= 0. \end{aligned}$$

The sequence (B_n) is Catalan-like if and only if $c \neq 1$, in which case we obtain

$$\begin{aligned} t_{2k} &= c^{2k-1} - c^{k-1} \quad (k \geq 1) \\ t_{2k+1} &= c^{2k} \quad (k \geq 0) \end{aligned}$$

with $B_{2n} = c^{\binom{n}{2}}$.

D. Finally, we look at the convolution recursion

$$B_{2n} = \frac{n+1}{2} \sum_k \binom{2n-1}{2k} \frac{B_{2k}}{k+1} \frac{B_{2n-2-2k}}{n-k}, \quad B_0 = 1,$$

$$B_{2n+1} = 0.$$

The sequence (B_{2n}) we obtain are the *Genocchi numbers* $B_{2n} = G_{2n}$

$$(G_n) = (1, 0, 1, 0, 3, 0, 17, 0, 155, 0, 2073, \dots).$$

The associated sequence $\tau = (t_k)$ is given by

$$\begin{aligned} t_{2k} &= k(k+1) \quad (k \geq 1) \\ t_{2k+1} &= (k+1)^2 \quad (k \geq 0). \end{aligned}$$

The generating function of the Genocchi numbers is given by

$$G(z) = \sum_{n \geq 0} \frac{G_n z^n}{n!} = (ztg(\frac{z}{2}))''.$$

8 Determinants of Hankel matrices

We now discuss the characterization of Catalan-like sequences by means of the determinants of the Hankel matrices H_n and $\widetilde{H}_n$, as introduced in Sect. 1.

Let (B_n) be the Catalan-like sequence associated with $\sigma = (s_k)$, $\tau = (t_k)$, and recursive matrix $A = (a_{n,k})$. We consider the first three Hankel matrices of order 0, 1 and 2:

$$H_n = (B_{i+j}), \quad \widetilde{H}_n = (B_{i+j+1}), \quad \widetilde{\widetilde{H}}_n = (B_{i+j+2})$$

for $0 \leq i, j \leq n$.

The key is the fundamental theorem of Sect. 2:

$$ATA^T = H, \tag{8.1}$$

where $T = \begin{pmatrix} T_0 & & & \\ & T_1 & & 0 \\ & 0 & T_2 & \\ & & & \ddots \end{pmatrix}$, $T_n = t_1 \dots t_n$, $T_0 = 1$, and $H = (B_{i+j})$.

We further consider the infinite Hankel matrices

$$\widetilde{H} = (B_{i+j+1}), \quad \widetilde{\widetilde{H}} = (B_{i+j+2}).$$

Denote by ΔA the shifted matrix

$$\Delta A = \begin{pmatrix} a_{10} & a_{11} & a_{12} & \dots \\ a_{20} & a_{21} & a_{22} & \dots \\ a_{30} & a_{31} & a_{32} & \dots \\ & \dots & & \end{pmatrix}$$

and by J the infinite matrix

$$J = \begin{pmatrix} s_0 & 1 & & & \\ t_1 & s_1 & 1 & & 0 \\ & t_2 & s_2 & 1 & \\ & 0 & \ddots & \ddots & \ddots \end{pmatrix}.$$

The recursion

$$a_{n,k} = a_{n-1,k-1} + s_k a_{n-1,k} + t_{k+1} a_{n-1,k+1}$$

can thus be written in matrix form

$$\Delta A \; = \; A \; J. \tag{8.2}$$

Looking at (8.1) we find that

$$(\Delta A) T A^T = \widetilde{H} \tag{8.3}$$

$$(\Delta A) T (\Delta A)^T = \widetilde{\widetilde{H}}. \tag{8.4}$$

Now we restrict the matrix equations to the rows and columns up to n. For an infinite matrix $M = (m_{ij})$ we denote by $M_n = (m_{ij})$, $0 \le i, j \le n$, its nth principal minor.

Since A_n is lower triangular with 1's on the main diagonal, we have $\det A_n = 1$ and obtain from (8.1) our first result, already mentioned in Sect. 2:

Result 1. $\det H_n = \Pi_n := T_0T_1\dots T_n$.

Looking at (8.2) we have

$$\det(\Delta A)_n = \det J_n$$

which gives, by (8.3),

Result 2. $\det \widetilde{H}_n = d_n\Pi_n$, *where* $d_n = \det J_n$.

From the definition of J we see that the d_n's are determined by the recursion:

$$\begin{aligned} d_{-1} &= 1, \quad d_0 = s_0 \\ d_n &= s_n d_{n-1} - t_n d_{n-2} \; (n \geq 1). \end{aligned} \tag{8.5}$$

Finally, we look at (8.4). Restricting to the nth principal minor this gives, for $i + j < 2n$,

$$\sum_{k=0}^{n} a_{i+1,k} T_k a_{j+1,k} = B_{i+j+2},$$

and, for $i = j = n$, since the index only runs up to n,

$$\sum_{k=0}^{n} a_{n+1,k} T_k a_{n+1,k} = B_{2n+2} - T_{n+1}.$$

Hence

$$(\Delta A)_n \begin{pmatrix} T_0 & & 0 \\ & \ddots & \\ 0 & & T_n \end{pmatrix} (\Delta A)_n^T = \begin{pmatrix} B_2 & B_3 & \dots & B_{n+2} \\ & & \dots & \\ B_{n+2} & B_{n+3} & \dots & B_{2n+2} - T_{n+1} \end{pmatrix},$$

and thus

$$(\det J_n)^2 \Pi_n = \det \widetilde{\widetilde{H}}_n - T_{n+1} \det \widetilde{\widetilde{H}}_{n+1}.$$

Solving this last recursion and taking d_n as in (8.5), we obtain

Result 3. $\det \widetilde{\widetilde{H}}_n = \Pi_{n+1} \sum_{j=-1}^{n} \frac{dj^2}{T_{j+1}}$. (8.6)

Examples. For $\tau \equiv 1$ we have $T_n = 1$ and $\Pi_n = 1$ for all n, and the formulae read:

$$\det H_n = 1, \;\; \det \widetilde{H}_n = d_n, \;\; \det \widetilde{\widetilde{H}}_n = \sum_{j=-1}^{n} d_j^2. \tag{8.7}$$

To apply (8.7) we have to solve (8.5):

i) $\sigma = (1, 2, 2, \dots)$, $B_n = C_n$. Here (8.5) yields $d_n = 1$ for all n and we have proved our initial characterization: the Catalan numbers C_n are the unique sequence with $\det H_n = \det \widetilde{H}_n = 1$ for all n. Furthermore, $\det \widetilde{\widetilde{H}}_n = n + 2$.

ii) $\sigma \equiv 1$, $B_n = M_n$ (Motzkin). We find (see [1]) that

$$\det \widetilde{H}_n = \begin{cases} 1 & n \equiv 0, 5 \pmod 6 \\ 0 & n \equiv 1, 4 \pmod 6 \\ -1 & n \equiv 2, 3 \pmod 6 \end{cases}.$$

Furthermore, $\det \widetilde{\widetilde{H}}_n = \lceil \frac{2n+4}{3} \rceil$.

iii) $\sigma = (0, 2, 2, 2, \dots)$, $B_n = F_n$ (Fine). Here $\det \widetilde{H}_n = -n$.

iv $\sigma = (3, 2, 2, \dots)$, $B_n = \binom{2n+1}{n}$. We obtain $\det \widetilde{H}_n = 2n + 3$.

v) $\sigma = (3, 3, 3, \dots)$, $B_n =$ hexagonal number. This gives $\det \widetilde{H}_n = f_{2n+4}$ (Fibonacci number).

Finally, we look at a few examples with $\tau \not\equiv 1$.

vi) $\sigma = (1, 2, 3, \dots)$, $\tau = (1, 2, 3, \dots)$, $B_n =$ Bell number. We have $T_n = n!$, and thus $\Pi_n = n!! := \prod_{k=0}^{n} (k!)$. Recursion (8.5) yields $d_n = 1$ for all n, and we obtain (see [3]):

$$\det H_n = n!!, \quad \det \widetilde{H}_n = n!!.$$

vii) $\sigma = (1, 3, 5, 7, \dots)$, $\tau = (1, 4, 9, 16, \dots)$, $B_n = n!$. Here $T_n = n!^2$, $\Pi_n = (n!!)^2$, and (8.5) gives $d_n = (n + 1)!$. Hence

$$\det H_n = (n!!)^2, \quad \det \widetilde{H}_n = (n!!)^2 (n + 1)!.$$

This survey would not be complete without a short discussion of the beautiful result of Gessel–Viennot [15] connecting determinants to non-intersecting path systems (see also [20]).

We first state the Gessel–Viennot result and then apply it to our situation. Let $G = (V, E)$ be a directed acyclic graph, where every edge $e \in E$ carries a (real) weight $w(e)$. We extend these weights to arbitrary directed paths P by setting

$$w(P) = \prod_{e \in P} w(e).$$

For $u, v \in V$ let $\mathcal{P}(u, v)$ be the set of directed paths from u to v.

Choose any two sets of $n+1$ vertices $\mathcal{U} = \{u_0, u_1, \dots, u_n\}$, $\mathcal{V} = \{v_0, v_1, \dots, v_n\}$, where the sets $\mathcal{U}$ and $\mathcal{V}$ need not be disjoint. A *path system* from $\mathcal{U}$ to $\mathcal{V}$ is a permutation σ of $\{0, 1, \dots, n\}$ together with $n + 1$ directed paths $P_{0,\sigma(0)} \in \mathcal{P}(u_0, v_{\sigma(0)})$, $\dots P_{n,\sigma(n)} \in \mathcal{P}(u_n, v_{\sigma(n)})$; the path system is said to be *vertex–disjoint* if two distinct paths never share a vertex.

Now associate to $\mathcal{U}$ and $\mathcal{V}$ the *path matrix* $M = (m_{ij})$ where

$$m_{ij} = \sum_{P \in \mathcal{P}(u_i, v_j)} w(P).$$

Here is the theorem of Gessel–Viennot:

$$\det M = \sum_{\mathcal{VD}} (\operatorname{sign} \sigma) w(P_{0,\sigma(0)}) \ldots w(P_{n,\sigma(n)}), \tag{8.8}$$

where the sum is extended over all vertex-disjoint path systems from $\mathcal{U}$ to $\mathcal{V}$.

In our situation we are given as usual $\sigma = (s_k)$, $\tau = (t_k)$ and convert the recursion

$$a_{n,k} = a_{n-1,k-1} + s_k a_{n-1,k} + t_{k+1} a_{n-1,k+1}$$

into a weighted graph as in Sect. 4.

In other words, we consider the lattice graph $G = G^{\sigma,\tau} = (V, E)$ in the upper half–plane, that is,

$$V = \{(x, y) : x, y \in \mathbb{Z}, y \geq 0\}$$

with the edges going east, north–east or south–east. The weights are

E : 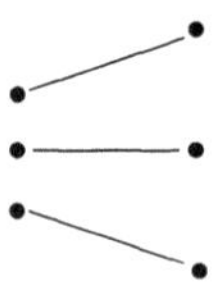 weight

north–east	1	
east	s_k	$y = k$
south–east	t_{k+1}	$y = k + 1$ to $y = k$.

Then, as in Sect. 4, we find (see the figure) that, for any vertices u, v of G on the x-axis which are n steps apart,

$$m(u, v) = \sum_{P \in \mathcal{P}(u,v)} w(P) = B_n.$$

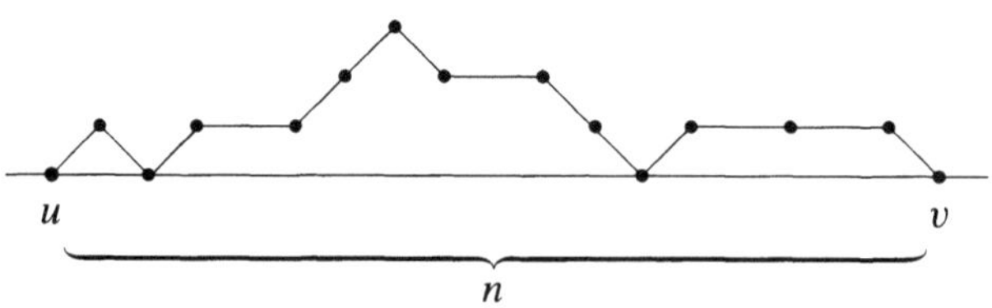

With the theorem of Gessel–Viennot we can now prove our results on the determinants of H_n, $\widetilde{H}_n$ and $\widetilde{\widetilde{H}}_n$ at one glance ("proof by picture").

Result 1.

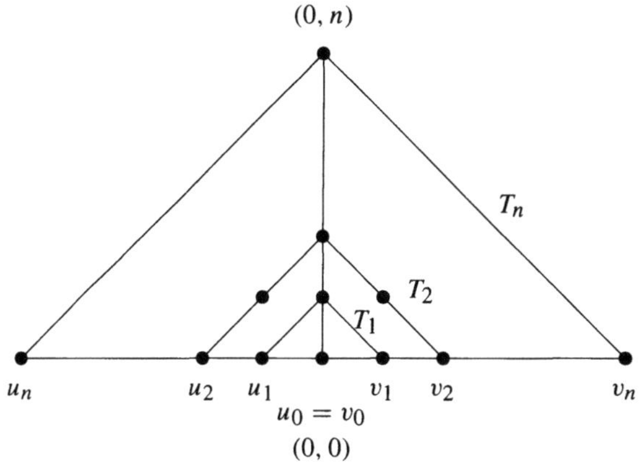

In this case $M = H_n$. There is clearly only one vertex-disjoint system with $\sigma = \text{id}$, as depicted in the figure. Hence $\det H_n = T_0 T_1 \ldots T_n = \Pi_n$.

Result 2.

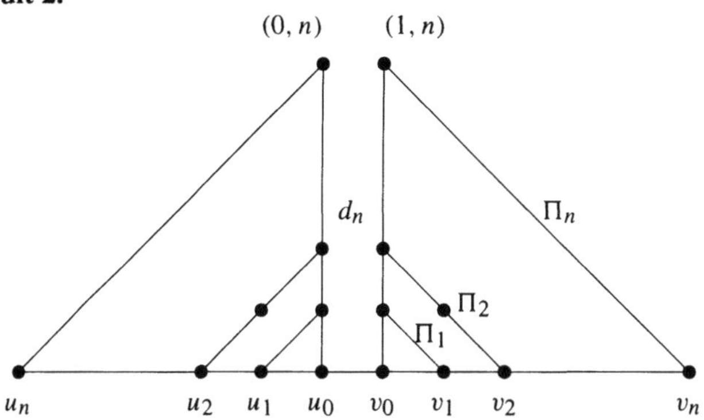

We have $M = \widetilde{H}_n$. The vertex-disjoint systems are forced to take the paths outside the strip between u_0 and v_0 as in Result 1. Hence

$$\det \widetilde{H}_n = d_n \Pi_n.$$

To compute d_n recursively, we classify the systems according to the top as in the following figure, and arrive precisely at the recursion (8.5).

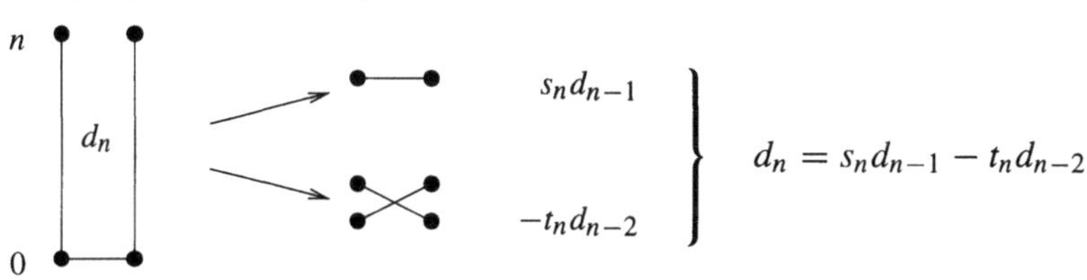

Note that we have sign $\sigma = -1$ in the second case, since the edges cross.

Result 3.

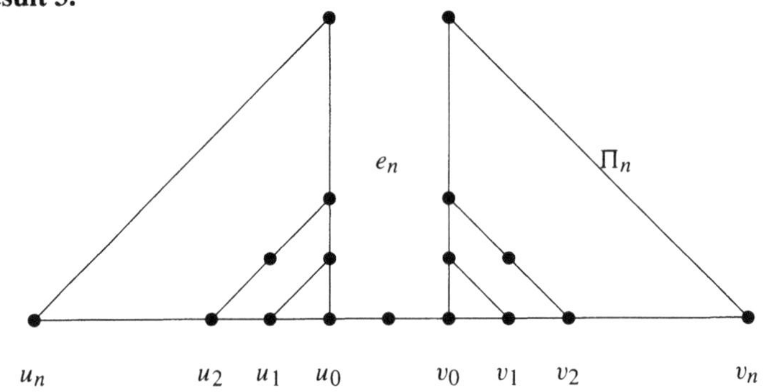

We have $M = \widetilde{\widetilde{H}}_n$, and it again suffices to consider the strip between u_0 and v_0, that is,

$$\det \widetilde{\widetilde{H}}_n = e_n \Pi_n.$$

Classifying the vertex-disjoint systems inside the strip as suggested by the following figure, we obtain (8.6).

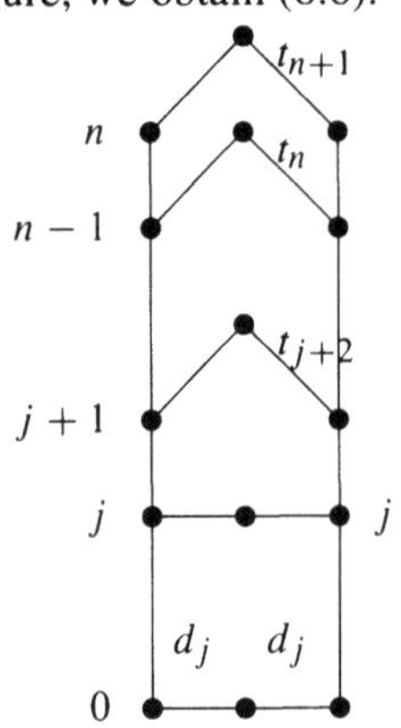

Indeed,

$$e_n = \sum_{j=-1}^{n} d_j^2 (t_{j+2} \cdots t_{n+1}) = T_{n+1} \sum_{j=-1}^{n} \frac{d_j^2}{T_{j+1}},$$

and thus

$$\det \widetilde{\widetilde{H}}_n = \Pi_{n+1} \sum_{j=-1}^{n} \frac{d_j^2}{T_{j+1}}.$$

9 Ballot numbers

We come to our final characterization of the sequence (C_n) of Catalan numbers, by means of a *ballot table*. We repeat the set-up. Consider the recursive system

$$\begin{aligned} &b_{0,0} = 1, \quad b_{0,k} = 0 \quad (k > 0) \\ &b_{n,k} = b_{n-1,k-1} + b_{n-1,k} + \ldots + b_{n-1,n-1} \ (n \geq 1). \end{aligned} \tag{9.1}$$

Then

$$b_{n,0} = C_n \quad \text{and} \quad \sum_{k=0}^{n} b_{n,k} = C_{n+1}. \tag{9.2}$$

(Of course, we still have to prove this.)

Thus the sequence (C_n) can be defined via (9.1). But, on the other hand, the sequences (C_n) and (C_{n+1}) are, as we know, also given via 3-term recursions

$$a_{n,k} = a_{n-1,k-1} + s_k a_{n-1,k} + t_{k+1} a_{n-1,k+1}, \tag{9.3}$$

namely, $\sigma = (1, 2, 2, \ldots)$, $\tau \equiv 1$, resp. $\sigma = (2, 2, 2, \ldots)$, $\tau \equiv 1$.

It is the purpose of this chapter to exhibit some general connections between ballot tables on the one hand and recursive matrices on the other.

We define a *generalized ballot table* as follows. Let $\mu = (m_0, m_1, m_2, \ldots)$ be a sequence of real numbers. We associate to μ the infinite lower triangular matrix $B^\mu = (b_{n,k})$ by setting

$$\begin{aligned} &b_{0,0} = 1, \quad b_{0,k} = 0 \quad (k > 0) \\ &b_{n,k} = b_{n-1,k-1} + m_k b_{n-1,k} + b_{n-1,k+1} + \ldots + b_{n-1,n-1} \quad (n \geq 1). \end{aligned} \tag{9.4}$$

B^μ is called the *ballot table* corresponding to μ. We are interested in the *sums* $S_n^\mu = \sum_{k=0}^{n} b_{n,k}$, and the elements in column *zero*, $Z_n^\mu = b_{n,0}$.

Hence for our example $\mu \equiv 1$ (still to be proved):

$$S_n^\mu = C_{n+1}, \ Z_n^\mu = C_n,$$

that is, the *sum* corresponds to $\sigma = (2, 2, 2, \ldots)$, $\tau \equiv 1$, and the zero-element to $\sigma = (1, 2, 2, \ldots)$, $\tau \equiv 1$.

Generalizing this example we use the following abbreviated notation:

$\mu \overset{S}{\longleftrightarrow} {\sigma \atop \tau}$ means that $S_n^\mu = B_n^{\sigma,\tau}$ for all n where S_n^μ and $B_n^{\sigma,\tau}$ are defined according to (9.4) and (9.3),

$\mu \overset{Z}{\longleftrightarrow} {\sigma \atop \tau}$ means that $Z_n^\mu = B_n^{\sigma,\tau}$ for all n.

The following theorem is basic [4]. *For any* $\alpha \in \mathbb{R}$,

$$\begin{aligned} &\textit{if} \quad \mu \overset{S}{\longleftrightarrow} {\sigma \atop \tau} \quad \textit{then} \quad (\alpha, \alpha, \mu) \overset{S}{\longleftrightarrow} {(\alpha + 1, \sigma) \atop (1, \tau)} \\ &\qquad\qquad\qquad\qquad\quad (\alpha, \mu) \overset{Z}{\longleftrightarrow} {(\alpha, \sigma) \atop (1, \tau)}. \end{aligned} \tag{9.5}$$

Iterating (9.5) we obtain as immediate consequences:

$$\mu = (m_0, m_0, m_1, m_1, m_2, m_2, \dots) \quad \overset{S}{\longleftrightarrow} \quad \begin{array}{l} \sigma = (m_0 + 1, m_1 + 1, m_2 + 1, \dots) \\ \tau = (1, 1, 1, \dots) \end{array}$$

and, in particular,

$$\mu = (a, a, s, s, s, s, \dots) \quad \overset{S}{\longleftrightarrow} \quad \begin{array}{l} \sigma = (a + 1, s + 1, s + 1, \dots) \\ \tau = (1, 1, 1, \dots). \end{array}$$

Furthermore,

$$\mu = (m_0, m_1, m_1, m_2, m_2, \dots) \quad \overset{Z}{\longleftrightarrow} \quad \begin{array}{l} \sigma = (m_0, m_1 + 1, m_2 + 1, \dots) \\ \tau = (1, 1, 1, \dots) \end{array}$$

and, in particular,

$$\mu = (a, s, s, s, \dots) \quad \overset{Z}{\longleftrightarrow} \quad \begin{array}{l} \sigma = (a, s + 1, s + 1, \dots) \\ \tau = (1, 1, 1, \dots). \end{array}$$

Looking at our list of Catalan-like numbers in Sect. 6(A), we arrive at the following interesting cases, where τ is always $\tau \equiv 1$.

$$\begin{array}{llllll}
\mu = (1, 1, 1, \dots) & \overset{S}{\longleftrightarrow} & \sigma = (2, 2, 2, \dots) & S_n = C_{n+1} & \text{(Catalan)} \\
\mu = (0, 0, 0, \dots) & \overset{S}{\longleftrightarrow} & \sigma = (1, 1, 1, \dots) & S_n = M_n & \text{(Motzkin)} \\
\mu = (1, 1, 1, \dots) & \overset{Z}{\longleftrightarrow} & \sigma = (1, 2, 2, \dots) & Z_n = C_n & \text{(Catalan)} \\
\mu = (0, 0, 0, \dots) & \overset{Z}{\longleftrightarrow} & \sigma = (0, 1, 1, \dots) & Z_n = R_n & \text{(Riordan)} \\
 & & & R_n = S_{n-1} - Z_{n-1} & \\
\mu = (0, 1, 1, \dots) & \overset{Z}{\longleftrightarrow} & \sigma = (0, 2, 2, \dots) & Z_n = F_n & \text{(Fine)} \\
 & & & F_n = S_{n-1} - Z_{n-1} & \\
\mu = (1, 0, 0, \dots) & \overset{Z}{\longleftrightarrow} & \sigma = (1, 1, 1, \dots) & Z_n = M_n & \text{(Motzkin)} \\
 & & & M_n = S_{n-1}. &
\end{array}$$

In particular, this finally proves our claim for $\mu \equiv 1$: $S_n = C_{n+1}$ and $Z_n = C_n$.

This last table, containing the Catalan, Motzkin, Riordan and Fine numbers, partly explains the ubiquity of these numbers in counting problems via the following approach.

Consider the usual ballot recursion (9.1)

$$b_{n,k} = b_{n-1,k-1} + b_{n-1,k} + \dots + b_{n-1,n-1}.$$

Suppose we are given sets $\mathcal{S}_{n,k}$ depending on two parameters n and k, $0 \le k \le n$, with $\mathcal{S}_n = \bigcup_{k=0}^{n} \mathcal{S}_{n,k}$, $|\mathcal{S}_0| = 1$. Assume that the following holds for all $n \ge 1$.

(A) There are maps $\varphi_{i,j} : \mathcal{S}_{n-1,i} \longrightarrow \mathcal{S}_{n,j}$ for all pairs (i, j) with $0 \le j \le i+1 \le n$.

(B) Any $P \in \mathcal{S}_{n,k}$ is the image under exactly one map $\varphi_{i,k}$; we denote this pre-image by ψP.

Then, clearly, the numbers $b_{n,k} = |\mathcal{S}_{n,k}|$ satisfy recursion (9.1), and hence $|\mathcal{S}_n| = C_{n+1}$.

Now we obtain further interpretations if the ballot sequence $\mu = (m_k)$ consists only of 0's and 1's, because $m_k = 1$, (resp. 0) simply corresponds to the fact that the map $\varphi_{k,k} : \mathcal{S}_{n-1,k} \longrightarrow \mathcal{S}_{n,k}$ is allowed (resp. forbidden). But, according to our list, these sequences μ lead precisely to our four number sequences (C_n), (M_n), (R_n) and (F_n).

For convenience, we list again the first few numbers:

$$
\begin{aligned}
C_n &= (1, 1, 2, 5, 14, 42, 132, \dots) \\
M_n &= (1, 1, 2, 4, 9, 21, 51, \dots) \\
R_n &= (1, 0, 1, 1, 3, 6, 15, \dots) \\
F_n &= (1, 0, 1, 2, 6, 18, 57, \dots).
\end{aligned}
$$

We illustrate this "ballot approach" to enumeration problems with three classical examples [28]. We just give the sets $\mathcal{S}_{n,k}$ and the maps $\varphi_{i,j}$ and ψ; the proofs are then straightforward.

Examples

(1) Non-crossing partitions. Let $\mathcal{S}_n$ be the set of non-crossing set partitions of $\{1, 2, \dots, n+1\}$, where we draw a partition on a horizontal line, with arcs (above the line) connecting the elements in a block. For example,

126 − 3 − 45 corresponds to 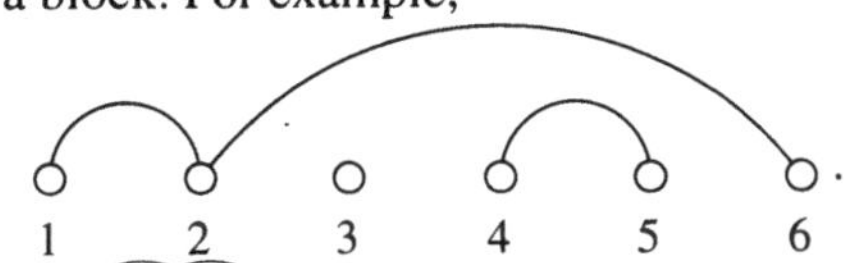

Non-crossing means that crossing arcs 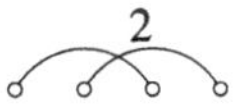 are forbidden.

For a partition P we denote by P_i the block containing i. Now we set

$$\mathcal{S}_{n,k} = \{P \in \mathcal{S}_n : |P_{n+1}| = k+1\}.$$

(A) The map $\varphi_{i,j} : \mathcal{S}_{n-1,i} \longrightarrow \mathcal{S}_{n,j}$ is defined as follows. Let $P \in \mathcal{S}_{n-1,i}$. Take the j smallest elements in P_n and merge them with $n+1$ to form a new block. This gives the partition $\varphi_{i,j}(P) \in \mathcal{S}_{n,j}$. For example

$$\varphi_{2,1}(126 - 3 - 45) = 26 - 3 - 45 - 17.$$

(B) Let $P \in \mathcal{S}_{n,k}$. Delete $n+1$ and merge $P_{n+1} \backslash n+1$ with P_n; this gives $\psi\pi \in \mathcal{S}_{n-1}$. For example,

$$\psi(126 - 3 - 45) = 3 - 1245.$$

Interpretating the various μ leading to our counting sequences we obtain the following well-known results:

C_n = # non-crossing partitions of $\{1, 2, \ldots, n\}$

M_{n-1} = # non-crossing partitions of $\{1, 2, \ldots, n\}$ without singleton blocks except possibly $\{n\}$

R_n = # non-crossing partitions of $\{1, 2, \ldots, n\}$ without singleton blocks

F_n = # non-crossing partitions of $\{1, 2, \ldots, n\}$ without singletons seen from above

M_n = # non-crossing partitions of $\{1, 2, \ldots, n\}$ without singletons seen from below but not from above ( is forbidden).

As an example, we list the $F_4 = 6$ partitions of $\{1, 2, 3, 4\}$ without singletons seen from above:

(2) Permutations. Let $\mathcal{S}_n$ be the set of permutations $\pi = \pi_1 \ldots \pi_{n+1}$ of $\{1, 2, \ldots, n+1\}$ with no subsequence $\pi_i \pi_j \pi_k$ $(i < j < k)$ such that $\pi_j < \pi_k < \pi_i$, For short we say that the *pattern* 312 is forbidden. Set $\mathcal{S}_{n,k} = \{\pi \in \mathcal{S}_n : \pi_1 = k + 1\}$. The maps are then:

(A) Let $\pi_1 = i+1, \pi_2\pi_3 \ldots \pi_n \in \mathcal{S}_{n-1,i}$. Then $\pi' = \pi'_1 \ldots \pi'_{n+1} = \varphi_{i,j}(\pi)$ is given as follows: $\pi'_1 = j + 1$, and for $k \geq 2$ we have $\pi'_k = \pi_{k-1}$ if $\pi_{k-1} < j + 1$, resp. $\pi'_k = \pi_{k-1} + 1$ if $\pi_{k-1} \geq j + 1$.

(B) For $\pi = k + 1, \pi_2 \ldots \pi_{n+1} \in \mathcal{S}_{n,k}$, set $\psi\pi = \pi'_1 \ldots \pi'_n \in \mathcal{S}_{n-1}$ where $\pi'_i = \pi_{i+1}$ if $\pi_{i+1} < k + 1$, resp. $\pi'_i = \pi_{i+1} - 1$ if $\pi_{i+1} > k + 1$.

This gives:

C_n = # permutations of $\{1, \ldots, n\}$ without a 312-pattern

M_{n-1} = # permutations of $\{1, \ldots, n\}$ without a 312-pattern such that, if $\pi_i < \pi_{i+1}$, then there exists $j > i + 1$ with $\pi_i < \pi_j < \pi_{i+1}$

R_n = # permutations of $\{1, \ldots, n\}$ as in the previous line and with $\pi_1 \geq 2$

F_n = # permutations of $\{1, \ldots, n\}$ without a 312-pattern and without a splitting element π_k, meaning $\pi_k = k$ and $\pi_i < \pi_k$ $(i < k)$ and $\pi_i > \pi_k$ $(i > k)$.

We list the $F_4 = 6$ permutations with this last property:

2143, 2341, 2431, 3241, 3421, 4321.

(3) Parallelogram animals. This example has been studied in many papers, e.g., [8, 10, 11, 17]. An animal P consists of rows $r_0, r_1, \ldots, r_t$ of lattice points such that

a) the points in row i have integral coordinates $(f_i, i), (f_i + 1, i), \ldots, (\ell_i, i)$ with $f_0 = 0$,

b) $f_i \le f_{i+1} \le \ell_i \le \ell_{i+1}$ for all i.

The number $n = \ell_t + t$ is called the *site perimeter* of P. The figure shows the five animals with site perimeter 2:

Let $\mathcal{S}_n$ be the set of parallelogram animals with site perimeters n, and

$$\mathcal{S}_{n,k} = \{P \in \mathcal{S}_n : \ell_0(P) = k\}.$$

The maps $\varphi_{i,j}$ and ψ are given as follows:

(A) Let $P \in \mathcal{S}_{n-1,i}$. If $j = i + 1$, add a point to the bottom line, moving P one step to the right. If $j \le i$, add a new bottom row of $j + 1$ points, moving P one step up.

(B) Let $P \in \mathcal{S}_{n,k}$. If $f_1 > 0$, delete the left–most point on the bottom line. If $f_1 = 0$, delete the bottom line.

The figure shows the operations:

We obtain:

C_{n+1} = # animals with site perimeter n

M_n = # animals with site perimeter n and $\ell_0 < \ell_1 < \ldots < \ell_t$ ("no walls")

R_{n+1} = # animals with site perimeter n, without walls and singleton rows

F_{n+1} = # animals with site perimeter n without singleton rows

M_{n+1} = # animals with site perimeter n, without walls, except singleton rows.

The figure shows the $F_4 = 6$ corresponding animals with $n = 3$:

There are numerous *weighted* versions (see [4, 8]). We discuss just one example for parallelogram animals. Consider the set $\mathcal{S}_n$ of animals with site perimeter n. For $P \in \mathcal{S}_n$ set

$$\begin{aligned} b(P) &= \ell_0 && \text{(base)} \\ w(P) &= \#\{i : \ell_i = \ell_{i+1}\} && \text{(wall)} \\ h(P) &= t && \text{(height)} \end{aligned}$$

and define the *weight* $W(P)$ of P

$$W(P) = \alpha^{b(P)} \beta^{w(P)} \gamma^{h(P)}.$$

Then the following can be proved [4]:

$$\sum_{P \in \mathcal{S}_n} W(P) = B_n^{\sigma,\tau} \text{ where } \begin{aligned} \sigma &= (\alpha + \beta\gamma, 1 + \beta\gamma, 1 + \beta\gamma, \dots) \\ \tau &= (\gamma, \gamma, \gamma, \dots). \end{aligned}$$

In particular, for $\alpha = 0$, $\beta = 1$, we obtain

$$\sum_{P \in \mathcal{S}_n, b(P)=0} \gamma^{h(P)} = B_n^{\sigma,\tau} \text{ with } \begin{aligned} \sigma &= (\gamma, \gamma + 1, \gamma + 1, \dots) \\ \tau &= (\gamma, \gamma, \gamma, \dots). \end{aligned}$$

These sequences σ, τ have already appeared in Sect. 7, and we obtain the Narayana polynomial

$$B_n^{\sigma,\tau} = \sum_{k=0}^{n-1} N(n,k) \gamma^{n-k},$$

and with $b(P) = 0$

$$\sum_{P \in \mathcal{S}_n} \gamma^{h(P)} = \frac{1}{n+1} \sum_{k=0}^{n} \binom{n+1}{k} \binom{n+1}{k+1} \gamma^k.$$

This gives the famous result (see, e.g., [8]) that the number of animals which fit within a rectangle with dimensions $a \times b$ (but not a smaller one) equals

$$\frac{1}{a+b-1} \binom{a+b-1}{a} \binom{a+b-1}{b}.$$

For $a = 2$, $b = 3$ we obtain the following six animals:

References

[1] Aigner, M. (1998): Motzkin numbers. European J. Combin. **19**, 663–675

[2] Aigner, M. (1999): Catalan-like numbers and determinants. J. Combin. Theory Ser. A **87**, 33–51

[3] Aigner, M. (1999): A characterization of the Bell numbers. Discrete Math. **205**, 207–210

[4] Aigner, M. (2000): Enumeration via ballot tables. Preprint

[5] Barnabei, M., Brini, A., Nicoletti, G. (1982): Recursive matrices and umbral calculus. J. Algebra **75**, 546–573

[6] Barnabei, M., Montefusco, L.B. (1998): Recursive properties of Toeplitz and Hurwitz matrices. Linear Alg. Appl. **274**, 367–388

[7] Bernhart, F.R. (1999): Catalan, Motzkin and Riordan numbers. Discrete Math. **204**, 73–112

[8] Bousquet-Mélou, M. (1994): Polyominoes and polygons. In: Barcelo, H., Kalai, G. (eds.) Jerusalem Combinatorics '93. (Contemporary Mathematics, vol. 178). American Mathematical Society, Providence, RI, pp. 55–70

[9] Chihara, T.S. (1978): An introduction to orthogonal polynomials. Gordon and Breach, New York

[10] Delest, M.-P., Fadou, J.M. (1993): Enumeration of skew Ferrers diagrams. Discrete Math. **112**, 65–79

[11] Delest, M.-P., Viennot, G. (1984): Algebraic languages and polynominoes enumeration. Theoret. Comput. Sci. **34**, 169–206

[12] Deutsch, E. (1999): Dyck path enumeration. Discrete Math. **204**, 167–202

[13] Donaghey, R., Shapiro, L.W. (1977): The Motzkin numbers. J. Combin. Theory Ser. A **23**, 291–301

[14] Flajolet, P. (1985): Combinatorial aspects of continued fractions. Discrete Math. **32**, 125–161

[15] Gessel, I.M., Viennot, G. (1985): Binomial determinants, paths, and hook length formulae. Adv. Math. **58**, 300–321

[16] Goulden, I., Jackson, D. (1983): Combinatorial enumeration. Wiley, New York

[17] Gouyou-Beauchamps, D., Viennot, G. (1988): Equivalence of the two-dimensional animal problem to a one-dimensional path problem. Adv. Appl. Math. **9**, 334–357

[18] Harary, F., Read, R. (1970): The enumeration of tree-like polyhexes. Proc. Edinburgh Math. Soc. **17**, 1–13

[19] Hofbauer, J., Fürlinger, J. (1985): q-Catalan numbers. J. Combin. Theory Ser. A **40**, 248–264

[20] Lindström, B. (1973): On the vector representation of induced matroids. Bull. London Math. Soc. **5**, 85–90

[21] Polýa, G. (1976): Problems and theorems in analysis, vol. II. Theory of functions, zeros, polynomials, determinants, number theory, geometry. Springer, Berlin

[22] Roman, S. (1984): The umbral calculus. Academic Press, New York

[23] Roman, S., Rota, G.-C. (1978): The umbral calculus. Adv. Math. **27**, 95–188

[24] Riordan, J. (1968): Combinatorial identities. Wiley, New York

[25] Shapiro, L.W., Getu, S., Woan, W.-J., Woodson, L.C. (1991): The Riordan group. Discrete Appl. Math. **34**, 229–239

[26] Sprugnoli, R. (1994): Riordan arrays and combinatorial sums. Discrete Math. **132**, 267–290

[27] Stanley, R. (1997): Enumerative combinatorics, vol. I. (Cambridge Studies in Advanced Mathematics, vol. 49). Cambridge University Press, Cambridge

[28] Stanley, R. (1999): Enumerative combinatorics, vol. 2. (Cambridge Studies in Advanced Mathematics, vol. 62). Cambridge University Press, Cambridge
[29] Viennot, G. (1984): Une théorie combinatoire des polynômes orthogonaux. Lecture Notes, Université de Quebec

Enumerative combinatorics on words

D. Perrin

1 Introduction

Generating series, also called generating functions, play an important role in combinatorial mathematics. Many enumeration problems can be solved by transferring the basic operations on sets into algebraic operations on formal series leading to a solution of an enumeration problem. The famous paper by Doubilet, Rota and Stanley, *The idea of generating function* [40], places the subject in a general mathematical frame-work allowing one to present in a unified way the diversity of generating functions, from the ordinary ones to the exponential or even Dirichlet.

Their place within the field of combinatorics on words is special. Indeed it was M. P. Schützenberger's point of view that sets of words can be considered as series in several noncommutating variables. The generating series of the set then appears as the image of a noncommutative series through a homomorphism. This gives rise to a rich domain in which an interplay between classical commutative algebra and combinatorics on words is present.

In these lectures, I will survey several aspects of generating functions on words. The emphasis is on the most elementary case corresponding to sets of words which can be defined using a finite automaton, usually called regular. The corresponding series are actually rational. Two special cases will be considered in turn. The first is the case of sets of words corresponding to leaves in a tree and usually called prefix codes. A recent result due to Frédérique Bassino, Marie-Pierre Béal and myself [10] is presented. It completely characterizes the generating series of regular prefix codes. The second is the case of sets of words considered up to a cyclic permutation, often called necklaces. The corresponding generating series are the zeta functions of symbolic dynamics.

A word on the terminology used here. We constantly use the term *regular* where a richer terminology is often used. In particular, what we call here a regular sequence is, in Eilenberg's terminology, an $\mathbb{N}$-rational sequence (see [22, 41] or [18]).

2 Regular sequences and automata

We consider the set A^* of all words on a given alphabet A. A subset of A^* is often called a *formal language*. For sets $X, Y \subset A^*$, we denote

$$\begin{aligned} X+Y &= X \cup Y, \\ XY &= \{xy \mid x \in X, y \in Y\}, \\ X^* &= \{x_1x_2\cdots x_n \mid x_i \in X, n \geq 0\}. \end{aligned}$$

We say that the pair (X, Y) is *unambiguous* if for each $z \in XY$ there is at most one pair $(x, y) \in X \times Y$ such that $z = xy$.

We say that a set of nonempty words X is a *code* if for each $x \in X^*$ there is at most one sequence $(x_1, x_2, \ldots, x_n)$ with $x_i \in X$ such that $x = x_1 x_2 \cdots x_n$ (one also says that X is *uniquely decipherable*). A particular case of a code is a *prefix code*. This is a set of words X such that no element of X is a prefix of another. It is easy to see that such a set is either reduced to the empty word or does not contain the empty word and is then a code.

The *length distribution* of a set of words X is the sequence $u_X = (u_n)_{n \geq 0}$ with

$$u_n = \mathrm{Card}(X \cap A^n).$$

We denote by u_X the formal series

$$u_X(z) = \sum_{n \geq 0} u_n z^n,$$

which is the ordinary generating series of the sequence u_X.

For example, the length distribution of $X = A^*$ is $u(z) = \frac{1}{1-kz}$ where $k = \mathrm{Card}(A)$.

The *entropy* of a formal language X is

$$h(X) = \log(1/\rho),$$

where ρ is the radius of convergence of the series $u_X(z)$. It is well-defined provided X is infinite and thus ρ is finite. If the alphabet A has k elements, then $h(X) \leq \log k$.

The following result relates the basic operations on sets with operations on series.

Proposition 1. *The following properties hold for any subsets X, Y of A^*.*

(i) *If $X \cap Y = \emptyset$, then $u_{X+Y} = u_X + u_Y$.*
(ii) *If the pair (X, Y) is unambiguous, then $u_{XY} = u_X u_Y$.*
(iii) *If X is a code, then $u_{X^*} = 1/(1 - u_X)$.*

Proof. The first two formulae are clear. If X is a code, every word in X^* has a unique decomposition as a product of words in X. This implies that

$$u_{X^n} = (u_X)^n$$

and thus,

$$u_{X^*} = 1 + u_X + \cdots + u_{X^n} + \cdots = 1/(1 - u_X). \qquad \square$$

Example 1. The set $X = \{b, ab\}$ is a prefix code. The series u_{X^*} is

$$u_{X^*}(z) = \frac{1}{1 - z - z^2}.$$

Let $(F_n)_{n\geq 0}$ be the sequence of Fibonacci numbers defined by $F_0 = 0$, $F_1 = 1$, and $F_{n+2} = F_{n+1} + F_n$. It follows from the recurrence relation that

$$\frac{z}{1 - z - z^2} = \sum_{n\geq 0} F_n z^n.$$

Consequently, $u_{X^*}(z) = \sum_{n\geq 0} F_{n+1} z^n$. It can also be proved by a combinatorial argument that the number of words of length n in X^* is F_{n+1}.

There are several variants of the generating series considered above. One may first define

$$p_X(z) = \sum_{n\geq 0} \frac{u_n}{k^n} z^n,$$

where $k = \mathrm{Card}(A)$. The coefficient of z^n in $p_X(z)$ is the probability for a word of length n to be in the set X. The relation between u_X and p_X is simple since $p_X(z) = u_X(z/k)$. Another variant of the generating series is the *exponential generating series* of the sequence $(u_n)_{n\geq 0}$ defined as

$$e(z) = \sum_{n\geq 0} \frac{u_n}{n!} z^n.$$

We will also use the zeta function of a sequence $(u_n)_{n\geq 1}$ defined as

$$\zeta(z) = \exp \sum_{n\geq 1} \frac{u_n}{n} z^n.$$

2.1 Regular sequences

We consider sequences of natural integers $s = (s_n)_{n\geq 0}$. We shall not distinguish between such a sequence and the formal series $s(z) = \sum_{n\geq 0} s_n z^n$.

We usually denote a vector indexed by elements of a set Q, also called a Q-vector, with boldface symbols. For $\mathbf{v} = (v_q)_{q\in Q}$ we say that v is nonnegative, denoted $\mathbf{v} \geq 0$, (resp., positive, denoted $\mathbf{v} > 0$) if $v_q \geq 0$ (resp., $v_q > 0$) for all $q \in Q$. The same conventions are used for matrices. A nonnegative $Q \times Q$-matrix M is said to be *irreducible* if, for all indices p, q, there is an integer m such that $(M^m)_{p,q} > 0$. The matrix is *primitive* if there is an integer m such that $M^m > 0$.

The *adjacency matrix* of a graph $G = (Q, E)$ is the $Q \times Q$-matrix M such that, for each $p, q \in Q$, the integer $M_{p,q}$ is the number of edges from p to q. The adjacency matrix of a graph G is irreducible iff the graph is strongly connected. It is primitive if, moreover, the g.c.d. of lengths of cycles in G is 1.

Let G be a finite graph and let I, T be two sets of vertices. For each $n \geq 0$, let s_n be the number of distinct paths of length n from a vertex of I to a vertex of T. The sequence $s = (s_n)_{n\geq 0}$ is called the sequence *recognized* by (G, I, T) or also by G if I and T are already specified. When $I = \{i\}$ and $T = \{t\}$, we simply write (G, i, t) instead of $(G, \{i\}, \{t\})$.

A sequence $s = (s_n)_{n\geq 0}$ of nonnegative integers is said to be *regular* if it is recognized by such a triple (G, I, T), where G is finite. We say that the triple (G, I, T) is a *representation* of the sequence s. The vertices of I are called *initial* and those of T *terminal*. Two representations are said to be *equivalent* if they recognize the same sequence.

A representation (G, I, T) is said to be *trim* if every vertex of G is on some path from I to T. It is clear that any representation is equivalent to a trim one.

A well-known result in theory of finite automata allows one to use a particular representation of any regular sequence s such that $s_0 = 0$. One can always choose in this case a representation (G, i, t) of s with a unique initial vertex i and a unique final vertex $t \neq i$ such that no edge is entering vertex i and no edge is going out of vertex t. Such a representation is called a *normalized representation* (see, for example, [37, p. 14]).

Let (G, i, t) be a trim normalized representation. If we merge the initial vertex i and the final vertex t in a single vertex still denoted by i, we obtain a new graph denoted by $\overline{G}$, which is strongly connected. The triple $(\overline{G}, i, i)$ is called the *closure* of (G, i, t).

Let s be a regular sequence such that $s_0 = 0$. The *star* s^* of the sequence s is defined by

$$s^*(z) = \frac{1}{1 - s(z)}.$$

Proposition 2. *If* (G, i, t) *is a normalized representation of* s*, its closure* $(\overline{G}, i, i)$ *recognizes the sequence* s^*.

Proof. The sequence s is the length distribution of the paths of first returns to vertex i in $\overline{G}$, that is, of finite paths going from i to i without going through vertex i. The length distribution of the set of all returns to i is thus $1 + s(z) + s^2(z) + \ldots = 1/(1 - s(z))$. □

An equivalent definition of regular sequences uses vectors instead of sets I, F. Let $\mathbf{i}$ be a Q-row vector of nonnegative integers and let $\mathbf{t}$ be a Q-column vector of nonnegative integers. We say that $(G, \mathbf{i}, \mathbf{t})$ recognizes the sequence $s = (s_n)_{n\geq 0}$ if, for each integer $n \geq 0$,

$$s_n = \mathbf{i}M^n\mathbf{t},$$

where M is the adjacency matrix of G. The proof that both definitions are equivalent follows from the fact that the family of regular sequences is closed under addition (see [22]). A triple $(G, \mathbf{i}, \mathbf{t})$ recognizing a sequence s is also called a representation of s and two representations are called equivalent if they recognize the same sequence.

A sequence $s = (s_n)_{n\geq 0}$ of nonnegative integers is *rational* if it satisfies a recurrence relation with integral coefficients. Equivalently, s is rational if there exist two polynomials $p(z)$, $q(z)$ with integral coefficients and with $q(0) = 1$ such that

$$s(z) = \frac{p(z)}{q(z)}.$$

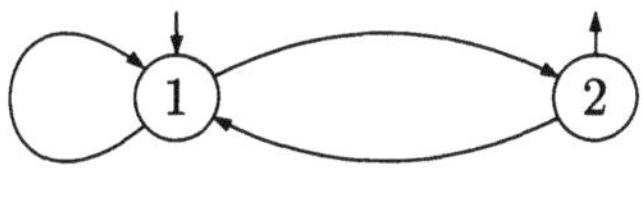

Fig. 1. The Fibonacci graph

For example, the sequence s defined by $s(z) = \frac{z}{1-z-z^2}$ is the sequence of Fibonacci numbers also defined by $s_0 = 0, s_1 = 1$ and $s_{n+1} = s_n + s_{n-1}$. It is recognized by the graph of Fig. 1 with $I = \{1\}$ and $T = \{2\}$.

Any regular sequence is rational. The converse is, however, not true (see Sect. 3.6).

A theorem of Soittola [41], also found independently in [27], characterizes those rational sequences which are regular. We say that a rational sequence has a *dominating root* either if it is a polynomial or if it has a real positive pole which is strictly smaller than the modulus of any other. A sequence r is a *merge* of the sequences r_i if there is an integer p such that

$$r(z) = \sum_{i=0}^{p-1} z^i r_i(z^p).$$

Theorem 1 (Soittola). *A sequence of nonnegative integers $r = (r_n)_{n\geq 0}$ is regular if and only if it is a merge of rational sequences having a dominating root.*

This result shows that whether a rational series is regular is decidable (see [41]). In the positive case, there is an algorithm computing a representation of the sequence.

2.2 Finite automata

We present here a brief introduction to the concepts used in automata theory. For a general reference, see [37] or [22].

An *automaton* over the alphabet A is composed of a set Q of *states*, a set $E \subset Q \times A \times Q$ of *edges* or *transitions* and two sets $I, T \subset Q$ of *initial* and *terminal* states.

A *path* in the automaton $\mathcal{A}$ is a sequence

$$(p_1, a_1, p_2), (p_2, a_2, p_3), \ldots, (p_n, a_n, p_{n+1})$$

of consecutive edges. Its label is the word $x = a_1 a_2 \cdots a_n$. A path is *successful* if it starts in an initial state and ends in a terminal state. The set *recognized* by the automaton is the set of labels of its successful paths.

An automaton is *deterministic* if, for each state p and each letter a, there is at most one edge which starts at p and is labeled by a. The term *right resolving* is also used.

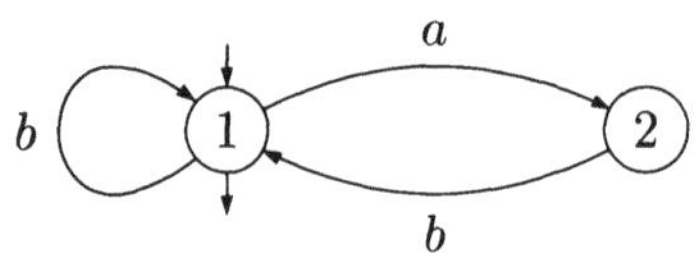

Fig. 2. Golden mean automaton

Example 2. Let $\mathcal{A}$ be the automaton given in Fig. 2 with 1 as unique initial and terminal state. It recognizes the set X^* where X is the prefix code $X = \{b, ab\}$.

A set of words X over A is *regular* if it can be recognized by a finite automaton.

It is a classical result that a set of words is regular iff it can be obtained by a finite number of operations union, product and star, starting from the finite sets.

The following result is also classical (see [22] for example).

Proposition 3. *Every regular set can be recognized by a finite deterministic automaton having a unique initial state.*

The following theorem is of fundamental importance. It belongs to the early folklore of automata theory.

Theorem 2. *The length distributions of regular sets are the regular sequences.*

Proof. Let X be a regular set. By Proposition 3, it can be recognized by a deterministic automaton $\mathcal{A}$. Since $\mathcal{A}$ is deterministic, there is at most one path with given label, origin and end. Thus the number of paths of length n from the initial state to a terminal state is equal to the number u_n of words of X of length n.

Conversely, let u be a regular sequence enumerating the paths in a graph G from I to T. We consider the graph G as an automaton with all edges with distinct labels. Let X be the set of labels of paths from I to T. The sequence u is the length distribution of the set X. □

Example 3. If $X = a^*b$, then

$$u_X(z) = \frac{z}{1-z}.$$

2.3 Beyond regular sequences

There are several natural classes of series beyond the rational ones. The algebraic series are those satisfying an algebraic equation. More generally, the hypergeometric series are those such that the quotient of two successive terms is given by a rational fraction (see [26]).

The class of algebraic series is linked with the class of context-free sets (see [23]). A typical example of a context-free set is the set of words on the binary alphabet $\{a, b\}$ having as many a's as b's. We compute below its length distribution which is an algebraic series.

Example 4. The set of words on $A = \{a, b\}$ having an equal number of occurrences of a and b is a submonoid of A^* generated by a prefix code D. Since any word of D^* of length $2n$ is obtained by choosing n positions among $2n$, we have

$$u_{D^*}(z) = \sum_{n \geq 0} \binom{2n}{n} z^{2n}.$$

By a simple application of the binomial formula, we obtain

$$u_{D^*}(z) = (1 - 4z^2)^{-\frac{1}{2}}.$$

Indeed, this follows from the simple identity

$$\binom{-\frac{1}{2}}{n} = \frac{1}{(-4)^n} \binom{2n}{n}.$$

We have $u_D(z) = 1 - 1/u_{D^*}(z)$ and thus

$$u_D(z) = 1 - \sqrt{1 - 4z^2}.$$

Thus $u_D(z)$ is an algebraic series, a solution of the equation

$$f^2 - 2f + 4z^2 = 0.$$

3 Enumeration on regular trees

We now turn to the study of generating sequences linked with trees. In fact, we do not enumerate trees but objects within a tree such as the nodes or the leaves at each level. This is actually equivalent to the enumeration of particular sets of words, namely, prefix-closed sets and prefix codes, as we shall see below (Sect. 4).

3.1 Graphs and trees

In this paper, we use directed multigraphs, i.e., graphs with possibly several edges with the same origin and the same end. We simply call them graphs in all that follows. We denote by $G = (Q, E)$ a graph with Q as set of vertices and E as set of edges. We also say that G is a graph on the set Q.

A *tree* T on a set of nodes N with *root* $r \in N$ is a function $T : N - \{r\} \longrightarrow N$ which associates to each node distinct from the root its father $T(n)$, in such a way that, for each node n, there is a nonnegative integer h such that $T^h(n) = r$. The integer h is the *height* of the node n.

A tree is *k-ary* if each node has at most k children. A node without children is called a *leaf*. A node which is not a leaf is called *internal*. A node n is a *descendant* of a node m if $m = T^h(n)$ for some $h \geq 0$. A k-ary tree is *complete* if all internal nodes have exactly k children and have at least one descendant which is a leaf.

For each node n of a tree T, the *subtree* rooted at n, denoted T_n is the tree obtained by restricting the set of nodes to the descendants of n.

Two trees S, T are isomorphic, denoted $S \equiv T$, if there is a map which transforms S into T by permuting the children of each node. Equivalently, $S \equiv T$ if there is a bijective map $f : N \to M$ from the set of nodes of S onto the set of nodes of T such that $f \circ S = T \circ f$. Such a map f is called an isomorphism.

If T is a tree with N as set of nodes, the *quotient graph* of T is the graph $G = (Q, E)$ where Q and E are defined as follows. The set Q is the quotient of N by the equivalence $n \equiv m$ if $T_n \equiv T_m$. Let $\bar{m}$ denote the class of a node m. The number of edges from $\bar{m}$ to $\bar{n}$ is the number of children of m equivalent to n.

Conversely, the set of paths in a graph with given origin is a tree. Indeed, let $G = (Q, E)$ be a graph. Let $r \in Q$ be a particular vertex and let N be the set of paths in G starting at r. The tree T having N as set of nodes and such that $T(p_0, p_1, \dots, p_n) = (p_0, p_1, \dots, p_{n-1})$ is called the *covering tree* of G starting at r.

Both constructions are mutually inverse in the sense that any tree T is isomorphic to the covering tree of its quotient graph starting at the image of the root.

Proposition 4. *Let T be a tree with root r. Let G be its quotient graph and let i be the vertex of G which is the class of the root of T. For each vertex q of G and for each $n \geq 0$, the number of paths of length n from i to q is equal to the number of nodes of T at height n in the class of q.*

A tree is said to be *regular* if it admits only a finite number of non-isomorphic subtrees, i.e., if its quotient graph is finite.

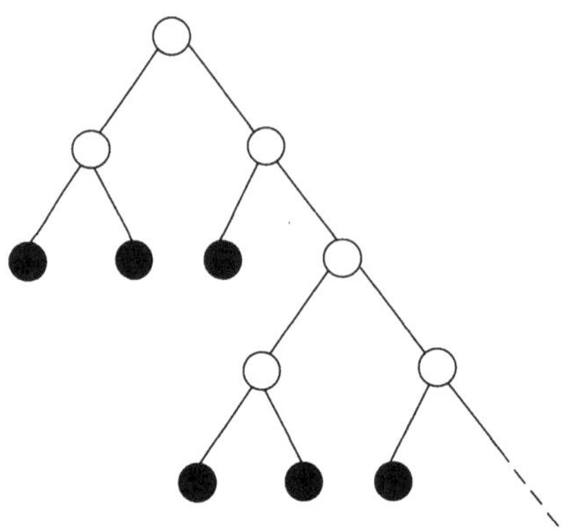

Fig. 3. A regular tree

For example, the infinite tree represented in Fig. 3 is a regular tree. Its quotient graph is represented in Fig. 4.

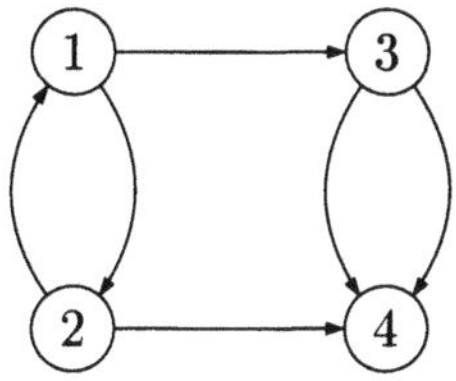

Fig. 4. And its quotient graph

3.2 Regular sequences and trees

If T is a tree, its *generating sequence of leaves* is the sequence of numbers $s = (s_n)_{n\geq 0}$, where s_n is the number of leaves at height n. We also simply say that s is the *generating sequence* of T.

The following result is a direct consequence of the definitions.

Theorem 3. *The generating sequence of a regular tree is a regular sequence.*

Proof. Let T be a regular tree and let G be its quotient graph. Since T is regular, G is finite. The leaves of T form an equivalence class t. By Proposition 4, the generating sequence of T is recognized by (G, i, t) where i is the class of the root of T. □

We say that a sequence $s = (s_n)_{n\geq 1}$ satisfies the Kraft inequality for the integer k if

$$\sum_{n\geq 0} s_n k^{-n} \leq 1,$$

i.e., in terms of the formal series $s(z) = \sum_{n\geq 0} s_n z^n$, if

$$s(1/k) \leq 1.$$

We say that s satisfies the strict Kraft inequality for k if $s(1/k) < 1$. The following result is well-known (see [4, p. 35], for example).

Theorem 4. *A sequence s is the generating sequence of a k-ary tree iff it satisfies the Kraft inequality for the integer k.*

We next consider the Kraft's equality case. If $s(1/k) = 1$, then any tree T having s as generating sequence is complete. The converse property is not true in general (see [22, p. 231]). However, it is a classical result that, when T is a complete regular tree, its generating sequence satisfies $s(1/k) = 1$ (see Proposition 8).

For the sake of a complete description of the construction described above in the proof of Theorem 4, we have to specify the choice made at each step among the leaves at height n. A possible policy is to choose to give as many children as possible to the nodes which are not leaves and of maximal height.

If we start with a finite sequence s satisfying Kraft's inequality, the above method builds a finite tree with generating sequence equal to s. It is not true that this incremental method gives a regular tree when we start with a regular sequence, as shown in the following example.

Let $s(z) = z^2/(1-2z^2)$. Since $s(1/2) = 1/2$, we may apply the Kraft construction to build a binary tree with length distribution s. The result is the tree $T(X)$ where X is the set of prefixes of the set

$$Y = \bigcup_{n\geq 0} 01^n0\{0,1\}^n.$$

which is not regular.

If s is a regular sequence such that $s_0 = 0$, there exists a regular tree T having s as generating sequence. Indeed, let (G, i, t) be a normalized representation of s. The generating sequence of the covering tree of G starting at i is s. Moreover, if s satisfies the Kraft inequality for an integer k, it is, however, not true that the regular covering tree obtained is k-ary, as shown in the following example.

Let s be the regular sequence recognized by the graph on the left of Fig. 5 with $i = 1$ and $t = 4$. We have $s(z) = 3z^2/(1-z^2)$. Furthermore $s(1/2) = 1$ and thus s satisfies Kraft's equality for $k = 2$. However there are four edges going out of vertex 2 and its regular covering tree starting at 1 is 4-ary. A solution for this example is given by the graph on the right of Fig. 5. It recognizes s and its covering tree starting at 1 is the regular binary tree of Fig. 3.

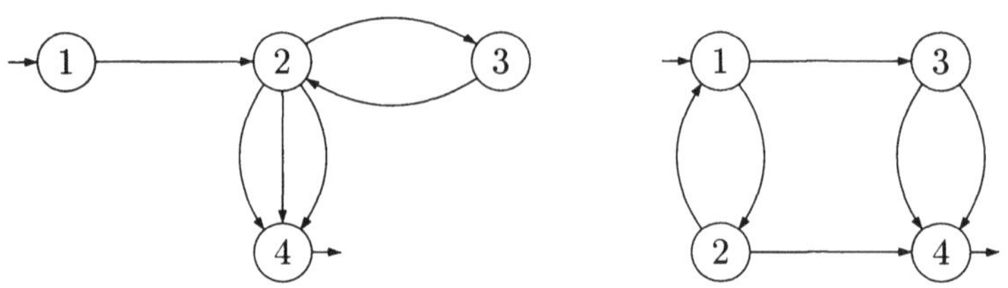

Fig. 5. Graphs recognizing $s(z) = 3z^2/(1-z^2)$

The aim of Sect. 3.5 is to build from a regular sequence s that satisfies the Kraft inequality for an integer k, a tree with generating sequence s which is both regular and k-ary.

3.3 Approximate eigenvectors

Let M be the adjacency matrix of a graph G. By the Perron–Frobenius theorem (see [25] for a general presentation and [30], [28] or [11] for the link with graphs and regular sequences), the nonnegative matrix M has a nonnegative real eigenvalue of maximal modulus denoted by λ, also called the spectral radius of the matrix.

When G is strongly connected, the matrix is irreducible and the Perron–Frobenius theorem asserts that the dimension of the eigenspace of the matrix M corresponding to λ is equal to one, and that there is a positive eigenvector associated to λ.

Let k be an integer. A *k-approximate* eigenvector of a nonnegative matrix M is, by definition, an integral vector $\mathbf{v} \geq 0$ such that

$$M\mathbf{v} \leq k\mathbf{v}.$$

We have the following result (see [30, p. 152]).

Proposition 5. *An irreducible nonnegative matrix M with spectral radius λ admits a positive k-approximate eigenvector iff $k \geq \lambda$.*

For a proof, see [30, p. 152]. When M is the adjacency matrix of a graph G, we also say that $\mathbf{v}$ is a k-approximate eigenvector of G. The computation of an approximate eigenvector can be obtained by the use of Franaszek's algorithm (see, for example, [30]). It can be shown that there exists a k-approximate eigenvector with elements bounded above by k^{2n} where n is the dimension of M [5]. Thus the size of the coefficients of a k-approximate eigenvector is bounded above by an exponential in n and can be, in the worst case, of this order of magnitude.

The following result is well-known. It links the radius of convergence of a sequence with the spectral radius of the associated matrix.

Proposition 6. *Let s be a regular sequence recognized by a trim representation (G, I, T). Let M be the adjacency matrix of G. The radius of convergence of s is the inverse of the maximal eigenvalue of M.*

Proof. The maximal eigenvalue λ of M is $\lambda = \limsup_{n \geq 0} \sqrt[n]{\|M^n\|}$, where $\| \ \|$ is any one of the equivalent matrix norms. Let ρ be the radius of convergence of s and, for each $p, q \in Q$, let ρ_{pq} be the radius of convergence of the sequence $u_{pq} = (M^n_{pq})_{n \geq 0}$. Then $1/\lambda = \min \rho_{pq}$. Since (G, I, T) is trim, we have $\rho_{pq} \geq \rho$ for all $p, q \in Q$. On the other hand, $\rho \geq \min \rho_{pq}$ since s is a sum of some of the sequences u_{pq}. Thus $\rho_s = \min \rho_{pq}$ which concludes the proof. □

As a consequence of this result, the radius of convergence ρ of a regular sequence s is a pole. Indeed, with the above notation, $s(z) = \mathbf{i}(I - Mz)^{-1}\mathbf{t}$. Then $\det(I - Mz)$ is a denominator of the rational fraction s, the poles of s are among the inverses of the eigenvalues of M. And since $1/\lambda$ is the radius of convergence of s, it has to be a pole of s. In particular, s diverges for $z = \rho$.

The following result, due to Berstel, is also well-known. It allows one to compute the radius of convergence of the star of a sequence.

Proposition 7. *Let s be a regular sequence. The radius of convergence of the series $s^*(z) = 1/(1 - s(z))$ is the unique real number r such that $s(r) = 1$.*

For a proof, see [22, pp. 211–214], [18, p. 82] or [11, p. 84]. As a consequence, we obtain the following result.

Proposition 8. *Let s be a regular sequence and let λ be the inverse of the radius of convergence of s^*. The sequence s satisfies the Kraft strict inequality $s(1/k) < 1$ (resp., equality $s(1/k) = 1$) if and only if $\lambda < k$ (resp., $\lambda = k$).*

We have thus proved the following result, which is the basis of the constructions of the subsequent sections.

Proposition 9. *Let s be a regular sequence satisfying Kraft's inequality $s(1/k) \leq 1$. Let (G, i, t) be a normalized representation of s and let $(\overline{G}, i, i)$ be the closure of (G, i, t). Then the adjacency matrix M of $\overline{G}$ admits a k-approximate eigenvector.*

In fact, under the hypothesis of Proposition 9, the graph G itself also admits a k-approximate eigenvector. Indeed, let $\overline{\mathbf{w}} = (\overline{w}_q)_{q \in Q-t}$ be a k-approximate eigenvector of $\overline{G}$. Then the vector $\mathbf{w} = (w_q)_{q \in Q}$ defined by $w_q = \overline{w}_q$ for $q \neq t$ and $w_t = \overline{w}_i$ is a k-approximate eigenvector of G. This is illustrated in the following example.

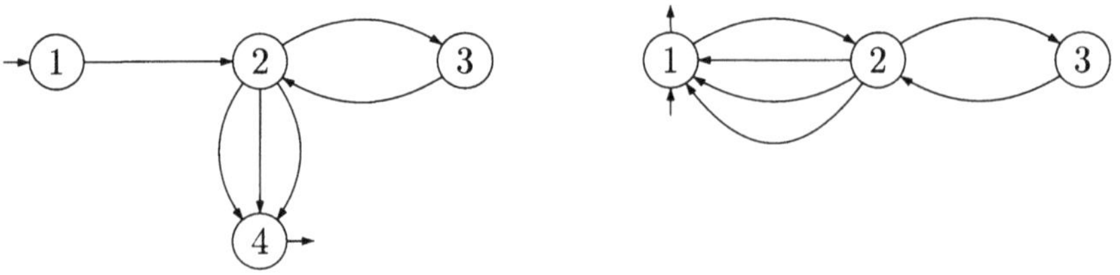

Fig. 6. The graphs G and $\overline{G}$

For example, we again consider $s(z) = 3z^2/(1 - z^2)$ (see Fig. 5). The sequence s is recognized by the normalized representation $(G, 1, 4)$ where G is the graph represented on the left of Fig. 6. The graph $\overline{G}$ is represented on the right. The vectors

$$\mathbf{w} = \begin{bmatrix} 3 \\ 2 \\ 1 \\ 3 \end{bmatrix}, \overline{\mathbf{w}} = \begin{bmatrix} 3 \\ 2 \\ 1 \end{bmatrix}$$

are 2-approximate eigenvectors of G and $\overline{G}$ respectively.

3.4 The multiset construction

In this section, we present the multiset construction. It can be considered as a version with multiplicities of the subset construction used in automata theory to replace a finite automaton by an equivalent deterministic one. We only use unlabeled graphs but the construction can be easily generalized to graphs with edges labeled by symbols from an alphabet.

Our construction is also linked with one used by D. Lind to build a positive matrix with given spectral radius (see [30], especially Lemma 11.1.9).

For convenience we use the term *multiset* of elements of a set Q as a synonym of Q-vector. If $\mathbf{u} = (u_q)_{q \in Q}$ is such a multiset, the coefficient u_q is also called the *multiplicity* of q. The *degree* of $\mathbf{u}$ is the sum $\sum_{q \in Q} u_q$ of all multiplicities.

We start with a triple $(G, \mathbf{i}, \mathbf{t})$ where $G = (Q, E)$ is a finite graph and $\mathbf{i}$ (resp., $\mathbf{t}$) is a row (resp., column) Q-vector. We denote by M the adjacency matrix of G.

Let m be a positive integer. We define another triple $(H, \mathbf{J}, \mathbf{X})$ which is said to be obtained by the *multiset construction*. The graph H is called an *extension* of the graph G. The extension is not unique and depends, as we shall see, on some arbitrary choices. The set S of vertices of H is formed of multisets of elements of Q of total degree at most m. Thus, an element of S is a nonnegative vector $\mathbf{u} = (u_q)_{q \in Q}$ with indices in Q such that $\sum_{q \in Q} u_q \leq m$. This condition ensures that H is a finite graph.

We now describe the set of edges of the graph H by defining its adjacency matrix N. Let U be the $S \times Q$-matrix defined by $U_{\mathbf{u},q} = u_q$. Then N is any nonnegative $S \times S$-matrix which satisfies

$$NU = UM.$$

Equivalently, for all $\mathbf{u} \in S$,

$$\sum_{\mathbf{v} \in S} N_{\mathbf{u},\mathbf{v}} \mathbf{v} = \mathbf{u}M.$$

Let us comment informally on the above formula. We can describe the construction of the graph H as a sequence of choices. If we reach a vertex $\mathbf{u}$ of H, we partition the multiset $\mathbf{u}M$ of vertices reachable from the vertices composing $\mathbf{u}$, into multisets of degree at most m to define the vertices reachable from $\mathbf{u}$ in H. The integer $N_{\mathbf{u},\mathbf{v}}$ is the multiplicity of $\mathbf{v}$ in the partition. The formula simply expresses the fact that the result is indeed a partition. In general, there are several possible partitions. The matrix U is called the *transfer matrix* of the extension.

Further, we define the S-row vector $\mathbf{J}$ and the S-column vector $\mathbf{X}$. Let $\mathbf{J}$ be the S-row vector such that $J_{\mathbf{i}} = 1$ and $J_{\mathbf{u}} = 0$ for $\mathbf{u} \neq \mathbf{i}$. Let $\mathbf{X}$ be the S-column vector such that $X_{\mathbf{u}} = \mathbf{u} \cdot \mathbf{t}$.

Thus

$$\mathbf{J}U = \mathbf{i}, \quad \mathbf{X} = U\mathbf{t}.$$

To avoid unnecessary complexity, we only keep in S the vertices reachable from $\mathbf{i}$. Thus, we replace the set S by the set of elements $\mathbf{u}$ of S such that there is a path from $\mathbf{i}$ to $\mathbf{u}$.

The number of multisets of degree at most m on a set Q with n elements is $\frac{n^{m+1}-1}{n-1}$. Thus the number of vertices of a multiset extension is of order n^m. It is polynomial in n if m is taken as a constant.

Fig. 7. The graphs G and H

For example, let G be the graph represented in Fig. 7 on the left. The graph H represented on the right is a multiset extension of G with

$$\mathbf{i} = \begin{bmatrix} 1 & 0 \end{bmatrix}, \quad \mathbf{j} = \begin{bmatrix} 0 \\ 1 \end{bmatrix}.$$

The matrices M, N and U are

$$M = \begin{bmatrix} 2 & 1 \\ 0 & 1 \end{bmatrix}, N = \begin{bmatrix} 1 & 1 \\ 0 & 2 \end{bmatrix}, U = \begin{bmatrix} 1 & 0 \\ 1 & 1 \end{bmatrix}, \mathbf{J} = \begin{bmatrix} 1 & 0 \end{bmatrix}, \mathbf{X} = \begin{bmatrix} 0 \\ 1 \end{bmatrix}.$$

In this case, the matrix U is invertible and the matrices M, N are conjugate.

The basic property of an extension is as follows.

Proposition 10. *Let H be an extension of G. The triple $(H, \mathbf{J}, \mathbf{X})$ is equivalent to $(G, \mathbf{i}, \mathbf{t})$.*

Proof. For each $n \geq 0$, we have

$$UM^n = N^n U.$$

Consequently, for each integer $n \geq 0$,

$$\begin{aligned} \mathbf{J}N^n\mathbf{X} &= \mathbf{J}N^n U\mathbf{t} \\ &= \mathbf{J}UM^n\mathbf{t} \\ &= \mathbf{i}M^n\mathbf{t}. \end{aligned}$$

This shows that $(H, \mathbf{J}, \mathbf{X})$ recognizes s. □

We will also make use of the following additional property of extensions.

Proposition 11. *Let H be an extension of G. Let M (resp., N) be the adjacency matrix of G (resp., H) and let U be the transfer matrix. If $\mathbf{w}$ is a k-approximate eigenvector of M, the vector $\mathbf{W} = U\mathbf{w}$ is a k-approximate eigenvector of N. If $\mathbf{w}$ is positive, then $\mathbf{W}$ is also positive.*

Proof. We have

$$NW = NUw = UMw \leq kUw = kW.$$

Since all rows of U are distinct from $\mathbf{0}$, the vector $\mathbf{W}$ is positive whenever $\mathbf{w}$ is positive. □

In the next section, we will choose a particular extension of the graph G called admissible and which is defined as follows. Let $\mathbf{w}$ be a positive Q-vector and let m be a positive integer. Let H be an extension of G, let U be the transfer matrix and let $\mathbf{W} = U\mathbf{w}$. We say that H is *admissible* with respect to $\mathbf{w}$ and m if, for each $\mathbf{u} \in S$, all but possibly one of the vertices $\mathbf{v}$ such that $(\mathbf{u}, \mathbf{v})$ is an edge of H satisfy $W_{\mathbf{v}} \equiv 0 \bmod m$.

Theorem 5. *For any graph G on Q, any positive Q-vector $\mathbf{w}$ and any integer $m > 0$, the graph G admits an admissible extension with respect to $\mathbf{w}$ and m.*

The proof relies on the following combinatorial lemma. This lemma is also used in a similar context by Adler et al. [1] and Marcus [34]. It is actually presented in [3] as a nice variant of the pigeon-hole principle.

Lemma 1. *Let $w_1, w_2, \ldots, w_m$ be positive integers. Then there is a non-empty subset $S \subset \{1, 2, \ldots, m\}$ such that $\sum_{q \in S} w_q$ is divisible by m.*

Proof. Either the partial sums $w_1, w_1 + w_2, w_1 + w_2 + w_3, \ldots, w_1 + w_2 + \cdots + w_m$ are all distinct (mod m), or two are congruent (mod m). In the former case, at least one partial sum must be congruent to 0 (mod m). In the latter, there are indices p, r, $1 \leq p < r \leq m$, such that

$$w_1 + w_2 + \cdots + w_p \equiv w_1 + w_2 + \cdots + w_r (\bmod m).$$

Hence $w_{p+1} + w_{p+2} + \cdots + w_r \equiv 0 \pmod m$. □

Proof of Theorem 5. We build progressively the set of edges of H. Let $\mathbf{u}$ be an element of S. We prove by induction on the degree $d(\mathbf{u}M) = \sum_{q \in Q} (\mathbf{u}M)_q$ of $\mathbf{u}M$ that there exist $\mathbf{v}_1, \ldots, \mathbf{v}_n \in S$ such that $\mathbf{u}M = \sum_{i=1}^{n} \mathbf{v}_i$ and $W_{\mathbf{v}_i} \equiv 0 \bmod m$ if $1 \leq i \leq n-1$. If $\mathbf{u}M \in S$, i.e., if $d(\mathbf{u}M) \leq m$, we choose $n = 1$ and $\mathbf{v}_1 = \mathbf{u}M$. Otherwise, there exists a decomposition $\mathbf{u}M = \mathbf{v} + \mathbf{u}'$ such that $d(\mathbf{v}) = m$. Let $w_1, w_2, \ldots, w_m$ be the sequence of integers formed by the w_q repeated v_q times. By Lemma 1 applied to the sequence of integers w_i, there is a decomposition $\mathbf{v} = \mathbf{v}' + \mathbf{r}$ with $\mathbf{v}' \neq 0$ such that $W_{\mathbf{v}'} \equiv 0 \bmod m$. We have $\mathbf{u}M = \mathbf{v}' + \mathbf{w}'$ with $\mathbf{w}' = \mathbf{r} + \mathbf{u}'$. Since $d(\mathbf{w}') < d(\mathbf{u}M)$, we can apply the induction hypothesis to $\mathbf{w}'$ to obtain the desired result. □

For an S-vector $\mathbf{W}$, we denote by $\lceil \frac{\mathbf{W}}{m} \rceil$ the S-vector $\mathbf{Z}$ such that, for each $\mathbf{u}$ in S,

$$Z_{\mathbf{u}} = \lceil \frac{W_{\mathbf{u}}}{m} \rceil.$$

Summing up the previous results, we obtain the following statement.

Proposition 12. *Let H be an admissible extension of G with respect to* $\mathbf{w}$ *and m. Let M (resp., N) be the adjacency matrix of G (resp., H), let U be the transfer matrix and let* $\mathbf{W} = U\mathbf{w}$. *If* $\mathbf{w}$ *is a positive k-approximate eigenvector of M, then* $\lceil \frac{\mathbf{W}}{m} \rceil$ *is a positive k-approximate eigenvector of N.*

Proof. By Proposition 11, the vector $\mathbf{W}$ is a positive k-approximate eigenvector of N. Thus

$$N\mathbf{W} \leq k\mathbf{W}.$$

Let $\mathbf{u}$ be an element of S. We have $W_{\mathbf{v}} \equiv 0 \bmod m$ for all indices $\mathbf{v}$ such that $N_{\mathbf{u},\mathbf{v}} > 0$ except possibly for an index $\mathbf{v}_0$. The previous inequality implies that

$$\sum_{\mathbf{v} \in S - \{\mathbf{v}_0\}} N_{\mathbf{u},\mathbf{v}} \frac{W_{\mathbf{v}}}{m} + N_{\mathbf{u},\mathbf{v}_0} \frac{W_{\mathbf{v}_0}}{m} \leq k \frac{W_{\mathbf{u}}}{m}.$$

Since $\frac{W_{\mathbf{v}}}{m}$ is a nonnegative integer for $\mathbf{v} \in Q - \{\mathbf{v}_0\}$, we get

$$\sum_{\mathbf{v} \in S - \{\mathbf{v}_0\}} N_{\mathbf{u},\mathbf{v}} \frac{W_{\mathbf{v}}}{m} + N_{\mathbf{u},\mathbf{v}_0} \lceil \frac{W_{\mathbf{v}_0}}{m} \rceil \leq k \lceil \frac{W_{\mathbf{u}}}{m} \rceil.$$

This proves that

$$N \lceil \frac{\mathbf{W}}{m} \rceil \leq k \lceil \frac{\mathbf{W}}{m} \rceil. \qquad \square$$

3.5 Generating sequences of leaves

In what follows, we show how the multiset construction allows one to prove the main result of [9] concerning the generating sequences of regular trees. We begin with the following lemma, which is also used in the next section. We use the term leaf for a vertex of a graph without outgoing edges.

Lemma 2. *Let G be a graph on a set Q of vertices. Let $i \in Q$ and $T \subset Q$. If G admits a positive k-approximate eigenvector* $\mathbf{w}$, *then there is a graph G' and a set of vertices I' of G' such that:*

1. *G' admits the k-approximate eigenvector* $\mathbf{w}'$ *with all components equal to* 1;
2. *the triple* $(G, i, \mathbf{w})$ *is equivalent to the triple* $(G', I', \mathbf{w}')$;
3. *if $w_p = 1$ for all $p \in T$, there is a set of vertices T' of G' such that the triple (G, i, T) is equivalent to the triple (G', I', T'); moreover, if T is the set of leaves of G, we can choose for T' the set of leaves of G'.*

We now state the main result of [9].

Theorem 6. *Let $s = (s_n)_{n \geq 0}$ be a regular sequence of nonnegative integers and let k be a positive integer such that $\sum_{n \geq 0} s_n k^{-n} \leq 1$. Then there is a k-ary rational tree having s as its generating sequence.*

Proof. We consider a regular sequence s and a positive integer k such that $\sum_{n\geq 0} s_n k^{-n} \leq 1$. Since the result holds trivially for $s(z) = 1$, we may suppose that $s_0 = 0$. Let (G, i, t) be a normalized representation of s and let $\overline{G}$ be the closure of G as defined at the beginning of Sect. 2.1. We denote by M (resp., $\overline{M}$) the adjacency matrix of G (resp., $\overline{G}$). Let $\overline{Q} = Q - \{t\}$ be the vertex set of $\overline{G}$. Let λ be the spectral radius of $\overline{M}$. By Proposition 8, the matrix $\overline{M}$ admits a positive k-approximate eigenvector $\overline{\mathbf{w}}$. By definition, we have $\overline{M}\overline{\mathbf{w}} \leq k\overline{\mathbf{w}}$.

Let $\mathbf{w}$ be the Q-vector defined by $w_q = \overline{w}_q$ for all $q \in \overline{Q}$ and $w_t = \overline{w}_i$. Then, since there is no edge going out of t in G, $\mathbf{w}$ is a positive k-approximate eigenvector of M. Let $\mathbf{t}$ be the Q-vector which is the characteristic vector of the vertex t. Let $m = w_i$.

By Theorem 5 there exists an admissible extension H of G with respect to $\mathbf{w}$ and m. Let U be the transfer matrix and let $\mathbf{W} = U\mathbf{w}$. Since $w_t \equiv 0 \bmod m$, we may choose H with the following additional property. For all $\mathbf{u} \in S$, either $u_t = 0$ or $\mathbf{u} = \mathbf{t}$.

According to Proposition 10, the sequence s is recognized by $(H, \mathbf{J}, \mathbf{X})$ where $\mathbf{J}$ is the characteristic row vector of $\mathbf{i}$ and $\mathbf{X}$ is the characteristic column vector of $\mathbf{t}$. This means that s is recognized by the normalized representation consisting of the graph H, the initial vertex $\mathbf{i}$, which we identify to i, and the terminal vertex $\mathbf{t}$, which we identify to t.

Let N be the adjacency matrix of H. By Proposition 12, the vector $\lceil \frac{\mathbf{W}}{m} \rceil$ is a positive k-approximate eigenvector of N. Note that $\lceil \frac{\mathbf{W}}{m} \rceil_i = \lceil \frac{\mathbf{W}}{m} \rceil_t = 1$.

We may now apply Lemma 2 to construct a triple (H', I', T') equivalent to (H, i, t). The set T' is the set of leaves of H'. Since $\lceil \frac{\mathbf{W}}{m} \rceil_i = 1$, I' is reduced to one vertex i'. Since H' admits a k-approximate eigenvector with all components equal to one, the graph H' is of outdegree at most k. Finally s is the generating sequence of the covering tree of H' starting at i'. This tree is k-ary and regular. □

We next consider the above constructions in the particular case of the equality in Kraft's inequality. In this case, the result is a complete k-ary tree. Indeed, by Proposition 8, the matrix $\overline{M}$ admits a positive integral eigenvector $\mathbf{w}$ for the eigenvalue k. We have, for all $p \in \overline{Q}$,

$$\sum_{q \in \overline{Q}} M_{p,q} w_q = k w_p.$$

As a consequence, for any $\mathbf{u} \neq \mathbf{t}$, we have

$$\sum_{\mathbf{v} \in S} N_{\mathbf{u},\mathbf{v}} W_{\mathbf{v}} = k W_{\mathbf{u}}.$$

Then the graph constructed in Lemma 2 is of constant outdegree k. Thus the k-ary tree obtained is complete.

We also consider the complexity of the construction used in the proof of Theorem 6. Let n be the number of vertices of the graph G giving a normalized representation of s. The size of the integer $m = w_i$ is exponential in n (see Sect. 3.3). Thus the number of vertices of the graph H is bounded by a double exponential in n. The final regular tree is the covering tree of a graph whose set of vertices has the same size in order of magnitude.

For example, let s be the sequence defined by

$$s(z) = \frac{z^2}{(1-z^2)} + \frac{z^2}{(1-5z^3)}.$$

Since $s(1/2) = 1$, it satisfies the Kraft equality for $k = 2$. The sequence s is recognized by (G, i, t) where $G = (Q, E)$ is the graph given in Fig. 8 with $Q = \{1, 2, 3, 4, 5, 6, 7\}, i = 1, t = 4$. The adjacency matrix of G admits the 2-approximate eigenvector represented in Fig. 8, where the coefficients of $\mathbf{w}$ are represented in squares beside the vertices. Thus $m = 3$.

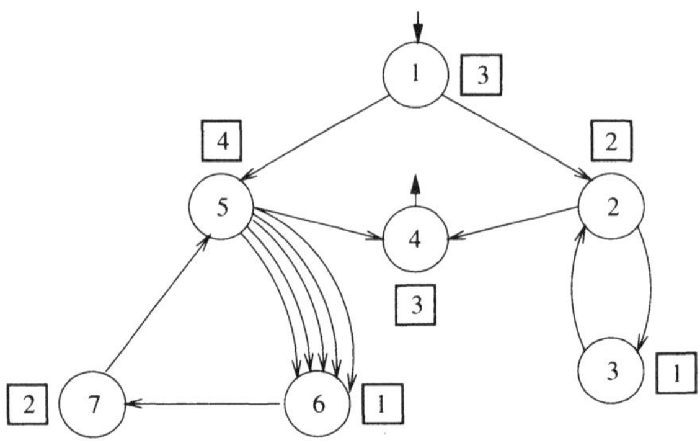

Fig. 8. A normalized representation of s

An admissible extension H of G with respect to $\mathbf{w}$ and m is given in Fig. 9. In this figure, each multiset of S is represented by a sequence of vertices with repetitions corresponding to the multiplicity. For example, the multiset $\mathbf{u} = (0, 0, 1, 0, 0, 2, 0)$ is represented by $(3, 6, 6)$. The sequence s is recognized by the normalized representation $(H, 1, 4)$, where the initial and final vertices are named as they appear on Fig. 9. The coefficients of $\mathbf{W}$ are represented in squares beside the vertices.

A regular binary tree T having s as generating sequence of leaves, is given in Fig. 10. In this figure, the nodes have been renumbered, with the children of a node with a given label represented only once. The leaves of the tree are indicated by black boxes. The tree itself is obtained from the graph of Fig. 9 by application of the construction of Lemma 2. For example, the vertex $(2, 5)$, which has coefficient 6 in $\mathbf{W}$, is split into two vertices named 2 and 3 in the tree.

This example was suggested to us by Christophe Reutenauer [38]. To check directly that the length distribution is equal to $s(z)$, one may compute from the graph

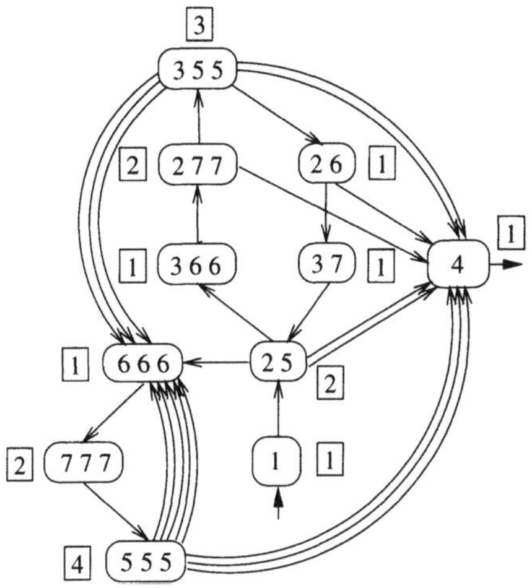

Fig. 9. An admissible extension H

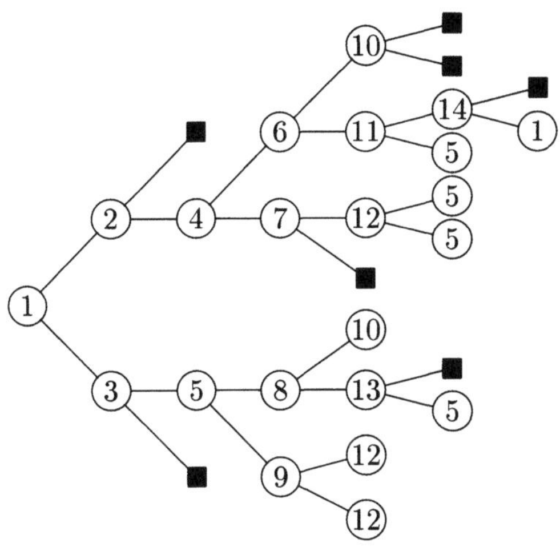

Fig. 10. A regular binary tree with length distribution s

the following regular expression of $s(z)$ and check by an elementary computation (possibly with the help of a symbolic computation system) that it is equal to $s(z)$:

$$s(z) = (z^6)^*(2z^2 + z^4 + 2z^5 + z^6 + (z^2 + 3z^5)(5z^3)^*3z^3). \tag{1}$$

(Note for a reader unfamiliar with regular expressions. The first factor $(z^6)^*$ corresponds to the vertex labeled 1 at level 6 of the tree. The term $2z^2 + z^4 + 2z^5 + z^6$ corresponds to the leaves reached by a path which does not use a vertex labeled 5.

The factor $(z^2+3z^5)(5z^3)^*$ corresponds to the paths from the root to a vertex labeled 5. Finally, the factor $3z^3$ corresponds to the direct paths from 5 to a leaf.)

This example shows an interesting feature of this problem. In fact, from the point of view of regular expressions, the difficult operation in this problem is the sum. It would be a simple matter to build a rational tree for each term of the sum in the expression (1) (see the example of Fig. 5). The difficulty would then be to merge these trees to obtain one corresponding to the sum.

A curious consequence of Theorem 6 is the following property of regular sequences.

Corollary 4. *Let $k \geq 2$ be an integer and let u be a regular sequence such that $u(1/k) \leq 1$ and $u(0) = 0$. Then there exist k regular sequences $u_1, \ldots, u_k$ such that $u_i(1/k) \leq 1$ and*

$$u(z) = \sum_{i=1}^{k} z u_i(z).$$

Proof. This is a simple consequence of Theorem 6. Indeed, if X is a regular prefix code on the k element alphabet A, then $X = \sum_{a \in A} aX_a$ where each X_a is a regular prefix code on the alphabet A. □

We know of no direct proof of this result.

3.6 Generating sequences of nodes

In this section, we consider the generating sequence of the set of all nodes in a tree instead of just the set of leaves. This is motivated by the fact that, in search trees, the information can be carried either by the leaves or by all the nodes of the tree. We will see that the complete characterization of the generating sequences of nodes in regular trees (Theorem 7) is more complicated than that for leaves.

Soittola (see [41, p. 104]) has characterized the series which are generating sequences of nodes in a regular tree. We characterize those that correspond to k-ary trees (Theorem 7). We also give a more direct construction in a particular case (Theorem 8).

Let T be a tree. The generating sequence of nodes of the tree T is the sequence $t = (t_n)_{n \geq 0}$, where t_n is the number of nodes of T at height n. The sequence t satisfies $t_0 \leq 1$ and, moreover, if T is a k-ary tree, the condition

$$t_n \leq k t_{n-1}$$

for all $n \geq 1$. If T is a regular tree, then t is a regular sequence. We now completely characterize the regular sequences t which are the generating sequences of nodes of a k-ary regular tree.

Theorem 7. *Let $t = (t_n)_{n \geq 0}$ be a regular sequence and let k be a positive integer. The sequence $(t_n)_{n \geq 0}$ is the generating sequence of nodes of a k-ary regular tree iff it satisfies the following conditions:*

(i) *the radius of convergence of t is strictly greater than $1/k$;*
(ii) *the sequence $s(z) = t(z)(kz - 1) + 1$ is regular.*

Proof. We first show that the conditions are necessary. Let $\overline{T}$ be the complete k-ary tree obtained by adding i new leaves to each node that has $k - i$ children. Since T is a regular tree, $\overline{T}$ is also regular.

Let s be the generating sequence of leaves of $\overline{T}$. Since $\overline{T}$ is complete, $s(1/k) = 1$. Since $kt_n = s_{n+1} + t_{n+1}$ for all $n \geq 0$, we have

$$1 - s(z) = t(z)(1 - kz).$$

Since s is a regular sequence, its radius of convergence is strictly larger than $1/k$ (see Sect. 3.3). Since the value of the derivative of s at $z = 1/k$ is $kt(1/k)$, the same holds for t. This proves the necessity of the conditions.

Conversely, if t satisfies the conditions of the theorem, the regular series $s(z) = t(z)(kz - 1) + 1$ satisfies $s(1/k) = 1$. Thus, by Theorem 6, s is the generating sequence of leaves of a complete k-ary regular tree. The internal nodes of this tree form a k-ary regular tree whose generating sequence of nodes is t. □

The sequence s defined by condition (ii) is rational as soon as t is regular and therefore rational. For a regular sequence t, condition (ii) is decidable in view of the theorem of Soittola (Theorem 1).

We may observe that condition (ii) of the theorem implies the non-negativity of the coefficients of the series s and thus the inequality $t_n \leq kt_{n-1}$ $\forall n \geq 1$. It also implies that $t_0 \leq 1$.

We now show that there are regular sequences t satisfying $t_n \leq kt_{n-1}$ for all $n \geq 1$ and condition (i) of the theorem for which the sequence $s(z) = t(z)(kz - 1) + 1$ is not regular. The example is based on an example of a rational sequence with nonnegative coefficients which is not regular (see [18, p. 95]). Let

$$r_n = b^{2n} \cos^2(n\theta)$$

with $\cos(\theta) = \frac{a}{b}$ where the integers a, b are such that $b \neq 2a$ and $0 < a < b$. The sequence r is rational, has nonnegative integer coefficients and is not regular. Its poles are $\frac{1}{b^2}$, $\frac{1}{b^2}e^{2i\theta}$ and $\frac{1}{b^2}e^{-2i\theta}$. We now define the sequence t as follows:

$$\begin{aligned} t_{2h} &= k^h, \\ t_{2h+1} &= k^h + r_h. \end{aligned}$$

We also assume that $b^2 < k$. By Soittola's theorem, the sequence t is regular since it is a merge of rational sequences having a dominating root. The radius of convergence of t is $\frac{1}{\sqrt{k}} > \frac{1}{k}$. Therefore the sequence t satisfies the first condition of Theorem 7. Let s be the sequence defined by $s(z) = t(z)(kz - 1) + 1$. If $h = 2p$ is even, then

$$\begin{aligned} s_h &= kt_{h-1} - t_h \\ &= kk^{p-1} + kr_{p-1} - k^p + 1 = kr_{p-1} + 1. \end{aligned}$$

Thus the sequence s is not regular.

The above example does not work for small values of k (the least value is $k = 10$). We do not know of similar examples for $2 \leq k \leq 9$.

Finally we describe a particular case of Theorem 7 in which there is a relatively simple method, based on the multiset construction, of building the regular tree with a given generating sequence of nodes. This avoids the use of Soittola's characterization which leads to a method of higher complexity.

A *primitive* representation of a regular sequence s is a representation $(G, \mathbf{i}, \mathbf{t})$ such that the adjacency matrix of G is primitive. The following result is proved in [8] with a different proof using the state-splitting method of symbolic dynamics. The proof given in [9] relies on a simpler construction.

Theorem 8. *Let $t = (t_n)_{n\geq 0}$ be a regular sequence and let k be a positive integer such that $t_0 = 1$ and $t_n \leq k t_{n-1}$ for all $n \geq 1$ and such that:*

(i) *the radius of convergence of t is strictly greater than $1/k$;*
(ii) *t has a primitive representation.*

Then $(t_n)_{n\geq 0}$ is the generating sequence of nodes by height of a k-ary regular tree.

The proof of this theorem which is given in [9] uses the multiset construction. It relies on the following lemma.

Lemma 3. *Let M be a primitive matrix with spectral radius λ. Let $\mathbf{v}$ be a nonnull and nonnegative integral vector and let k be an integer such that $\lambda < k$. Then there is a positive integer n such that $M^n\mathbf{v}$ is a positive k-approximate eigenvector of M.*

Proof. For a primitive matrix M with spectral radius λ, it is known that the sequence $((\frac{M}{\lambda})^n)_{n\geq 0}$ converges to $\mathbf{r}.\mathbf{l}$ where $\mathbf{r}$ is a positive right eigenvector and $\mathbf{l}$ a positive left eigenvector of M for the eigenvalue λ with $\mathbf{l} \cdot \mathbf{r} = 1$ (see, for example, [30, p. 130]). Thus $(\frac{M^n}{\lambda^n}\mathbf{v})_{n\geq 0}$ converges to $\mathbf{r}.\mathbf{l}.\mathbf{v}$ which is equal to $\rho\mathbf{r}$ where ρ is a nonnegative real number. Since $M\mathbf{r} = \lambda\mathbf{r}$, we get, for a large enough integer n,

$$M\frac{M^n}{\lambda^n}\mathbf{v} \leq k\frac{M^n}{\lambda^n}\mathbf{v}$$

or equivalently $MM^n\mathbf{v} \leq kM^n\mathbf{v}$. If n is large enough, we also have $M^n\mathbf{v} > 0$ since M is primitive. □

The proof of Theorem 8 uses a shift of indices of the sequence to obtain a new sequence to which a simple application of the multiset construction can be applied. We illustrate it in an example.

Let t be the series recognized by the graph G of Fig. 11 with

$$\mathbf{i} = \begin{bmatrix} 1 & 0 & 0 \end{bmatrix} \quad \text{and} \quad \mathbf{t} = \begin{bmatrix} 1 \\ 1 \\ 0 \end{bmatrix}.$$

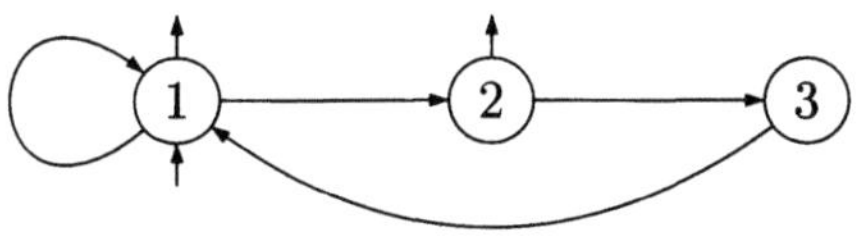

Fig. 11. A primitive representation G of t

The adjacency matrix M of G is the primitive matrix

$$M = \begin{bmatrix} 1 & 1 & 0 \\ 0 & 0 & 1 \\ 1 & 0 & 0 \end{bmatrix}.$$

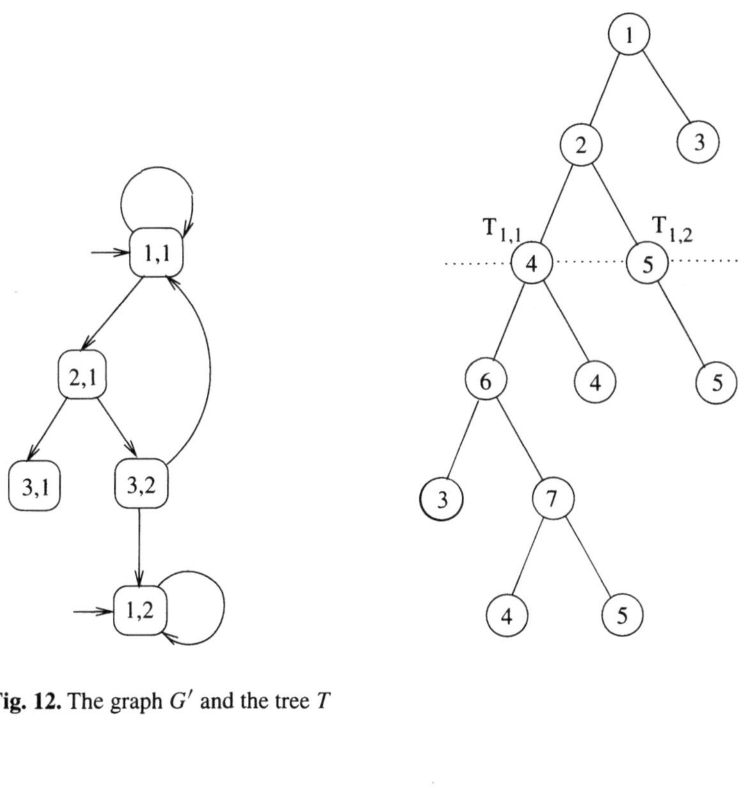

Fig. 12. The graph G' and the tree T

Its spectral radius is less than 2. The hypotheses of Theorem 8 are thus satisfied. We have

$$M^2\mathbf{t} = \begin{bmatrix} 2 \\ 1 \\ 2 \end{bmatrix} \quad \text{and} \quad M^3\mathbf{t} = \begin{bmatrix} 3 \\ 2 \\ 2 \end{bmatrix}.$$

Since $M^3\mathbf{t} \leq 2M^2\mathbf{t}$, the vector $\mathbf{W} = M^2\mathbf{t}$ is an approximate eigenvector of M (the existence of such a vector is asserted by Lemma 3). Let $\mathbf{w} = M^2\mathbf{t}$.

Applying Lemma 2, we obtain from G the graph G' represented on the left side of Fig. 12 . Moreover, $(G, \mathbf{i}, \mathbf{w})$ is equivalent to $(G', I', \mathbf{w}')$ where I' is the set of initial vertices indicated in Fig. 12 and $\mathbf{w}$ is the vector with all components equal to 1. The covering trees $T_{1,1}$ and $T_{1,2}$ of G' starting at the vertices of I' give, with the appropriate shift of indices, the binary regular tree T represented on the right side of Fig. 12 (the nodes of the tree have been renumbered).

4 Generating sequences of prefix codes

There is a close connection between trees and prefix codes or prefix-closed sets of words. We present below the translation of some of the notions and results seen before in terms of prefix codes.

4.1 Trees and prefix codes

Let R be a set of words in the alphabet $A = \{0, 1, \dots, k-1\}$. The set R is said to be *prefix-closed* if any prefix of an element of R is also in R. The set X of words which are not proper prefices of words in R is a prefix code, called the prefix code associated to R.

When R is prefix-closed, we can build a tree $T(R)$ as follows. The set of nodes is R, the root is the empty word ϵ and $T(a_1a_2\cdots a_n) = a_1a_2\cdots a_{n-1}$. The leaves of T form a prefix code which is the prefix code associated to R. The generating sequence of T is the generating sequence of X.

For example, let $R = \{\epsilon, 0, 1, 10, 11\}$. The tree $T(R)$ is represented in Fig. 13. The associated prefix code is $X = \{0, 10, 11\}$.

Let X be a prefix code on an alphabet with k symbols. It is clear that its length distribution $u = (u_n)_{n\geq 1}$ satisfies Kraft's inequality

$$\sum_{n\geq 1} u_n k^{-n} \leq 1,$$

or, equivalently, $u(1/k) \leq 1$. The number $u(1/k)$ can actually be interpreted as the probability that a sufficiently long word has a prefix in X.

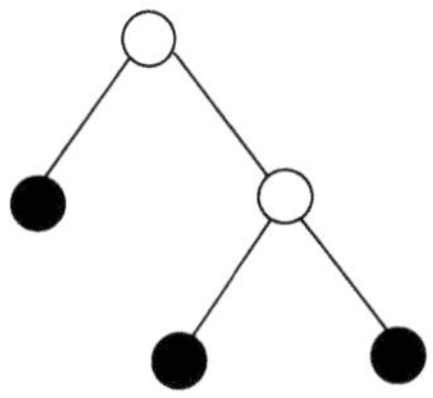

Fig. 13. The tree $T(X)$

There is also a connection with the notion of entropy. In fact, if X is a prefix code, the entropy of X^* is equal to $\log(1/\rho)$ where ρ is the solution of the equation $u_X(\rho) = 1$. Thus Kraft's inequality expresses the fact that $h(X^*) \leq \log k$.

Conversely, Kraft–McMillan's theorem states that, for any such sequence $u = (u_n)_{n\geq 1}$, there exists a prefix code X on a k-symbol alphabet such that $u = u_X$.

The equality case in Kraft's inequality corresponds to a particular class of prefix codes often called *complete*. A prefix code X on the alphabet A is complete if any word in A either has a prefix in X or is a prefix of a word of X.

Theorem 6 shows that the generating sequences of regular prefix codes are exactly the regular sequences satisfying Kraft's inequality.

4.2 Bifix codes

We investigate here the length distributions of a particular class of prefix codes, called bifix. Several other classes of prefix codes could give rise to a similar studies (for a description to these classes, see [21]).

The definition of a suffix code is the analog of that of prefix code. It is a set of words X such that no element of X is a suffix of another. The notion of a complete suffix code is also analogous. A *bifix code* is a set X of words which is both a prefix and a suffix code.

Any set of words of fixed length is obviously a bifix code but there are more complicated examples.

Example 5. The set

$$X = \{aaa, aaba, aabb, ab, baa, baba, babb, bba, bbb\}$$

is a complete prefix code, pictured in Fig. 14. It is also a complete suffix code as one may check by reading its words backwards.

Surprisingly, it is an open problem to characterize the length distributions of bifix codes. The following simple example shows that they are more constrained than those of prefix codes.

Example 6. The sequence $u(z) = z + 2z^2$ is not realizable as the length distribution of a bifix code on a binary alphabet although $u(1/2) = 1$. Indeed, one of the symbols has to be in X, say a. Then bb is the only word of length 2 that can be added.

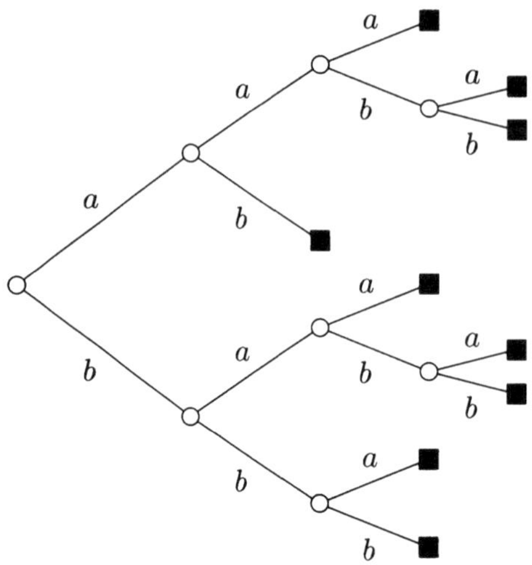

Fig. 14. The bifix code X

The following nice partial result is due to Ahlswede, Balkenhol and Khachatrian [2]. We state the result for a binary alphabet. It can be readily generalized to k symbols but this presents less interest.

Theorem 9. *For any integer sequence u such that*

$$u(1/2) \leq 1/2,$$

there is a bifix code X such that $u = u_X$.

Proof. The proof is by induction. We suppose that we have already built a bifix code X formed of words of length at most $n-1$ with length distribution $(u_1, u_2, \ldots, u_{n-1})$. We have

$$\sum_{i=1}^{n} u_i 2^{-i} \leq 1/2,$$

and thus

$$2 \sum_{i=1}^{n} u_i 2^{n-i} \leq 2^n.$$

Finally, we obtain

$$u_n \leq 2^n - 2 \sum_{i=1}^{n-1} u_i 2^{n-i}.$$

The expression on the right-hand side is at most equal to the number of elements of the set $A^n - XA^* - A^*X$. Thus, we can choose u_n words of length n which do not have a prefix or a suffix in X. The result follows by induction. □

The authors of [2] formulate the interesting conjecture that Theorem 9 is still true if the hypothesis $u(1/2) \leq 1/2$ is replaced by $u(1/2) \leq 3/4$.

There are known additional conditions imposed on length distributions of bifix codes. For example, one has the following result, originally due to Schützenberger (see [16]).

Theorem 10. *If X is a finite complete bifix code on k symbols, then $u_X(1/k) = 1$ and $\frac{1}{k}u'_X(1/k)$ is an integer.*

The number $\frac{1}{k}u'_X(1/k)$ can be interpreted as the average length of a word of X. Indeed,

$$zu'_X(z) = \sum_{x \in X} |x|z^{|x|}.$$

Example 7. For the bifix code of Example 5, we have

$$u_X(z) = z^2 + 4z^3 + 4z^4$$

and thus

$$u'_X(z) = 2z + 12z^2 + 16z^3.$$

Hence $\frac{1}{2}u'_X(1/2) = 3$.

The conditions of Theorem 10 show directly that the sequence of Example 6 is not realizable. Indeed, it satisfies the first condition but not the second. The conditions of Theorem 10 are not sufficient. Indeed, if $u(z) = z + 4z^3$, then we have $u(1/2) = 1$ and $u'(1/2) = 4$ although it is clearly impossible that $u = u_X$ for a bifix code X.

Recently, Ye and Yeung [44] have made some progress on this problem. In particular, they are able to prove that Theorem 9 still holds when $u(1/2) \leq 5/8$, provided $u_1 = 1$.

5 Zeta functions, subshifts of finite type and circular codes

In this section, we present a number of results on interrelated objects which are connected with cyclic permutation of words. The link with enumerative combinatorics was developed in Lothaire's volume [31] and later in R. Stanley's book [43]. We begin with notions classical in symbolic dynamics (see [30] or [28] for a general reference; see [15] or [24] for the link with finite automata).

5.1 Subshifts of finite type

A *subshift* is a set of bi-infinite words on a finite alphabet A which avoids a given set F of forbidden words. It is a topological space as a closed subset of the space $A^{\mathbb{Z}}$ of functions from $\mathbb{Z}$ into the set A. The *full shift* on A is the set of all bi-infinite words on A. It corresponds to the case $F = \emptyset$.

A *sofic* subshift is the set of bi-infinite labels of paths in a finite automaton. A sofic subshift is called *irreducible* if the automaton can be chosen to be strongly connected. A *subshift of finite type* is the set of bi-infinite words avoiding a finite set of finite words. Any subshift of finite type is sofic but the converse is not true. The *edge shift* of a finite graph G is the set S_G of bi-infinite paths in G (viewed as bi-infinite sequences of edges). It is a subshift of finite type.

The *shift* σ is the function on a subshift S which maps a point x to the point $y = \sigma(x)$ whose ith coordinate is $y_i = x_{i+1}$.

A *morphism* from a subshift S into a subshift T is a function $f : S \to T$ which is continuous and invariant under the shift. A bijective morphism is called a *conjugacy*. Any subshift of finite type is conjugate to some edge shift.

The *entropy* $h(S)$ of a subshift S is the entropy of the formal language formed by the finite blocks occurring in words of S. It can be shown that the entropy is a topological invariant, in the sense that two conjugate subshifts have the same entropy.

While the entropy is a measure of the number of forbidden words, it is possible to study the number of minimal forbidden words. This gives rise to another invariant of subshifts [13, 14].

An integer p is a *period* of a point $x = (a_n)_{n\in\mathbb{Z}}$ if $a_{n+p} = a_n$ for all $n \in \mathbb{Z}$. Equivalently, p is a period of x if $\sigma^p(x) = x$. The *zeta function* of a subshift S is defined as the series

$$\zeta(S) = \exp \sum_{n\geq 1} \frac{p_n}{n} z^n$$

where p_n is the number of words with period n in S. It is also a topological invariant, since a point of period n is mapped by a conjugacy to a point of the same period.

The following result, due to Bowen and Lanford [19], is classical (see [30]).

Proposition 13. *Let G be a finite graph and let M be the adjacency matrix of G. Then*

$$\zeta(S_G) = \det(I - Mz)^{-1}.$$

Proof. First we have, for each $n \geq 1$,

$$\mathrm{Tr}(M^n) = p_n$$

since the coefficient (i, j) of M^n is the number of paths from i to j. Thus

$$\begin{aligned}\zeta(S_G) &= \exp\sum_{n\geq 1}\frac{p_n}{n}z^n\\ &= \exp\sum_{n\geq 1}\frac{\mathrm{Tr}(M^n)}{n}z^n\\ &= \exp\mathrm{Tr}(\log(I-Mz)^{-1})\\ &= \det(I-Mz)^{-1}\end{aligned}$$

since, by the formula of Jacobi, $\exp\mathrm{Tr} = \det\exp$. □

Example 8. Let S be the edge shift of the graph G of Fig. 15. We have

$$M = \begin{bmatrix} 1 & 1 & 0\\ 0 & 0 & 1\\ 1 & 0 & 0\end{bmatrix}.$$

Consequently,

$$\zeta(S) = \frac{1}{1-z-z^3}.$$

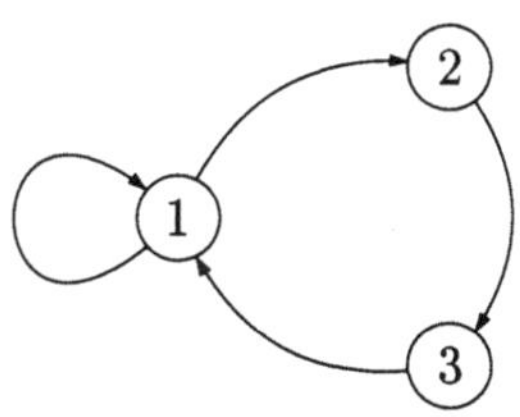

Fig. 15. A subshift of finite type

Let S be a subshift of finite type and let p_n be the number of points with period n. Let q_n be the number of points with least period n. Since q_n is a multiple of n, we also write $q_n = nl_n$. We then have the following formula expressing the zeta function as an infinite product using the integers l_n as exponents:

$$\zeta(S) = \Pi_{n\geq 1}(1-z^n)^{-l_n},$$

as one may verify using $p_n = \sum_{d|n} dl_d$ and the definition of $\zeta(S)$.

A classical result, related with what follows, is the following statement, known as Krieger's embedding theorem.

Theorem 11. *Let S, T be two subshifts of finite type. There exists an injective morphism $f : S \to T$ with $f(S) \neq T$ iff*

1. $h(S) < h(T)$;
2. *for each $n \geq 1$, $q_n(S) \leq q_n(T)$ where $q_n(S)$ (resp., $q_n(T)$) is the number of points of S (resp., T) of least period n.*

The following result is the basis of many applications of symbolic dynamics to coding. It is due to Adler, Coppersmith and Hassner [1].

Theorem 12. *If S is an irreducible subshift of finite type such that $h(S) \geq \log k$, then S is conjugate to a subshift of finite type S_G where the graph G has outdegree at least k.*

The proof is based on a state-splitting algorithm using approximate eigenvectors and Lemma 1. This result is part of a number of constructions leading to sliding block codes used in magnetic recording (see [35], [11] or [30]). At the same time it gives the following result.

Theorem 13. *If S is a subshift of finite type such that $h(S) \leq \log k$, then there is a graph G of outdegree at most k such that S is conjugate to S_G.*

There is a connection between this theorem and Theorem 6. Indeed let u be a regular sequence of integers such that $u(1/k) \leq 1$. Let G be a normalized graph recognizing u (in the sense of Sect. 2.1). Let $\bar{G}$ be the graph obtained by merging the initial and terminal vertex. Then $h(S_{\bar{G}}) \leq \log k$. We can apply Theorem 13 to obtain a graph H with outdegree at most k such that S_G and S_H are conjugate. This gives the conclusion of Theorem 6 provided the initial-terminal vertex did not split in the construction. The following examples show both cases (for details, see [7] and [8]).

Example 9. Let G be the graph of Fig. 5. The splitting of vertex 2 gives a graph of outdegree 2. A normalization gives the automaton on the right.

Example 10. The sequence of the example given in Fig. 6 is recognized by a graph G such that $\bar{G}$ has three cycles of length 2. The solution as a binary tree has only two cycles of length 2 and thus could not be obtained by state-splitting.

5.2 Circular codes

A *circular word*, or *necklace*, is the equivalence class of a word under cyclic permutation. For a word w, we denote by $\bar{w}$ the circular word represented by w.

Let X be a set of words and let $w = x_1 x_2 \cdots x_n$ with $x_i \in X$. The set of cyclic permutations of the sequence $(x_1, x_2, \ldots, x_n)$ is called a *factorization* of the circular word $\bar{w}$.

A *circular code* is a set X of words such that the factorization of circular words is unique.

Example 11. The set $X = \{a, aba\}$ is a circular code. Indeed, the position of the symbols b determines uniquely the occurrences of aba.

Example 12. The set $X = \{ab, ba\}$ is not a circular code. Indeed, the circular word $\bar{w}$ for $w = abab$ has two factorizations namely (ab, ab) and (ba, ba).

The following characterization is useful (see [16]).

Proposition 14. *A set X is a circular code if and only if it is a code and, for all* $u, v \in A^*$,

$$uv, vu \in X^* \Rightarrow u, v \in X^*.$$

Example 13. This gives another way to prove that the set $X = \{ab, ba\}$ is not a circular code. Indeed, otherwise we would have $a, b \in X^*$ which is a contradiction.

Let X be a finite code. The *flower automaton* of X, denoted $\mathcal{A}_X$, is the following automaton. Its set of states is

$$Q = \{(u, v) \in A^+ \times A^+ \mid uv \in X\} \cup (1, 1).$$

The transitions are of the form $(u, av) \xrightarrow{a} (ua, v)$ or $(1, 1) \xrightarrow{a} (a, v)$ or $(u, a) \xrightarrow{a} (1, 1)$. The unique initial and final state is $(1, 1)$.

Example 14. The flower automaton of the circular code $\{a, aba\}$ is pictured in Fig. 16.

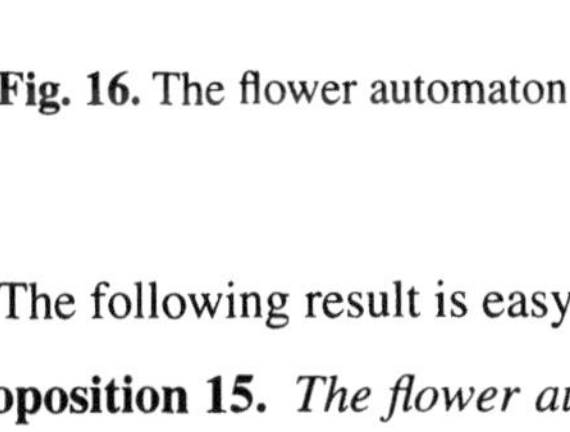

Fig. 16. The flower automaton of $\{a, aba\}$

The following result is easy to prove.

Proposition 15. *The flower automaton* $\mathcal{A}_X$ *recognizes* X^*. *The code X is circular iff, for each word w, there is at most one cycle with label w.*

We now study the length distributions of circular codes. Let X be a circular code and let $u(z) = (u_n)_{n\geq 1}$ be its length distribution. For each $n \geq 1$, let p_n be the number of words w of length n such that $\bar{w}$ has a factorization in words of X.

Proposition 16. *The sequences* $(p_n)_{n\geq 1}$ *and* $(u_n)_{n\geq 1}$ *are related by*

$$\exp \sum_{n\geq 1} \frac{p_n}{n} z^n = \frac{1}{1 - u(z)}. \tag{2}$$

Proof. Each p_n depends only on the first n terms of the sequence (u_n). It is therefore possible to assume that the sequence (u_n) is finite, i.e., that the code X is finite. Let $\mathcal{A}$ be the flower automaton of X. Let S be the subshift of finite type associated with the graph of $\mathcal{A}$. Then p_n is the number of elements of period n in S. Indeed, each word w such that $\bar{w}$ has a factorization is counted exactly once as the label of a cycle in $\mathcal{A}$. We also have that

$$\det(I - Mz) = 1 - u(z).$$

Thus, the result follows from Proposition 13. □

The explicit relation between the numbers u_n and p_n is as follows. For each $i \geq 1$, let $u^{(i)} = (u_n^{(i)})_{n\geq 1}$ be the length distribution of X^i. Equivalently, $u_n^{(i)}$ is the coefficient of degree n of $u(z)^i$. Then, for each $n \geq 1$,

$$p_n = \sum_{i=1}^{n} \frac{n}{i} u_n^{(i)}.$$

We also have, for each $n \geq 1$, that

$$p_n = nu_n + \sum_{i=1}^{n-1} p_i u_{n-i}. \tag{3}$$

This formula can be easily deduced from formula (2) by taking the logarithmic derivative of each side of the formula. It shows directly that, for any sequence $(u_n)_{n\geq 1}$ of nonnegative integers, the sequence p_n defined by formula (2) consists of nonnegative integers.

Formula (3) is known as Newton's formula in the field of symmetric functions. In fact, the numbers u_n can be considered, up to sign, as elementary symmetric functions and the p_n as the sums of powers (see [32]). The link between Witt vectors and symmetric functions was established in [42].

Let $p_n = \sum_{d|n} dl_d$. Then l_n is the number of nonperiodic circular words of length n with a factorization. In terms of generating series, we have

$$\exp \sum_{n\geq 1} \frac{p_n}{n} z^n = \prod_{n\geq 1} (1 - z^n)^{-l_n}. \tag{4}$$

Putting formulae (2) and (4) together, we obtain

$$\frac{1}{1 - u(z)} = \prod_{n\geq 1} (1 - z^n)^{-l_n}. \tag{5}$$

For any sequence $(u_n)_{n\geq 1}$ of nonnegative integers, the sequence $l = (l_n)_{n\geq 1}$ thus defined consists of nonnegative integers. This can be proved either by a direct computation or by a combinatorial argument since any sequence u of nonnegative integers is the length distribution of a circular code on a large enough alphabet. We write $l = \phi(u)$ and we say that l is the *ϕ-transform* of the sequence u.

We denote by $\varphi_n(k)$ the number of nonperiodic circular words of length n on k symbols. The numbers $\varphi_n(k)$ are called the *Witt numbers*. It is clear that the sequence $(\varphi_n(k))_{n\geq 1}$ is the ϕ-transform of the sequence $(k^n)_{n\geq 1}$.

The corresponding particular case of identity (5)

$$1 - kz = \prod_{n\geq 1}(1 - z^n)^{\varphi_n(k)}$$

is known as the *cyclotomic identity.*

The following array displays a tabulation of the Witt numbers for small values of n and k.

n	$\varphi_n(2)$	$\varphi_n(3)$	$\varphi_n(4)$
1	2	3	4
2	1	3	6
3	2	8	20
4	3	18	60
5	6	48	204
6	9	116	670
7	18	312	2340
8	30	810	8160
9	56	2184	29120
10	99	5880	104754

The value $\varphi_3(4) = 20$ is famous because of the genetic code: there are precisely 20 amino-acids coded by words of length 3 over the 4-symbol alphabet A,C,G,U.

For any sequence $a = (a_n)_{n\geq 1}$, let

$$p_n = \sum_{d|n} d a_d^{n/d}.$$

The pair (a, p) is called a *Witt vector* (see [29] or [36]). The numbers p_n are the *ghost components*. In terms of generating series, one has

$$\exp \sum_{n\geq 1} \frac{p_n}{n} z^n = \prod_{n\geq 1}(1 - a_n z^n)^{-1}.$$

The following result is due to Schützenberger (see [16]).

Theorem 14. *Let $u = (u_n)_{n\geq 1}$ be a sequence of nonnegative integers and let $l = (l_n)_{n\geq 1}$ be the ϕ-transform of u. The sequence $(u_n)_{n\geq 1}$ is the length distribution of a circular code on k symbols iff, for all $n \geq 1$,*

$$l_n \leq \varphi_n(k).$$

Several complements to Theorem 14 appear in [6]. In particular, the relation with Kraft's inequality is studied. The equality case in Kraft's inequality is characterized in terms of the sequence of inequalities above.

There is a connection between Theorem 14 and Krieger's embedding theorem (Theorem 11), in the sense that Theorem 14 gives a simple proof of Theorem 11 in a particular case. We next consider the particular case of a subshift of finite type, which is called a *renewal system.*

A renewal system S is the edge shift of a graph G made up of cycles sharing exactly one vertex. Such a graph is determined by the sequence $u = (u_i)_{1\leq i\leq n}$ where u_i is the number of loops of length i. Let T_k be the full shift on k symbols. Suppose that the pair formed by S and T_k satisfies the hypotheses of Krieger's theorem. The number $q_n(S)$ of points of least period n is nl_n where $l = (l_n)_{n\geq 1}$ is the ϕ-transform of the sequence u and $q_n(T_k) = n\varphi_n(k)$. Thus, the sequence u satisfies the hypotheses of Theorem 14. Consequently, there is a circular code X such that $u_X = u$. The flower automaton of X defines an embedding of S_G into the full shift on k symbols. This gives an alternative proof of Krieger's theorem in this case.

It would be interesting to have a proof of Krieger's theorem along the same lines in the general case.

To close this section, we mention the following open problem: if the sequence u is regular and satisfies the inequalities

$$l_n \leq \varphi_n(k) \qquad (n \geq 1),$$

where $l = \phi(u)$, does there exist a regular circular code on k symbols such that $u = u_X$?

5.3 Zeta functions

Theorem 13 admits the following generalization due to Reutenauer [39].

Theorem 15. *The zeta function of a sofic subshift is regular.*

We have seen already (Theorem 13) that the zeta function of a subshift of finite type is a rational fraction, and indeed the inverse of a polynomial. The stronger statement that it is regular follows from the following formula which allows us to compute $\det(I - Mz)$ when M is the adjacency matrix of a $n \times n$ graph G. We have

$$\det(I - Mz) = (1 - v_1(z)) \cdots (1 - v_n(z)),$$

where $v_i(z)$ is the length distribution of the set of first returns to state i using only states $\{i, i + 1, \ldots, n\}$ (see [12]).

The proof that the zeta function of a sofic subshift is rational is a result of Manning [33] and of Bowen [20]. For an exposition, see [30] or [12]. A generalization appears in [17].

Acknowledgements. This paper is based on several texts written together with Frédérique Bassino and Marie-Pierre Béal (in particular [9] and [10]). I warmly thank them for agreeing to the use of this material here.

The link between length distributions of circular codes and symmetric functions was brought to my attention by Jacques Désarménien and Jean-Yves Thibon. This connection seems to open promising perspectives for the future.

References

[1] Adler, R.L., Coppersmith, D., Hassner, M. (1983): Algorithms for sliding block codes. An application of symbolic dynamics to information theory. IEEE Trans. Inform. Theory **29**, 5–22

[2] Ahlswede, R., Balkenhol, B., Khachatrian, L. (1997): Some properties of fix-free codes. Preprint **97-39**. Fakultät für Mathematik, Universität Bielefeld

[3] Aigner, M., Ziegler, G.M. (1998): Proofs from The Book. Springer, Berlin

[4] Ash, R.B. (1990): Information theory. Dover, New York

[5] Ashley, J.J. (1998): A linear bound for sliding-block decoder window size. IEEE Trans. Inform. Theory **34**, 389–399

[6] Bassino, F. (1999): Generating functions of circular codes. Adv. in Appl. Math. **22**, 1–24

[7] Bassino, F., Béal, M.-P., Perrin, D. (1997): Enumerative sequences of leaves in rational trees. In: Degano, P. et al. (eds.) Automata, Languages and Programming. (Lecture Notes in Computer Science, vol. 1256), Springer, Berlin, pp. 76–86

[8] Bassino, F., Béal, M.-P., Perrin, D. (1999): Enumerative sequences of leaves and nodes in rational trees. Theoret. Comput. Sci. **221**, 41–60

[9] Bassino, F., Béal, M.-P., Perrin, D. (1999): Length distributions and regular sequences, In: Rosenthal, J., Marcus, B. (eds.) Codes, Systems and Graphical Models. (IMA Volumes in Mathematics and its Applications, vol.123). Springer, New York

[10] Bassino, F., Béal, M.-P., Perrin, D. (2000): A finite state version of the Kraft-McMillan theorem. SIAM J. Comput. **30**, 1211–1230

[11] Béal, M.-P. (1993): Codage symbolique. Masson, Paris

[12] Béal, M.-P. (1995): Puissance extérieure d'un automate déterministe, application au calcul de la fonction zêta d'un système sofique. RAIRO Inform. Théor. Appl. **29**, 85–103

[13] Béal, M.-P., Mignosi, F., Restivo, A. (1996): Minimal forbidden words and symbolic dynamics. In: Puech, C., Reischuk, R. (eds.) STACS96. (Lecture Notes in Computer Science, vol. 1046). Springer, Berlin, pp. 555–566

[14] Béal, M.-P., Mignosi, F., Restivo, A., Sciortino, M. (2000): Forbidden words in symbolic dynamics. Adv. in Appl. Math. **25**, 163–193

[15] Béal, M.-P., Perrin, D. (1997): Symbolic dynamics and finite automata. In: Rosenberg, G., Salomaa, A. (eds.) Handbook of Formal Languages, vol. 2. Springer, Berlin, pp. 463–505

[16] Berstel, J., Perrin, D. (1985): Theory of codes. Academic Press, Orlando, FL

[17] Berstel, J., Reutenauer, C. (1990): Zeta functions of formal languages. Trans. Amer. Math. Soc. **321**, 533–546
[18] Berstel, J., Reutenauer, C. (1988): Rational series and their languages. Springer, Berlin
[19] Bowen, R., Lanford, O.E. (1970): Zeta functions of restrictions of the shift transformation. In: Chern, S.-S., Smale, S. (eds.) Global Analysis. (Proceedings of Symposia in Pure Mathematics, vol. 14). American Mathematical Society, Providence, RI, pp. 43–49
[20] Bowen, R. (1978): On Axiom A diffeomorphisms. (Regional Conferenc Series in Mathematics, no. 35). American Mathematical Society, Providence, RI
[21] Bruyère, V., Latteux, M. (1996): Variable-length maximal codes. In: Meyer auf der Heide, F., Monien, B. (eds.) Automata, Languages and Programming. (Lecture Notes in Computer Science, vol. 1099). Springer, Berlin, pp. 24–47
[22] Eilenberg, S. (1974): Automata, languages and machines, vol. A. Academic Press, New York
[23] Flajolet, P. (1987): Analytic models and ambiguity of context-free languages. Theoret. Comput. Sci. **49**, 283–309
[24] Forney, G.D., Marcus, B.H., Sindhushayana, N.T., Trott, M. (1995): Multilingual dictionary: system theory, coding theory, symbolic dynamics, and automata theory. In: Calderbank, R. (ed.) Different Aspects of Coding Theory. (Proceedings of Symposia in Applied Mathematics, vol. 50). American Mathematical Society, Providence, RI, pp. 109–138
[25] Gantmatcher, F.R. (1959): Theory of matrices, vol. 2. Chelsea, New York
[26] Graham, R.L., Knuth, D., Patachnik, O. (1989): Concrete mathematics. A foundation for computer science. Addison Wesley, Reading, MA
[27] Katayama, T., Okamoto, M., Enomoto, H. (1978): Characterization of the structure-generating functions of regular sets and the DOL growth functions. Information and Control **36**, 85–101
[28] Kitchens, B.P. (1998): Symbolic dynamics. One-sided, two-sided and countable state Markov shifts. Springer, Berlin
[29] Lang, S. (1984): Algebra, 2nd edition. Addison Wesley, Reading, MA
[30] Lind, D.A., Marcus, B.H. (1995): An introduction to symbolic dynamics and coding. Cambridge University Press, Cambridge
[31] Lothaire, M. (1997): Combinatorics on words. Cambridge University Press, Cambridge
[32] Macdonald, I.G. (1995): Symmetric functions and Hall polynomials, 2nd edition. Clarendon Press, Oxford
[33] Manning, A. (1971): Axiom A diffeomorphisms have rational zeta functions. Bull. London Math. Soc. **3**, 215–220
[34] Marcus, B.H. (1979): Factors and extensions of full shifts. Monatsh. Math. **88**, 239–247
[35] Marcus, B.H., Roth, R.M., Siegel, P.H. (1998): Constrained systems and coding for recording channels. In: Pless, V.S. et al. (eds.) Handbook of Coding Theory, vol. II. North Holland, Amsterdam, pp. 1635–1764
[36] Metropolis, N., Rota, G.-C. (1983): Witt vectors and the algebra of necklaces. Adv. Math. **50**, 95–125
[37] Perrin, D. (1990): Finite Automata. In: van Leeuwen, J. (ed.) Handbook of Theoretical Computer Science, vol. B. Elsevier, Amsterdam, pp. 1–57
[38] Reutenauer, C. (1997): Personal communication
[39] Reutenauer, C. (1997): **N**-rationality of zeta functions. Adv. in Appl. Math. **29**, 1–17
[40] Rota, G.-C. (1975): Finite operator calculus. Academic Press, New York
[41] Salomaa, A., Soittola, M. (1978): Automata-theoretic properties of formal power series. Springer, New York
[42] Scharf, T., Thibon, J.-Y. (1996): On Witt vectors and symmetric functions. Algebra Colloq. **3**, 231–238

[43] Stanley, R.P. (1997): Enumerative combinatorics, vol. 1. Cambridge University Press, Cambridge,
[44] Ye, C., Yeung, R.W. (2000): Basic properties of fix-free codes. IEEE Trans. Inform. Theory, submitted

Algebraic Combinatorics and Theoretical Computer Science

Alphabet splitting

A. Lascoux

I met Gian-Carlo in 1976, landing in MIT trying to establish Western connections. I was already working with Marcel-Paul Schützenberger who was one of his great friends and I needed no further introduction. Moreover, the few dollars that I had in my pocket were forcing me to eat fast, and this was contrary to Gian-Carlo's sense of hospitality. The outcome was that I appeared several times on his list of professional expenses (section: restaurants). It was still an epoch in which tax inspectors readily accepted supporting combinatorics. This is no longer the case since combinatorics received the imprimatur of the Bourbaki Seminar. Let me however point out that meals were followed by long discussions about the comparative merits of algebraic structures, Gian Carlo for his part tirelessly asking me to repeat the definition of λ-rings that he copied each time in his black notebook with a new illustrative example.

Gian-Carlo thought of himself as an epigraphist of the richness of the past and an advocate of the algebraic structures which permit their integration into up-to-date research.

For example, the inclusion-exclusion principle leads to the study of ordered structures and to their Möbius functions. Invariant theory opens onto umbral calculus, differential calculus has a discrete counterpart, etc.

However, it does not suffice to highlight algebraic structures in order to give them life; one has also to market them. This entails showing that the energy spent in learning them is compensated by new light shed on classical domains, connections between different fields and the creation of new mathematical objects – one cannot always grind the same grain, one also has to sow!

Umbral calculus [13], Möbius functions [23], Hopf algebras [19], Baxter algebras [25], etc., all of them can be phrased in a few words.

For example, umbral calculus is the art of raising indices and lowering exponents, Möbius inversion consists of inverting triangular matrices with unit diagonals, and so on.

But these notions are not as straightforward as it seems at first glance. How to compute on ordered sets when one cannot represent them and experiment on them as soon as they become sufficiently voluminous? Even in the case of such a classical object as the symmetric group, though one can describe its Möbius function (with respect to the Ehresmann–Bruhat order), one does not know much about its intervals. One conjectures that Kazhdan–Lusztig depends only on the isomorphism type of the interval, but one does not know how to compute these polynomials in general, though they are the first level in the description of intervals.

As concerns raising indices, nothing prevents us from doing so, but one wants to do it *without loss of information*.

It is clear, for example, that in a linear identity one is allowed to replace a_i, $i = 1, 2, \ldots$ by $a^1, a^2, \ldots$, supposing that a is an indeterminate or a transcendental element. But why is it that inside *straightening* relations, so dear to Rota [21], between minors of matrices, one can restrict to Vandermonde matrices

$$\begin{bmatrix} a^1 & a^2 & a^3 & \cdots \\ b^1 & b^2 & b^3 & \cdots \\ \cdot & \cdot & \cdot & \cdots \end{bmatrix}$$

instead of having to take a generic one

$$\begin{bmatrix} a_1 & a_2 & a_3 & \cdots \\ b_1 & b_2 & b_3 & \cdots \\ \cdot & \cdot & \cdot & \cdots \end{bmatrix} \quad ?$$

Umbral calculus rules are simple, but it is far less elementary to determine the limits of their domain of validity [23]. I shall approach this territory through the simplest way I know, that is, addition.

Addition of what? Addition of alphabets, i.e., a disjoint union of sets of indeterminates:

$$(\mathbb{A} = \{a\},\ \mathbb{B} = \{b\}) \quad \mapsto \quad \mathbb{A} + \mathbb{B} := \{a\} \cup \{b\}.$$

Restrict for the moment to finite sets. Taking an extra indeterminate z, one can write generating functions

$$\lambda_z(\mathbb{A}) := \prod_{a \in \mathbb{A}} (1 + za), \quad \sigma_z(\mathbb{A}) := \prod_{a \in \mathbb{A}} \frac{1}{1 - za}. \tag{1}$$

the expansion of which defines *complete functions* $S^i(\mathbb{A})$ and *elementary functions* $\Lambda^i(\mathbb{A})$

$$\lambda_z(\mathbb{A}) = \sum z^i \, \Lambda^i(\mathbb{A}), \quad \sigma(\mathbb{A}) = \sum z^i \, S^i(\mathbb{A}). \tag{2}$$

Addition of alphabets translates into a product of generating functions

$$\lambda_z(\mathbb{A} + \mathbb{B}) = \lambda_z(\mathbb{A}) \, \lambda_z(\mathbb{B}), \quad \sigma_z(\mathbb{A} + \mathbb{B}) = \sigma_z(\mathbb{A}) \, \sigma_z(\mathbb{B}). \tag{3}$$

One also needs *power sums* ψ_i:

$$\psi_i(\mathbb{A}) = \sum_{a \in \mathbb{A}} a^i, \quad \sum_{i=1}^{\infty} \psi_i(\mathbb{A}) \frac{z^i}{i} = \log(\sigma_z(\mathbb{A})). \tag{4}$$

However, having learnt addition, one knows how to multiply by a positive integer

$$\mathbb{A} = \{a\} \to 2\mathbb{A} := \{a'\} \cup \{a''\}, \quad 3\mathbb{A} := \{a'\} \cup \{a''\} \cup \{a'''\}, \ \ldots$$

that is, one knows how to double, triple, ..., each letter (erasing the diacritical signs at the final stage)

$$\lambda_z(2\mathbb{A}) = \lambda_z(\mathbb{A})^2, \quad \lambda_z(3\mathbb{A}) = \lambda_z(\mathbb{A})^3, \ldots .$$

These tools are sufficient to invert formal series $f(z) = z + f_1 z^2 + f_2 z^3 + \cdots$, that is, to find $g(z) = z + g_1 z^2 + g_2 z^3 + \cdots$ such that

$$f(g(z)) = z \quad \& \quad g(f(z)) = z.$$

This problem was solved by Lagrange, and it involves calculating successive derivatives of powers of f; these operations are easily expressed in terms of alphabets.

Indeed, writing f as $f = z\sigma_{-z}(\mathbb{A})$ (every formal series factorizes formally!), one finds that the powers of g are

$$g^k(z) = k\, z^k \sum \frac{z^i}{i+k}\, \Lambda^i\big((i+k)\mathbb{A}\big), \tag{5}$$

and only require multiplication of alphabets by integers if $k \in \mathbb{Z}$ (formulae are still valid for $k \in \mathbb{C}$; the limit case $k = 0$ gives the logarithm).

The theory of symmetric functions gives, for every choice of pairs of adjoint bases (with respect to the natural scalar product), an expansion of $\Lambda^i((i+k)\mathbb{A})$ in terms of the coefficients of f or $1/f$. The expansions below appear in the classical literature, many mathematicians having wanted to leave their own interpretation of Lagrange inversion. We refer to Vincent Prosper's thesis [9] and to its implementation [10].

In short, the compact notation $\Lambda^i((i+k)\mathbb{A})$ *contains* the following formulae (see the comments about symmetric functions at the end):

$$\begin{aligned}
\Lambda^i((i+k)\mathbb{A}) &= (-1)^i \textstyle\sum_{|J|=i} \psi_J(-(i+k))S^J(\mathbb{A}) && \text{(a)}\\
&= \textstyle\sum_{|J|=i} \psi_J(i+k)\Lambda^J(\mathbb{A}) && \text{(b)}\\
&= \textstyle\sum_{|J|=i} S_J(i+k)S_{J^\sim}(\mathbb{A}) && \text{(c)}\\
&= \textstyle\sum_{|J|=i} \Lambda^J(i+k)\psi_J(\mathbb{A}) && \text{(d)}\\
&= (-1)^i \textstyle\sum_{|J|=i} \frac{(-i-k)^{\ell(J)}}{m_1!m_2!\cdots}(\frac{\psi_1}{1})^{m_1}(\frac{\psi_2}{2})^{m_2}(\frac{\psi_3}{3})^{m_3}\cdots && \text{(e)}
\end{aligned} \tag{6}$$

are the expansions in terms of the basis of products of complete functions S^J, products of elementary functions Λ^J, Schur functions S_J, monomial functions ψ_J, products of power sums ψ^J, noting in (6e) the partitions given exponentially: $J = 1^{m_1}2^{m_2}\cdots$.

Let us emphasize once more that all these expansions are a consequence of the operation $\mathbb{A} \to k\,\mathbb{A}$, by use of (1), ... ,(4).

Let us go back to Gian-Carlo's opus. I say that the fundamental article "Finite Operator Calculus" [13] rests upon Lagrange inversion, and eventually, upon multiplication of alphabets by integers.

To state the problem in a few words: one wants to deform the derivative

$$D := \frac{d}{dx} \to Q = D + a_1 D^2 + a_2 D^2 + \cdots$$

while preserving properties like the existence of a distinguished basis $\{x^n\} : Dx^n = nx^{n-1}$, Taylor formula, etc.; in other words, the tools of classical analysis in one variable.

Let us introduce an alphabet $\mathbb{A}$ (which will remain formal) and write

$$Q = D\sigma_{-D}(\mathbb{A}) = D - S^1(\mathbb{A})D^2 + S^2(\mathbb{A})D^2 - S^3(\mathbb{A})D^3 + \cdots . \tag{7}$$

One wants polynomials $P_n, n = 1, 2, \ldots, P_0 := 1$, such that

$$Q(P_n) = nP_{n-1} \quad \& \quad P_n(0) = 0,\ n \geq 1. \tag{8}$$

Rota's theorem ([17, Theorem 4]), is

$$P_n = x\,\lambda_D(n\mathbb{A})\,x^{n-1}, \qquad n \geq 1. \tag{9}$$

Proof. f' being the derivative of a function $f(x)$ of x, one has $f(D)x = x\,f(D) + f'(D)$. Taking f such that $f(D) = \lambda_D(-\mathbb{A})$, one has $Q = D\,f = f\,D$ and

$$\begin{aligned}
Q\,P_n &= f\,DP_n = f\,Dxf^{-n}\,x^{n-1} \\
&= f\,f^{-n}\,x^{n-1} + fx\,Df^{-n}x^{n-1} \\
&= f^{-n+1}\,xx^{n-2} + fxf^{-n}\,Dx^{n-1} \\
&= (1-n)f'\,f^{-n}\,x^{n-2} + xf^{-n+1}x^{n-2} \\
&\quad + xf'\,f^{-n}\,(n-1)x^{n-2} + xf^{-n+1}(n-1)x^{n-2} \\
&= n\,P_{n-1} \qquad \square
\end{aligned}$$

Rota's theorem has many corollaries which are extensions of classical analysis:

- existence of a reproducing kernel

$$\sum \frac{P_n(y)}{n!}\,Q^n(f(x)) = f(x+y); \tag{10}$$

- existence of a Cauchy formula

$$\sum \frac{P_n(y)}{n!}\,x^n\sigma_{-x}(n\mathbb{A}) = \exp(xy); \tag{11}$$

- existence of a binomiality property

$$P_n(x+y) = \sum \binom{n}{i} P_i(x)P_{n-i}(y). \tag{12}$$

These formulae can be interpreted as consequences of replacing the usual exponential by $\sum x^n P_n(x)/n!$. They may appear much more surprising once specialized.

For example, let $Q = D(1-D)$, i.e., $\mathbb{A}$ is such that $S^1(\mathbb{A}) = 1$, $S^i(\mathbb{A}) = 0$, $i > 1$. Then $\Lambda^i(n\mathbb{A}) = \binom{n+i-1}{i}$,

$$P_n = \sum \frac{(n-i)\cdots(n+i-1)}{i!}\,x^{n-i}, \tag{13}$$

and the Cauchy formula gives the expansion of exp,

$$\exp(xy) = \sum \frac{P_n(y)}{n!} x^n (1-x)^n, \tag{14}$$

used by Schur [24] to compute $\sin(\pi x)$.

Abel polynomials occur for

$$Q = D + aD^2/1! + a^2D^3/2! + \cdots,$$

that is, for Q such that $Q(f(x)) = \frac{d}{dx} f(x+a)$.

The corresponding "Abel alphabet" is such that $S^i(\mathbb{A}) = (-a)^i/i!$, $\Lambda^i(n\mathbb{A}) = S^i(n\mathbb{A}) = (-na)^i/i!$, and finally

$$P_n(x) = x\,(x-na)^{n-1}. \tag{15}$$

In this case, power sums are simpler since

$$\psi_1(n\mathbb{A}) = -na \quad \& \quad \psi_i(n\mathbb{A}) = 0,\ i > 1.$$

One may notice that dimensions $\dim(J)$, J a partition, of representations of the symmetric group are given by the specialization $S^i(\mathbb{A}) = 1/i!$. Determinants in the $S^i(n\mathbb{A})$ can therefore be computed in terms of these dimensions (for which one has many combinatorial interpretations). For example, the evaluation of Schur functions in a multiple of the Abel alphabet, for $J = [j_1, \ldots, j_n]$, $0 \le j_1 \le \cdots \le j_n$, is:

$$S_J(n\mathbb{A}) = \frac{(-a\,n)^{|J|}}{|J|!} \dim(J) = (-a\,n)^{|J|} \det \left| \frac{1}{(j_i + k - i)!} \right|_{1 \le i,k \le n}.$$

As a last example of deforming the derivative, leading to the Bernoulli numbers:

$$Q = D \frac{D}{\exp(D)-1} = D\,(1 - \frac{1}{2}D + \frac{B_2}{2!}D^2 + \frac{B_4}{4!}D^4 + \cdots) \tag{16}$$

Thus

$$S^i(\mathbb{A}) = B_i/i! \quad \& \quad \Lambda^i(\mathbb{A}) = (-1)^i/(i+1)!, \quad i \ge 0.$$

The elementary functions are the same as in Abel's case (for $a = 1$), up to a shift of indices which has consequences, which (in medical terminology) are termed drastic:

$$P_n(x) = x \left(\frac{\exp(D)-1}{D} \right)^n x^{n-1} = x(1 + D/2! + D^2/3! + \cdots)^n x^{n-1}. \tag{17}$$

Compact notations for coefficients of powers of series ease computations. Moreover, since these notations refer to the symmetric functions, they allow one to benefit from the full machinery that one can find in such good treatises of symmetric functions as [8] (see also [7]).

In the present case of the Bernoulli alphabet, the identity $\frac{d}{dx}\exp(x) = \exp(x)$ implies $\psi_i(\mathbb{A}) = -S^i(A)\ \forall i > 1$, and therefore the Newton relations between complete functions S^i and power sums ψ_k,

$$nS^n = \psi_1 S^{n-1} + \psi_2 S^{n-2} + \cdots \psi_n S^0,$$

give the following recursion (Gessel told me that this is classical)

$$\begin{aligned}-(n+1)S^n(\mathbb{A}) &= \frac{-n-1}{n!}B_n \\ &= S_2(\mathbb{A})S_{n-2}(\mathbb{A}) + S_4(\mathbb{A})S_{n-4}(\mathbb{A}) + \cdots + S_{n-2}(\mathbb{A})S_2(\mathbb{A}).\end{aligned} \tag{18}$$

One can compare the underlying algorithm to that of Augusta Ada Byron, Lady Lovelace,

$$\frac{2x-1}{2(2x+1)} = B_2\frac{2x}{2!} + B_4\frac{2x(2x-1)(2x-2)}{4!} + B_6\frac{2x\cdots(2x-4)}{6!} + \cdots. \tag{19}$$

To close this chapter on Bernoulli numbers, let me express them in terms of Schubert polynomials [6] specialized in $x_1 = x_2 = \cdots = 1$. For example,

$$(-1)^{n-1}B_{2n} = \frac{Y_{[2,(3,0)^{n-1}]}}{2^n(n-1)!(2^{2n}-1)} = \frac{n\,Y_{[1,(2,0)^{n-1}]}}{2^{2n}(2^{2n}-1)} \tag{20}$$

$$B_{12} = Y_{[2,3,0,3,0,3,0,3,0,3]}\frac{1}{31449600} = Y_{[1,2,0,2,0,2,0,2,0,2]}\frac{1}{1397760}.$$

One can reduce the determinants expressing the above Schubert polynomials, and finally one obtains the Bernoulli numbers as minors of the Pascal matrix $\left|\binom{i}{2j-i}\right|$, up to factors $2\times(2^{2n}-1)$. For example, for $n=6$, the product of $B_{12} = -\frac{691}{2730}$ by $2\times(2^{2n}-1)$ $(=8190)$ is equal to the determinant (=2073) of the following matrix (writing '·' for '0', and with the full shifted Pascal matrix from which it is extracted written on the right):

$$\begin{bmatrix} 1 & 1 & \cdot & \cdot & \cdot & \cdot \\ \cdot & 1 & 1 & \cdot & \cdot & \cdot \\ \cdot & \cdot & 3 & 1 & \cdot & \cdot \\ \cdot & \cdot & 1 & 6 & 1 & \cdot \\ \cdot & \cdot & \cdot & 5 & 10 & 1 \\ \cdot & \cdot & \cdot & 1 & 15 & 15 \end{bmatrix} \leftarrow \begin{bmatrix} 1 & 1 & \cdot & \cdot & \cdot & \cdot & \cdot & \cdot & \cdot & \cdot & \cdot \\ \cdot & 1 & 2 & 1 & \cdot & \cdot & \cdot & \cdot & \cdot & \cdot & \cdot \\ \cdot & \cdot & 1 & 3 & 3 & 1 & \cdot & \cdot & \cdot & \cdot & \cdot \\ \cdot & \cdot & \cdot & 1 & 4 & 6 & 4 & 1 & \cdot & \cdot & \cdot \\ \cdot & \cdot & \cdot & \cdot & 1 & 5 & 10 & 10 & 5 & 1 & \cdot \\ \cdot & \cdot & \cdot & \cdot & \cdot & 1 & 6 & 15 & 20 & 15 & \cdot \end{bmatrix}$$

Combinatorialists tirelessly point out that they practised q-calculus long before specialists of quantum groups took it over.

They introduced a *q-derivative* $D_q : f \mapsto \frac{f(qx)-f(x)}{qx-x}$. In the same manner as before, it can be deformed

$$D_q \mapsto Q = D_q\sigma_{-D_q}(\mathbb{A}) = D_q - S^1(\mathbb{A})D_q^2 + S^2(\mathbb{A})D_q^2 - S^3(\mathbb{A})D_q^3 + \cdots. \tag{21}$$

Since $D_q(x^n) = [n]\,x^{n-1}$, with $[n] := 1+q+\cdots+q^{n-1}$, one looks for polynomials P_n^q such that

$$Q(P_n^q) = [n]P_{n-1}^q \quad \& \quad P_n^q(0) = 0,\ n \geq 1. \tag{22}$$

Actually one has to replace the relation $P_n = x\lambda_D(n\mathbb{A})x^{n-1}$ by an equivalent one to be able to q-ify it, and one finds that

$$P_n^q = \left(1 + \sum (n - |J|) \frac{(n-1)\cdots(n-\ell(J)+1)}{\prod m_i!} \Lambda^J(\mathbb{A})\, D_q^{|J|}\right) x^n, \tag{23}$$

sum over all partitions $J \neq 0$, $J = 1^{m_1}2^{m_2}3^{m_3}\cdots$, $\ell(J) = m_1 + m_2 + \cdots$, $|J| = m_1 + 2m_2 + 3m_3 + \cdots$, with $\Lambda^J := (\Lambda^1)^{m_1}(\Lambda^2)^{m_2}(\Lambda^3)^{m_3}\cdots$.

The q-integers only appear in the action of D_q and not in the coefficients of the powers of D_q. Since P_n^q is obtained from x^n and not x^{n-1}, formula (23) is not a straightforward consequence of (9). For more about extensions of the umbral calculus, see [12].

I now have to argue that addition and multiplication by integers do not cover all of mathematics. Indeed, one must have recourse to rational numbers, or even complex numbers.

To state it differently: we met the doubling of alphabets $\mathbb{A} = \{a\} \mapsto 2\mathbb{A} := \{a'\} \cup \{a''\}$, but the inverse operation is required:

$$\mathbb{A} \mapsto \frac{1}{2}\mathbb{A} = \mathbb{B}, \quad \mathbb{B} = \{b\} \mapsto \{b'\} \cup \{b''\} = \mathbb{A}.$$

Realizing that at the level of power sums one has

$$\psi_i(2\mathbb{A}) = \sum (a')^i + \sum (a'')^i = 2\psi_i(\mathbb{A}),$$

one sees that $\mathbb{B} = k\,\mathbb{A}$, $k \in \mathbb{C}$, can be defined by the equations

$$\psi_i(\mathbb{B}) := k\,\psi_i(\mathbb{A}),\ i \geq 1. \tag{24}$$

It is however more illuminating to discover that every polynomial can play the rôle of an alphabet, that is, power sums ψ_i are operators on the ring of polynomials ($\alpha \in \mathbb{C}$, u monomial):

$$P = \sum_{\alpha,u} \alpha\, u \Rightarrow \psi_i(P) = \sum_{\alpha,u} \alpha\, u^i. \tag{25}$$

As the ring of symmetric polynomials $\mathfrak{S}ym$ is generated by the ψ_i, $i = 1, 2, \ldots$, formula (25) transforms any symmetric polynomial into an operator on the ring of polynomials. Thus, the generating functions of the operators Λ^i or S^i are

$$P = \sum_{\alpha,u} \alpha\, u \mapsto \lambda_z(P) = \prod (1 + zu)^\alpha, \tag{26}$$

$$P = \sum_{\alpha,u} \alpha\, u \mapsto \sigma_z(P) = \prod (1 - zu)^{-\alpha}. \tag{27}$$

Each formula (25), (26), (27), as one wishes, characterizes the structure of the λ-ring of the ring of polynomials. Grothendieck had choosen *lambda operations*, that is, the *exterior powers* Λ^i of classes of vector bundles, the S^i's being the *symmetric powers*. Algebraic topologists, for their part, prefer *Adams operations* ψ_i (see the remarks about λ-rings at the end).

Notice the different rôles played by constants α and monomials u:

$$\begin{aligned} &\psi_i(\alpha) = \alpha, \quad S^i(\alpha) = \tbinom{\alpha+i-1}{i}, \quad \Lambda^i(\alpha) = \tbinom{\alpha}{i} \\ &\psi_i(u) = u^i, \quad S^i(u) = u^i, \qquad \Lambda^i(u) = 0,\ i > 1, \quad \Lambda^1(u) = u. \end{aligned} \tag{28}$$

Computers cannot distinguish indeterminates α from monomials u. It is more correct to characterize "monomials" as *rank 1 elements* (i.e., elements $x \neq 0$ such that $\Lambda^i(x) = 0\ \forall i > 1$), and "constants" as elements invariant under all the ψ_i's (Rota [17] called them *elements of binomial type*).

One should not restrict to polynomials, but extend to rational functions or formal series, keeping the definition

$$\psi_i\left(\frac{\sum \alpha u}{\sum \beta v}\right) = \frac{\sum \alpha u^i}{\sum \beta v^i}, \tag{29}$$

with summations finite or not.

With some precautions, one can just as well use "Laurent monomials" with exponents in $\mathbb{Z}$ rather than in $\mathbb{N}$.

If q is a rank 1 element, one can now write $\frac{1}{1-q} = 1 + q + q^2 + \cdots$, i.e., $\frac{1}{1-q}$ is the infinite alphabet $\{1, q, q^2, \dots\}$ and $\psi_i(\frac{1}{1-q}) = \frac{1}{1-q^i}$, $i \geq 1$. By the way, we learn from Cauchy that

$$S_i(\frac{1}{1-q}) = \frac{1}{(1-q)\cdots(1-q^i)} \tag{30}$$

and therefore

$$\sigma_x(\frac{1}{1-q}) = \sum \frac{x^i}{(1-q)\cdots(1-q^i)} \tag{31}$$

is the q-exponential, the equality $\frac{1}{1-q} = 1+q+q^2+\cdots$ leading to the factorization

$$\sigma_x(\frac{1}{1-q}) = \prod_{i=1}^{\infty} \frac{1}{1-xq^i} \tag{32}$$

which turns the q-exponential into an object rather easier to manipulate than the classical exponential [18].

Combining elements of different types in a λ-ring, one can cover various expansions with the "+" symbol. For example, let x be of binomial type, q be of rank 1, and y be such that $y(1-q)$ is of rank 1. The q-Charlier polynomials are defined by:

$$\mathrm{Ch}_n(x) := n!\, \Lambda^n(x-y). \tag{33}$$

To get their explicit expression, it is preferable to introduce $z := y(1-q)$. By combining (3), (28) and (30), it follows that

$$\Lambda^n(x - \frac{z}{1-q}) = \sum \Lambda^{n-i}(x)\Lambda^i(\frac{-z}{1-q}) = \sum \Lambda^{n-i}(x)(-z)^i S^i(\frac{1}{1-q}),$$

$$\mathrm{Ch}_n(x) = n! \sum \frac{x\cdots(x-n+i+1)}{(n-i)!} \frac{(-y)^i(1-q)^i}{(1-q)\cdots(1-q^i)}. \tag{34}$$

For $y=1, q\to 1$, one recovers the classical polynomials (see [13, p. 64]).

We can now go back to the umbral calculus, having in mind that several lifts are possible for a_i:

$$a_i \mapsto S_i(\mathbb{A}),\ a_i \mapsto \psi_i(\mathbb{A}),\ a_i \mapsto i!S_i(\mathbb{A}), \dots .$$

This is where λ-wisdom is not enough, and some help is needed to allow experimentation, e.g., from ACE [1], and its library SFA [10] devoted to λ-rings.

Let us take, for example, the identity

$$\exp(bx) = \sum \frac{b(ak+b)^{k-1}}{k!}(x\exp(-ax))^k, \tag{35}$$

that Riordan [11] gives as an application of Lagrange inversion. How should we interpret it in a λ-ring ? One can introduce two alphabets $\mathbb{A}$, $\mathbb{B}$, replace $\exp(ax)$ by $\sigma_x(\mathbb{A})$, $\exp(bx)$ by $\sigma_x(\mathbb{B})$ and look for the coefficients c_k of the expansion

$$\sigma_x(\mathbb{B}) = \sum c_k(\frac{x}{\sigma_x(\mathbb{A})})^k = \sum c_k x^k\, \lambda_{-x}(k\mathbb{A}). \tag{36}$$

The first terms are

$$\begin{aligned}1 + xS^1(\mathbb{B}) + x^2S^2(\mathbb{B}) + \cdots = 1 &+ xS^1(\mathbb{B})\lambda_{-x}(\mathbb{A}) + x^2(S^2(\mathbb{B})\\ &+ S^1(\mathbb{B})S^1(\mathbb{A}))\lambda_{-x}(2\mathbb{A}) + x^3(S^3(\mathbb{B}) + 2S^2(\mathbb{B})S^1(\mathbb{A})\\ &+ 2S^1(\mathbb{B})S^2(\mathbb{A}) + S^1(\mathbb{B})S_{11}(\mathbb{A}))\lambda_{-x}(3\mathbb{A}) + \cdots .\end{aligned}$$

One sees that the c_k's are linear in the $S_i(\mathbb{B})$'s. One can therefore restrict to the case $\mathbb{B} := b$ of cardinality 1. By homogeneity, one can even put $b=1$. Finally one has to solve

$$\frac{1}{1-x} = \sum c_k x^k\, \lambda_{-x}(k\mathbb{A}),$$

i.e.,

$$1 = (1-x)\sum c_k x^k\, \lambda_{-x}(k\mathbb{A}) = \sum c_k x^k\, \lambda_{-x}(k\mathbb{A}+1);$$

the solution of this system combining addition and multiplication is:

$$c_k = \frac{1}{k} S^{k-1}(k\mathbb{A}+2),\ k \geq 1. \tag{37}$$

For example, $c_3 = \frac{1}{3} S_2(2\mathbb{A}+2) = 2S^2(\mathbb{A}) + S_{11}(\mathbb{A}) + 2S^1(\mathbb{A}) + 1$ and, if we go back to a homogeneous expression of degree 3,

$$c_3 = 2S^2(\mathbb{A})S^1(\mathbb{B}) + S_{11}(\mathbb{A})S^1(\mathbb{B}) + 2S^1(\mathbb{A})S^2(\mathbb{B}) + S^3(\mathbb{B}).$$

Riordan's identity can be found in the specialization $S^i(\mathbb{A}) = a^i/i!$, $S^i(\mathbb{B}) = b^i/i!$, since then

$$c_3 \mapsto 2\frac{a^2}{2}\frac{b}{1} + \frac{a^2}{2}\frac{b}{1} + 2\frac{a}{1}\frac{b^2}{2} + \frac{b^3}{3!} = \frac{b(3a+b)^2}{3!}.$$

One may ponder over the fact that the generalization of Riordan's identity is proved by replacing $\exp(-bx)$ by $1 - bx$. This "umbra" looks too paltry to shelter the exponential, in that it uses only formula (25) and it uncovers the integer 2 which was not present in the formulation of (25).

We could give numerous classical examples which can be decoded in terms of multiples of alphabets (cf. the article of Brenti [2]). The split alphabets promised by the title are more difficult to find, and I shall end with a single example, that of Jack polynomials indexed by partitions of length 1 [8]. Indeed their generating function is

$$\sum \frac{z^k}{\alpha^k k!} J_k^\alpha(x_1, \ldots, x_n) = \prod (1 - zx_i)^{-\frac{1}{\alpha}} = \sigma_z(\frac{1}{\alpha}X), \tag{38}$$

the last expression based on the assumption that the x_i's are rank 1-elements, with sum X. An example of computations related to Jack polynomials and using λ-techniques can be found in [5].

It will be much easier for me to give a longer list of examples when λ-rings are taught at school.

1 The Cauchy formula

The space of symmetric functions $\mathfrak{Sym}$ in indeterminates $x_1, \ldots, x_n$ has a long history. Newton showed that it is a space of polynomials in $\Lambda^1, \ldots, \Lambda^n$ or $\psi_1, \ldots, \psi_n$. A great many of the classical problems in this field amount to changing bases and linear algebra.

Kostka understood that there exists a canonical scalar product for which Schur functions are an orthonormal basis; power sums, an orthogonal basis; monomial functions are adjoint to products of complete functions, etc. All these statements can be summarized by the existence of a *Cauchy kernel*

$$K(X,Y) := \prod_{i,j} \frac{1}{1 - x_i y_j}$$

(by introducing a second set of indeterminates $y_1, \dots, y_n$ that one can suppose to be of the same cardinality). This kernel defines the scalar product, that is, every expansion "separating variables"

$$K(X, Y) = \sum U_J(X)V_J(Y)$$

provides a pair of adjoint bases $\{U_J\}$, $\{V_J\}$ of $\mathfrak{S}ym$.

Thus expansions like

$$K(X, Y) = \prod_i \sum_k y_i^k S^k(X) \quad \& \quad K(X, Y) = \prod_{i,j} \left(1 + x_i y_j + (x_i y_j)^2 + \cdots\right)$$

show that the monomial basis is adjoint to the basis of products of complete functions, and that the basis of power sums is orthogonal. Diagonalizing $K(X, Y)$ is less straightforward and can be deduced from the Binet–Cauchy theorem expressing minors of a product of matrices (in that case the matrix $\left[\frac{1}{1-x_i y_j}\right]$ is the product of two Vandermonde matrices; cf. [8]).

The above three expansions, combined with the involution $X \mapsto -X$ and the symmetry in X and Y, give formulae $(6a), \dots, (6e)$, when Y is specialized to the integer $i + k$.

By deforming the Cauchy kernel by introducing one or two parameters, one generalizes Schur functions into *Hall–Littlewood polynomials* or *Macdonald polynomials*. However, these polynomials constitute an orthogonal basis, not an orthonormal one, and one needs to introduce extra conditions to characterize them (contrary to Schur functions which are the only orthonormal basis of $\mathfrak{S}ym$ over $\mathbb{Z}$).

1.1 λ-rings

λ-rings are rings with operators λ^i, $i \in \mathbb{N}$, satisfying three axioms expressing their compatibility with ring operations (cf. [4]), addition, $\lambda^i(x + y)$; multiplication, $\lambda^i(xy)$; the third axiom stating a universality property with respect to composition $x \mapsto \lambda^i(\lambda^j(x))$ (called *plethysm* by Littlewood and combinatorialists after him).

Instead of taking λ^i, one can use the S^i defined by $S^i(x) := (-1)^i \lambda^i(-x)$, or the ψ_k's defined by the generating function (4) (replacing A by x).

Our present knowledge of plethysm is too scanty to allow wide use in explicit computations. There remains only addition (this is an avatar of the Chu–Vandermonde formula)

$$S^k(x + y) = \sum_{i+j=k} S^i(x)S^j(y)$$

and multiplication, which is given by the Cauchy formula

$$S^k(xy) = \sum_{J:|J|=k} S_J(x)S_J(y)$$

(with the $S_J(x)$'s defined as determinants in the $S^j(x)$'s).

In short one can say that Cauchy's formula generates the majority of identities in the theory of λ-rings, as well as in the theory of classical symmetric functions.

We restricted our constructions to rings of polynomials, which renders axiomatics simpler. The basic objects are monomials u and scalars α (here taken in $\mathbb{C}$). The three axioms can be written at a stroke in formula (25)

$$\psi_k\left(\sum \alpha\, u\right) = \sum \alpha\, u^k, \quad k > 0.$$

In a general λ-ring, one has elements more general than "polynomials".

1.2 Schubert polynomials

These polynomials generalize Schur functions and constitute a linear basis of the ring of polynomials in $x_1, x_2, \ldots$, stable for divided differences. Their combinatorics involves permutations rather than partitions as in the case of symmetric functions.

Following Newton, for every i one defines the *i-th divided difference* ∂_i as the operator (written on the right)

$$f\,\partial_i = \frac{f(\ldots x_i, x_{i+1} \ldots) - f(\ldots x_{i+1}, x_i \ldots)}{x_i - x_{i+1}}.$$

For every k and every $\nu := [\nu_1, \ldots, \nu_k]$, $\nu_1 \geq \cdots \geq \nu_k \geq 0$, one puts

$$Y_\nu := x^\nu$$

and one recursively defines *Schubert polynomials* by

$$Y_J\,\partial_i = Y_{[\ldots, j_{i+1}, j_i - 1, \ldots]} \quad \text{if} \quad j_i > j_{i+1} \tag{39}$$

(since $\partial_i^2 = 0$, then $Y_J\,\partial_i = 0$ if $j_i \leq j_{i+1}$). In other terms, Schubert polynomials are the images of *dominant* monomials under iterated divided differences.

When J is of the type $j_1 \leq j_2 \leq \cdots \leq j_n$, $j_{n+1} = 0 = j_{n+2} = \cdots$, then Y_J is equal to the Schur function of index $[j_1, \ldots, j_n]$ in the variables $x_1, \ldots, x_n$. Binomial determinants are obtained by applying (39) to Schur functions. Thus

$$Y_{2345} \xrightarrow{\partial_4\partial_5} Y_{234003} \xrightarrow{\partial_3} Y_{230303}$$

$$Y_{1234} \xrightarrow{\partial_4\partial_5} Y_{123002} \xrightarrow{\partial_3} Y_{120202}$$

Techniques of divided differences replace manipulations of determinants and provide identities on binomial determinants which are specializations of Schubert polynomials ($x_i = 1$ for all i). In this way, one obtains the determinants that we gave in relation with Bernoulli numbers. Gessel and Viennot [3] prefer to use a combinatorics of non-intersecting paths to study binomial determinants.

As in the case of symmetric functions, the full theory can be condensed into the existence of a Cauchy kernel

$$K_n(X, Y) := \prod_{i,j\,:\,i+j \leq n} (x_i - y_j)$$

which generalizes the *resultant* $\prod_{i,j}(x_i - y_j)$ (this last one being diagonal in the basis of Schur functions).

Let us remark that it is equivalent to consider $\prod_{i,j}(x_i - y_j)$ or $\prod_{i,j} 1/(1 - x_i y_j)$. This is not so in the non-symmetric case; the expansion of $\prod_{i,j:i+j\leq n} 1/(1 - x_i y_j)$ involves *key polynomials* (associated to *isobaric divided differences*) and not Schubert polynomials.

References

[1] ACE (Veigneau, S. et al.) (1998): An algebraic environment for the computer algebra system MAPLE; http://phalanstere.univ-mlv.fr/~ace (1998)

[2] Brenti, F. (1999): A class of q-symmetric functions arising from plethysm. Preprint

[3] Gessel, I., Viennot, X. (1985): Binomial determinants, paths, and hook length formulae. Adv. Math. **58**, 300–321

[4] Knutson, D. (1973): λ-rings and the representation theory of the symmetric group. (Lecture Notes in Mathematics, vol. 308). Springer, Berlin

[5] Lascoux, A., Lassalle, M. (2000): Une identité remarquable en théorie des partitions. Math. Ann. To appear

[6] Lascoux, A., Schützenberger, M.P. (1982): Polynômes de Schubert. C. R. Acad. Sci. Paris Sér. I Math. **294**, 447–450

[7] Lascoux, A., Schützenberger, M.P. (1985): Formulaire raisonné de fonctions symétriques. Université de Paris VII, Paris

[8] Macdonald, I.G. (1995): Symmetric functions and Hall polynomials, 2nd edition. Clarendon Press, Oxford

[9] Prosper, V. (1999): Combinatoire des polynômes multivariés, Thèse, Université de Marne-la-Vallée, Marne-la-Vallée; http://schubert.univ-mlv.fr/~vince/vpthesis.html

[10] Prosper, V. (2000): SFA, a package on symmetric functions considered as operators over the ring of polynomials for the computer algebra system MAPLE. J. Symbolic Comput. **29**, 83–94

[11] Riordan, J. (1968): Combinatorial identities. Wiley, New York

[12] Roman, S. (1985): More on the umbral calculus, with emphasis on the q-umbral calculus. J. Math. Anal. Appl. **107**, 222–254

[13] Rota, G.-C., Kahaner, D., Odlyzko, A. (1973): Finite operator calculus. J. Math. Anal. Appl. **42**, 684–760

[14] Rota, G.-C. (1964): On the foundations of combinatorial theory. I. Theory of Möbius functions. Z. Wahrsch. Verw. Gebiete **2**, 340–368

[15] Rota, G.-C. (1964): The number of partitions of a set. Amer. Math. Monthly **71**, 498–504

[16] Rota, G.-C. (1969): Baxter algebras and combinatorial identities. I, II. Bull. Amer. Math. Soc. **75**, 325–329, 330–334

[17] Mullin, R., Rota, G.-C. (1970): On the foundations of combinatorial theory. III. Theory of binomial enumeration. In: Harris, B. (ed.) Graph Theory and its Applications. Academic Press, New York, pp. 167–213

[18] Goldman, J., Rota, G.-C. (1970): On the foundations of combinatorial theory. IV. Finite vector spaces and Eulerian generating functions. Stud. Appl. Math. **49**, 239–258

[19] Rota, G.-C. (1978): Hopf Algebra methods in combinatorics. In: Problèmes combinatoires et théorie des graphes. (Colloques Internationaux CNRS, vol. 260). Éditions du Centre National de la Recherche Scientifique, Paris, pp. 363–365

[20] Rota, G.-C., Joni, S.A. (1979): Coalgebras and bialgebras in combinatorics. Stud. Appl. Math. **61**, 93–139
[21] Rota, G.-C., Kung, J.P. (1984): The invariant theory of binary forms. Bull. Amer. Math. Soc. (N.S.) **10**, 27–85
[22] Rota, G.-C., Chen, W.Y.C. (1992): q-analogs of the inclusion-exclusion principle and permutations with restricted position. Discrete Math. **104**, 7–22
[23] Rota, G.-C. (1996): Report on the present state of combinatorics. Discrete Math. **153**, 289–303
[24] Schur, I. (1981): Bermerkungen sur Theorie der beschränkten Bilinearformen mit unendlich vielen Veränderlichen. J. Reine Angew. Math. **140**, 1–28

Some operations on the family of equivalence relations

T. Britz, M. Mainetti, L. Pezzoli

1 Introduction

Throughout the history of mathematics, the notion of an equivalence relation has played a fundamental role. It dates back at least to the time when the natural numbers first were introduced: a non-negative integer may be thought of as a representative of the equivalence class of sets with the same cardinality. To express such a simple and "obvious" fact with equivalence relations may seem unnecessarily cumbersome. Nothing is further from the truth. Equivalence relations play a decisive role as building elements in every area of mathematics. For instance, algebra is firmly founded on equivalence relations: groups theory, rings theory, modules and fields would basically be impossible to define and use without equivalence relations.

Equivalence relations are so ubiquitous in everyday life that we often forget about their proactive existence. Much is still unknown about equivalence relations. Were this situation remedied, the theory of equivalence relations could initiate a chain reaction generating new insights and discoveries in many fields dependent upon it.

This paper springs from a simple acknowledgement: the only operations on the family of equivalence relations fully studied, understood and deployed are the binary join $\vee$ and meet $\wedge$ operations. Very little is known about the possible classification of all operations on equivalence relations.

Some open questions are which other natural operations (binary or n-ary) may be defined on the family of equivalence relations on a set, how they relate to the existing operations, which properties they possess and whether or not they may lead to a fuller understanding and classification of the algebra of equivalence relations. The present paper presents some new ideas and indicates how they may be used to define a class of operations.

In addition to this fascinating topic, we expand the classification given by Jónsson [8] on the type of a pair of equivalence relations. A pair (R, T) of equivalence relations is said to be of type n if their join $R \vee T$ can be expressed as a sequence of n compositions of R and T. We refine such a classification with the introduction of intermediate types, called n weak, n strong and $n.5$. An extensive discussion is dedicated to this topic, with several examples.

2 Preliminaries

Given sets S and S', we will call any subset $R \subseteq S \times S'$ a *relation* on $S \times S'$. The converse relation R^{-1} of a relation R is the set $R^{-1} = \{(b, a) : (a, b) \in R\}$. One may interpret relations as the edge set of a directed, bipartite graph having components

S and S'. A binary operation $\circ$ is defined on any pair of relations $R \subseteq S \times S'$ and $R' \subseteq S' \times S''$ by

$$R \circ R' = \{(a, c) : (a, b) \in R, (b, c) \in R' \text{ for some } b \in S'\}.$$

We will omit the $\circ$ and write RT. Also, we will write $\alpha_1 R_1 \alpha_2 R_2 \cdots R_{n-1}\alpha_n$ instead of $(\alpha_1, \alpha_2) \in R_1, (\alpha_2, \alpha_3) \in R_2, \ldots, (\alpha_{n-1}, \alpha_n) \in R_{n-1}$.

2.1 Equivalence relations

Now let R be a relation on the set $S \times S$. R is called *reflexive* if aRa for all $a \in S$, *symmetric* if $R^{-1} = R$, and *transitive* if $RR \subseteq R$. A relation $R \subseteq S \times S$ that is reflexive, symmetric and transitive is called an *equivalence relation*. Any equivalence relation R on the set S partitions S into *equivalence classes*, the family of which we will denote by S_R or S/R. This partitioning is unique, so we have an isomorphism $R_\pi \mapsto \pi_R$ between the family of equivalence relations and the family $\Pi[S]$ of partitions of S. Endowed with the partial order of refinement, that is, $\pi \leq \pi'$ if and only if every block of π is contained in a block of π', $\Pi[S]$ is seen to be a lattice with unique minimal and maximal elements, namely $\hat{0}$, the partition composed of one-element blocks, and $\hat{1}$, the partition with the one block S. Re-using the notation of meet $\vee$ and join $\wedge$, we obtain from $\Pi[S]$ the binary operations on the family of equivalence relations on S:

$$R \wedge T = R_{\pi_R \wedge \pi_T} \text{ and } R \vee T = R_{\pi_R \vee \pi_T}.$$

These have the following specific descriptions.

Proposition 1.

$$R \wedge T = R \cap T \tag{1}$$

$$\begin{aligned} R \vee T = \; & R \cup RT \cup RTR \cup \cdots \\ & \cup T \cup TR \cup TRT \cup \cdots . \end{aligned} \tag{2}$$

Proof. (1) $R \cap T$ is the largest relation contained in both R and T. As $R \cap T$ is an equivalence relation, $R \cap T$ is the largest equivalence relation contained in both R and T.

(2) $R \vee T$ is the smallest equivalence relation containing both R and T. From the transitivity of $R \vee T$, one obtains the expression (2). □

Two equivalence relations R and T are said to *commute* if and only if $RT = TR$. Commuting equivalence relations can be characterized by various means, among others by

Theorem 1. *Let R and T be equivalence relations on a set S. The following statements are then equivalent:*

1. *R and T commute;*

2. $R \vee T = RT$;
3. RT *is an equivalence relation.*

Proof. 1 $\Rightarrow$ 2: Since R and T commute, (2) reduces to $R \vee T = RT$.

2 $\Rightarrow$ 3: $R \vee T$ is an equivalence relation.

3 $\Rightarrow$ 1: R, T and RT are equivalence relations and therefore symmetric. Hence, $TR = (R^{-1}T^{-1})^{-1} = (RT)^{-1} = RT$. □

Two equivalence relations R and T are *independent* if and only if $\varrho \cap \tau$ is non-empty for all classes $\varrho \in S_R$ and $\tau \in S_T$. The following characterization in terms of the equivalence classes of R and T is due to P. Dubreil and M.-L. Dubreil-Jacotin [3].

Theorem 2. *Two equivalence relations R and T commute if and only if, for any equivalence class $C \in S_{R \vee T}$, the restrictions $R_{|C}$, $T_{|C}$ are independent.*

This theorem will, in a rephrased form, be proved in Sect. 4.

2.2 Finite types

Let R and T be two relations on S. If there is an n such that

$$R \vee T = \overbrace{RTR\ldots}^{n+1} \cup \overbrace{TRT\ldots}^{n+1},$$

then (R, T) is said to be of *type* $n.5$, or just of *type* n. (R, T) is said to be of

strong type n if $RTR \cdots = TRT \ldots$ and

weak type n if $RTR \cdots \subseteq TRT \ldots$ or $TRT \cdots \subseteq RTR \ldots$.

Denote the families of pairs (R, T) of type $n.5$, weak type n, and strong type n by $\mathcal{H}_n$, $\mathcal{W}_n$ and $\mathcal{S}_n$, respectively. Evidently, a pair of strong type n is also of weak type n, and a pair of weak type n is also of type $n.5$. A pair of type $n.5$ is also of strong type $n+1$. These remarks are contained in the following result.

Proposition 2. *The families $\mathcal{H}_n$, $\mathcal{W}_n$ and $\mathcal{S}_n$ form the sequence*

$$\begin{aligned} \emptyset \neq \mathcal{S}_0 \subsetneq \mathcal{W}_0 \subsetneq \mathcal{H}_0 \subsetneq \mathcal{S}_1 &= \mathcal{W}_1 \subsetneq \mathcal{H}_1 \subsetneq \mathcal{S}_2 \subsetneq \mathcal{W}_2 \subsetneq \mathcal{H}_2 \subsetneq \mathcal{S}_3 \\ &= \mathcal{W}_3 \subsetneq \mathcal{H}_3 \subsetneq \mathcal{S}_4 \cdots . \end{aligned}$$

Proof. Figure 1 shows elements of $\mathcal{S}_n$, $\mathcal{W}_n$ (n even) and $\mathcal{H}_n$, for $n = 0$, which are not of lesser type. The rows and columns represent the equivalence classes of R and T, respectively. The non-empty intersections between these are denoted by the black squares. Corresponding examples for all $n \geq 1$ are given in Fig. 2. To conclude the proof, we must show that $\mathcal{S}_n = \mathcal{W}_n$ for all odd n. Let (R, T) be of weak type n, n odd. Then, without loss of generality, $R \vee T = RTR \cdots T$. As R, T and $R \vee T$ are symmetric, $TRT \cdots R = (R^{-1}T^{-1}R^{-1} \cdots T^{-1})^{-1} = (RTR \cdots T)^{-1} = (R \vee T)^{-1} = R \vee T = RTR \cdots T$. Thus, $\mathcal{W}_n \subseteq \mathcal{S}_n$ for all odd n. □

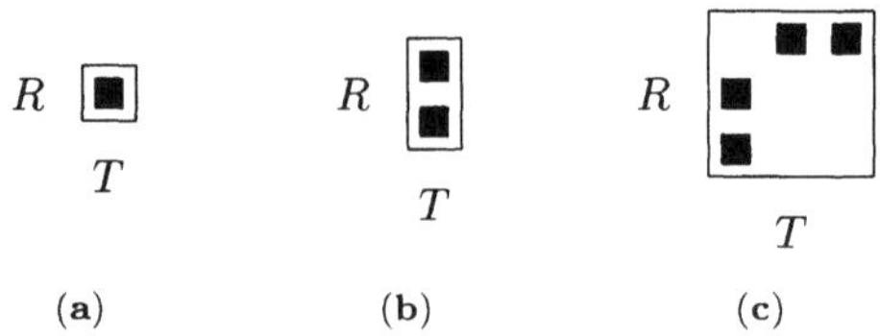

Fig. 1. Examples of the different types for $n = 0$: **(a)** strong **(b)** weak **(c)** 0.5

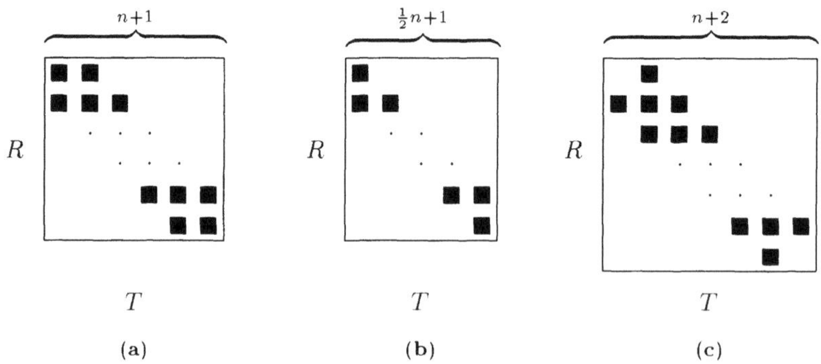

Fig. 2. Examples of the different types for any $n \geq 1$: **(a)** strong **(b)** weak, n even **(c)** $n.5$

M.-L. Dubreil-Jacotin and her husband P. Dubreil investigated pairs of types 1 and 2 [2, 3]. The originator of the term 'type', B. Jónsson, investigated pairs of types 1, 2 and 3 in relation to embeddings of lattices [8]. Also, G. Hutchinson studied pairs of type n in [7]. The pairs of strong type 1 are evidently the pairs of commuting equivalence relations. In Sect. 4, we shall give structural characterizations of the pairs of type 1.5, strong type 2 and weak type 2.

3 $R - T$ compatible relations

Let R, T be two equivalence relations on a set S. We will assume in what follows that there are finitely many classes of R and T, that is, that S_R and S_T are finite sets. For any $\alpha \in S$, $R(\alpha)$ and $T(\alpha)$ will indicate the classes of R and T containing α, respectively. An equivalence relation E on S is said to be $R - T$ *compatible* if the following conditions are satisfied:

$$\forall \alpha, \alpha', \alpha'' \in S, \alpha E \alpha' R \alpha'' : \; R(\alpha) \cap T(\alpha'') \cap E(\alpha'') \neq \emptyset; \tag{3}$$

$$\forall \alpha, \alpha', \alpha'' \in S, \alpha E \alpha' T \alpha'' : \; T(\alpha) \cap R(\alpha'') \cap E(\alpha'') \neq \emptyset; \tag{4}$$

$$R \wedge T \subseteq E \;\; \text{(or, equivalently, } \pi_{R \wedge T} \leq \pi_E). \tag{5}$$

An example is shown in Fig. 3, which also illustrates the structure of the $R - T$ compatible relations. The rows correspond to the classes of R, the columns correspond to the classes of T and the boxes correspond to the classes of E. Assume that α and α' are in the same class of E and that α' and α'' are in the same class of R (see Fig. 3(a)). By (3), there exists an element β which belongs to the same class of R as α and the same classes of T and E as α'' (see Fig. 3(b)). Now, α'' and β are in the same class of E, and β and α are in the same class of R, so by applying (3) to α'', β, and α, we find an element β' belonging to the same class of R as α'' and the same classes of T and E as α. Similarly, there exists an element β'' belonging to the same class of R as β and the same classes of T and E as α'. By transitivity, α, α', β, and β' belong to the same class of E (see Fig. 3(c)).

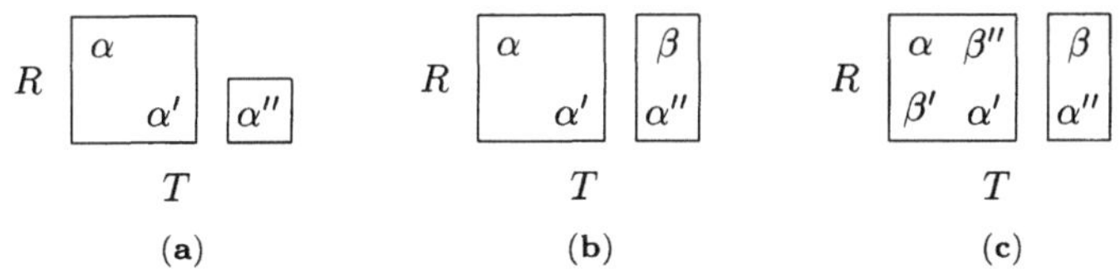

Fig. 3. The structure of $R - T$ compatible relations

These arguments can then be generalized to give a structural characterization of $R - T$ compatible equivalence relations. Define for all classes $C \in S_E$

$$\begin{aligned} R_C &= \{\varrho \in S_R : \varrho \cap C \neq \emptyset\}, \\ T_C &= \{\tau \in S_T : \tau \cap C \neq \emptyset\} \text{ and} \\ (R, T)_C &= \{(\varrho, \tau) \in S_R \times S_R : \varrho \cap \tau \cap C \neq \emptyset\}. \end{aligned}$$

Theorem 3. *Let E be an equivalence relation such that $R \wedge T \subseteq E$. Then E is $R - T$ compatible if and only if the following conditions hold:*

1. $\forall C \in S_E : (R, T)_C = R_C \times T_C$;
2. $\forall C, C' \in S_E : R_C, R_{C'}$ $(T_C, T_{C'})$ *are either disjoint or identical.*

In looser terms, condition 1 states that the classes of E are rectangular, as demonstrated in Fig. 3(c). Condition 2 states that any two classes of E that occupy common rows (columns) occupy the same rows (columns), as demonstrated in Fig. 3(b).

Proof. Assume E is $R - T$ compatible. Let $C \in S_E$ be given and choose $(\varrho, \tau) \in R_C \times T_C$. By definition, there exist elements $\alpha \in \varrho \cap C$ and $\alpha' \in \tau \cap C$. Since $\alpha E \alpha' R \alpha'$, (3) implies that there exists $\beta \in \varrho \cap \tau$ such that $\beta E \alpha'$. As $\beta \in C$, the pair (ϱ, τ) is an element of $(R, T)_C$, proving condition 1. Let $C, C' \in S_E$ be given. Assume that there exists $\varrho \in R_C \cap R_{C'}$, and choose any $\varrho' \in R_C$. Then there exist elements $\alpha \in \varrho' \cap C$, $\alpha' \in \varrho \cap C$ and $\alpha'' \in \varrho \cap C'$, so there is a $\beta \in \varrho' \cap T(\alpha'')$ such that $\beta E \alpha''$. Hence $\varrho' \in R_{C'}$. Similarly, $R_{C'} \subseteq R_C$, so $R_{C'} = R_C$. Conversely, assume that conditions 1 and 2 hold and that $\alpha E \alpha' R \alpha''$. As $R(\alpha') \in R_{E(\alpha)} \cap R_{E(\alpha'')}$, condition 2 implies $R_{E(\alpha)} = R_{E(\alpha'')}$, so there is an element $\beta \in R(\alpha) \cap E(\alpha'')$. By condition 1, we may find an element $\beta' \in E(\alpha'') \cap R(\beta) \cap T(\alpha'')$. As $\beta' \in R(\alpha) \cap T(\alpha')$ and $\beta' E \alpha''$, we have shown (3). Condition (4) is proved similarly, and E is $R - T$ compatible. □

From condition 2 we immediately obtain

Corollary 1. *Let E be $R - T$ compatible. Then $E \subseteq RT \cap TR$.*

The following result concerns the structure of the family of $R - T$ compatible equivalence relations.

Proposition 3. *The family of $R - T$ compatible equivalence relations forms a sublattice of the equivalence relations on S, in particular, a sublattice of the interval $[R \wedge T; R \vee T]$.*

Proof. Let E_1, E_2 be $R - T$ compatible equivalence relations. $E_1 \wedge E_2$ is easily seen to be $R - T$ compatible. Assume $\alpha(E_1 \vee E_2 \alpha' T)\alpha''$. Since S_R and S_T are finite, $E_1 \vee E_2$ is of finite type N for some N, so there exist $\alpha_1, \ldots, \alpha_{N-1}$ such that $\alpha E_1 \alpha_1 E_2 \cdots \alpha_{N-1} E_{j_{N-1}} \alpha' T \alpha''$ where $j_{N-1} \in \{1, 2\}$ (see Fig. 4(a)). By the $R - T$ compatibility of E_1 and E_2, we inductively find $\beta_1, \ldots, \beta_N$ such that $\alpha_{N-i} T \beta_i R \beta_{i-1}$, and $\beta_i E_{j_i} \beta_{i-1}$, where $\beta_0 = \alpha''$, $\alpha_N = \alpha'$, $\alpha_0 = \alpha$, and $j_i \in \{1, 2\}$ (see Fig. 4(b)). Then $\beta_N(E_1 \vee E_2)\alpha''$ and, by the transitivity of R, $\beta_N R \alpha''$, so $E_1 \vee E_2$ satisfies (4). Similarly, $E_1 \vee E_2$ satisfies (3). Since $E_1 \vee E_2$ trivially satisfies (5), $E_1 \vee E_2$ is $R - T$ compatible. Let E be an $R - T$ compatible equivalence relation. By (5), $R \wedge T \subseteq E$. By the previous corollary, $E \subseteq RT \cap TR \subseteq R \vee T$. □

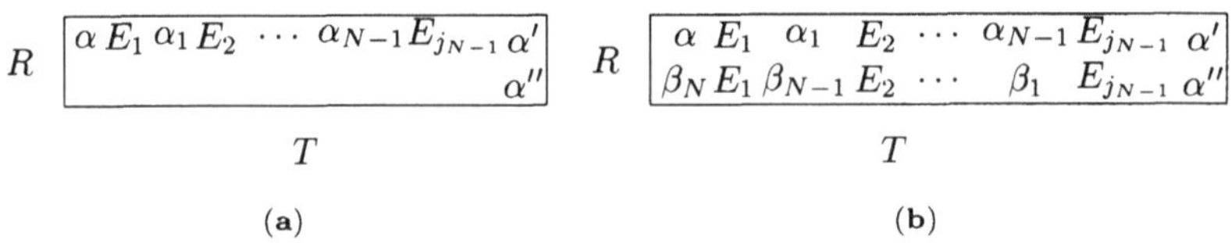

Fig. 4. $E_1 \vee E_2$ satisfies (4)

As $R \wedge T$ is $R - T$ compatible, the family of $R - T$ compatible equivalence relation is non-empty. Proposition 3 then guarantees that the following definition is well-defined and introduces a binary operation on the family of equivalence relations.

Definition 1. Let $R * T$ denote the largest $R - T$ compatible equivalence relation.

We will now look at the classes of $R * T$. Set $F = S/R * T$ and define the relations R^F and T^F on F by

$$R^F = \{(\phi, \phi') \in F \times F : R_\phi \cap R_{\phi'} \neq \emptyset\}$$
$$T^F = \{(\phi, \phi') \in F \times F : T_\phi \cap T_{\phi'} \neq \emptyset\}.$$

Lemma 1. *R^F and T^F are equivalence relations on $F \times F$.*

Proof. Theorem 3 implies that $R^F = \{(\phi, \phi') \in F \times F : R_\phi = R_{\phi'}\}$ and $T^F = \{(\phi, \phi') \in F \times F : T_\phi = T_{\phi'}\}$ and the proof follows immediately. □

We will call the structure $(F; R^F, T^F)$ the *shape* of $(S; R, T)$.

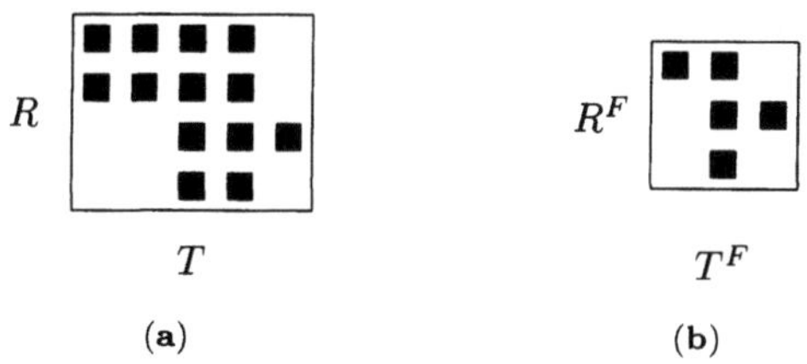

Fig. 5. (a) $(S; R, T)$ and **(b)** its shape

To simplify the notation, we will write F_{R^F} and F_{T^F} as F_R and F_T. $(S; R, T)$ and its shape are in a sense isomorphic structures; to be more precise:

Proposition 4. *Every structure $(S; R, T)$ such that $R \wedge T = \hat{0}$ is determined, up to isomorphism, by:*

- *its shape $(F; R^F, T^F)$,*
- *a function $r : F_R \to \mathbb{N}$, and*
- *a function $c : F_T \to \mathbb{N}$.*

Proof. This follows from Lemma 1 with the functions r and c given by $r(\varrho) = |R_\phi|$ for some $\phi \in \varrho$ and $c(\tau) = |T_\phi|$ for some $\phi \in \tau$. □

The value $r(\varrho)$ is the number of rows in R that are represented by ϱ (τ). c is the function, analog as to r, on the set of columns in T. Bearing Theorem 3 in mind, we obtain

Corollary 2. *Let (R, T) be of type n. If (R, T) is of weak type 1 or less, then (R^F, T^F) is a row and column disjoint union of element sets. Otherwise, (R, T) and (R^F, T^F) are of the same types.*

We can say more about these shapes. Define the functions $f : F_R \to \mathcal{P}(F_T)$ and $g : F_T \to \mathcal{P}(F_R)$ by

$$f(\varrho) = \{\tau \in F_T : \tau \cap \varrho \neq \emptyset\}; \qquad g(\tau) = \{\varrho \in F_R : \varrho \cap \tau \neq \emptyset\}.$$

Proposition 5. *The following statements are equivalent:*

1. *$(F; R, T)$ is a shape;*
2. *$R * T = \hat{0}$;*
3. *f and g are injective.*

Proof. $1 \Rightarrow 2$: Let $(F; R, T)$ be the shape of, say, $(S; R', T')$. Assume $R * T \neq \hat{0}$. Then $(F/R * T; R^{F/R*T}, T^{F/R*T}) \neq (F; R, T)$ is the shape of $(S; R', T')$, contradicting the maximality of $R' * T'$.

$2 \Rightarrow 3$: Assume that $f(\varrho) = f(\varrho')$ for some $\varrho, \varrho' \in F_R$, $\varrho \neq \varrho'$. Then E defined by

$$\alpha E \alpha' \iff \alpha = \alpha' \; or \; (\{R(\alpha), R(\alpha')\} = \{\varrho, \varrho'\} \; and \; T(\alpha) = T(\alpha'))$$

is $R-T$ compatible and not contained in $R*T$, contradicting the maximality of $R*T$.

$3 \Rightarrow 2 \Rightarrow 1$: If f and g are injective, then $R * T = \hat{0}$ by Theorem 3. Then $F = F/R * T$, so $(F; R, T)$ is its own shape. □

We observe that if any of these statements is true, then $R \wedge T = \hat{0}$. Furthermore, Proposition 5 informs us of how the $*$ operation may be performed. Given equivalence relations R and T on S, f_S (g_S) is injective if and only if no two rows (columns) of $(S; R, T)$ are identical. Proceeding iteratively to delete any row or column that appears more than once, until there no longer are recurring rows or columns, we obtain the shape $(F = S/R * T; R^F, T^F)$ of $(S; R, T)$ (see Fig. 5).

4 Some applications

We will now look at some applications of the preceding section. The first is a restatement of P. Dubreil and M.-L. Dubreil-Jacotin's result, Theorem 2. The condition in Theorem 2 is that the classes of $R \vee T$ are rectangular and mutually row and column disjoint. By Theorem 3, this is equivalent to the statement that $R \vee T$ is $R - T$ compatible. In general, $R * T \leq R \vee T$ so we may restate Theorem 2 as follows.

Theorem 4. *Equivalence relations R and T commute if and only if $R \vee T = R * T$.*

Proof. Assume R and T commute. By Theorem 1, $R \vee T = RT$, so RT is an equivalence relation and, trivially, $R - T$ compatible. By Proposition 3, $R \vee T = RT \subseteq R * T \subseteq R \vee T$, so $R \vee T = R * T$. Conversely, assume $R \vee T = R * T$. By Corollary 1, $RT \cup RT \subseteq R \vee T = R * T \subseteq RT \cap TR$, so $RT = TR$. □

We will now show some properties of pairs (R, T) of types n using results from the previous section. The first results are structural characterizations of type 1.5, strong type 2 and weak type 2. We will assume that $R \vee T = \hat{1}$. In general, a pair (R, T) is of the stated type if and only if the stated conditions hold for all restrictions $(R_{|C}, T_{|C})$, $C \in S_{R \vee T}$.

Proposition 6. *Let R and T be equivalence relations on S such that $R \vee T = \hat{1}$. Then (R, T) is of type 1.5 if and only if $f(F_R)$ is a complete chain of $\mathcal{P}(F_T)$.*

The stated condition is also valid if we put $g(F_T)$ in place of $f(F_R)$. In terms of illustrations, the condition states that the shape of $(S; R, T)$ is a square triangle, for example, , after suitable row and column permutations.

Proof. By Corollary 2, (R, T) is of type 1.5 if and only if (R^F, T^F) is of type 1.5. Assume that (R, T) is of type 1.5 and let $f, f' \in f(F_R)$. If $f \nsubseteq f'$ and $f' \nsubseteq f$, then we may find $\phi, \phi' \in F$ such that $T^F(\phi) \in f - f'$ and $T^F(\phi') \in f' - f$. But then $(\phi, \phi') \notin R^F T^F \cup T^F R^F$, so (R^F, T^F) is not of type 1.5. Hence, (F_R) is a chain of $\mathcal{P}(F_T)$. By Proposition 5, g is injective, so the chain is complete. Conversely, suppose $f(F_R)$ is a complete chain of $\mathcal{P}(F_T)$, and let $\phi, \phi' \in F$. Without loss of generality, $f(R^F(\phi)) \subseteq f(R^F(\phi'))$, so $\phi T^F R^F \phi'$. Hence, (R^F, T^F) and (R, T) are of type 1.5. □

Proposition 7. *Let R and T be equivalence relations on S such that $R \vee T = \hat{1}$. Then (R, T) is of strong type 2 if and only if conditions 1 and 2 below hold. (R, T) is of weak type 2 if and only if either 1 or 2 holds.*

1. *Any two elements of $f(F_R)$ intersect non-trivially.*
2. *Any two elements of $g(F_T)$ intersect non-trivially.*

Proof. Assume that (R, T) is of strong type 2 and let $f, f' \in f(F_R)$. If $\{f, f'\}$ is not a chain, then we may find $\phi, \phi' \in F$ such that $T^F(\phi) \in f - f'$ and $T^F(\phi') \in f' - f$. As (R, T) is of strong type 2, $\phi R^F \phi_1 T^F \phi_2 R^F \phi'$ for some $\phi_1, \phi_2 \in F$. Hence, $T^F(\phi_1) \in f \cap f'$. Similarly, we may show that any $g, g' \in g(F_T)$ intersect non-trivially. Conversely, assume that conditions 1 and 2 hold and let $\phi, \phi' \in F$. As $f(R^F(\phi)) \cap f(R^F(\phi'))$ and $g(T^F(\phi)) \cap g(T^F(\phi'))$ are non-empty, $\phi R^F T^F R^F \phi'$ and $\phi T^F R^F T^F \phi'$. Hence, (R, T) is of strong type 2. Similar arguments show the case of weak type 2. □

For higher types, necessary and sufficient conditions, similar to those in Propositions 6 and 7, are fairly complicated so we will omit further structural characterizations. We shall now turn our attention to the representations of (R, T). Since S_R and S_T are finite, any representation of a pair (R, T) contains a finite number of squares. Let S_n, W_n and H_n denote the minimal number of squares in a representation of (R, T), where (R, T) is of strong type n, weak type n and type $n.5$, respectively, and not of lesser type. Pairs such as these may be said to be of *pure* type. The following theorem describes these minimal representations.

Theorem 5. *For all $n \geq 1$, $S_n = 2n + 2$. For all odd $n \geq 1$, $W_n = 2n + 2$ and $H_n = n + 2$. For all even $n \geq 1$, $W_n = n + 2$ and $H_n = 2n + 4$. Figure 6 shows representations with these minimal numbers of squares. Furthermore, the representations in (a), (b), and (c) are unique up to row and column permutations and transposition.*

Proof. It is clear that any representation of (R, T), where (R, T) is of pure weak type n, must contain a row and column permutation of the representation Fig. 6(b). Since Fig. 6(b) represents a pair of pure weak type n, this is a unique minimal representation. Similarly, Fig. 6(c) is a unique minimal representation. Let (R, T) then be of pure strong type n. We may then find a pair of squares, a and b, that are connected by paths $RT \cdots$ and $TR \cdots$ of length $n + 1$ and that cannot be connected

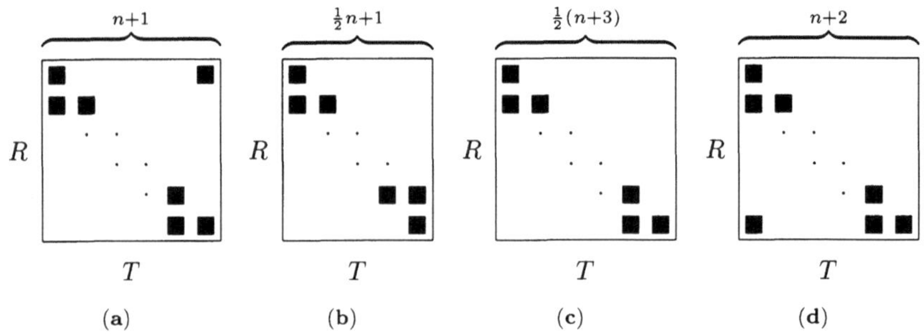

Fig. 6. Minimal representations for each pure type n: **(a)** strong n **(b)** weak, n even **(c)** $n.5$, n odd **(d)** $n.5$, n even

by any path of shorter length. Let c be a square on one path and d be a square on the other path, such that c and d are not endpoints a, b. Assume that c and d are in the same row. As the arguments are similar in all cases, we will only illustrate one case, namely, the following:

$$a \overbrace{\cdots TR}^{k} eTc \overbrace{RT \cdots}^{n-k} b \qquad a \overbrace{\cdots RT}^{m} \; fRd \overbrace{TR \cdots}^{n-m} b$$

for some $k, m \leq n$ and where e and f immediately precede c and d on each path. As $cRdRf$ we have the path

$$a \overbrace{\cdots RT}^{m} d \overbrace{RT \cdots}^{n-k} b.$$

If $k \geq m$, then this path has length $m + n - k < n + 1$, contradicting the choice of a and b. Similarly, $m \ngeq k$. In all other cases, we also find contradictions, so c and d cannot lie in the same row. Likewise, we may show that c and d cannot lie in the same column. Such a pair of paths is a unique, minimal representation of a pair of pure strong type n. Figure 6(a) shows such a pair of paths.

Assume now that there is a representation of a pair of pure type $n.5$, for some even n, that contains fewer than $2n+4$ squares. It must contain two paths $aRT \cdots Rc$ and $bTR \cdots Td$ of length $n+1$ such that there are no paths $aTR \cdots Tc$, $bRT \cdots Rd$ of length $n+1$. Since there are fewer than $2n+4$ squares, the paths share a subpath, say $e \cdots f$ of length m. Let k and l denote the lengths of the paths $aRT \cdots e$ and $bTR \cdots e$, respectively. As $k-l$ is odd, we may assume that $k > l$. Then the distance from a to d along these paths is $k+m+(n+1-m-l) = n+1+(k-l) > n+1$, so there must be another path (see Fig. 7) utilizing some further $p-1$ squares. The paths from a to c and from b to d containing the new path are longer than n. By inspecting the squares that join the paths, we obtain the inequalities

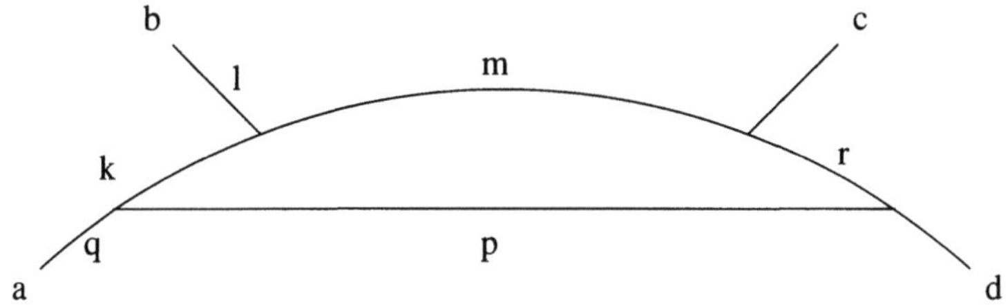

Fig. 7. Some paths in the representation

$$q + p + r + (n + 1 - m - k) - 1 \geq n + 1 \Leftrightarrow p \geq m + 1 + (k - q - r)$$
$$l + (k - q) - 1 + p + (n + 1 - m - l - r) \geq n + 1$$
$$\Leftrightarrow p \geq m + 1 - (k - q - r)$$

Adding the latter pair of inequalities we obtain $p \geq m + 1$, so the number of squares in the representation is at least $2n + 4 + p - 1 - m \geq 2n + 4$.

Finally, we note that $\mathcal{W}_n = \mathcal{S}_n$ for all odd n, according to Proposition 2. □

We remark that Fig. 6(b) is the same as Fig. 2(b). Figure 6(a) shows a row and column permutation of the representation obtained by removing all squares except the end squares of the diagonal from Fig. 2(a). Similarly, Fig. 6(d) shows a row and column permutation of the figure obtained by removing all squares except the top square of the diagonal from Fig. 2(c). The following figure shows a minimal representation for a pair of pure type $n.5$, for any even $n > 2$, which cannot be obtained from Fig. 6(d) by row and column permutations and transposition.

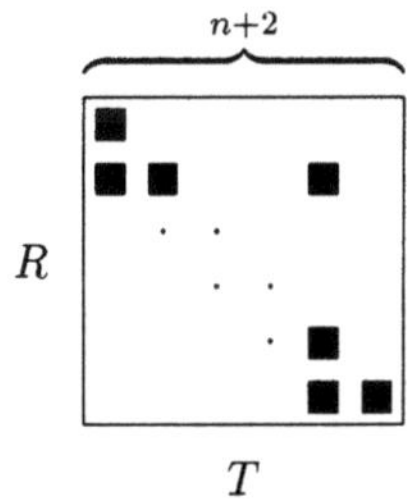

Fig. 8. An alternative to Fig. 6(d), for $n > 2$

5 Generalizing $*$ to n-ary operations

Let equivalence relations $R_1, \ldots, R_n, E$ on S be given. We shall then say that E is $(R_1, \ldots, R_n)$ *p-compatible*, for some $1 \leq p < n$, if and only if the following conditions are satisfied:

$$\forall \alpha, \alpha', \alpha'' \in S, \sigma \in \Sigma_n : \alpha E \alpha' R_{\sigma(j)} \alpha'' \ \ for\ all\ j \leq p \implies$$
$$\bigcap_{j \leq p} R_{\sigma(j)}(\alpha) \cap \bigcap_{j > p} R_{\sigma(j)}(\alpha'') \cap E(\alpha'') \neq \emptyset; \quad (6)$$
$$\bigwedge_{i \leq n} R_i \subseteq E. \quad (7)$$

Using Proposition 1 inductively, one obtains the expression

$$\bigvee_{i \leq n} R_i = \bigcup_{\substack{m \geq 1 \\ i_1, \dots, i_m \leq n}} R_{i_1} \cdots R_{i_m}. \quad (8)$$

We will now generalize Theorem 3 to obtain a structural characterization of those equivalence relations which are $(R_1, \dots, R_n)$ p-compatible. Let $C \in S_E$ be a class of E. For all $1 \leq i_1 < \cdots < i_m \leq n$ for $m \leq n$ define

$$R_{(i_1, \dots, i_m), C} = \{(\varrho_1, \dots, \varrho_m) \in \prod_{j \leq m} S_{R_{i_j}} : \varrho_1 \cap \cdots \cap \varrho_m \cap C \neq \emptyset\}.$$

Lemma 2. *Let E be $(R_1, \dots, R_n)$ p-compatible. Then, for all classes $C \in S_E$ and $1 \leq i_1 < \cdots < i_{p+1} \leq n$,*

$$R_{(i_1, \dots, i_{p+1}), C} = \prod_{j \leq p+1} R_{i_j, C}.$$

Proof. Let $(\varrho_1, \dots, \varrho_{p+1}) \in \prod_{j \leq p+1} R_{i_j, C}$. For each $k \leq p+1$ find an element $\alpha_k \in \varrho_k \cap C$. We now proceed by induction on k to find elements $\beta_k \in \varrho_1 \cap \cdots \cap \varrho_{k+1} \cap C$ for all $k \leq p$. For $k = 1$ set $\beta_0 = \alpha_1$. Assume then that we have found $\beta_{k-1} \in \varrho_1 \cap \cdots \cap \varrho_k \cap C$ for some $k \leq p$. Now, $\beta_{k-1} E \alpha_{k+1} R_{i_j} \alpha_{k+1}$ for all $j = 1, \dots, \widehat{k+1}, \dots, p+1$, so by (6) there exists an element β_k which is contained in

$$\bigcap_{\substack{j \leq p+1 \\ j \neq k+1}} R_{i_j}(\beta_{k-1}) \cap R_{i_{k+1}}(\alpha_{k+1}) \cap \bigcap_{j > p+1} R_{i_j}(\alpha_{k+1}) \cap C$$
$$= \varrho_1 \cap \cdots \cap \varrho_k \cap \bigcap_{\substack{j \leq p+1 \\ j > k+1}} R_{i_j}(\beta_{k-1}) \cap \varrho_{k+1} \cap \bigcap_{j > p+1} R_{i_j}(\alpha_{k+1}) \cap C$$
$$\subseteq \varrho_1 \cap \cdots \cap \varrho_{k+1} \cap C,$$

so $(\varrho_1, \dots, \varrho_{k+1}) \in R_{(i_1, \dots, i_{k+1}), C}$. □

This may be strengthened.

Lemma 3. *Let E be $(R_1, \dots, R_n)$ p-compatible. Then, for all classes $C \in S_E$,*

$$R_{(1, \dots, n), C} = \prod_{i \leq n} R_{i, C}.$$

This means that the classes of E are n-boxes, regardless of the value of p.

Proof. Let $(\varrho_1, \ldots, \varrho_n) \in \prod_{i \leq n} R_{i,C}$. As in the proof of Lemma 2, we proceed by induction on k. For $k = p + 1$, we may apply Lemma 2 to find an element $\alpha_{p+1} \in \varrho_1 \cap \cdots \cap \varrho_{p+1} \cap C$. Assume then that α_k is contained in $\varrho_1 \cap \cdots \cap \varrho_k \cap C$ for some $k \geq p+1$. By Lemma 2, there exists an element $\alpha \in \varrho_{k-p+1} \cap \cdots \cap \varrho_{k+1} \cap C$. Now, $\alpha E \alpha_k R_j \alpha_k$ for all $j = k - p + 1, \ldots, k + 1$, so by (6) there is an element α_{k+1} contained in

$$\bigcap_{j=k-p+1,\ldots,k+1} R_j(\alpha) \cap \bigcap_{\substack{j \leq k-p \\ j > k+1}} R_j(\alpha_k) \cap C$$
$$= \varrho_{k-p+1} \cap \cdots \cap \varrho_{k+1} \cap \varrho_1 \cap \cdots \cap \varrho_{k-p} \cap \bigcap_{j>k+1} R_j(\alpha_k) \cap C$$
$$\subseteq \varrho_1 \cap \cdots \cap \varrho_{k+1} \cap C,$$

so $(\varrho_1, \ldots, \varrho_{k+1}) \in R_{(1,\ldots,k+1),C}$. □

We may now state the following theorem.

Theorem 6. *Let E be an equivalence relation on S such that $\wedge R_i \subseteq E$. Then E is $(R_1, \ldots, R_n)$ p-compatible if and only if the following conditions hold:*

1. $\forall C \in S_E : R_{(1,\ldots,n),C} = \prod_{i \leq n} R_{i,C}$;
2. $\forall C, C' \in S_E, 1 \leq i_1 < \cdots < i_p \leq n : R_{(i_1,\ldots,i_p),C}, R_{(i_1,\ldots,i_p),C'}$ *are disjoint or identical.*

Proof. If E is $(R_1, \ldots, R_n)$ p-compatible, then Lemma 3 implies condition 1. The arguments in the rest of the proof are as in the proof of Theorem 3. □

This immediately implies

Corollary 3. *Let E be $(R_1, \ldots, R_n)$ p-compatible for some p. Then E is contained in $\bigcap_{\sigma \in \Sigma_n} R_{\sigma(1)} \cdots R_{\sigma(n)}$.*

Theorem 6 may be used to describe the families of $(R_1, \ldots, R_n)$ p-compatible equivalence relations. In order to obtain this description, we need the following

Lemma 4. *If E is $(R_1, \ldots, R_n)$ p-compatible, then E is $(R_1, \ldots, R_n)$ p'-compatible for all $p' \geq p$.*

Proof. Let E be $(R_1, \ldots, R_n)$ p-compatible and let $p' \geq p$ be given. Further, let $C, C' \in S_E$ and $1 \leq i_1 < \cdots < i_{p'} \leq n$ be given and assume that there exists an element $(\varrho_1, \ldots, \varrho_{p'}) \in R_{(i_1,\ldots,i_{p'}),C} \cap R_{(i_1,\ldots,i_{p'}),C'}$. Choose any $(\varrho'_1, \ldots, \varrho'_{p'}) \in R_{(i_1,\ldots,i_{p'}),C}$. Then, for all $1 \leq j_1 < \cdots < j_p \leq p'$ the sets $\varrho_{j_1} \cap \cdots \cap \varrho_{j_p} \cap C$, $\varrho_{j_1} \cap \cdots \cap \varrho_{j_p} \cap C'$ and $\varrho'_{j_1} \cap \cdots \cap \varrho'_{j_p} \cap C$ are non-empty, so, by 6.2, $\varrho'_{j_1} \cap \cdots \cap \varrho'_{j_p} \cap C'$ is also non-empty. Theorem 6 then implies that $\varrho'_1 \cap \cdots \cap \varrho'_{p'} \cap C'$ is non-empty, so $R_{(i_1,\ldots,i_{p'}),C} \subseteq R_{(i_1,\ldots,i_{p'}),C'}$. Similarly, $R_{(i_1,\ldots,i_{p'}),C'} \subseteq R_{(i_1,\ldots,i_{p'}),C}$, so, by Theorem 6, E is $(R_1, \ldots, R_n)$ p'-compatible. □

We now extend the notion of $(R_1, \ldots, R_n)$ p-compatibility to all values $p = 0, \ldots, n$. From (6) and (8), it is reasonable to say that E is $(R_1, \ldots, R_n)$ 0-compatible if and only if $E = \wedge R_i$, and $(R_1, \ldots, R_n)$ n-compatible if and only if E contains $\wedge R_i$ and is contained in $\vee R_i$. Now, for all $p = 0, \ldots, n$ define $\mathcal{F}_p$ to be the family of $(R_1, \ldots, R_n)$ p-compatible equivalence relations on S. Thus, $\mathcal{F}_0$ is the singleton set $\{\wedge R_i\}$, and $\mathcal{F}_n$ is the interval $[\wedge R_i; \vee R_i]$ in $\Pi[S]$. Using this notation, we may state the following nice proposition concerning the structure of the family $\{\mathcal{F}_p,\ p = 0, \ldots, n\}$.

Proposition 8. *The families $\mathcal{F}_0, \ldots, \mathcal{F}_n$ and the lattice of partitions of S, $\Pi[S]$, form a linear order of sublattices:*

$$\{\wedge R_i\} = \mathcal{F}_0 \subseteq \mathcal{F}_1 \subseteq \cdots \subseteq \mathcal{F}_n = [\wedge R_i; \vee R_i] \subseteq \Pi[S].$$

Proof. By Lemma 4, Corollary 3 and (8), $\mathcal{F}_p \subseteq \mathcal{F}_{p'}$ for all $0 \leq p < p' \leq n$. By the arguments in the proof of Proposition 3, we may prove that each $\mathcal{F}_p$ is a sublattice of $\Pi[S]$. To conclude the proof, we remark that, if $\mathcal{L}'$ and $\mathcal{L}''$ are sublattices of a lattice $\mathcal{L}$ and $\mathcal{L}' \subseteq \mathcal{L}''$, then $\mathcal{L}'$ is a sublattice of $\mathcal{L}''$. □

We may then generalize the $*$ operation as follows.

Definition 2. Let $R_1, \ldots, R_n$ be equivalence relations on S. For each $p = 0, \ldots, n$ let $T^p(R_1, \ldots, R_n)$ denote the largest $(R_1, \ldots, R_n)$ p-compatible relation on S.

By Proposition 8, $\wedge R_i = T^0(R_i) \leq T^1(R_i) \leq \cdots \leq T^n(R_i) = \vee R_i$. Figure 9 demonstrates that there need not be equality anywhere:

$$T^0(R, T, U) < T^1(R, T, U) < T^2(R, T, U) < T^3(R, T, U).$$

Fig. 9. **(a)** R,T and U **(b)** $T^1(R, T, U)$ **(c)** $T^2(R, T, U)$ **(d)** $T^3(R, T, U)$

Given certain equalities in the above equation, one might wish to investigate the properties of the relations $R_1, \ldots, R_n$. A simple case is described by

Proposition 9. *The following statements are equivalent:*

1. $T^n(R_1, \ldots, R_n) = T^p(R_1, \ldots, R_n)$ *for some* $0 < p < n$;
2. $T^n(R_1, \ldots, R_n) = T^p(R_1, \ldots, R_n)$ *for all* $0 < p \leq n$;
3. $\forall j, j'$: R_j *and* $R_{j'}$ *commute.*

Loosely speaking, the above statements describe the situations in which the classes of $\vee R_i$ are n-boxes, disjoint in R_i for all $i \leq n$.

Proof. $2 \Rightarrow 1$: Trivial.

$1 \Rightarrow 3$: Immediate from Theorem 6.

$3 \Rightarrow 2$: Assume that all pairs $R_j, R_{j'}$ commute. Since any permutation of n elements can be expressed as a composition of transpositions (j, j'), $j, j' = 1, \ldots, n$, the relations $R_{\sigma(1)} R_{\sigma(2)} \cdots R_{\sigma(n)}$ are equal for all $\sigma \in \Sigma_n$. Hence, (8) implies that R_i satisfies both condition 6.1 and condition 6.2 for all $p \leq n$, so we may conclude that $\vee R_i$ is maximally $(R_1, \ldots, R_n)$ p-compatible for all $p \geq 1$. □

As a special case of this result, we again obtain Theorem 2:

$$T^2(R, T) = T^1(R, T) \iff R \text{ and } T \text{ commute.}$$

References

[1] Birkhoff, G. (1967): Lattice theory. 3rd ed. (AMS Colloqium Publications, vol. 25). American Mathematical Society, Providence, RI

[2] Dubreil, P. (1950): Relations binaires et applications. C. R. Acad. Sci. Paris **230**, 1028–1030

[3] Dubreil, P., Dubreil-Jacotin, M.-L. (1939): Théorie algébrique des relations d'équivalence. J. Math. Pures Appl. (9) **18**, 63–95

[4] Finberg, D., Mainetti, M., Rota, G.-C. (1996): The logic of commuting equivalence relations. In: Ursini, A., Aglianò, P. (eds.) Logic and algebra (Pontignano, 1994). (Lecture Notes in Pure and Appl. Math. vol. 180). Dekker, New York, pp. 69–96

[5] Grätzer, G. (1978): General lattice theory. Birkhäuser, Basel

[6] Haiman, M. (1984): The theory of linear lattices. Ph.D. thesis, Massachusetts Institute of Technology

[7] Hutchinson, G. (1973): Recursively unsolvable word problems of modular lattices and diagram-chasing. J. Algebra **26**, 385–399

[8] Jónsson, B. (1953): On the representation of lattices. Math. Scand. **1**, 193–206

[9] Ore, Ø. (1942): Theory of equivalence relations. Duke Math. J. **9**, 573–627

Solving linear recurrences using functionals

A. Di Bucchianico, G. Soto y Koelemeijer

1 Introduction

Many counting problems (especially those related to lattice path counting) give rise to recurrence relations. Several techniques are available for solving recurrence relations (see, e.g., [3]). A technique not mentioned in [3] is the umbral calculus. The main contribution in this respect is due to Niederhausen [4–10]. Niederhausen finds polynomial solutions to recurrence relations. In fact, if we have a recurrence relation in two variables, then his approach interprets one variable as the index of a polynomial sequence and the other variable as the argument of a polynomial. Moreover, he assumes that the solutions can be represented as so-called Sheffer sequences (for definitions, see Sect. 2). This assumption is true for a large class of recurrences. The umbral calculus provides several explicit formulas for calculations with these Sheffer sequences. These formulas are expressed in terms of linear operators on the vector space of polynomials. The recurrence relation is then translated into an equation involving the so-called delta operator of the Sheffer sequence. If this equation can be solved explicitly, then the basic sequence can be calculated by the transfer formula from umbral calculus (which, in some sense, is equivalent to Lagrange inversion). The deep part of the work of Niederhausen is that he has extended this basic scheme to allow for certain classes of initial values (see references mentioned above). Related to the work of Niederhausen are two papers by Watanabe [18, 19]. In particular, [18] contains a multivariate extension of the Niederhausen approach in case of the ballot problem.

Other umbral calculus based approaches to solving linear recurrences can be found in a paper by Roman et al. [16] ((see also [15, Sect. 5.6]) and two papers by Razpet [12, 13]). The Roman paper addresses a generalization of the well-known three-term recurrence relation satisfied by orthogonal polynomials. The solution is given without any explanation on how it is obtained. Verification is by direct computation using standard umbral calculus techniques. Although the paper is written in terms of functionals, it is easily seen that his method is essentially equivalent to the Niederhausen approach (cf. [1, Sect. 3.7] and [2]). The approach in the Razpet papers, however, are completely different. Not only is the subject different (recurrence relations based on lattice path counting), the approach is also different. The solutions in [12] are found by considering generic solutions that are of the form $\frac{1}{i!}\left\langle (1+t)^j \mid s_i \right\rangle$, where $(s_n)_{n\in\mathbb{N}}$ is a Sheffer sequence. In [13] this pattern is extended to one dimension higher in order to solve a linear recursion with three arguments. Unfortunately, Razpet gives no explanation for the several special choices he makes, nor does he relate his approach to that of Niederhausen.

In this paper we explain the true nature of the Razpet approach. We provide a general framework for his approach that includes the Niederhausen framework. The generic form of our solutions is $\frac{1}{i!}\langle h^j / s_i \rangle$, where h is a formal power series with $h(0) = 1$ and $h'(0) \neq 0$ and $(s_n)_{n \in \mathbb{N}}$ is a Sheffer sequence. The main advantage of our method is that it allows to postpone the actual choice of h and the delta operator f of the Sheffer sequence as long as possible. I.e., the redundancy in the generic form may be exploited to incorporate insights that can only be seen while solving the recurrence relation at hand. It must be noted, however, that our results are restricted to the general form of solutions. The next step (which is beyond the scope of this paper) is to include the deep results of Niederhausen for handling various types of boundary condition and initial values. Our approach is illustrated by solving recurrences explicitly. We also show that several calculations of Razpet can be simplified by using basic theorems from umbral calculus.

This paper is partially based on [17].

2 Prerequisites

In this section we review basic results of umbral calculus that we need in this paper. For proofs we refer to [15]. This section (except perhaps for Theorem 11) may be skipped by readers familiar with the umbral calculus in terms of functionals.

Let $\mathcal{F}$ denote the algebra of formal power series in the variable t over the field $\mathbb{C}$. Thus an element of $\mathcal{F}$ has the form

$$f(t) = \sum_{k=0}^{\infty} a_k t^k, \qquad a_k \in \mathbb{C}. \tag{1}$$

Addition and multiplication are defined in the usual way. If $a_0 \neq 0$, then f has a multiplicative inverse, denoted by f^{-1}. Such power series are called **invertible series**. If $a_0 = 0$ and $a_1 \neq 0$, then f will be called a **delta series**. A delta series $f(t)$ has a compositional inverse, denoted by $\overline{f}(t)$, such that $\overline{f}(f(t)) = f(\overline{f}(t)) = t$.

The formal derivative of the series (1) is defined by

$$D_t f(t) = \sum_{k=1}^{\infty} k\, a_k t^{k-1}. \tag{2}$$

We denote by $\mathcal{P}$ the vector space of polynomials in one variable with complex coefficients. A **polynomial sequence** is a sequence in $\mathcal{P}$ such that $\deg(p_n) = n$ for all $n \in \mathbb{N}$. Let $\mathcal{P}^*$ be the vector space of all linear functionals on $\mathcal{P}$. The action of the functional $L \in \mathcal{P}^*$ on the polynomial $p \in \mathcal{P}$ will be denoted by

$$\langle L \mid p \rangle . \tag{3}$$

Each formal power series

$$f(t) = \sum_{k=0}^{\infty} \frac{c_k}{k!} t^k \tag{4}$$

defines a linear functional on $\mathcal{P}$ if we set

$$\langle f \mid x^n \rangle = c_n \qquad \text{for } n \geq 0. \tag{5}$$

In particular,

$$\left\langle t^k \mid x^n \right\rangle = n!\, \delta_{n,k}. \tag{6}$$

It is easy to show that any linear functional $L \in \mathcal{P}^*$ can be represented in this way. It is clear that for all $f \in \mathcal{F}$

$$f(t) = \sum_{k=0}^{\infty} \frac{\langle f \mid x^k \rangle}{k!} t^k. \tag{7}$$

Theorem 1. *If f and g are in $\mathcal{F}$, then*

$$\langle f\, g \mid x^n \rangle = \sum_{k=0}^{n} \binom{n}{k} \left\langle f \mid x^k \right\rangle \left\langle g \mid x^{n-k} \right\rangle.$$

Theorem 2. *If f is in $\mathcal{F}$, then, for all $p \in \mathcal{P}$,*

$$\langle f \mid x\, p(x) \rangle = \langle D_t\, f(t) \mid p \rangle\,.$$

We now give some examples.

Example 1. 1. The **evaluation functional** is the invertible functional e^{yt} and

$$\left\langle e^{yt} \mid p \right\rangle = p(y).$$

2. The **Abel functional** is the delta functional te^{yt} and

$$\left\langle te^{yt} \mid p \right\rangle = p'(y).$$

Remark 1. It is clear that for any polynomial p

$$\left\langle t^k \mid p \right\rangle = p^{(k)}(0), \tag{8}$$

where $p^{(k)}$ is the kth derivative.

Taking the above remark further, we may also associate operators with formal power series in the following way. Each power series of the form (4) defines a linear operator on $\mathcal{P}$ if we set

$$f\, x^n = \sum_{k=0}^{n} \binom{n}{k} c_k\, x^{n-k} \text{ for } n \geq 0. \tag{9}$$

The next theorem links the two possible actions of power series.

Theorem 3. *If f and g are power series and $p \in \mathcal{P}$, then*

$$\langle f\, g \mid p \rangle = \langle g \mid f\, p \rangle .$$

We now are ready to introduce the important concept of Sheffer sequences.

Theorem 4. *Let f be a delta series and let g be an invertible series. Then there exists a unique sequence $(s_n)_{n \in \mathbb{N}}$ in $\mathcal{P}$ satisfying the orthogonality conditions*

$$\left\langle g\, f^k \mid s_n \right\rangle = n!\, \delta_{n,k}, \tag{10}$$

for all $n, k \geq 0$.

Definition 1. The sequence $(s_n)_{n \in \mathbb{N}}$ in Theorem 4 is called the **Sheffer sequence** for the ordered pair (g, f).

Definition 2. The Sheffer sequence for the ordered pair $(1, f)$ is called the **associated sequence** (or basic sequence) for f.

Theorem 5. *A polynomial sequence $(p_n)_{n \in \mathbb{N}}$ is an associated sequence if and only if, for all $n \in \mathbb{N}$,*

$$p_n(x + y) = \sum_{k=0}^{n} \binom{n}{k} p_k(x)\, p_{n-k}(y). \tag{11}$$

Polynomial sequences satisfying (11) are called **sequences of binomial type**. If $(p_n)_{n \in \mathbb{N}}$ is of binomial type, then $(p_n/n!)_{n \in \mathbb{N}}$ satisfies (11) with the binomial coefficient deleted. Such sequences are called polynomials of convolution type in [1]. As a consequence, the literature on umbral calculus has two slightly different notions of Sheffer sequences, that differ by a factor $n!$. This also affects all formulas in this section. In this paper we strictly follow the conventions adopted in [15].

Theorem 6. *A polynomial sequence $(s_n)_{n \in \mathbb{N}}$ is a Sheffer sequence if and only if there exists an associated sequence $(p_n)_{n \in \mathbb{N}}$ such that, for all $n \in \mathbb{N}$,*

$$s_n(x + y) = \sum_{k=0}^{n} \binom{n}{k} s_k(x)\, p_{n-k}(y).$$

Corollary 1. *A polynomial sequence $(s_n)_{n \in \mathbb{N}}$ is Sheffer if and only if there exist a sequence c_n of constants with $c_0 \neq 0$ and an associated sequence $(p_n)_{n \in \mathbb{N}}$ such that*

$$s_n = \sum_{k=0}^{n} \binom{n}{k} c_{n-k}\, p_k .$$

Yet another useful characterization of Sheffer sequences is given by the following theorem.

Theorem 7. *A polynomial sequence* $(s_n)_{n\in\mathbb{N}}$ *is a Sheffer sequence if and only if there exists a delta series* f *such that*

$$f\, s_n = n\, s_{n-1}$$

for $n \geq 1$. *Moreover, there exists an invertible series* g *such that* $(g\, s_n)_{n\in\mathbb{N}}$ *is the associated sequence of* f.

The so-called transfer formulas yield explicit expressions for the associated sequence of a delta series f. We only need the following form.

Theorem 8 (Transfer formula). *Let* $(p_n)_{n\in\mathbb{N}}$ *be the associated sequence of a delta series* f. *Then*

$$p_n(x) = x \left(\frac{f(t)}{t}\right)^{-n} x^{n-1}. \qquad (12)$$

The following theorem shows how to expand functionals. This theorem is a powerful extension of Taylor's formula.

Theorem 9 (Functional expansion theorem). *Let* $(s_n)_{n\in\mathbb{N}}$ *be the Sheffer sequence for* (g, f). *For any* $h \in \mathcal{F}$ *we have*

$$h = \sum_{k=0}^{\infty} \frac{\langle h \mid s_k \rangle}{k!}\, g\, f^k.$$

The next theorem shows the structure of the generating function of Sheffer sequences.

Theorem 10. 1. *Let* $(p_n)_{n\in\mathbb{N}}$ *be the associated sequence for the delta series* f. *The generating function of* $(p_n)_{n\in\mathbb{N}}$ *is*

$$e^{y\overline{f}(t)} = \sum_{k=0}^{\infty} p_k(y)\, \frac{t^k}{k!},$$

where $\overline{f}$ *is the compositional inverse of* f.

2. *Let* $(s_n)_{n\in\mathbb{N}}$ *be the Sheffer sequence for* (g, f). *The generating function of* $(s_n)_{n\in\mathbb{N}}$ *is*

$$\frac{1}{g(\overline{f}(t))}\, e^{y\overline{f}(t)} = \sum_{k=0}^{\infty} s_k(y)\, \frac{t^k}{k!}.$$

It follows by expansion of the generating function of Theorem 10 that each associated sequence (and hence, by Theorem 5, also each sequence of binomial type) may be written in the form $n! \sum_{k=0}^{n} c_n^{k*} \frac{x^k}{k!}$, where c_n^{k*} denotes the nth element of the k-fold convolution of the sequence $(c_n)_{n\in\mathbb{N}}$ of coefficients. In fact, a stronger result is true.

Theorem 11 ([1, Theorems 2.3.10 and 2.4.12]).

a) *Let* $(p_n)_{n\in\mathbb{N}}$ *be a sequence of polynomials of binomial type. A polynomial sequence* $(q_n)_{n\in\mathbb{N}}$ *is of binomial type if and only if there exists a sequence of constants* $(c_n)_{n\in\mathbb{N}}$ *with* $c_0 = 0$ *and* $c_1 \neq 0$ *such that*

$$q_n = n! \sum_{k=0}^{n} c_n^{k*} \frac{p_k}{k!}.$$

b) *Let* $(s_n)_{n\in\mathbb{N}}$ *be a Sheffer sequence. A polynomial sequence* $(r_n)_{n\in\mathbb{N}}$ *is a Sheffer sequence if and only if there exists a sequence of constants* $(c_n)_{n\in\mathbb{N}}$ *with* $c_0 = 0$ *and* $c_1 \neq 0$ *such that*

$$r_n = n! \sum_{k=0}^{n} c_n^{k*} \frac{s_k}{k!}.$$

3 Main results

In this section we explain the basic mechanism behind the Razpet approach to solving recurrences. We start by studying properties of the generic form of the solutions, which are of the form

$$\frac{1}{i!}\left\langle h^j \mid s_i \right\rangle,$$

where h is a power series with $h(0) = 1$ and $h'(0) \neq 0$ and $(s_n)_{n\in\mathbb{N}}$ is a Sheffer sequence. After that we illustrate the general form of the Razpet approach by solving the recurrence relation

$$r(i, j) = r(i, j-1) + a\, r(i-1, j) + b\, r(i-1, j+1)$$

with boundary conditions $r(i, -1) = 0$ for $i \geq 1$ and $r(0, j) = 1$ for $j \geq -1$. This recurrence occurs, e.g., when one counts lattice paths from $(0, 0)$ to (i, j) with admissible steps $(0, 1)$, $(1, 0)$, and $(1, 1)$, where the latter two steps receive weights a and b, respectively. As an extra advantage of our generalization, we find several simplifications of the original computations by Razpet in [12]. We obtain these simplifications by applying the functional expansion theorem (Theorem 9). The same recurrence relation with other boundary conditions has been studied in detail in [4].

The next two theorems show that, like Niederhausen, we may assume that the solution of the recurrence relation can be represented as a Sheffer polynomial. However, our candidate solutions are represented in a redundant way, which we will exploit later on in order to simplify calculations.

In the next lemma real powers of formal power series occur. We refer to [11] for details.

Lemma 1. *Let* h *be a formal power series with* $h(0) = 1$ *and* $h'(0) \neq 0$. *Then* $(\langle h^x \mid t^n \rangle)_{n\in\mathbb{N}}$ *is a sequence of polynomials of binomial type.*

Proof. First note that $\langle h^x | t^0 \rangle = h(0)^x = 1$. By Theorem 1, we have

$$\left\langle h^{x+y} \,\middle|\, t^n \right\rangle = \sum_{k=0}^{n} \binom{n}{k} \left\langle h^x \,\middle|\, t^k \right\rangle \left\langle h^y \,\middle|\, t^{n-k} \right\rangle.$$

Since h has a nonzero linear term $-it$ follows that $\langle h^x \mid t^n \rangle$ is a polynomial in x of degree exactly n. Hence, $(\langle h^x \mid t^n \rangle)_{n\in\mathbb{N}}$ is a sequence of polynomials of binomial type. □

Theorem 12. *Let h be a formal power series with $h(0) = 1$ and $h'(0) \neq 0$ and let $(p_n)_{n\in\mathbb{N}}$ be a sequence of polynomials of binomial type. Then the sequence $(\langle h^x \mid p_n \rangle)_{n\in\mathbb{N}}$ is of binomial type.*

Proof. First note that $\langle h^x \mid p_0 \rangle = \langle h^x \mid 1 \rangle = h(0)^x = 1$. It follows from Theorem 11a) that there exists a sequence of constants $(c_n)_{n\in\mathbb{N}}$ with $c_0 = 0$ and $c_1 \neq 0$ such that $p_n(t) = n! \sum_{k=0}^{n} c_n^{k*} \frac{t^n}{n!}$. Hence,

$$\left\langle h^x \mid p_n \right\rangle = \left\langle h^x \,\middle|\, n! \sum_{k=0}^{n} c_n^{k*} \frac{t^k}{k!} \right\rangle = n! \sum_{k=0}^{n} c_n^{k*} \frac{\left\langle h^x \mid t^k \right\rangle}{k!}.$$

Since $c_0 = 0$, we have that $c_n^{n*} = c_1^n \neq 0$. Thus $\langle h^x \mid p_n \rangle$ is a polynomial of degree n exactly. Combining Lemma 1 with Theorem 11a) yields that $(\langle h^x \mid p_n \rangle)_{n\in\mathbb{N}}$ is of binomial type. □

Theorem 13. *Let h be a formal power series with $h(0) = 1$ and $h'(0) \neq 0$ and let $(s_n)_{n\in\mathbb{N}}$ be a Sheffer sequence of polynomials. Then the sequence $(\langle h^x \mid s_n \rangle)_{n\in\mathbb{N}}$ is a Sheffer sequence.*

Proof. It follows from Corollary 1 that $(s_n)_{n\in\mathbb{N}}$ is a Sheffer sequence if and only if there exists a sequence of constants $(c_n)_{n\in\mathbb{N}}$ with $c_0 \neq 0$ and a sequence of binomial type $(p_n)_{n\in\mathbb{N}}$ such that $s_n = \sum_{k=0}^{n} \binom{n}{k} c_k\, p_{n-k}$. Hence,

$$\left\langle h^x \mid s_n \right\rangle = \left\langle h^x \,\middle|\, \sum_{k=0}^{n} \binom{n}{k} c_k\, p_{n-k} \right\rangle = \sum_{k=0}^{n} \binom{n}{k} c_k \left\langle h^x \mid p_{n-k} \right\rangle.$$

Combining this with Theorem 12, we obtain that $(\langle h^x \mid s_n \rangle)_{n\in\mathbb{N}}$ is a Sheffer sequence. □

Remark 2. Alternative proofs for Theorems 12 and 13 are possible by using the concept of umbral operator (cf. [15, Sect. 3.4]). In fact, Lemma 1 implies that the linear operator that sends x^n to $\langle h^x \mid t^n \rangle$ is an umbral operator. Theorem 12 is then a consequence of the fact that the image under an umbral operator of a sequence of binomial type is of binomial type. In turn, Theorem 13 follows by combining this use of umbral operators with the concept of Sheffer operator (cf. [15, Sect. 3.5]).

We now apply the above results to solve the recurrence relation

$$r(i,j) = r(i,j-1) + a\,r(i-1,j) + b\,r(i-1,j+1) \tag{13}$$

with boundary conditions $r(i,-1) = 0$ for $i \geq 1$ and $r(0,j) = 1$ for $j \geq -1$.

The first step is to substitute the generic solution $r(i,j) = \frac{1}{i!}\left\langle h^j \mid s_i \right\rangle$, where h is a formal power series with $h(0) = 1$ and $h'(0) \neq 0$ and $(s_n)_{n\in\mathbb{N}}$ is the Sheffer sequence for (g, f). In order to obtain information from this substitution, we must apply certain rules so that we have s_i on the right-hand side of each bracket. Conversely, these rules indicate which kind of recurrences can be handled by the method proposed in this paper.

In the following we fix i and j (but note that the roles of i and j may be interchanged by relabeling). The basic substitution rules are:

1. $a\,r(i,j) \to \frac{1}{i!}\left\langle a\,h^j \mid s_i \right\rangle$. This is just linearity of the functional h^j.
2. $r(i-k,j) \to \frac{1}{i!}\left\langle h^j f^k \mid s_i \right\rangle$. Thus shifts in the "$i$" variable are interpreted as actions of the delta operator of the Sheffer sequence (cf. Theorem 7).
3. $j\,r(i,j) \to \frac{1}{i!}\left\langle \frac{h}{h'} D_t h^j \mid s_i \right\rangle$. Thus multiplication by the "j" variable is taken care of by differentiating h.

The last substitution rule depends on the choice of umbral calculus, i.e., the choice of the constant $k!$ in (4). Every choice of constant yields a different umbral calculus with a different "differentiation operator" (e.g., in this way it is possible to do q-analysis). See [15, Chap. 6] for more details.

It is interesting to compare our substitutions with the Niederhausen approach. Niederhausen interprets $r(i,j)$ as $\frac{1}{i!}\,s_i(j)$, where $(s_n)_{n\in\mathbb{N}}$ is Sheffer. Niederhausen needs an extra substitution rule $r(i,j-\ell) = \left(f\,E^\ell\,s_i\right)(j)$, where E denotes the shift operator. In our terminology, if $h = e^t$, then $\frac{i}{i!}(E^{-\ell} s_i(j) = \left\langle h^{j-\ell} \mid s_i \right\rangle$.

In the case of recurrence (13), the substitutions yield for all $i \in \mathbb{N}$ that

$$\left\langle h^j \mid s_i \right\rangle = \left\langle h^{j-1} \mid s_i \right\rangle + \left\langle a\,f\,h^j \mid s_i \right\rangle + \left\langle b\,f\,h^{j+1} \mid s_i \right\rangle.$$

Since $(s_n)_{n\in\mathbb{N}}$ is a basis for $\mathcal{P}$ and functionals are determined by their action on a basis, we obtain the relation

$$h^j = h^{j-1} + a\,f\,h^j + b\,f\,h^{j+1}. \tag{14}$$

If $a + b\,h(0) \neq 0$, then, since h is invertible, (14) simplifies to

$$f = \frac{h-1}{h\,(a+b\,h)} \tag{15}$$

For more complex recurrences, the relation between f and h may be more complicated. Lagrange inversion may be used with good success in such cases (see, e.g., [7]).

Note that our proposed solution $\left\langle h^j \mid s_i \right\rangle$ seemingly has three parameters: f, g, and h. However, (14) shows that h determines f uniquely. Conversely, given f, any h that satisfies (14) yields a solution to (13). The next lemma shows that there is an additional implicit restriction to admissible power series h.

Lemma 2. *Let h be an invertible series and let f be as in (14). Assume that $a + bh(0) \neq 0$. Then f is a delta series if and only if $h(0) = 1$ and $h'(0) \neq 0$.*

Proof. Write (15) in the form $f = (h-1)\,h^{-1}\,(a+bh)^{-1}$. Since both h and $a+b\,h$ are invertible, we conclude that $f(0) = 0$ if and only if $h(0) = 1$. Moreover, the chain rule yields that $f'(0) \neq 0$ if and only if $h'(0) \neq 0$. This proves the results, since f is a delta series if and only if $f(0) = 0$ and $f'(0) \neq 0$. □

The boundary conditions of (13) are rather special. The next lemma shows that the "horizontal" boundary conditions ($r(i,-1) = 0$ for $i = 1, 2, \ldots$ and $r(0,-1) = 1$) and the invertible series g of the Sheffer sequence determine each other uniquely.

Lemma 3. *Let h be a formal power series with $h(0) = 1$ and $h'(0) \neq 0$, and $(s_n)_{n\in\mathbb{N}}$ be the Sheffer sequence for (f, g). Let $r(i,j) = \frac{1}{i!}\left\langle h^j \mid s_i \right\rangle$ be a solution to (13) with boundary conditions $r(i,\alpha) = c_i$ for $i = 0, 1, \ldots$ for some fixed $\alpha \geq -1$. Then the invertible series associated to $(s_n)_{n\in\mathbb{N}}$ satisfies*

$$g = \frac{h^\alpha}{\sum_{i=0}^\infty c_i\, f^i}.$$

Proof. "⇒" First note that $c_0 = r(0,\alpha) = 1$. By the functional expansion theorem (Theorem 9), we have $h^\alpha = \sum_{i=0}^\infty \frac{\langle h^\alpha \mid s_i\rangle}{i!}\, g\, f^i$, which immediately yields the desired result by simple rearrangements.

"⇐" By Theorems 3 and 7, we have

$$\frac{1}{k!}\left\langle h^\alpha \mid s_k \right\rangle = \frac{1}{k!}\left\langle h^\alpha \,\middle|\, g^{-1} p_k \right\rangle = \frac{1}{k!}\left\langle h^\alpha g^{-1} \mid p_k \right\rangle = \frac{1}{k!}\left\langle \sum_{i=0}^\infty c_i\, f^i \,\middle|\, p_k \right\rangle = c_k.$$

This completes the proof. □

The other boundary conditions are satisfied automatically.

Lemma 4. *Let h be an invertible formal power series with $h(0) = 1$ and $h'(0) \neq 0$, and let $(s_n)_{n\in\mathbb{N}}$ be the Sheffer sequence for (g, f). If $r(i,j) = \frac{1}{i!}\left\langle h^j \mid s_i \right\rangle$, then $r(0,j) = 1/g_0$ for all j.*

Proof. It follows from Theorems 3 and 7 that $r(0,j) = \left\langle h^j \mid s_0 \right\rangle = \left\langle hg^{-1} \mid 1 \right\rangle = \left\langle g^{-1} \mid 1 \right\rangle = 1/g_0$. □

Corollary 2. *Let h be a formal power series with $h(0) = 1$ and $h'(0) \neq 0$ and let $(s_n)_{n\in\mathbb{N}}$ be the Sheffer sequence for (g, f). If $r(i,j) = \frac{1}{i!}\left\langle h^j \mid s_i \right\rangle$ satisfies (13), then $r(0,j) = 1$ for all j.*

Proof. By Lemma 4, we only need to prove that $g_0 = 1$. However, this follows immediately from Lemma 3, since $h(0) = 1$ and $c_0 = r(0,-1) = 1$. □

At this point it is important to note that our strategy is to postpone the actual choice of f and of h as long as possible. The reason is that, depending on the actual recurrence, we may obtain simplifications of calculations at the final stage of the computations. This is impossible to see when starting to solve a recurrence.

Now that we have translated all boundary conditions into conditions on the power series involved, the time has come to do the real calculations. We need to perform two calculations:

- explicit calculation of $(s_n)_{n\in\mathbb{N}}$,
- calculate the action of powers of h on $(s_n)_{n\in\mathbb{N}}$ to find $r(i, j)$.

We perform the first calculation by applying the transfer formula (Theorem 8). The second calculation follows from the defining property (5). It seems sensible to try some standard choices for f or h and see whether the associated h or f is manageable. It is here that we see the main advantage of our approach. In the case at hand (i.e., (14)), it is not advisable to fix f and then determine h, since there is a quadratic term in h. Rather we should try to find an h such that the corresponding f has the property that positive powers of $t/f(t)$ are easy to calculate. Therefore we wish to have a simple numerator in (15), which prompts us to choose $h - 1 = t$ as a sensible choice. In view of (15), we make the extra assumption $a \neq -b$. Hence, our delta series becomes

$$f = \frac{t}{(1+t)\,((a+b)+bt)}.$$

It follows from Lemma 3 with $\alpha = -1$ and $c_i = 0$ for $i \geq 1$ that $g = h^{-1}$. The transfer formula (Theorem 8), together with Theorem 7, yields that the associated sequence of f equals

$$p_n(x) = x\ ((1+t)\,(a+b(1+t)))^n\ x^{n-1},$$

and since $g = h^{-1}$, the Sheffer sequence for (g, f) equals

$$s_n = (1+t)\,p_n.$$

Finally, the solution to (13) is

$$r(i, j) = \frac{1}{i!}\left\langle (1+t)^{j+1} \,\middle|\, x\ ((1+t)\,(a+b(1+t)))^i\ x^{n-1} \right\rangle,$$

which, by Theorem 2, can be rewritten as

$$r(i, j) = \frac{j+1}{i!}\left\langle (1+t)^{i+j}\ ((a+b)+bt)^i \,\middle|\, x^{i-1} \right\rangle.$$

The binomial formula together with (6) finally yield

$$r(i, j) = \frac{j+1}{i}\sum_{k=0}^{i-1}\binom{i+j}{k}\binom{i}{k+1}(a+b)^{k+1}\,b^{i-1-k}.$$

It is very well possible that, by applying different choices for h, we may obtain new identities by equating the solutions $\frac{1}{i!}\left\langle h^j \mid s_i \right\rangle$.

It is also possible to solve linear recurrences of the form (13) where the coefficients a and b are power series interpreted as linear operators (cf. [7]).

4 Multivariate results

In this section we briefly indicate how to generalize the results of the previous sections to higher dimensions. Several multivariate settings for the umbral calculus exist. We follow the approach of [14]. In fact, [14] describes a much more general setting than we need. However, the restricted setting in Sect. 8 of that paper suffices for us. The setting in that section is the algebra of formal power series in n variables and the vector space of polynomials in n variables. Briefly speaking, a delta set is an n-tuple of formal power series in n variables with additional conditions to allow a compositional inverse. In order to obtain analogues of the important transfer formula, we need a further restriction to diagonal delta sets. This is no problem for us since we only need simple choices for the power series involved. The generic form of a solution for a recurrence in $2n$ variables is

$$r(i_1, \ldots, i_n, j_1, \ldots, j_n) = \frac{1}{i_1! \ldots i_n!} \left\langle h_1^{j_1} \ldots h_n^{j_n} \mid s_{i_1,\ldots,i_n}(t_1, \ldots, t_n) \right\rangle . \tag{16}$$

Here we assume that the multivariate Sheffer sequence is such that it spans the vector space of polynomials in n variables (Roman calls this a strong sequence) and the corresponding delta set is diagonal. The proofs of Sect. 3 can be adapted without too much difficulty to this multivariate setting by using Remark 2 and the results of [14]. Note that in the generic form (16) there is much freedom in partitioning the variables.

In case there are an odd number of variables in the recursion, we may apply the procedure followed in [13]. There a generating function is defined with respect to one of the variables. This yields a recurrence relation for the generating function. Solving this generating function while treating the dummy variable of the generating function as a parameter, we obtain a recurrence in one variable less. Hence, this is a recurrence in an even number of variables.

5 Final remarks

We provided a general setting for the Razpet approach to solving linear recurrences using umbral calculus. The Niederhausen approach appears as a special case. The advantage of our general setting is the flexibility of choosing the parameters (f and h) of the solution. We also showed how to simplify calculations using the functional expansion theorem. Finally, we briefly mentioned how to adapt our approach to deal with recurrences in more than two variables.

As said before, Niederhausen has developed several powerful ways to handle more complicated boundary conditions and initial values than the simple boundary conditions that we discussed. It would therefore be of interest to extend his results to our setting.

Acknowledgements. We are grateful to Katherine Humphreys for several useful remarks that improved an earlier version of this paper.

References

[1] Di Bucchianico, A. (1997): Probabilistic and analytical aspects of the umbral calculus. (CWI Tract., vol. 119). Centrum voor Wiskunde en Informatica, Amsterdam

[2] Di Bucchianico, A., Loeb, D.E. (1998): Natural exponential families and umbral calculus. In: Sagan, B., Stanley, R.P. (eds.) Mathematical Essays in Honor of Gian-Carlo Rota. Birkhäuser Boston, Boston, MA, pp. 43–60

[3] Mohanty, S.G. (1979): Lattice path counting and applications. Academic Press, New York

[4] Niederhausen, H. (1979): Lattice paths with three step directions. Congr. Numer. **24**, 753–774

[5] Niederhausen, H. (1980): Linear recurrences under side conditions. European J. Combin. **1**, 353–368

[6] Niederhausen, H. (1980): Sheffer polynomials and linear recurrences. Congr. Numer. **29**, 689–698

[7] Niederhausen, H. (1985): A formula for explicit solutions of certain linear recursions on polynomial sequences. Congr. Numer. **49**, 87–98

[8] Niederhausen, H. (1996): Symmetric Sheffer sequences and their applications to lattice path counting. J. Statist. Plann. Inference **54**, 87–100

[9] Niederhausen, H. (1997); Lattice path enumeration and umbral calculus. In: Balakrishnan, N. (ed.) Advances in Combinatorial Methods and Applications to Probability and Statistics. Birkhäuser Boston, Boston, MA, pp. 87–100

[10] Niederhausen, H. (1999): Recursive initial value problems for Sheffer sequences. Discrete Math. **204**, 319–327

[11] Niven, I. (1969): Formal power series. Amer. Math. Monthly **76**, 871–889

[12] Razpet, M. (1990): An application of the umbral calculus. J. Math. Anal. Appl. **149**, 1–16

[13] Razpet, M. (1990): A new class of polynomials with applications. J. Math. Anal. Appl. **150**, 85–99

[14] Roman, S.M. (1979): The algebra of formal series. III. Several variables. J. Approx. Theory **26**, 340–381

[15] Roman, S.M. (1984): The umbral calculus. Academic Press, New York

[16] Roman, S.M., De Land, P.N., Shiflett, R.C., Schultz, H.S. (1983): The umbral calculus and the solution to certain recurrence relations. J. Combin. Inform. System Sci. **8**, 235–240

[17] Soto y Koelemeijer, G. (1998): Recurrences and umbral calculus. Master's thesis, Delft University of Technology

[18] Watanabe, T. (1986): On a generalization of polynomials in the ballot problem. J. Statist. Plann. Inference **14**, 143–152

[19] Watanabe, T. (1987): On a determinant sequence in the lattice path counting. J. Math. Anal. Appl. **123**, 401–414

Polynomiality of the q, t-Kostka revisited

A. M. Garsia*, M. Zabrocki

1 Introduction

The polynomiality problem for the q, t-Kostka coefficients [11] was posed by Macdonald in the fall 1988 meeting of the Lotharingian seminar. It remained open for quite a few years, when suddenly, in 1996, several proofs of varied difficulty appeared in a period of only a few months. At the present there are three basically different approaches to proving the polynomiality of the q, t-Kostka coefficients:

(1) via plethystic formulas (Garsia-Tesler [4], Garsia-Remmel [3]);
(2) via vanishing properties (Sahi [13], Knop [7, 8]);
(3) via Rodrigues formulas (Lapointe–Vinet [10], Kirillov–Noumi [6]).

Each of these approaches has its own special advantages. The plethystic approach led to very efficient algorithms for computing these coefficients and ultimately produced remarkably simple explicit formulas [2]. The vanishing properties approach led to the discovery of basic non-symmetric variants of the Macdonald polynomials with remarkable combinatorial implications that still remain to be fully explored. The approach via Rodrigues formulas stems from a pioneering paper of Lapointe–Vinet [9] on Jack polynomials. Although originally it was based on deep affine Hecke algebra identities, eventually the idea led to some of the most elementary proofs of the polynomiality result (see [10] and [6]). In particular it produced a family of symmetric function operators $\{B_k^{q,t}\}_{k=1,2,\ldots}$ which permitted the construction of the Macdonald "integral forms" $J_\mu(x; q, t)$, one part at a time, starting from 1, according to an identity of the form

$$J_{\mu'}(x; q, t) = B_{\mu_1}^{q,t} B_{\mu_2}^{q,t} \cdots B_{\mu_k}^{q,t} \mathbf{1}.$$

Our main contribution here is a remarkably simple argument which shows that families of operators $B_k^{q,t}$, yielding such a formula, may be obtained by q-twisting in a minor way *any* sequence of operators $\{B_k^t\}_{k=1,2,\ldots}$ which yields the analogous formula

$$Q_{\mu'}(x; t) = B_{\mu_1}^t B_{\mu_2}^t \cdots B_{\mu_k}^t \mathbf{1}$$

for the Hall–Littlewood polynomial $Q_\mu(x; q, t)$. As a byproduct we obtain that the polynomiality of the q, t-Kostka is an immediate consequence of the polynomiality of the the Kostka–Foulkes coefficients. What is surprising is that this fact was missed for so many years by researchers in this area. What might be even more surprising is that we obtain a remarkably general result by further simplifying some of the

* Supported by NSF

arguments used in [10] and [6]. To give a more precise description of our results we need some notation.

We shall deal with identities in the algebra Λ of symmetric functions in a finite or infinite alphabet $X = \{x_1, x_2, x_3, \dots\}$ with coefficients in the field of rational functions $\mathbb{Q}(q,t)$. We denote by $\Lambda_{\mathbb{Z}[q,t]}$ the algebra of symmetric functions in X with coefficients in $\mathbb{Z}[q,t]$

An essential notational tool in our presentation is the notion of *plethystic substitution* and we need to recall its definition. Briefly, if $E = E(t_1, t_2, t_3, \dots)$ is a given formal series in the variables $t_1, t_2, t_3, \dots$ (which may include the parameters q, t) and $f \in \Lambda$ has been expressed in terms of the power basis in the form

$$F = Q(p_1, p_2, p_3, \dots),$$

then the *plethystic substitution* of E in F, denoted $F[E]$, is defined simply by setting

$$F[E] = Q(p_1, p_2, p_3, \dots)\Big|_{p_k \to E(t_1^k, t_2^k, t_3^k, \dots)}. \tag{1.1}$$

This operation is easily programmed in any symbolic manipulation software which includes a symmetric function package. We shall adopt the convention that, inside the plethystic brackets [ ], X and X_n respectively stand for $x_1 + x_2 + x_3 + \cdots$ and $x_1 + x_2 + \cdots + x_n$. A similar notation is adopted when we work with any other alphabet. Note that if $P \in \Lambda$ then $P[X_n]$ simply means $P(x_1, x_2, \dots, x_n)$. Plethystic substitution need not be restricted to symmetric polynomials. In fact, the substitution in (I.1) makes sense even when Q is a formal power series. In this vein, we shall systematically use the symbol $\Omega[X]$ to represent the symmetric function series

$$\Omega[X] = \prod_i \frac{1}{1 - x_i} = \exp\Big(\sum_{k \geq 1} \frac{p_k[X]}{k}\Big). \tag{1.2}$$

This notation is particularly convenient since, for any two alphabets X, Y from (1.1) and (1.2), we easily derive that

$$\Omega[X + Y] = \Omega[X]\Omega[Y], \quad \Omega[X - Y] = \Omega[X]/\Omega[Y]. \tag{1.3}$$

In particular we see that the Macdonald kernel $\Omega_{q,t}(x_1, x_2, \dots, x_n; y_1, y_2, \dots, y_k)$ (see [12, p. 309, (2.5)]) may be expressed in the compact form

$$\Omega_{q,t}(x_1, x_2, \dots, x_n; y_1, y_2, \dots, y_k) = \Omega\Big[X_n Y_k \frac{1-t}{1-q}\Big] \tag{1.4}$$

We represent partitions here by their French Ferrers diagram, that is, with rows decreasing from bottom to top. To express the fact that a certain lattice cell s belongs to the Ferrers diagram of a partition λ we shall simply write $s \in \lambda$. More generally we shall always identify partitions with their corresponding diagrams. The number of rows of the Ferrers diagram of a partition μ will be called the *length* of μ and denoted $l(\mu)$. We shall often make use of the operation of prepending a column of length k to the diagram of a partition μ of length $\leq k$; for convenience we shall denote the resulting partition by the symbol $\mu + 1^k$.

Let us recall that the *integral forms* $J_\lambda(x; q, t)$ of the Macdonald polynomials $P_\lambda(x; q, t)$ and $Q_\lambda(x; q, t)$ are defined (see [12, p. 352]) by setting

$$J_\lambda(x; q, t) = h_\lambda(q, t) P_\lambda(x; q, t) = h'_\lambda(q, t) Q_\lambda(x; q, t) \tag{1.5}$$

with

$$h_\lambda(q, t) = \prod_{s\in\lambda} \left(1 - q^{a_\lambda(s)} t^{l_\mu(s)+1}\right), \quad h'_\lambda(q, t) = \prod_{s\in\lambda} \left(1 - q^{a_\lambda(s)+1} t^{l_\mu(s)}\right) \tag{1.6}$$

where, for a cell $s \in \lambda$, $a_\lambda(s)$ and $l_\lambda(s)$ represent the *arm* and the *leg* of s in λ, that is, the number of cells of λ that are respectively strictly *EAST* and *NORTH* of s. Recall that Garsia–Haiman [1] introduced the modified versions $H_\mu[X; q, t\,]$ of the integral forms by setting

$$H_\mu[X; q, t\,] = J_\mu[\frac{X}{1-t}; q, t\,]. \tag{1.7}$$

These polynomials offer direct access to the Macdonald q, t-Kostka coefficients because their Schur function expansion reduces to

$$H_\mu[X; q, t\,] = \sum_\lambda S_\lambda[X] K_{\lambda\mu}(q, t).$$

This follows immediately from the definition of $K_{\lambda\mu}(q, t)$, given in [12, p. 354, (8.11)], as the coefficients appearing in the expansion

$$J_\mu[X; q, t\,] = \sum_\lambda S_\lambda[X(1-t)]\, K_{\lambda\mu}(q, t), \tag{1.8}$$

It is also shown in [12, p. 354, (8.12)] that we have

$$J_\mu[X; 0, t\,] = Q_\mu[X; t\,] = \sum_\lambda S_\lambda[X(1-t)]\, K_{\lambda\mu}(t), \tag{1.9}$$

where $Q_\lambda[X; t\,]$ is the Hall–Littlewood polynomial and $K_{\lambda\mu}(t)$ denotes the corresponding *Kostka–Foulkes* coefficient. We shall also set

$$H_\mu[X; t\,] = Q_\lambda[\frac{X}{1-t}; t\,] = \sum_\lambda S_\lambda[X] K_{\lambda\mu}(t). \tag{1.10}$$

It turns out that the polynomiality of the $K_{\lambda\mu}(q, t)$ is an immediate consequence of the following general result first proved in [14].

Theorem 1.1. *For any linear operator V acting on Λ and $P \in \Lambda$, set*

$$\widetilde{V}^q P[X] = V^Y P[qX + (1-q)Y]\,\big|_{Y=X} \tag{1.11}$$

where V^Y is simply V acting on polynomials in the Y variables. This given, if $G_k = G_k(X,t)$ is any linear operator on Λ with the property that

$$G_k H_\mu[X;t] = H_{\mu+1^k}[X;t] \quad \textit{for all } \mu \textit{ of length} \le k, \tag{1.12}$$

then $\widetilde{G_k}^q$ has the property

$$\widetilde{G_k}^q H_\mu[X;q,t] = H_{\mu+1^k}[X;q,t] \textit{for all } \mu \textit{ of length} \le k. \tag{1.13}$$

In particular, the modified Macdonald polynomial $H_\mu[X;q,t]$ may be obtained from the "Rodriguez" formula:

$$H_\mu[X;q,t] = \widetilde{G_{\mu'_1}}^q \widetilde{G_{\mu'_2}}^q \cdots \widetilde{G_{\mu'_h}}^q \mathbf{1} \tag{1.14}$$

where $\mu' = (\mu'_1, \mu'_2, \ldots, \mu'_h)$ denotes the conjugate of μ.

Our main contribution here is a simple, direct and elementary proof of this result which only uses identities given in the original paper of Macdonald.

Note that, since the Kostka–Foulkes matrix $K(t) = \|K_{\lambda\mu}(t)\|_{\lambda\mu}$ is unitriangular, it follows that its inverse $H(t) = K(t)^{-1}$ has entries in $\mathbb{Z}[t]$. This implies that the "trivial" operator $TG_k = TG_k(X;t)$ defined by setting for the $\{H_\mu[X;t]\}_\mu$ basis

$$TG_k H_\mu[X;t] = \begin{cases} H_{\mu+1^k}[X;t] & \text{if } \mathrm{l}(\mu) \le \mathrm{k} \\ 0 & \text{otherwise,} \end{cases} \tag{1.15}$$

acts integrally on the Schur basis. This given, we see that the desired result

$$K_{\lambda\mu}(q,t) \in \mathbb{Z}[q,t] \tag{1.16}$$

is an immediate consequence of (1.14) with $G = TG$.

This paper is divided into two sections. In the first section we prove Theorem 1.1 and, in the second, we give a number of applications, including the explicit derivation of the action of a variant of the operator TG on the monomial basis.

2 Rodrigues operators for the integral forms

We shall start by proving a result analogous to Theorem 1.1 for the Macdonald integral forms. To this end we shall need a number of auxiliary results.

Proposition 2.1. *If $V(X;t)$ is a linear operator on symmetric functions in X that depends only on t and we set*

$$\widetilde{V}^q(X;t)P[X] = V(Z;t)^Z P\big[qX + (1-q)Z\big]\Big|_{Z=X}, \tag{2.1}$$

then

$$\frac{\widetilde{V}^q(X;t)\Omega\left[XY\frac{1-t}{1-q}\right]}{\Omega\left[XY\frac{1-t}{1-q}\right]} = \frac{V(X;t)^X\Omega[XY(1-t)]}{\Omega[XY(1-t)]}; \tag{2.2}$$

in particular this ratio is independent of q.

Proof. We have

$$\begin{aligned}\widetilde{V}^q(X;t)\Omega\left[XY\frac{1-t}{1-q}\right] &= V(Z;t)^Z\Omega\left[\left(qX+(1-q)Z\right)Y\frac{1-t}{1-q}\right]\Big|_{Z=X}\\ &= \Omega\left[qXY\frac{1-t}{1-q}\right]V(X;t)^X\Omega\left[\left(1-q\right)XY\frac{1-t}{1-q}\right]\\ &= \Omega\left[XY\frac{1-t}{1-q}\right]\Omega\left[(q-1)XY\frac{1-t}{1-q}\right]V(X;t)^X\Omega\left[XY(1-t)\right]\\ &= \Omega\left[XY\frac{1-t}{1-q}\right]\Omega\left[XY(t-1)\right]V(X;t)^X\Omega\left[XY(1-t)\right].\end{aligned}$$

This proves (2.2). $\square$

Recall that, for an alphabet $Y_k = \{y_1, y_2, \cdots, y_k\}$ and an interval $I \subseteq [1,k] = \{1,2,\ldots k\}$, Macdonald sets (see [12, p. 315, (3.5)])

$$A_I[Y_k;t] = t^{\binom{|I|}{2}} \prod_{\substack{i\in I\\ j\in[1,k]-I}} \frac{ty_i - y_j}{y_i - y_j}.$$

Note also that, for any symmetric polynomial $P[Y_k]$, the operator T_I^q which replaces y_i by qy_i for every $i \in I$ may be written in the form

$$T_I^q P[Y_k] = P[Y_k + (q-1)Y_I] \tag{2.3}$$

where

$$Y_I = \sum_{i\in I} y_i .$$

This given, we have the following identity due to Kirillov–Noumi [6].

Proposition 2.2. *Let $M^{(Y_k)}(u)$ denote the Macdonald operator acting on polynomials in the alphabet Y_k, that is,*

$$M^{(Y_k)}(u) = \sum_{r=0}^{k} u^r \sum_{\substack{I\subseteq[1,k]\\ |I|=r}} A_I[Y_k;t]T_I^q .$$

Then, for $k \leq n$,

$$M^{(Y_k)}(-1)\Omega\left[X_nY_k\frac{1-t}{1-q}\right] = y_1y_2\cdots y_k \sum_{l(\mu)\leq k} J_{\mu+1^k}[X_n;q,t]\,\frac{P_\mu[Y_k;q,t]}{h'_\mu(q,t)}. \tag{2.4}$$

Proof. We start with the Macdonald *Cauchy*-identity from equation (4.13) in [12],

$$\Omega\big[X_n Y_k \frac{1-t}{1-q}\big] = \sum_{l(\lambda)\le k} Q_\lambda[X_n; q, t\,],\ P_\lambda[Y_k; q, t\,],$$

the summation being only over partitions λ of length $\le k$ because P_λ vanishes when evaluated on an alphabet whose cardinality is smaller than the length of λ. Applying $M^{(Y_k)}(u)$ to both sides of this identity and using Theorem (4.15) of [12, p. 324] we get

$$M^{(Y_k)}(u)\Omega\big[X_n Y_k \frac{1-t}{1-q}\big]$$
$$= \sum_{l(\lambda)\le k} Q_\lambda[X_n; q, t\,]\Big(\prod_{i=1}^{k}(1+u\,t^{k-i}q^{\lambda_i})\Big) P_\lambda[Y_k; q, t\,].$$

Now note that, if $l(\lambda) < k$, then the term corresponding to λ in this sum will vanish if we set $u = -1$. We thus obtain

$$M^{(Y_k)}(-1)\Omega\big[X_n Y_k \frac{1-t}{1-q}\big]$$
$$= \sum_{l(\lambda)=k} Q_\lambda[X_n; q, t\,]\Big(\prod_{i=1}^{k}(1-t^{k-i}q^{\lambda_i})\Big) P_\lambda[Y_k; q, t\,]. \quad (2.5)$$

Now it follows from Theorem (4.17) of [12, p. 325] that, if $l(\lambda) = k$, then

$$P_\lambda[Y_k; q, t\,] = y_1 y_2 \cdots y_k\, P_\mu[Y_k; q, t\,].$$

with

$$\mu = (\lambda_1 - 1, \lambda_2 - 1, \ldots, \lambda_k - 1).$$

Thus we may rewrite (2.5) in the form

$$M^{(Y_k)}(-1)\Omega\big[X_n Y_k \frac{1-t}{1-q}\big]$$
$$= y_1 y_2 \cdots y_k \sum_{l(\mu)\le k} Q_{\mu+1^k}[X_n; q, t\,]\Big(\prod_{i=1}^{k}(1-t^{k-i}q^{\mu_i+1})\Big) P_\mu[Y_k; q, t\,]. \quad (2.6)$$

Now from (2.5) and (2.6) we derive the fact that

$$Q_{\mu+1^k}[X_n; q, t\,]\Big(\prod_{i=1}^{k}(1-t^{k-i}q^{\mu_i+1})\Big) = \frac{J_{\mu+1^k}[X_n; q, t\,]}{h'_\mu(q, t)}.$$

Thus the desired identity in (2.4) immediately follows upon using this in (2.6). □

Proposition 2.3.

$$\frac{M^{(Y_k)}(u)\Omega\left[X_nY_k\frac{1-t}{1-q}\right]}{\Omega\left[X_nY_k\frac{1-t}{1-q}\right]}=\sum_{r=0}^{k}u^r\sum_{\substack{I\subseteq[1,k]\\|I|=r}}A_I[Y_k;t]\;\Omega\left[X_nY_I(t-1)\right]\;;\tag{2.7}$$

in particular this ratio is independent of q.

Proof.

$$\begin{aligned}M^{(Y_k)}(u)\Omega\left[X_nY_k\frac{1-t}{1-q}\right]&\\&=\sum_{r=0}^{k}u^r\sum_{\substack{I\subseteq[1,k]\\|I|=r}}A_I[Y_k;t]\Omega\left[X_n\big(Y_k+(q-1)Y_I\big)\frac{1-t}{1-q}\right]\\&=\Omega\left[X_nY_k\frac{1-t}{1-q}\right]\sum_{r=0}^{k}u^r\sum_{\substack{I\subseteq[1,k]\\|I|=r}}A_I[Y_k;t]\,\Omega\left[X_nY_I(t-1)\right]\end{aligned}$$

and this proves (2.7). □

Theorem 2.1. *If $B_k(X;t)$ is any operator with the property that*

$$B_k(X;t)J_\mu[X;0,t]=J_{\mu+1^k}[X;0,t]\quad \textit{for all } l(\mu)\le k,\tag{2.8}$$

then

$$\widetilde{B_k}^q(X;t)J_\mu[X;q,t]=J_{\mu+1^k}[X;q,t]\quad \textit{for all } l(\mu)\le k.\tag{2.9}$$

Proof. Suppose we show that

$$y_1y_2\cdots y_k\widetilde{B_k}^q(X_n;t)^{X_n}\Omega\left[X_nY_k\frac{1-t}{1-q}\right]=M^{(Y_k)}(-1)\Omega\left[X_nY_k\frac{1-t}{1-q}\right].\tag{2.10}$$

Then by combining this with Proposition 2.2 we get

$$\begin{aligned}y_1y_2\cdots y_k\sum_{l(\mu)\le k}\widetilde{B_k}^q(X_n;t)^{X_n}J_\mu[X_n;q,t]\frac{P_\mu[Y_k;q,t]}{h'_\mu(q,t)}&\\=y_1y_2\cdots y_k\sum_{l(\mu)\le k}J_{\mu+1^k}[X_n;q,t]\frac{P_\mu[Y_k;q,t]}{h'_\mu(q,t)}&\end{aligned}$$

and (2.9) then follows by equating coefficients of $P_\mu[Y_k;q,t]$. To show (2.10) we need only verify that

$$y_1y_2\cdots y_k\frac{\widetilde{B_k}^q(X_n;t)^{X_n}\Omega\left[X_nY_k\frac{1-t}{1-q}\right]}{\Omega\left[X_nY_k\frac{1-t}{1-q}\right]}=\frac{M^{(Y_k)}(-1)\Omega\left[X_nY_k\frac{1-t}{1-q}\right]}{\Omega\left[X_nY_k\frac{1-t}{1-q}\right]},\tag{2.11}$$

and, since we have shown that both sides of this equation are independent of q, we need only verify this equality at $q = 0$. However, the hypothesis in (2.8) yields that

$$
\begin{aligned}
y_1 y_2 \cdots y_k \frac{\widetilde{B_k}^q (X_n; t)^{X_n} \Omega\left[X_n Y_k \frac{1-t}{1-q}\right]}{\Omega\left[X_n Y_k \frac{1-t}{1-q}\right]}\Bigg|_{q=0}
&= \frac{y_1 y_2 \cdots y_k}{\Omega[X_n Y_k (1-t)]} \sum_{l(\mu)\le k} B_k(X_n; t)^{X_n} J_\mu[X_n; 0, t] \frac{P_\mu[Y_k; 0, t]}{h'_\mu(0,t)} \\
&= \frac{y_1 y_2 \cdots y_k}{\Omega[X_n Y_k (1-t)]} \sum_{l(\mu)\le k} J_{\mu+1^k}[X_n; 0, t] \frac{P_\mu[Y_k; 0, t]}{h'_\mu(0,t)}
\end{aligned}
$$

and again by Proposition 2.3 we see that this is precisely

$$
\frac{M^{(Y_k)}(-1)\Omega\left[X_n Y_k \frac{1-t}{1-q}\right]}{\Omega\left[X_n Y_k \frac{1-t}{1-q}\right]}\Bigg|_{q=0}.
$$

This completes the proof of (2.9). □

Before we can proceed with the proof of Theorem 2.1 we need one more auxiliary result. To begin with, it will be convenient to consider the substitution $X \to X/(1-t)$ as a linear operator on symmetric functions. More precisely, for any symmetric polynomial P and any alphabet X, we set

$$
F^t P[X] = P\big[X(1-t)\big].
$$

Now a somewhat surprising development is that the operation $V \to \widetilde{V}^q$ defined in (1.11) commutes with conjugation by F^t. In fact, we may state

Proposition 2.4. *For any linear operator V acting on Λ and any polynomial $P \in \Lambda$ we have*

$$
F^t\, \widetilde{V}^q\, {F^t}^{-1} P[X] = \widetilde{(F^t\, V {F^t}^{-1})}^q\, P[X]. \tag{2.12}
$$

Proof. It is sufficient to prove (2.12) for the Schur basis. Note that, for any partition λ, the addition formula for Schur functions gives

$$
\begin{aligned}
\widetilde{(F^t V {F^t}^{-1})}^q\, S_\lambda[X] &= (F^t\, V {F^t}^{-1})^Y S_\lambda[qX + (1-q)Y]\Big|_{Y=X} \\
&= \sum_{\mu\subseteq\lambda} S_{\lambda/\mu}[qX]\, (F^t\, V {F^t}^{-1})^Y S_\mu[(1-q)Y]\Big|_{Y=X} \qquad (2.13)\\
&= \sum_{\mu\subseteq\lambda} S_{\lambda/\mu}[qX]\, (F^t\, V {F^t}^{-1})^X S_\mu[(1-q)X].
\end{aligned}
$$

In the same vein we see that the left-hand side of (2.12), for $P = S_\lambda$, gives

$$\begin{aligned} F^t\, \widetilde{V}^q F^{t^{-1}} S_\lambda[X] &= F^t\, \widetilde{V}^q S_\lambda[\frac{X}{1-t}] \\ &= F^t\, V^Y S_\lambda \left[\frac{qX+(1-q)Y}{1-t}\right] |_{Y=X} \\ &= F^t \sum_{\mu\subseteq\lambda} S_{\lambda/\mu}[\frac{qX}{1-t}]\, V^Y S_\mu[\frac{(1-q)Y}{1-t}]\Big|_{Y=X} \\ &= \sum_{\mu\subseteq\lambda} S_{\lambda/\mu}[qX]\, F^t\, V^X S_\mu[\frac{(1-q)X}{1-t}], \end{aligned}$$

and it is easily seen that this is another way to write the last expression in (2.13). □

We are now in a position to give our

Proof of Theorem 1.1. By assumption we have

$$G_k H_\mu[X;t] = H_{\mu+1^k}[X;t] \quad \text{for all } \mu \text{ of length} \le k. \tag{2.14}$$

Thus from (1.10) we derive that

$$G_k F^{t^{-1}} Q_\mu[X;t] = F^{t^{-1}} Q_{\mu+1^k}[X;t].$$

Now this, using (1.9), may be rewritten as

$$F^t G_k F^{t^{-1}} J_\mu[X;0,t] = J_{\mu+1^k}[X;0,t] \quad \text{for all } \mu \text{ of length} \le k.$$

In other words the operator

$$B_k = B_k(X;t) = F^t\, G_k F^{t^{-1}}$$

satisfies the hypothesis of Theorem 2.1. It then follows that

$$(\widetilde{F^t G_k F^{t^{-1}}})^q\, J_\mu[X;q,t] = J_{\mu+1^k}[X;q,t] \quad \text{for all } \mu \text{ of length} \le k.$$

But Proposition 2.4 yields

$$F^t \widetilde{G}_k^{\,q} F^{t^{-1}} J_\mu[X;q,t] = J_{\mu+1^k}[X;q,t] \quad \text{for all } \mu \text{ of length} \le k$$

and (1.7) shows that this is just another way of writing (1.13), completing the proof of Theorem 1.1. □

Remark 2.1. We should note that any symmetric function operator $V(q,t) = V(x;q,t)$ of the form

$$V(q,t) = \sum_{I\subseteq[1,k]} a_I(x,t)\, T_I^q \tag{2.15}$$

(in particular the Macdonald operator) satisfies the identity

$$V(q,t) = \widetilde{V(0,t)}^{q}. \tag{2.16}$$

In fact, since for any $P \in \Lambda$ we have

$$V(q,t)P[X] = \sum_{I \subseteq [1,k]} a_I(x,t)\, P[X + (q-1)X_I],$$

then

$$V(0,t)\, P[X] = \sum_{I \subseteq [1,k]} a_I(x,t)\, P[X - X_I].$$

Thus

$$\begin{aligned} \widetilde{V(0,t)}^{q} P[X] &= V(0,t)^Y P[qX + (1-q)Y]\Big|_{Y=X} \\ &= \sum_{I \subseteq [1,k]} a_I(y,t)\, P[qX + (1-q)(Y - Y_I)]\Big|_{Y=X} \\ &= \sum_{I \subseteq [1,k]} a_I(x,t)\, P[qX + (1-q)(X - X_I)] \\ &= \sum_{I \subseteq [1,k]} a_I(x,t)\, P[X + (q-1)X_I] \end{aligned}$$

which is (2.16).

We then see that Propositions 2.1 and 2.3 are both particular cases of the following general fact.

Proposition 2.5. *If $V(q,t) = V(x;q,t)$ is an operator on Λ with the property*

$$V(q,t) = \widetilde{V(0,t)}^{q},$$

then

$$\frac{V(q,t)^X \Omega\big[XY\frac{1-t}{1-q}\big]}{\Omega\big[XY\frac{1-t}{1-q}\big]} = \frac{V(0,t)^X \Omega\big[XY(1-t)\big]}{\Omega\big[XY(1-t)\big]};$$

in particular this ratio is independent of q.

The proof follows exactly the same steps we used to prove Proposition 2.1.

3 Applications

It is shown in [12, p. 324, (4.14)] that

$$P_\mu[X;t,t] = S_\mu[X]. \tag{3.1}$$

Thus (I.5) gives

$$J_\mu[X; t, t] = \prod_{s\in\mu}(1 - t^{h_\mu(s)})S_\mu[X], \tag{3.2}$$

where $h_\mu(s) = a_\mu(s) + l_\mu(s) + 1$ denotes the hook-length corresponding to the cell s in μ. This given, Theorem 2.1 has the following immediate corollary.

Theorem 3.1. *If B_k is any operator on Λ with the property*

$$B_k(X; t)J_\mu[X; 0, t] = J_{\mu+1^k}[X; 0, t] \quad \textit{for all } l(\mu) \leq k, \tag{3.3}$$

then the operator $\widetilde{B_k}^t$ defined by setting for $P \in \Lambda$

$$\widetilde{B_k}^t(X; t)P[X] = B_k(Y; t)^Y P[tX + (1 - t)Y]\big|_{Y=X} \tag{3.4}$$

has the property

$$\widetilde{B_k}^t(X; t)S_\mu[X] = \Big(\prod_{i=1}^{k}\left(1 - t^{k+1-i+\mu_i}\right)\Big) S_{\mu+1^k}[X] \quad \textit{for all } l(\lambda) \leq k. \tag{3.5}$$

Proof. This follows by setting $q = t$ in (2.9), using formula (3.2) and canceling the common factor. □

This result has the following converse.

Theorem 3.2. *If $B_k(X; t)$ is any operator with the property in (3.5), then the operator $\widetilde{B_k}^q(X, t)$ defined by setting for $P \in \Lambda$,*

$$\widetilde{B_k}^q(X; t)P[X] = B_k(Y; t)^Y P[qX + (1 - q)Y]\big|_{Y=X} \tag{3.6}$$

has the property

$$\widetilde{B_k}^q(X; t)J_\mu[X; q, t] = J_{\mu+1^k}[X; q, t] \quad \textit{for all } l(\mu) \leq k, \tag{3.7}$$

Proof. The argument follows almost verbatim what we did to prove Theorem 2.1 except that at one point we must set $q = t$ rather than $q = 0$. □

Of course we may produce a polynomiality proof based on this theorem; however all our attempts yielded a more complicated argument than that based on Theorem 1.1.

For representation-theoretical reasons Garsia–Haiman were led to consider the polynomials

$$\widetilde{H}_\mu[X; q, t] = \sum_\lambda S_\lambda[X]\tilde{K}_{\lambda\mu}(q, t) \tag{3.8}$$

where

$$\tilde{K}_{\lambda\mu}(q, t) = t^{n(\mu)}K_{\lambda\mu}(q, 1/t) \tag{3.9}$$

with

$$n(\mu) = \sum_i (i-1)\mu_i. \tag{3.10}$$

Note that setting $q = 0$ in (3.8) gives

$$\widetilde{H}_\mu[X; 0, t] = \widetilde{H}_\mu[X; t] = \sum_\lambda S_\lambda[X]\tilde{K}_{\lambda\mu}(t)$$

where

$$\tilde{K}_{\lambda\mu}(t) = t^{n(\mu)} K_{\lambda\mu}(1/t)$$

is the so-called *cocharge* Kostka–Foulkes polynomial. Now it turns out that a result analogous to Theorem 1.1 also holds for the basis $\{\widetilde{H}_\mu[X; q, t]\}_\mu$.

Theorem 3.3. *If $\mathcal{H}_k = \mathcal{H}_k(X, t)$ is any linear operator on Λ with the property that*

$$\mathcal{H}_k \widetilde{H}_\mu[X; t] = \widetilde{H}_{\mu+1^k}[X; t] \quad \textit{for all } \mu \textit{ of length} \leq k, \tag{3.11}$$

then the operator $\widetilde{\mathcal{H}_k}^q$ defined by (1.11) *has the property*

$$\widetilde{\mathcal{H}_k}^q \widetilde{H}_\mu[X; q, t] = \widetilde{H}_{\mu+1^k}[X; q, t] \quad \textit{for all } \mu \textit{ of length} \leq k. \tag{3.12}$$

In particular, the modified Macdonald polynomial $\widetilde{H}_\mu[X; q, t]$ may be obtained from the "Rodrigues" formula

$$\widetilde{H}_\mu[X; q, t] = \widetilde{\mathcal{H}_{\mu'_1}}^q \widetilde{\mathcal{H}_{\mu'_2}}^q \cdots \widetilde{\mathcal{H}_{\mu'_h}}^q \mathbf{1}\widetilde{\mathcal{H}_{\mu'_h}}^q \mathbf{1}\widetilde{\mathcal{H}_{\mu'_h}}^q \mathbf{1} \tag{3.13}$$

where $\mu' = (\mu'_1, \mu'_2, \ldots, \mu'_h)$ denotes the conjugate of μ.

Proof. Note that setting $t = 1/t$ in (3.11) gives

$$t^{-n(\mu)}\mathcal{H}_k(X; 1/t)H_\mu[X; t] = t^{-n(\mu+1^k)} H_{\mu+1^k}[X; t] \quad \text{for all } \mu \text{ of length} \leq k$$

and this, by using (3.10), may be rewritten as

$$t^{\binom{k}{2}}\mathcal{H}_k(X; 1/t)H_\mu[X; t] = H_{\mu+1^k}[X; t] \quad \text{for all } \mu \text{ of length} \leq k.$$

Thus we may apply Theorem 2.1 and derive the fact that

$$t^{\binom{k}{2}}\mathcal{H}_k(Y; 1/t)^Y H_\mu[qX + (1-q)Y; q, t]\big|_{Y=X} = H_{\mu+1^k}[X; q, t] \quad \text{for all } \mu \text{ of length} \leq k.$$

Setting $t = 1/t$ and using (3.9) transforms this back to (3.12). □

Some interesting developments follow by combining the present identities with one of the simplest of the "Rodrigues" operators introduced by Lapointe–Vinet in [10]. To see how this comes about we need to recall this beautiful result.

Theorem 3.4. *The operator*

$$LV_k(X_n;q,t) = \frac{1}{(\frac{1}{q};\frac{1}{t})_{n-k}} M^{(X_n)}\Big(-\frac{1}{qt^{n-k-1}}\Big)\underline{e_k[X_n]}, \tag{3.14}$$

where $\underline{e_k[X_n]}$ *denotes multiplication by* $e_k[X_n]$, *has the property*

$$LV_k(X_n;q,t)J_\mu[X_n;q,t] = J_{\mu+1^k}[X_n;q,t] \quad \textit{for all } l(\mu) \le k.$$

Proof. The argument is so elementary that it might as well be reproduced here. It is shown by Macdonald in [12, p. 340, (6.24), (iv)] that

$$e_k[X_n]P_\mu[X_n;q,t] = P_{\mu+1^k}[X_n;q,t] + \sum_{\substack{\lambda/\mu\in V_k\\ \lambda\neq\mu+1^k}} P_\lambda[X_n;q,t]\Psi_{\lambda\mu}(q,t)$$

where $\lambda/\mu \in V_k$ indicates that the sum is over partitions λ such that λ/μ is a vertical k-strip. Now, applying the Macdonald operator $M^{(X_n)}(u)$ to both sides and using [12, p. 324, (4.15)] we get, for $l(\mu) \le k$,

$$\begin{aligned} &M^{(X_n)}(u)e_k[X_n]P_\mu[X_n;q,t] \\ &\quad = \prod_{i=1}^{k}\big(1+ut^{n-i}q^{\mu_i+1}\big)\prod_{i=k+1}^{n}\big(1+ut^{n-i}\big)P_{\mu+1^k}[X_n;q,t] \\ &\qquad + \sum_{\substack{\lambda/\mu\in V_k\\ \lambda\neq\mu+1^k}}\prod_{i=1}^{n}\big(1+ut^{n-i}q^{\lambda_i}\big)P_\lambda[X_n;q,t]\Psi_{\lambda\mu}(q,t). \end{aligned} \tag{3.15}$$

Now 1^k is the shortest vertical k-strip that may be added to μ, and any other will spill a cell at height $k+1$. Thus each term of the sum in (3.15) will contain the factor

$$\big(1+ut^{n-k-1}q\big)$$

which vanishes when we set $u = -1/q\,t^{n-k-1}$. Thus (3.15) gives

$$\begin{aligned} &\frac{1}{\big(\frac{1}{q};\frac{1}{t}\big)_{n-k}} M^{(X_n)}\Big(-\frac{1}{qt^{n-k-1}}\Big)e_k[X_n]P_\mu[X_n;q,t] \\ &\qquad\qquad = \prod_{i=1}^{k}\big(1-t^{k+1-i}q^{\mu_i}\big)P_{\mu+1^k}[X_n;q,t] \end{aligned}$$

and this is easily converted to (3.14) by means of (1.5). □

To state and prove our next result we need some notation. To begin, if $\sigma = (\sigma_1, \sigma_2, \dots, \sigma_n)$ is a permutation in the symmetric group $\mathcal{S}_n$ and $a = (a_1, a_2, \dots, a_n)$ is a given vector, we set

$$\sigma a = (a_{\sigma_1}, a_{\sigma_2}, \dots, a_{\sigma_n});$$

we also let Supp(a) denote the *support* of a, that is, the set of elements

$$\mathrm{Supp}(a) = \{i : a_i \neq 0\}.$$

Next, for any two vectors $a = (a_1, a_2, \dots, a_n)$ and $b = (b_1, b_2, \dots, b_n)$, we shall write $a \approx b$ if and only if the components of b are a rearrangement of the components of a. More precisely, we set $a \approx b$ if and only if for some $\sigma \in \mathcal{S}_n$ we have

$$b = \sigma a.$$

With this notation the monomial symmetric function m_λ may be represented by the sum

$$m_\lambda[X_n] = \sum_{p \approx \lambda} x^p. \tag{3.16}$$

Finally, we shall generically denote by $\epsilon = (\epsilon_1, \epsilon_2, \dots, \epsilon_n)$ the indicator vector of a subset of $\{1, 2, \dots, n\}$. Thus setting $|\epsilon| = k$ will mean that ϵ represents a subset of cardinality k. In particular, the elementary symmetric function e_k may be represented by the sum

$$e_k[X_n] = \sum_{|\epsilon|=k} x^\epsilon. \tag{3.17}$$

It turns out that, not withstanding the presence of terms $1/q$ in (3.14), we can evaluate the limit of the operator $LV_k(X_n; q, t)$ as $q \to 0$. We next state the following surprising corollary of the Lapointe–Vinet result.

Theorem 3.5. *Let $TLV_k(t)$ be the operator defined by setting for the monomial basis*

$$TLV_k(t)\, m_\lambda = 0 \quad \textit{when } l(\lambda) > k \tag{3.18}$$

and

$$TLV_k(t)\, m_\lambda = \sum_{|\epsilon|=k} \sum_{p \approx \lambda} c_{\epsilon,p}(t)\, S_{p+\epsilon} \quad \textit{when } l(\lambda) \leq k, \tag{3.19}$$

where $S_{p+\epsilon}$ denotes the corresponding signed Schur function and

$$c_{\epsilon,p}(t) = \begin{cases} (-1)^{n-k} t^{\binom{n-k}{2}} \prod_{\epsilon_i=0} \left(-t^{k+1-i}\right) \prod_{\epsilon_i+p_i=1} \left(1 - t^{k+1-i}\right) \\ \qquad\qquad \textit{if } \mathrm{Supp}(p) \subseteq \mathrm{Supp}(\epsilon) \\ 0 \qquad\qquad \textit{otherwise.} \end{cases} \tag{3.20}$$

Then we also have

$$TLV_k Q_\mu[X;t] = \begin{cases} Q_{\mu+1^k}[X;t] & \text{if } l(\mu) \le k \\ 0 & \text{otherwise.} \end{cases} \tag{3.21}$$

Proof. The original definition of the Macdonald operator, given in [12, p. 315, (3.2)] may be written in the form

$$M^{(X_n)}(u) = \frac{1}{\Delta_n(x)} \sum_{\sigma \in S_n} \operatorname{sign}(\sigma) x^{\sigma\delta} \prod_{i=1}^{n} \left(1 + ut^{n-\sigma_i} T_i^q\right) \tag{3.22}$$

where $\Delta_n(x)$ denotes the Vandermonde determinant in the variables $x_1, \ldots, x_n$, T_i^q denotes the operator that replaces x_i by qx_i and, for convenience, we have set

$$\delta = (n-1, n-2, \ldots, 1, 0).$$

This given, taking account of (3.16) and (3.17), we may write the action of the Lapointe–Vinet operator LV_k on the monomial basis in the form

$$LV_k\, m_\lambda[X_n] = \frac{1}{(\frac{1}{q}; \frac{1}{t})_{n-k}} \sum_{|\epsilon|=k} \sum_{p \approx \lambda} \frac{1}{\Delta_n(x)} \sum_{\sigma \in S_n} \operatorname{sign}(\sigma) x^{\sigma\delta} \prod_{i=1}^{n} \left(1 - t^{k+1-\sigma_i} q^{p_{\sigma_i} + \epsilon_{\sigma_i} - 1}\right) x^{\sigma(p+\epsilon)}. \tag{3.23}$$

Since we may set

$$\frac{1}{\Delta_n(x)} \sum_{\sigma \in S_n} \operatorname{sign}(\sigma) x^{\sigma(\delta+p+\epsilon)} = S_{p+\epsilon}[X_n],$$

we see that (3.23) reduces to

$$LV_k(q,t)\, m_\lambda[X_n] = \sum_{|\epsilon|=k} \sum_{p \approx \lambda} c_{\epsilon,p}(q,t)\; S_{p+\epsilon}[X_n] \tag{3.24}$$

with

$$c_{\epsilon,p}(q,t) = \frac{1}{(\frac{1}{q}; \frac{1}{t})_{n-k}} \prod_{i=1}^{n} \left(1 - t^{k+1-i} q^{p_i+\epsilon_i-1}\right). \tag{3.25}$$

Our next step is to evaluate $c_{\epsilon,p}(q,t)$ at $q=0$. To this end it is convenient to rewrite (3.25) in the form

$$c_{\epsilon,p}(q,t) = \frac{(-1)^{n-k} t^{\binom{n-k}{2}}}{(q;t)_{n-k}} \frac{q^{n-k}}{q^{\#\{i : \epsilon_i+p_i=0\}}} \prod_{\epsilon_i+p_i>1} \left(1 - t^{k+1-i} q^{p_i+\epsilon_i-1}\right) \times \prod_{\epsilon_i+p_i=1} \left(1 - t^{k+1-i}\right) \prod_{\epsilon_i+p_i=0} \left(q - t^{k+1-i}\right). \tag{3.26}$$

Now note that, since $|\epsilon| = k$, we necessarily have that

$$\#\{i : \epsilon_i + p_i = 0\} \leq n - k$$

with equality only if

$$\mathrm{Supp}(p) \subseteq \mathrm{Supp}(\epsilon). \tag{3.27}$$

Thus it follows that

$$\frac{q^{n-k}}{q^{\#\{i : \epsilon_i + p_i = 0\}}}\Bigg|_{q=0} = 0$$

when (3.27) fails to hold. Since we have the obvious evaluations

$$\begin{aligned} (q;t)_{n-k}\Big|_{q=0} &= 1 \\ \prod_{\epsilon_i + p_i > 1} (1 - t^{k+1-i} q^{p_i + \epsilon_i - 1})\Big|_{q=0} &= 1 \\ \prod_{\epsilon_i + p_i = 0} (q - t^{k+1-i})\Big|_{q=0} &= \prod_{\epsilon_i + p_i = 0} (-t^{k+1-i}), \end{aligned}$$

we see that from (3.26) we derive the fact that

$$c_{\epsilon,p}(0,t) = c_{\epsilon,p}(t),$$

with $c_{\epsilon,p}(t)$ precisely as defined in (3.20). Thus from (3.24) we get

$$LV_k(0,t)\, m_\lambda[X_n] = \sum_{|\epsilon|=k} \sum_{p \approx \lambda} c_{\epsilon,p}(0,t)\ S_{p+\epsilon}[X_n] = TLV_k(t)\, m_\lambda[X_n].$$

To show (3.21) note that from (1.8) we get

$$LV_k(q,t) J_\mu[X_n; q,t] = \sum_\lambda \Big(LV_k(q,t) S_\lambda[X_n(1-t)] \Big) K_{\lambda\mu}(q,t).$$

Thus the polynomiality of the $K_{\lambda\mu}(q,t)$ insures that we can safely set $q = 0$ here and obtain

$$\begin{aligned} LV_k(q,t) J_\mu[X_n; q,t]\Big|_{q=0} &= \sum_\lambda \Big(LV_k(0,t) S_\lambda[X_n(1-t)] \Big) K_{\lambda\mu}(0,t) \\ &= \sum_\lambda \Big(TLV_k(t) S_\lambda[X_n(1-t)] \Big) K_{\lambda\mu}(t) \\ &= TLV_k(t) Q_\mu[X_n; t]. \end{aligned}$$

Thus when $l(\mu) \leq k$ the Lapointe–Vinet result (Theorem 3.5) yields

$$TLV_k(t) Q_\mu[X;t] = J_{\mu+1^k}[X;0,t] = Q_{\mu+1^k}[X;t].$$

This proves the first alternative in (3.21). To show the second we note that [12, p. 209, (2.6)] implies that we have an expansion of the form

$$J_\mu[X;t] = \sum_{\lambda \leq \mu} m_\lambda[X]\, \xi_{\lambda\mu}(t). \tag{3.28}$$

Thus when $l(\mu) > k$ we shall have $l(\lambda) > k$ for all the summands in (3.28) and (3.18) which then gives

$$TLV_k(t)\, J_\mu[X_n;t] = 0.$$

This completes our proof. □

We note that the Lapointe–Vinet result has one further curious consequence.

Theorem 3.6. *Let $W_k = W_k(X_n;t)$ be the operator defined by setting, for every $P \in \Lambda$,*

$$W_k P[X] = \frac{1}{(\frac{1}{t};\frac{1}{t})_{n-k}} \sum_{r=0}^{n} \left(\frac{-1}{t^{n-k}}\right)^r \sum_{\substack{I \subseteq [1,n] \\ |I|=r}} B_I(x;t)\, P[X - X_I] \tag{3.29}$$

with

$$B_I(x;t) = \frac{1}{\Delta_n(x)} T_I^t\, \Delta_n(x) e_k[X_n]. \tag{3.30}$$

Then

$$\widetilde{W_k}^q\, J_\mu[X_n;q,t] = J_{\mu+1^k}[X_n;q,t] \quad \textit{for all } \mu \textit{ of length} \leq k. \tag{3.31}$$

Proof. Using formula (3.5) of [12, p. 316], we may write the Lapointe–Vinet result (for $l(\mu) \leq k$) in the form

$$J_{\mu+1^k}[X_n;q,t] = \frac{1}{(\frac{1}{q};\frac{1}{t})_{n-k}} \sum_{r=0}^{n} \left(\frac{-1}{q\, t^{n-k-1}}\right)^r \sum_{\substack{I \subseteq [1,n] \\ |I|=r}} \frac{T_I^t\, \Delta_n(x)}{\Delta_n(x)}\, T_I^q e_k[X_n] J_\mu[X_n;q,t]$$

where, for each $i \in I$, T_I^t replaces x_i by tx_i. This given, we may set $q = t$ and obtain

$$J_{\mu+1^k}[X_n; t, t] = \frac{1}{(\frac{1}{t}; \frac{1}{t})_{n-k}} \sum_{r=0}^{n} \left(\frac{-1}{t^{n-k}}\right)^r \sum_{\substack{I \subseteq [1,n] \\ |I|=r}} B_I(x; t)\ J_\mu[X_n + (t-1)X_I; t, t\,] = \widetilde{W}_k^{\,t}\ J_\mu[X_n; t, t\,]. \tag{3.32}$$

Since

$$J_\mu[X_n; t, t\,] = h_\mu(t) S_\mu[X_n],$$

we derive from (3.32) the fact that the operator W_k satisfies the hypothesis of Theorem 3.2; thus (3.31) is an immediate consequence of Theorem 3.2. □

We conclude with a final application of Theorem 1.1.

Theorem 3.7. *Let $G_k(X; t)$ be any operator with the property that*

$$G_k(X; t) H_\mu[X; t] = H_{\mu+1^k}[X; t] \quad \textit{for all } \mu \textit{ of length} \le k. \tag{3.33}$$

Then the operator

$$\widetilde{G}_k^{\,t}(X; q) = \widetilde{G}_k^{\,q}(X; t)\Big|_{t=q}$$

has the property

$$\omega \widetilde{G}_k^{\,t}(X; q) \omega H_\mu[X; q, t\,] = H_{(k,\mu)}[X; q, t\,] \quad \textit{for all } \mu \textit{ with } \mu_1 \le k. \tag{3.34}$$

Proof. It follows from the Macdonald duality formula [12, p. 327, (5.1)] that the polynomial $H_\mu[X; q, t\,]$ satisfies the identity

$$H_{\mu'}[X; q, t\,] = \omega\, H_\mu[X; t, q\,]. \tag{3.35}$$

Now, assuming (3.33), it follows from Theorem 1.1 that

$$\widetilde{G}_k^{\,q}(X; t)\, H_{\mu'}[X; q, t\,] = H_{\mu'+1^k}[X; q, t\,] \quad \text{for all } \mu' \text{ of length} \le k.$$

Interchanging q and t we get

$$\widetilde{G}_k^{\,t}(X; q)\, H_{\mu'}[X; t, q\,] = H_{\mu'+1^k}[X; t, q\,] \quad \text{for all } \mu' \text{ of length} \le k.$$

and two applications of (3.35) then give

$$\widetilde{G}_k^{\,t}(X; q)\, \omega H_\mu[X; t, q\,] = \omega H_{(k,\mu)}[X; q, t\,] \quad \text{for all } \mu \text{ with } \mu_1 \le k,$$

which is simply another way of writing formula (3.34). □

The simplicity of our operator $\omega \widetilde{G}_k^{\,t}(X; q)\, \omega$ and our proof of (3.34) should be contrasted with the complexity of the developments in the recent Kajihara–Noumi paper [5].

References

[1] Garsia, A.M., Haiman, M. (1996): Some natural bigraded S_n-modules and q, t-Kostka coefficients. The Foata Festschrift. Electron. J. Combin. **3**, no. 2, Research Paper 24, approx. 60 pp. (electronic)

[2] Garsia, A.M., Haiman, M., Tesler, G. (1999): Explicit plethystic formulas for Macdonald q, t-Kostka coefficients. The Andrews Festschrift. Sém. Lothar. Combin. **42**, B42m, 45 pp. (electronic) (Website: http://www.emis.de/journals/SLC/)

[3] Garsia, A.M., Remmel, J. (1998): Plethystic formulas and positivity for q, t-Kostka coefficients. In: Sagan, B.E., Stanley, R. (eds.) Mathematical Essays in Honor of Gian-Carlo Rota. (Progress in Mathematics, vol. 161), Birkhäuser Boston, Boston, MA

[4] Garsia, A.M., Tesler, G. (1996): Plethystic formulas for Macdonald q, t-Kostka coefficients. Adv. Math. **123**, 144–222

[5] Kajihara, Y., Noumi, M. (2000): Raising operators of row type for Macdonald polynomials. Compositio Math. **120**, 119–136

[6] Kirillov, A., Noumi, M. (1996): q-difference raising operators for Macdonald polynomials and the integrality of transition coefficients. In: van Diejen, J.F., Vinet, L. (eds.) Algebraic Methods and q-Special Functions. (CRM Proceedings and Lecture Notes, vol. 22). American Mathematical Society, Providence, RI, pp. 227–243 (Also q-alg/9605005. 4 May 96)

[7] Knop, F. (1997): Integrality of two variable Kostka functions. J. Reine Angew. Math. **482**, 177–189

[8] Knop, F. (1997): Symmetric and non-symmetric quantum Capelli polynomials. Comment. Math. Helv. **72**, 84–100

[9] Lapointe, L., Vinet, L. (1995): Rodrigues formulas for the Jack polynomials and the Macdonald-Stanley conjecture. Internat. Math. Res. Notices **1995**, 419–424

[10] Lapointe, L., Vinet, L. (1997): Rodrigues formulas for the Macdonald polynomials. Adv. Math. **130**, 261–279

[11] Macdonald, I.G. (1988): A new class of symmetric functions. Actes du 20e Séminaire Lotharingien. (Publication de l'Institute de Recherche Mathématique Avancée, vol. 372). Université Louis Pasteur, Strasbourg, pp. 131–171

[12] Macdonald, I.G. (1995): Symmetric functions and Hall polynomials, 2nd edition. Clarendon Press, Oxford.

[13] Sahi, S. (1996): Interpolation, integrality, and a generalization of Macdonald's polynomials. Internat. Math. Res. Notices, **1996**, 457–471

[14] Zabrocki, M. (2000): q-Analogs of symmetric function operators. Preprint

A combinatorial approach to the theory of PI-algebras and exponential growth

A. Giambruno

1 Introduction

The theory of PI-algebras has been developed in the last decade mainly through combinatorial methods pertaining to the representation theory of the symmetric group in characteristic zero. These methods together with the study of the asymptotic behavior of a numerical sequence related to a PI-algebra have lead to the notion of the *PI-exponent* of an algebra. Through the scale determined by the exponent many results have been obtained in the last few years shedding new light on the understanding of the polynomial identities of an algebra.

Throughout F will be a field of characteristic zero. If $X = \{x_1, x_2, \dots\}$ is a countable set, we write $F\langle X\rangle$ for the free associative algebra on X over F. Recall that a polynomial $f(x_1, \dots, x_n) \in F\langle X\rangle$ is an identity for an F-algebra A if $f(a_1, \dots, a_n) = 0$ for all $a_1, \dots, a_n \in A$. It is easy to check that the polynomial identities of a given algebra A form a T-ideal $\mathrm{Id}(A)$ of $F\langle X\rangle$ i.e., a two-sided ideal which is invariant under all endomorphisms of $F\langle X\rangle$. In case $\mathrm{Id}(A) \neq 0$ and, so, A satisfies a nontrivial identity, then we say that A is a PI-algebra. By making use of a Vandermonde determinant argument, it is easy to see that a polynomial is an identity for an algebra if and only if all its homogeneous components are still identities for A. Moreover, if $\mathrm{char} F = 0$, by the well-known linearization process every polynomial f is equivalent to a finite set of multilinear polynomials. Hence $\mathrm{Id}(A)$ is completely determined by all the multilinear polynomials it contains.

For $n \geq 1$ let $V_n = \mathrm{Span}\{x_{\sigma(1)} \cdots x_{\sigma(n)} | \sigma \in S_n\}$ be the space of multilinear polynomials in the variables $x_1, \dots, x_n$ where S_n is the symmetric group of degree n. We write $c_n(A) = \dim_F V_n/(V_n \cap \mathrm{Id}(A))$ and we say that $c_n(A)$ is the nth codimension of A.

In the language of varieties of algebras, let var(A) denote the variety of algebras generated by A. Then $V_n/(V_n \cap \mathrm{Id}(A))$ describes all the multilinear elements of degree n of the relatively free algebra of var(A) of countable rank.

The asymptotic behavior of the sequence $\{c_n(A)\}_{n\geq 1}$ allows us to obtain important information on the identities of the algebra A. It is clear that A is a PI-algebra if and only if $c_n(A) < n!$ for some $n \geq 1$. Actually much more can be said: in [32] Regev proved that A is a PI-algebra if and only if the sequence of codimensions is exponentially bounded. Hence there exist constants α, β such that $c_n(A) \leq \alpha\beta^n$, for all $n \geq 1$.

It was recently proved that, for any PI-algebra A, $\lim_{n\to\infty} \sqrt[n]{c_n(A)}$ exists and is an integer. This integer, denoted Exp(A), is called the PI-exponent of the algebra A. We shall give below an account of the known results about the exponent and we shall characterize the corresponding T-ideals for certain classes of algebras. The main tool is the representation theory of the symmetric group S_n and some of the results will be expressed in the language of Young diagrams.

If A is an algebra with involution $*$ (e.g., the algebra of $n \times n$ matrices over F), then one can define similar codimensions in the free algebra with involution $F\langle X, *\rangle$. Several results can be obtained in this framework and, apart from their intrinsic interest, the study of the $*$-codimensions can help to give a better understanding of the ordinary case. The main tool for studying the $*$-identities of an algebra is the representation theory of the hyperoctahedral group $\mathbf{Z}_2 \wr S_n$.

2 The PI-exponent of an algebra

Throughout A will be a PI-algebra over the field F of characteristic zero. The Grassmann algebra G plays an important role in the theory of PI-algebras. Recall that G is the algebra generated by a countable dimensional vector space with basis $\{e_1, e_2, \dots\}$. subject to the relations $e_i e_j = -e_j e_i$, for $i, j = 1, 2, \dots$. G has a natural $\mathbf{Z}_2$-grading $G = G^{(0)} \oplus G^{(1)}$ where $G^{(0)}$ and $G^{(1)}$ are the spaces generated by the monomials in the e_i's of even degree and odd degree respectively.

In [22] Kemer characterized the T-ideals $\mathrm{Id}(A)$ for which the sequence of codimensions $\{c_n(A)\}_{n\geq 1}$ is polynomially bounded. He showed that $\mathrm{Id}(A)$ has this property if and only if $\mathrm{Id}(A) \not\subseteq \mathrm{Id}(G)$ and $\mathrm{Id}(A) \not\subseteq \mathrm{Id}(UT_2)$ where $UT_2 = UT_2(F)$ is the algebra of 2×2 upper triangular matrices over F. The sequences of codimensions for the algebras G and UT_2 are known: in [24] it was shown that $c_n(G) = 2^{n-1}$. Also, it follows from [26] that $c_n(UT_2) = 2 + 2^{n-1}(n-2)$. Hence, as a consequence of Kemer's result (see also [18]) it follows that there exists no PI-algebra with growth of codimensions intermediate between polynomial and exponential.

In the 1980s it was conjectured by Amitsur that, for any PI-algebra A, $\lim_{n\to\infty} \sqrt[n]{c_n(A)}$ exists and is an integer. This conjecture has been recently verified.

Theorem 1 ([15], [16]). *For any PI-algebra A, $\lim_{n\to\infty} \sqrt[n]{c_n(A)}$ exists and is an integer.*

Define $\lim_{n\to\infty} \sqrt[n]{c_n(A)} = \mathrm{Exp}(A)$ and call $\mathrm{Exp}(A)$ the *PI-exponent* of the algebra A. Thus any algebra of exponential growth has exponent $d \geq 2$.

An explicit method for computing $\mathrm{Exp}(A)$ for any given PI-algebra A was given in [15], [16]. This method relies on a well-known result of Kemer reducing the study of the identities of an algebra to that of the identities of the Grassmann envelope of a suitable finite dimensional $\mathbf{Z}_2$-graded algebra. Recall that if $B = B^{(0)} + B^{(1)}$ is a $\mathbf{Z}_2$-graded algebra over F, then $G(B) = B^{(0)} \otimes G^{(0)} + B^{(1)} \otimes G^{(1)}$ is the Grassmann envelope of B. Kemer's theorem is the following.

Theorem 2 ([23, Theorem 2.3]). *If A is a PI-algebra, then there exists a finite-dimensional $\mathbf{Z}_2$-graded algebra B such that* $\mathrm{Id}(A) = \mathrm{Id}(G(B))$.

The method of [15] and [16] reads as follows: let A be a PI-algebra over F and suppose that F is algebraically closed. Let $\mathrm{Id}(A) = \mathrm{Id}(G(B))$ with $B = B^{(0)} + B^{(1)}$

a finite dimensional $\mathbf{Z}_2$-graded algebra over F. Let $B = C + J$ be the Wedderburn decomposition of B (cf. [8, Theorem 72.19]) where C a maximal semisimple $\mathbf{Z}_2$-graded subalgebra and J is the Jacobson radical of B. It is well-known that $C = C_1 \oplus \cdots \oplus C_k$ where $C_1, \ldots, C_k$ are simple subalgebras homogeneous in the $\mathbf{Z}_2$-grading.

Consider all possible products of the type

$$D_1 J D_2 J \ldots J D_h \neq 0 \tag{1}$$

where $D_1, \ldots, D_h$ are distinct algebras among $C_1, \ldots, C_k$ and define

$$p = \dim(D_1 \oplus \cdots \oplus D_h).$$

If d is the maximal value of p where $D_1, \ldots, D_h$ satisfy (1), then $d = \operatorname{Exp}(G(B)) = \operatorname{Exp}(A)$.

From the above description it is clear that over an algebraically closed field the PI-exponent of a finite dimensional algebra A equals the dimension of a suitable semisimple subalgebra of A. Since the codimensions of an algebra do not change upon extension of the base field one can easily get the following result. For an algebra B we write $Z(B)$ for the center of B.

Theorem 3 ([15, Corollary 3, 4]). *Let A be a finite dimensional algebra over the field F.*

1) *If A is semisimple, then* $\operatorname{Exp}(A) = \dim_{Z(B)} B$ *where B is a simple subalgebra of A of greatest dimension.*
2) *A is central simple over F if and only if* $\operatorname{Exp}(A) = \dim_F A$.

Proof. Let $A = A_1 \oplus \cdots \oplus A_t$ be the decomposition of A into simple subalgebras. If $\bar{F}$ is the algebraic closure of F, then $\bar{A} = A \otimes_F \bar{F} \cong \oplus_i (A_i \otimes_F \bar{F}) \cong \oplus_i (B_{i1} \oplus \cdots \oplus B_{is_i})$ where $B_{i1} \cong \ldots \cong B_{is_i}$ are simple algebras central over $\bar{F}$ and $s_i = [Z(A_i) : F]$. By the description before the theorem, $\operatorname{Exp}(A) = \operatorname{Exp}(\bar{A}) = \max_i\{\dim_{\bar{F}} B_{i1}\}$ and the conclusion of the first part follows since $(\dim_{Z(A_i)} A_i)[Z(A_i) : F] = \dim_F A_i = \dim_{\bar{F}} A_i \otimes_F \bar{F} = (\dim_{\bar{F}} B_{i1})[Z(A_i) : F]$.

Suppose now that $\operatorname{Exp}(A) = \dim_F A = \dim_{\bar{F}} \bar{A}$. If $\bar{A}$ is nilpotent then $\operatorname{Exp}(A) = 0$. Hence $\bar{A}$ contains a semisimple subalgebra B. But then, by the description of the exponent, $\dim_{\bar{F}} \bar{A} = \operatorname{Exp}(A) \leq \dim_{\bar{F}} B$. Thus $\bar{A} = B$ is semisimple and by the hypothesis and the above we see that A must be central simple over F. □

An important class of T-ideals is given by the so-called verbally prime T-ideals. They were introduced by Kemer who showed that they are the basic building blocks for the study of the identities of arbitrary PI-algebras (see [23]). Recall that a T-ideal I is verbally prime if, whenever $f, g \in F\langle X\rangle$ are polynomials in distinct variables and $fg \in I$, then either $f \in I$ or $g \in I$. It turns out that if I is a verbally prime T-ideal then $I = \operatorname{Id}(A)$ where A is one of the following algebras: $F\langle X\rangle$, $M_k(F)$, $M_k(G)$

and $M_{k,l}(G)$ $(0 < l \leq [k/2])$ where

$$M_{k,l}(G) = \begin{array}{c} \\ k \\ l \end{array} \begin{array}{c} \begin{array}{cc} k & l \end{array} \\ \begin{pmatrix} G^{(0)} & G^{(1)} \\ G^{(1)} & G^{(0)} \end{pmatrix} \end{array}$$

and $M_k(F)$, $M_k(G)$ are the algebras of $k \times k$ matrices over F and G respectively.

As a consequence of the description of the PI-exponent we can now easily compute $\operatorname{Exp}(A)$ for any of the PI-algebras above.

Corollary 1 ([6],[33]).

1) $\operatorname{Exp}(M_k(F)) = k^2$.
2) $\operatorname{Exp}(M_k(G)) = 2k^2$.
3) $\operatorname{Exp}(M_{k,l}(G)) = (k+l)^2$.

Proof. Part 1) is part of Theorem 3 above. To prove 2), notice that $M_k(G) \cong M_k(F) \otimes G^{(0)} + cM_k(F) \otimes G^{(1)}$ is the Grassmann envelope of $M_k(D)$ where $D = F + cF, c^2 = 1$, with grading $(M_k(F), cM_k(F))$. Hence $\operatorname{Exp}(M_k(G)) = \dim M_k(F) + \dim M_k(F) = 2k^2$.

We prove 3) by noting that $M_{k,l}(G) \cong M_{k,l}^{(0)}(F) \otimes G^{(0)} + M_{k,l}^{(1)}(F) \otimes G^{(1)}$ is the Grassmann envelope of $M_{k,l}(F)$ with grading $(M_{k,l}^{(0)}(F), M_{k,l}^{(1)}(F))$. Hence $\operatorname{Exp}(M_{k,l}(G)) = \dim M_{k,l}^{(0)}(F) + \dim M_{k,l}^{(1)}(F) = k^2 + l^2 + 2kl = (k+l)^2$. □

One last application of the above method for computing $\operatorname{Exp}(A)$ is given for the algebra $UT_m = UT_m(F)$ of $m \times m$ upper triangular matrices over F. If the e_{ij}'s are the usual matrix units, write $UT_m = F_1 \oplus \cdots \oplus F_m + J$ where for every i, $F_i = Fe_{ii}$ and J is the ideal of strictly upper triangular matrices. Clearly $F_1 J F_2 \cdots J F_m \neq 0$ which implies the following.

Corollary 2. $\operatorname{Exp}(UT_m) = m$.

3 Algebras of low PI-exponent

A first application of the description of the exponent deals with algebras A such that $\operatorname{Exp}(A) \leq 1$, i.e., with polynomial growth of codimensions.

A remark is in order. If A is such an algebra then there exists a finite dimensional algebra B such that $\operatorname{Id}(A) = \operatorname{Id}(B)$.

In fact, since by [24] $c_n(G) = 2^{n-1}$, then, for n large, $c_n(A) < c_n(G)$. This implies that $\operatorname{Id}(A) \not\subseteq \operatorname{Id}(G)$ and, by a theorem of Kemer [23, Theorem 2.3] there exists a finite dimensional algebra B such that $\operatorname{Id}(A) = \operatorname{Id}(B)$. Hence, when studying T-ideals $\operatorname{Id}(A)$ with polynomial growth of codimensions, we may assume that A is finite dimensional. We shall next determine the structure of any finite dimensional algebra of PI-exponent at most one in case F is algebraically closed.

Theorem 4 ([18, Theorem 2]). *Let A be a finite dimensional algebra over an algebraically closed field F. Then the sequence of codimensions $\{c_n(A)\}_{n\geq 1}$ is polynomially bounded if and only if*

1) $A = A_0 \oplus A_1 \oplus \cdots \oplus A_m$, *a vector space direct sum of F-algebras where, for $i = 1, \ldots, m$, $A_i = B_i + J_i$, $B_i \cong F$, J_i is a nilpotent ideal of A_i and $A_0, J_1, \ldots, J_m$ are nilpotent right ideals of A;*
2) *for all $i, k \in \{1, \ldots, m\}, i \neq k$, $A_i A_k = 0$ and $B_i A_0 = 0$.*

Proof. Let $A = B + J$ be the Wedderburn decomposition of A where B is a semisimple subalgebra of A and $J = J(A)$ its Jacobson radical. Write $B = B_1 \oplus \cdots \oplus B_m$ with $B_1, \ldots, B_m$ simple F-algebras. Since $c_n(A)$ is polynomially bounded, by the description of the exponent, $B_i J B_k = 0$ for all $i \neq k$ and $\dim_F B_i = 1$, $i, k = 1, \ldots, m$.

Let $e = e_1 + \cdots + e_m$ be the decomposition of the unit element of B into orthogonal central (in B) idempotents; thus $e_i B = B_i \cong F$. Define, for all $i = 1, \ldots, m$, $J_i = e_i J$ and $J_0 = \{x \in J | Bx = 0\}$. It is easy to show that $A = B + J = (B_1 + J_1) \oplus \cdots \oplus (B_m + J_m) \oplus J_0$; let $A_i = B_i + J_i$ and $A_0 = J_0$.

For $i \neq k \in \{1, \ldots, m\}$, $A_i A_k = (B_i + J_i)(B_k + J_k) = 0$ since $e_i e_k = 0$ and $B_i J B_k = 0$. Also, for $i \neq 0$, $B_i A_0 = 0$.

Conversely, let A satisfy 1) and 2). From the relations $A_i A_k = 0$ and $B_i A_0 = 0$ it follows that $J = A_0 + J_1 + \cdots + J_m$ is a nilpotent two-sided ideal of A and $A = B_1 \oplus \cdots B_m \oplus J$ where $B_i \cong F$ for all i. Since from the defining relations $A_i A_k = 0$ and $B_i A_0 = 0$, it follows that $B_i J B_k = 0$ for all $i \neq k$, then $\mathrm{Exp}(A) \leq 1$.
□

Through the description of $\mathrm{Exp}(A)$ one can easily get Kemer's result characterizing algebras whose codimension sequence is polynomially bounded.

Theorem 5 ([22]). *Let A be a* PI-*algebra. Then* $\mathrm{Exp}(A) \leq 1$ *if and only if* $\mathrm{Id}(A) \not\subseteq \mathrm{Id}(G)$ *and* $\mathrm{Id}(A) \not\subseteq \mathrm{Id}(UT_2)$.

Proof. Since $\mathrm{Exp}(G) = \mathrm{Exp}(UT_2) = 2$, one direction is clear. Now suppose that $\mathrm{Id}(A) \not\subseteq \mathrm{Id}(G)$ and $\mathrm{Id}(A) \not\subseteq \mathrm{Id}(UT_2)$ and that $\mathrm{Exp}(A) > 1$. As mentioned above we may assume that $A = G(B)$ where B is a finite dimensional $\mathbf{Z}_2$-graded algebra. Also, by standard arguments we may assume that F is algebraically closed. Let $B = B_1 \oplus \cdots \oplus B_k + J$ be the Wedderburn decomposition of B where J is the Jacobson radical of B and $B_1, \ldots, B_k$ are simple subalgebras homogeneous in the $\mathbf{Z}_2$-grading.

Recall that a simple finite dimensional $\mathbf{Z}_2$-graded algebra over F is isomorphic to one of the following algebras:

i) $M_{a,b}(F) = \begin{pmatrix} A_{11} & A_{12} \\ A_{21} & A_{22} \end{pmatrix}$ where $A_{11}, A_{12}, A_{21}, A_{22}$ are any $a \times a$, $a \times b$, $b \times a$ and $b \times b$ matrices respectively, $a > 0$, $b \geq 0$, with grading

$$M_{a,b}^{(0)}(F)=\begin{pmatrix}A_{11} & 0\\ 0 & A_{22}\end{pmatrix},\quad M_{a,b}^{(1)}(F)=\begin{pmatrix}0 & A_{12}\\ A_{21} & 0\end{pmatrix};$$

ii) $M_N(F)\oplus cM_N(F)$ where $c^2=1$, with grading $M_N(F)^{(0)}=M_N(F)$, $M_N(F)^{(1)}=cM_N(F)$.

Now, if B contains one simple component of type i) with $a+b\geq 2$ or of type ii) with $N\geq 1$, then we would get that $G(B)$ contains a subalgebra isomorphic to either UT_2 or G, a contradiction.

Therefore, since $\mathrm{Exp}(G(B))>1$, we see that there exist $B_i, B_l, i\neq l$, such that $B_iJB_l\neq 0$ and $B_i\cong B_l\cong F$. Let $1_1, 1_2$ be the unit elements of B_i and B_l respectively. Then $1_1(j_0+j_1)1_2\neq 0$ for some $j_0\in J^{(0)}, j_1\in J^{(1)}$ where $J=J^{(0)}+J^{(1)}$ is the induced $\mathbf{Z}_2$-grading on J. Suppose for instance that $1_1j_01_2\neq 0$. If we set $1_1=u_{11}$, $1_2=u_{22}$, $1_1j_01_2=u_{12}$, then u_{11}, u_{22}, u_{12} generate a $\mathbf{Z}_2$-graded subalgebra V of B and $\mathrm{Id}(G(V))=\mathrm{Id}(UT_2)$. Similarly if $1_1j_11_2\neq 0$. Thus we see that $\mathrm{Id}(G(B))\subseteq\mathrm{Id}(UT_2)$ in this case also. □

Another description of PI-algebras of exponent ≤ 1 can be given in the language of S_n-characters. We need to introduce some terminology.

The symmetric group S_n acts on the left on the space V_n by

$$\sigma f(x_1,\dots,x_n)=f(x_{\sigma(1)},\dots,x_{\sigma(n)}),\ \sigma\in S_n,\ f(x_1,\dots,x_n)\in V_n.$$

This action induces a structure of left S_n-module on $\frac{V_n}{V_n\cap Id(A)}$ and we write $\chi_n(A)$ for its S_n-character; $\chi_n(A)$ is called the nth cocharacter of A.

It is well-known that there is a one-to-one correspondence between partitions of n and irreducible S_n-characters. Let χ_λ denote the irreducible S_n-character associated to the partition $\lambda=(\lambda_1,\dots,\lambda_t)\vdash n$ (we write $|\lambda|=n$) and let

$$\chi_n(A)=\sum_{\lambda\vdash n}m_\lambda\chi_\lambda$$

where $m_\lambda\geq 0$ are the corresponding multiplicities.

We write $\chi_\lambda(1)=d_\lambda$ for the degree of the irreducible S_n-character χ_λ and, if T_λ is a tableau of shape λ, we let e_{T_λ} be the corresponding essential idempotent of FS_n. We have (see [18])

Theorem 6 ([22]). *Let A be an algebra over the field F. Then $\{c_n(A)\}_{n\geq 1}$ is polynomially bounded if and only if*

$$\chi_n(A)=\sum_{\substack{\lambda\vdash n\\ |\lambda|-\lambda_1<q}}m_\lambda\chi_\lambda$$

for some constant q (in case $\dim_F A<\infty$, q can be taxen to be the nilpotency index of $J(A)$).

Proof. We clearly may assume that F is algebraically closed and suppose first that the codimensions of A are polynomially bounded. We may assume that A is finite dimensional over F and let $A = B_1 \oplus \cdots \oplus B_k + J$. Let λ be a partition of n such that $|\lambda| - \lambda_1 \geq q$ and suppose by contradiction that $m_\lambda \neq 0$. Then there exists a tableau T_λ such that $e_{T_\lambda}(x_1, \ldots, x_n) \notin Id(A)$. Let $\lambda' = (\lambda'_1, \ldots, \lambda'_t)$ be the conjugate partition of λ. Then $e_{T_\lambda}(x_1, \ldots, x_n)$ is a linear combination of polynomials each alternating on t disjoint sets of $\lambda'_1, \ldots, \lambda'_t$ variables, respectively. We shall reach a contradiction by proving that each such polynomial f vanishes in A.

Fix a basis of A which is the union of bases of $B_1, \ldots, B_m$ and J respectively. Since $B_i B_k = B_i J B_k = 0$ for all $i \neq k$, in order to get a nonzero value of f we must replace all the variables with elements of J and of one simple component, say, B_i. Also, since $\dim B_i = 1$, we can substitute at most one element of B_i in each alternating set. Hence we can substitute in all at most $t = \lambda_1$ elements from B_i. It follows that, in order to get a nonzero value, we must substitute at least $|\lambda| - \lambda_1 \geq q$ elements from J. Since $J^q = 0$ we get that $f \equiv 0$ and with this contradiction the proof of the first part of the theorem is complete.

Suppose now that $\chi_n(A) = \sum_{\lambda \vdash n} m_\lambda \chi_\lambda$ and $m_\lambda = 0$ whenever $|\lambda| - \lambda_1 \geq q$. By [5] the multiplicities m_λ are polynomially bounded; hence

$$c_n(A) = \sum_{\substack{\lambda \vdash n \\ |\lambda| - \lambda_1 < q}} m_\lambda d_\lambda \leq C n^t \sum_{\substack{\lambda \vdash n \\ |\lambda| - \lambda_1 < q}} d_\lambda$$

and the proof now follows from the hook formula for the degrees d_λ. □

The previous theorem says that A has polynomial growth of codimensions if and only if all the irreducible characters appearing with non-zero multiplicity in $\chi_n(A)$ have associated diagram with at most $q - 1$ boxes below the first row where $J^q = 0$.

One can go one step further and try to characterize algebras of PI-exponent two. To this end we introduce the following two algebras:

$$UT_2(G^{(0)}, G) = \begin{pmatrix} G^{(0)} & G \\ 0 & G \end{pmatrix}, \quad UT_2(G, G^{(0)}) = \begin{pmatrix} G & G \\ 0 & G^{(0)} \end{pmatrix}.$$

It is not hard to check that the two algebras above have distinct ideals of identities. Also $\operatorname{Exp}(UT_2(G^{(0)}, G)) = \operatorname{Exp}(UT_2(G^{(0)}, G)) = 3$.

By making use of the above algebras one can get a characterization of algebras of PI-exponent greater that 2.

Theorem 7 ([21, Theorem 1]). *Let A be a PI-algebra. Then* $\operatorname{Exp}(A) > 2$ *if and only if* $\operatorname{Id}(A) \subseteq \operatorname{Id}(B)$ *where B is one of the algebras* $UT_3(F)$, $M_2(F)$, $M_{1,1}(G)$, $UT_2(G^{(0)}, G)$, $UT_2(G, G^{(0)})$.

Proof. Suppose $\operatorname{Exp}(A) = p > 2$. As in the proof of the previous theorem, we may assume that $A = G(B)$ where B is a finite dimensional $\mathbf{Z}_2$-graded algebra and F is algebraically closed.

Let $B = B_1 \oplus \cdots \oplus B_k + J$ where $B_1, \ldots, B_k$ are simple subalgebras homogeneous in the $\mathbf{Z}_2$-grading. For each $i = 1, \ldots, k$, let $B_i = B_i^{(0)} + B_i^{(1)}$ and $J = J^{(0)} + J^{(1)}$ be the induced $\mathbf{Z}_2$-grading.

Now, if B contains one simple component of type $M_{a,b}(F)$ with $a + b \geq 2$ or of type $M_N(F) \oplus cM_N(F)$ with $N \geq 2$, then we would get that $G(B)$ contains an algebra isomorphic to either $M_2(F)$ or $M_{1,1}(G)$ and we would be done.

Therefore, since $p > 2$, we may assume that one of the following possibilities occurs:

1) for some $i \neq l$, $B_i J B_l \neq 0$ where $B_i \cong F + cF, c^2 = 1$ and $B_l \cong F$;
2) for some $i \neq l$, $B_i J B_l \neq 0$ where $B_i \cong F$ and $B_l \cong F + cF, c^2 = 1$;
3) there exist B_i, B_l, B_m distinct such that $B_i J B_l J B_m \neq 0$ and $B_i \cong B_l \cong B_m \cong F$.

Suppose 1) holds. Then there exists $a + cb \in B_i$ such that $(a + cb) j 1_3 \neq 0$ where 1_3 is the unit element of B_l and $j \in J$ is homogeneous. Also we may assume that $(a + cb) j_0 1_3 \neq 0$ for some $j_0 \in J^{(0)}$. Write $a + cb = u_{11}(a + b) + u_{22}(a - b)$ where $u_{11} = \frac{1+c}{2}, u_{22} = \frac{1-c}{2}$ and $1 = 1_{B_i}$ is the unit element of B_i. Set $u_{33} = 1_3$.

First consider the case when $j_0 u_{33}$ and $c j_0 u_{33}$ are linearly dependent over F. Since $c^2 = 1$ it follows that $c j_0 u_{33} = \pm j_0 u_{33}$.

Suppose $c j_0 u_{33} = j_0 u_{33}$. Then $u_{11} j_0 u_{33} = j_0 u_{33}$ and $u_{22} j_0 u_{33} = 0$. If we set $u_{13} = j_0 u_{33}$, then the u_{hk}'s behave like the corresponding matrix units of 3×3 matrices and the algebra generated by $u_{11}, u_{22}, u_{33}, u_{13}$ over F is isomorphic to the $\mathbf{Z}_2$-graded algebra $D = \begin{pmatrix} F & 0 & F \\ 0 & F & 0 \\ 0 & 0 & F \end{pmatrix}$ with grading

$$D^{(0)} = \left\{ \begin{pmatrix} \lambda & 0 & \nu \\ 0 & \lambda & 0 \\ 0 & 0 & \mu \end{pmatrix} \right\}, \quad D^{(1)} = \left\{ \begin{pmatrix} \lambda & 0 & \nu \\ 0 & -\lambda & 0 \\ 0 & 0 & 0 \end{pmatrix} \right\}.$$

Clearly,

$$G(D) = \left\{ \begin{pmatrix} a+b & 0 & z \\ 0 & a-b & 0 \\ 0 & 0 & t \end{pmatrix} \right\}$$

where $a, t \in G_0, b \in G_1$ and $z \in G$, and it is easy to check that $G(D) \cong UT_2(G, G^{(0)})$. Hence $\mathrm{Id}(A) \subseteq \mathrm{Id}(UT_2(G, G^{(0)}))$ and we are done in this case.

It can be shown that the same conclusion holds in all the remaining cases. Also in case 2) the same procedure as above shows that $\mathrm{Id}(A) \subseteq \mathrm{Id}(UT_2(G^{(0)}, G))$.

Finally suppose that 3) holds. Then there exist $j_0, j_0' \in J^{(0)}, j_1, j_1' \in J^{(1)}$ such that $1_1(j_0 + j_1) 1_2 (j_0' + j_1') 1_3 \neq 0$ where $1_1, 1_2, 1_3$ are the unit elements of B_i, B_l, B_m

respectively. In this case at least one of the products $1_1 j_r 1_2 j'_s 1_3$, $r, s \in \{0, 1\}$, is nonzero. Then, for fixed r and s, set $u_{11} = 1_1$, $u_{22} = 1_2$, $u_{33} = 1_3$, $u_{12} = 1_1 j_r 1_2$, $u_{23} = 1_2 j'_s 1_3$, $u_{13} = 1_1 j_r 1_2 j'_s 1_3$ and let D_{rs} be the $\mathbf{Z}_2$-graded subalgebra of B generated by $u_{11}, u_{22}, u_{33}, u_{12}, u_{23}, u_{13}$. By taking the Grassmann envelope of the algebra D_{rs}, we see that $G(B)$ must contain at least one of the following four algebras denoted E_1, E_2, E_3, E_4 respectively:

$$UT_3(F), \begin{pmatrix} G^{(0)} & G^{(0)} & G^{(1)} \\ 0 & G^{(0)} & G^{(1)} \\ 0 & 0 & G^{(0)} \end{pmatrix}, \begin{pmatrix} G^{(0)} & G^{(1)} & G^{(1)} \\ 0 & G^{(0)} & G^{(0)} \\ 0 & 0 & G^{(0)} \end{pmatrix}, \begin{pmatrix} G^{(0)} & G^{(1)} & G^{(0)} \\ 0 & G^{(0)} & G^{(1)} \\ 0 & 0 & G^{(0)} \end{pmatrix}.$$

It is easy to check that each E_i satisfies the identity $[x_1, x_2][x_3, x_4][x_5, x_6] \equiv 0$ and so all the identities of $UT_3(F)$ (cf. [29]). On the other hand, each one of the algebras E_2, E_3, E_4 has a subalgebra isomorphic to $UT_3(F)$. For instance, in the case of E_4 this subalgebra is generated by $e_{11}, e_{22}, e_{33}, xe_{12}, ye_{23}$ and xye_{13} where x and y are two distinct generators of G. Hence $\mathrm{Id}(A) \subseteq \mathrm{Id}(UT_3(F))$ and we are done.

From the description of the exponent it follows that

$$\mathrm{Exp}(UT_2(G^{(0)}, G)) = \mathrm{Exp}(UT_2(G, G^{(0)})) = \mathrm{Exp}(UT_3(F)) = 3$$

and $\mathrm{Exp}(M_2(F)) = \mathrm{Exp}(M_{1,1}(G)) = 4$. Hence, if $\mathrm{Id}(A) \subseteq \mathrm{Id}(B)$ where B is one of the five algebras above, then $\mathrm{Exp}(A) > 2$. □

Putting together the previous two results we get

Corollary 3. *Let A be a PI-algebra. Then* $\mathrm{Exp}(A) = 2$ *if and only if* $\mathrm{Id}(A) \not\subseteq \mathrm{Id}(B)$ *where B is one of the algebras $UT_3(F)$, $M_2(F)$, $M_{1,1}(G)$, $UT_2(G^{(0)}, G)$, $UT_2(G, G^{(0)})$ and either* $\mathrm{Id}(A) \subseteq \mathrm{Id}(G)$ *or* $\mathrm{Id}(A) \subseteq \mathrm{Id}(UT_2(F))$.

4 Upper block-triangular matrices and minimal PI-algebras

An interesting and useful class of finite dimensional algebras is given by the upper block-triangular matrix algebras.

Let $d_1, \ldots, d_m$ be positive integers. Then

$$UT(d_1, \ldots, d_m) = \begin{pmatrix} M_{d_1}(F) & & * \\ 0 & \ddots & \\ \vdots & & \\ 0 & \cdots\ 0 & M_{d_m}(F) \end{pmatrix}$$

where, for $i = 1, \ldots, m$, $M_{d_i}(F)$ is the algebra of $d_i \times d_i$ matrices over F, denotes the algebra of all upper block-triangular matrices over F. The Wedderburn decomposition of this algebra is readily written as

$$UT(d_1, \ldots, d_m) = B_1 \oplus \cdots \oplus B_m + J$$

where, for $i = 1, \dots, m$, $B_i \cong M_{d_i}(F)$ and the Jacobson radical J consists of all strictly upper block-triangular matrices. Since $B_1 J \cdots J B_m \neq 0$ it follows that $\mathrm{Exp}(UT(d_1, \dots, d_m)) = d_1^2 + \cdots + d_m^2$.

The relevance of these algebras is given by the following

Theorem 8 ([20, Lemma 2]). *Let F be an algebraically closed field of characteristic zero and let A be a finite dimensional algebra over F with* $\mathrm{Exp}(A) = d \geq 2$. *Then there exists a subalgebra of A isomorphic to $UT(d_1, \dots, d_m)$ and $d_1^2 + \cdots + d_m^2 = d$.*

Proof. Let $A = B_1 \oplus \cdots \oplus B_t + J$ where $B_1, \dots, B_t$ are simple subalgebras of A. Since $\mathrm{Exp}(A) = d$, we may assume that $B_1 J B_2 J \cdots J B_m \neq 0$ for some $m \leq t$ and $\dim B_1 + \cdots + \dim B_m = d$. Since F is algebraically closed, for every i, $B_i \cong M_{d_i}(F)$ for a suitable d_i.

For convenience we identify $B_1 \oplus \cdots \oplus B_m$ with a subalgebra of $M_q(F)$ of block-diagonal matrices where $q = d_1 + \cdots + d_m$. In fact if we set

$$q_0 = 0, q_1 = d_1, \dots, q_k = d_1 + \cdots + d_k, \dots, q_m = q_1 + \cdots + d_m = q,$$

then B_j is the linear span of all matrix units $e_{\alpha\beta}$ with $q_{j-1} + 1 \leq \alpha, \beta \leq q_j$, $j = 1, \dots, m$.

Since $B_1 J B_2 J \cdots J B_m \neq 0$ there exist $u_i \in J$ and $e_{\alpha_j \beta_j} \in B_j$ such that

$$e_{\alpha_1\beta_1} u_1 e_{\alpha_2\beta_2} u_2 \cdots u_{m-1} e_{\alpha_m\beta_m} \neq 0 \tag{1}$$

and

$$q_{i-1} + 1 \leq \alpha_i, \beta_i \leq q_i, i = 1, \dots, m.$$

In order to prove the theorem it is enough to define elements x_{ij} in the algebra generated by $B_1, \dots, B_m, u_1, \dots, u_{m-1}$ and show that they form a basis of a subalgebra C of A with the same structural constants as $UT(d_1, \dots, d_m)$.

If $q_{k-1} + 1 \leq i, j \leq q_k$ for some $1 \leq k \leq m$, then define $x_{ij} = e_{ij} \in B_k$. Now let $1 \leq i < j \leq q$ and

$$q_{k-1} + 1 \leq i \leq q_k, \quad q_{k+s-1} + 1 \leq j \leq q_{k+s}$$

where $1 \leq k \leq m, s \geq 1$ and $k + s \leq m$. Then define

$$x_{ij} = e_{i\alpha_k} e_{\alpha_k\beta_k} u_k e_{\alpha_{k+1}\beta_{k+1}} u_{k+1} \cdots u_{k+s-1} e_{\alpha_{k+s}\beta_{k+s}} e_{\beta_{k+s} j}.$$

It is not hard to show that these elements have the required properties and the theorem is proved. □

The T-ideal of an upper block-triangular matrix algebra has some special properties that we shall now describe. First we need to compare the identities of two such algebras.

Theorem 9 ([20, Corollary 1]). *If $(d_1, \dots, d_m)$ and $(q_1, \dots, q_s)$ are two distinct sequences of positive integers, then*

$$\mathrm{Id}(UT(d_1, \dots, d_m)) \neq \mathrm{Id}(UT(q_1, \dots, q_s)).$$

Recall that if I is a T-ideal of the free algebra $F\langle X\rangle$ then $I = \mathrm{Id}(A)$ for some PI-algebra A. Then in this case we write $\mathrm{Exp}(I) = \mathrm{Exp}(A)$. If A and B are two PI-algebras and $\mathrm{Id}(A) = \mathrm{Id}(B)$, we say that A and B are PI-equivalent.

Among algebras of some fixed PI-exponent a prominent role is played by the so-called minimal algebras. Let A be a PI-algebra and $\mathrm{Exp}(A) = d$. We say that A is *minimal of exponent* d if, for every T-ideal I such that $I \underset{\neq}{\supset} \mathrm{Id}(A)$, we have that $\mathrm{Exp}(I) < d$.

From Theorem 5 it follows that the algebras G and UT_2 are the only two minimal algebras of PI-exponent 2 (up to equivalence).

A family of minimal PI-algebras was given by Drensky in [9] and [10]. He proved that, if A is an algebra such that $\mathrm{Id}(A)$ is generated by either the polynomial $[x_1, x_2] \cdots [x_{2n-1}, x_{2n}]$ or $[x_1, x_2, x_3] \cdots [x_{3n-2}, x_{3n-1}, x_{3n}]$, then A is minimal. Drensky also conjectured that an algebra A is minimal if and only if its T-ideal of identities $\mathrm{Id}(A)$ is a product of verbally prime T-ideals.

The connection between minimal algebras and upper block-triangular matrix algebras is given by the following.

Theorem 10 ([20, Theorem 1]). *Let A be a finitely generated algebra over the field F of characteristic zero. Then A is minimal of exponent $d \geq 2$ if and only if $\mathrm{Id}(A) = \mathrm{Id}(UT(d_1, \dots, d_m))$ for some $d_1, \dots, d_m$ with $d_1^2 + \cdots + d_m^2 = d$.*

By using a well-known result of Lewin [27] it can be proved that the T-ideal of identities of an upper block-triangular matrix algebra coincides with the product of the T-ideals of identities of the block matrices. From the previous theorem a proof of Drensky's conjecture in the case of finitely generated algebras follows easily.

Theorem 11 ([20, Corollary 4]). *Let A be a finitely generated PI-algebra and $\mathrm{Exp}(A) = d \geq 2$. Then A is minimal of exponent d if and only if $\mathrm{Id}(A) = I_1 \cdots I_k$ where $I_1, \dots, I_k$ are verbally prime T-ideals.*

5 Algebras with involution and *-exponent

An interesting class of associative algebras is given by the algebras with involution. Throughout F will be a field of characteristic zero and A an F-algebra with involution $*$. Hence $*$ is an anti-automorphism of A of order two.

We let $F\langle X, *\rangle = F\langle x_1, x_1^*, x_2, x_2^*, \dots\rangle$ be the free algebra with involution $*$ of countable rank. Recall that a polynomial $f(x_1, x_1^*, \dots, x_n, x_n^*) \in F\langle X, *\rangle$ is a $*$-identity for the algebra A if $f(a_1, a_1^*, \dots, a_n, a_n^*) = 0$ for all $a_1, \dots, a_n \in A$. The set $\mathrm{Id}(A, *)$ of all $*$-identities of A is a $*$-T-ideal of $F\langle X, *\rangle$, i.e., an ideal invariant under all endomorphisms of $F\langle X, *\rangle$ commuting with the involution.

The connection between algebras with involution satisfying a *-identity and PI-algebras, was given by Amitsur in [1] and [2]. He showed that, if A satisfies a nontrivial *-identity, then A is a PI-algebra.

An important example of a PI-algebra with involution is given by the algebra of $k \times k$ matrices over F. It turns out that, if $*$ is an involution on $M_k(F)$, then $\mathrm{Id}(M_k(F), *) = \mathrm{Id}(M_k(F), t)$ or $\mathrm{Id}(M_k(F), s)$ where t and s are the transpose and the symplectic involution on $M_k(F)$, respectively. Hence, when studying the *-identities of $M_k(F)$, we may consider only these two involutions.

In characteristic zero since every *-identity is equivalent to a system of multilinear ones, we may restrict ourselves to multilinear polynomials. For $n \geq 1$, let $V_n(*)$ be the space of multilinear $*$-polynomials in $x_1, x_1^*, \ldots, x_n, x_n^*$. For a PI-algebra A with involution, we write $c_n(A, *) = \dim V_n(*)/(V_n(*) \cap \mathrm{Id}(A, *))$ and we say that $c_n(A, *)$ is the nth *-codimension of A.

In [14] it was noted that, as in the ordinary (non-involution) case, if A satisfies a nontrivial $*$-identity, then the sequence $\{c_n(A, *)\}_{n \geq 1}$ is exponentially bounded; moreover an explicit bound was found in [3].

The asymptotic behavior of $c_n(M_k(F), *)$ was determined in [4] for both $* = t$ and $* = s$. It turns out that $c_n(M_k(F), *) \underset{n \to \infty}{\simeq} Cn^t k^{2n}$ for some explicit constants C and t. This result was achieved by proving that the $*$-trace codimensions and the $*$-codimensions are asymptotically equal and then combining a result of Loday and Procesi [28] on the $*$-trace identities of $M_k(F)$ with some asymptotic computations of Regev [34].

For general algebras, in [13] and [30] it was shown that there are no PI-algebras with involution with intermediate growth of the *-codimensions. In [17] the authors determined the exponential behavior of the sequence $c_n(A, *)$ for any finite dimensional algebra A. The theorem is the following.

Theorem 12 ([17]). *If A is a finite dimensional algebra with involution, then the limit $\lim_{n \to \infty} \sqrt[n]{c_n(A, *)}$ exists and is an integer.*

Define $\lim_{n \to \infty} \sqrt[n]{c_n(A, *)} = \mathrm{Exp}(A, *)$. As in the ordinary (non-involution) case $\mathrm{Exp}(A, *)$ (called the **-PI-exponent* of A) can be explicitly computed.

Recall that an algebra with involution A is $*$-simple if A has no proper $*$-invariant ideals (i.e., ideals I such that $I^* = I$). It is well-known and easy to prove that, if A is $*$-simple, then either A is simple or $A \cong A_1 \oplus A_1^{\mathrm{op}}$ where A_1 is a simple homomorphic image of A and $*$ on $A_1 \oplus A_1^{\mathrm{op}}$ is the exchange involution $(a, b)^* = (b, a)$ (cf. [35, Proposition 2.13.24]).

Suppose now that F is algebraically closed and A is a finite dimensional F-algebra. By [18, Theorem 4] there exists a maximal semisimple subalgebra B of A which is invariant under $*$. Let $A = B + J$ be the Wedderburn decomposition of A with $B = B^*$. Then $\mathrm{Exp}(A, *) = \max_i \dim_F (C_1^{(i)} + \cdots + C_{t_i}^{(i)})$ where $C_1^{(i)}, \ldots, C_{t_i}^{(i)}$ are distinct $*$-simple subalgebras of B and $C_1^{(i)} J C_2^{(i)} J \cdots J C_{t_i}^{(i)} \neq 0$.

For an algebra with involution A over F we let $Z = Z(A)$ be the center of A and $Z^+ = Z(A)^+ = Z \cap A^+$ be the symmetric center of A. The next result gives an exact estimate of $\mathrm{Exp}(A, *)$ in the case of simple or semisimple algebras.

Theorem 13 ([20, Corollary 2,3]). *Let A be a finite dimensional algebra with involution over F.*

1) *If A is semisimple, then* $\mathrm{Exp}(A, *) = \dim_{Z(B)^+} B$ *where B is a $*$-simple subalgebra of A of greatest dimension.*
2) *A is $*$-simple and $F = Z^+$ if and only if* $\mathrm{Exp}(A, *) = \dim_F A$.

It is worth mentioning that, in the light of the previous result, the equality $\mathrm{Exp}(M_k(F), *) = \mathrm{Exp}(M_k(F)) = \dim_F M_k(F)$ obtained in [4] is actually characteristic of finite dimensional simple algebras with involution of the first kind, i.e., $Z^+ = Z$.

From the description of the *-PI-exponent we can also derive the analogue of Theorem 4.

We first need to introduce some terminology. For every $i = 1, 2, \ldots$, let $s_i = x_i + x_i^*$ and $k_i = x_i - x_i^*$. Then $V_n(*)$ can be viewed as a space of polynomials in symmetric and skew variables.

Now, for $r = 0, \ldots, n$, let

$$\begin{aligned} V_{r,n-r} = \ & \mathrm{Span}\{w_{\sigma(1)} \cdots w_{\sigma(n)} | \sigma \in S_n, \ w_i = s_i \\ & \text{for } i = 1, \ldots, r \text{ and } w_i = k_i \text{ for } i = r+1, \ldots, n\}. \end{aligned}$$

Thus $V_{r,n-r}$ is the space of multilinear polynomials in $s_1, \ldots, s_r, k_{r+1}, \ldots, k_n$. It is clear that in order to study $V_n(*) \cap \mathrm{Id}(A, *)$ it is enough to study $V_{r,n-r} \cap \mathrm{Id}(A, *)$ for all r. Define

$$c_{r,n-r}(A, *) = \dim_F \frac{V_{r,n-r}}{V_{r,n-r} \cap \mathrm{Id}(A, *)}.$$

It is not hard to show (see [11, Theorem 1.3]) that the codimensions $c_n(A, *)$ and $c_{r,n-r}(A, *)$ are related by the following equality:

$$c_n(A, *) = \sum_{r=0}^{n} \binom{n}{r} c_{r,n-r}(A, *).$$

We can now prove the characterization of finite dimensional algebras with polynomial growth of $*$-codimensions.

Theorem 14 ([18, Theorem 6]). *Let A be a finite dimensional algebra with involution over an algebraically closed field F. Then the sequence of $*$-codimensions $\{c_n(A, *)\}_{n \geq 1}$ is polynomially bounded if and only if*

1) *the sequence of codimensions $\{c_n(A)\}_{n \geq 1}$ is polynomially bounded;*
2) *A=B+J, where B is a maximal semisimple subalgebra of A and $b = b^*$ for all $b \in B$.*

Proof. By [14, Lemma 4.4], for all n, $c_n(A) \le c_n(A,*) \le \alpha n^t$ for some constants α, t, and the sequence of codimensions is polynomially bounded.

From the description of the $*$-exponent it follows that $A = B + J$, where $B = B^* = B_1 \oplus \cdots \oplus B_m$ and $B_i = B_i^* \cong F$, $B_i J B_k = 0$ for all $i \neq k$. It is clear then that $*$ is the identity map on B.

Conversely, suppose that the sequence of codimensions is polynomially bounded and $A = B + J$ with B semisimple and $*$ is the identity on B. In this case, if $a \in A$, write $a = b + j$, $b \in B$, $j \in J$. Then $a - a^* = j - j^* \in J$ and J contains the set of skew elements of A.

Notice that if $f(x_1, \dots, x_n) \in V_n \cap \mathrm{Id}(A)$ then, for every $r = 0, \dots, n$,

$$f(s_1, \dots, s_r, k_{r+1}, \dots, k_n) \in V_{r,n-r} \cap \mathrm{Id}(A,*).$$

Hence $c_{r,n-r}(A,*) \le c_n(A) \le \alpha n^t$, for some α, t and for all r. Let $J^q = 0$. Since J contains all skew elements of A, then, for all $r \le n-q$, $V_{r,n-r} \cap \mathrm{Id}(A,*) = V_{r,n-r}$, and $c_{r,n-r}(A,*) = 0$ follows. But then for all n we obtain

$$c_n(A,*) = \sum_{r=0}^{n} \binom{n}{r} c_{r,n-r}(A,*) \le \alpha n^t \sum_{r=n-q+1}^{n} \binom{n}{r}$$

$$= \alpha n^t \sum_{r=0}^{q-1} \binom{n}{r} \le \alpha n^{t+q}$$

and $c_n(A,*)$ is polynomially bounded. □

6 $*$-cocharacters and polynomial growth

Recall that, if $\mathbf{Z}_2 = \{1, *\}$ is the multiplicative group of order 2, then the hyperoctahedral group of degree n is the wreath product $H_n = \mathbf{Z}_2 \wr S_n = \{(a_1, \dots, a_n; \sigma) | a_i \in \mathbf{Z}_2 = \{1, *\}, \sigma \in S_n\}$; hence H_n is the semidirect product of $\mathbf{Z}_2^n$ and S_n where S_n acts on $\mathbf{Z}_2^n$ by permuting the coordinates.

If we view $V_n(*)$ as the space of multilinear polynomials in the symmetric variables s_i and the skew variables k_i ($i = 1, \dots, n$), then $V_n(*)$ has a natural structure of left H_n-module induced by defining for $h = (a_1, \dots, a_n; \sigma) \in H_n$, $hs_i = s_{\sigma(i)}$ and $hk_i = k_{\sigma(i)}^{a_{\sigma(i)}} = \pm k_{\sigma(i)}$ (see [14]).

If A is an algebra with involution, then the space $V_n(*) \cap \mathrm{Id}(A,*)$ is invariant under the H_n action; this in turn induces a structure of left H_n-module on $V_n(*)/(V_n(*) \cap \mathrm{Id}(A,*)) = V_n(A,*)$; let $\chi_n(A,*)$ be its character. It is well-known that the irreducible H_n-characters are in one-to-one correspondence with the pairs of partitions (λ, μ) for which $\lambda \vdash r$, $\mu \vdash n - r$ and $r = 0, \dots, n$. If $\chi_{\lambda,\mu}$ denotes the H_n-character corresponding to (λ, μ), then the character $\chi_n(A,*)$ has the decomposition

$$\chi_n(A,*) = \sum_{|\lambda|+|\mu|=n} m_{\lambda,\mu}\chi_{\lambda,\mu}$$

where $m_{\lambda,\mu} \geq 0$ are the corresponding multiplicities.

There is an easy way to relate H_n-representations and $S_r \times S_{n-r}$-representations, $r = 0, 1, \ldots, n$: recall that $V_{r,n-r}$ is the space of multilinear polynomials in $s_1, \ldots, s_r, k_{r+1}, \ldots, k_n$. We write $s_1 = y_1, \ldots, s_r = y_r, k_{r+1} = z_1, \ldots, k_n = z_{n-r}$ and we view $V_{r,n-r}$ as the space of multilinear polynomials in the symmetric variables $y_1, \ldots, y_r$ and in the skew variables $z_1, \ldots, z_{n-r}$. We let $S_r \times S_{n-r}$ act on the left on $V_{r,n-r}$ as follows: if $f(y_1, \ldots, y_r, z_1, \ldots, z_{n-r}) \in V_{r,n-r}$, then

$$(\sigma,\tau)f = f(y_{\sigma(1)}, \ldots, y_{\sigma(r)}, z_{\tau(1)}, \ldots, z_{\tau(n-r)}),$$

for $(\sigma,\tau) \in S_r \times S_{n-r}$.

Since T-ideals are invariant under permutations of symmetric (respectively, skew) variables, we see that $V_{r,n-r} \cap \mathrm{Id}(A,*)$ is invariant under this action and so

$$V_{r,n-r}(A,*) = V_{r,n-r}/(V_{r,n-r} \cap \mathrm{Id}(A,*))$$

has an induced structure of left $S_r \times S_{n-r}$-module; we write $\psi_{r,n-r}(A,*)$ for its character. By complete reducibility we decompose

$$\psi_{r,n-r}(A,*) = \sum_{\substack{\lambda\vdash r\\ \mu\vdash n-r}} m_{\lambda,\mu}(\chi_\lambda \otimes \chi_\mu)$$

where χ_λ (respectively χ_μ) denotes the usual S_r-character (respectively S_{n-r}-character), $\chi_\lambda \otimes \chi_\mu$ is the irreducible $S_r \times S_{n-r}$-character associated to the pair (λ,μ) and $m_{\lambda,\mu} \geq 0$ is the corresponding multiplicity.

The relation between the H_n-character of $V_n(A,*)$ and the $S_r \times S_{n-r}$- character of $V_{r,n-r}(A,*)$ is given in [11, Theorem 1.3] as follows: if A is an algebra with involution, then, for all $r \leq n$,

$$\chi_n(A,*) = \sum_{r=0}^{n} \sum_{\substack{\lambda\vdash r\\ \mu\vdash n-r}} m_{\lambda,\mu}\chi_{\lambda,\mu}, \text{ and } \psi_{r,n-r}(A,*) = \sum_{\substack{\lambda\vdash r\\ \mu\vdash n-r}} m_{\lambda,\mu}(\chi_\lambda \otimes \chi_\mu).$$

A characterization of algebras with involution whose $*$-codimension sequence is polynomially bounded can be given in the language of H_n-characters.

Let $G_2 = F \oplus F$ be the 2-dimensional algebra with exchange involution $(a,b)^* = (b,a)$. It is not difficult to see that a basis of $\mathrm{Id}(G_2,*)$ is given by the $*$-polynomials $[y_1,y_2]$, $[y_1,z_2]$ and $[z_1,z_2]$. Also one can determine the decomposition of the nth $*$-cocharacter of G_2. In fact (see [12]) we have that, for every $n \geq 1$,

$$\chi_n(G_2,*) = \sum_{r=0}^{n} \chi_{(r),(n-r)}.$$

Hence $c_n(G_2,*) = 2^n$.

Theorem 15 ([12, Theorem 2]). *Let A be an algebra with involution. Then* $\mathrm{Id}(A,*) \not\subseteq \mathrm{Id}(G_2,*)$ *if and only if* $z^d \in \mathrm{Id}(A,*)$, *for some* $d \geq 1$, *where z is a skew variable.*

Proof. Since $z^d \notin \mathrm{Id}(G_2,*)$, one implication is obvious. Suppose now that

$$\mathrm{Id}(A,*) \not\subseteq \mathrm{Id}(G_2,*)$$

and let $f = f(y_1,\dots,y_r,z_1,\dots,z_{n-r}) \in \mathrm{Id}(A,*)$, $f \notin \mathrm{Id}(G_2,*)$. By the standard multilinearization process we clearly may assume that f is multilinear; hence f does not vanish on a basis of G_2. Let $a=(1,1)$, $b=(1,-1)$; then $\{a\}$, $\{b\}$ are bases of the subspace of symmetric and skew elements of G_2 respectively and $b^2=a$. Since $f \notin \mathrm{Id}(G_2,*)$, then

$$0 \neq f(a,\dots,a,b,\dots,b) = f(b^2,\dots,b^2,b,\dots,b) = \alpha b^{n+r}$$

where $\alpha \neq 0$ is the sum of all the coefficients of f. Since z^2 is a symmetric variable, it follows that $f(z^2,\dots,z^2,z,\dots,z) = \alpha z^{n+r} \in \mathrm{Id}(A,*)$ and, since $\alpha \neq 0$, we get $z^{n+r} \in \mathrm{Id}(A,*)$. □

Lemma 1. *Let A be such that $c_n(A,*) \leq \alpha n^t$ for some α. Then there exists a constant β such that*

$$\chi_n(A,*) = \sum_{|\lambda|+|\mu|=n} m_{\lambda,\mu}\chi_{\lambda,\mu} \tag{1}$$

where $m_{\lambda,\mu}=0$ if either $|\lambda|-\lambda_1>\beta$ or $|\mu|-\mu_1>\beta$.

Proof. We decompose the ordinary n^{th} cocharacter of A into irreducible S_n-characters

$$\chi_n(A) = \sum_{\nu\vdash n} m_\nu\chi_\nu \tag{2}$$

where, for $\nu \vdash n$, m_ν is the multiplicity of χ_ν in $\chi_n(A)$. By [14, Lemma 4.4] $c_n(A) \leq c_n(A,*) \leq \alpha n^t$; hence by Theorem 6 there exists a constant β such that, in (2), $m_\nu = 0$ whenever $|\nu|-\nu_1>\beta$.

Suppose by contradiction that there exist $\lambda \vdash r$, $\mu \vdash n-r$ such that $|\lambda|-\lambda_1>\beta$ (or $|\mu|-\mu_1>\beta$) and $m_{\lambda,\mu}\neq 0$. Then there exist tableaux T_λ, T_μ such that $e_{T_\lambda}e_{T_\mu}$ has a nontrivial action on $V_{r,n-r}(A,*)$. This says that there exists a nontrivial polynomial $f \in e_{T_\lambda}e_{T_\mu}V_{r,n-r}$ such that $f=f(y_1,\dots,y_r,z_1,\dots,z_{n-r})$ is not a $*$-identity of A. In particular this implies that the polynomial $f(x_1,\dots,x_r,x_{r+1},\dots,x_n)$ is not a ordinary identity of A. Notice that, by the above, $F(S_r\otimes S_{n-r})f(x_1,\dots,x_n)$ is an irreducible $S_r\otimes S_{n-r}$-module where S_r acts on $x_1,\dots,x_r$ and S_{n-r} acts on $x_{r+1},\dots,x_n$.

Let $M = F(S_r\otimes S_{n-r})f \uparrow S_n$ be the S_n-module which is obtained by inducing $F(S_r\otimes S_{n-r})f$ up to S_n. If $\chi^{S_n}(M)$ denotes the S_n-character of M, then we can decompose

$$\chi^{S_n}(M) = \sum_{\nu\vdash n} m'_\nu\chi_\nu. \tag{3}$$

Since $|\lambda| - \lambda_1 \geq \beta$, (or $|\mu| - \mu_1 \geq \beta$) by the Littlewood–Richardson rule [25, Theorem 2.8.13], we see that $m'_\nu = 0$ for all $\nu \vdash n$ such that $|\nu| - \nu_1 \leq \beta$.

Since $M \subseteq V_n(A)$, by comparing (2) and (3) we see that $m'_\nu \leq m_\nu$ for all $\nu \vdash n$. Hence $m'_\nu = 0$ for all $\nu \vdash n$ and this contradicts $M \neq 0$. This completes the proof of the lemma. □

For a partition $\lambda \vdash n$, let $d_\lambda = \chi_\lambda(1)$ be the degree of the irreducible S_n-character χ_λ. If $d_{\lambda,\mu} = \chi_{\lambda,\mu}(1)$ is the degree of the irreducible H_n-character, then it is well-known that $d_{\lambda,\mu} = \binom{n}{r} d_\lambda d_\mu$ where $\lambda \vdash r, \mu \vdash n-r$.

Lemma 2. *Let A be such that $c_n(A,*) \leq \alpha n^t$ for some α. Then there exists a constant γ such that, in (1), $m_{\lambda,\mu} = 0$ if $|\lambda| > \gamma$ and $|\mu| > \gamma$.*

Proof. Let $n > 2t$. If r is such that $r > t$ and $n - r > t$, then $\binom{n}{r} \geq \binom{n}{t+1}$. Since $\binom{n}{t+1}$ is a polynomial in n of degree $t+1$ and $c_n(A,*) \leq \alpha n^t$, there exists N such that, for $n > N$, $\binom{n}{r} \geq \binom{n}{t+1} > c_n(A,*)$. Let $\gamma = \max\{N, t\}$ and suppose that $r = |\lambda| > \gamma$ and $n - r = |\mu| > \gamma$. In case $m_{\lambda,\mu} \neq 0$, we get

$$c_n(A,*) \geq d_{\lambda,\mu} = \binom{n}{r} d_\lambda d_\mu \geq \binom{n}{r} > c_n(A,*),$$

a contradiction. Hence $m_{\lambda,\mu} = 0$ and the lemma is proved. □

Theorem 16 ([12, Theorem 3]). *Let A be an algebra with involution. Then $c_n(A,*) \leq \alpha n^t$ for some α, t if and only if there exists a constant δ such that*

$$\chi_n(A,*) = \sum_{|\lambda|+|\mu|=n} m_{\lambda,\mu}\chi_{\lambda,\mu}$$

and $m_{\lambda,\mu} = 0$ whenever either $|\lambda| - \lambda_1 > \delta$ or $|\mu| > \delta$.

Proof. Since the $*$-codimensions of G_2 have exponential growth, we see that $\mathrm{Id}(A,*) \not\subseteq \mathrm{Id}(G_2,*)$. Hence, by Theorem 15, there exists an integer $d \geq 1$ such that $z^d \in \mathrm{Id}(A,*)$. Let $\delta = d(\beta + \gamma + 1)$ where β and γ are the integers determined in Lemma 1 and Lemma 2 respectively.

Let μ be such that $|\mu| > \delta$ and suppose that, for some λ, $\chi_{\lambda,\mu}$ appears with nonzero multiplicity in $\chi_n(A,*)$, i.e., $m_{\lambda,\mu} \neq 0$. But then by Lemma 2 we get $|\lambda| \leq \gamma$ and by Lemma 1 we get $|\mu| - \mu_1 \leq \beta$.

Let T_λ and T_μ be any two tableaux of shapes λ and μ respectively. For any polynomial $f \in e_{T_\lambda} e_{T_\mu} V_{r,n-r}$ we denote by g the polynomial obtained from f by identifying all the skew variables corresponding to the first row of μ (denote this variable by z). Recall that since $\mathrm{char} F = 0$, f and g are equivalent $*$-identities; it follows that in order to prove that f vanishes in A it is enough to show that so does g.

Since $n = |\lambda| + |\mu| > \delta = d(\beta + \gamma + 1)$ and $n - \mu_1 = |\lambda| + |\mu| - \mu_1 \leq \beta + \gamma$, then $\mu_1 \geq (d-1)(\beta + \gamma + 1) + 1$; from this it follows that every monomial of g contains z^d as a submonomial. Since z^d vanishes on A, we see that $g \in \mathrm{Id}(A,*)$ and so $m_{\lambda,\mu} = 0$, a contradiction. This completes the proof of the theorem. □

We remark that the above theorem says that, if $\chi_{\lambda,\mu}$ appears with nonzero multiplicity in $\chi_n(A,*)$, then the number of boxes of μ and the number of boxes of λ below the first row must be bounded by a constant.

Another characterization of algebras with involution having polynomial growth of $*$-codimensions can be given by means of G_2 and M where M is an algebra introduced and extensively studied in [30]: M is the 4-dimensional algebra over F with involution $*$ generated by $\{a, b, c, c^*\}$ satisfying the relations: $a = a^* = a^2, b = b^* = b^2, ac = cb = c, c^*a = bc^* = c^*$ and all the other products are zero. These algebras play the role of the infinite dimensional Grassmann algebra and of the algebra of 2×2 upper triangular matrices in the general theory of PI-algebras.

The final result is the following.

Theorem 17 ([13, Theorem 4]). *Let A be an algebra with involution. Then the sequence of $*$-codimensions $\{c_n(A,*)\}_{n\geq 1}$ is polynomially bounded if and only if* $\mathrm{Id}(A,*) \not\subseteq \mathrm{Id}(G_2,*)$ *and* $\mathrm{Id}(A,*) \not\subseteq \mathrm{Id}(M,*)$.

References

[1] Amitsur, S.A. (1968): Rings with involution. Israel J. Math. **6**, 99–106

[2] Amitsur, S.A. (1968): Identities in rings with involution. Israel J. Math. **7**, 63–68

[3] Bahturin, Y., Giambruno, A., Zaicev, M. (1999): G-identities on associative algebras. Proc. Amer. Math. Soc. **127**, 63–69

[4] Berele, A., Giambruno, A., Regev, A. (1996): Involution codimensions and trace codimensions of matrices are asymptotically equal. Israel J. Math. **96**, 49–62

[5] Berele, A., Regev, A. (1983): Applications of hook Young diagrams to P.I. algebras. J. Algebra **82**, 559–567

[6] Berele, A., Regev, A. (1995): On the codimensions of the verbally prime P.I. algebras. Israel J. Math. **91**, 239–247

[7] Berele, A., Regev, A. (1998): Codimensions of products and intersections of verbally prime T-ideals. Israel J. Math. **103**, 17–28

[8] Curtis, C.W., Reiner, I. (1962): Representation theory of finite groups and associative algebras. Wiley, New York

[9] Drensky, V. (1987): Extremal varieties of algebras. I. Serdica **13**, 320–332 (Russian)

[10] Drensky, V. (1988): Extremal varieties of algebras. II. Serdica **14**, 20–27 (Russian)

[11] Drensky, V., Giambruno, A. (1994): Cocharacters, codimensions and Hilbert series of the polynomial identities for 2×2 matrices with involution. Canad. J. Math. **46**, 718–733

[12] Giambruno, A., Mishchenko, S. (2000): Polynomial growth of the $*$-codimensions and Young diagrams. Comm. Algebra, to appear

[13] Giambruno, A., Mishchenko, S. (2000): On star-varieties with almost polynomial growth. Algebra Colloq., to appear

[14] Giambruno, A., Regev, A. (1985): Wreath products and P.I. algebras. J. Pure Appl. Algebra **35**, 133–149

[15] Giambruno, A., Zaicev, M. (1998): On codimension growth of finitely generated associative algebras. Adv. Math. **140**, 145–155

[16] Giambruno, A., Zaicev, M. (1999): Exponential codimension growth of PI-algebras: an exact estimate. Adv. Math. **142**, 221–243

[17] Giambruno, A., Zaicev, M. (1999): Involution codimensions of finite dimensional algebras and exponential growth. J. Algebra **222**, 471–484

[18] Giambruno, A., Zaicev, M. (2001): A characterization of algebras with polynomial growth of the codimensions. Proc. Amer. Math. Soc. **129**, 59–67

[19] Giambruno, A., Zaicev, M. (2000): Minimal varieties of of algebras of exponential growth. Electron. Res. Announc. Amer. Math. Soc. **6**, 40–44

[20] Giambruno, A., Zaicev, M. (2000): Minimal varieties of exponential growth. Preprint

[21] Giambruno, A., Zaicev, M. (2000): A characterization of varieties of associative algebras of exponent two. Serdica Math. J, **26**, 245–252

[22] Kemer, A.R. (1978): T-ideals with power growth of the codimensions are Specht. Sibirsk. Mat. Zh. **19**, 54–69 (Russian) [English transl: Siberian Math. J. **19**, 37–48 (1978)]

[23] Kemer, A.R. (1991): Ideals of identities of associative algebras. (Translations of Mathematical Monographs, vol. 87). American Mathematical Society, Providence, RI

[24] Krakowsky, A., Regev, A. (1973): The polynomial identities of the Grassmann algebra. Trans. Amer. Math. Soc. **181**, 429–438

[25] James, G., Kerber, A. (1981): The representation theory of the symmetric group. (Encyclopedia of Mathematics and its Applications, vol. 16). Addison-Wesley, London

[26] Latyshev, V.N. (1977): Complexity of nonmatrix varieties of associative algebras. Algebra i Logika **16**, 149–183 (Russian) [English transl.: Algebra and Logic **16** , 98–122 (1977)

[27] Lewin, J. (1974): A matrix representation for associative algebras. I. Trans. Amer. Math. Soc. **188**, 293–308

[28] Loday, J.L., Procesi, C. (1988): Homology of symplectic and orthogonal algebras. Adv. Math. **69**, 93–108

[29] Malcev, J.N. (1971): A basis for the identities of the algebra of upper triangular matrices. Algebra i Logika **10**, 393–400 (Russian) [English transl.: Algebra and Logic **10**, 242–247 (1971)]

[30] Mishchenko, S., Valenti, A. (2000): A star-variety with almost polynomial growth. J. Algebra **223**, 66–84

[31] Procesi, C. (1967): Non-commutative affine rings. Atti Accad. Naz. Lincei, Mem. Cl. Sci. Fis. Mat. Natur. Sez. I **8**, 239–255

[32] Regev, A. (1972): Existence of identities in $A \otimes B$. Israel J. Math. **11**, 131–152

[33] Regev, A. (1984): Codimensions and trace codimensions of matrices are asymptotically equal. Israel J. Math. **47**, 246–250

[34] Regev, A. (1981): Asymptotic values for degrees associated with strips of Young diagrams. Adv. Math. **41**, 115–136

[35] Rowen, L.H. (1988): Ring theory, vols. I, II. Academic Press, Boston, MA

On the permanent of certain circulant matrices

B. Codenotti, G. Resta

1 Introduction

The computation of the permanent of a matrix seems to be a very hard task, even for sparse (0, 1)-matrices. A number of results show that it is extremely unlikely that there is a polynomial time algorithm for computing the permanent.[1] The best known algorithm is due to Ryser [22] and takes $O(n\,2^n)$ operations, where n is the matrix size.

Even more striking is the fact that the computation of the permanent is essentially as difficult when restricted to the class of (0, 1)-matrices with only three nonzero entries per row and column [6]. It is thus natural to ask which additional constraints could make the permanent easier to compute, e.g., for which classes of matrices with three nonzeros per row it is possible to design algorithms significantly faster than Ryser's method.

An interesting and extensively studied class of permanents is that of permanents of circulant matrices, whose structural and computational properties have been analyzed by many authors (see [19, 20, 4, 1]). Some structure of circulant permanents has been exposed by investigations on the recurrences they satisfy [19] and by the employment of spectral techniques which yield asymptotic estimates [20]. These results suggest that there could be room for tractability; however they must be contrasted with an intrinsic limitation, expressed by the presence, in a generic circulant, of large arbitrary submatrices (see Sect. 4). This limitation vanishes for very sparse circulant matrices [4, 1]; in this case we present in Sect. 5 a very efficient computational approach, based upon recent results by Galluccio and Loebl [10, 11]. This approach has its historical roots in the work of Kac and Ward [13] and Kasteleyn [14–16] on the Ising model and the dimer problem. These papers revealed the crucial role of the genus of a graph in determining the feasibility of computing permanents via determinants.

This last issue has been of major importance since 1913 when, in a seminal paper [21], Pólya raised the question of *convertibility*, i.e., the transformation of the permanent of a given matrix into the determinant of a slightly modified matrix (see the next section for the appropriate definitions). We will see that the notion of convertibility (which turns out to be extremely significant in the context of matrices associated with planar graphs) can be viewed as a special case of expressibility as

[1] Technically, Valiant proved that computing the permanent of a (0, 1)-matrix is $\#P$-complete (see [23, 24]). The class $\#P$ contains those functions that can be computed in polynomial time by a counting (nondeterministic) Turing machine, and the $\#P$-complete problems represent the hardest problems within the class. More recently, several authors have found even stronger negative results [6, 9].

a linear combination of a limited number of determinants (where the appropriate context is that of graphs of bounded genus).

The original contribution of this paper consists of a polynomial time algorithm for computing the permanent of (0, 1)-circulant matrices with three ones per row, a result which should be contrasted with the above mentioned hardness of the problem for the permanent of (0, 1)-matrices with three arbitrarily placed ones per row.

The rest of the paper is organized as follows.

In Sect. 2 we give the main definitions and notation used throughout the paper. In Sect. 3 we recall the crucial role of determinant and permanent in algebraic complexity. In Sect. 4 we show that a circulant graph on n vertices with enough nonzero entries contains an arbitrary subgraph of size proportional to $\sqrt{n}$. In Sect. 5 we continue the work begun in [4] and in [1] on permanents of sparse circulants, and provide a framework for their polynomial time computation. Finally, in Sect. 6 we sketch some further results.

2 Definitions and notation

Let Σ be the set of all permutations of the first n integers, and $A = (a_{i,j})$ be an $n \times n$ matrix.

The permanent of A is defined as

$$\sum_{\sigma \in \Sigma} \prod_{i=1}^{n} a_{i,\sigma(i)},$$

where $\sigma = (\sigma_1, \ldots, \sigma_n)$.

The determinant of A is defined as

$$\sum_{\sigma \in \Sigma} \operatorname{sign}(\sigma) \prod_{i=1}^{n} a_{i,\sigma(i)},$$

where the *sign* of a permutation σ is $+$ or $-$ depending on the parity of the number $s(\sigma)$ of exchanges necessary to take $(1, 2, \ldots, n)$ into σ, i.e., $\operatorname{sign}(\sigma) = (-1)^{s(\sigma)}$.

We will denote the permanent and the determinant of a square matrix A as $\operatorname{per}(A)$ and $\det(A)$, respectively.

A (0, 1)-matrix A is said to be *convertible* if there exists a $(-1, 1)$-matrix X such that $\operatorname{per}(A) = \det(A \star X)$, where $\star$ denotes the elementwise product, i.e., the (i, j)-th entry of the matrix $A \star X$ is $a_{i,j} x_{i,j}$. As we will see in the next sections, convertibility plays a central role, both computationally and mathematically, in the investigations on the hardness of computing permanents. (For more on convertibility, the reader is encouraged to consult [3, 18].)

A *cycle cover* in a digraph with n vertices is a set of n edges forming a set of oriented and disjoint cycles. A *matching* in a graph is a set of pairwise node disjoint edges, and a *perfect matching* (or 1-factor) is a matching such that each node of the graph is incident to exactly one of the edges of the matching.

The permanent of a (0, 1)-matrix has an interpretation in terms of both the digraph and the bipartite graph that can be associated with the matrix. More precisely, if A is an $n \times n$ (0, 1)-matrix, we denote by $D(A)$ the digraph whose adjacency matrix is A and by $G[A]$ the $2n$-node bipartite graph associated with A in the natural way. Then the permanent of A is equal to the number of cycle covers in $D(A)$ as well as to the number of perfect matchings in $G[A]$.

An $n \times n$ circulant matrix $A = (a_{ij})$ is a matrix fully determined by its first row, where each of the other rows is obtained by a cyclic shift of the previous one, i.e., $a_{ij} = \alpha_{1+(n+j-i) \bmod n}$. Thus A is fully described by the n-tuple $[\alpha_1, \alpha_2, \ldots, \alpha_n]$.

Let P_n denote the $n \times n$ (0, 1)-matrix with 1's only in positions $(i, i+1)$, $i = 1, 2, \ldots, n-1$, and $(n, 1)$. Any (0, 1)-circulant matrix can be written as $P^{t_1} + P^{t_2} + \ldots + P^{t_k}$, where $P = P_n$ and $0 \leq t_1 < t_2 < \ldots < t_k < n$.

A graph which admits a circulant adjacency matrix will be called a *circulant* graph.

We now show how a graph is related to the determinant of its adjacency matrix.

Let Y be an $n \times n$ matrix of variables $\{y_{ij} : 1 \leq i, j \leq n\}$. Let G be a digraph on n vertices, in which each arc (i, j) is assigned a weight y_{ij}.

Since each permutation can be decomposed into cycles, and since it *covers* all the indices, there exists a direct correspondence between permutations and cycle covers in the graph corresponding to the matrix Y. Thus

$$\det(Y) = \sum_{c \in CC} (-1)^{\sharp(c)} \Pi(c), \tag{1}$$

where CC denotes the set of cycle covers, $\sharp(c)$ the number of even length cycles belonging to the cover c, and $\Pi(c)$ the product of the elements of Y corresponding to the edges in c. Y is called the *weighted* adjacency matrix of G.

3 Role of permanents (and determinants) in algebraic complexity

The goal of this section is to introduce the reader to the most important computational properties of permanents and determinants. Our exposition will mainly follow Valiant (see [23] and [24]).

Reducibility and completeness are the key notions in computational complexity, where complete problems are the hardest in a class of interreducible problems. Valiant has provided an algebraic framework in which:

- the class of polynomials under investigation contains those polynomials whose coefficients can be described by a short (i.e., polynomial size) formula;
- reducibility takes the form of a transformation (called projection) of a polynomial onto another by means of substitutions for indeterminates of constants or other indeterminates;
- completeness by projections within the above class of polynomials gives rise to a class containing several natural and important problems, e.g., the permanent.

In the following, we will see that determinant and permanent play a fundamental role in algebraic complexity. Indeed the question of whether or not a polynomial can be efficiently computed turns out to be equivalent to the question of whether of not it is a projection of a moderate size determinant, while the question of algebraic intractability can be cast in terms of intractability of the permanent.

Definition 1. Let X be a set of indeterminates, and A a set of polynomials. A function $\sigma : X \to A$ is called a *substitution*. If A only consists of constants and indeterminates, then σ is called a *simple substitution*. The application of a substitution σ to a polynomial p is denoted by p^σ.

Let $q \in \mathbf{F}[y_1, \ldots, y_n]$ and $p \in \mathbf{F}[x_1, \ldots, x_m]$. We say that q is a *projection* of p if $q = p^\sigma$.

We will deal with infinite families of polynomials, defined as

$$P = \{P_j \, : \, P_j \in \mathbf{F}[x_1, \ldots, x_j], \; j = 1, 2, \ldots\}.$$

We will say that a function $t(n)$ is *p-bounded* if there exist two constants h and k such that $t(n) \leq k + n^h$, for any n.

Definition 2. Let P and Q be two families of polynomials, and let t be a function mapping integers onto integers. Q is called a *t-projection* of P if, for any i, there exist $j \leq t(i)$ and σ such that $Q_i = P_j^\sigma$. If $t(n)$ is a p-bounded function, then the t-projection is called a *p-projection*.

When dealing with polynomials, we speak of formulae describing them, and of their size (or length). A formula F is an algebraic expression formed from constants, indeterminates and compositions of formulae by the operators $\{+, \times\}$. The size $|F|$ of the formula F is the number of compositions needed in its construction.

It is possible to show that each polynomial p described by a formula of size $|p|$ is the projection of the determinant of a $(|p|+2) \times (|p|+2)$ matrix. The computation of the determinant can thus be considered as *universal*, in the sense that any formula corresponding to a polynomial can be expressed as the determinant of a matrix with a number of rows only slightly larger than the size of the formula.

We can now state the universality result.

Theorem 1 ([24]). *Any polynomial $p \in \mathbf{F}[x_1, \ldots, x_k]$ is the projection of the determinant of an $s \times s$ matrix, where $s = |p| + 2$.*

It is clear that the permanent also enjoys the property expressed by Theorem 1, since it differs from the determinant only in the signs in the summation.

The following definition formalizes the notion of a family of polynomials whose coefficients can be described by a formula of size bounded by a polynomial. We will see that this notion captures the structure of several important families of polynomials which do not have a polynomial-size formula.

Definition 3. A family of polynomials P over $\mathbf{F}$ is *p-definable* if and only if one of the following two conditions is satisfied:

1. There exist a family Q over $\mathbf{F}$ and a p-bounded function t such that, for all i, we have

$$|Q_i| \le t(i) \text{ and } P_i = \sum_{\sigma} Q_i^{\sigma} \prod_{\sigma(x_k)=1} x_k, \tag{2}$$

where the summation spans all the 2^j substitutions $\{x_1, \ldots, x_j\} \to \{0,1\}^j$, $j \le i$, and where $Q_i \in \mathbf{F}[x_1, \ldots, x_i]$. We will say that P is *p-defined* by Q.
2. P is the p-projection of a p-definable family.

The following theorem by Valiant shows that the permanent is p-definable.

Theorem 2. *The permanent is p-defined by the family Q, where*

$$Q_{n\times n} = \left(\prod_{i=1}^{n}\sum_{j=1}^{n} y_{ij}\right) \prod_{\substack{i=k \\ \text{or } j=h}} (1 - y_{ij}y_{kh}) \tag{3}$$

Proof. First of all note that $Q_{n\times n}$ can be computed by a formula of size $O(n^3)$, i.e., the family Q has p-bounded formula size, and thus the p-definition is legal.

To see that Q p-defines the permanent, note that, if n entries of a matrix A are multiplied, then their product contributes to $\text{per}(A)$ only if all the rows and columns contain exactly one of the selected entries. If we consider the two factors whose product is equal to $Q_{n\times n}$ in (3), we see that the first vanishes if there exists a row whose entries are all zero, while the second ensures that no two entries are chosen from the same row or column. □

We can define a notion of completeness with respect to p-definability.

Definition 4 (Completeness). Let P be a p-definable family over $\mathbf{F}$. We say that P is *complete* over $\mathbf{F}$ if any p-definable family over $\mathbf{F}$ is a p-projection of P.

We now prove that the permanent is complete.

Theorem 3 **([24]).** *Let* $\mathbf{F}$ *be a field of characteristic different from 2. Then the permanent is complete over* $\mathbf{F}$.

Proof. We show that an arbitrary p-definable family P is a p-projection of the permanent, which is p-definable by Theorem 2. Let P be the p-projection of $\tilde{P}$, where $\tilde{P}$ is defined as

$$\tilde{P}_i[x_1, \ldots, x_i] = \sum_{\sigma\in\{0,1\}^j} Q_i^{\sigma} \prod_{\sigma(x_k)=1} x_k,$$

where the summation spans all the 2^j substitutions of x_k with 0 or 1, for $k = 1, \ldots, j$.

We now design a graph G'' whose adjacency matrix has permanent equal to $\tilde{P}_i[x_1, \ldots, x_i]$.

Let f be a minimum size formula for Q_i; we build a graph G' corresponding to f by a procedure similar to the one used in [24] to prove the universality of the determinant.

In other words, the universality result allows us to consider a graph G' such that the determinant of its weighted adjacency matrix is equal to Q_i. We want to transform G' into a new graph G'' whose weighted adjacency matrix has permanent equal to $\tilde{P}_i$.

In order to achieve this, we consider, for each k, the edges labeled x_k in G'. These edges are replaced by a global structure which interconnects them, as shown in Fig. 1, and which is based upon a subgraph V with 4 nodes, of which we have represented only two distinct nodes a and b. All the edges used to interconnect the copies of V to G' have unit weight, except for the edge which connects the most 'external' copies of V, to which the weight x_k is assigned. This substitution operation is executed for each k, grouping together the edges of weight x_k. The detail of the substitution for a single edge is shown in Fig. 2.

The correct transformation of G' is obtained by selecting the following adjacency matrix for V, where node a (b) corresponds to row 1 (4) and the two intermediate nodes to rows 2 and 3:

$$V = \begin{pmatrix} 0 & 1 & -1 & -1 \\ 1 & -1 & 1 & 1 \\ 0 & 1 & 1 & 2 \\ 0 & 1 & 3 & 0 \end{pmatrix}.$$

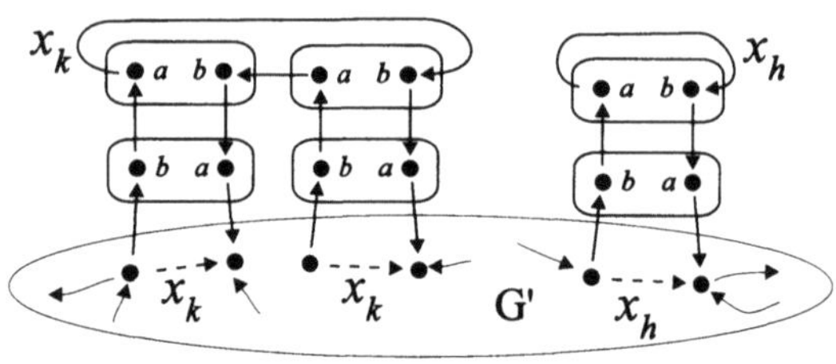

Fig. 1. The generation of G''. All the edges with the same weight are connected by a global structure. Only nodes a and b are shown in V

If we denote by $B[r|c]$ the matrix obtained from B by deleting row r and column c, it is immediate to check that $\mathrm{per}(V) = \mathrm{per}(V[1|1]) = \mathrm{per}(V[4|4]) = 0$, while $\mathrm{per}(V[1|4]) = \mathrm{per}(V[4|1]) = 4$. It follows that a cycle entering from node a and exiting from b, or vice versa, will receive from V a multiplicative contribution of 4, while if V is crossed in another way it will receive a null contribution. The global effect is that either all the (substituted) edges x_k of G' and the edge x_k (added in G'') belong to a cycle cover or their contribution will be null.

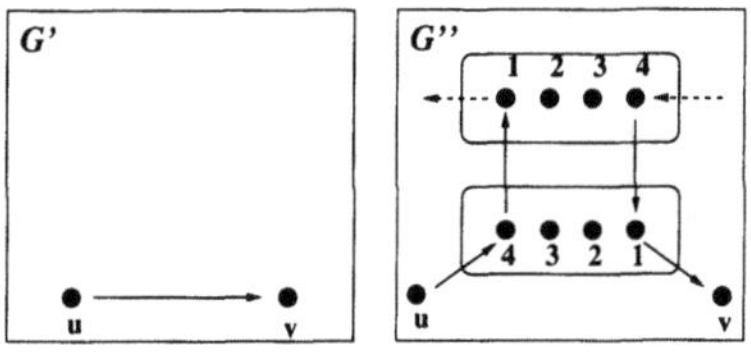

Fig. 2. The replacement of an edge by a suitable subgraph

In this way, we simulate the product between Q_i^σ and those x_k's for which $\sigma(x_k) = 1$, so that the permanent of the matrix corresponding to G'' is equal to $\tilde{P}_i \cdot 4^J$, where J is the number of graphs V used in the process. Giving weight $(2^{-1})^{2J}$ to the edge going from t to s in G'', we obtain the desired result. □

These results have shown that all the polynomials with short formula size can be efficiently computed by reduction to a determinant, and that all the polynomials whose coefficients can be specified by a short formula reduce by projections to the permanent, which is complete for such a class of polynomials.

Therefore it turns out that, despite their closeness, permanent and determinant play very different roles in algebraic complexity, the first being paradigmatic of computational intractability, and the latter of tractability. Thus the classical mathematical question of the reduction of the permanent to the determinant, first raised by Pólya [21], becomes of central importance in computational complexity.

This question will be the guiding thread of the investigation of permanents of circulant matrices in the subsequent sections. The permanent of a circulant matrix seems to be easier to compute than an arbitrary permanent, since it has a great deal of structure (see, e.g., [8, 19, 20]). However no efficient algorithm has been discovered yet; here we describe some partial progress on the issue of its existence.

4 Circulants vs arbitrary graphs

The purpose of this section is to study the presence of arbitrary subgraphs in circulant graphs. Before stating the result, we need to review some properties of certain sequences of integers.

Theorem 4 ([2, 12]). *If m is a power of a prime, then there exist m integers $a_1, \dots, a_m$ such that $1 \le a_1 < a_2 < \dots < a_m \le m^2 - 1$, and all the sums $a_i + a_j$ are distinct, $1 \le i \le j \le m$.*

Corollary 1. *For any n, there exist at least $M(n) = \lfloor \sqrt{n}/2 \rfloor$ integers $1 \le a_1 < \dots < a_m \le n$, such that all the differences $a_j - a_i$ are distinct, $1 \le i \le j \le m$.*

Proof. First note that, if all the sums $a_i + a_j$ are distinct, then all the differences $a_j - a_i$ are also distinct, $1 \le i \le j \le m$. In fact, $a_j - a_i = a_h - a_k$ implies $a_j + a_k = a_h + a_i$.

If $\lfloor\sqrt{n}\rfloor$ is a power of a prime, then there exist $m = \lfloor\sqrt{n}\rfloor \geq M(n)$ numbers $1 \leq a_1 < \ldots < a_m \leq m^2 - 1 \leq n$ whose differences are distinct and we are done. If $\lfloor\sqrt{n}\rfloor$ is not a power of a prime, we consider the two consecutive powers of 2 such that $2^h < \lfloor\sqrt{n}\rfloor < 2^{h+1}$. Letting $m = 2^h \geq \lfloor\sqrt{n}/2\rfloor = M(n)$ we obtain the result. □

We can now prove a structural property of circulant matrices, which hints at a possible obstacle to the efficient computation of permanents of circulant matrices with enough nonzero entries.

Theorem 5. *Let A be an $n \times n$ circulant matrix with first row $[\alpha_1, \alpha_2, \ldots, \alpha_n]$. A has a principal submatrix M of size $m = \theta(\sqrt{n})$ whose off-diagonal entries are distinct elements of the first row of A. More precisely, we have*

$$\lfloor\sqrt{n}/(2\sqrt{2})\rfloor \leq m \leq \lceil\sqrt{n} + 1\rceil.$$

Proof. The ij-th entry of the circulant matrix A can be defined as $a_{ij} = \alpha_{1+(n+j-i)\bmod n}$.

The upper bound on m is trivial, since there are $m^2 - m$ off-diagonal entries in M and $n - 1$ in A. Hence we must have $m^2 - m \leq n - 1$, from which we obtain $m \leq \frac{1}{2} + \sqrt{n - \frac{3}{4}} < \lceil\sqrt{n} + 1\rceil$.

For the lower bound we want to determine a set of indices $S = \{s_1 < s_2 < \ldots < s_m\}$ such that the principal submatrix $M = A(S)$ does not contain repeated elements, except for the diagonal which is constant and equal to the diagonal of A by construction.

Let us assume, for simplicity, that $s_m < n/2$. This implies that the entries of the strictly upper triangular submatrix U of M must be chosen from $\alpha_2, \alpha_3, \ldots, \alpha_{s_m}$, while those below the diagonal from $\alpha_{n+2-s_m}, \ldots, \alpha_n$. This ensures that entries above the main diagonal do not overlap with entries below the main diagonal.

Two entries of U, say u_{ij} and u_{hk}, with $(i, j) \neq (h, k)$, take the same values if and only if $\alpha_{s_j - s_i + 1} = \alpha_{s_k - s_h + 1}$. This implies that $u_{ij} \neq u_{hk}$ if and only if $s_j - s_i \neq s_k - s_h$. The same condition holds for the entries below the diagonal of M.

By Corollary 1 there exists a sequence of $M(n/2) = \lfloor\sqrt{n/2}/2\rfloor$ indices that satisfies $s_j - s_i \neq s_k - s_h$ and such that $s_m < n/2$. □

The above result shows that a circulant graph contains an arbitrary subgraph of size $\theta(\sqrt{n})$. Thus it would be rather surprising to come up with algorithms for computing permanents of arbitrary circulants with running times less than $T(\sqrt{n})$, where $T(n)$ is the worst case running time of the best available algorithm for the general case.

5 Permanents of very sparse circulant matrices

Building upon the convertibility of matrices corresponding to planar graphs, in [1] it was shown that the permanent of (0, 1)-circulant matrices with 3 ones in each row can be computed in $O(n^3 2^{O(\sqrt{n})})$ time. Here we take advantage of a recent result by

Galluccio and Loebl [10] to show that the problem can in fact be solved in polynomial time.

Galluccio and Loebl built upon previous work by Kasteleyn (see [14–16]) to show that the generating function of the perfect matchings of a graph of genus g may be written as a linear combination of 4^g Pfaffians.

In the following we give some definitions taken from [10, 11] and then show how the results in [10] lead to a polynomial time algorithm for computing $\mathrm{per}(I+P^i+P^j)$. (Recall that P is the circulant matrix whose first row is all zero except for its second entry, which is equal to 1.) It is also useful to recall that, for the $n \times n$ matrix $A = I + P^i + P^j$, where n is a prime number greater than 2, the bipartite graph $G[A]$ is a cycle of length $2n$ with n additional chords, whose length d is a function of n, i, and j [4].

Definition 5. The *genus* g of a surface in $\mathbf{R}^3$ is the maximum number of nonintersecting closed curves which one can draw on the surface without disconnecting it. (Equivalently, we can say that a surface has genus g if it is topologically equivalent to a sphere with g "handles".)

For instance, a sphere and a plane have genus 0, while a torus has genus 1.

Definition 6. A graph has *genus* g if it can be embedded in a surface of genus g, but not in a surface of genus $g - 1$.

According to this definition, it is clear that all planar graphs have genus 0.

We now define the generating function of the perfect matchings of a graph.

Definition 7. Given a graph $G = (V, E)$, we associate with each edge e of G a variable x_e and we let $x = \{x_e \mid e \in E\}$. For each $M \subset E$, $x(M)$ will denote the product of the variables associated with the edges of M.

An *orientation* of a graph $G = (V, E)$ is a digraph $D = (V, A)$ obtained from G by fixing an orientation for each edge of G.

Let C be a cycle in a graph G, and let D be an orientation of G. C is said to be *clockwise odd* in D if it has an odd number of edges directed (according to D) in agreement with the clockwise orientation.

The *generating function* of the perfect matchings of G is the polynomial $\mathcal{P}(G, x)$ which is equal to the sum of $x(P)$ over all perfect matchings P of G.

The following two definitions, which we take from [10], provide a standard way to embed a graph in a surface of genus g. We first describe a surface of genus g as a polygon connected by $2g$ "bridges", which is equivalent to a sphere with g handles. Then we define a sort of normal form for a graph embedding, i.e., we introduce the notion of g-graph, which is required in order to apply the main result of Galluccio and Loebl.

Definition 8. A surface S_g of genus g consists of a base B_0 and $2g$ bridges B^i_j, $i = 1, \ldots, g$, $j = 1, 2$, where:

1. B_0 is a convex $4g$−gon with vertices $a_1, \ldots, a_{4g}$ numbered clockwise.

2. B_1^i, $i = 1, \dots, g$, is a 4−gon with vertices x_1^i, x_2^i, x_3^i, x_4^i numbered clockwise. It is glued with B_0 so that side $[x_1^i, x_2^i]$ of B_1^i is identified with side $[a_{4(i-1)+1}, a_{4(i-1)+2}]$ of B_0, and $[x_3^i, x_4^i]$ of B_1^i is identified with $[a_{4(i-1)+3}, a_{4(i-1)+4}]$ of B_0.
3. B_2^i, $i = 1, \dots, g$, is a 4-gon with vertices y_1^i, y_2^i, y_3^i, y_4^i numbered clockwise. It is glued with B_0 so that side $[y_1^i, y_2^i]$ of B_2^i is identified with side $[a_{4(i-1)+2}, a_{4(i-1)+3}]$ of B_0 and $[y_3^i, y_4^i]$ of B_1^i is identified with side $[a_{4(i-1)+4}, a_{4(i-1)+5 \pmod{4g}}]$ of B_0.

Definition 9. We say that a graph G is a *g-graph* if it can be embedded in S_g so that all the vertices belong to the base B_0, and the embedding is obtained using at most one bridge for each edge of G. The set of edges embedded entirely in the base will be denoted by E_0, and the set of edges embedded in the bridge B_j^i by E_j^i. We say that a g-graph is *proper* if it satisfies the following additional conditions.

1. The outer face of $G_0 = (V, E_0)$ is a cycle and is embedded in the boundary of B_0.
2. If $e \in E_1^i$, then e is embedded entirely in B_1^i, one of its endpoints lies on side $[x_1^i, x_2^i]$ and the other one lies on $[x_3^i, x_4^i]$. The same holds for edges in E_2^i.
3. Each vertex of G is incident with at most one edge e such that $e \notin E_0$.
4. G_0 has a perfect matching.

Note that, for a graph G embedded in a generic surface of genus g, it is not immediate whether it can be embedded in S_g as a proper g-graph and, if so, how. When there is an embedding in S_g, we will see later how to transform G into a proper g-graph G' such that $\mathcal{P}(G, x) = \mathcal{P}(G', x')$, by adding to G suitable auxiliary vertices and edges.

We have seen that a proper g-graph G is composed of a planar part, which contains all the vertices, and an arbitrary number of extra edges, each embedded in one of the $2g$ bridges. The following definition makes clear that, by taking advantage of the structure of S_g, it is possible to add to the planar part of G the edges on any single bridge, and still obtain a planar subgraph of G.

Definition 10 ([10]). Let G be a proper g-graph and let $G_j^i = (V, E_0 \cup E_j^i)$. We draw $B_0 \cup B_j^i$ on the plane as follows: B_0 is unchanged, and the side $[x_1^i, x_4^i]$ if $j = 1$ (or $[y_1^i, y_4^i]$ if $j = 2$) of B_j^i is drawn so that it belongs to the external boundary of $B_0 \cup B_j^i$. In this way we obtain a planar embedding of G_j^i, which we call a *planar projection* of E_j^i outside B_0.

It is well-known that, for a planar graph G, it is possible to choose an orientation of its edges so that each inner face of each 2-connected component[2] of G is clockwise odd. This property allows us to evaluate $\mathcal{P}(G, x)$.

[2] A 2-connected component of a graph is a connected subgraph which cannot be disconnected by deleting only one of its edges.

Definition 11. Let $G = (V, E)$ be a graph on $2n$ vertices, and let D be an orientation of G. We denote by $A(D)$ the skew-symmetric matrix with rows and columns indexed by V, where $a_{vw} = x_{vw}$ if $(v, w) \in D$, $a_{vw} = -x_{vw}$ if $(w, v) \in D$, and $a_{vw} = 0$ otherwise. The *Pfaffian* of the skew-symmetric matrix $A(D)$ is defined as

$$\mathrm{Pf}(A(D)) = \sum_P s^*(P) a_{i_1 j_1} \dots a_{i_n j_n},$$

where $P = \{\{i_1, j_1\}, \dots, \{i_n, j_n\}\}$ is a partition of the set $\{1, \dots, 2n\}$ into pairs satisfying $i_k < j_k$, $k = 1, \dots, n$, and $s^*(P)$ is equal to the sign of the permutation $i_1 j_1 \dots i_n j_n$ of $12 \dots (2n-1)(2n)$.

Each nonzero term of the expansion of the Pfaffian of $A(D)$ is either equal to $x(P)$ or to $-x(P)$, where P denotes a perfect matching of G. If $s(D, P)$ indicates the sign of the term $x(P)$, then we have $\mathrm{Pf}(A(D)) = \sum_P s(D, P) x(P)$.

We can now state a classical result connecting Pfaffian and determinant (see, e.g., [17]).

Theorem 6. *Let G be a graph and let D be an orientation of G. Then* $[\mathit{Pf}(A(D))]^2 = \det(A(D))$.

A graph G is called *Pfaffian* if it has a *Pfaffian orientation*, i.e., an orientation D such that $\mathcal{P}(G, x)^2 = [\mathrm{Pf}(A(D))]^2 = \det(A(D))$.

The planar part of a g-graph has a Pfaffian orientation; adding to it the edges of a single bridge, we can determine an orientation for these edges which is still Pfaffian, since the resulting graph is still planar (see Definition 10). Considering for each bridge its Pfaffian orientation together with the opposite orientation, and combining in all possible ways these orientations for the $2g$ bridges, we obtain the so-called $2^{2g} = 4^g$ *relevant orientations* for G. We now describe more precisely the above process, again following [10].

Definition 12. Let $G = (V, E)$ be a proper g-graph. A Pfaffian orientation D_0 of G_0 such that each inner face of each 2-connected component of G_0 is clockwise odd in D_0 is called a *basic orientation* of G_0.

Definition 13. Let $G = (V, E)$ be a proper g-graph, and D_0 a basic orientation of G_0. We define the orientation D^i_j of each G^i_j as follows: let G^i_j be embedded in the plane by the planar projection of E^i_j outside B_0, and complete the basic orientation D_0 of G_0 to an orientation of G^i_j so that each inner face of each 2-connected component of G^i_j is clockwise odd.

The orientation $-D^i_j$ is defined by reversing the orientation D^i_j of G^i_j.

Definition 14. Let G be a proper g-graph, $g \geq 1$. We say that an orientation D of G is *relevant* if it is equal to the basic orientation D_0 on G_0, and to D^i_j or $-D^i_j$ on E^i_j.

We define the *type* $r(D) \in \{+1, -1\}^{2g}$ of an orientation D as follows: the value of $r(D)$ in position $2i + j$ is given by the sign of D^{i+1}_j in D, for $i = 0, \dots, g-1$ and $j = 1, 2$.

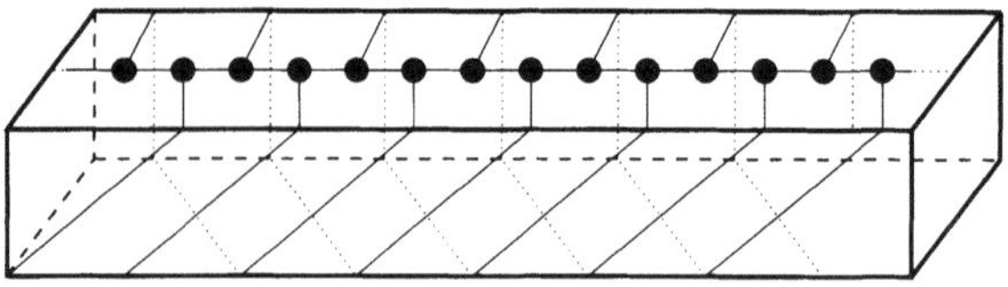

Fig. 3. Embedding the graph corresponding to $I + P + P^3$ in the torus. We show a fragment (without the wrap-around) of a square-section torus. Along the horizontal direction, we embed all the vertices and the edges that connect them in a cycle (which corresponds to two of the 3 "diagonals", e.g., $I + P$). The other edges, corresponding to P^3, are arranged in order to form a path which crosses the torus in a spiral-like fashion

We can now state the main result of Galluccio and Loebl [10].

Theorem 7 **([10]).** *Let G be a proper g-graph. Then*

$$\mathcal{P}(G, x) = s_0 \sum_{i=1}^{4^g} c(r(D_i))\mathrm{Pf}(A(D_i)),$$

where

- D_i, $i = 1, \ldots, 4^g$, *are the relevant orientations of G;*
- $c(r(D))$ *is equal to the products of* c_i, $i = 0, \ldots, g-1$, *where* $c_i = c(r_{2i+1}, r_{2i+2})$, $c(1, 1) = c(1, -1) = c(-1, 1) = 1/2$, *and* $c(-1, -1) = -1/2$;
- s_0 *is the sign of the term* $x(M)$ *for an arbitrary perfect matching M of G with respect to an arbitrary relevant orientation D.*

Given a graph embedded in S_g, we can make it a proper g-graph by adding some auxiliary vertices and edges. Given an edge $e = uv$ whose embedding uses k bridges (thus violating the definition of proper g-graph), we replace it with a path $P_e = (u, e_1, v_1, \ldots, v_{2k}, e_{2k+1}, v)$. We embed the additional vertices $v_1, \ldots, v_{2k}$ so that each new edge uses at most one bridge. With $x'_{e_1} = x_e$ and $x'_{e_i} = 1$ for $i > 1$, it is easy to see that the new graph G' obtained in this way satisfies $\mathcal{P}(G', x') = \mathcal{P}(G, x)$. Finally, we add edges e so that the outer face of the planar part of G' is a cycle, and we let $x'_e = 0$, so that these new arcs do not modify the value of $\mathcal{P}(G', x')$.

We now show how to apply Theorem 7 to the computation of $\mathrm{per}(M_n^d)$, where $M_n^d = I_n + P_n + P_n^d$, and n is prime.

The first step towards the application of Theorem 7 consists of determining the value of the genus of the bipartite graph associated with M_n^d.

Theorem 8. *The bipartite graph G associated with* M_n^d *has genus 1.*

Proof. For n prime, the graph G can be described as a cycle of length $2n$, with the addition of n chords of constant length. Such a graph can easily be embedded in a torus: it is sufficient to embed the $2n$ cycle in the external maximal circumference of the torus (see Fig. 3). The additional chords will then become mutually parallel and fold on the smaller circumference in a spiral-like fashion.

We will see in the next section that the same result can be proved for n composite.

□

In general, the transition from the embedding in the torus to an embedding in S_1 seems to be nontrivial and nonunique. An embedding of the graph corresponding to M_5^3 is shown in Fig. 4; this embedding has been chosen since it can be easily generalized to one for M_n^d. The $2n$ cycle is broken and all but one of its edges constitute the planar part of the embedding. The n chords are divided in two sets, of cardinalities $n-d+1$ and $d-1$. The embedding of the first set of edges can be done using only one bridge, while for the latter we use both bridges.

We now show how to transform the graph G into the proper 1-graph G'.

The basic orientation of the planar part of G' is described by matrix A, while matrices B and C show the corresponding orientations of the edges on the two bridges. The auxiliary edges (dotted in Fig. 4) are added only to ensure that the planar part of G' contains a cycle, and are only needed to define the orientation of the rest of the graph. Since these edges must not contribute to the count of perfect matchings, we have set to zero the corresponding entries in A, B and C.

The columns (rows) of A, B and C are indexed by the vertices $2, 4, \ldots, 16$ $(1, 3, \ldots, 15)$. The adjacency matrices of the different parts of G' are

$$A' = \begin{pmatrix} 0 & A \\ -A^T & 0 \end{pmatrix}, \quad B' = \begin{pmatrix} 0 & B \\ -B^T & 0 \end{pmatrix}, \quad C' = \begin{pmatrix} 0 & C \\ -C^T & 0 \end{pmatrix}.$$

However we can actually work with A, B and C, since we are interested in Pfaffians and determinants, and can thus take advantage of the equality $\det(A' \pm B' \pm C') = \det(A \pm B \pm C)^2$.

We have

$$A = \left(\begin{array}{ccccc|cc|c} 1 & 1 & & & & & & \\ & 1 & 1 & & & & & \\ & & 1 & 1 & & & & \\ & & & 1 & 1 & & & \\ \hline 1 & & & & & & & 1 \\ \hline & & & & & -1 & & \\ & & & & & & 1 & \\ \hline & & & & 1 & & & \end{array}\right), \quad B = \left(\begin{array}{ccccc|cc|c} & & & & & & & \\ & & & & & & & \\ & & & & & & & \\ & & & & & & & \\ \hline & & & & & & & \\ \hline 1 & & & & & & & \\ & 1 & & & & & & \\ \hline & & & & & & & -1 \end{array}\right),$$

$$C = \left(\begin{array}{ccc|cc|c}
 & -1 & & & & \\
 & & -1 & & & \\
 & & & -1 & & \\
 & & & & -1 & \\
\hline
-1 & & & & & \\
\hline
 & & & & & \\
\hline
 & & & & &
\end{array}\right).$$

The above embedding of G' can be generalized, and it is easy to discover the structure of the matrices A, B and C corresponding to M^d (see below). The signs of the entries of matrix C depend on the parity of d, as we show below, where we use the notation C_{odd} and C_{even} to describe C in the two cases. We have

$$A = \left(\begin{array}{cccc|c|c}
1 & 1 & & & & \\
 & \ddots & \ddots & & & \\
 & & 1 & 1 & & \\
\hline
1 & & & & & 1 \\
\hline
 & & & & Y_{d-1} & \\
\hline
 & & & 1 & &
\end{array}\right), \quad
B = \left(\begin{array}{c|c|c}
 & & \\
\hline
 & & \\
\hline
I_{d-1} & & \\
\hline
 & & -1
\end{array}\right).$$

$$C_{\text{odd}} = \left(\begin{array}{c|c|c}
-I_{n-d} & & \\
 & Y_{d-1} & \\
\hline
-1 & & \\
\hline
 & & \\
\hline
 & &
\end{array}\right), \quad
C_{\text{even}} = \left(\begin{array}{c|c|c}
I_{n-d} & & \\
 & -Y_{d-1} & \\
\hline
1 & & \\
\hline
 & & \\
\hline
 & &
\end{array}\right)$$

where Y_k is a $k \times k$ diagonal matrix whose entries are $(-1, 1, -1, \ldots)$.

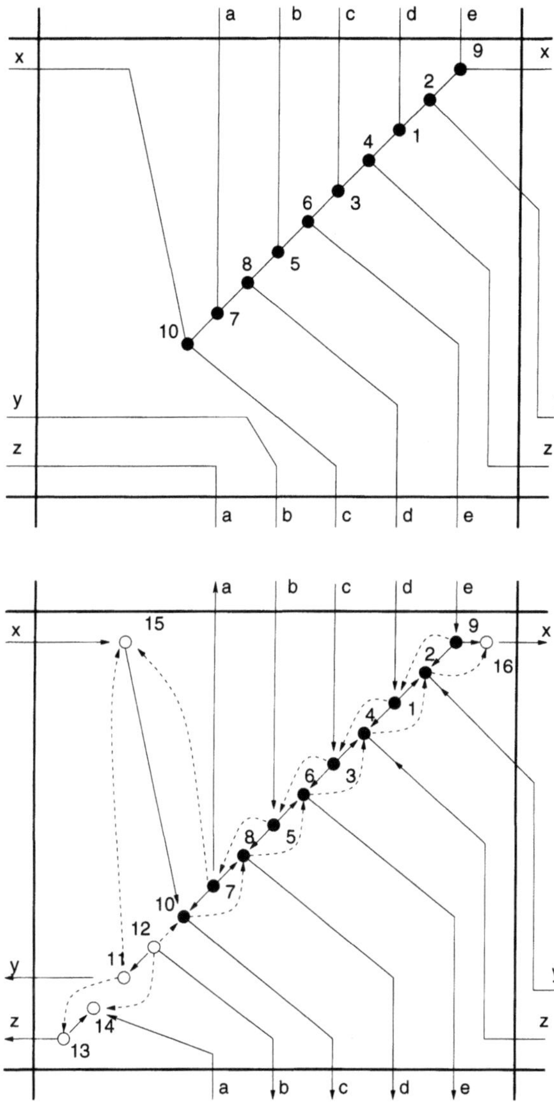

Fig. 4. Embedding of the bipartite graph G associated with $I_5 + P_5 + P_5^3$. Above we show the embedding of G in S_1, below the embedding of the corresponding proper 1-graph. Auxiliary vertices are white, and auxiliary edges are dotted. The two bridges connect the opposite side of the central square, which contains the planar part of the graph. Basic orientations for all the edges are shown

According to Theorem 7, we can then write

$$\begin{aligned}\mathrm{per}(M^d) = \frac{1}{2}|\det(A+B+C) + \det(A+B-C)\\ + \det(A-B+C) - \det(A-B-C)|.\end{aligned}$$

Table 1. We show, for n prime, $11 \leq n \leq 61$, the values of the 4 determinants which contribute to the permanent of $I + P + P^k$, for $k = 6, 7$. In particular we have $d_1 = \det(A + B + C)$, $d_2 = \det(A + B - C)$, $d_3 = \det(A - B + C)$, $d_4 = \det(A - B - C)$, and $\text{per}(|A + B + C|) = |d_1 + d_2 + d_3 - d_4|/2$

n	d_1	d_2	d_3	d_4	$\text{per}(I + P + P^6)$
11	3	1	199	-199	201
13	159	131	27	-53	185
17	717	103	137	-613	785
19	573	229	761	-647	1105
23	2487	2209	2347	-1151	4097
29	10443	18503	24071	-3481	28249
31	46875	20089	20089	-20089	53571
37	193143	187591	297851	-67489	373037
41	878223	306353	988511	-571787	1372437
43	1760853	1124107	1874801	-516689	2638225
47	6052569	1754323	7748327	-3980243	9767731
53	58853217	17683663	43574057	-20084033	70097485
59	353797869	152350273	408073147	-95822609	505021949
61	774569463	185499049	713175401	-280330381	976787147

n	d_1	d_2	d_3	d_4	$\text{per}(I + P + P^7)$
11	67	69	23	-23	91
13	1	3	521	-521	523
17	409	309	307	-137	581
19	191	1371	1483	-229	1637
23	599	2073	3313	-2209	4097
29	3481	10443	24071	-18503	28249
31	46439	18789	5953	-38069	54625
37	297851	193143	67489	-187591	373037
41	988511	878223	306353	-571787	1372437
43	1205291	1479117	1205291	-1205291	2547495
47	7145599	3878913	1352191	-6450751	9413727
53	36419269	37973337	20449733	-34871351	64856845
59	139606391	291647859	245155739	-230894141	453652065
61	313849027	514885143	410510603	-487297037	863270905

In Table 1 we report the values of these determinants and permanents for $d = 6, 7$ and for several values of n.

A careful analysis of the procedure leading to the matrices A, B and C shows that it is possible to project the sign changes expressed by these matrices into the original matrix, and thus to obtain the permanent of the matrix M as a linear combination of determinants of 4 matrices which differ from M only in the sign of some entries, and not in size or zero-nonzero pattern.

Following [10], we can achieve this by explicitly considering the 4 relevant orientations of G', i.e., $(A \pm B \pm C)$. Then, for each such orientation D' and for each edge e of G which was changed into a path P_e of odd length in G', we orient e in G in agreement with the direction in D' of an odd number of edges of P_e. We then orient all untouched edges of G as in D'.

Applying this procedure to the above matrices A, B and C, we readily obtain the following result.

Theorem 9. *Let Q^k be the $n \times n$ Toeplitz matrix whose entries are all zero, except for a diagonal of ones starting from the $(k+1)$-th entry of the first row, and let $\hat{Q}^k = (Q^{n-k})^T$. Note that $P^k = Q^k + \hat{Q}^k$. Then the following result holds:*

$$\begin{aligned}\operatorname{per}(I + P + P^j) = {} & \frac{\det(I - Q + \hat{Q} + Q^j - \hat{Q}^j) + \det(I - P + P^j)}{2} + \\ & + \frac{\det(I - Q + \hat{Q} - Q^j + \hat{Q}^j) - \det(I - P - P^j)}{2},\end{aligned}$$

for any prime number $n > 3$ and any j, $1 < j < n/2$.

Note that, because of the symmetries of the problem, for n prime and for any matrix $A = I + P^i + P^j$, there exists a matrix $B = I + P + P^k$ with $k < n/2$ and such that $\operatorname{per}(A) = \operatorname{per}(B)$ (see [4, 1]). Therefore Theorem 9 applies to all matrices of the form $I + P^i + P^j$.

We can now take advantage of the result of Dietrich and Osborne [7] on the computation of the determinant of Toeplitz matrices with $O(n \log^2 n)$ arithmetic operations, to show that the permanent of a prime size circulant matrix with three ones per row can be computed in $O(n \log^2 n)$ time.

6 Further results and conclusions

In order to extend the results of the previous section to the case of n composite, we analyze the structure of the bipartite graph $G[A]$, where $A = I_n + {P_n}^i + {P_n}^j$, for n composite. Without loss of generality, we may assume that n, i, and j are mutually prime, i.e., $c = (n, i, j) = 1$. Otherwise the problem of computing $\operatorname{per}(A)$ can be reduced to the evaluation of $\operatorname{per}(I_{n/c} + P_{n/c}^{i/c} + P_{n/c}^{j/c})^c$ since, as proved in [4], $G[I_n + P^i + P^j]$ is isomorphic to c disjoint copies of $G[I_{n/c} + P^{i/c} + P^{j/c}]$.

We label the left and right vertices with $0, 1, \ldots, n-1$ and $0', 1', \ldots, (n-1)'$, respectively. The $3n$ edges of $G[A]$ are given by

$$\{a \leftrightarrow (a+k)' : a \in 0, 1, \ldots, n-1, \text{ and } k \in \{0, i, j\}\},$$

where $a + k$ is computed modulo n.

We assume that $j > i$, and let $t = j - i$ and $r = (n, t)$. The following facts hold (for the proofs, see [5]).

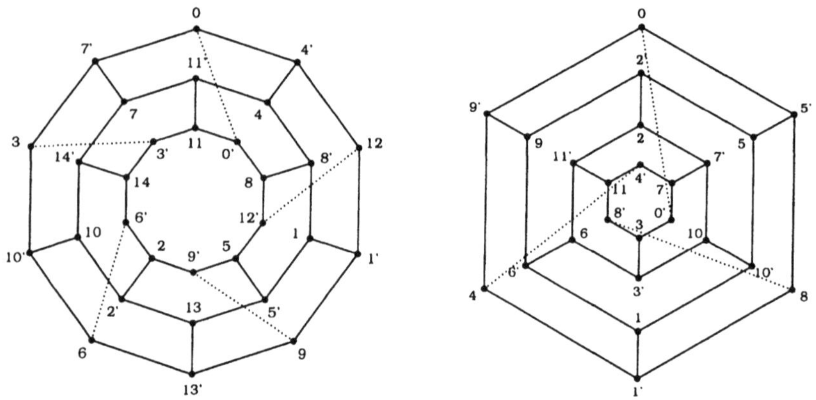

$G[I_{15} + P_{15}^4 + P_{15}^7]$ $G[I_{12} + P_{12}^5 + P_{12}^9]$

Fig. 5. We give drawings of the graphs $G[I + P^4 + P^7]$ (for $n = 15$) and $G[I + P^5 + P^9]$ (for $n = 12$) which illustrate the properties stated by Lemmas 3 and 4. We have $r = (15, 7 - 4) = 3$ and $r = (12, 9 - 5) = 4$, respectively. The $15/3 = 5$ and $12/4 = 3$ edges which should be drawn "around" the smaller circumference of the torus are dotted

Lemma 3. *If we remove from $G[A]$ the edges of type $a \leftrightarrow a'$, for $a = 0, \ldots, n-1$, then the remaining graph, which contains $2n$ vertices and $2n$ edges, can be partitioned into r independent cycles, each having $2n/r$ vertices.*

Lemma 4. *The following facts hold.*

(a) *For each cycle C_a, there are no $b \leftrightarrow b'$ edges that connect vertices of C_a to other vertices of C_a.*
(b) *If an edge of type $c \leftrightarrow c'$ connects two cycles C_a, C_b, then n/r such edges connect C_a and C_b. If $r = 2$, then the number of connecting edges is $2n/r = n$.*
(c) *If $r > 2$, then, for each pair of interconnected cycles C_a and C_b, the subgraph of $G[A]$ induced by the vertices of C_a and C_b is planar.*

An illustration of Lemmas 3 and 4 is shown in Fig. 5.

Lemmas 3 and 4 imply that $G[A]$ can be embedded in the torus, i.e., its genus is one, and thus per(A) can be obtained as a linear combination of four determinants of matrices with the same nonzero pattern as A, as in the case of n prime.

The problem of computing the permanent of circulants with more than three ones per row appears to be more complicated, since the genus of the corresponding graph does not grow linearly with the number of diagonals in the matrix. For example, the genus of $G[I + P + P^j + P^{j\pm 1}]$ is still 1, as illustrated in Fig. 6, while, in the general case ($A = I + P^i + P^j + P^k$), the genus of $G[A]$ is much larger.

The issues discussed in this section will be the subject of a forthcoming paper [5], where we provide a complete description of the computation of circulant permanents via determinants.

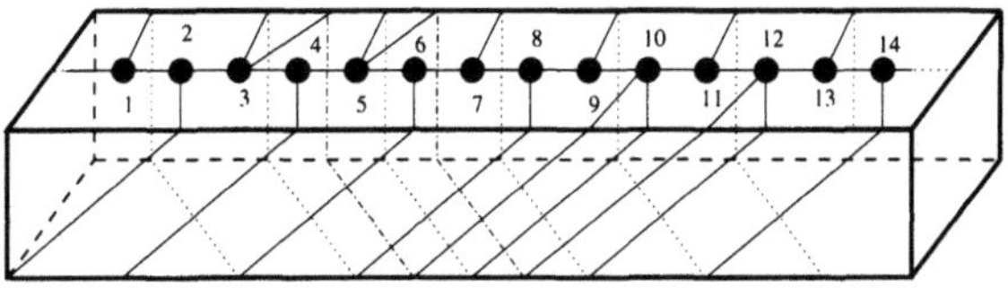

Fig. 6. Embedding the graph corresponding to $G[I + P + P^3 + P^4]$ in the torus. (Compare with Fig. 3.) To improve readibility, we only show edges (3, 10) and (5, 12) among those related to P^4

References

[1] Bernasconi, A., Codenotti, B., Crespi, V., Resta, G. (1999): How fast can one compute the permanent of circulant matrices? Linear Algebra Appl. **292**, 15–37

[2] Bose, R.C., Chowla, S. (1962): Theorems in the additive theory of numbers. Comment. Math. Helv. **37**, 141–147

[3] Brualdi, R.A, Shader, B.L. (1995): Matrices of sign-solvable linear systems. Cambridge University Press, Cambridge

[4] Codenotti, B., Crespi, V., Resta, G. (1997): On the permanent of certain (0,1) Toeplitz matrices. Linear Algebra Appl. **267**, 65–100

[5] Codenotti, B., Resta, G. (2000): Circulant permanents. In preparation

[6] Dagum, P., Luby, M., Mihail, M., Vazirani, U. (1988): Polytopes, permanents, and graphs with large factors. In: Proceedings of the 29th IEEE Symposium on Foundations of Computer Science. Institute of Electrical and Electronics Engineers. Washington, DC, pp. 412–421

[7] Dietrich, C.R., Osborne, M.R. (1996): $O(n \log^2 n)$ determinant computation of a Toeplitz matrix and fast variance estimation. Appl. Math. Lett. **9**, 29–31

[8] Elspas, B., Turner, J. (1970): Graphs with circulant adjacency matrices. J. Combin. Theory **9**, 297–307

[9] Feige, U., Lund, C. (1992): On the hardness of computing the permanent of random matrices. In: Proceedings of the Twenty-Fourth Annual ACM Symposium on the Theory of Computing. Association for Computing Machinery, New York, pp. 643–654

[10] Galluccio, A., Loebl, M. (1999): On the theory of Pfaffian orientations. I. Perfect matchings and permanents. Electron. J. Combin. **6**, R6

[11] Galluccio, A., Loebl, M. (1999): On the theory of Pfaffian orientations. II. T-joins, k-cuts, and duality of enumeration. Electron. J. Combin. **6**, R7

[12] Halberstam, H., Roth, K.F. (1966): Sequences. Clarendon Press, Oxford

[13] Kac, M., Ward, J.C. (1952): A combinatorial solution of the two-dimensional Ising model. Phys. Rev. **88**, 1332–1337

[14] Kasteleyn, P.W. (1961): The statistics of dimers on a lattice. I. The number of dimer arrangements on a quadratic lattice. Physica **27**, 1209–1225

[15] Kasteleyn, P.W. (1963): Dimer statistics and phase transitions. J. Math. Phys. **4**, 287–293

[16] Kasteleyn, P.W. (1967): Graph theory and crystal physics. In: Harary, F. (ed.). Graph Theory and Theoretical Physics. Academic Press, London, pp. 43–110

[17] Lovasz, L., Plummer, M.D. (1986): Matching theory. (Annals of Discrete Mathematics, vol. 29). North-Holland, Amsterdam

[18] McCuaig, W., Robertson, N., Seymour, P.D., Thomas, R. (1997): Permanents, Pfaffian orientations, and even directed circuits (extended abstract). Proceedings of the Twenty-Ninth ACM Symposium on the Theory of Computing. Association for Computing Machinery, New York, pp. 402–405

[19] Minc, H. (1978): Permanents. (Encyclopedia of Mathematics and its Applications, vol. 6). Addison-Wesley, Reading, MA

[20] Minc, H. (1987): Permanental compounds and permanents of (0, 1) circulants. Linear Algebra Appl. **86**, 11–42

[21] Polya, G. (1913): Aufgabe 424. Arch. Math. Phys. (3), **20**, 271

[22] Ryser, H.J. (1963): Combinatorial mathematics. (The Carus Mathematical Monographs no. 14). Mathematical Association of America, Buffalo, NY

[23] Valiant, L.G. (1979): The complexity of computing the permanent. Theoret. Comput. Sci. **8**, 189–201

[24] Valiant, L.G. (1979): Completeness classes in algebra. In: Conference Record of the Eleventh ACM Symposium on Theory of Computing. Association for Computing Machinery, New York, pp. 249–261

Episturmian words and morphisms (results and conjectures)

J. Justin

1 Introduction

The theory of Sturmian words is well developed and is an active subject of research (survey in [2]).

Among several generalizations to larger alphabets (billiard sequences, sequences with low complexity, ...) one is particularly interesting. It was introduced by Rauzy [11] and studied by Arnoux and Rauzy [1], mainly for 3-letter alphabets. These authors study these words by way of their Rauzy factor graphs which are rather simple and also by coding them with interval exchange transformations.

With Droubay and Pirillo we have in [6] begun a study of a slightly more general class of infinite words that we call episturmian. Our study makes use of the palindrome closure operation introduced by de Luca [5] in relation with Sturmian words.

Here we recall some results of [6] essentially about the standard episturmian words and morphisms.

Then we present results and conjectures of research in progress about all episturmian words and morphisms, shifts, frequencies of letters.

When applied to Sturmian words this allows us to rediscover, in part, known results in what we hope is an enlightening and unifying way. This is the case, for instance, for return words [13], fractional powers [12], palindromic factors [7], fixed points of morphisms [9]. We also mention [4], an extension of Fine and Wilf's theorem for three periods which deals with episturmian words on three letters, generalizing a characterization of Sturmian words in [5].

2 Notations

(We will not review common notation in the theory of words).

- The alphabet will always be called A and will be finite.
- If $u \in A^* \cup A^\omega$, $\mathrm{Alph}(u)$ is the set of letters occurring in u.
- If $s \in A^\omega$, $\mathrm{Ult}(s)$ is the set of letters occurring infinitely many times in s.
- $F(s)$ is the set of factors of s.
- A factor v of s is *right* (resp. *left*)-*special* if, for at least 2 different letters x, y, vx and vy (resp. xv and yv) belong to $F(s)$.
- Infinite words here are right-infinite, but 2-sided infinite words will also occur in Theorem 10.
- $\widetilde{u}$ is the *reversal* of u.

- The *right palindromic closure* of u is the shortest palindrome having prefix u. It is $u^{(+)} = vw\tilde{v}$ where $vw = u$ with w the longest palindromic suffix of u.
- Sturmian words (classical definition with slope and intercept).

3 Episturmian words

Definition 1. An infinite word t is *episturmian* if the set $F(t)$ of its factors is closed under reversal and, for all $n \in \mathbb{N}$, t has at most one factor of length n which is right special.

Definition 2. An episturmian word s is *standard* if all the left-special factors of s are prefixes of s.

Episturmian words are uniformly recurrent. Two episturmian words are *equivalent* if they have the same set of factors. In any equivalence class of episturmian words, exactly one is standard unless these words are periodic (in this case there is a finite number of standard episturmian words in the class).

In [6], the standard episturmian words were introduced first, extending to an arbitrary alphabet a construction of de Luca [5] for standard Sturmian words. This is

Theorem 1. *An infinite word s is standard episturmian if and only if, whenever u is a prefix of s, then $u^{(+)}$ is also a prefix of s.*

Corollary 1. *There exists a bijection which associates to any standard episturmian word $s \in A^{\omega}$ an infinite word $\Delta(s) = x_1x_2x_3\cdots$, $x_i \in A$, such that the infinite sequence of all palindromic prefixes of s, $u_1 = \epsilon, u_2, u_3, \ldots$ is given by $u_{n+1} = (u_nx_n)^{(+)}$, for all $n \geq 1$.*

Hereafter the notations for $\Delta(s)$ and the palindromic prefixes will be retained. $\Delta(s)$ is called the *directive word of s*.

Theorem 2. *A standard episturmian word s is periodic if and only if $|\mathrm{Ult}(\Delta(s))| = 1$.*

Remark 1. When $A = \{a, b\}$ and s is not periodic we obtain exactly the Sturmian words. Writing $\Delta(s) = a^{d_1}b^{d_2}a^{d_3}b^{d_4}\cdots$, $(d_1, d_2, d_3, d_4, \ldots)$ is the "directive sequence" of s and coincides, grosso modo, with the sequence of the partial quotients of the continued fraction expansion of the slope of s [2].

Remark 2. Sturmian words have two properties which cannot, it seems, be generalized to episturmian words:

a) the continued fraction of the slope introduces a simple relation between the structure of the word and the frequencies of its letters;
b) Sturmian words are "balanced"; this is a very useful characteristic property.

Indeed these properties have some analogues for episturmian words; in particular these words have a balance property which, when $|A| = 2$, coincides with the balance property of Sturmian words but unhappily is not characteristic for $|A| > 2$.

However it is possible to say many things about episturmian words without the help of these properties. In our opinion this may allow us to attain a simpler understanding of the theory of Sturmian words itself.

4 Standard episturmian morphisms

For any $a \neq b$ in A define endomorphisms ψ_a, θ_{ab} of A^* by:

$$\psi_a(a) = a, \quad \psi_a(x) = ax \quad \text{for } x \neq a;$$
$$\theta_{ab}(a) = b, \quad \theta_{ab}(b) = a, \quad \theta_{ab}(x) = \text{otherwise}.$$

A morphism φ will be called *standard episturmian* if it belongs to the monoid of morphisms, $\mathcal{S}$, generated by all ψ_a , θ_{ab}, $a, b \in A$.

Theorem 3. *Let $s \in A^\omega$ and $\Delta = x_1 x_2 \cdots$, $x_i \in A$. Then s is standard episturmian with directive word Δ if and only if there exists an infinite sequence of infinite words, $s_0 = s, s_1, s_2, \ldots$ such that, for all $i \in \mathbb{N}$,*

$$s_i = \psi_{x_{i+1}}(s_{i+1}).$$

Corollary 2. *If s is standard episturmian then, for any $\varphi \in \mathcal{S}$, $s' = \varphi(s)$ is standard episturmian. Moreover, the directive words $\Delta(s)$, $\Delta(s')$ satisfy $\Delta(s') = w\pi(\Delta(s))$ for some $w \in A^*$ and some permutation π of A.*

Now let $\text{Alph}(s) = B \subseteq A$ with $s \in A^\omega$ standard episturmian. We say that s (or any equivalent infinite word) is *B-strict* if $\text{Ult}(\Delta(s)) = B$ (this means that for any right-special factor u of s, $uB \subseteq F(s)$, or equivalently that the complexity function of s is asymptotically $(|B| - 1)n + k$ for some k). We also can say "strict" when the alphabet is not specified.

The next theorem is important as, together with Corollary 2, it characterizes standard episturmian morphisms of A^*.

Theorem 4. *If there exist two A-strict standard episturmian words s, s' such that $s' = \varphi(s)$ for some endomorphism φ of A^*, then $\varphi \in \mathcal{S}$.*

Corollary 3. *An A-strict standard episturmian word s is the fixed point of a morphism if and only if its directive word $\Delta(s)$ is periodic.*

Example. The Rauzy word [10], r, sometimes called the Tribonacci word as it is the analogue for three letters of the Fibonacci word,

$$r = abacabaabacabab \cdots$$

has directive word $\Delta(r) = (abc)^\omega$. So r is a fixed point of the morphism $\psi_a\psi_b\psi_c$, but, as $(bca)^\omega$ is $\Delta(r)$ apart from a permutation of the alphabet, we even have $r = \rho(r)$ where $\rho = \psi_a\theta_{ac}\theta_{ab}$, i.e., $\rho(a) = ab$, $\rho(b) = ac$, $\rho(c) = a$.

5 Morphisms of episturmian words

Now for $a \in A$ define $\overline{\psi}_a$ by $\overline{\psi}_a(x) = \widetilde{\psi_a(x)}$ for $x \in A$. The *episturmian morphisms* of A^* are the elements of the monoid $\mathcal{E}$ generated by all $\overline{\psi}_a$, ψ_a, θ_{ab}, $a \neq b \in A$.

Now let $\overline{A} = \{\overline{a} \mid a \in A\}$. A letter $\overline{x}$ is considered as x with *spin* 1 while x itself is considered as x with spin 0. An *infinite spinned word* is an element of $(A \cup \overline{A})^\omega$. We generally write $\check{\Delta} = \check{x}_1\check{x}_2\check{x}_3\cdots$, $x_i \in A$ for a spinned word when the spins are not explicitly specified.

Theorem 5. *A recurrent infinite word t is episturmian if and only if there exist a spinned word $\check{\Delta} = \check{x}_1\check{x}_2\check{x}_3\cdots$, $x_i \in A$, and an infinite sequence $t_0 = t$, t_1, t_2, $\ldots$ of infinite words such that, for $i \in \mathbb{N}$, $t_i = \psi_{x_{i+1}}(t_{i+1})$ if $\check{x}_{i+1} = x_{i+1}$ and $t_i = \overline{\psi}_{x_{i+1}}(t_{i+1})$ if $\check{x}_{i+1} = \overline{x}_{i+1}$.*

We will say that $\check{\Delta}$ is a *spinned directive word* for t or *directs* t.

Corollary 4. *If t is episturmian, then, for any $\varphi \in \mathcal{E}$, $\varphi(t)$ is episturmian. Moreover, if t is directed by $\check{\Delta}$, then $\varphi(t)$ is directed by $f\pi(\check{\Delta})$ for some $f \in (A \cup \overline{A})^*$ and some permutation π of A (the sequence of spins in $\check{\Delta}$ remaining unchanged).*

Theorem 6 (generalizes Theorem 4)**.** *If t and t' are A-strict episturmian words with $t' = \varphi(t)$ for some endomorphism φ of A^*, then $\varphi \in \mathcal{E}$.*

Remark 3. Let $B = \mathrm{Ult}(\Delta(s))$ with $s \in A^\omega$ standard episturmian. Then any word in the class of s is the image under some morphism of $\mathcal{E}$ of some B-strict episturmian word. For instance, for $|A| = 3$, the episturmian words on A are the Arnoux–Rauzy words [1] (case $|B| = 3$) and the images of Sturmian words (case $|B| = 2$) and the images of a^ω, $a \in A$, (case $|B| = 1$) under morphisms of $\mathcal{E}$.

The next corollary characterizes fixed points of morphisms.

Corollary 5. *An A-strict episturmian word is a fixed point of a morphism if and only if it has a spinned directive word $\check{\Delta}$ which is periodic, i.e., $\check{\Delta} = f^\omega$ for some $f \in (A \cup \overline{A})^*$.*

Example. The Fibonacci word $f = abaaba\cdots$ has directive word $\Delta(f) = (ab)^\omega$. Also af is directed by $(a\overline{b})^\omega$ and is a fixed point of $\psi_a\overline{\psi}_b$. Similarly bf, directed by $(\overline{a}b)^\omega$, is a fixed point of $\overline{\psi}_a\psi_b$.

More generally (see [3] for the Sturmian case), we have

Corollary 6. *If a standard episturmian word s is a fixed point of a morphism, then, for any $x \in \mathrm{Alph}(s)$, xs is also a fixed point of a morphism.*

In general the spinned word directing an episturmian word is not unique. For instance, the Fibonacci word f is directed by $\Delta(f) = (ab)^\omega$ but also by $\overline{a}\overline{b}(a\overline{b})^\omega$ and by infinitely many others.

In the opposite direction we have

Theorem 7. *Let $\check{\Delta}$ be an infinite spinned word with $|\mathrm{Ult}(\Delta)| \geq 2$. Then:*

1) *If all spins of $\check{\Delta}$ are ultimately 0, then there is exactly one infinite word directed by $\check{\Delta}$. It can be obtained by a shift of the standard episturmian word whose directive word is Δ.*

2) *If some letter of* Ult(Δ) *has spin 0 in* $\check{\Delta}$ *infinitely many times, then exactly one infinite word is directed by* $\check{\Delta}$ *and this word is episturmian.*
3) *If all spins of* $\check{\Delta}$ *are ultimately 1, then there are exactly* $|\,\mathrm{Ult}(\Delta)|$ *episturmian words directed by* $\check{\Delta}$.

We omit for simplicity the case $|\mathrm{Ult}(\Delta)| = 1$ which leads to a periodic episturmian word but also, possibly, to some non-recurrent words.

6 Shifts and morphisms

Let s be standard episturmian with directive word $\Delta(s) = x_1x_2\cdots$. Define $\mu_n = \psi_{x_1}\psi_{x_2}\cdots\psi_{x_n}$, $n \geq 1$, μ_0 to be the identity and $h_n = \mu_n(x_{n+1})$. Then the h_n are prefixes of s and are related to the palindromic prefixes u_n by $u_{n+1} = h_{n-1}u_n$. Also set $\sigma_0 = 0$ and $\sigma_n = |h_{n-1}|$, $n \geq 1$.

Theorem 8. *With notation as above, let* $\check{\Delta}$ *be* Δ *with spin 1 exactly at positions* m_1, m_2, $\ldots$, m_p *(that is, for* x_{m_1}, x_{m_2}, $\ldots$, x_{m_p}*). Then the (unique) episturmian word directed by* $\check{\Delta}$ *is* s *shifted by* $\sigma_{m_1} + \sigma_{m_2} + \cdots \sigma_{m_p}$ *positions.*

This theorem gives an interpretation of the spins when only a finite number of them are 1.

Moreover $\sigma_0, \sigma_1, \sigma_2, \cdots$ can be considered as a system of numeration and every positive integer $< \sigma_p$ can be written as a sum of some of the $\sigma_q, q < p$. Consequently any positive shift can be obtained in this way. In general this representation is not unique.

Now, if t, t' are infinite words with $t' \in A^n t$, we say that t' is a *negative shift* of t.

Theorem 9. *Let* s *be standard episturmian with directive word* $\Delta = x_1x_2\cdots$. *For any* $c \in$ Ult(Δ)*, let* $\check{\Delta}$ *be* Δ *with spins 1 except when* $x_i = c$*. Then the (unique) episturmian word directed by* $\check{\Delta}$ *is* cs.

Corollary 7. *If the episturmian word* t *is a negative shift of the standard episturmian word* s*, then it is directed by some* $\check{\Delta}$ *where all spins are ultimately 1 except for exactly one letter, all of whose spins are ultimately 0. Moreover, all episturmian negative shifts of* s *can be obtained in this way.*

Theorem 10. *Let* s *be standard episturmian with directive word* $\Delta = x_1x_2\cdots$, *let* $\overline{\Delta} = \overline{x}_1\overline{x}_2\cdots$ *and let* t *be any one of the words directed by* $\overline{\Delta}$. *Then* $\tilde{t}s$ *is a 2-sided episturmian word equivalent to* s.

Examples. In case s is standard Sturmian on $\{a, b\}$ we get $t_1 = abs$, $t_2 = bas$.

The situation for $|A| > 2$ is less simple. For instance, for the Rauzy word $r = abac\cdots$, we get three words (which are fixed points of $\overline{\psi}_a\overline{\psi}_b\overline{\psi}_c$), $t_1 = abacababa\cdots$, $t_2 = bacabaaba\cdots$, $t_3 = cabaabaca\cdots$, whence three (episturmian) extensions to the left of the Rauzy word.

Theorem 11. *With s standard episturmian and Δ its directive word as previously, let $\check{\Delta}$, having infinitely many spins 0 and 1, be Δ with spins 1 exactly at positions $i_1 < i_2 < i_3 < \cdots$ and let $\check{\Delta}_p$ be Δ with spin 1 at positions $i_1, i_2, \cdots, i_p$ only. Let t_p be directed by $\check{\Delta}_p$, i.e., t_p is s shifted by $\sigma_{i_1} + \sigma_{i_2} + \cdots \sigma_{i_p}$ positions. Then the episturmian word directed by $\check{\Delta}$ is the limit of the t_p for p approaching infinity.*

This allows us, at least formally, to give a meaning to the infinite sum $\sum \sigma_{i_j}$.

In case s is standard Sturmian with slope α, this works: the numbers $\alpha \sum_{j \leq p} \sigma_{i_j}$ mod 1 and their limit are the intercepts of the words t_p above and of their limit.

This also seems to work when s is generated by a morphism because, perhaps under some assumptions, the calculation of the sums is possible.

We conjecture that, in the general case, there is a notion of intercept for episturmian words. This conjecture is also suggested by the construction of (strict) episturmian words from interval exchanges: all words in an equivalence class are obtained by starting from all points on the unit circle, that is, from all real numbers modulo 1.

7 Frequencies of letters (in progress)

It seems that the existence of, and some properties of frequencies of letters in an episturmian word s can follow from ergodic properties of the interval exchange transformation used in [1] for constructing it. Our aim (hope?) is to get information about frequencies more directly by using the properties of episturmian words given here and in [6]. The first step is to prove the existence of frequencies with the following strong definition:

For any $x \in A$, and for any factors $v_1, v_2, \ldots$ of s with increasing lengths, the frequencies of x in the v_i have a limit independent of the v_i.

Assuming this and supposing for simplicity that $A = \{a, b, c\}$, let α, β, γ be the frequencies of the letters. Then we make the following conjectures.

Conjecture 1. The frequencies α, β, γ are linearly independent over the rationals if and only if s is A-strict.

Conjecture 2. For $\xi \in [0, 1]$ irrational, there exists, apart from a relabeling of the letters, exactly one class of A-strict episturmian words in which a letter has frequency ξ.

These two conjectures are apparently correct when the word is generated by a morphism.

Lastly, as $\alpha + \beta + \gamma = 1$, we can represent (α, β, γ) by a point M inside an equilateral triangle ABC, with α, β, γ barycentric (or triangular) coordinates of M relatively to A, B, C.

Not every point inside ABC gives the frequencies of an episturmian word. Indeed the first letter of the standard episturmian word s always occurs in any factor xy $(x, y \in A)$ of s. So its frequency is $\geq 1/2$. So the median triangle $A'B'C'$ (i.e., A'

is the midpoint of BC and so on) is forbidden. But as $s = \psi_{x_1}(s_1)$ we can say the same thing for the frequencies in s_1. This prohibits new triangles for $M(\alpha, \beta, \gamma)$. Continuing infinitely we get that the allowed domain for M is a kind of fractal object recalling the Sierpinski gasket and apparently having measure 0 (see remarks at the end of [1]).

Remark 4. When we delete only the interior of the forbidden triangles $A'B'C'$ and so on, the set of the remaining $M(\alpha, \beta, \gamma)$ corresponds to all episturmian words on A. When we also delete the vertices of these triangles, we exclude the periodic episturmian words and when we finally also delete the boundaries of the triangles, the remaining $M(\alpha, \beta, \gamma)$ correspond to the A-strict episturmian words.

References

[1] Arnoux, P., Rauzy, G. (1991): Représentation géometrique des suites de complexité $2n + 1$. Bull. Soc. Math. France **119**, 199–215

[2] Berstel, J., Séébold, P. (2000): Sturmian words. In: Lothaire, M., Algebraic Combinatorics on Words. (Preliminary Version (8 Nov. 2000)). Institut Gaspard Monge, Université Marne-la-Vallée. To appear, pp. 41–98

[3] Berstel, J., Séébold, P. (1994): A remark on morphic Sturmian words. Marne-la-Vallée Inform. Théor. Appl. **28**, 255–263

[4] Castelli, M.G., Mignosi, F., Restivo, A. (1999): Fine and Wilf's theorem for three periods and a generalization of Sturmian words. Theoret. Comput. Sci. **218**, 83–94

[5] de Luca, A. (1997): Sturmian words: structure, combinatorics, and their arithmetics. Theoret. Comput. Sci. **183**, 45–82

[6] Droubay, X., Justin, X., Pirillo, G. (2000): Episturmian words and some constructions of de Luca and Rauzy. Theoret. Comput. Sci., to appear. See also: Palindrome factors and a characterization by de Luca of standard Sturmian infinite words. LIAFA 98-06, Laboratoire d'Informatique et d'Algorithmique Fondements et Applications, Université de Paris VII, Paris

[7] Droubay, X., Pirillo, G. (1999): Palindromes and Sturmian words. Theoret. Comput. Sci. **223**, 73–85

[8] Mignosi, F., Seebold, P.: (1993): Morphismes Sturmiens et règles de Rauzy. J. Théor. Nombres Bordeaux **5**, 221–233

[9] Parvaix, B. (1997): Propriétés d'invariance des mots sturmiens. J. Théor. Nombres Bordeaux **9**, 351–369

[10] Rauzy, G. (1982): Nombres algébriques et substitutions. Bull. Soc. Math. France **110**, 147–178

[11] Rauzy, G. (1985): Mots infinis en arithmétique. In: Nivat, M., Perrin, D. (eds.) Automata on Infinite Words. (Lecture Notes in Computer Science, vol. 194). Springer, Berlin, pp. 165–171

[12] Vandeth, D. (2000): Sturmian words and words with a critical exponent. Theoret. Comput. Sci. **242**, 283–300

[13] Vuillon, L. (1999): A characterization of Sturmian words by return words. LIAFA 99/21, Laboratoire d'Informatique et d'Algorthmique Fondements et Applications, Université de Paris VII, Paris

A curious characteristic property of standard Sturmian words

G. Pirillo

1 Introduction

Sturmian words have been studied for a very long time (Bernoulli [1], Christoffel [3], Markoff [18], Morse and Hedlund [19], Coven and Hedlund [5], Lunnon and Pleasants [17], ...). They are infinite words that have exactly $n + 1$ factors of length n, for each $n \geq 0$. Thus they are written in a binary alphabet. Recently a natural and promising generalization to an arbitrary alphabet was presented in [9].

Sturmian words are related to number theory (continued fractions [11]), theory of algorithms [15] and even mathematical education [14] and are extensively studied.

They have several properties: for instance, like the famous Thue–Morse sequence [26], they are uniformly recurrent and they are exactly those among aperiodic infinite words which have minimal complexity [4, 19].

In the literature they have many names and are defined in several ways (cutting sequences, mechanical words, . . .). For these reasons, it *is difficult to trace back many of the properties of Sturmian words because of the scattered origins, terminology and notation* (see [2, p. 42] for an updated survey on the subject).

A Sturmian word is *standard* if and only if its prefixes are exactly its left special factors if and only if it is invariant under all decimations [12] (see also [22]). Standard sturmian words represent real numbers in a natural way.

The object of the present paper is to prove that w is a palindrome prefix of a standard Sturmian word if and only if wab and wba are conjugate. In a forthcoming survey paper we will discuss the characterization of standard Sturmian words.

2 Preliminaries

For a set A (*alphabet*) whose elements are called *letters*, the free monoid A^* generated by A is the set of the (finite) words on A. The *empty word* is denoted by ϵ. If $u = u(1)u(2)\cdots u(m)$, $u(i) \in A$, is a word, its *length* is $|u| = m$. Also $|u|_x$, $x \in A$, is the number of occurrences of x in u. The *reversal* of u, i.e., the word $u(m)u(m-1)\cdots u(2)u(1)$, is denoted by $\widetilde{u}$. A word equal to its reversal is a *palindrome*. The words u and v are conjugate and we write $u \sim v$ if there exist words s and t such that $u = st$ and $v = ts$.

In the same way an *infinite word* s is an infinite sequence

$$s = s(1)s(2)\cdots s(i)\cdots, \quad s(i) \in A, \quad i \in \mathbb{N}_+ = \mathbb{N} \setminus \{0\}.$$

The set of infinite words on A is A^ω. We say that $s \in A^\omega$ is *ultimately periodic* if it can be written as $s = uvvv\cdots = uv^\omega$ for $u, v \in A^*$ (*periodic* if $u = \epsilon$).

For a finite or infinite word t, the word $w = t(i)t(i+1)\cdots t(j)$ is a *factor* of t (*proper* factor if $w \neq t$). It is a *prefix* of t if $i = 1$. If $t \in A^*$ and $|t| = m$, then w is a *suffix* of t if $j = m$. A word $u \in \{a,b\}^* \cup \{a,b\}^\omega$ is *balanced* if $||w|_a - |w'|_a| \leq 1$ whenever w, w' are factors of u with $|w| = |w'|$. Let $m \in A^* \cup A^\omega$. Then $F(m)$ denotes the set of factors of m and the *alphabet* of m is $\mathrm{Alph}(m) = F(m) \cap A$.

Let s be an infinite word; then a factor u of s is *right* (resp., *left*) *special* in s if there exist $x, y \in A$, $x \neq y$, such that $ux, uy \in F(s)$ (resp., $xu, yu \in F(s)$).

An infinite word s has property Pi if every left-most occurrence of a palindrome in s is a central factor of a palindrome prefix of s.

3 The Fibonacci word

Let $\varphi : \{a,b\}^* \to \{a,b\}^*$ be the morphism given by $\varphi(a) = ab$, $\varphi(b) = a$. Let $f_0 = b$ and, for $n \geq 0$,

$$f_{n+1} = \varphi(f_n).$$

For $n \geq 2$, let $g_n = f_{n-2}f_{n-1}$ and let h_n be the longest common prefix of f_n and g_n. For example we have: $f_1 = a$, $f_2 = ab$, $f_3 = aba$, $f_4 = abaab$, $g_2 = ba$, $g_3 = aab$, $g_4 = ababa$, $h_2 = \epsilon$, $h_3 = a$, $h_4 = aba$.

Notice that, for $n \geq 0$, $|f_n|$ is the nth element of the sequence of Fibonacci numbers F_n. We have $F_0 = 1$, $F_1 = 1$ and, for each $n \geq 2$, $F_n = F_{n-1} + F_{n-2}$; for $0 \leq n \leq 13$ the Fibonacci numbers F_n are 1, 1, 2, 3, 5, 8, 13, 21, 34, 55, 89, 144, 233, 377. For each $n \geq 2$, $f_n = f_{n-1}f_{n-2}$ which implies that, for each $n \geq 1$, f_n is a prefix of f_{n+1}.

Hence, there exists a unique infinite word, namely, the Fibonacci word denoted by f, such that, for each $n \geq 1$, f_n is a prefix of f, and we have

$$f = abaababaabaababaababaabaababaabaababaababa \ldots.$$

Some well-known facts concerning f are collected in Lemma 1.

Lemma 1. *For each $n \geq 2$,*

i) *(near-commutative property) $f_{n+2} = f_{n+1}f_n = f_n g_{n+1} = h_{n+2}xy$ and $g_{n+2} = f_n f_{n+1} = f_{n+1}g_n = h_{n+2}yx$, where $x, y \in \{a,b\}$, $x \neq y$, and*

$$xy = \begin{cases} ab & \text{if } n \text{ is even} \\ ba & \text{if } n \text{ is odd;} \end{cases}$$

ii) *the words $h_n ab$ and $h_n ba$ are conjugate, i.e., $f_n \sim g_n$ (for instance, $aba \sim aab$, $abaab \sim ababa$, $abaababa \sim abaabaab$, and so on).*

4 Sturmian words

In this section we give the most common definition and some known results.

Definition 1. An infinite word $s = s_1s_2s_3 \cdots$, $s_i \in \{a, b\}$, is *Sturmian* if there exist real numbers $\alpha, \rho \in [0, 1]$ such that either, for all i,

$$s(i) = a \text{ if } \lfloor \rho + (n+1)\alpha \rfloor = \lfloor \rho + n\alpha \rfloor, \quad s(i) = b \text{ otherwise,}$$

or, for all i,

$$s(i) = a \text{ if } \lceil \rho + (n+1)\alpha \rceil = \lceil \rho + n\alpha \rceil, \quad s(i) = b \text{ otherwise.}$$

The infinite word s is *proper* Sturmian if α is irrational, *periodic* Sturmian if α is rational and *standard* Sturmian if $\rho = 0$.

Proposition 1. *An infinite word is Sturmian if and only if for every n it has exactly $n + 1$ factors of length n.*

Proposition 2. *An infinite word is Sturmian if and only if it contains exactly one palindrome word of even length and two palindrome words of odd length for each non-negative integer.*

Proposition 3. *An infinite non-ultimately-periodic word u is Sturmian if and only if it is balanced.*

Proposition 4. *A Sturmian word is standard if and only if its prefixes are exactly its left special factors.*

Proposition 5. *An infinite non-ultimately-periodic word on a two letter alphabet is standard Sturmian if and only if it has property Pi.*

5 Results and proofs

The Fibonacci word, related to the golden ratio, is a very particular example of a standard Sturmian word. We are convinced that, if one looks **carefully** at the properties of this precious example, one can discover relevant general properties of standard Sturmian words (see, for instance, [20, 21]). An easy property of its palindrome prefixes (see Lemma 1, ii)), suggested to us the following

Theorem 1. *A word w is a palindrome prefix of a standard Sturmian word if and only if wab and wba are conjugate.*

Proof. *("only if" part.)* It is possible to say that this part is well-known. In any case, for a complete proof, one can use notions defined in [6] (PAL, PER, Stand, ...).

("if" part.) This part seems to be unknown and, our knowledge, it has never been mentioned in the literature.

Let w be written on an arbitrary alphabet and suppose that $wab \sim wba$. If $|w| = 0$ or $|w| = 1$ the statement is easily verified. So suppose that $|w| > 1$. We know that, for some words u, v, we have $wab = uv$ and $wba = vu$. If $|u| = 1$, then $u = a$ and $w = a^{|w|}$ which is a palindrome prefix of a standard Sturmian word. So suppose that $|u| \geq 2$. If $|v| = 1$, then $v = b$ and $w = b^{|w|}$ which is a palindrome prefix of a standard Sturmian word.

So we can suppose, $|u| \geq 2$ and $|v| \geq 2$. If, without loss of generality, $|u| \leq |v|$ for instance, we easily see that u is a prefix of v and so w is a fractional power of both u and v.

Put $p = |u|$ and $q = |v|$ and $d = \mathrm{GCD}\,(p, q)$. If d were greater than 1, then $|w| = p + q - 2 \geq p + q - d$ and, by a result of Fine and Wilf [16], w would have period d and consequently $u = h^{p/d}$ and $v = h^{q/d}$ for some non-empty word h which is clearly impossible as u has suffix ba and v has suffix ab. So $d = 1$.

Now we show that $\mathrm{Alph}(w) \subseteq \{a, b\}$. If this is not the case, suppose that there is a letter $c \in \mathrm{Alph}(w)$, $c \neq a$, $c \neq b$ and w is one of the words of minimal length such that $wab \sim wba$. Suppose that $|u| < |v|$ and put $q = kp + r$, $0 \leq r < p$. We can write $v = u^k v_1$ with v_1 a prefix of u of length r. Also put $v_1 = v'ab$. By the conjugation relation of wab and wba we get $uv'ba = v'abu$. As $v'abu \sim= uv'ab$, we get with $w' = uv'$, $w'ab \sim w'ba$. As $\mathrm{Alph}(w'ab) = \mathrm{Alph}(wab)$ and $|w'| < |w|$ we get a contradiction. Consequently $\mathrm{Alph}(wab) = \{a, b\}$.

As $d = 1$, by a theorem of [6], w is a palindrome prefix of some standard Sturmian word.

Remark 1. We presented the proof of the above theorem on several occasions and, in particular, at two meetings in March and September 1999 before illustrating it at the VI Incontro Italiano di Combinatoria Algebrica, International Conference dedicated to the memory of Gian-Carlo Rota (Maratea, Italy, 5–10 October 1999). We always pointed out that the "if" part of the theorem was, to our knowledge, never mentioned in the literature.

Remark 2. Aldo de Luca, in Firenze in November 1999, informed us that he found an alternative proof of the above theorem and on 10 January 2000 he sent it to us [7].

Below we reproduce the content of the proof of de Luca in its entirety.

We denote by PER the set of all palindrome prefixes of some standard Sturmian word. Let Σ be the set of all words w satisfying the equation:

$$w = Cxy = AB$$

where x and y are distinct letters and A, B, C are palindromes. It is well-known [6] that

$$\mathrm{PER}\,\{ab, ba\} = \Sigma. \tag{1}$$

A word $w \in PER$ if and only if wab and wba are conjugate.

We prove the "if part" since the "only if part" is trivial. From the hypothesis the words wab and baw are conjugate, so that we can write

$$wab = \alpha\beta, \ baw = \beta\alpha. \tag{2}$$

The proof is by induction on the length $|w|$ of w. If $|w| = 0$, then the result is trivial since $w = \epsilon$. Let us then suppose that $|w| > 0$. Since

$$wab\alpha = \alpha baw,$$

we conclude that $ab\alpha$ and αba are conjugate. Moreover, we have

$$\alpha ba = \gamma\delta, \ ab\alpha = \delta\gamma, \ w = (\gamma\delta)^n\gamma \tag{3}$$

for a suitable $n \geq 0$. We first suppose that $|\alpha| < |w|$ so that by induction $\alpha \in$ PER. Since α is a palindrome and $(\alpha ba)^{\sim} = ab\alpha$ it follows from (3) that γ, δ and w are palindromes.

From (2), since α and w are palindromes, it follows that

$$baw = \beta^{\sim}\alpha = \beta\alpha$$

so that β is a palindrome. Thus $wab \in \Sigma$ and $w \in$ PER.

We now suppose that $|\alpha| = |w|$. In this case $\beta = ab = ba$ which is absurd. If $|\alpha| = |w| + 1$, then $\beta = b$, $\alpha = wa = aw$, so that w and α are powers of the letter a and the result follows trivially. If $|\alpha| = |w| + 2$, then $\beta = \epsilon$ so that $wab = baw$. By solving this equation we obtain $w = (ba)^n b$, i.e., w (and then α and β) are palindromes, so that $wab \in \Sigma$ which concludes the proof. □

Acknowledgement. We wish to thank Aldo de Luca for giving us his authorization to use his proof in this paper.

Remark 3. We would like to end this paper with some quotations by G.-C. Rota:

In mathematics, as anywhere today, it is becoming more difficult to tell the truth [13] (see also [23]). The truth offends.... The ideal of all science, not only of mathematics, is to do away with any kind of synthetic a posteriori statement and to leave only analytic trivialities in its wake. Science may be defined as the transformation of synthetic facts of nature into analytic statements of reason ... hic Rhodus, hic salta *[24] (see also [23]). Hardy was right after all: mathematicians are out to debunk the fakery that lies concealed underneath every logically correct proof. But they will not admit that their task is one of debunking; they will rather pretend that they are busy proving new theorems and stating new conjectures in compliance with the canons of present day logic [24] (see also [25]).*

References

[1] Bernoulli, J. (1772): Sur une nouvelle espèce de calcul. In: Recueil pour les Astronomes. Tome 1. Berlin, pp. 255–284

[2] Berstel, J., Séébold, P. (2000): Sturmian words. In: Lothaire, M., Algebraic Combinatorics on Words. (Preliminay Version (8 Nov. 2000)). Insitut Gaspard Monge, Université Marne-la-Vallée. To appear, pp. 41–98
[3] Christoffel, E.B. (1875): Observatio arithmetica. Mat. Ann. **6**, 145–152
[4] Coven, E.M. (1974): Sequences with minimal block growth. II. Math. Systems Theory **8**, 376–382
[5] Coven, E.M., Hedlund, G.A. (1973): Sequences with minimal block growth. Math. Systems Theory **7**, 138–153
[6] de Luca, A. (1997): Sturmian words: structure, combinatorics, and their arithmetics. Theoret. Comput. Sci. **183**, 45–82
[7] de Luca, A. (2000): Personal communication, e-mail of 10 January 2000
[8] de Luca, A., Mignosi, F. (1994): Some combinatorial properties of Sturmian words. Theoret. Comput. Sci. **136**, 361–385
[9] Droubay, X., Justin, J., Pirillo, G. : Episturmian words and some constructions of de Luca and Rauzy. Theoret. Comput. Sci., to appear
[10] Droubay, X., Pirillo, G. (1999): Palindromes and Sturmian words. Theoret. Comput. Sci. **223**, 73–85
[11] Hardy, G. H., Wright, E.M. (1971): An introduction to the theory of numbers. 4th edition. Clarendon Press, Oxford
[12] Justin, J., Pirillo, G. (1997): Decimations and Sturmian words. RAIRO Inform. Théor. Appl. **31**, 271–290
[13] Kac, M., Rota G.-C., Schwartz J. T. (1992): Discrete thoughts. Birkhäuser Boston, Boston, MA
[14] Knuth, D.E. (1972): Sequences with precisely $k + 1$ k-blocks: Solution of problem E2307. Amer. Math. Monthly **79**, 773–774
[15] Knuth, D.E. (1973): The art of computer programming. Vol. 3: Sorting and searching. Addison-Wesley, Reading, MA
[16] Lothaire, M. (1983): Combinatorics on words. Addison-Wesley, Reading, MA
[17] Lunnon, W.F., Pleasants, P.A.B. (1992): Characterization of two-distance sequences. J. Austral. Math. Soc. Ser. A **53**, 198–218
[18] Markoff, A. (1882): Sur une question de Jean Bernoulli. Math. Ann. **19**, 27–36
[19] Morse, M., Hedlund, G.A. (1940): Symbolic dynamics II. Sturmian trajectories. Amer. J. Math. **62**, 1–42
[20] Pirillo, G. (1999) From the Fibonacci word to Sturmian words. Publ. Math. Debrecen, **54**, Suppl., 961–971
[21] Pirillo, G. (1999): A new characteristic property of the palindrome prefixes of a standard Sturmian word. Sém. Lothar. Combin. **43**, Art B43f (electronic)
[22] Rauzy, G. (1985): Mots infinis en arithmétique. In: Nivat, M., Perrin, D. (eds.), Automata on Infinite Words. (Lecture Notes in Computer Science vol 194). Springer, Berlin, pp. 165–171
[23] Rota G.-C. (1993): Pensieri discreti. Garzanti, Milano
[24] Rota G.-C. (1997): Indiscrete thoughts. Birkhäuser Boston, Boston, MA
[25] Rota G.-C. (1999): Lezioni napoletane. La Città del Sole, Napoli
[26] Thue, A. (1906): Über unendliche Zeichenreihen. Skr. Norske Vid. Selsk. Kristiania I. Mat. Naturv. Klasse I, **1906** (7), 1–22

FSC
www.fsc.org
MIX
Papier aus verantwortungsvollen Quellen
Paper from responsible sources
FSC® C105338

If you have any concerns about our products,
you can contact us on
ProductSafety@springernature.com

In case Publisher is established outside the EU,
the EU authorized representative is:
Springer Nature Customer Service Center GmbH
Europaplatz 3, 69115 Heidelberg, Germany

Printed by Libri Plureos GmbH
in Hamburg, Germany